Clinical Cardiogenetics

H.F. Baars · J.J. van der Smagt
P.A.F.M. Doevendans
(Editors)

Clinical Cardiogenetics

Springer

Editors
H.F. Baars
Department of Cardiology
TweeSteden Hospital
Tilburg
The Netherlands
and
Department of Cardiology
Heart and Lung Center Utrecht
University Medical Centre Utrecht
The Netherlands

J.J. van der Smagt
Clinical Geneticist
University Medical Centre
Utrecht
The Netherlands

P.A. Doevendans
Department of Cardiology
University Medical Center Utrecht
Utrecht
The Netherlands

ISBN: 978-1-84996-470-8   e-ISBN: 978-1-84996-471-5
DOI: 10.1007/978-1-84996-471-5
Springer London Dordrecht Heidelberg New York

British Library Cataloguing in Publication Data
A catalogue record for this book is available from the British Library

Library of Congress Control Number: 2010937629

© Springer-Verlag London Limited 2011
Apart from any fair dealing for the purposes of research or private study, or criticism or review, as permitted under the Copyright, Designs and Patents Act 1988, this publication may only be reproduced, stored or transmitted, in any form or by any means, with the prior permission in writing of the publishers, or in the case of reprographic reproduction in accordance with the terms of licenses issued by the Copyright Licensing Agency. Enquiries concerning reproduction outside those terms should be sent to the publishers.
The use of registered names, trademarks, etc., in this publication does not imply, even in the absence of a specific statement, that such names are exempt from the relevant laws and regulations and therefore free for general use.
Product liability: The publishers cannot guarantee the accuracy of any information about dosage and application contained in this book. In every individual case the user must check such information by consulting the relevant literature.

Cover design: eStudioCalamar, Figueres/Berlin

Printed on acid-free paper

Springer is part of Springer Science+Business Media (www.springer.com)

Preface

Dear colleague,

With this comprehensive textbook on genetic aspects of cardiovascular disease for clinicians, we aim to provide a tool for clinicians, which simply has not been available before. Despite the fact that there is increasing awareness of genetic aspects of cardiovascular disease among both cardiologists and their patients, such a textbook has been missing. There was a small booklet *Cardiovascular Genetics for Clinicians*, edited by P.A. Doevendans and A.A.M. Wilde, published in 2001 by Kluwer Academic Publishers (ISBN 1-4020-0097-9), nowadays largely outdated. But 2 years after the American Heart Association Journals launched *Circulation: Cardiovascular Genetics* – a journal entirely devoted to genetic aspects of cardiovascular disease, we felt the absence of a comprehensive textbook for clinicians a serious omission. Of course, all medical and genetic facts presented in this textbook can be found elsewhere in academic journals. But as far as we know, this textbook is the first effort to compile the massive amount of knowledge available in a single easy-to-read (which does not requiring prior expert genetic knowledge) textbook with useful clinical handles for cardiologists.

In the Netherlands, over the last two decades, outclinics for cardiogenetic disorders have been established in all academic teaching hospitals. In these outclinics, cardiologists, clinical geneticists, molecular geneticists, genetic laboratories, genetic nurses, and social workers closely cooperate in order to provide the most optimal care for individuals, with cardiac disease with probable genetic cause, and their family members. However, the prevalence of genetic cardiomyopathies and primary electrical diseases is sufficiently high, that it is neither desirable nor possible that the care for this group of patients and their relatives remains solely restricted to a small number of academic centres. Therefore, basic knowledge on the genetic aspects of these types of disorders and on what to do with as yet asymptomatic at risk family members should be considered to be a prerequisite for all practising cardiologists. In 2009, in the Netherlands, a first, largely consensus-based guideline on surveillance in hypertrophic cardiomyopathy and how to go about screening of relatives of index patients has been approved by the national professional organisations of both cardiologists and clinical geneticists. If only one lesson could be learned from this guideline, it would be that simply treating and surveilling your patient, whilst completely ignoring the genetic aspects, is no longer considered adequate care.

Many cardiologists feel uncomfortable about this, as they have not been specifically trained in genetics let alone in explaining genetics to their patients; they fear genetic discrimination and not completely without reason, which makes it even more difficult to discuss the topic with their patients. Furthermore, they rightfully claim

Contributors

H. F. Baars, MD Department of Cardiology, TweeSteden Hospital, Tilburg and Department of cardiology, Heart and Lung Center Utrecht, University Medical Centre Utrecht, The Netherlands

A. F. Baas, MD, PhD Division of Biomedical Genetics, Department of Medical Genetics, University Medical Center Utrecht, The Netherlands

A. Bakker, MD Department of Vascular Medicine, Academic Medical Center, University of Amsterdam, Amsterdam, The Netherlands

K. D. Becker Pathway Genomics Corporation, San Diego, California, USA

B. Benito, MD Electrophysiology Research Program, Research Center, Montreal Heart Institute, Montreal, QC, Canada

M. Borggrefe, MD First Department of Medicine – Cardiology, University Medical Centre Mannheim, Mannheim, Germany

M. L. Bots Julius Center for Health Sciences and Primary Care, University Medical Center Utrecht, Utrecht, The Netherlands

J. Brugada, MD, PhD Hospital Clínic, Universitat de Barcelona Villarroel, Barcelona, Spain

P. Brugada, MD, PhD Heart Rhythm Management Centre, Cardiovascular Institute, UZ Brussel, VUB Brussels, Belgium

R. Brugada, MD, PhD Cardiovascular Genetics Center, Universitat de Girona, Spain

K. Caliskan, MD Department of Cardiology, Erasmus Medical Centre, Rotterdam, The Netherlands

L. Carrier, MD, PhD Department of Experimental Pharmacology and Toxicology, Cardiovascular Research Center, University Medical Center Hamburg-Eppendorf, Hamburg, Germany and Inserm, University Pierre et Marie Curie, Paris, France

H. ten Cate, MD, PhD Laboratory for Clinical Thrombosis and Haemostasis, Department of Internal Medicine, Cardiovascular Research Institute Maastricht, Maastricht University Medical Center, Maastricht, The Netherlands

I. Christiaans, MD PhD Department of Clinical Genetics and Department of Cardiology, Academic Medical Centre, Amsterdam, The Netherlands

R. Schimpf, MD First Department of Medicine – Cardiology,
University Medical Centre Mannheim, Mannheim, Germany

J.-J. Schott, PhD L'institut du thorax, Inserm U915, Faculte de Medecine,
Nantes, France

H. Schunkert Universität zu Lübeck, Medizinische Klinik II,
Ratzeburger Allee 160, Lübeck, Germany

C. E. Seidman Howard Hughes Institute, Harvard Medical School,
Boston, MA, USA

J. Seidman Howard Hughes Institute, Harvard Medical School,
Boston, MA, USA

J. J. van der Smagt, MD Clinical Geneticist, University Medical Centre,
Utrecht, The Netherlands

H. J. M. Smeets, MD, PhD Department of Population Genetics,
Genomics and Bioinformatics, University of Maastricht, The Netherlands

G. A. Somsen, MD, PhD Department of Cardiology,
Onze Lieve Vrouwe Gasthuis, Amsterdam, The Netherlands
Cardiologie Centra Nederland / Cardiology Centers of The Netherlands,
Amsterdam, The Netherlands

J. P. van Tintelen, MD, PhD Clinical Geneticist, University Medical Centre,
Groningen, The Netherlands

I. I. Tulevski, MD, PhD RESULT Cardiology Centers of the Netherlands, The
Netherlands

C. Veltmann, MD First Department of Medicine – Cardiology,
University Medical Centre Mannheim, Mannheim, Germany

M. de Visser, MD, PhD Department of Neurology, Academic Medical Centre,
Amsterdam, The Netherlands

C. van der Werf, MD Department of Cardiology, Heart Failure
Research Centre, Academic Medical Centre, Amsterdam, The Netherlands

A. A. M. Wilde, MD, PhD Department of Cardiology, Heart Failure Research
Centre, Academic Medical Centre, Amsterdam, The Netherlands

C. Wolpert, MD Department of Medicine – Cardiology,
Klinikum Ludwigsburg, Ludwigsburg, Germany

M. H. Zafarmand Department of Cardiology,
University Medical Center Utrecht, Utrecht, The Netherlands and
Julius Center for Health Sciences and Primary Care,
University Medical Center Utrecht, Utrecht, The Netherlands

Abbreviations

ACS	Acute coronary syndrome
ADP	Adenosine diphosphate
(A)MI	(Acute) myocardial infarction
Bp	Base pair
CAD	Coronary artery disease
CHD	Coronary heart disease
EPCR	Endothelial protein C receptor
F	Coagulation factor
GP	Glycoprotein
HMWK	High molecular weight kininogen
HVR	Hypervariable region
PAI-1	Plasminogen activator inhibitor-1
PAR	Protease-activated receptor
PCI	Percutaneous coronary intervention
PlA	Platelet antigen
(s)TM	(Soluble) Thrombomodulin
STE-ACS	ST-segment elevation ACS
TAFI	Thrombin activatable fibrinolysis inhibitor
TF	Tissue factor
tPA	Tissue type plasminogen activator
uPA	Urokinase plasminogen activatot
VNTR	Variable number of tandem repeats
vWF	von Willebrand factor
vWD	von Willebrand disease

Part I

Genetics

Introduction to Molecular Genetics

M.M.A.M. Mannens and H.J.M. Smeets

1.1 Introduction

Genetic knowledge is being rapidly introduced into clinical medicine. Knowledge on genes and gene defects, gene expression, and gene products have been gathered as a result of the human genome project and large-scale resequencing projects at a rapid pace.[1,2] The genetic cause of the vast majority of important monogenic disorders is nowadays known, and more rare disorders are being unraveled quickly. Developments in genomics and sequencing technologies enable molecular geneticists with an accelerated and detailed characterization of genetic defects, genetic predisposition, and/or genetic background of individual patients. This also provides extensive knowledge on the molecular pathophysiology.

The introduction of genetic tests for heritable cardiac abnormalities is of a fairly recent nature. Disorders, like hereditary cardiac arrhythmia syndromes or inherited cardiomyopathies, have been genetically unraveled in the last 2 decades, and research is ongoing to improve DNA-diagnostics.[3,4] Genetic testing offers many opportunities, but also a considerable number of risks and uncertainties. Therefore, introduction in the clinic has to be performed with great care. Not every test that can be done should be done. It is evident that genetic testing must be beneficial for the patient and his or her family. If he or she is affected, then the test can be performed to confirm the diagnosis and in some instances to predict prognosis and adjust treatment. It is clear that such genetic tests affect not only the patient involved, but also will be of concern to relatives and future offspring. In asymptomatic relatives of patients, it may be possible to determine their genetic status and predict what the chances will be of developing symptoms in the years to follow. Especially this kind of predictive genetic testing is becoming more and more important in the field of genetic cardiovascular diseases, enabling the discrimination between carriers and noncarriers of a specific genetic risk. It is obvious that in this area of predictive medicine, ethical and social (health insurance) aspects play important roles as well. Therefore, these investigations should be embedded in a multidisciplinary approach of cardiologists, clinical geneticists, laboratory specialists and social workers, psychologists, and ethicists. Society should define the outlines, within which genetic testing should be performed, preventing the social and economical discrimination of individuals based on their genetic burden.

1.1.1 DNA, What Is It, Where Is It, and How Are Proteins Made?

DNA (deoxyribonucleic acid) is an antiparallel dimer of nucleic acid strands (Fig. 1.1a–c). It is composed of nucleotides (base + deoxyribose + phosphate group). These nucleotides are polymerized through a phophodiester linkage. Base pairing of nucleotides are possible between adenine (A) and thymine (T) or cytosine (C) and guanine (G). CG bonds are stronger than AT bonds because the CG base pairs form three hydrogen bonds and the AT base pairs only two. The DNA *double helix* structure was originally published by Watson and

M.M.A.M. Mannens (✉)
Department of Clinical Genetics, Academic Medical Center,
P.O. Box 22700 1100 DE, Amsterdam, The Netherlands
e-mail: m.a.mannens@amc.uva.nl

Crick and is composed of two strands wound around each other in a helical structure and antiparallel, that is, the strands run in opposite directions (5'–3' versus 3'–5'). DNA does not form a perfect helix since the sugar phosphate backbones are slightly offset from the center of the helix. This creates major (relative open) grooves and minor (relatively closed) grooves (Fig. 1.1c). DNA in eukaryotic cells is located in the nucleus. The mitochondria (mainly involved in the energy supply for the cell) have multiple copies of their own small circular DNA molecules. Since the total DNA content in the human cell is about 2 m long and the nucleus is about 6×6^{-10} m, it is obvious that DNA has to be packed very condensed. DNA is packed in chromosomes, a complex of DNA and histone proteins. The human cell has a diploid genome consisting of 46 *chromosomes*, which means that all chromosomes are present in two copies (homologous pairs) with the exception of the sex chromosomes, which can be present as XX (female) or XY (male). During gametogenesis (forming of oocyte or sperm cell), the chromosomes become haploid (only one copy per cell) and genetic material between both homologous chromosomes is exchanged (*recombination* Fig. 1.2). This means that genetic information (such as disease causing mutations) can switch from one chromosome to the other homologue. This may sometimes be a problem in indirect DNA-diagnostics using markers in the vicinity of the gene defect. Of course *replication* of DNA is needed for cell division. A complex of enzymes unwinds the DNA at many positions and replicates the two strands of the DNA helix resulting in two copies of the original DNA molecule. Since DNA molecules are between 48,000,000 and 240,000,000 nucleotides long, mistakes in replication are likely, leading to potential mutations. The cell, however, has a number of mechanisms (e.g., proofreading activity of enzymes, mismatch repair, or excision repair) that correct such replication errors. However, still one mismatch out of 10^9 nucleotides remains.

The genetic code lies in the genes in the DNA molecules. These genes code for the specific amino acid sequence of proteins. *Messenger RNA* (mRNA) is needed as an intermediate between DNA and the protein since DNA is localized in the nucleus and proteins are synthesized on *ribosomes* outside the nucleus. Eukaryotic genes consist of coding sequences (exons), noncoding sequences (*introns*), and *regulatory sequences*

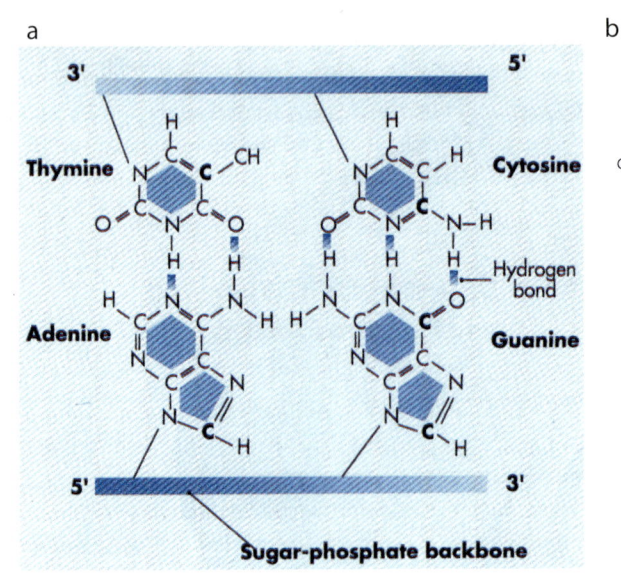

Fig. 1.1 Structure of DNA. a base (C, T, A, or G) combined with a deoxyribose and a phosphate group is called a nucleotide. These nucleotides are polymerized through phophodiester linkage. DNA is read from the 5' to the 3' end. (**a**) The four bases that make-up the actual DNA code, Adenine always pairs with Thymine with two hydrogen bonds and Cytosine always pairs with Guanine using three hydrogen bonds()s. (**b**) Chemical structure of DNA, showing the sugar backbone and the polymerization through phosphodiester linkage, as DNA is read in 5' to 3' direction, the code of this stretch of DNA would read:GATC. (**c**) DNA double helix istructure, the base pairs in the middle are aligned around the helical axis. The major and minor grooves are indicated. Adapted with permission from Jorde, Carey, Bamshad, White, Medical Genetics third edition, Mosby Elsevier 2006

(promoter sequence, enhancer, and repressor) that start, enhance, or repress gene expression. Only 2% of the DNA actually codes for the exons in approximately 20,000–25,000 human genes, 25% of the DNA codes for introns. The remainder of the DNA is, for instance, needed for regulation of gene expression, maintenance of chromosomes, syntheses of rRNAs (ribosomal RNAs) or tRNAs (transfer RNAs), or segregation of chromosomes during cell division. Part of the genome might be a coincidental result of evolution and might not have any function at all, but new functions of DNA sequences are still revealed. Recently, microRNAs have been discovered that do not code for a protein but have proven to be regulators of gene expression. Between humans, the genetic variation is less than 0.1%.

The enzyme *RNA polymerase* can use a single-stranded DNA molecule as a template and it copies the genetic code in the genes in a single-stranded mRNA molecule. The chemical structure of mRNA is similar to DNA with the exception that ribose is used instead of deoxyribose for the sugar backbone and that the base *uracil* is used instead of thymine. During a process called *transcription*, a pre-mRNA is (Fig. 1.3) synthesized according to the DNA code. This pre-mRNA consists of introns and exons. The introns are removed by a process called splicing leading to a smaller mRNA that contains only the genetic code for a protein. Transcription is very efficient since sometimes many different mRNA's encoding different proteins can be synthesized from a single gene.

These mRNAs are transported from the nucleus to the *ribosomes* in the cytosol. Ribosomes are cellular complexes consisting of proteins and ribosomal RNAs (rRNAs). In the ribosomes, the mRNA code is used as a template to synthesize proteins (a process called *translation*) (Fig. 1.4). mRNA, ribosomes, transfer RNAs (tRNA), and amino acids are the key components in this process. tRNAs function as amino acid carriers and as recognition molecules that identify the mRNA nucleotide sequence and translate that sequence into the amino acid sequence of proteins. Each tRNA recognizes one of the 20 specific amino acids (Table 1.1). A second tRNA recognition site binds to a specific nucleotide triplet (a combination of three sequential nucleotides in the mRNA [called a codon] that encodes for a specific amino acid, see Fig. 1.4). By repeating this process for all codons and linking subsequent amino acids to the previous one, the sequence of the genetic code is

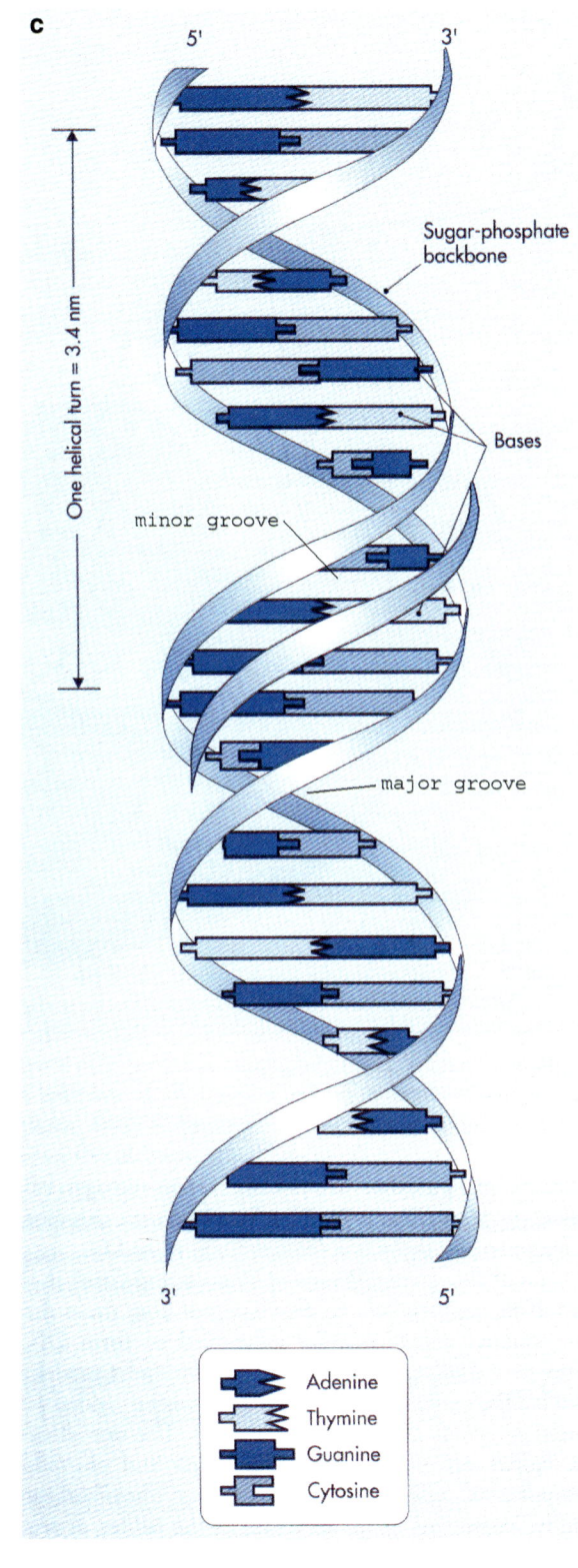

Fig. 1.1 (continued)

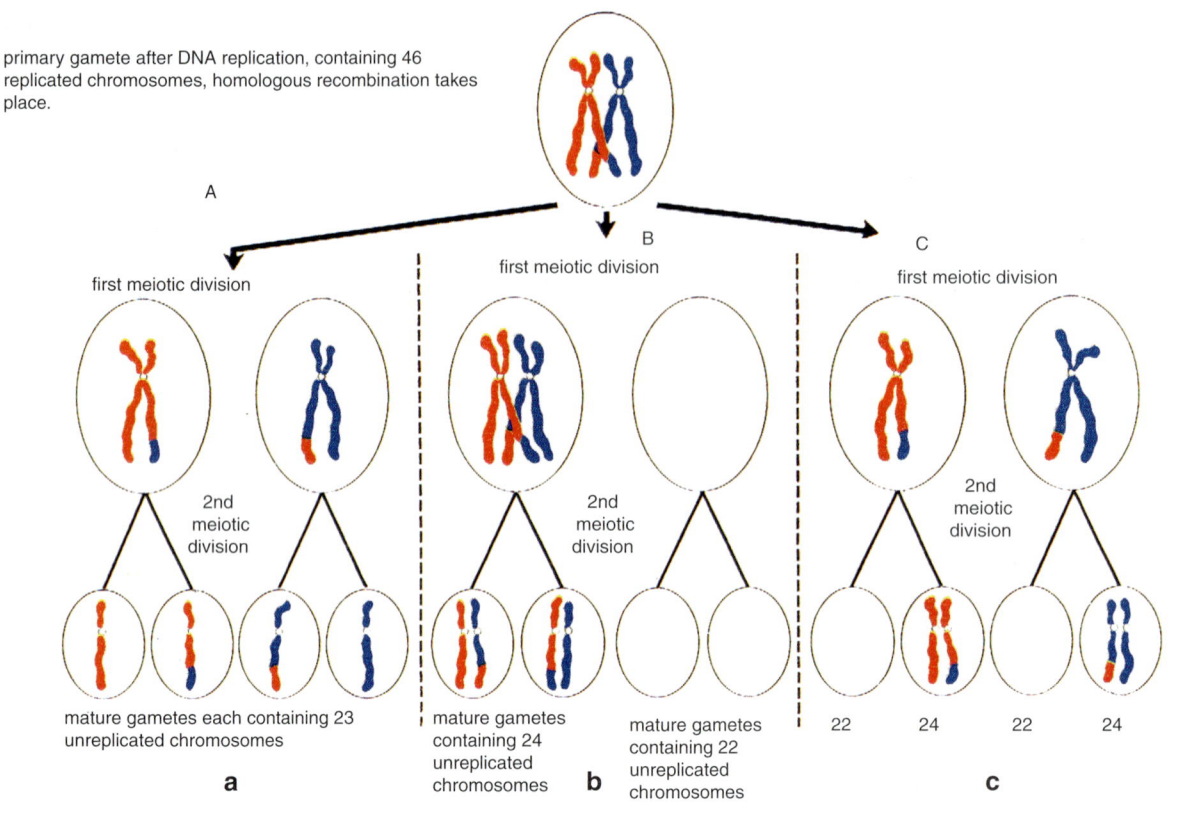

Fig. 1.2 Meiosis: (**a**) demonstrates the normal stages of meiosis(after division one the cell contains 23 replicated chromosomes-22 autosomes and 1 sex-chromosome). (**b**) Demonstrates nondisjunction in meiosis 1(the most frequent cause of for instance Down syndrome). (**c**) Demonstrates nondisjunction in meiosis2. Appreciate the effect of recombination in the mature gamete, in this way each grandparent contributes to both copies of each autosome of his/her grandchild. Adapted from Langman Inleiding tot de embryologie Bohn Scheltema & Holkema 9ᵉ revised edition druk 1982

translated into protein. Translation is also a very efficient process since many ribosomes (polysome) translate a single mRNA. Both transcription and translation are initiated and terminated by specific sequences in the genetic code.

Mutations are changes in the DNA that can lead to disease. Mutations can occur due to chemical modifications such as smoking, UV light (sun), radioactivity, and chemical instability of the DNA. Mutations can also be the result of replication errors. The effect can be a somatic nonhereditary disease (e.g., cancer) or a hereditary disease (e.g., hereditary cancer or for instance a hereditary cardiac condition), provided the mutation is transmitted to the next generation because the mutation is present in or has newly occurred in the gametes. When a single nucleotide is changed (called a *point mutation*) the effect may be alteration of the amino acid code. Besides, such a mutation can affect a regulatory sequence or lead to a premature stop codon. Also one or more nucleotides can be deleted (called a *deletion*) or added (*insertion*). Deletions and insertions of triplets of nucleotides within the reading frame will delete or add the associated amino acids. As nucleotides are read in triplets (called a codon, each codon coding for a specific amino acid) any deletion or insertion that cannot be divided by three will lead to a disturbance of the reading frame of the genetic code (and will therefore cause a frameshift mutation). This leads to a completely different protein (or more often to loss of the protein), as sooner or later a premature stop codon will be introduced. Mutations can cause a loss of function or gain of (an abnormal) function of a protein. A 50% reduced dosage of a protein may not be enough for normal function (this is called *haploinsufficiency*) and may thus cause disease. Some mutations can have a *dominant negative* effect on protein function which means that the mutation leads to an altered protein that interferes with the function of the wild type (=normal) protein, which is produced from the unmutated copy of the gene. Finally, some mutations

1 Introduction to Molecular Genetics

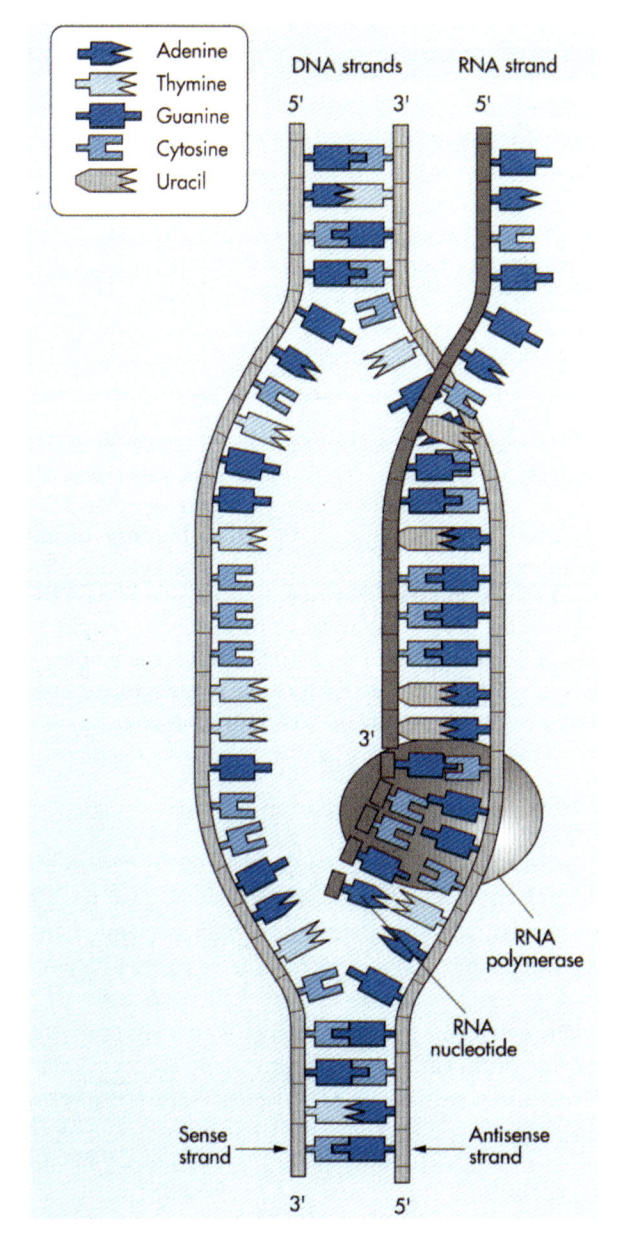

Fig. 1.3 Visualizes the transcription of the DNA code into messenger RNA. This messenger RNA contains both exons and introns. It has to be processed, and the introns (nonprotein coding parts) of the gene have to be removed (a process called splicing), before a mature messenger RNA leaves the nucleus to the cytosol, where it will be translated into protein. Reprinted with permission Jorde, Carey, Bamshad, White, Medical Genetics third edition, Mosby Elsevier 2006

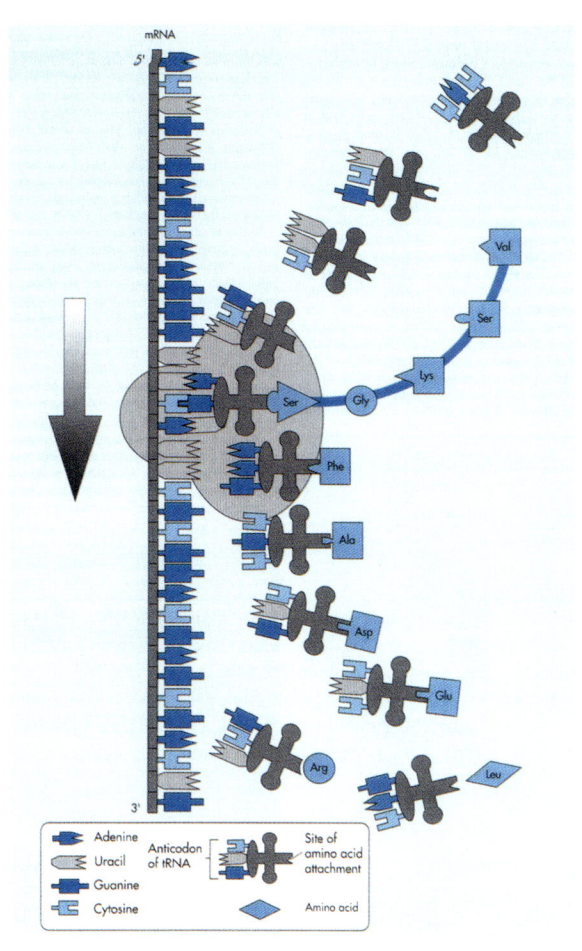

Fig. 1.4 Translation. The genetic code of the mRNA determines the order in which tRNAs are recognized and thus the order of the amino acids in the protein. The light gray oval structure depicts the ribosome, where translation takes place and on which the proteins are formed. Reprinted with permission Jorde, Carey, Bamshad, White, Medical Genetics third edition, Mosby Elsevier 2006

can be lethal leading to a nonviable embryo; such mutations are always de novo, or the transmitting parent has been rescued by the fact that the mutation is not present in all of his or her cells (a situation that is called *mosaicism*). Mutations can vary in size from one nucleotide up to complete extra chromosomes (as for instance in the case of Down's syndrome). Especially for small missense mutations only changing a single amino acid in the protein, the effect of the mutation may be difficult to predict. The location of such a mutation in the protein will to a large extent determine the effect of the mutation on protein function (e.g., is it within an important domain of the protein or is the specific amino acid residue very conserved during evolution, an indication that it may be important for normal function and that changing this residue may be not tolerated). Frameshift mutations at the extreme end of a gene often are less devastating for instance because only a small part of the protein is altered. In addition, a

Table 1.1 key to the genetic code

	T		C		A		G	
T	TTT	F	CTT	L	ATT	I	GTT	V
	TTC	L	CTC	L	ATC	I	GTC	V
	TTA	S	CTA	L	ATA	I	GTA	V
	TTG	L	CTG	L	ATG	M	GTG	V
C	TCT	S	CCT	P	ACT	T	GCT	A
	TCC	S	CCC	P	ATT	T	GCC	A
	TCA	S	CCA	P	ATA	T	GCA	A
	TCG	S	CCG	P	ACG	T	GCG	A
A	TAT	Y	CAT	H	AAT	N	GAT	D
	TAC	Y	CAC	H	AAC	N	GAC	D
	TAA	X	CAA	Q	AAA	K	GAA	E
	TAG	X	CAG	Q	AAG	K	GAG	E
G	TGT	C	CGT	R	AGT	S	GGT	G
	TGC	C	CGC	R	AGC	S	GGC	G
	TGA	X	CGA	R	AGA	R	GGA	G
	TGG	W	CGG	R	AGG	R	GGG	G

Top row depicts the first base of each codon at the DNA level (at the RNA level thymine T is replaced by uracil U
The most left column shows the second base of each codon. Amino acids are indicated by their single letter codes(three-letter codes are also often used): *A* alanine(ala), *C* cysteine(cys), *D* aspartic acid (asp), *E* glutamic acid (glu), *F* phenylalanine(phe), *G* glycine (gly), *H* histidine (his), *I* isoleucine (ile), *K* Lysine (lys), *L* Leucine(leu), *M* Methionine (met), *N* asparagine (asn), *P* proline (pro),*Q* glutamine (gln), *R* arginine (arg), *S* serine (ser), *T* threonine (thr), *V* valine (val), *W* tryptophan (try), *Y* tyrosine (tyr), *X* stop codon
As you can see in the table each amino acid can be coded by different codons

kind of quality control mechanism called *nonsense mediated messenger –RNA decay* largely prevents the expression of truncated erroneous proteins.

1.1.1.1 Genotyping in Mendelian and Non-Mendelian Disorders

Genetic diseases can be divided in *Mendelian* and *non-Mendelian* diseases.[5] The first group is caused by defects in *autosomal* or *X-chromosomal* genes that have a dominant or recessive manifestation. In case of a *dominant* disorder (Fig. 1.5), the disease will already become manifest if only one *allele* (one of both copies of a gene) is affected. In case of *recessive* segregation, both homologous alleles must be affected to develop the disease. Dominant and recessive are no absolute terms. Some carriers of dominant mutations remain healthy, which is called *non-penetrance*.[6] This can be disease, age- and sex-dependent. A *penetrance* of 90% means that 90% of the mutation carriers of a specific gene defect will be affected (at a defined age). Moreover, manifestations of genetic disease often vary among mutation carriers even within a single family. This is called *variable expression*, probably caused by other genetic and environmental factors. One has to realize that not every disease case within a single family has to have a genetic cause. Nongenetic cases mimicking genetic disease are called *phenocopies*, and this can be a problem especially in disorders with a commonly occurring environmentally induced counterpart, like for instance left ventricular hypertrophy due to untreated hypertension in a family with hypertrophic cardiomyopathy caused by a sarcomere gene mutation.

Sometimes mutations in the same gene can lead to different disorders, as do for instance mutations in the SCN5A gene involved in Long QT syndrome type 3, Brugada syndrome, or isolated conduction defects.[7–9] Some specific neuromuscular disorders, like myotonic dystrophy (in which about 30% of patients die from cardiac arrythmias), become more severe in subsequent generations. This phenomenon is called *anticipation* and the molecular basis is an unstable stretch of small DNA repeats, which in most cases increase in size when transmitted to the next generation.[5] The size of the unstable DNA fragment is in general related to disease severity and age of onset.

1 Introduction to Molecular Genetics

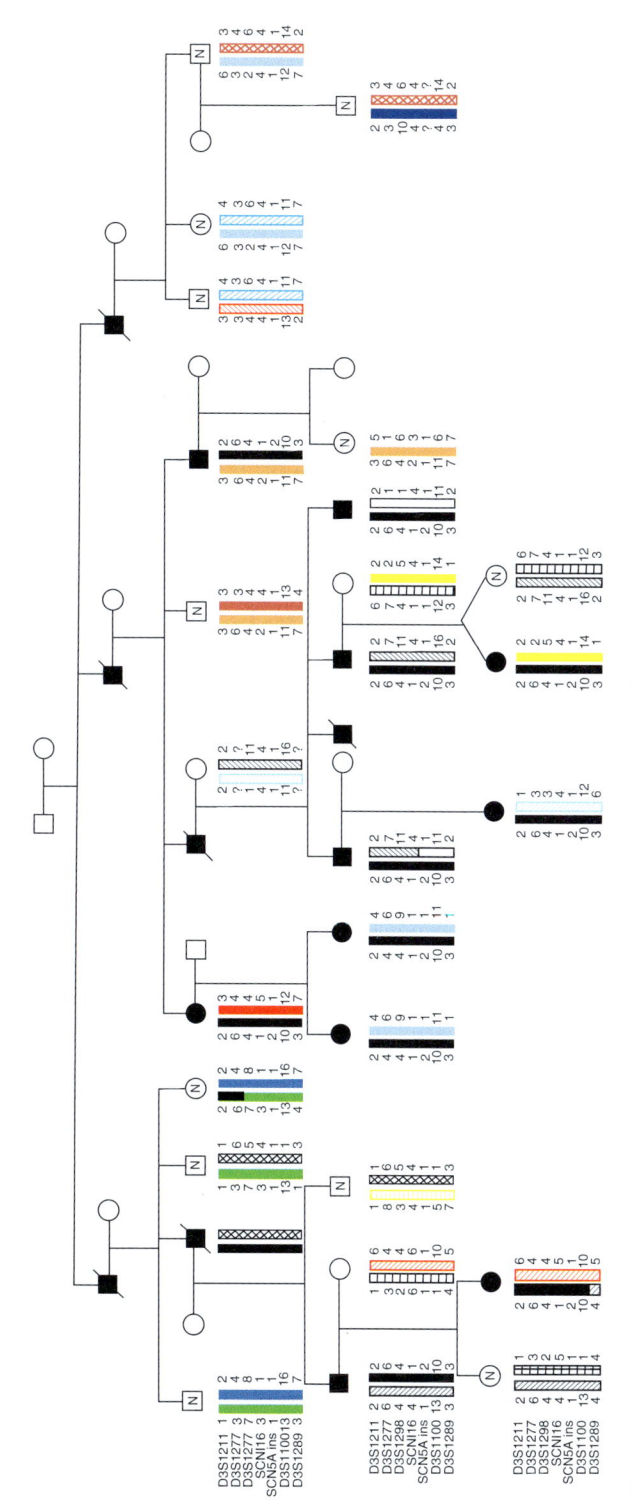

Fig. 1.5 Linkage analyses in a family presenting with "Brugada" syndrome and LQT3. Affected persons are indicated with filled symbols. *N* not affected. Deceased persons are indicated with a crossed line. The disease is showing evident autosomal dominant transmission with an average half of the offspring of affected individuals inheriting the disease. There is male to male transmission, thus excluding X-linked dominant disease. The haplotype (haplotype = a stretch of DNA on a single chromosome as defined by the alleles of a set of linked genetic markers that are usually inherited together, thus without recombination) indicated with a black bar segregates with the disease. Marker alleles are numbered for each tested marker on chromosome 3

Carriers of recessive disorders are usually without symptoms, thus including most females in X-linked recessive disorders. However, in some disorders it is possible to identify them by biochemical testing, reduced enzymatic activity, or by expression of minimal signs (for instance as a marginally increased QT-interval in carriers of the recessive Jervell-Lange-Nielsen syndrome mutations [Jervell-Lange-Nielsen syndrome is a combination of long QT syndromes and deafness]). Finally, some disorders, like inherited cancers can be dominant at the family level and recessive at the gene level. In that case a mutation in a specific gene segregates in the family but a *second hit*, destroying the other wild-type (=normal) allele of the gene, has to occur in order for disease to arise. If the chance of this second hit occurring is very high, than a dominant segregation pattern will be observed. For these reasons terms like dominant and recessive segregation should be carefully defined. This also applies to X-chromosomal diseases, which are predominantly expressed in males, because males only have one X-chromosome, whereas females have two. Dominant and recessive are terms describing genetic diseases and not genes or mutations themselves. For mutations in a specific gene, both dominant and recessive disease can occur, either leading to the same or a different phenotype. Single gene mutations in the KCNQ1 gene lead to dominant Romano-Ward syndrome, whereas mutations affecting both copies of the same gene cause the recessive Jervell-Lange-Nielsen syndrome.[10] The latter may also result from mutations in the KCNE1 gene, demonstrating the *genetic heterogeneity* of this disorder. Often the nomenclature and classification of genetic disorders have to be adjusted at the moment the genes causing the disorders have been identified.

In addition to Mendelian inheritance, other segregation patterns exist for monogenic disorders. Some diseases show a *maternal segregation pattern* and are caused by defects in the *mitochondrial DNA*, which is, for example, the case for a number of isolated or rare syndromic dilated or hypertrophic cardiomyopathies.[11] On rare occasions gene defects only become manifest when transmitted by either the mother or the father. This phenomenon is called *genomic imprinting*[12] and results from the fact that for a limited number of genes only the paternal or maternal copy is expressed. A defect in such an imprinted gene will only be detected, if the mutation is present in the active (expressed) allele of the gene. When looking at phenomena like reduced penetrance and variable expression, it is immediately clear that clinical manifestation of monogenic disorders is not constant within families. No clear discrimination between *monogenic disorders* with one major gene defect and modifying factors, and complex genetic disorders with more than one minor gene defects and contributing environmental factors, exists. *Polygenic diseases* are caused by more than one gene, which often contribute in a quantitative manner to the clinical manifestations and are considered risk factors, as the manifestations may vary, depending on the genetic background. Complex genetic disorders are polygenic disorders, which can often be influenced by the environment to a relatively large extent. These disorders are much more frequent than monogenic disorders and the role of the environment is often stronger, which means that a disease-state does not always have a major genetic cause.[13] The underlying genetic risk factors can be identified in large-scale population studies in which common genetic variants (frequency > 1%) are being tested across the entire genome (*whole genome association (WGA) studies*) for being more or less frequent in the patient population compared to a matched control population. These variants can be the functional causative factor or they can be merely markers that such a contributing factor is located in their immediate vicinity. The success of these studies largely depends on the size and composition of the population and the number of genetic factors involved. If the size of the population is too small, the number of genetic factors too large, and the effect of the genetic factor too small, then it is unlikely that any meaningful association can be found. However, many loci and factors are currently being identified, which usually have only a small to moderate effect on the manifestation of the complex genetic disorder. This implies that these factors may allow for the identification of the underlying pathophysiological processes, but by themselves only have a limited value in predicting the risk of developing the disorder under investigation in a specific individual. Apart from these common variants, it is anticipated that many different, rare variants in candidate genes can also explain part of the genetic risk of complex diseases. These cannot be assessed by WGA, but possibly by sequencing all or large numbers of candidate genes or complete exomes (all exons in the genome), using massive parallel or next-generation sequencing approaches. These rare variants contributing to complex genetic diseases will be found in the near future.

1.1.1.2 Genetic Heterogeneity of Monogenic Disorders

Diseases like familial hypertrophic cardiomyopathy or congenital Long QT syndrome are genetically heterogeneous monogenic disorders. The diseases are caused by single gene defects in patients, but defects in a large number of different genes can lead to the same or very similar disease and the severity of clinical manifestation may depend on additional mutations in other genes and/or environmental factors. Familial hypertrophic cardiomyopathy is mainly a disease of the cardiac sarcomere and many mutations in many sarcomeric genes have been detected in patients.[4,14] At this point, a complete screening of all sarcomere genes would be technically possible, but very difficult to implement in a routine diagnostic setting. It is also not necessary for those families in which hypertrophy can be clearly demonstrated and progression of disease is relatively benign. However, testing can be essential in those families in which sudden cardiac death (SCD) occurs, especially in those cases where hypertrophy is mild, and individuals at risk for dangerous arrhythmias may not be so easy to detect by nongenetic means. At this moment candidate genes are routinely analyzed sequentially, which is laborious and time-consuming, but the pace of technical developments in this field will allow for more optimal diagnostic strategies in the near future, making analysis of a large number of involved genes in parallel feasible. Such protocols will become available to diagnostic laboratories in the years to come, and can be considered a first step toward the personal genome (see also section alternative technologies).

When no good phenotypic selection criteria for which genes to test first exist, the frequency with which causative mutations in the different genes are found can be used to establish *a mutation screening strategy*, starting with the most commonly mutated gene. For familial hypertrophic cardiomyopathy, one generally starts with the beta-MHC gene (MYH7) (±35% of the mutations), followed by the myosin binding protein-gene (MYBPC3) (±15%) and the Troponin T gene (TNNT2) (±3–5%). A bias exists in these percentages as some genes have been more frequently screened in patients than others. Moreover, these percentages can be different in different populations. For instance in the Netherlands a single mutation in the myosin binding protein gene accounts for almost 25% of hypertrophic cardiomyopathy families. It should be stressed that the chances of rapidly identifying a mutation in a patient increases if many affected family members are available for testing, and genetic markers can be used to identify the locus involved (many candidate loci can then be easily excluded using polymorphic genetic markers). In some families it is also possible to perform DNA-diagnostics, if the underlying genetic defect is not located in any of the known genes. A prerequisite is that the size of the family is such that segregation of the disorder with a specific locus can be proven with high likelihood (as a general rule of thumb one would need >10 patients with unambiguous genetic status consenting to participate in case of an autosomal dominant disorder). It should be noted that careful clinical examination is a prerequisite as any false diagnosis will lead to a wrong conclusion with respect to the gene or genetic locus involved. Also carriers of the mutation that are not yet expressing the disease as a result of their young age may confuse such a genetic undertaking. The combination of clinical and genetic data can not only be of use in care for individual patients and their families, but is also important for extending the knowledge on the disease (frequency of non-penetrant carriers, number of de novo *(=new or spontaneous)* mutations), and for instance for the development of clinical selection criteria for the gene involved, like in the congenital Long QT syndrome, and for establishing a more exact genotype-based prognosis. *Genotype–phenotype correlations* are important for counseling but as long as not all contributing factors to the phenotype are known such knowledge should be used with great care in individual cases. Congenital Long QT syndrome is most often caused by mutations in cardiac potassium or sodium channel genes.[3,15] Based on genotype–phenotype correlations, it is possible to select the most likely gene involved using clinical information such as the trigger of cardiac events (physical stress and diving as a trigger for arrythmia point to the KCNQ1 gene and acoustic stimuli to KCNH2 gene). The characteristic repolarization pattern on the ECG also gives information on which gene is most likely to be involved.[16] The identification of the causative mutation has to confirm this. This is specifically important, because for these disorders gene-specific prophylactic therapies exists and can save lives, if carriers are identified early. It can be expected that gene-specific therapy may evolve to mutation specific therapy in the future. Genetic testing can provide for these disorders a clear contribution to patient management, either by

analyses are necessary but linkage of a marker with a disease locus is considered to be proven if the chance of linkage as compared to chance segregation is 1,000:1. Linkage is excluded at odds of chance to linkage of 100:1. These likelihoods are given in logarithms or LOD scores (log of the odds). A lod score of 3 means a chance of 1,000 to 1 odds for linkage as compared to chance. Lod scores of 3 can only be achieved in large families (roughly more than ten informative meioses, i.e., heterozygotes for whom the segregation of the homologous chromosomes in relation to the disease can be determined). This means that the genetic status of at least 11 individuals should be known without doubt. Smaller families are often studied to give an indication of the gene involved or exclude genes not involved, but conclusive DNA-diagnosis by linkage is not possible in such smaller families. Figure 1.5 gives an example of a linkage study. Polymorphic markers are numbered according to their length. The "black bar" combination of polymorphic markers (haplotype) is segregating with the disease locus, a combination of LQT3 and "Brugada" syndrome segregating within a single family. These markers on chromosome 3 segregate with the disease giving a lod score of 6.5. Analysis of the relevant gene (a sodium channel SCN5A) revealed a three nucleotide insertion (TGA) at position 5537, causing the disease.

1.1.1.5 Direct DNA-Diagnostics: Screening for Small Mutations

Sanger Sequencing

The sequence of a particular DNA fragment can be determined by using a primer sequence (short oligonucleotide complementary to a small DNA fragment adjacent to the DNA fragment to be sequenced).[5] Using this primer as a starting point, the sequence of a single-stranded DNA fragment can be artificially copied by an enzyme called DNA-polymerase and the addition of the chemical components of DNA (the nucleotides adenine, cytosine, thymine, and guanine). The newly synthesized strands are chemically modified by the addition of small amounts of dideoxynucleotides that are labeled with fluorochromes or radioactive isotopes. The dideoxynucleotides represent all four existing nucleotides (A, T, C, or G) that are added in four distinct reactions or are labeled differently so they can be recognized as such. Incorporation of a dideoxynucleotide stops the synthesis of DNA. This occurs randomly, but with a predetermined frequency. Addition of modified adenine for instance reveals the position of all adenines in the genetic code by creating partially synthesized DNA fragments that stop at all positions of adinine. By doing so with all four modified nucleotides and by size separating fragments with a resolution of one nucleotide by high voltage electrophoresis, the genetic code can be read simply by determining the length of the all fragments. The nucleotide present in a particular fragment comes before that of a longer fragment. Figure 1.8 shows an example of such a sequence reaction. The procedure is nowadays fully automated and the dideoxynucleotides are labeled with fluorochromes of four different colors. For capillary sequencers, the separated fragments are detected with laser technology (colored peaks). By comparing the known sequence with the sequence in patients, a mutation can be found. Although the sequencing is fully automated, the quality control of the sequence and comparison to the wild type (normal) sequence is not completely automated; therefore, sequencing an entire gene for mutations is still quite laborious. In the case shown (Fig. 1.8), a DNA mutation was found leading to a change from amino acid glutamine to a stop at position 356. The result is an incomplete nonfunctional ion channel protein (KCNQ1) that causes the long QT syndrome type 1 (Romano-Ward syndrome) in this patient.

Gene Scanning Methods: Alternatives to Sequencing Entire Candidate Genes

As an alternative to the laborious sequencing of entire genes many clinical laboratories use a *gene scanning method* to quickly go through a gene for possible aberrations; subsequently they only sequence those parts of the gene that demonstrated an abnormal pattern using the scanning technique. Many different techniques exist, each with their pros and cons, and differing sensitivity and specificity. Most of these techniques are either based on the fact that most mutations will cause a conformational change of the single-stranded DNA and therefore altered characteristics at gel electrophoresis (SSCP single-stranded conformation polymorphism), or on the difference in stability between heteroduplexes and homoduplexes (DGGE denaturing

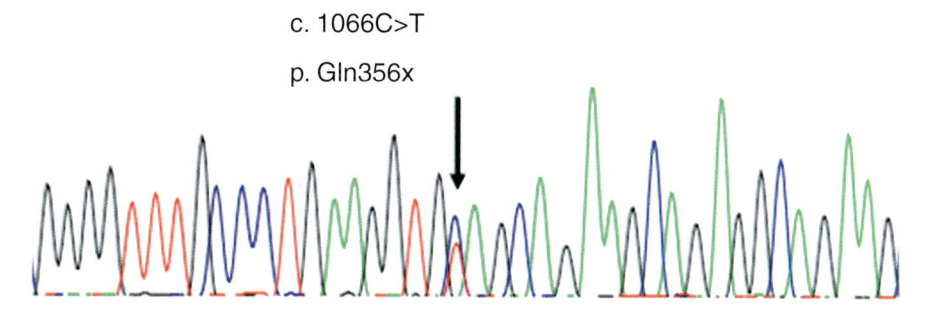

GGGGTTTGCCCTGAAGGTG ^C_T AGCAGAAGCAGAGGCAGAAG

c. 1066C>T

p. Gln356x

Fig. 1.8 Sequence analysis. A DNA heterozygous C to T mutation was found at position 1066, leading to a change from a codon (CAG) for amino acid glutamine at position 356 in the protein to a stop codon (TAG) at position 356. The result is an incomplete nonfunctional ion channel protein (KCNQ1) that causes the long QT syndrome type 1 (Romano-Ward syndrome) in this patient

gradient gel electrophoresis, DHPLC denaturing high-performance liquid chromatography, HRMC high resolution melting curve analysis).

These latter methods detect DNA consisting of mismatched mutant and wild-type DNA strands.[5] These so-called heteroduplexes are less stable than homoduplexes (consisting of either two mutated or two wild type strands) and can be, respectively, detected by an altered melting profile, using denaturing agents and HPLC, saturating dyes and heating or denaturing agents and electrophoresis. The methods are relatively easy to perform and detect up to 100% of mutations. These approaches are illustrated by an example of DGGE (Fig. 1.9). In this technique, the DNA is denatured (made single stranded) and then allowed to reanneal (become double stranded again), in case of a heterozygous mutation different types of molecules will be formed: homoduplexes when two wild type (normal strands) reanneal or when two mutated strands reanneal, and heteroduplexes when a wild type strand reanneals with a mutated strand.

The newly formed double-stranded DNA molecules demonstrate changes in electrophoretic mobility on a gel with a gradient of denaturing agents. Because of the mutation, DNA fragments will melt at different positions in the denaturing gel and thus get stuck at different positions (Fig. 1.9). Formally, these techniques have been designed to detect heterozygous mutations as heteroduplexes are easiest to identify, but often homoduplexes consisting of two mutated strands show a pattern that can also be discerned from the wild-type homoduplex (see Fig. 1.9), thus allowing for the detection of homozygous mutations. Figure 1.9b. shows a N543H heterozygous mutation in the LDL receptor gene causing familial hypercholesterolemia compared to control DNA. The aberrant homo- and heteroduplexes are clearly visible.

An alternative technology to DGGE is denaturing high-performance liquid chromatography (DHPLC).[19] In contrast to DGGE, DHPLC uses chromatography instead of electrophoresis to detect aberrant DNA fragments such as insertions, deletions, and also single base substitutions. The sensitivity of the technology is comparable to sequencing, although a sequence reaction will still be needed to determine the actual sequence change. The detection level of DHPLC is even higher than in sequencing (1% of a mutation in a pool of wild type (=normal) molecules is detectable[20]). This may be of importance when looking for mosaic mutations that are present in only a fraction of the cells.

New Technologies Emerging on the Horizon

Many alternatives techniques exist or emerge that fall outside the scope of this book. Most promising are those techniques that allow screening of multiple genes in parallel for each patient, allowing a more rapid identification of the underlying gene defect, especially for those genes in which mutations are rarely found, and yielding a more complete picture of all mutations and variants involved. This is of importance for both

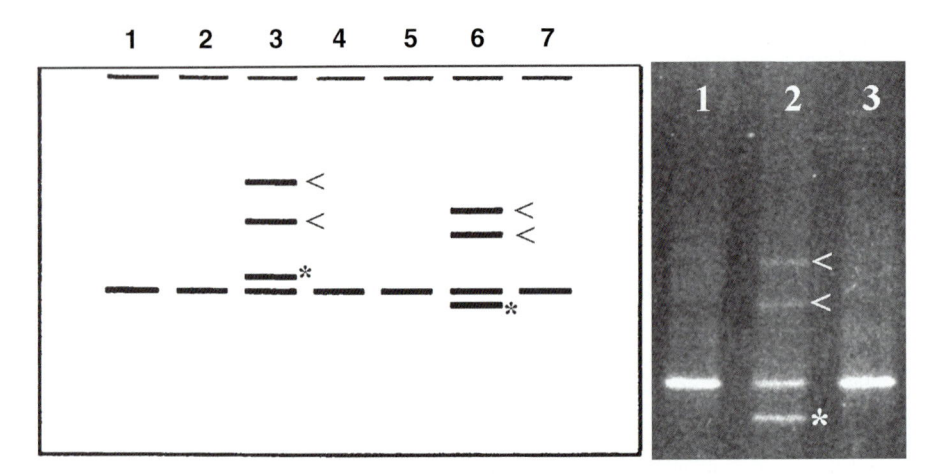

Fig. 1.9 The left-hand part shows a theoretical DGGE analysis. Mutants are seen in lane 3 and 6, revealing homo- and heteroduplexes (heteroduplexes indicated with < and mutated homoduplexes indicated with *). The heteroduplexes are less stable as a result of the mismatch so they denature more easily and get stuck higher in the gel at a lower concentration of denaturing agents than the homoduplexes. Besides heteroduplexes, both gels show an abnormal homoduplex different from the homoduplex present in the loaded controls, corresponding to the homoduplex formed by two mutated strands. The right-hand part shows a N543H heterozygous mutation in the LDL receptor gene (lane 2) causing familial hypercholesterolemia compared to control DNA (lane 1 and 3). The aberrant homo- and heteroduplexes are clearly visible on this DGGE gel. (*Reprinted with permission Jorde, Carey, Bamshad, White, Medical Genetics third edition, Mosby Elsevier 2006)

diagnostics and prognostics. Mass spectrometry is an example that can be used to quickly screen large numbers of samples for known mutations. By using microarray or DNA-microchip approaches the entire sequence of multiple genes can be rebuilt by overlapping oligonucleotides and the hybridization pattern of the DNA of the patient resolves its sequence rapidly.[21] Next generation sequencing (also called massive parallel sequencing) will enable us to screen multiple genes in multiple patients in one short experiment (400–600 Mb per run). Both approaches still require prior amplification or capture of the candidate genes involved for which a variety of enrichment strategies are available. So far these non-PCR-based enrichment strategies have not reach the required coverage for diagnostic applications. However, developments in sequencing technology proceed in a very rapid pace and it can be expected that within 3–5 years a complete human genome can be determined within a week for less than $1,000 without any prior handling or enrichment. However, in general practice, many of these techniques are not yet or only just available or applicable. They have been successful in research, but they have to be validated for diagnostic purposes. These approaches will yield massive numbers of data and variants in patients, which have to be filtered and broken down to the pathological variants, which are responsible for or involved in the disease manifestations. It will require a major effort from bioinformatics and functional approaches to cope with and interpret these huge amounts of sequence variants. ICT solutions are required to deal with the immense data sets generated by these technologies.

In case of quantitative differences, like deletions and duplications, real-time PCR or MLPA has been proven very valuable since these techniques can provide information on gene (or chromosome) copy number within hours. Sometimes, analyses at the RNA or protein level are necessary to predict the result of a DNA chance or as alternative in cases where no DNA alteration can be found. A major breakthrough is also expected in the field of copy number variations from the microarray or DNA-microchip and next generation sequencing approaches mentioned above. These technologies can be used to screen the complete sequence of multiple genes for small mutations or to analyse large numbers of clones (cDNA or genomic clones) for major rearrangements. In a different application oligonucleotide clones are attached on the array and the expression level of those genes can be determined in RNA samples from normal or affected tissue.[22] Specific expression profiles may exist for specific gene defects or pathogenic states, allowing a refined characterization of the underlying genetic cause or disease state in the patient. As the microarray and next generation sequencing technology approach can also be used to determine large numbers of risk factors, it is clear that the time where a more complete knowledge of genetic defects and predisposition of patients emerge is nearby. It is, however, also clear that interpreting this information and explaining it to patients will be the major bottleneck in the years to come. It will be important to focus on relevant information and established knowledge in health care. Scientists, clinicians, and society will have to collaborate closely to make the genomic revolution a success, and that will be of benefit to mankind.

References

1. International Human Genome Consortium. Initial sequencing and analysis of the human genome. *Nature*. 2001;409:860-921.
2. Venter JC, Adams MD, Myers EM, et al. The sequence of the human genome. *Science*. 2001;291:1304-1351.
3. Van Langen IM, Hofman N, Wilde AAM. Family and population strategies for screening and counselling of inherited cardiac arrhythmias. *Ann Med*. 2004;36(Suppl 1):116-124.
4. Morita H, Seidman J, Seidman CE. Genetic causes of human heart failure. *J Clin Invest*. 2005;115:518-526.
5. Strachan T, Read AP. *Human Molecular Genetics*. 2nd ed. Oxford: Bios Scientific Publishers; 1999.
6. Priori SG, Napolitano C, Schwartz PJ. Low penetrance in the Long-QT syndrome. Clinical impact. *Circulation*. 1999;99:529-533.
7. Priori SG. Long QT and Brugada syndromes: from genetics to clinical management. *J Cardiovasc Electrophysiol*. 2000;11:1174-1178.
8. Bezzina C, Veldkamp MW, Van den Berg MP, et al. A single Na$^+$ channel mutation causing both long-QT and Brugada syndromes. *Circ Res*. 1999;85:1206-1213.
9. Bezzina CR, Rook MB, Wilde AAM. Cardiac sodium channel and inherited arrhythmia syndromes. *Cardiovasc Res*. 2001;49:257-271.
10. Neyroud N, Tesson F, Denjoy I, et al. A novel mutation in the potassium channel gene KVLQT1 causes the Jervell and Lange-Nielsen cardioauditory syndrome. *Nat Genet*. 1997;15:186-189.
11. Towbin JA, Lipshultz SE. Genetics of neonatal cardiomyopathy. *Curr Opin Cardiol*. 1999;14:250-262.
12. Reik W, Walter J. Genomic imprinting: parental influence on the genome. *Nat Rev Genet*. 2001;2:21-32.
13. Haines Jl, Pericak-Vance MA, eds. *Approaches to Gene Mapping in Complex Human Diseases*. New York: Wiley-Liss; 1998.
14. Bonne G, Carrier L, Richard P, Hainque B, Schwartz K. Familial hypertrophic cardiomyopathy. From mutations to functional defects. *Circ Res*. 1998;83:580-593.
15. Nicol RL, Frey N, Olson EN. From the sarcomere to the nucleus: role of genetics and signaling in structural heart disease. *Annu Rev Genomics Hum Genet*. 2000;01:179-223.
16. Wilde AAM, Roden DM. Predicting the long-QT genotype from clinical data. From sense to science. *Circulation*. 2000;102:2796-2798.
17. Hawkins JR. *Finding Mutations, the Basics*. Oxford: Oxford University Press; 1997.
18. DeCoo IFM, Gussinklo T, Arts PJW, van Oost BA, Smeets HJM. A PCR test for progressive external ophthalmoplegia and Kearns-Sayre syndrome on DNA from blood samples. *J Neur Sci*. 1997;149:37-40.
19. Underhill PA, Jin L, Lin AA, et al. Detection of numerous Y chromosome biallelic polymorphisms by denaturing high-performance liquid chromatography. *Genome Res*. 1997;10:996-1005.
20. van den Bosch BJC, de Coo RFM, Scholte HR, et al. Mutation analysis of the entire mitochondrial genome using denaturing high performance liquid chromatography. *Nucl Acids Res*. 2000;28:e89-96.
21. Lipshutz RJ, Fodor SPA, Gingeras TR, Lockhart DJ. High density synthetic oligonucleotide arrays. *Nat Genet*. 1992;1(Suppl):20-25.
22. Young RA. Biomedical discovery with DNA arrays. *Cell*. 2000;102:9-15.

Clinical Genetics

J.J. van der Smagt, Peter J. van Tintelen, and Carlo L.M. Marcelis

2.1 Introduction

Clinical geneticists are medical doctors that combine general knowledge of medicine with specific expertise in genetics, genetic diagnosis, and genetic aspects of disease. They are specifically trained in communicating the implications of genetic information and genetic disease to patients and their families. In addition, they are trained in syndrome diagnosis and clinical dysmorphology.

In some countries, clinical genetics has evolved as a recognized medical specialty, whereas in other countries genetic services are provided by traditional specialists with specific additional training in genetics. Usually, only university hospitals and sometimes large regional hospitals have departments of clinical genetics. This centralized approach may limit access to genetic services, but facilitates the availability of staff and resources that are necessary for conducting family investigations, that are usually beyond the scope of the individual medical specialist. By working on a consultant basis in different hospitals and by employing trained genetic nurses for specific tasks, clinical geneticists try to optimize access to genetic services.

With the very rapid advances in genetic knowledge, and more specifically genetic knowledge of human disease, many physicians will not find the time to keep up with the pace. Even though new generations of doctors are much better trained, both in genetics and in quickly acquiring adequate information from (internet) databases, cooperation between "organ specialists" and "genetic specialists" seems to be the best option for the near future. Importantly, many genetic diseases are relatively rare, so that individual specialists will only encounter patients with a specific genetic disease on an occasional basis. This makes it difficult for them to obtain sufficient experience in providing adequate genetic information, in addressing the genetic questions of both patients and their families, and eventually in interpreting the often complex genetic test results.

Over the last one and a half decade, genetics has become increasingly important to the field of cardiology.[1,2] There is increasing awareness that some cardiac disorders occur in families and that important genetic factors play a role in disease causation. This holds true for not only rare monogenic disorders such as different types of cardiomyopathy, congenital long QT syndromes, and catecholaminergic polymorphic ventricular tachycardia, but also for more common complex disorders such as cardiovascular disease, hypertension, and diabetes. In the latter group of disorders many different additive genetic and environmental contributions, each of relatively small effect size, are hypothesized to be causing the disease. Important progress has been made in understanding the molecular background predisposing to different types of cardiac disease.

Meanwhile, in clinical genetic practice, focus has shifted from primarily reproductive issues (parents wanting to know the risk of recurrence after the birth of a child with a mental handicap or serious congenital abnormality, for example, a congenital heart defect) to include the assessment of risk of genetic disease, occurring later in life, in individuals with a positive family history. This started in neurology with individuals at risk for mostly untreatable neurodegenerative disease, like Huntington's chorea, wanting to know their genetic status in order to make future plans. Subsequently genetic diagnosis entered the field of oncology, where

J.J. van der Smagt (✉)
University Medical Centre Utrecht,
Heidelberglaan 100,
3584 CX Utrecht, The Netherlands
e-mail: j.j.vandersmagt@umcutrecht.nl

it has become an increasingly important tool in identifying individuals at high risk of getting cancer. Of course, in the field of oncology genetic testing has important medical implications, as individuals at risk may opt for increased cancer surveillance and preventive treatment strategiesmay be devised, based on genetic information.

Cardiology is a third discipline of medicine where large-scale so-called presymptomatic testing of healthy at risk individuals has become available for some of the primary electric heart diseases and cardiomyopathies. An important driving force behind this development is the "sudden cardiac death phenotype," motivating both families and their treating cardiologists to undertake steps to identify individuals at increased risk of life-threatening arrhythmias at an early stage. Although for most disease entities family studies have not yet actually been proven to be beneficial, identifying those individuals at risk seems a logical first step in the development of preventive strategies. However, genetics of cardiac disease may be complicated, for example, by *genetic heterogeneity* (different genetic causes result in clinically identical disease) and the fact that test results may be difficult to interpret. Cooperation between cardiologists and clinical geneticists is, therefore, of great importance.

In this chapter basic concepts in genetics and important issues that have to be considered in case of genetic testing are discussed.

2.1.1 The Clinical Genetic Intake

For those cardiologists involved in caring for families with genetic cardiac disorders it is important to gain some experience in constructing pedigrees and recording family histories.

2.1.1.1 Family History

History taking will be more time-consuming than usual as, besides the regular cardiac anamnesis, detailed information on several family members has to be obtained.[3] Usually, information on three (sometimes four) generations is considered sufficient. Whenever possible, information should be collected on first degree relatives (parents, siblings, and children), second degree relatives (grandparents, uncles/aunts, and nephews/nieces), and third degree relatives (first cousins). On average, they share 50%, 25%, and 12.5% of their DNA with the index patient. Information on past generations may be sparse or even misleading as many conditions could not be correctly diagnosed in the past, whereas in contrast, younger generations will be less informative as they may not have lived long enough yet for disease symptoms to become manifest. Therefore, information on more distant relatives, like first cousins, from the same generation as the index patient may prove essential.

The reliability of the information obtained through family history taking will vary from case to case. In general, accuracy decreases with the decreasing degree of relationship. As a general rule it is wise to confirm important information by checking medical records, whenever possible. If this involves family members, their written consent to retrieve these records will be required.

While taking a family history it is important to be as specific as possible. People may leave out vital information when they do not think that it is important. Possible cardiac events should be specifically asked for, and approximate ages at which they occurred should be recorded. Of course, also the circumstances in which the event took place have to be noted. Depending on the nature of the condition under investigation it may be necessary to ask for specific events, like diving or swimming accidents in case of suspected long QT syndrome type 1. It is useful to keep in mind that syncope resulting from arrhythmias may in the past have been diagnosed as seizures or epilepsy.

If family members are under cardiac surveillance elsewhere it is prudent to record this, and if individuals are deceased as a result of a possible cardiac event always inquire whether autopsy has been performed. Information on consanguinity is often not readily volunteered, and should be specifically asked for. Depending on the nature of the disorder under investigation it may also be important to inquire about medical conditions not specifically involving the heart. For example, when investigating a family with possible autosomal dominant dilated cardiomyopathy, it would be prudent to also ask for signs of skeletal muscle disease in family members.

2.1.2 Pedigree Construction

Drawing a *pedigree* is a helpful tool in assessing any familial disorder. Presenting family history information in a pedigree allows to quickly visualize family

2 Clinical Genetics

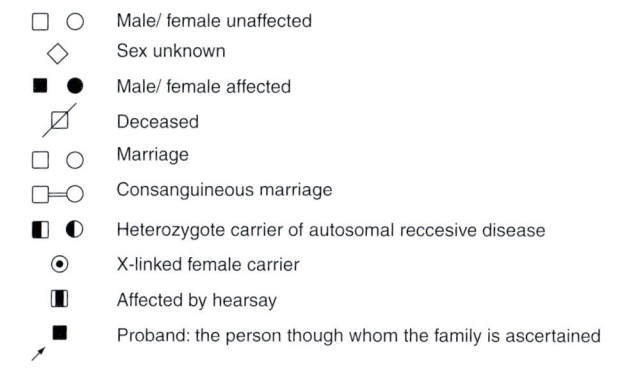

Fig. 2.1 Symbols used to denote individuals in a pedigree

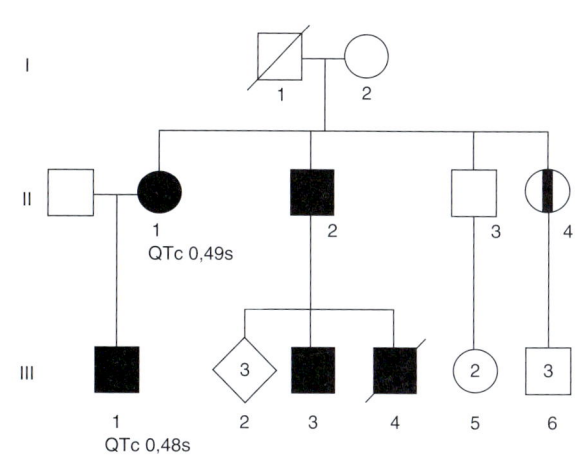

- ☐ ○ Male/ female unaffected
- ◇ Sex unknown
- ■ ● Male/ female affected
- ⌀ Deceased
- ☐―○ Marriage
- ☐═○ Consanguineous marriage
- ◨ ◐ Heterozygote carrier of autosomal reccesive disease
- ⊙ X-linked female carrier
- ▯ Affected by hearsay
- ↗■ Proband: the person though whom the family is ascertained

Fig. 2.2 Example of a small pedigree like one could draw up, while taking a family history, during consultation of a family suspected of long QT syndrome type 1. Footnotes with this pedigree could be: III-1 index patient (27-10-2000), syncope while playing soccer, spontaneous recovery, QTc 0.48 s, repolarization pattern compatible with lQT type I. I-1: no medical information, died in unilateral car accident at age 32. II-1: (02-3-1975), no symptoms, QTc 0,49 sec. II-2 (10-5-1977) known with seizures as a child. II-4: (13-6-1979)said to have fainted during exercise more than one occasion. III-4 sudden death, while swimming at age 12yrs. No other persons known with seizures ,syncope or sudden death known in the family

structure and asses the possible inheritance patterns.[4,5] In addition, a drawn pedigree will make it more easy to see which, and how many, family members are at risk for cardiac disease, and who should be contacted. The symbols commonly used for pedigree construction are represented in Fig.2.1.

Nowadays, different software packages exist for pedigree construction. These packages have the advantage that it is easier to update pedigrees and that pedigrees can be more easily added to other digital medical files. Frequently, the software also offers options that are valuable for doing genetic research.

However, the great advantage of pen and paper is that the pedigree can be constructed while taking the family history, thus ensuring that no important family members are overlooked.

A few tips and tricks (see Fig. 2.2):

- Start drawing your pedigree on a separate sheet of paper. Start with your index patient in the middle of the paper and go from there.
- Add a date to your pedigree.
- Numbering: by convention generations are denoted by a Roman numeral, whereas individuals within a generation are identified by an Arabic numeral. In this way each individual can be identified unambiguously by combining the two numbers (II-3, III-1, etc.). Additional information on a specific individual can be added in a footnote referring to this identification number.
- The most important clinical information can be directly added to the pedigree (see Fig. 2.2).
- Record approximate dates (e.g., birth year, or 5-year interval), not ages. Add age at time of death.
- Especially in case of a suspected autosomal recessive disorder, names and places of birth of all grandparents of the index patient should be recorded (usually in a footnote). Consanguinity is unlikely when paternal and maternal grandparents come from very different areas. If birthdates are also available this could facilitate genealogical studies in search of consanguinity.
- Levels of evidence: for individuals that are probably affected based on heteroanamnestic information, but whose medical records have not yet been checked, the symbol "affected by hearsay" (see Fig.2.1) can be used.
- For counseling reasons add information on both sides of the family. Unexpected additional pathology may be of importance to your patient and his or her offspring.

2.1.3 Basic Concepts in Inherited Disease

A single copy of the human genome contains over three billion base pairs and is estimated to contain

20,000–25,000 protein coding genes.[4] Genes are transcribed into messenger RNA in the nucleus. Subsequently, the noncoding parts of genes (introns) are spliced out to form the mature messenger RNA, which is in turn translated into protein in the cytosol. Proteins consist of chains of amino acids. Each amino acid is coded by one or more combinations of three nucleotides in the DNA.

Less than 3% of DNA is protein coding. The remainder codes for RNA-genes, contains regulatory sequences, or consists of DNA of undetermined function, sometimes misleadingly referred to as "junk DNA."

DNA is stored on 23 chromosome pairs (Fig. 2.3), present in the nucleus of each cell; 22 pairs of autosomes and one pair of sex chromosomes. During gametogenesis (the production of oocytes and sperm cells) meiosis takes place ensuring that only one copy of each pair is transmitted to the offspring. Since chromosomes are present in pairs, humans are diploid organisms. They have two complete copies of DNA, one copy contributed by the father and one by the mother. Therefore, each gene at each locus is present in two copies. These are usually referred to as the two alleles of that specific gene.

The exception to this rule are the sex chromosomes, as males have only one X-chromosome and one Y-chromosome, the first being inherited from the mother, the latter from the father. Thus, males have only a single copy of most X-linked genes. In addition to the nuclear DNA, small circular DNA molecules are present in the mitochondria in the cytoplasm. Many copies of this mtDNA will be present per cell. The mtDNA is exclusively inherited from the mother. Oocytes may contain up to 100,000 copies of mtDNA. MtDNA only codes for 37 genes, all involved in mitochondrial function.

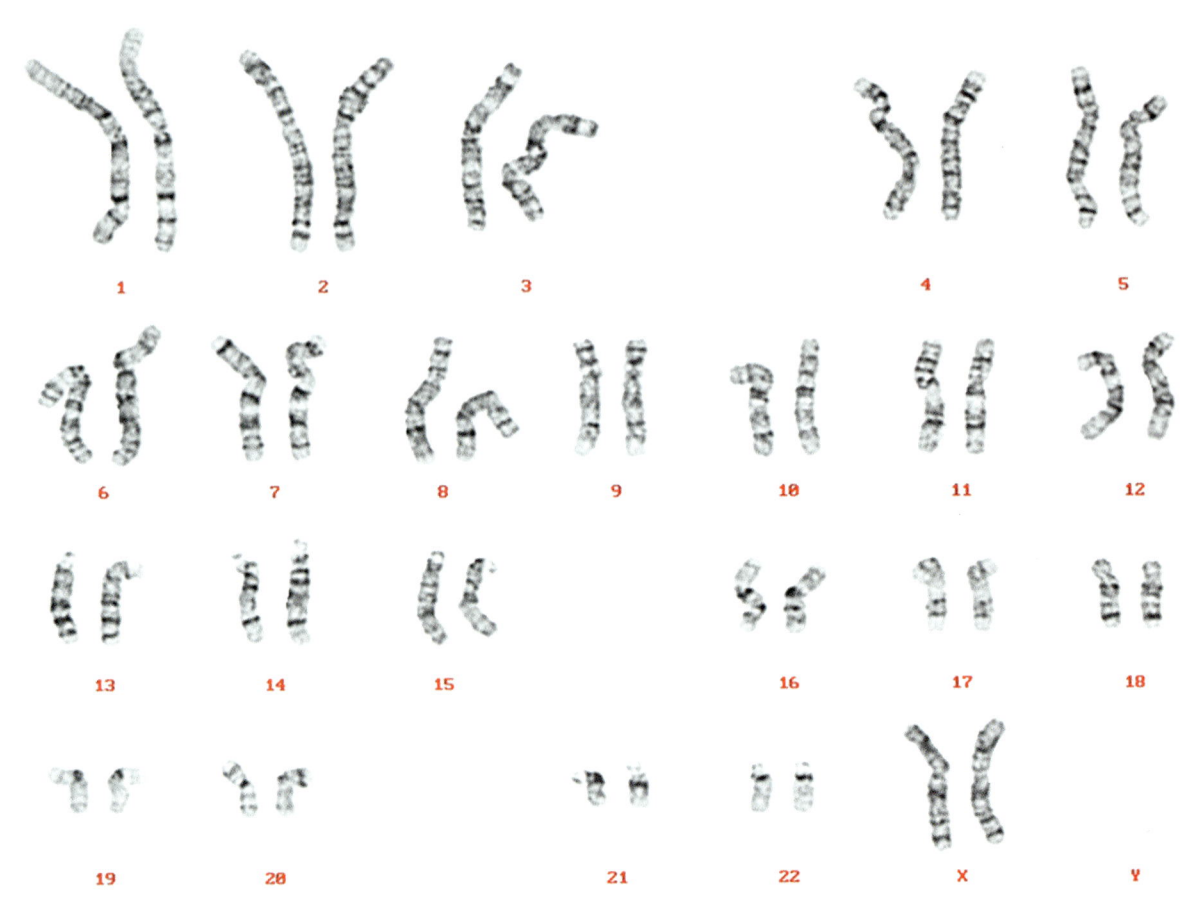

Fig. 2.3 Normal female karyogram (46, XX): the way the chromosomes are shown, when DNA is visualized through a light microscope

2.1.4 Mitosis and Meiosis

Two types of cell divisions exist: mitosis and meiosis. Mitosis ensures the equal distribution of the 46 chromosomes over both daughter cells. In order to accomplish this, first the DNA on each chromosome has to be replicated. At cell division each chromosome consists of two identical DNA-chromatids (sister chromatids), held together at a single spot: the centromere. To ensure orderly division, the DNA in the chromosome has to be neatly packaged (a process called condensation). This is when chromosomes actually become visible through a microscope. Prior to cell division a bipolar mitotic spindle develops, the completely condensed chromosomes move to the equator of the cell, the nuclear membrane dissolves and microtubular structures develop reaching from both poles of the spindle to the centromere of each chromosome. Subsequently, the centromeres divide and the sister chromatids are pulled to opposite poles of the dividing cell. Cell division results in two daughter cells, each with 46 unreplicated chromosomes and exactly the same nuclear genetic information as the original cell (Fig. 2.4).

Meiosis is a specialized cell division that is necessary to finish the process of gametogenesis. The goal is to produce gametes that contain only 23 unreplicated chromosomes. The vital steps of meiosis are outlined in Fig. 2.5. One of the hallmarks of meiosis is that both replicated chromosomes of each pair come in close apposition to each other and actually exchange genetic material before meiosis takes place. This more or less random process is called homologous *recombination*.

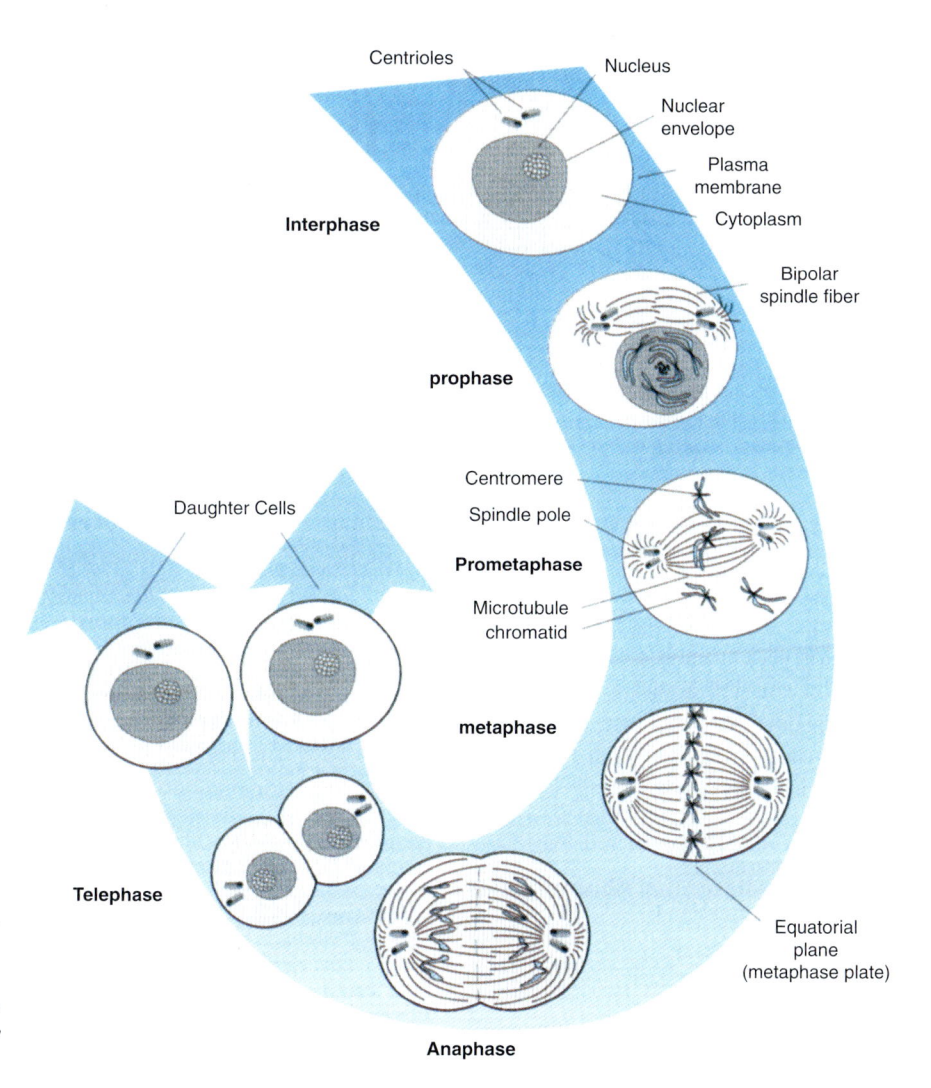

Fig. 2.4 Different stages of mitosis, leading to two daughter cells with exactly the same nuclear DnA content. (Reprinted with permission Jorde, Carey, Bamshad, White, Medical Genetics third edition, Mosby Elsevier 2006)

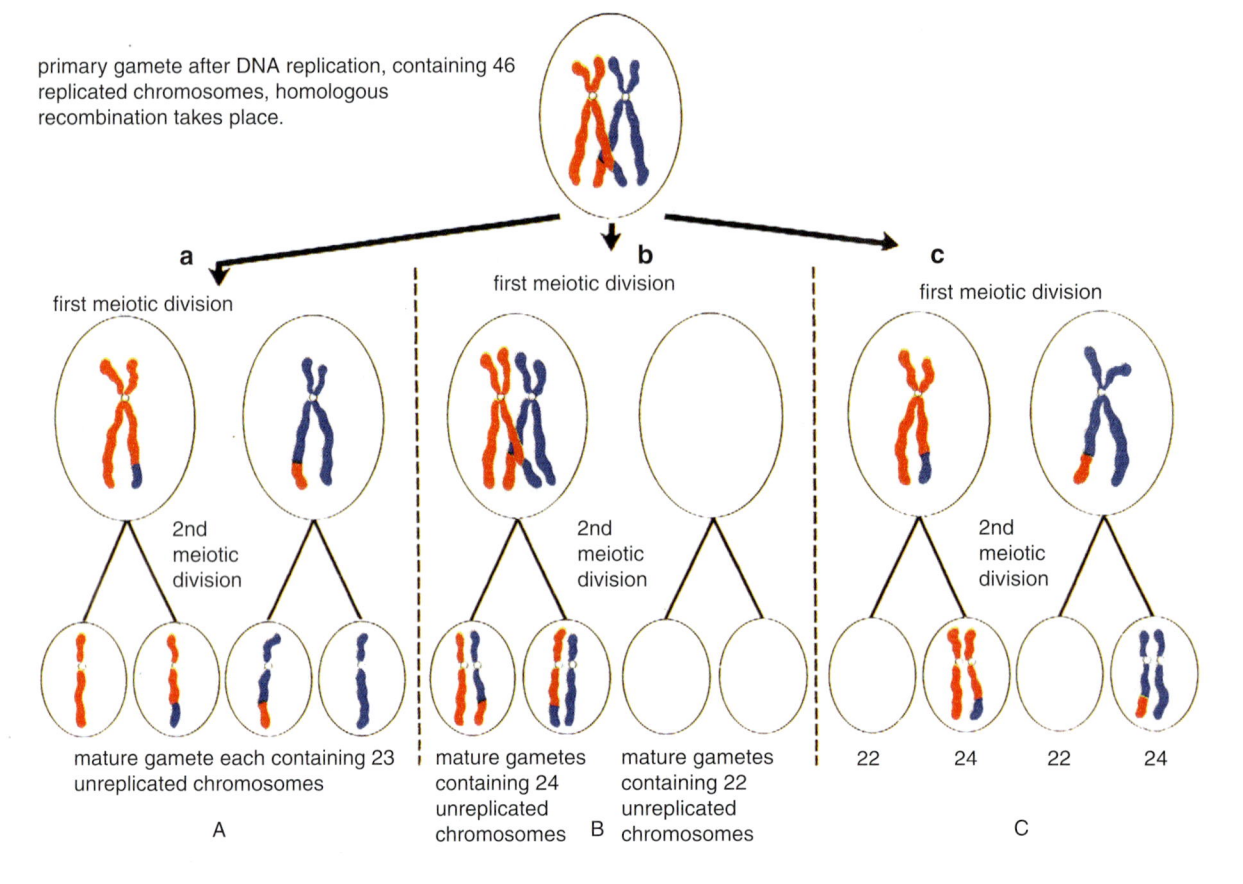

Fig. 2.5 Meiosis: (**a**) demonstrates the normal stages of meiosis (after division one the cell contains 23 replicated chromosomes – 22 autosomes and 1 sex-chromosome). (**b**) Demonstrates nondisjunction in meiosis 1 (the most frequent cause of for instance Down syndrome). (**c**) Demonstrates nondisjunction in meiosis 2. Appreciate the effect of recombination in the mature gamete, in this way each grandparent contributes to both copies of each autosome of his/her grandchild (**b**). (Adapted from Langman Inleiding tot de embryologie Bohn Scheltema & Holkema 9ᵉ herziene druk 1982)

Recombination ensures that each individual is able to produce an infinite number of genetically different offspring. Apart from ensuring genetic diversity, recombination is also necessary for proper segregation of the homologous chromosomes during meiosis I. During male meiosis the X and Y are able to function as a chromosome pair, thus ensuring proper segregation of sex chromosomes. They can recombine at the tip of their short arms.

2.1.4.1 Chromosomal Abnormalities

Mutations may affect single genes, but also the genomic architecture at a larger scale can be affected. Such aberrations, when visible through a microscope, are called chromosomal abnormalities. Humans have 22 pairs of autosomes and one pair of sex chromosomes. Abnormalities can be divided in numerical (any deviation from 46 chromosomes) and structural defects (abnormal chromosomes). A whole set of 23 extra chromosomes is called triploidy. It results from fertilization or meiotic error. Children with triploidy die before or immediately after birth. A single extra chromosome is called a trisomy. They most often result from meiotic error. Only three autosomal trisomies are potentially viable: trisomy 21 (Down syndrome), trisomy 18, and trisomy 13. All three have a high chance of being associated with heart defects.

In structural chromosome abnormalities, a distinction is made between balanced and unbalanced defects. In balanced defects chromosome parts are displaced but there is no visible extra or missing chromosome material. Balanced rearrangements are most often not associated with an abnormal phenotype, but they may predispose to unbalanced offspring. Unbalanced chromosome abnormalities have a very high risk of being associated with mental retardation and birth defects. As heart development is a very complex process, probably involving hundreds of genes, chances are that this process will be disturbed in one way or another in case of a visible chromosomal abnormality. Indeed, heart defects are very frequent in children with structural chromosomal abnormalities.

Smaller abnormalities are not readily visible through the microscope and will be missed unless specific techniques are applied to detect them. Still, so called microdeletions may contain a large number of genes and are often associated with heart defects. Examples of microdeletion syndromes associated with heart defects are the 22q11.2 (velo-cardio-facial/DiGeorge) deletion syndrome, Williams–Beuren syndrome, 1p36 deletion syndrome and Wolf–Hirschhorn syndrome.

As a general rule, regardless whether a visible chromosome abnormality or a microdeletion is involved, the resulting heart defect will not occur as an isolated feature. Often associated birth defects, developmental delay and/or abnormal growth will be present. Therefore, it is in this category of heart defect patients with additional anomalies that a chromosome abnormality has to be considered.

In contrast to mutation analysis, chromosome analysis requires dividing cells for the chromosomes to become visible through a microscope. Usually, white blood cells or cultured fibroblasts are used for chromosome analysis.

2.1.4.2 Inheritance Patterns

A genetic component plays a role in many diseases. Usually the genetic contribution to disease is appreciated when either a clear pattern of inheritance or significant familial clustering of a disease is noted.[5]

Classical genetic disease follows a recognizable Mendelian inheritance pattern. These disorders are called monogenic disorders as a mutation at a single locus conveys a very strong risk of getting the disease. Sometimes, indeed everybody with a specific mutation develops the disease (this is called complete *penetrance*). In that case the influence of environmental factors or contributions at other genetic loci seems negligible. In practice, however, most monogenic diseases display considerable variation in disease manifestation, severity, and age at onset (clinical variability), even within a single family (where every affected person has the same mutation). Especially in autosomal dominant disease, the chance of developing clinical manifestations of disease when a specific pathogenic mutation is present is often far less than 100% (*incomplete or reduced penetrance*). However, such clinically unaffected mutation carriers may foster severely affected children when they transmit the mutation to their offspring. So, even in so-called monogenic diseases many other genetic and nongenetic factors can usually modify clinical outcome.

Whereas monogenic diseases are often relatively rare, there is a clear genetic contribution to many common disorders such as coronary artery disease, hypertension, and hypercholesterolemia. In the vast majority of patients, these diseases are explained by the combined additive effect of unfavorable genetic variants at multiple different loci and environmental factors (anything nongenetic), eventually causing disease. Polygenic disease, multifactorial disease, and *complex genetic disease* are all terms used to denote this category of diseases. When looking at pedigrees apparently nonrandom clustering within the family can often be noted, however, without a clear Mendelian inheritance pattern. Very common complex disorders may mimic autosomal dominant disease, whereas in less common disorders like, for example, congenital heart defects, a genetic contribution is very likely although the majority of cases will present as sporadic cases without a positive family history. Importantly, frequent complex genetic diseases may have less common monogenic subtypes. FH (familial hypercholesterolemia) as a result of mutations in the LDL receptor or MODY (maturity onset diabetes in the young), are examples of monogenic subtypes of diseases that most often have a complex etiology.

2.1.5 Single Gene Disorders: Mendelian Inheritance

In *single gene disorders*, inheritance patterns can be explained in terms of Mendelian inheritance. Of importance in the first place is whether the causative gene is on one of the autosomes or on one of the sex chromosomes, more specifically on the X-chromosome (the Y-chromosome contains very few disease-related genes and will not be discussed further).

The second distinction to be made is whether mutations in the gene follow a *dominant* or *recessive* mode of inheritance.

2.1.5.1 Autosomal Dominant Inheritance

Autosomal dominant disease is caused by dominant mutations on one of the autosomes. Dominant mutations already cause disease when only one of both alleles is mutated. Most individuals with dominant disease are *heterozygous* for the mutation (they have one mutated and one normal allele). Heterozygous carriers of such a mutation have a high risk of clinically expressing disease symptoms. It is the most common form of inheritance in monogenic cardiac disease. It is characterized by (see Fig. 2.6):

- Equal chance of males and females to be affected.
- Individuals in more than one generation are usually affected (unless a new mutation has occurred).
- Father-to-son transmission can occur
- On average 50% of offspring will be affected (assuming complete penetrance)

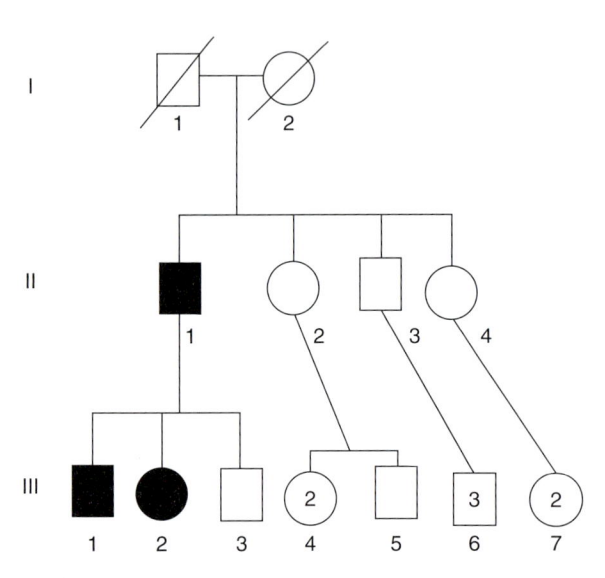

Fig. 2.6 Example of a small autosomal dominant pedigree, the observed male to male transmission (II-1 > III-1) excludes X-linked dominant inheritance. If we assume this disorder has full penetrance a de novo mutation must have occurred in II-1

Although this inheritance pattern is rather straightforward, in practice precise predictions are often complicated by issues of penetrance and *variable expression* (see paragraph *on penetrance and variable expressivity*).

2.1.5.2 Autosomal Recessive Inheritance

In recessive inheritance disease occurs only when both alleles of the same gene are mutated.

Affected patients carry mutations on both the paternal and maternal allele of a disease gene. New mutations are rarely encountered. Therefore, it is reasonable to assume that both healthy parents are heterozygous carriers of one mutation. These healthy individuals are often called "carriers." It is reasonable to assume that each person is carrier of one or more disease-associated *autosomal recessive* mutations.

Affected patients can be *homozygous* (the same mutation on both alleles of the gene) or *compound heterozygous* (different mutations on the two alleles of the gene) for the mutation. If *consanguinity* is involved, a single mutation that was present in the common ancestor is transmitted to the patient by both parents, leading to the homozygous state.

In most cases, but not always autosomal recessive conditions are limited to a single sibship (see Fig. 2.7). If vertical transmission of an autosomal recessive disease occurs this is called "pseudodominance." Pseudodominance can be encountered in case of consanguinity in multiple generations or in case of a very high population frequency of healthy heterozygous carriers. Autosomal recessive inheritance is characterized by:

- Equal chance of males and females to be affected.
- Parents of patients are usually healthy carriers.
- The chance that the next child (a sib) will be affected is 25%.
- Affected individuals are usually limited to a single sibship.
- The presence of consanguinity in the parents favors, but does not prove, the autosomal recessive inheritance mode.

2.1.5.3 X-Linked Recessive Inheritance

X-linked disorders are caused by mutations on the X-chromosome. The X-chromosome does not contain

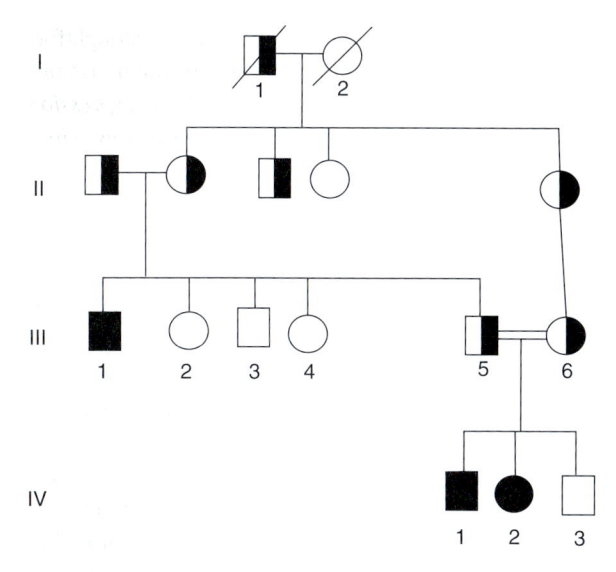

Fig. 2.7 Example of an autosomal recessive pedigree, illustrating the role of consanguinity in AR disease. III-5 and III-6 are first cousins. IV-1 and IV-2 inherited both their mutated alleles from a single heterozygous grandparent (in this case I-1). In this pedigree heterozygous carriers are indicated, usually heterozygous carriers for AR disorders can only be unambiguously identified by DNA analysis

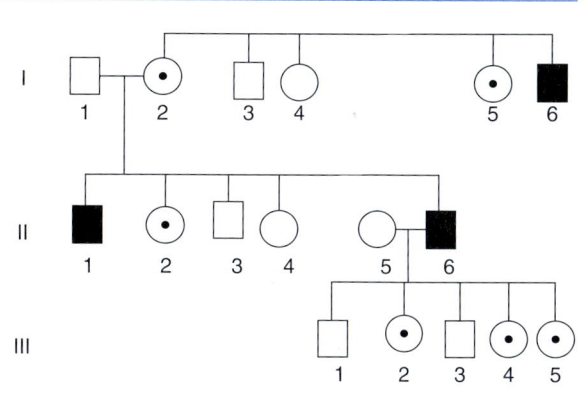

Fig. 2.8 Example of an X-linked recessive disorder, the disease is transmitted via apparently healthy heterozygous females, only hemizygous males manifest the disease. All daughters of an affected male will be carriers. Carrier females are indicated with dots within circles, usually DNA analysis will be required to unambiguously identify carrier females

"female specific" genes. As females have two X-chromosomes and males only one, in X-linked disorders, usually, a difference in disease expression will be noted between males and females.

In X-linked recessive inheritance (see Fig. 2.8):

- No male to male transmission occurs.
- Heterozygous females are healthy.
- All daughters of affected males will be healthy carriers.
- Sons of carrier women have a 50% chance of being affected.
- Daughters of carrier women have a 50% chance of being a healthy carrier.

It should be noted that in female somatic cells only one X-chromosome is active. The other X-chromosome is inactivated. This process of *X-inactivation* (called *Lyonization*) is random, occurs early in embryogenesis and remains fixed, so that daughter cells will have the same X-chromosome inactivated as the cell they were derived from. Usually, in a given female tissue, approximately half of the cells will express the paternal X-chromosome and the other half of the cells, the maternal X-chromosome. However, for a variety of reasons, significant deviations of this equal distribution of active X-chromosomes may occur (called *skewing of X-inactivation*). Naturally, this may influence disease expression in case of X-linked disease. For example, if the X-chromosome containing an X-linked recessive mutation is expressed in over 90% of cells in a given tissue, disease may develop a female like it does in males.

2.1.5.4 X-Linked Dominant Inheritance

In X-linked dominant disorders, heterozygous females are most likely to be affected. However, on average these heterozygous females are often less severely affected than hemizygous males. Exceptions, however, do exist.

Some X-linked dominant disorders may be lethal in hemizygous males like, for instance, the Oculo-Facio-Cardio-Dental syndrome that is associated with congenital heart defects. Hemizygous males will miscarry, leading to a reduced chance of male offspring in affected females.

The characteristics of X-linked dominant inheritance are (see Fig. 2.9):

- No male-to-male transmission occurs.
- Heterozygous females are affected.
- All daughters of an affected male will also be affected.
- Affected females have a 50% chance of having affected children.

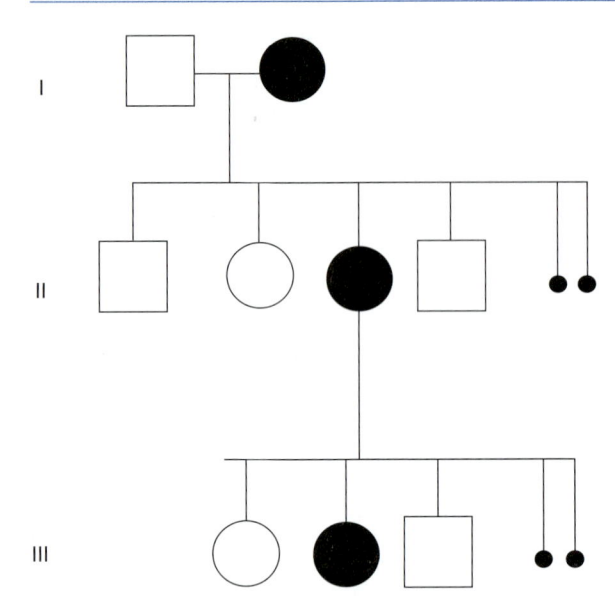

Fig. 2.9 Example pedigree of an X-linked dominant disorder with early lethality in males. Affected males that are conceived will miscarry (leading to skewed sex ratios in offspring). The black dots in the pedigree represent miscarriages. Based on such a limited pedigree, definite distinction from autosomal dominant inheritance would be impossible

Especially in X-linked disorders the distinction between dominant and recessive disease may be blurred, with some heterozygous females being not affected at all, whilst others are affected to the same degree as hemizygous men. In cardiogenetics several X-linked disorders are known where heterozygous females may be asymptomatic, but also run a high risk of significant disease, like for instance in Fabry disease. In Duchenne Muscular dystrophy, considered to be an X-linked recessive disorder, females only very rarely develop severe skeletal muscle weakness, but they are at considerable risk for dilation of the left ventricle and should be monitored by a cardiologist.

2.1.6 Non-Mendelian Inheritance

In fact, any deviation from the classical rules of Mendel could be categorized under the heading of non-Mendelian inheritance. Such deviations can, for example, result from genome disorders (de novo deletions or duplications of larger stretches of DNA or even whole chromosomes), epigenetic factors (these are factors not changing the actual DNA-code, but change the way in which specific genes are expressed) and unstable mutations (trinucleotide repeat mutations such as in myotonic dystrophy, that may expand over generations and lead to a more severe phenotype in subsequent generations). However, for sake of brevity only multifactorial inheritance and maternal (mitochondrial) inheritance will be briefly discussed.

2.1.6.1 Multifactorial Inheritance

Although genetic factors very often contribute to disease, most of the time this will not be in a monogenic fashion. The majority of disorders are caused by a complex interplay of multiple unfavorable genetic variants at different loci in combination with environmental (nongenetic factors). The genetic variants involved may each by themselves have a limited effect. It is the additive effect of multiple factors that eventually will lead to disease, hence the name *multifactorial inheritance*. In this paragraph no distinction is made between multifactorial inheritance and polygenic inheritance (no important environmental contribution). In general practice, such a distinction is most often of no importance unless specific environmental factors can be identified that can be influenced. Hereditability is a measure used to indicate the contribution of inherited factors to a multifactorial phenotype. In animal studies heritability can be calculated, as both environment and genetic composition of the animals can be controlled. In man hereditability can only be estimated indirectly.

In multifactorial inheritance, sometimes clustering of a condition within a family may be observed that cannot be easily explained by chance. Especially in common diseases like diabetes or hypertension (when the underlying genetic variants occur at high frequency in the population) this clustering may mimic Mendelian inheritance. However, in more rare disorders like, for example, in congenital heart disease, an identified patient may well be the only affected one in the family. Still, family members will be at increased risk of a congenital heart defect.

Many continuous traits like, for example, blood pressure, can be explained in terms of the additive effect of multiple deleterious or protective genetic and environmental factors. In case of hypertension the sum of all these factors would be defined as disease liability, which is distributed as a Gaussian curve in the population. At the right side of the curve (highest liability) those with hypertension are found. Their close relatives who will share many of the predisposing genetic (and possibly also environmental factors) with the hypertensive patient, will usually have a higher than average disease liability; however, they may not meet with the clinical criteria for hypertension. For discontinuous traits like, for example, congenital heart defects, a threshold model has been proposed (Fig. 2.10). If disease susceptibility exceeds the threshold level disease will arise. Again, the liability of close relatives of a patient with a heart defect will be, on average, closer to the disease threshold than that of unrelated individuals, but most of them will not exceed the threshold and therefore will have anatomically normal hearts.

It is important to realize that some disorders are more multifactorial than others. Sometimes mutations at a single locus will not be sufficient to cause disease, but have a strong effect. If a mutation in such a *major gene* is present, little else has to go wrong for disease to occur. Therefore, strict separation between Mendelian and multifactorial disease is artificial. Indeed, genes that are involved in rare monogenetic variants of a disease may also play a role in the more common multifactorial forms of the disease.

The following characteristics can be applied to multifactorial inheritance:

- Familial clustering may occur, but usually no Mendelian inheritance pattern can be identified.
- Recurrence risks for family members are in general lower than in monogenic disease.
- Risk of disease rapidly falls with decreasing degree of relationship to the index patient.
- Risk may be higher for relatives of severely affected patients.
- Risk estimations are usually based on empirical (observational) data.
- These risks are not fixed risks, like in Mendelian disease. New disease cases in a family may indicate a higher genetic load, and therefore a higher risk for relatives.

At this point in time the use of predictive genetic testing in multifactorial disease is limited, as usually only a small part of morbidity can be explained by the genetics variants that have thus far been identified for these disorders. As these variants have by themselves only a small effect, the odds of getting the disease, once an unfavorable variant has been identified, are small. Still, commercial genetic tests, supplying risk profiles for many common conditions, based on genetic profiles, are readily available via the internet. Such risk predictions are imprecise and differ substantially between different test providers.

Although exceptions may exist, predictive genetic testing in multifactorial disease is not likely to play a role of importance in genetic counseling in the near future. In contrast, genetic tests for common disorders may play a role in clinical practice in the near future, for example, in risk stratification, and in identifying groups that are eligible for specific treatments.

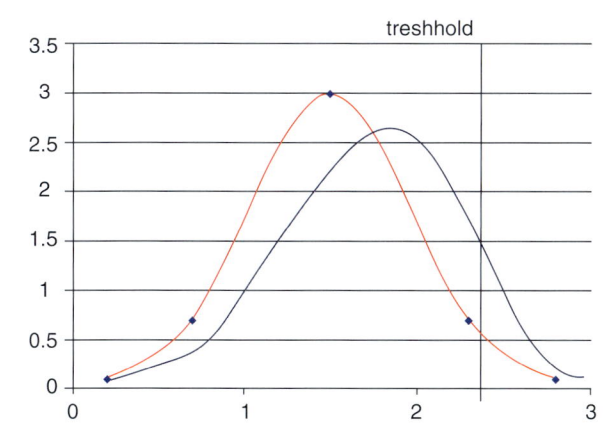

Fig. 2.10 Example of a liability distribution of a discontinuous multifactorial trait (a congenital heart defect) in a given population. The red curve is for the general population. The area under the red curve to the right of the threshold represents the proportion of individuals with CHD in the general population. The blue curve is for first degree relatives of a patient with CHD. Since CHD is a discontinuous trait (it is either present or absent) a threshold is introduced. Everybody with a liability exceeding the threshold will have CHD. Liability for CHD will be determined by the additive effect of unfavorable genetic and environmental factors. As a result of shared unfavorable factors the liability curve for first degree relatives has shifted to the right explaining the fact that a larger proportion of first degree relatives will be affected with CHD in comparison to the general population, whereas the majority of relatives has no CHD, as their liability does not exceed the threshold

2.1.6.2 Maternal (Mitochondrial) Inheritance

Mitochondria are present in most cells in different numbers, and are the principle providers of energy by means of the respiratory chain. Mitochondria contain small circular DNA molecules of their own (*mtDNA*). These molecules are only 16,569 base pairs in length and code for only 37 genes. Thirteen polypeptides of the respiratory chain are encoded by the mitochondrial DNA, whereas the remainder (the majority) are encoded by the nuclear DNA. The rest of the mitochondrial genes play a role in mitochondrial translation (transfer RNA's and ribosomal RNA's).

Somatic cells typically contain 1000–10,000 mtDNA molecules (two to ten molecules per mitochondrion). Mitochondrial DNA-replication is under nuclear control and suited to meet with the energy requirement of the cell. It is not associated with cell division like the nuclear DNA. When a cell divides, mitochondria randomly segregate to daughter cells within the cytoplasm. Oocytes may contain up to 100,000 copies of mtDNA, whereas sperm cells usually contain only a few hundreds. Moreover, these paternal copies do not enter the oocyte at fertilization. Therefore, the paternal contribution to the mtDNA is negligible, and mtDNA is inherited exclusively via the mother, hence the concept of *maternal inheritance*.

Whereas nuclear genes are present in two copies per cell, mitochondrial genes are present in thousands of copies. In maternally inherited disease, in a specific tissue, a significant part of the mtDNA copies may carry a similar mtDNA mutation, whereas the remainder of the copies is normal (wild type). This phenomenon is called *heteroplasmy*. Here, again, a threshold is important, that is determined by the specific energy requirement of the tissue. If the percentage of mutated mtDNA becomes so high that the energy requirement cannot be fulfilled, this may result in mitochondrial disease. If a mutation is present in all mtDNA molecules in a specific tissue, this is called *homoplasmy*. The mechanism that leads to homoplasmy of certain mtDNA mutations is not yet fully understood.

Mitochondrial DNA differs in many aspects from nuclear DNA. In contrast to nuclear DNA, most of the mtDNA codes for genes. Therefore, any random mutation in the mtDNA is much more likely to disrupt an actual gene, than is the case in the nuclear genome. DNA-repair mechanisms to repair acquired DNA-damage, as are present in the nucleus, are lacking, leading to accumulation of mtDNA mutations, for example, in aging. On the other hand, since mtDNA genes are present in hundreds to thousands of copies per cell, acquired mutations rarely lead to recognizable mitochondrial disease. Only a minute fraction of mtDNA mutations will become "fixed" and will subsequently be transmitted to offspring

It is important to realize that maternal inheritance is not equivalent to mitochondrial disease. As most of the proteins active in the mitochondrion are encoded by nuclear genes, mitochondrial diseases may be inherited in other fashions, most often in an autosomal recessive manner.

Mitochondrial disease affects many tissues, although tissues with the highest energy requirements (muscle, brain) are most often involved. Cardiac muscle may be involved in different mitochondrial conditions. Sometimes a cardiomyopathy may be the first or most prominent manifestation of a mutation in the mtDNA.

The following characteristics apply to maternal inheritance

- Men and women are affected with similar frequencies, however only females transmit the disease to offspring.
- Phenotypes may be extremely variable (and unpredictable) as a result of different levels of heteroplasmy in different tissues.
- The percentage of mutated mtDNA in one specific tissue may not accurately predict the level of heteroplasmy in other tissues. This is a major problem for example in prenatal diagnosis.
- Affected females are likely to transmit mutated mtDNA to all of their offspring, but non-penetrance will result if the threshold for disease expression is not reached.

2.1.7 New (De Novo) Mutations

Mutations can occur at any time both during gametogenesis or regular cell-division. If a detected mutation is present in neither of the parents (that is, if it is absent in the blood of both parents) the mutation is called "*de novo*." De novo mutations may have arisen in the sperm or egg cell, or may even have occurred after conception. Mutation rates in genes (the number of mutations per gene per generation) are on average very low, in the order of 10^{-5}–10^{-7}. Therefore, if in an isolated patient a de novo mutation in a candidate gene for the disorder is being found, it is usually regarded as a pathogenic mutation.

It should be realized that most new mutations will go unnoticed. When they are situated in noncoding DNA or in recessive genes they will have no immediate effect, whereas new mutations in important dominant genes may be lethal, and may therefore not be ascertained.

2.1.8 Mosaicism

When mutations (or chromosomal abnormalities) arise shortly after conception, *mosaicism* may result. Mosaicism is defined as the presence of genetically different cell populations (usually an abnormal and a normal cell line) within a single individual. The importance of mosaicism in relation to cardiac disease is that (at least in theory) mutations that are not detected in the blood of the affected individual may be present in the heart. Preliminary observations suggest that this may be important in some types of congenital heart disease.

Germline mosaicism is a special type of mosaicism, where a population of precursor spermatocytes or oocytes carries a specific mutation that is not detected in other tissues. As a result of germline mosaicism a healthy (apparently noncarrier) parent may unexpectedly transmit a disease mutation to several offspring. The classic observation of germline mosaicism is in Duchenne muscular dystrophy, where apparently noncarrier females may give birth to more than one affected son with exactly the same dystrophin mutation. However, germline mosaicism may occur in any disorder including cardiac disorders and, therefore, it should be considered a possibility in any apparently de novo mutational event.

2.1.9 On Penetrance and Variable Expressivity

The *penetrance* of a specific mutation refers to its ability to cause a disease phenotype. In monogenic disease mutations may show 100% penetrance. For instance, most dystrophin mutation will cause Duchenne muscular dystrophy in all hemizygous males. However, especially in autosomal dominant disease, penetrance is often reduced; for example, not everybody with the mutation actually becomes ill. Whether or not disease symptoms develop may be dependent on a constellation of other genetic (genetic background) or environmental factors, such as lifestyle. Disease penetrance is not necessarily identical to having actual clinical complaints. Especially in cardiogenetics, many clinically asymptomatic individuals with, for example, a cardiomyopathy or long QT syndrome may have easily noticeable abnormalities on ECG or echocardiography. Such individuals may not realize their genetic status, but they cannot be regarded as true non-penetrants. Usually, they should be under cardiac surveillance and often preventive treatment will be indicated. Penetrance, in this way, is to some extent dependent on how well individuals have been examined for disease symptoms. If true non-penetrance occurs, genetic diagnosis may be the only means to identify individuals that may transmit the disease to their offspring. For decisions with respect to patient care it is more useful to look at penetrance of specific phenotypic traits, for example, the chance of a ventricular arrhythmia in case of a *KCNQ1* mutation in long QT syndrome type 1.

In congenital heart disease penetrance is fixed, as the disease is either present or not. In diseases that manifest themselves later in life this is not true. For instance, in an autosomal dominant inherited cardiomyopathy penetrance at the age of 10 may be low, whereas at the age of 60 most individuals with the genetic defect will have developed disease manifestations. In this case there is *age-dependent penetrance*. Of course, this will influence risk estimations based on clinical observation. At the age of 10 a child of a cardiomyopathy patient from this family may still have an almost 50% chance of having inherited the familial mutation despite a normal cardiac evaluation, while at the age of 60 a normal cardiac evaluation severely reduces the chance of the mutation being present. If sound scientific data are available on penetrance these can be used in genetic counseling and decision making. However, unfortunately this is often not the case.

Variable expression is used to indicate the presence of variation in disease symptoms and severity in individuals with a similar mutation. For example, in desmin myopathy some individuals may mainly suffer from skeletal myopathy, whereas in others from the same family cardiac manifestations may be the principle determinant of the disease.

2.2 Genotype–Phenotype Correlations

This term refers to the extent to which it is possible to predict a phenotype (i.e., clinical disease manifestation) given a specific genotype and vice versa. In an era where presymptomatic genetic testing becomes more and more customary, this is an issue of great importance. If it were possible to predict phenotype based on genotype with great accuracy, this would lend additional legitimacy to genetic testing, especially if early intervention would change disease course. Indeed, there have been claims that, for example, hypertrophic cardiomyopathy caused by mutations in the gene encoding cardiac Troponin T (*TNNT2*) has a higher potential for malignant arrhythmias than mutations in some other genes.[6,7] Also within a given gene, some mutations may have a stronger pathogenic effect than others.

Without doubt significant genotype–phenotype correlations do exist, but it is prudent to regard such claims with caution, as some of them may also be the result of selection and *publication bias*. From a clinical point of view it is obvious that, if intrafamilial (where every affected individual has the same mutation) variation in disease severity and penetrance is considerable, little can be expected of phenotype predictions based on the presence of this family-specific mutation alone. As a result of the difficulty in establishing straightforward genotype–phenotype correlations, the role of genetic information in cardiac risk stratification protocols has been limited thus far.

The reverse situation needs also to be considered. Clinical information on history and clinical data such as, for example, T-wave morphology in patients suspected of having a form of long QT syndrome is very helpful in selecting the genes that should be analyzed first.[8] In the long QT syndromes genotype–phenotype correlations can be used in practice: clinical parameters suggest a specific genotype and subsequently, genotype-specific therapy can be instituted. Accurate clinical information may improve the yield of genetic testing and may decrease costs and time needed for these analyses.

2.3 Basic Concepts in Population Genetics

Population genetics studies genetic variation and genetic disease in the context of populations. Here, a population is defined as the group of individuals that are likely to get offspring together, and the genetic diversity that is contained within this group. Populations are not only delimited by geographical boundaries such as borders, rivers, mountains, islands, but also by religious, ethnic, and cultural differences.

Some insights from population genetics are important to the field of clinical genetics and necessary for understanding genetic phenomena that are relevant to clinical practice like, for instance, *founder effects*. Two important population genetic "laws" predict the distribution of neutral genetic variation (i.e., the Hardy–Weinberg equilibrium) and the frequency of disease mutations (mutation-selection equilibrium), respectively.

2.3.1 Hardy–Weinberg Equilibrium

The *Hardy–Weinberg equilibrium* predicts that the relative frequency of different genotypes at a locus within a population remains the same over generations. For an autosomal gene G with two alleles A and a with an allele frequency of p and q, respectively, the possible genotypes AA, Aa, and aa will occur with a frequency of p^2, $2pq$, and q^2. As there are only two alleles for G, $p+q=1$.

However, for the Hardy–Weinberg equilibrium to be true, many assumptions have to be made. The population has to be infinitely large, there has to be random mating with respect to G, there has to be no selection against any of the G genotypes, no new mutations occur in G, and there is no migration introducing G alleles into, or removing G alleles from the population. Clearly no situation in real life will ever satisfy all these criteria.

The Hardy–Weinberg equilibrium is a neutral equilibrium. Small deviations from the expected genotype frequencies occur by chance (genetic drift) and over multiple generations a significant difference in genotype frequency (when compared to the original equilibrium) may become apparent. There is no driving force correcting such chance deviations. As a matter of fact a new Hardy–Weinberg equilibrium is established with each generation.

In real life, new mutations do occur and often selection does exist against disease-associated alleles, causing them to disappear from the gene pool. However, mutation rates for recessive disorders are extremely small and selection pressure is low, as selection works only against the homozygous affected. Therefore, in autosomal recessive disorders, the

Hardy–Weinberg equilibrium can be used to calculate carrier frequencies for recessive disorders if the frequency of the disorder in the population (q^2) is known. Because of the limitations mentioned above, such calculations have to be regarded as estimates and interpreted with caution.

2.3.2 Mutation–Selection Equilibrium

To understand the dynamics of disease causing (not neutral) alleles, another equilibrium is of importance: the *mutation-selection equilibrium*. New disease alleles will arise with a given frequency as a result of new mutations, but when diseased individuals are less likely to reproduce, they also disappear again from the gene pool. Therefore, the equilibrium that predicts the frequency of disease alleles is a function of the mutation rate, the reproductive fitness, and the mode of inheritance of the disease.

The easiest example is a severe congenital heart defect as a result of a new autosomal dominant mutation. If this heart defect is lethal, reproductive fitness is nil, and the population frequency of the mutated autosomal dominant gene would be identical to the mutation frequency. In, for example, long QT syndrome type 1, most mutation carriers, however, will reproduce, but reproductive fitness is somewhat reduced as a result of some affected individuals dying from arrhythmias at a young age.[9] Here, the actual frequency of the disease allele is much larger than the mutation frequency, as most disease alleles will be inherited. Still, if no new mutations would occur, the disease would eventually die out as a result of reduced fitness.

Mutation-selection equilibrium is more stable than the Hardy–Weinberg equilibrium. If for some reason more new mutations arise than expected, selective pressure increases as well since there are more affected individuals to target, moving the equilibrium again in the direction of the original state. However, if reproductive fitness increases significantly as a result of improved therapies, eventually a new equilibrium with a higher population frequency of the mutated allele will be established.

2.4 Founder Mutations

If a population descends from a relatively limited number of ancestors, the genetic variation is largely dependent on the variation that was present in this small group of ancestors. If by chance a rare disease allele was present in one of these "founders," this disease allele may achieve an unusual high frequency in this founder population, which is not found in other populations. This is especially true if selection against the mutation is small, so that the mutation is not easily eliminated from the gene pool.

For example, in the Netherlands over 20% of hypertrophic cardiomyopathy is caused by a single c.2373_2374insG mutation in the MYBPC3 gene.[10] In order to prove that this is indeed a founder mutation and not a mutation that has occurred de novo more than once, it was established that the mutation in each patient lies on an identical genetic marker background (haplotype), which must have been present in the founder. If the mutation had occurred many times de novo it would have been expected to be associated with different haplotypes.

Founder effects, like the one described above, can help explain why certain diseases are more frequent in some populations than others. Moreover, it is important to be aware of these mechanisms as they can aid in devising efficient strategies for molecular diagnosis in specific populations.

2.5 Genetic Isolates

Genetic isolates are small, closed communities within a larger population where people tend to marry among each other. Consanguinity is more likely and even if this is not the case, genetic variation within an isolate is much more limited, because of the absence of new genes contributing to the gene pool. As a result some genetic diseases may have a much higher frequency within an isolate than in the population as whole, while in contrast other genetic diseases may be virtually absent. Therefore, it may be important to realize whether or not a specific patient comes from a genetic isolate.

2.5.1 Consanguinity

Consanguineous marriages are very common in some cultures and unusual in others.[11] Marriages between first cousins are most frequent. They share 12.5% of their DNA, derived from their common ancestor. In some cultures uncles are allowed to marry their nieces. Such second degree relatives share 25% of their DNA. This situation, from a genetic point of view, is no different from double first cousins that have all four grandparents in common and, therefore, also share 25% of their DNA.

Consanguinity may have significant social and economical advantages, especially in low-income societies. However, the genetic risks cannot be ignored, but they are highly dependent on the degree of relationship. The problem with consanguinity arises from the reduction to homozygosity in offspring of consanguineous parents. If both parents carry the same recessive mutation in their shared DNA, there is a 25% risk of the mutation being homozygous in each child. Therefore, consanguinity mainly increases the likelihood of autosomal recessive disease. The chance that a recessive disorder is caused by consanguinity increases with decreasing frequency of the disorder. In other words, the relative risk increase as a result of consanguinity is highest for the rarest recessive disorders. For example, thus far a rare form of catecholaminergic polymorphic ventricular tachycardia (CPVT) as a result of an autosomal recessive mutation in the CASQ2 gene has only been found in consanguineous families.[12] In addition, one also has to be aware of the fact that autosomal dominantly inherited disease may also run in consanguineous families. If offspring has inherited both affected alleles, the clinical picture is often severe and lethal at an early age. Examples have been found long QT syndrome and hypertrophic cardiomyopathy.[13, 14]

If consanguinity occurs frequently within a population, the population becomes inbred. In such a population, for any genetic locus, the frequency of heterozygotes will be lower than expected under Hardy–Weinberg equilibrium (because of reduced random mating). This will lead to overestimation of carrier frequencies.

In multifactorial disease consanguinity may play a role as well, although less conspicuous than in autosomal recessive disease. Shared predisposing genetic variants, present in heterozygous form in the parents, have a 25% chance of being present in homozygous form in the offspring, thus increasing the likelihood for multifactorial disorders.

Information on consanguinity is not always volunteered and should be specifically asked for. Sometimes consanguinity is present, but not known to the family.

Most individuals have little information on relatives dating further than three generations back. If ancestors from both parents are from the same small isolated community, consanguinity may still be suspected. When of importance, genealogical studies may be used to substantiate this.

2.5.2 Genetic Testing

Any test to identify a genetic disease can be considered a genetic test. Genetic testing using DNA analysis is available for an increasing number of cardiac diseases and conditions that are associated with cardiovascular disease in a wider context. Two important differences between genetic DNA tests, when compared to other diagnostic tests, need mentioning. First, DNA tests usually have health implications that last a life time, while the genetic defect in itself is not amenable to treatment. Second, the implications of genetic test results often are not limited to the patient in front of you, but also are of concern to (future) family members. The family and not the individual patient could be regarded as the "diagnostic unit" in genetic disease. As a result of these notions DNA testing is, and should only be offered as part of a genetic counseling procedure in order to assure that patients fully understand the scope of the tests that are being performed. This is especially true for monogenic disease and tests for very high risk genes.

2.5.3 Genetic Counseling

Genetic counseling is a two-way communication process aiming at helping patients with genetic disease or at (perceived) increased risk of genetic disease, and their relatives, to understand the genetic risk and decide on a suitable course of action.[15] Genetic counseling is offered by trained medical or paramedical professionals. Its goals are:

- To help patients and their family members comprehend medical facts (diagnosis, symptoms, complications, course, variation, and management)
- To help patients and their family members understand the basic facts of the genetic contribution to

their disorder, where this is relevant for communicating risks to specific family members and recurrence risks in (future) children
- To make them understand the options available to deal with risks and recurrence risks (preventive treatments, lifestyle adjustments, reproductive options, prenatal diagnosis)
- To help counselees choose a suitable course of action in view of their individual risk of disease, goals, personal and cultural values, religious beliefs, and facilitate this course of action
- Support counselees in making the best possible adjustment to their disease condition or to their increased risk of genetic disease

Most common counseling situations for cardiac disorders can be grouped into one of the three categories mentioned below. All three have their own dynamics and major issues:

- Parents who have a child with a congenital heart defect, a syndrome that has important cardiovascular implications, or other cardiac disease. They want to be informed about prognosis, recurrence risk to other children, and the possibility of prenatal diagnosis.
- Patients that have a cardiac defect or cardiac disease themselves and have questions about genetic aspects and prognosis. They may also be concerned about risk to family members, most often (future) children and/or sibs.
- Those who have been referred because of a positive family history for cardiac disease or suspicion thereof, or a family history for sudden cardiac death at a young age. They come for information on their personal risk, questions about the usefulness of presymptomatic cardiac evaluation and, if possible, they may opt for presymptomatic genetic testing.

In the counseling process different stages can be discerned:

- *Diagnostic phase.* For meaningful genetic counseling, the genetic diagnosis has to be as precise as possible. As we are considering conditions that often run in families, medical information from other affected family members is necessary to arrive at the correct family diagnosis. In addition to a history, family history, physical examination, and cardiac investigations it may therefore be necessary to collect medical records or pathology reports of relatives of the counselee. If genetic testing has previously been performed in the family, it is vital that this information is retrieved.
- *Informative counseling.* During this stage, information on the condition, natural course, complications, prognosis, genetics, management, and treatment options are communicated with the counselee. Although standardized written information can be very helpful, it cannot fully replace counseling. Information should be tailored to the specific needs and knowledge of the counselee, and the counselor should verify whether vital information is understood.
- *Supportive counseling.* During this phase, the emphasis is on the decision-making process and coping. Counselees and their partners have to make choices with regard to, for example, reproductive issues or presymptomatic genetic testing. Choices have to be reinforced and counselees and their relatives have to be supported in coping with the genetic disease that runs in their family, and eventually with the outcome of genetic testing. The results of genetic testing in one individual often alter disease risks for relatives. It may be necessary to inform relatives about the (newly identified) genetic disease in their family, and the fact that they themselves are, or are not, at risk to be genetically predisposed to this disease. Counselees should be aided in communicating such difficult news to their relatives.
- *Follow-up.* People should be encouraged to return if important questions remain or if new issues arise. Ideal and easy solutions for problems related to inherited disease are rarely, if ever, available. Therefore, coping with genetic information may take quite some time, and what seems to be the best course of action may change over time. Moreover, new information on other affected persons in the family may alter earlier assumptions with respect to the genetic contribution to the disease. Such information may therefore change the conclusions of the genetic analysis, especially if risk estimates were based on empirical data.

It is customary that the conclusions of the genetic analysis are offered to the counselees in writing, to facilitate later decisions being made based on adequate medical and genetic information.

Dealing with unfavorable genetic information may be very distressing for the ones involved. In clinical genetics departments specially trained social workers and/or psychologists are available to offer supportive care whenever needed.

Some paradigms are inherently associated with the genetic counseling process:

- Nondirectiveness. Historically, nondirectiveness is an important hallmark of genetic counseling. The counselor provides adequate information and support. The counselee decides. This notion stems from time that genetic counseling was mainly concerned with reproductive issues. Naturally, counselors should have no say in the reproductive decisions made by their clients. Also, in presymptomatic testing of late onset neurodegenerative disease, where medical interventions to change disease course are virtually absent, maximum nondirectiveness should be applied in counseling.
- However, with a changing focus in medical genetics to disorders that are, at least to some extent, amenable to early intervention or preventive treatment, the applicability of nondirective genetic counseling becomes less obvious. For example, in long QT syndrome type 1, where β-blocker therapy has been proven to be effective in symptomatic patients, nondirective counseling seems less indicated.[16] In practice, in cardiogenetics, a balance that respects both patient autonomy, and assures that the appropriate medical decisions are made, should be sought for.
- Informed consent. Informed consent is not unique to clinical genetics or genetic counseling. However, some institutions will require written informed consent prior to DNA-testing, especially if presymptomatic testing of apparently healthy individuals is concerned. This is no rule of thumb and may vary based on individual insights and local differences in the medico-legal situation.
- Privacy issues. These are also not unique to genetic medicine, but may be more urgent in this discipline. Genetic information may have a huge impact on insurability and career options. The extent to which this is true is largely dependent on legislation dealing with genetic discrimination, which varies between countries. However, a danger of discrimination on genetic grounds always exists. Therefore, maximum confidentiality of genetic information should be assured. Providing genetic information to third parties, without written permission from the individual involved, would be defendable only in case of a medical emergency. In contrast, genetic information is much harder to keep confidential because DNA is shared by relatives that are likely to benefit significantly from this information. When appropriate, permission to use genetic information for the benefit of relatives should be actively acquired by the genetic counselor. Especially in families that communicate insufficiently, clinical geneticists may encounter problems with confidentiality and find themselves confronted with conflicting duties.

2.6 Predictive Testing and the Dynamics of Family Studies

Predictive or presymptomatic testing is performed on symptom free or perceived symptom free individuals in order to find out whether or not they have inherited the predisposition for a genetic disease. Often predictive testing takes place in the context of family studies. In family studies specific individuals are targeted for evaluation based on a positive family history for genetic disease. Both predictive testing and family studies are unique features of clinical genetics practice.

2.6.1 Predictive DNA Testing

Predictive DNA testing is usually performed for monogenic disorders with important health risks. Demonstrating that an individual has not inherited the family specific mutation usually reduces risks to population level, and also risks for offspring will be normalized. However, if the mutation is indeed identified, this does not automatically mean that the individual will get the disease. In many cardiac disorders penetrance is significantly reduced. In general, presence of the familial mutation will not allow for predictions on severity of the disease or age of onset,

Most genetic cardiac disorders show significant locus heterogeneity, that is, many different genes are associated with a similar phenotype. Besides, molecular heterogeneity (the number of different mutations in a gene) is immense. Therefore, as a paradigm, predictive genetic testing in a family is only possible if a causative family specific mutation has been identified in the index patient.

It is important to emphasize that predictive testing does not necessarily involve DNA testing. A cardiologist performing echocardiography in a symptom free sib of a hypertrophic cardiomyopathy patient is involved in both a family study and predictive testing. The detection of, even a very mild, hypertrophy of the interventricular septum, that as yet does not need treatment, will have serious consequences for this person. The adverse consequences (see next paragraph) of predictive testing based on concealed cardiac symptoms are no different from those associated with predictive DNA testing. Therefore, in a case like this, the same standards of genetic counseling should be applied prior to echocardiography.

In families where DNA-studies have been unsuccessful, family studies will have to rely solely on phenotype and therefore on cardiac evaluation. An important difference between family studies based on phenotype and those based on genotype arises when non-penetrance or age-dependent penetrance occurs. In that case, of course, a genetic test will be more sensitive to demonstrate the predisposition especially in young individuals. In conditions with age-dependent penetrance it may be prudent to re-evaluate individuals with a 50% prior chance of having inherited the genetic defect after a couple of years.

2.6.2 Adverse Consequences of Predictive Testing

Predictive testing may offer important medical and psychosocial benefits to the individuals tested. However, it should be realized that, in contrast to this, predictive testing can also have negative psychological and socio-economic repercussions.[17] Individuals may perceive themselves as less healthy, even when no disease symptoms can yet be demonstrated. Coming to terms with knowledge about one's own genetic predisposition, feelings of guilt toward children that are now also at increased risk, forced lifestyle changes and difficulty with choices regarding, for example, reproductive issues may cause a lot of distress and anxiety. Importantly, knowing that one has the predisposition for a serious late onset disorder is likely to complicate qualifying for, for example, life or health insurance or might interfere with career options. Last but not least, predictive testing can complicate family relationships, especially if some family members want to be tested while others decline testing. Test results of one person may also yield risk information with regard to other family members that may not want to know this. Therefore, predictive testing should not be embarked on without giving these issues serious thought. Opting for predictive testing should be a well-considered and autonomous decision of the individual involved. Pressure on individuals to undergo testing, for instance, by insurance companies or employers would be absolutely unethical.

2.6.3 Predictive Testing in Minors

Minors cannot make their own well-informed decisions with regard to predictive testing. It is a paradigm in clinical genetics not to perform predictive genetic testing in minors if there is no direct and important medical benefit.[18] Late onset disorders, or disorders that are not amenable to preventive treatment, are not to be tested in healthy children.[18] In some countries predictive genetic testing in minors is subject to specific restrictive legislation.

However, in many cardiac disorders like, for example, long QT syndromes, preventive therapy should be instituted at an early age. In such cases, postponing testing until children can make their own autonomous decisions is often not a realistic option. Thus, predictive genetic testing of minors can certainly be indicated. In the Netherlands, for example, predictive genetic testing of minors for cardiac disorders is performed in centers for cardiogenetics, according to a protocol that also involves participation of a psychologist or specialized social worker. It should be noted that parents who have their children tested for heritable arrhythmia syndromes are likely to experience major distress and anxiety.[19] This may influence the handling of their children and moreover parental anxiety is likely to lead to anxiety in children.

Although, as a rule of thumb, predictive testing in children is only performed if treatment or surveillance is possible and necessary, there may be exceptions to this rule that have to judged on a case-by-case basis.

The bottom line is that testing has to be in the interest of the child. For example, should a child from a hypertrophic cardiomyopathy family be talented enough to seriously pursue a professional career in sports, it would be unfair to postpone testing, thereby depriving the child from the possibility to choose another career at an earlier stage.

2.6.3.1 Conducting Family Studies

The way individuals are selected for evaluation in a family study primarily depends on the mode of inheritance of the disease. Cardiogenetic family studies most often involve autosomal dominant conditions, in which affected individuals are likely to occur in several generations and both males and females may be affected. Family studies are conducted using the "cascade method." As soon as a new disease carrier has been identified, his or her first degree relatives become the next targets for study. When, parents of a disease carrier are deceased it will often be difficult to determine whether the condition has been inherited from the mother or from the father. The possibility also remains that the disease predisposition was inherited from neither parent, but resulted from a de novo mutational event. A decision will have to be made whether to stop here or to pursue the family study further to aunts, uncles, and often first cousins at both sides. This decision depends in part on the medical information available on the parents and more distant relatives. Moreover, the magnitude of the risk for severe events that is associated with the familial disease, knowledge on the frequency of the familial character of the disease, and the availability of therapies that influence this risk are important issues when deciding how far family studies should be pursued.

The major justification for family studies is to unambiguously identify those individuals that run an increased risk of disease, in order to institute preventive therapies or closely monitor these individuals and enroll them in risk stratification protocols.

However, sometimes the targeted family members themselves may not be at high risk for serious disease anymore. Contacting them may still be justified if there is a considerable chance that the predisposition to a treatable disease has been transmitted to their children. This may, for example, be the case in elderly individuals from long QT syndrome families that never experienced arrhythmias themselves. Demonstrating the predisposition in them will not necessarily lead to treatment, but exclusion of the predisposition will render further testing unnecessary for all of his or her children. The medical benefits for elderly tested individuals may be limited, but also the socio-economic dangers of predictive testing may be less urgent in older individuals, as they will usually already have insurance and careers.

In case of a disorder that is not amenable to treatment, only reproductive counseling can be offered to family members that turn out to have the genetic predisposition. For personal reasons family members may opt for predictive testing. Uncertainty regarding genetic status may by itself be a major cause of distress and anxiety. However, if no clear medical benefits are to be expected, family studies should only be initiated on specific demand of the relatives themselves.

2.6.4 Contacting Family Members

It is not customary for clinical geneticists to directly contact family members of their patients with genetic disease. Usually a patient is supplied with short written information for family members on the disease, management, genetic aspects, and the risks involved. It will be pointed out to the patient which family members should be contacted and should receive this written information. Subsequently, the family members themselves are left to decide whether or not they want to act upon it. This strategy, using these family letters guarantees "maximum nondirectiveness" toward family members and also, in a way, deals with issues of patient confidentiality, as the patient agrees to distribute general information on the disease within the family. An inventory of responses to these family letters demonstrated that this is an effective means to inform relatives.[20] Still, especially when potentially dangerous arrhythmia syndromes are involved, one could challenge the passive nature of this strategy. Is it correct to leave the responsibility to actually distribute the written information solely with the patient? Should clinical geneticists not at least check whether the information has been received by family members and has been understood? This issue has not yet been settled, but alternatives are certainly conceivable. For instance, in the Netherlands in familial hypercholesterolemia (FH) family members are actively visited by genetic field workers.

2.6.5 Prenatal Diagnosis

Prenatal diagnosis can be requested for a number of different reasons. Termination of pregnancy may be the ultimate consequence once it has been established that the fetus has a very serious debilitating genetic disorder. However, the goal of prenatal diagnosis may also be to aid in planning peripartum medical interventions, or help parents to emotionally prepare for the birth of a child with a birth defect. Parents with a previous child with a congenital heart defect will qualify for specialized ultrasound in subsequent pregnancies. Depending on the severity and type of heart defect that is detected at ultrasound, parents may decide to terminate the pregnancy or to deliver in a center where appropriate neonatal intensive care is available. On rare occasions even fetal therapy can be applied; for instance some fetal tachyarrhythmias can be treated by putting the mother on medication.

Prenatal diagnosis can be divided in invasive diagnosis and noninvasive imaging studies, mainly prenatal ultrasound. Invasive prenatal diagnosis involves obtaining chorionic villi (placental cells), amniocytes (fetal cells present in amniotic fluid), or rarely cord blood, for genetic and, sometimes, protein or metabolite studies. The invasive procedures are associated with a small albeit significant risk of pregnancy loss. Therefore, these should be undertaken only if the prenatal diagnosis will have medical consequences. Like in predictive genetic testing, prenatal DNA-diagnosis for cardiac disorders will only be possible if the family-specific mutation has been identified beforehand.

Except for ultrasound diagnosis in pregnancies of couples to whom an earlier child with a congenital heart defect has been born, requests for prenatal diagnosis are infrequent in cardiogenetic practice. However, requests for prenatal diagnosis should always be taken seriously and the reasons should be explored. Frequently, other issues like feelings of guilt, fear of disapproval from friends or relatives, uncertainty about postnatal follow-up, and so on may be found to underlie such requests.

Preimplantation genetic diagnosis (PGD) is a technique in which in vitro fertilization (IVF) is combined with genetic diagnosis prior to implantation of the embryo into the womb. As genetic diagnosis has to be performed on one or two embryonal cells instead of millions of white blood cells, PGD is technically much more demanding. PGD may be an alternative to couples that are opposed to pregnancy termination, but would not be able to reproduce knowing that their child is at high risk of serious genetic disease. Success rates of PGD are limited by the limitations of the IVF procedure and the fact that after genetic testing fewer viable embryos may be left for implantation. PGD has been performed for a limited number of disorders that may have major cardiac consequences like, for instance, Marfan syndrome or myotonic dystrophy.[21, 22]

Besides prenatal diagnosis, which is performed in selected cases because of increased risk of genetic disease, also prenatal screening programs exists. In principle all pregnant women are eligible for prenatal screening programs that may be differently set-up in different countries. In most western countries nowadays prenatal ultrasound screening is offered to pregnant women at around 20 weeks of gestation. As congenital heart defects occur at high frequency in the general population, many more heart defects will be found by chance during ultrasound screening than by using other methods of prenatal diagnosis, even if the sensitivity of ultrasound screening may be relatively poor.

2.6.6 Population Genetic Screening Issues

In screening for genetic disease whole populations, or specific subpopulations, are targeted.

Screening programs for inherited disease should meet with the same standards as any screening program, the only difference being that the outcome of screening may also have bearing on family members.

Above all, screening must produce a useful outcome, usually a form of early preventive treatment. If surveillance is the only screening outcome it must be well established that this surveillance leads to important health benefits. In genetic screening, apart from identifying individuals who are likely to develop the disease, a goal of the program could be the identification of carriers of frequent autosomal recessive disease, like cystic fibrosis or thalassemia. Here, the useful outcome would be the possibility of preconception counseling, and the possible prevention of the unexpected birth of an affected child. This situation will not play a significant role in cardiogenetics as the only frequent autosomal recessive disorder associated with major cardiac problems, hemochromatosis, is a late onset disorder where preconception counseling would probably be without benefit.

The ethical framework for population screening requires that participation is voluntary, that easy understandable information is supplied to subjects about the goals of the program, that programs respect the privacy and autonomy of subjects, that screening results are confidential and that no pressure is exerted to follow a specific course of action based on screening results.

Organization and timing of screening are directly related to the goals of the program and to the question how subjects can be best recruited in order to maximize uptake of the program. For instance, mutations in long QT genes may be responsible for up to 10% of cases of sudden infant death syndrome.[23] This has led some professionals to propose ECG screening programs for neonates.[24] In developed countries, where programs for neonatal screening have already been implemented, it would seem logical to use the existing infrastructure for organizing such a screening program.

2.6.7 Interpreting Genetic Test Results

Although nowadays most genetic tests are based on direct mutation testing, interpretation of the results is not always straightforward. Without going into great depths on this subject, it may be appropriate to spend a few lines on this subject. Mutations can basically have effects in three different ways. They can cause loss of normal protein function. This is called haploinsufficiency. They can cause gain or change of normal protein function, or they can make the protein become toxic, if normal metabolism is disturbed. For example, loss of function mutations in the SCN5a gene causes Brugada syndrome and progressive conduction disease, whereas gain of function mutations in the same gene underlies long QT syndrome type 3.

If *nonsense mutations* (leading to a stop codon) or frameshift *mutations* (leading to disruption of the reading frame, which usually causes a premature stop) occur, one can be confident that this will lead to haploinsufficiency, unless the truncation is very close to the C-terminus of the gene. Actually, as a result of a process called nonsense-mediated messenger RNA decay, only very little truncated protein will be produced. Most splice mutations (especially those disturbing the readingframe) and larger rearrangements of genes will also lead to haploinsufficiency. Therefore, in most cases when such classes of mutations are detected in a candidate gene, one may assume that the mutation is causative for the disease phenotype.

Missense mutations (mutations changing only one amino acid in the protein) may lead to both loss of protein function or gain/change of function. Especially in case of structural proteins, where different protein molecules act together to form a structure, missense mutations may be more deleterious than truncating mutations, as the mutated proteins are incorporated into the structure and disrupt it. This is called a dominant negative effect.

However, many of the missense mutations detected may actually be rare variants without significant effect on protein function. Therefore, if a new missense mutation in a candidate gene for a specific disorder is identified it will often be difficult to predict whether or not it is the actual causative mutation. Such mutations of unknown pathogenicity are called *unclassified variants* (UVs). Unfortunately, in clinical practice UVs occur rather frequently and cannot always be satisfactorily resolved.

Absence of the same UV in a large number of healthy controls is compatible with pathogenicity, but merely confirms that the UV is indeed rare. One would need to demonstrate that the UV is significantly more common in diseased than in healthy persons in order to prove the pathogenic potential. This is hardly ever possible in clinical practice. Proving that a UV always cosegregates with the disease phenotype in a family can provide definite evidence for pathogenicity, but only if the family is large enough (about ten informative meiosis would be required). This is often not the case. Absence of the UV in an affected family member is, of course, strong evidence against the UV being the causal mutation. Importantly, if family history is negative, testing the healthy parents for presence of the UV can give vital information. If the UV turns out to be a de novo mutation, this can be regarded as very strong evidence in favor of pathogenicity. As mutation frequencies are exceedingly low, the chance that a new mutation would occur in the studied candidate gene just by coincidence is negligible.

If the amino acid change is likely to change the three-dimensional structure of the protein, or if it is located in a known binding site or important functional domain of the protein, and if it involves an evolutionary conserved residue, these are all indications that the UV is likely to be pathogenic, but none of these arguments does provide conclusive evidence. An amino acid residue is considered evolutionary conserved when all other species have the same amino acid at this position in the protein; in that case often other proteins of the same protein family also have the same amino acid at this position.

In a research setting it would often be possible to introduce the UV into a model system and observe its effect, but of course this is not possible in clinical practice.

Assumptions made regarding the pathogenic potential of missense mutations are therefore often provisional. It is important to realize this when using genetic information in clinical practice. Over-interpreting misses variants for pathogenic mutations is harmful in several ways, as on the one hand some individuals without a genetic predisposition to the disease will be stigmatized and unnecessarily kept under surveillance, while on the other hand, the actual causative mutation will go unnoticed and individuals may be released from surveillance, based on incorrect genetic information. The fact that a specific missense mutation has been published as a pathogenic mutation in the literature cannot always be regarded as sufficient evidence (one has to go back to the original publications and weigh the evidence).

2.6.8 The Cardiogenetics Outclinic

Since the care for individuals with genetic cardiac disease and their relatives requires both cardiologic and genetic expertise, in the Netherlands multidisciplinary outpatient clinics for cardiogenetics have been set up. In these outpatient clinics that are operating within the university hospitals, cardiologists, pediatric cardiologists, clinical geneticists, molecular geneticists, genetic nurses, psychologists, and/or social workers co-operate to provide integrated health care for this specific patient group. This is of benefit to patients because the number of hospital visits can usually be reduced, and also to health care providers because of easier communication. Besides, from a data collection and research point of view centralization of patients with inherited cardiac disease also has obvious advantages. It will be immediately clear that most of the regular care for this patient group will have to remain with cardiologists working in regional or local hospitals. With an estimated prevalence of 1 in 500 for hypertrophic cardiomyopathy, it would not only be unnecessary to follow all these patients in outclinics for cardiogenetics, but it would also be impossible. This implicates that a general awareness of the genetic aspects of cardiac disease among cardiologists is needed.

References

1. Hofman N, van Langen I, Wilde AM. Genetic testing in cardiovascular diseases. *Curr Opin Cardiol.* 2010;25: e-published.
2. Cowan J, Morales A, Dagua J, Hershberger RE. Genetic testing and henetic counseling in cardiovascular genetic medicine: overview and preliminary recommendations. *Congestive Heart Failure.* 2008;14:97-105.
3. Morales A, Cowan J, Dagua J, Hershberger RE. Family history: an essential tool for cardiovascular genetic medicine. *Congestive Heart Failure.* 2008;14:37-45.
4. Kingston HM. Genetic assessment and pedigree analysis. In: Rimoin DL, Connor JM, Pyeritz RE, Korf BR, eds. *Emery and Rimoin's Principles and Practice of Medical Genetics.* 4th ed. London: Churchill Livingstone; 2002.
5. Turnpenny P, Ellard S. *Emery's Elements of Medical Genetics.* 13th ed. London: Churchill Livingstone; 2008.
6. McKenna WJ, Thierfelder L, Suk HJ, Anan R, O'Donoghue A, et al. Mutations in the genes for cardiac troponin T and alpha-tropomyosin in hypertrophic cardiomyopathy. *N Engl J Med.* 1995;332:1058-1064.
7. Gimeno JR, Monserrat L, Pérez-Sánchez I, et al. Hypertrophic cardiomyopathy. A study of the troponin-T gene in 127 spanish families. *Rev Esp Cardiol.* 2009;62:1473-1477.
8. Van Langen IM, Birnie E, Alders M, Jongbloed RJ, Le Marec H, Wilde AA. The use of genotype-phenotype correlations in mutation analysis for the long QT syndrome. *J Med Genet.* 2003;40:141-145.
9. Schwartz PJ, Priori SG, Spazzolini C, et al. Genotype-phenotype correlation in the long-QT syndrome: gene-specific triggers for life-threatening arrhythmias. *Circulation.* 2001;103:89-95.
10. Alders M, Jongbloed R, Deelen W, et al. The 2373insG mutation in the MYBPC3 gene is a founder mutation, which accounts for nearly one-fourth of the HCM cases in the Netherlands. *Eur Heart J.* 2003;24:1848-1853.
11. Othman H, Saadat M. Prevalence of consanguineous marriages in Syria. *J Biosoc Sci.* 2009;41:685-692.
12. Lahat H, Pras E, Olender T, et al. A missense mutation in a highly conserved region of CASQ2 is associated with autosomal recessive catecholamine-induced polymorphic ventricular tachycardia in Bedouin families from Israel. *Am J Hum Genet.* 2001;69:1378-1384.

13. Hoorntje T, Alders M, van Tintelen P, et al. Homozygous premature truncation of the HERG protein: the human HERG knockout. *Circulation*. 1999;100:1264-1267.
14. Zahka K, Kalidas K, Simpson MA, et al. Homozygous mutation of MYBPC3 associated with severe infantile hypertrophic cardiomyopathy at high frequency among the Amish. *Heart*. 2008;94:1326-1330.
15. Resta R, Biesecker BB, Bennett RL, et al. A new definition of Genetic Counseling: National Society of Genetic Counselors' Task Force report. *J Genet Couns*. 2006;15:77-83.
16. Vincent GM, Schwartz PJ, Denjoy I, et al. High efficacy of beta-blockers in long-QT syndrome type 1: contribution of noncompliance and QT-prolonging drugs to the occurrence of beta-blocker treatment "failures". *Circulation*. 2009;119:215-221.
17. Hendriks KS, Hendriks MM, Birnie E, et al. Familial disease with a risk of sudden death: a longitudinal study of the psychological consequences of predictive testing for long QT syndrome. *Heart Rhythm*. 2008;5:719-724.
18. European Society of Human Genetics. Genetic testing in asymptomatic minors: recommendations of the European Society of Human Genetics. *Eur J Hum Genet*. 2009;17:720-721.
19. Hendriks KS, Grosfeld FJ, Wilde AA, et al. High distress in parents whose children undergo predictive testing for long QT syndrome. *Community Genet*. 2005;8:103-113.
20. van der Roest WP, Pennings JM, Bakker M, van den Berg MP, van Tintelen JP. Family letters are an effective way to inform relatives about inherited cardiac disease. *Am J Med Genet A*. 2009;149A:357-363.
21. Piyamongkol W, Harper JC, Sherlock JK, et al. A successful strategy for preimplantation genetic diagnosis of myotonic dystrophy using multiplex fluorescent PCR. *Prenat Diagn*. 2001;21:223-232.
22. Spits C, De Rycke M, Verpoest W, et al. Preimplantation genetic diagnosis for Marfan syndrome. *Fertil Steril*. 2006;86:310-320.
23. Arnestad M, Crotti L, Rognum TO, et al. Prevalence of long-QT syndrome gene variants in sudden infant death syndrome. *Circulation*. 2007;115:361-367.
24. Berul CI, Perry JC. Contribution of long-QT syndrome genes to sudden infant death syndrome: is it time to consider newborn electrocardiographic screening? *Circulation*. 2007;115:294-296.

Part II

Cardiomyopathy

Hypertrophic Cardiomyopathy

Imke Christiaans and Lucie Carrier

3.1 Introduction

Hypertrophic cardiomyopathy (HCM) has been described in medical literature for several centuries. The resurgence of anatomy in the Renaissance allowed further study of the disease, and dissections of victims of sudden death revealed bulky hearts.[1] Nowadays, HCM is still a major cause of sudden cardiac death (SCD) in the young, and the most common monogenetic heart disease.[2–4] This chapter discusses not only the epidemiology, diagnosis, pathophysiology, and therapy of HCM but also deals with topics specific for cardiogenetic diseases like the genetic background, risk stratification for SCD, and the screening strategies in HCM families. It is intended to be of help for all involved in the care for HCM patients and their families.

3.2 Prevalence and Diagnosis

Hypertrophic cardiomyopathy is a relatively common autosomal dominant disease affecting at least 1 in 500 persons in the general population worldwide.[5–9] The clinical diagnosis is made when there is a hypertrophied (often asymmetric), nondilated left ventricle on echocardiography (left ventricle wall thickness ≥ 15 mm) in the absence of other cardiac or systemic diseases that may cause cardiac hypertrophy, such as aortic valve stenosis and arterial hypertension.[10–14] Because of the hereditary nature of the disease, relatives of a patient can also be affected. In a patient's relative the diagnosis can be made if certain ECG abnormalities and/or a left ventricle wall thickness ≥ 13 mm are present.[15]

Considering the prevalence of clinically diagnosed HCM is at least 1 in 500, in the Netherlands with a population of 16.4 million[16], at least 32,800 people must have HCM and many more may be at risk of developing HCM, particularly mutation carrying relatives who have not developed left ventricular hypertrophy (LVH) yet. In a country like China the number of individuals with HCM would be more than 2.6 million. Many of these HCM patients are unaware of their disease and/or unaware of its hereditary nature. It is estimated that in the Netherlands only 10–20% of HCM mutation carriers have been identified and in other countries this number is probably much smaller; many individuals are therefore still unaware of their risk of HCM and associated SCD.

3.3 Pathophysiology and Natural History

The British pathologist Robert Donald Teare is assumed to be the first to describe hypertrophic cardiomyopathy in 1958.[17] Although bulky hearts and hypertrophy in sudden death victims have been described centuries earlier[1] and subaortic stenosis (a former name for hypertrophic cardiomyopathy) was described earlier in literature,[18,19] Teare was the first to describe the asymmetrical appearance of hypertrophy and its familial nature. He also described a disordered arrangement of muscle fibers at microscopic examination of the hearts of his cases, now known as myocyte disarray

I. Christiaans (✉)
Department of Clinical Genetics, Academic Medical Centre, Amsterdam, The Netherlands
Department of Cardiology, Academic Medical Centre, Amsterdam, The Netherlands
e-mail: i.christiaans@amc.uva.nl

(Fig. 3.1).[17] With electron microscopy one can also notice a disordered arrangement of the myofilaments as well. Myocardial disarray is not confined to the thickened parts of the left ventricle; left ventricle regions with normal thickness can also be disorganized.[20] Fibrosis is another feature of HCM visualized by microscopy or MRI. Both fibrosis and myocyte disarray are thought to be related to ventricular arrhythmias.[21] Besides hypertrophy, fibrosis, and disarray, abnormalities in the intramyocardial small vessels are another pathological finding in HCM. Vessels may show a thickening of the vessel wall and a decrease in luminal size.[22]

The diagnosis of HCM is most often made in adulthood, but HCM can present at any age. HCM is clinically heterogeneous; patients are often asymptomatic and diagnosed in routine screening, but HCM can also be a very disabling disease giving rise to dyspnea, exertional angina, palpitations, and (pre)syncope. The anatomic changes in HCM can be a substrate for arrhythmias, which may lead to palpitations, syncope, and even sudden cardiac death.

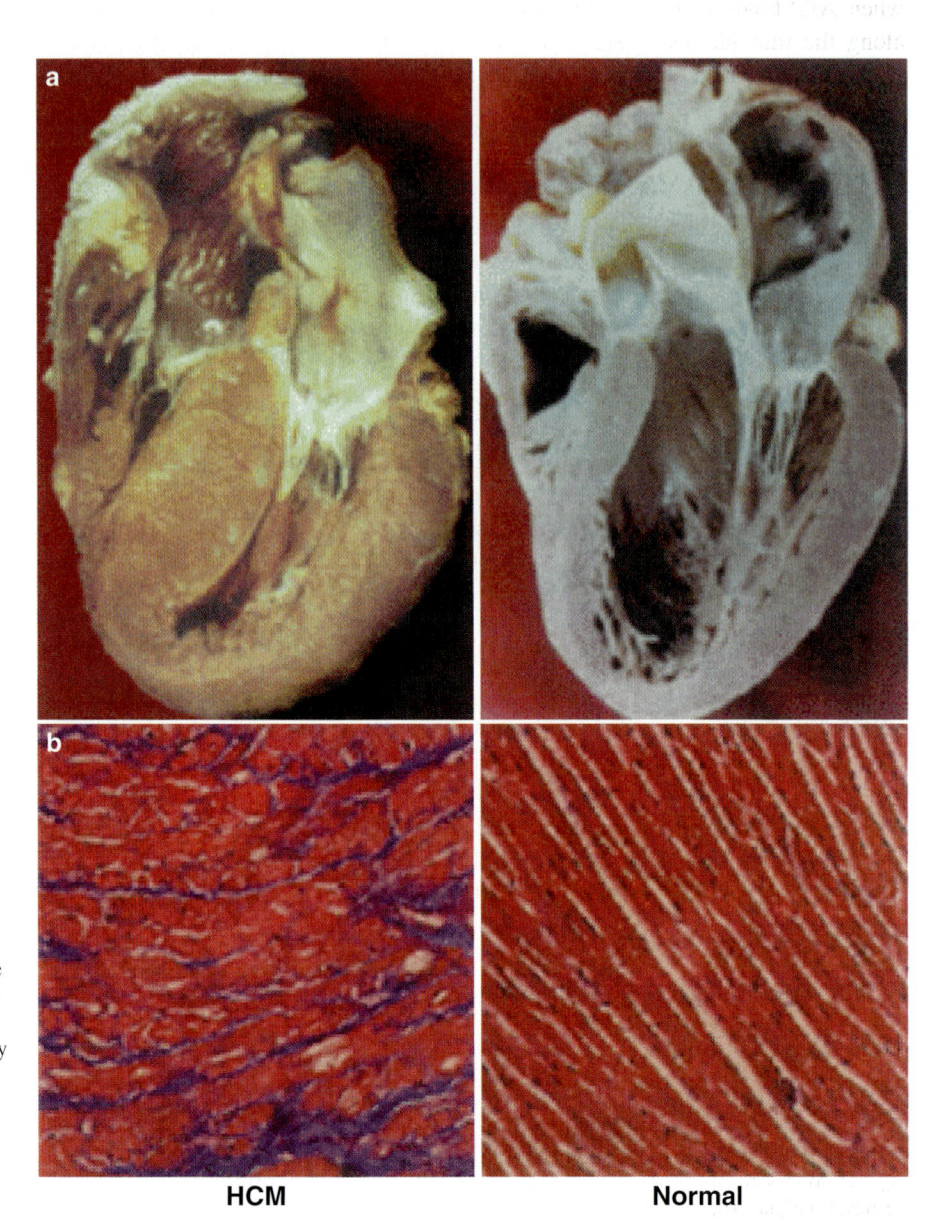

Fig. 3.1 (**a**) Normal heart and heart with hypertrophic cardiomyopathy. (**b**) Heart muscle on microscopy with structured fiber pattern in the normal heart and myocardial disarray in the heart with hypertrophic cardiomyopathy (Derived from Hypertrophic cardiomyopathy: from gene defect to clinical disease. Chung, MW, Tsoutsman, T, Semsarian C. *Cell Res.* 2003;13:9–20)

Early pathogenesis in HCM starts in the functional unit of contraction within the cardiomyocytes. The sarcomere is a protein complex divided into thick and thin myofilaments and proteins involved in the cytoarchitecture of the sarcomeres, like proteins in the Z-disc connecting the thin myofilaments of the sarcomere (Fig. 3.2). In response to electrical depolarization of the cardiomyocyte, intracellular levels of calcium rise. Calcium binds to the troponin complex of the thin filament and releases the inhibition of interactions between the thick filament (myosin) and thin filament (actin) by troponin I. The myosin head can then bind to actin and when ATP binds to myosin the myosin head moves along the thin filament. ATP hydrolysis occurs and force is generated. During relaxation, calcium is removed from the cytosol.[23] Calcium is not only a key molecule regulating both cardiac contraction and relaxation but also seems to play an important role in the early pathogenesis of HCM. It is hypothesized that the genetic defect in a gene encoding a sarcomeric protein disrupts normal contraction and relaxation of the sarcomere and calcium accumulates within the sarcomere. This leads to a reduction of calcium reuptake and eventually to reduced stores in the sarcoplasmic reticulum and increased calcium sensitivity. Myocytes in mice with HCM also display an inefficient use of ATP. This in combination with the increased calcium sensitivity triggers a remodeling process moderated by several transcription factors resulting in hypertrophy of cardiomyocytes. The increased mass of cardiomyocytes and inefficient use of ATP lead to an increased energy demand. When this energy demand cannot be met ischemia can result in premature myocyte death and replacement fibrosis.[23,24]

The hallmark of HCM, left ventricular hypertrophy (LVH), can appear at virtually any age and increases or decreases dynamically throughout life.[25,26] Studies suggest that extreme LVH occurs more frequently in the

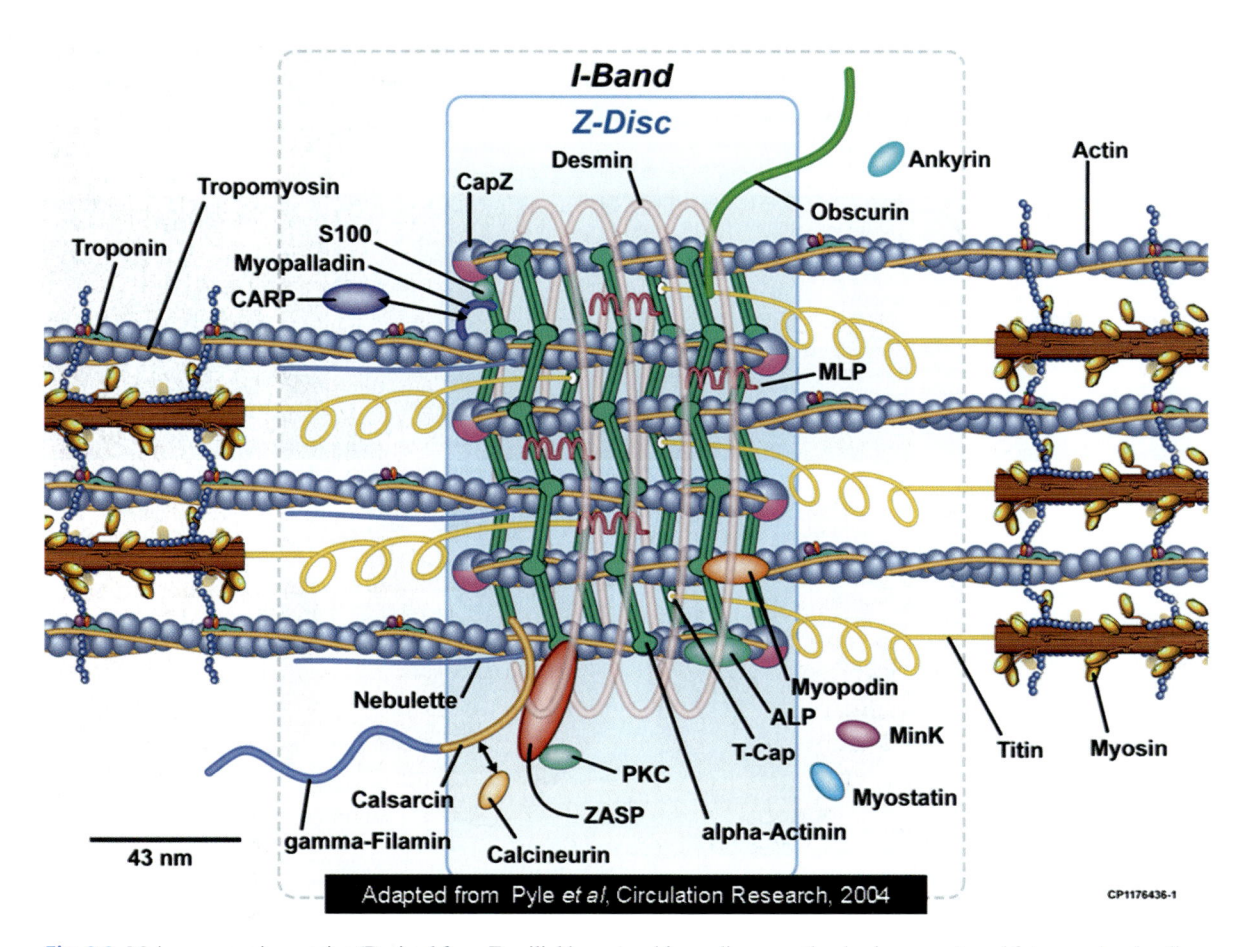

Fig. 3.2 Main sarcomeric proteins (Derived from Familial hypertrophic cardiomyopathy: basic concepts and future molecular diagnostics. Rodrigues JE, McCudden CR, Willis MS. *Clin Biochem*. 2009;42:755–765)

young, that is, in puberty or early adulthood.[27,28] A modestly decreasing magnitude of LVH with increasing age has also been found; however, this inverse relation was largely gender-specific for women.[29] These studies suggest that LVH is a dynamic feature of HCM, although all studies were cross-sectional in design.

There might be two possible explanations for the observation that left ventricular wall thickness is generally milder in patients of more advanced age than in the young. Firstly, young patients with marked LVH have a higher rate of premature cardiac death. This means that patients with extreme LVH are underrepresented in the subgroup of older patients (the so-called healthy survivors phenomenon). Secondly, progression to "end-stage" HCM may account in part for this.[29] About 5–10% of HCM patients progress to this end-stage, burn-out phase, which is characterized by systolic dysfunction, dilation of the left ventricle, and wall thinning, resembling the features of dilated cardiomyopathy.[26] Patients with this so-called dilated-hypokinetic evolution are younger at first evaluation, and more often have a family history of HCM or SCD than patients who do not develop these features.[30] Most of the reported patients with a dilated-hypokinetic evolution of HCM were 60 years or older.[25]

The most common location for LVH is the anterior part of the interventricular septum, giving rise to an asymmetrical appearance of hypertrophy. The asymmetrical septal hypertrophy can be divided into two subtypes: the sigmoid subtype with an ovoid LV cavity and a pronounced basal septal bulge and the reverse curvature subtype with a crescent shaped LV cavity and midseptal hypertrophy.[31] Other forms of hypertrophy, for example, concentric or neutral hypertrophy and apical hypertrophy can also be found in HCM (Fig. 3.3). The latter form seems to occur more frequently in East Asian patients.[32,33]

Not only can HCM present at any age, the clinical course is also variable. Patients can remain asymptomatic throughout life, but the disease can also give rise to heart failure and other adverse advents like sudden unexpected death and embolic stroke. First symptoms are often exertional dyspnea, disability (often associated with chest pain), dizziness, and (pre)

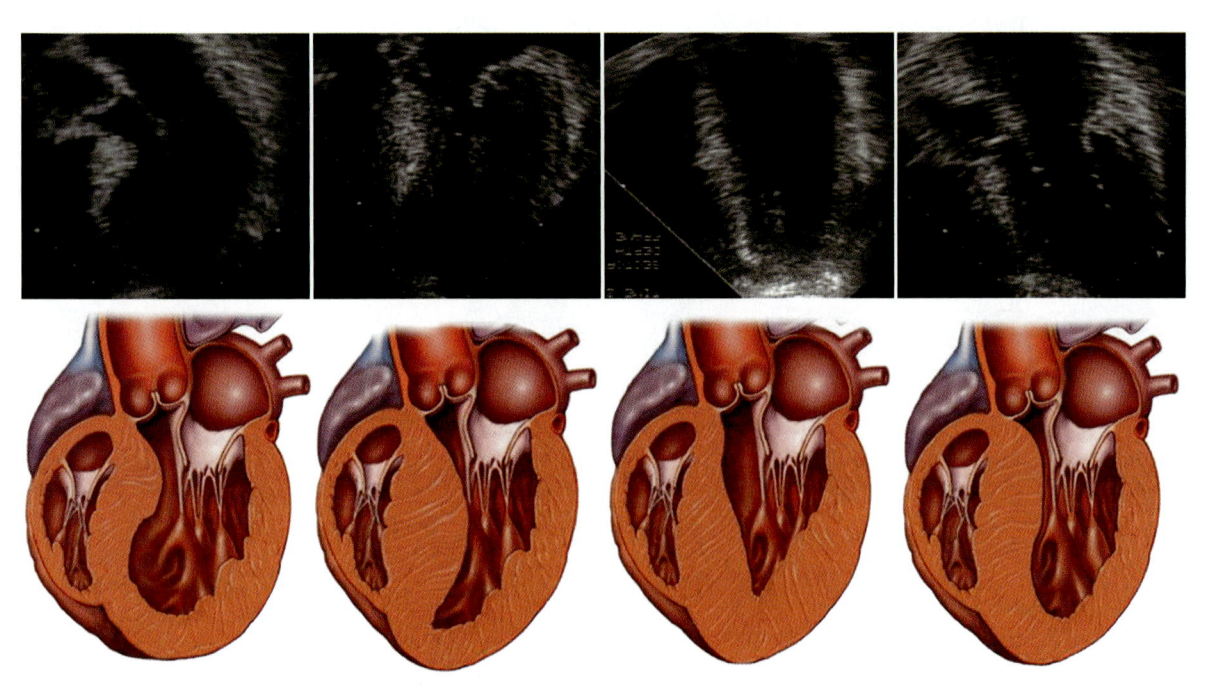

Fig. 3.3 Hypertrophic cardiomyopathy septal morphological subtypes based on standard echocardiography long-axis views taken at end-diastole (*top*). From *left* to *right*, sigmoid, reverse curve, apical, and neutral subtypes (Adapted from Echocardiography-guided genetic testing in hypertrophic cardiomyopathy: septal morphological features predict the presence of myofilament mutations. Binder J, Ommen SR, Gersh BJ, Van Driest SL, Tajik AJ, Nishimura RA, Ackerman MJ. *Mayo Clin Proc.* 2006;81: 459–467)

syncope. These symptoms are primarily caused by diastolic dysfunction with impaired filling of the left ventricle due to abnormal relaxation and increased stiffness of the myocardial wall, leading in turn to elevated left atrial and left ventricular end-diastolic pressures, pulmonary congestion, and impaired exercise performance.[34,35] Systolic function is often spared, but can arise in patients with the so called "end stage" form of HCM. Symptoms may become more disabling in the presence of other associated pathophysiological mechanisms, like myocardial ischemia, left ventricular outflow tract obstruction (LVOTO), and atrial fibrillation.

In the early descriptions of HCM, patients almost always had left ventricular outflow tract obstruction and the disease was also called HOCM (Hypertrophic Obstructive Cardiomyopathy). Nowadays, obstruction is still a frequent finding, but only a minority (20–30%) of HCM patients have LVOTO, which is in HCM often caused by systolic anterior motion (SAM) of the mitral valve leaflefts (Fig. 3.4).[31,36–39] In general LVOTO is defined as a peak gradient of 30 mmHg or more identified by Doppler echocardiography under basal (resting) conditions,[31,36–39] but the degree of LVOTO can vary in time and some patients only have provocable obstruction. The presence of LVOTO is a strong predictor of disease progression, symptoms of heart failure, and death due to heart failure and stroke.[38]

Atrial fibrillation is another frequent symptom in HCM (about 20%). It is associated with advanced age, congestive symptoms, and an increased left atrial size at diagnosis.[40] HCM patients with atrial fibrillation have an increased risk of heart failure-related death, besides the risk of cerebrovascular accidents.[36,38–40]

3.4 Disease Penetrance

HCM has long been regarded as a disease that mainly affects young people. It was thought that penetrance, that is, the presence of left ventricular hypertrophy, was complete at approximately 20 years of age.[41] Nowadays, it is more and more recognized that not only symptoms but also hypertrophy can develop at any point in life. Looking at symptoms of the disease, in an unselected patient group 65% of patients aged 75 years or older had no or only mild symptoms.[42] In this study the probability of survival for 5, 10, and 15 years for patients diagnosed with HCM after the age of 50 did not differ significantly from the (matched) general population. These data show that patients can have normal survival and that the disease may not give symptoms until late in life, or even may give no symptoms at all.

Although the abovementioned study is performed in an unselected population of HCM patients and based on the presence of symptoms, disease penetrance is best investigated in a population with proven mutation carriers. Studies in mutation carriers show that disease penetrance is incomplete in adult life, but reaches 95–100% after the age of 50 years.[25,43–45] The relationship between disease penetrance and the genotype (defined as the mutated gene or a specific mutation within a gene) is still not completely resolved. Studies including

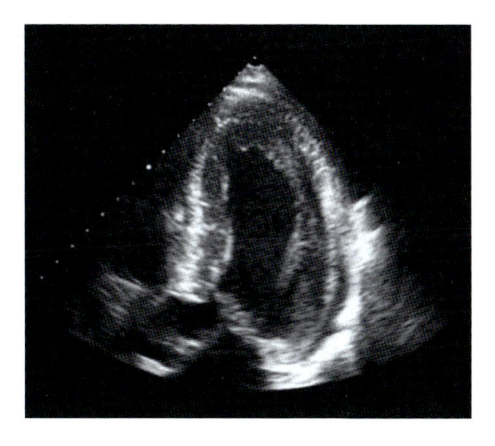

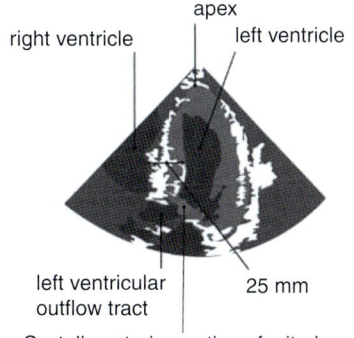

Fig. 3.4 Apical four chamber echocardiography in a HCM patient. Red *arrow* indicates the hypertrophic interventricular septum (>25 mm) and the white *arrow* indicates systolic anterior motion of a mitral valve leaflet in the left ventricular outflow tract

a large number of mutation carriers show no difference in disease penetrance between carriers of different mutated genes.[43,46]

3.5 Therapy

Management in HCM patients is directed toward control of symptoms, risk stratification for and prevention of SCD (see Sect. 5.6), and screening of relatives (see Sect. 5.8). Symptoms of dyspnea, angina, syncope, and fatigue are appraised and treated (Fig. 3.5).[34] Prophylactic pharmacological treatment in asymptomatic patients has not been proved to be effective in preventing progression of the disease. Drugs are often the only available treatment modality for symptomatic patients without LVOTO. Negative inotropic agents like beta-blockers and calcium antagonists have been shown to relieve symptoms. Beta-blockers decrease the heart rate, which results in a prolongation of the diastole and relaxation phase of the heart and an increase in passive ventricular filling. Besides, beta-blockers decrease left ventricular contractility and myocardial oxygen demand and can possibly reduce ischemia in myocardial microvessels. Verapamil, a calcium antagonist, also

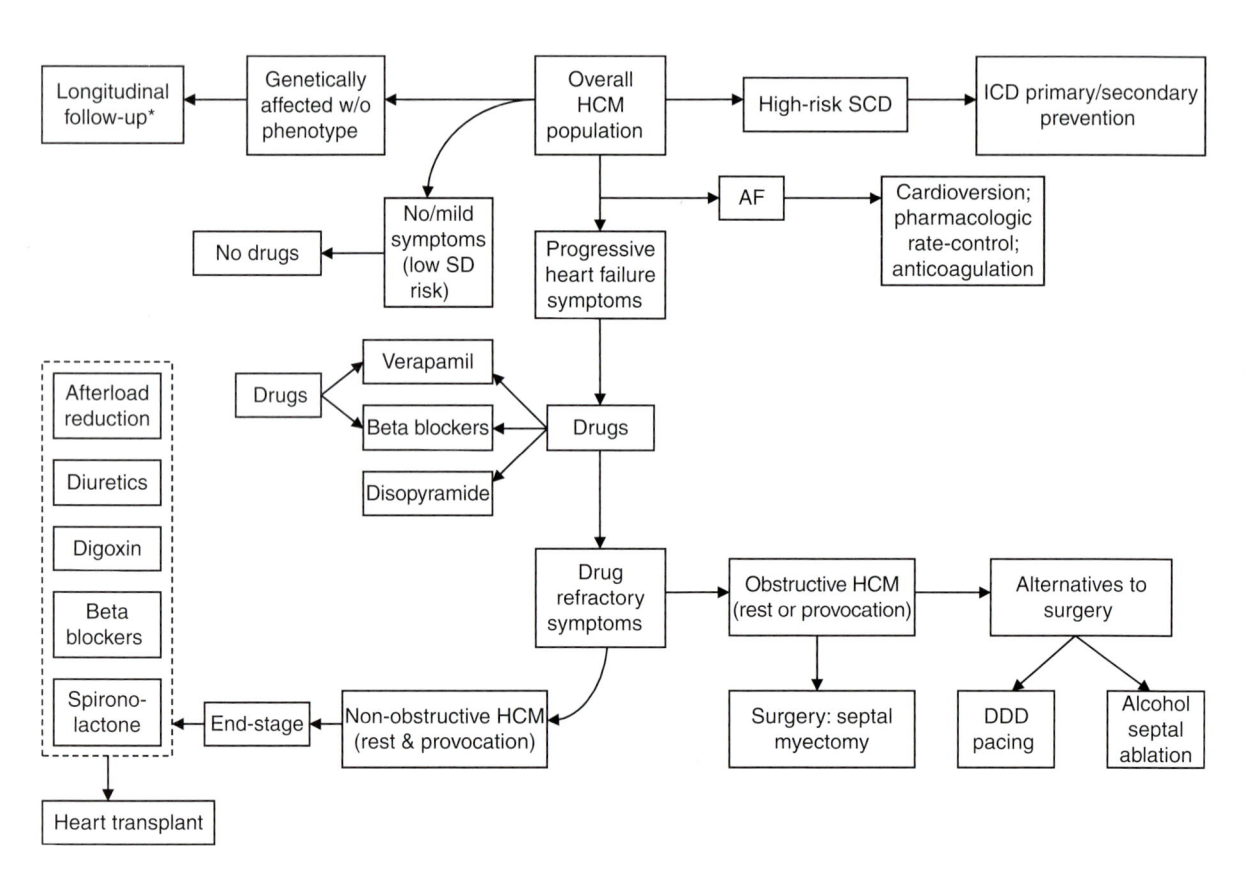

Fig. 3.5 Treatment strategies for patient sugroups within the broad clinical spectrum of hypertrophic cardiomyopathy (HCM). *AF* atrial fibrillation, *DDD* dual-chamber, *ICD* implantable cardioverter defibrillator, *SD* sudden death (Derived from American College of Cardiology/European Society of Cardiology clinical expert consensus document on hypertrophic cardiomyopathy. A report of the American College of Cardiology Foundation Task Force on Clinical Expert Consensus Documents and the European Society of Cardiology Committee for Practice Guidelines. Maron BJ et al.; Task Force on Clinical Expert Consensus Documents. American College of Cardiology; Committee for Practice Guidelines. European Society of Cardiology. *J Am Coll Cardiol.* 2003 Nov 5;42(9):1687–713)

has favorable effects on symptoms by improving ventricular relaxation and filling. In patients with progressive heart failure symptoms with LVOTO, disopyramide, a negative inotropic and type I-A antiarrhythmic drug can be used. It can decrease systolic anterior movement of the mitral valve, LVOTO, and mitral regurgitation. In end-stage HCM, load reducing drugs (ACE inhibitors, angiotensin-II receptor blockers, diuretics, digitalis, beta-blockers, or spironolactone) can be used to alleviate symptoms from systolic failure.[34]

For HCM patients with LVOTO, other treatment modalities are available if pharmacological treatment cannot alleviate symptoms. Ventricular septal myectomy (Morrow operation) is the next preferable treatment.[47] Septal myectomy has been associated with a considerable risk of mortality and morbidity. Nowadays, isolated myectomy has low operative mortality and brings effective long lasting improvement of symptoms.[48] An alternative, more recently developed treatment modality for symptomatic LVOTO is percutaneous alcohol septal ablation. The introduction of alcohol in the septal perforator branch of the left anterior descending coronary artery mimics septal myectomy.[49] Although both percutaneous alcohol septal ablation and septal myectomy can alleviate symptoms, the latter therapy is considered the gold standard because of a lower rate of procedural complications and more available knowledge on the long-term outcomes.[48,50]

Indirect evidence suggests an association between exercise and sudden cardiac death. Intense physical activity (e.g., sprinting) or systematic isometric exercise (e.g., heavy lifting) should be discouraged. (Young) HCM patients should be advised to avoid intense competitive sports and professional athletic careers. Bacterial endocarditis prophylaxis is recommended in HCM patients with LVOTO in resting or exercise conditions.[34]

3.6 Sudden Cardiac Death and Risk Stratification

One of the most used definitions of sudden cardiac death is: "Natural death due to cardiac causes, heralded by abrupt loss of consciousness within 1 hour of the onset of acute symptoms; preexisting heart diseases may have been known to be present, but the time and mode of death are unexpected."[51] In the past HCM was seen as a disease with an ominous prognosis, because of severe symptoms and high rates of SCD (3–6%/year).[26,51–56] These data, however, came from highly selected patient populations of severely affected patients treated in tertiary referral centers. These populations underrepresented clinically stable and asymptomatic patients.[57,58] Recent reports with less referral bias indicate overall annual mortality rates from HCM of 1–2% (SCD and heart failure-related death). Annual mortality from SCD alone in HCM patients is 0.4–1%.[59–65]

Although the overall risk of SCD for HCM patients is small in absolute terms, a small subset of patients is at much higher risk of SCD. Much would be gained in terms of treatment benefits and avoidance of diagnostic and treatment burden if this subset could be detected at an early stage. The ACC/ESC Consensus Report suggests that for this purpose all HCM patients should be identified and undergo cardiological diagnostics, to evaluate clinical signs of HCM and to estimate the risk of SCD.[34] Currently, six major risk factors for SCD have been identified which are mainly based on the outcome of various cardiological diagnostic tests: (1) prior cardiac arrest (ventricular fibrillation) or spontaneous sustained ventricular tachycardia, (2) a family history of premature sudden death, (3) unexplained syncope, (4) left ventricular wall thickness ≥ 30 mm, (5) abnormal exercise blood pressure, and (6) nonsustained ventricular tachycardia. Possible minor risk factors for SCD include: (1) atrial fibrillation, (2) myocardial ischemia, (3) left ventricular outflow tract obstruction, (4) a high-risk mutation, and (5) intense (competitive) physical exertion.[34] Recently, myocardial fibrosis assessed by magnetic resonance imaging is being evaluated as a possible risk factor for SCD. Associations with nonsustained ventricular tachycardia and other risk factors have been found, but a prognostic study evaluating fibrosis as an independent predictor for SCD is still lacking.[66–68]

HCM patients, particularly those under the age of 60 years, should undergo a yearly clinical assessment including risk stratification based on the presence of the six major risk factors for SCD: careful personal and family history, 12-lead ECG, echocardiography, Holter recording, and exercise test.[34] The presence of two major risk factors is associated with an estimated annual risk of SCD of 4–5%.[52] According to international guidelines this risk is regarded sufficiently high to justify primary prevention of SCD by means of an implantable cardioverter defibrillator (ICD). In patients with a prior cardiac arrest or spontaneous sustained ventricular tachycardia, an ICD should be implanted for secondary prevention.[34,69,70] Recent guidelines on arrhythmia and the prevention of SCD also support ICD implantation in HCM patients with only one of the six major risk factors and who are considered to be at high risk for SCD by their physician.[69,70] At present the implantable cardioverter defibrillator (ICD) appears to be an effective prophylactic treatment modality.[71,72] However, due to the low positive predictive value of the major risk factors, the decision to implant an ICD in HCM patients with only one major risk factor for SCD remains controversial.

3.7 Genetics of HCM

HCM is inherited as an autosomal dominant trait. In more than half of the HCM patients the disease causing mutation can be identified currently.[46,73–79] Mutations can be located in many genes, but are most often found in the genes encoding sarcomeric proteins (Table 3.1, Fig. 3.2). Sarcomeric genes can be divided into genes encoding for myofilament proteins [13,14,46,74,75,78–83] and genes encoding for Z-disc proteins.[84–89]

Most HCM patients are heterozygous for the mutation, but in 3–5% of cases, patients carry two mutations in the same gene (different alleles – compound heterozygous or homozygous–) or in different genes (digenic). This is generally associated with a more severe phenotype with younger age of onset (often < 10 years) and more adverse events suggesting a gene-dosage effect.[79,90–93] The two most frequently mutated genes are *MYBPC3* and *MYH7*, encoding the sarcomeric proteins cardiac myosin-binding protein C and beta myosin heavy chain, respectively (Table 3.1). Both proteins are major components of the sarcomeric thick filament. In contrast to *MYH7* and most of the other genes associated with HCM, 70% of *MYBPC3* mutations are nonsense or frameshift and should result in truncated proteins[94,95], suggesting haploinsufficiency. In the other genes missense mutations are most frequent, which create a mutant protein that interferes with normal function (dominant negative effect).

Since the discovery of the first genes for HCM, many papers on genotype–phenotype correlations have been written. At first, specific mutations, mainly in the *MYH7* gene, were described that were associated with a "malignant" phenotype (decreased survival).[96–98] So-called "benign" mutations were reported in families with normal longevity, as well.[96,97,99–104] These suggested "malignant" and "benign" mutations have been contradicted in many subsequent studies. Nowadays, it is believed that, possibly apart from rare exceptions, there are no clear genotype–phenotype relations with respect to magnitude of left ventricular hypertrophy and incidence of sudden cardiac death.[46,96,100,102,105–107] Moreover, genetic studies have revealed that not all mutation carriers are clinically affected, using standard echocardiography. This suggests the existence of modifier genes, which modulate the phenotypic expression of the disease. Associations have been found with polymorphisms in genes for the angiotensin II type 1 and type 2 receptors and in the promoter region of the calmodulin III gene.[108–110]

Non-sarcomeric genes have been associated with specific phenotypes which include, besides HCM, almost always a distinct noncardiac syndromic phenotype, like the PTPN11 gene in Noonan syndrome and the LAMP2 gene in Danon disease. However, mutations in the GLA gene, associated with Fabry disease, can give rise to HCM without further symptoms of Fabry disease, especially in women.[111] PRKAG2 is the other non-sarcomeric gene, which presents with an exclusively cardiac phenotype. The cardiac phenotype is distinct and includes besides HCM electrical pre-excitation (Table 3.1).[112]

De novo mutations and germline mosaicism occur very rarely in HCM.[113–117] Because most mutations are private, many of the identified mutations are novel. In certain countries/populations, however, founder mutations have been identified, in which haplotype analysis suggests a common ancestor. These founder mutations often comprise a large part (10–25%) of the detected mutations in these countries. Founder mutations have been found in the Netherlands,[118,119] South Africa,[120] Finland,[121] Italy,[74] Japan[25], and in the Amish population of the USA.[122]

Like in other genetic diseases, identified mutations in HCM patients can be pathogenic (disease causing), silent polymorphisms, or unclassified variants of which

Table 3.1 Genes associated with hypertrophic cardiomyopathy (HCM) and their detection rate [46,74,75,78,79,84–89]

Gene	Name	Detection rate
Sarcomeric		
Myofilament		
MYBPC3	Myosin-binding protein C	13–32%
MYH7	Beta myosin heavy chain	4–25%
TNNT2	Troponin T2	0.5–7%
TNNI3	Cardiac troponin I	<5%
MYL2	Myosin light chain 2	<5%
MYL3	Myosin light chain 3	<1%
TPM1	Alpha tropomyosin	<1%
ACTC	Alpha actin	<1%
TNNC1	Troponin C	<1%
Z-disc		
ACTN2	Alpha-2 actinin	
CSRP3	Cysteine- and glycine-rich protein 3	4–5%
LBD3 (or ZASP)	Lim domain-binding 3	
TCAP	Titin-cap (Telethonin)	
VCL	Vinculin	
TTN	Titin	
MYOZ2	Myozenin 2	<1%
Non-sarcomeric[a]		*Phenotype*
PRKAG2	AMP-activated protein kinase gamma 2	LVH/pre-excitation (Wolf-Parkinson-White syndrome)/conduction disturbances
LAMP2	Lysosome-associated membrane protein 2	Danon disease
GLA	Alpha galactosidase	Fabry disease
PTPN11	Protein-tyrosine phosphatase non-receptor-type 11	Noonan, Leopard, CFC syndrome
KRAS2	Kirsten rat sarcoma viral oncogen homolog	Noonan, Leopard, CFC syndrome
SOS1	Son of sevenless homolog 1	Noonan syndrome
BRAF1	V-RAF murine sarcoma viral oncogen homolog B1	CFC syndrome
MAP2K1	Mitogen-activated protein kinase kinase 1	CFC syndrome
MAP2K2	Mitogen-activated protein kinase kinase 2	CFC syndrome
HRAS	Harvey rat sarcoma viral oncogene homolog	Costello syndrome
GAA	Glucosidase alpha acid	Pompe disease
GDE	Glycogen debrancher enzyme	Glycogen storage disorder III
FXN	Frataxin	Friedreich ataxia
TTR	Transthyretin	Amyloidosis I
Mitochondrial DNA		LVH "plus"

[a]Because of specific phenotype mutation detection rate is not provided

the pathogenic effect is still unclear. Nonsense or frameshift mutations are most often pathogenic because they are predicted to result in a C-terminal truncated protein, that is likely to be nonfunctional. Moreover, due to the presence of two quality controls, the nonsense-mediated mRNA decay degrading nonsense (truncated) mRNAs[123] and the ubiquitin-proteasome degrading aberrant proteins;[124] truncating mutations most often do not result in the formation of protein at all and therefore lead to haploinsufficiency of the protein encoded by the mutated allele in the cells. Missense mutations create a mutant protein that either interferes with normal function (dominant negative effect) or assumes a new function. Sometimes, however, it remains unclear if a missense mutation results in a protein with no or abnormal function.

The amino acid substitution in missense mutations can give some indications for pathogenicity. Missense mutations at codons conserved between species and/or isoforms are more likely to be pathogenic than mutations at poorly conserved regions. Different in-silico methods have been developed to assess not only conservation, but also changes in protein structure, chemical and biophysical characteristics, and interactions. These in-silico methods to define pathogenicity have to be validated by comparison to a gold standard. Gold standards can be functional assays, frequently found mutations (e.g., founder mutations), or segregation with the phenotype. All of these potential standards, however, have their own strengths and weaknesses.[125,126] Unclassified variants and even polymorphisms in HCM-associated genes and other genes (e.g., polymorphisms in the renin-angiotensin–aldosterone system[127,128]) may exhibit phenotype modifying effects. In most countries, analysis of uncertain variants and modifiers in HCM patients is currently performed in research setting only and not used in clinical decision making.

3.8 Screening of Relatives: Genetic Counseling and Testing

Consensus documents on HCM encourage screening of relatives because of the risk of HCM-associated sudden cardiac death.[14] Identification of a disease causing mutation in a HCM patient (the proband) implies the opportunity of screening by means of predictive DNA testing in relatives. DNA testing for HCM, and especially predictive DNA testing in relatives, is not common practice in most countries, because health insurance does not cover the costs of DNA testing, and/or because genetic counseling is unavailable. DNA testing is therefore often only available in a research setting. Instead, most of these countries use clinical modalities such as echocardiography and ECG to screen relatives on the presence of disease. In the Netherlands, diagnostic DNA testing is covered by standard health insurance and DNA testing for HCM has been increasing since 1996 with specialized multidisciplinary cardiogenetics outpatient clinics now present in all eight university hospitals. Most mutations (80%) in the Netherlands are found in the MYBPC3 gene, mainly because of three founder mutations (c.2373_2374insG, c.2827C>T, and c.2864_2865delCT).

In the Netherlands, there is much experience with family screening for HCM and other cardiogenetic diseases as specialized outpatient clinics have been developed in the early stages of DNA testing (see Chap. 25). Here, systematic screening of relatives in families with a disease causing mutation, so-called cascade screening, starts with the proband. Probands in whom a disease causing mutation is detected receive a family letter to inform their direct relatives about the possibilities of DNA testing and its pros and cons. Predictive DNA testing in relatives >10 years can only be performed after genetic counseling. A genetic counseling session involves one session with a genetic counselor (clinical geneticist or genetic consultant) and a cardiologist and takes on average between 30 and 45 min. An additional session with a psychologist or social worker is offered to all counselees but is only mandatory before DNA testing when children are involved. The DNA test result can be communicated to the counselee face to face at the outpatient clinic, by telephone, or by mail, according to the counselee's preference. All counselees receive a letter with their DNA test result and its implications for the counselee and his or her relatives.[129]

Relatives without the familial HCM mutation can be discharged from cardiological follow-up. Relatives who carry the familial HCM mutation are, according to international guidelines, like affected HCM patients with or without mutation, advised to undergo regular cardiological evaluation to evaluate the presence of left ventricular hypertrophy and risk factors for sudden cardiac death.[34] During follow-up left ventricular hypertrophy (manifest disease) can present at any age. The presence of risk factors for SCD in mutation carriers is associated with an increased risk of SCD if HCM is manifest (i.e., if left ventricular hypertrophy is present). In mutation carrying relatives (still) without manifest HCM, risk stratification is advised; yet it is unclear if the major risk factors established for affected HCM patients are also valid for mutation carrying relatives without manifest disease and if the presence of these risk factors is associated with an increased risk of SCD.

If no mutation can be detected in a proband with HCM, first degree relatives still have a risk to develop hypertrophic cardiomyopathy. DNA testing in these relatives is impossible; they are all advised to undergo regular cardiological evaluations (annual between 12 and 18 years and once every 5 years >18 years) assessing the presence of hypertrophy or ECG abnormalities directing toward HCM.[34]

3.9 Summary

Hypertrophic cardiomyopathy (HCM) is the most common monogenetic heart disease affecting 1 in 500 people worldwide. Hallmark of the disease is left ventricular hypertrophy in the absence of cardiac or systemic disease that may cause hypertrophy. The disease can present at any age and is highly variable. Patients can remain asymptomatic throughout their life, but HCM is also associated with adverse clinical events, like heart failure, stroke, and sudden cardiac death. Therapy is mainly directed toward relieving symptoms of heart failure and left ventricular outflow tract obstruction. Risk stratification with clinical risk markers can identify patients at high risk for sudden cardiac death. In these patients, prevention of sudden cardiac death is effective with an implantable cardioverter defibrillator.

Because of the hereditary nature of the disease, first degree relatives are advised to undergo periodic cardiological evaluation for the presence of left ventricular hypertrophy. In about half of the HCM patients a disease causing mutation can be detected in one of the genes encoding for sarcomeric proteins. Detection of a disease causing mutation allows predictive genetic testing in relatives, and can thus better identify the relatives at risk for HCM and associated death. Although there is no evidence of a clear benefit of early pharmacological treatment in mutation carrying relatives, risk stratification for sudden cardiac death is also warranted in them and can save lives.

References

1. Coats CJ, Hollman A. Hypertrophic cardiomyopathy: lessons from history. *Heart.* 2008;94:1258-1263.
2. Basso C, Calabrese F, Corrado D, Thiene G. Postmortem diagnosis in sudden cardiac death victims: macroscopic, microscopic and molecular findings. *Cardiovasc Res.* 2001; 50:290-300.
3. Drory Y, Turetz Y, Hiss Y, et al. Sudden unexpected death in persons less than 40 years of age. *Am J Cardiol.* 1991; 68:1388-1392.
4. Maron BJ. Sudden death in young athletes. *N Engl J Med.* 2003;349:1064-1075.
5. Hada Y, Sakamoto T, Amano K, et al. Prevalence of hypertrophic cardiomyopathy in a population of adult Japanese workers as detected by echocardiographic screening. *Am J Cardiol.* 1987;59:183-184.
6. Maron BJ, Peterson EE, Maron MS, Peterson JE. Prevalence of hypertrophic cardiomyopathy in an outpatient population referred for echocardiographic study. *Am J Cardiol.* 1994; 73:577-580.
7. Maron BJ, Gardin JM, Flack JM, Gidding SS, Kurosaki TT, Bild DE. Prevalence of hypertrophic cardiomyopathy in a general population of young adults. Echocardiographic analysis of 4111 subjects in the CARDIA Study. Coronary Artery Risk Development in (Young) Adults. *Circulation.* 1995;92:785-789.
8. Morita H, Larson MG, Barr SC, et al. Single-gene mutations and increased left ventricular wall thickness in the community: the Framingham Heart Study. *Circulation.* 2006; 113:2697-2705.
9. Zou Y, Song L, Wang Z, et al. Prevalence of idiopathic hypertrophic cardiomyopathy in China: a population-based echocardiographic analysis of 8080 adults. *Am J Med.* 2004; 116:14-18.
10. Klues HG, Schiffers A, Maron BJ. Phenotypic spectrum and patterns of left ventricular hypertrophy in hypertrophic cardiomyopathy: morphologic observations and significance as assessed by two-dimensional echocardiography in 600 patients. *J Am Coll Cardiol.* 1995;26:1699-1708.
11. Maron BJ, Spirito P, Wesley Y, Arce J. Development and progression of left ventricular hypertrophy in children with hypertrophic cardiomyopathy. *N Engl J Med.* 1986;315:610-614.
12. Wigle ED, Sasson Z, Henderson MA, et al. Hypertrophic cardiomyopathy. The importance of the site and the extent of hypertrophy. A review. *Prog Cardiovasc Dis.* 1985; 28:1-83.
13. Elliott P, Andersson B, Arbustini E, et al. Classification of the cardiomyopathies: a position statement from the European Society of Cardiology Working Group on Myocardial and Pericardial Diseases. *Eur Heart J.* 2008;29: 270-276.
14. Maron BJ, Towbin JA, Thiene G, et al. Contemporary definitions and classification of the cardiomyopathies: an American Heart Association Scientific Statement from the Council on Clinical Cardiology, Heart Failure and Transplantation Committee; Quality of Care and Outcomes Research and Functional Genomics and Translational Biology Interdisciplinary Working Groups; and Council on Epidemiology and Prevention. *Circulation.* 2006;113:1807-1816.
15. McKenna WJ, Spirito P, Desnos M, Dubourg O, Komajda M. Experience from clinical genetics in hypertrophic cardiomyopathy: proposal for new diagnostic criteria in adult members of affected families. *Heart.* 1997;77:130-132.
16. Central Bureau of Statistics. October. 2008. Ref Type: Generic
17. Teare D. Asymmetrical hypertrophy of the heart in young adults. *Br Heart J.* 1958;20:1-8.

18. Chevers N. Observations on the diseases of the orifice and valves of the aorta. *Guys Hosp Rep.* 1842;7:387-442.
19. Vulpian A. Contribution à l'étude des rétrécissements de l'orifice ventriculo-aortique. *Arch Physiol.* 1868;3:456-457.
20. Maron BJ, Wolfson JK, Roberts WC. Relation between extent of cardiac muscle cell disorganization and left ventricular wall thickness in hypertrophic cardiomyopathy. *Am J Cardiol.* 1992;70:785-790.
21. Varnava AM, Elliott PM, Mahon N, Davies MJ, McKenna WJ. Relation between myocyte disarray and outcome in hypertrophic cardiomyopathy. *Am J Cardiol.* 2001;88:275-279.
22. Maron BJ, Wolfson JK, Epstein SE, Roberts WC. Intramural ("small vessel") coronary artery disease in hypertrophic cardiomyopathy. *J Am Coll Cardiol.* 1986;8:545-557.
23. Ahmad F, Seidman JG, Seidman CE. The genetic basis for cardiac remodeling. *Annu Rev Genomics Hum Genet.* 2005;6:185-216.
24. Tsoutsman T, Lam L, Semsarian C. Genes, calcium and modifying factors in hypertrophic cardiomyopathy. *Clin Exp Pharmacol Physiol.* 2006;33:139-145.
25. Kubo T, Kitaoka H, Okawa M, et al. Lifelong left ventricular remodeling of hypertrophic cardiomyopathy caused by a founder frameshift deletion mutation in the cardiac Myosin-binding protein C gene among Japanese. *J Am Coll Cardiol.* 2005;46:1737-1743.
26. Maron BJ. Hypertrophic cardiomyopathy: a systematic review. *JAMA.* 2002;287:1308-1320.
27. Elliott PM, Gimeno Blanes JR, Mahon NG, Poloniecki JD, McKenna WJ. Relation between severity of left-ventricular hypertrophy and prognosis in patients with hypertrophic cardiomyopathy. *Lancet.* 2001;357:420-424.
28. Maron BJ, Piccininno M, Casey SA, Bernabo P, Spirito P. Relation of extreme left ventricular hypertrophy to age in hypertrophic cardiomyopathy. *Am J Cardiol.* 2003;91:626-628.
29. Maron BJ, Casey SA, Hurrell DG, Aeppli DM. Relation of left ventricular thickness to age and gender in hypertrophic cardiomyopathy. *Am J Cardiol.* 2003;91:1195-1198.
30. Biagini E, Coccolo F, Ferlito M, et al. Dilated-hypokinetic evolution of hypertrophic cardiomyopathy: prevalence, incidence, risk factors, and prognostic implications in pediatric and adult patients. *J Am Coll Cardiol.* 2005;46:1543-1550.
31. Binder J, Ommen SR, Gersh BJ, et al. Echocardiography-guided genetic testing in hypertrophic cardiomyopathy: septal morphological features predict the presence of myofilament mutations. *Mayo Clin Proc.* 2006;81:459-467.
32. Kitaoka H, Doi Y, Casey SA, Hitomi N, Furuno T, Maron BJ. Comparison of prevalence of apical hypertrophic cardiomyopathy in Japan and the United States. *Am J Cardiol.* 2003;92:1183-1186.
33. Sakamoto T, Tei C, Murayama M, Ichiyasu H, Hada Y. Giant T wave inversion as a manifestation of asymmetrical apical hypertrophy (AAH) of the left ventricle. Echocardiographic and ultrasono-cardiotomographic study. *Jpn Heart J.* 1976;17:611-629.
34. Maron BJ, McKenna WJ, Danielson GK, et al. American College of Cardiology/European Society of Cardiology clinical expert consensus document on hypertrophic cardiomyopathy. A report of the American College of Cardiology Foundation Task Force on Clinical Expert Consensus Documents and the European Society of Cardiology Committee for Practice Guidelines. *J Am Coll Cardiol.* 2003;42:1687-1713.
35. Sanderson JE, Gibson DG, Brown DJ, Goodwin JF. Left ventricular filling in hypertrophic cardiomyopathy. An angiographic study. *Br Heart J.* 1977;39:661-670.
36. Autore C, Bernabo P, Barilla CS, Bruzzi P, Spirito P. The prognostic importance of left ventricular outflow obstruction in hypertrophic cardiomyopathy varies in relation to the severity of symptoms. *J Am Coll Cardiol.* 2005;45:1076-1080.
37. Elliott PM, Gimeno JR, Tome MT, et al. Left ventricular outflow tract obstruction and sudden death risk in patients with hypertrophic cardiomyopathy. *Eur Heart J.* 2006;27:1933-1941.
38. Maron MS, Olivotto I, Betocchi S, et al. Effect of left ventricular outflow tract obstruction on clinical outcome in hypertrophic cardiomyopathy. *N Engl J Med.* 2003;348:295-303.
39. Olivotto I, Maron BJ, Montereggi A, Mazzuoli F, Dolara A, Cecchi F. Prognostic value of systemic blood pressure response during exercise in a community-based patient population with hypertrophic cardiomyopathy. *J Am Coll Cardiol.* 1999;33:2044-2051.
40. Olivotto I, Cecchi F, Casey SA, Dolara A, Traverse JH, Maron BJ. Impact of atrial fibrillation on the clinical course of hypertrophic cardiomyopathy. *Circulation.* 2001;104:2517-2524.
41. Charron P, Komajda M. Molecular genetics in hypertrophic cardiomyopathy: towards individualized management of the disease. *Expert Rev Mol Diagn.* 2006;6:65-78.
42. Maron BJ, Casey SA, Hauser RG, Aeppli DM. Clinical course of hypertrophic cardiomyopathy with survival to advanced age. *J Am Coll Cardiol.* 2003;42:882-888.
43. Charron P, Carrier L, Dubourg O, et al. Penetrance of familial hypertrophic cardiomyopathy. *Genet Couns.* 1997;8:107-114.
44. Maron BJ, Niimura H, Casey SA, et al. Development of left ventricular hypertrophy in adults in hypertrophic cardiomyopathy caused by cardiac myosin-binding protein C gene mutations. *J Am Coll Cardiol.* 2001;38:315-321.
45. Niimura H, Bachinski LL, Sangwatanaroj S, et al. Mutations in the gene for cardiac myosin-binding protein C and late-onset familial hypertrophic cardiomyopathy. *N Engl J Med.* 1998;338:1248-1257.
46. Van Driest SL, Ommen SR, Tajik AJ, Gersh BJ, Ackerman MJ. Sarcomeric genotyping in hypertrophic cardiomyopathy. *Mayo Clin Proc.* 2005;80:463-469.
47. Morrow AG, Reitz BA, Epstein SE, et al. Operative treatment in hypertrophic subaortic stenosis. Techniques, and the results of pre and postoperative assessments in 83 patients. *Circulation.* 1975;52:88-102.
48. Brown ML, Schaff HV. Surgical management of obstructive hypertrophic cardiomyopathy: the gold standard. *Expert Rev Cardiovasc Ther.* 2008;6:715-722.
49. Faber L, Meissner A, Ziemssen P, Seggewiss H. Percutaneous transluminal septal myocardial ablation for hypertrophic obstructive cardiomyopathy: long term follow up of the first series of 25 patients. *Heart.* 2000;83:326-331.
50. Sorajja P, Valeti U, Nishimura RA, et al. Outcome of alcohol septal ablation for obstructive hypertrophic cardiomyopathy. *Circulation.* 2008;118:131-139.

51. Priori SG, Aliot E, Blomstrom-Lundqvist C, et al. Task Force on Sudden Cardiac Death of the European Society of Cardiology. *Eur Heart J.* 2001;22:1374-1450.
52. Elliott PM, Poloniecki J, Dickie S, et al. Sudden death in hypertrophic cardiomyopathy: identification of high risk patients. *J Am Coll Cardiol.* 2000;36:2212-2218.
53. Maki S, Ikeda H, Muro A, et al. Predictors of sudden cardiac death in hypertrophic cardiomyopathy. *Am J Cardiol.* 1998;82:774-778.
54. McKenna W, Deanfield J, Faruqui A, England D, Oakley C, Goodwin J. Prognosis in hypertrophic cardiomyopathy: role of age and clinical, electrocardiographic and hemodynamic features. *Am J Cardiol.* 1981;47:532-538.
55. Swan DA, Bell B, Oakley CM, Goodwin J. Analysis of symptomatic course and prognosis and treatment of hypertrophic obstructive cardiomyopathy. *Br Heart J.* 1971;33:671-685.
56. Koga Y, Itaya K, Toshima H. Prognosis in hypertrophic cardiomyopathy. *Am Heart J.* 1984;108:351-359.
57. Maron BJ, Spirito P. Impact of patient selection biases on the perception of hypertrophic cardiomyopathy and its natural history. *Am J Cardiol.* 1993;72:970-972.
58. Elliott PM, Gimeno JR, Thaman R, et al. Historical trends in reported survival rates in patients with hypertrophic cardiomyopathy. *Heart.* 2006;22:1374-1450.
59. Cecchi F, Olivotto I, Montereggi A, Squillatini G, Dolara A, Maron BJ. Prognostic value of non-sustained ventricular tachycardia and the potential role of amiodarone treatment in hypertrophic cardiomyopathy: assessment in an unselected non-referral based patient population. *Heart.* 1998;79:331-336.
60. Adabag AS, Casey SA, Kuskowski MA, Zenovich AG, Maron BJ. Spectrum and prognostic significance of arrhythmias on ambulatory Holter electrocardiogram in hypertrophic cardiomyopathy. *J Am Coll Cardiol.* 2005;45:697-704.
61. Maron BJ, Casey SA, Poliac LC, Gohman TE, Almquist AK, Aeppli DM. Clinical course of hypertrophic cardiomyopathy in a regional United States cohort. *JAMA.* 1999;281:650-655.
62. Kofflard MJ, Waldstein DJ, Vos J, Ten Cate FJ. Prognosis in hypertrophic cardiomyopathy observed in a large clinic population. *Am J Cardiol.* 1993;72:939-943.
63. Cannan CR, Reeder GS, Bailey KR, Melton LJI, Gersh BJ. Natural history of hypertrophic cardiomyopathy. A population-based study, 1976 through 1990. *Circulation.* 1995;92:2488-2495.
64. Kyriakidis M, Triposkiadis F, Anastasakis A, et al. Hypertrophic cardiomyopathy in Greece: clinical course and outcome. *Chest.* 1998;114:1091-1096.
65. Maron BJ, Olivotto I, Spirito P, et al. Epidemiology of hypertrophic cardiomyopathy-related death: revisited in a large non-referral-based patient population. *Circulation.* 2000;102:858-864.
66. Leonardi S, Raineri C, De Ferrari GM, et al. Usefulness of cardiac magnetic resonance in assessing the risk of ventricular arrhythmias and sudden death in patients with hypertrophic cardiomyopathy. *Eur Heart J.* 2009;30(16):2003-2010.
67. Moon JC, McKenna WJ, McCrohon JA, Elliott PM, Smith GC, Pennell DJ. Toward clinical risk assessment in hypertrophic cardiomyopathy with gadolinium cardiovascular magnetic resonance. *J Am Coll Cardiol.* 2003;41:1561-1567.
68. Adabag AS, Maron BJ, Appelbaum E, et al. Occurrence and frequency of arrhythmias in hypertrophic cardiomyopathy in relation to delayed enhancement on cardiovascular magnetic resonance. *J Am Coll Cardiol.* 2008;51:1369-1374.
69. Epstein AE, Dimarco JP, Ellenbogen KA, et al. Guidelines for Device-Based Therapy of Cardiac Rhythm Abnormalities: Executive Summary. A Report of the American College of Cardiology/American Heart Association Task Force on Practice Guidelines (Writing Committee to Revise the ACC/AHA/NASPE 2002 Guideline Update for Implantation of Cardiac Pacemakers and Antiarrhythmia Devices). *Circulation.* 2008;117:e350-e408.
70. Zipes DP, Camm AJ, Borggrefe M, et al. Guidelines for Management of Patients With Ventricular Arrhythmias and the Prevention of Sudden Cardiac Death: a report of the American College of Cardiology/American Heart Association Task Force and the European Society of Cardiology Committee for Practice Guidelines (writing committee to develop Guidelines for Management of Patients With Ventricular Arrhythmias and the Prevention of Sudden Cardiac Death): developed in collaboration with the European Heart Rhythm Association and the Heart Rhythm Society. *Circulation.* 2006;114:e385-e484.
71. Maron BJ, Shen WK, Link MS, et al. Efficacy of implantable cardioverter-defibrillators for the prevention of sudden death in patients with hypertrophic cardiomyopathy. *N Engl J Med.* 2000;342:365-373.
72. Maron BJ, Spirito P, Shen WK, et al. Implantable Cardioverter-defibrillators and prevention of sudden cardiac death in hypertrophic cardiomyopathy. *JAMA.* 2007;298:405-412.
73. Ackerman MJ, VanDriest SL, Ommen SR, et al. Prevalence and age-dependence of malignant mutations in the beta-myosin heavy chain and troponin T genes in hypertrophic cardiomyopathy: a comprehensive outpatient perspective. *J Am Coll Cardiol.* 2002;39:2042-2048.
74. Girolami F, Olivotto I, Passerini I, et al. A molecular screening strategy based on beta-myosin heavy chain, cardiac myosin binding protein C and troponin T genes in Italian patients with hypertrophic cardiomyopathy. *J Cardiovasc Med (Hagerstown).* 2006;7:601-607.
75. Ingles J, Doolan A, Chiu C, Seidman J, Seidman C, Semsarian C. Compound and double mutations in patients with hypertrophic cardiomyopathy: implications for genetic testing and counselling. *J Med Genet.* 2005;42:e59.
76. Van Driest SL, Vasile VC, Ommen SR, et al. Myosin binding protein C mutations and compound heterozygosity in hypertrophic cardiomyopathy. *J Am Coll Cardiol.* 2004;44:1903-1910.
77. Van Driest SL, Ommen SR, Tajik AJ, Gersh BJ, Ackerman MJ. Yield of genetic testing in hypertrophic cardiomyopathy. *Mayo Clin Proc.* 2005;80:739-744.
78. Erdmann J, Daehmlow S, Wischke S, et al. Mutation spectrum in a large cohort of unrelated consecutive patients with hypertrophic cardiomyopathy. *Clin Genet.* 2003;64:339-349.
79. Richard P, Charron P, Carrier L, et al. Hypertrophic cardiomyopathy: distribution of disease genes, spectrum of

mutations, and implications for a molecular diagnosis strategy. *Circulation.* 2003;107:2227-2232.
80. Bonne G, Carrier L, Bercovici J, et al. Cardiac myosin binding protein-C gene splice acceptor site mutation is associated with familial hypertrophic cardiomyopathy. *Nat Genet.* 1995;11:438-440.
81. Geisterfer-Lowrance AA, Kass S, Tanigawa G, et al. A molecular basis for familial hypertrophic cardiomyopathy: a beta cardiac myosin heavy chain gene missense mutation. *Cell.* 1990;62:999-1006.
82. Watkins H, Conner D, Thierfelder L, et al. Mutations in the cardiac myosin binding protein-C gene on chromosome 11 cause familial hypertrophic cardiomyopathy. *Nat Genet.* 1995;11:434-437.
83. Watkins H, McKenna WJ, Thierfelder L, et al. Mutations in the Genes for Cardiac Troponin T and {alpha}-Tropomyosin in Hypertrophic Cardiomyopathy. *N Engl J Med.* 1995;332: 1058-1065.
84. Posch MG, Thiemann L, Tomasov P, et al. Sequence analysis of myozenin 2 in 438 European patients with familial hypertrophic cardiomyopathy. *Med Sci Monit.* 2008;14: CR372-CR374.
85. Osio A, Tan L, Chen SN, et al. Myozenin 2 is a novel gene for human hypertrophic cardiomyopathy. *Circ Res.* 2007;100:766-768.
86. Vasile VC, Will ML, Ommen SR, Edwards WD, Olson TM, Ackerman MJ. Identification of a metavinculin missense mutation, R975W, associated with both hypertrophic and dilated cardiomyopathy. *Mol Genet Metab.* 2006;87:169-174.
87. Theis JL, Bos JM, Bartleson VB, et al. Echocardiographic-determined septal morphology in Z-disc hypertrophic cardiomyopathy. *Biochem Biophys Res Commun.* 2006;351: 896-902.
88. Bos JM, Poley RN, Ny M, et al. Genotype-phenotype relationships involving hypertrophic cardiomyopathy-associated mutations in titin, muscle LIM protein, and telethonin. *Mol Genet Metab.* 2006;88:78-85.
89. Hayashi T, Arimura T, Itoh-Satoh M, et al. Tcap gene mutations in hypertrophic cardiomyopathy and dilated cardiomyopathy. *J Am Coll Cardiol.* 2004;44:2192-2201.
90. Ho CY, Lever HM, DeSanctis R, Farver CF, Seidman JG, Seidman CE. Homozygous mutation in cardiac troponin T: implications for hypertrophic cardiomyopathy. *Circulation.* 2000;102:1950-1955.
91. Lekanne Deprez RH, Muurling-Vlietman JJ, Hruda J, et al. Two cases of severe neonatal hypertrophic cardiomyopathy caused by compound heterozygous mutations in the MYBPC3 gene. *J Med Genet.* 2006;43:829-832.
92. Richard P, Isnard R, Carrier L, et al. Double heterozygosity for mutations in the beta-myosin heavy chain and in the cardiac myosin binding protein C genes in a family with hypertrophic cardiomyopathy. *J Med Genet.* 1999;36:542-545.
93. Richard P, Charron P, Leclercq C, et al. Homozygotes for a R869G mutation in the beta -myosin heavy chain gene have a severe form of familial hypertrophic cardiomyopathy. *J Mol Cell Cardiol.* 2000;32:1575-1583.
94. Carrier L, Bonne G, Bahrend E, et al. Organization and sequence of human cardiac myosin binding protein C gene (MYBPC3) and identification of mutations predicted to produce truncated proteins in familial hypertrophic cardiomyopathy. *Circ Res.* 1997;80:427-434.
95. Richard P, Villard E, Charron P, Isnard R. The genetic bases of cardiomyopathies. *J Am Coll Cardiol.* 2006;48:A79-A89.
96. Marian AJ, Mares A Jr, Kelly DP, et al. Sudden cardiac death in hypertrophic cardiomyopathy. Variability in phenotypic expression of beta-myosin heavy chain mutations. *Eur Heart J.* 1995;16:368-376.
97. Anan R, Greve G, Thierfelder L, et al. Prognostic implications of novel beta cardiac myosin heavy chain gene mutations that cause familial hypertrophic cardiomyopathy. *J Clin Invest.* 1994;93:280-285.
98. Hwang TH, Lee WH, Kimura A, et al. Early expression of a malignant phenotype of familial hypertrophic cardiomyopathy associated with a Gly716Arg myosin heavy chain mutation in a Korean family. *Am J Cardiol.* 1998;82: 1509-1513.
99. Marian AJ, Roberts R. Molecular genetic basis of hypertrophic cardiomyopathy: genetic markers for sudden cardiac death. *J Cardiovasc Electrophysiol.* 1998;9:88-99.
100. Watkins H, Rosenzweig A, Hwang DS, et al. Characteristics and prognostic implications of myosin missense mutations in familial hypertrophic cardiomyopathy. *N Engl J Med.* 1992;326:1108-1114.
101. Roberts R, Sigwart U. New concepts in hypertrophic cardiomyopathies, part II. *Circulation.* 2001;104:2249-2252.
102. Fananapazir L, Epstein ND. Genotype-phenotype correlations in hypertrophic cardiomyopathy. Insights provided by comparisons of kindreds with distinct and identical beta-myosin heavy chain gene mutations. *Circulation.* 1994; 89:22-32.
103. Consevage MW, Salada GC, Baylen BG, Ladda RL, Rogan PK. A new missense mutation, Arg719Gln, in the beta-cardiac heavy chain myosin gene of patients with familial hypertrophic cardiomyopathy. *Hum Mol Genet.* 1994;3: 1025-1026.
104. Coviello DA, Maron BJ, Spirito P, et al. Clinical features of hypertrophic cardiomyopathy caused by mutation of a "hot spot" in the alpha-tropomyosin gene. *J Am Coll Cardiol.* 1997;29:635-640.
105. Havndrup O, Bundgaard H, Andersen PS, et al. The Val606Met mutation in the cardiac beta-myosin heavy chain gene in patients with familial hypertrophic cardiomyopathy is associated with a high risk of sudden death at young age. *Am J Cardiol.* 2001;87:1315-1317.
106. Epstein ND, Cohn GM, Cyran F, Fananapazir L. Differences in clinical expression of hypertrophic cardiomyopathy associated with two distinct mutations in the beta-myosin heavy chain gene. A 908Leu-Val mutation and a 403Arg-Gln mutation. *Circulation.* 1992;86:345-352.
107. Van Driest SL, Ackerman MJ, Ommen SR, et al. Prevalence and severity of "benign" mutations in the beta-myosin heavy chain, cardiac troponin T, and alpha-tropomyosin genes in hypertrophic cardiomyopathy. *Circulation.* 2002;106:3085-3090.
108. Deinum J, van Gool JM, Kofflard MJ, Ten Cate FJ, Danser AH. Angiotensin II type 2 receptors and cardiac hypertrophy in women with hypertrophic cardiomyopathy. *Hypertension.* 2001;38:1278-1281.
109. Osterop AP, Kofflard MJ, Sandkuijl LA, et al. AT1 receptor A/C1166 polymorphism contributes to cardiac hypertrophy in subjects with hypertrophic cardiomyopathy. *Hypertension.* 1998;32:825-830.

110. Friedrich FW, Bausero P, Sun Y, et al. A new polymorphism in human calmodulin III gene promoter is a potential modifier gene for familial hypertrophic cardiomyopathy. *Eur Heart J*. 2009;30(13):1648-1655.
111. Monserrat L, Gimeno-Blanes JR, Marin F, et al. Prevalence of Fabry Disease in a Cohort of 508 Unrelated Patients With Hypertrophic Cardiomyopathy. *J Am Coll Cardiol*. 2007;50:2399-2403.
112. Gollob MH, Green MS, Tang AS, Roberts R. PRKAG2 cardiac syndrome: familial ventricular preexcitation, conduction system disease, and cardiac hypertrophy. *Curr Opin Cardiol*. 2002;17:229-234.
113. Rai TS, Ahmad S, Bahl A, et al. Genotype phenotype correlations of cardiac beta-myosin heavy chain mutations in Indian patients with hypertrophic and dilated cardiomyopathy. *Mol Cell Biochem*. 2009;321:189-196.
114. Forissier JF, Richard P, Briault S, et al. First description of germline mosaicism in familial hypertrophic cardiomyopathy. *J Med Genet*. 2000;37:132-134.
115. Watkins H, Thierfelder L, Hwang DS, McKenna W, Seidman JG, Seidman CE. Sporadic hypertrophic cardiomyopathy due to de novo myosin mutations. *J Clin Invest*. 1992;90:1666-1671.
116. Watkins H, Anan R, Coviello DA, Spirito P, Seidman JG, Seidman CE. A de novo mutation in alpha-tropomyosin that causes hypertrophic cardiomyopathy. *Circulation*. 1995;91:2302-2305.
117. Cuda G, Perrotti N, Perticone F, Mattioli PL. A previously undescribed de novo insertion-deletion mutation in the beta myosin heavy chain gene in a kindred with familial hypertrophic cardiomyopathy. *Heart*. 1996;76:451-452.
118. Alders M, Jongbloed R, Deelen W, et al. The 2373insG mutation in the MYBPC3 gene is a founder mutation, which accounts for nearly one-fourth of the HCM cases in the Netherlands. *Eur Heart J*. 2003;24:1848-1853.
119. Michels M, Solima OII, Kofflard MJ, et al. Diastolic abnormalities as the first feature of hypertrophic cardiomyopathy in dutch myosin-binding protein c founder mutations. *J Am Coll Cardiol Img*. 2009;2:58-64.
120. Moolman-Smook JC, De Lange WJ, Bruwer EC, Brink PA, Corfield VA. The origins of hypertrophic cardiomyopathy-causing mutations in two South African subpopulations: a unique profile of both independent and founder events. *Am J Hum Genet*. 1999;65:1308-1320.
121. Jaaskelainen P, Miettinen R, Karkkainen P, Toivonen L, Laakso M, Kuusisto J. Genetics of hypertrophic cardiomyopathy in eastern Finland: few founder mutations with benign or intermediary phenotypes. *Ann Med*. 2004;36:23-32.
122. Zahka K, Kalidas K, Simpson MA, et al. Homozygous mutation of MYBPC3 associated with severe infantile hypertrophic cardiomyopathy at high frequency among the Amish. *Heart*. 2008;94:1326-1330.
123. Garneau NL, Wilusz J, Wilusz CJ. The highways and byways of mRNA decay. *Nat Rev Mol Cell Biol*. 2007;8:113-126.
124. Mearini G, Schlossarek S, Willis MS, Carrier L. The ubiquitin-proteasome system in cardiac dysfunction. *Biochim Biophys Acta*. 2008;1782:749-763.
125. Chan PA, Duraisamy S, Miller PJ, et al. Interpreting missense variants: comparing computational methods in human disease genes CDKN2A, MLH1, MSH2, MECP2, and tyrosinase (TYR). *Hum Mutat*. 2007;28:683-693.
126. Goldgar DE, Easton DF, Byrnes GB, Spurdle AB, Iversen ES, Greenblatt MS. Genetic evidence and integration of various data sources for classifying uncertain variants into a single model. *Hum Mutat*. 2008;29:1265-1272.
127. Ortlepp JR, Vosberg HP, Reith S, et al. Genetic polymorphisms in the renin-angiotensin-aldosterone system associated with expression of left ventricular hypertrophy in hypertrophic cardiomyopathy: a study of five polymorphic genes in a family with a disease causing mutation in the myosin binding protein C gene. *Heart*. 2002;87:270-275.
128. Perkins MJ, Van Driest SL, Ellsworth EG, et al. Gene-specific modifying effects of pro-LVH polymorphisms involving the renin-angiotensin-aldosterone system among 389 unrelated patients with hypertrophic cardiomyopathy. *Eur Heart J*. 2005;26:2457-2462.
129. van Langen I, Hofman N, Tan HL, Wilde AA. Family and population strategies for screening and counselling of inherited cardiac arrhythmias. *Ann Med*. 2004;36(Suppl 1):116-124.

Familial Dilated Cardiomyopathy

4

G. Aernout Somsen, G. Kees Hovingh, I.I. Tulevski, Jon Seidman, and Christine E. Seidman

4.1 Introduction

Dilated cardiomyopathy (DCM) is a disorder of the ventricular myocardium characterized by normal or thinned walls, enlarged chamber volumes, and diminished systolic function. Clinically, DCM gives rise to fatigue, shortness of breath, and increased morbidity and mortality. DCM occurs in response to underlying pathologies including valvular dysfunction, hypertension, or myocarditis, or as an idiopathic disorder of the myocardium. Among patients with idiopathic DCM, approximately 25–30% have affected first-degree family members, implying a genetic etiology.[16] However, a pathogenic mutation is found in a minority of cases of familial DCM upon molecular analyses. A number of candidate genes have been identified, and these encode for proteins of the myocyte contractile apparatus, the myocyte cytoskeleton, and nuclear envelope, as well as proteins involved in calcium homeostasis. In addition, the function of a number of proteins encoded by candidate genes is still unknown. To date, over 20 genes have been shown to play a pivotal role in the origin of DCM. This diversity of genetic etiologies in DCM explains why patients sometimes exhibit additional clinical manifestations, including defects in the conduction system resulting in arrhythmias, skeletal muscle abnormalities, deafness, and endocrinologic disease.

Current molecular testing strategies allow us to identify mutations in DCM patients. In addition to confirmation of the diagnosis in these patients with clinical signs of the disease, genetic testing also holds great value in the identification of carriers of the disease-causing mutation in family members. Molecular screening of these individuals does not only result in a potential strategy to diminish disease-related complications (arrhythmias and thromboembolism) in a very early phase, but will also greatly enhance our knowledge about the exact role of a mutated gene in the course of disease.

The list of genes is not deemed to be complete and more genes involved will be identified in the future. Knowledge of the full repertoire of key-role genes is indispensable for our understanding of pathways involved in the process of the failing heart.

4.2 Epidemiology and Prevalence

DCM (without coronary artery disease) has a prevalence of 40 per 100,000 in the USA and a reported incidence of 5–8 in 100,000 in the USA and Europe.[15]

Among 1,230 patients that were categorized as having primary or secondary DCM, 50% were defined as idiopathic (IDC) (Table 4.1)[13] Before the widespread application of echocardiography, assessment of the prevalence of DCM among family members was troublesome, likely resulting in an underestimation of the heritability. In recent years, however, more accurate numbers have been published. In their study, Grunig and colleagues showed that 35% of the cases of IDC were familial DCM, indicating that a substantial proportion of all DCM patients may have a genetic etiology.[16] While the prevalence and incidence of genetic disorders may vary between populations and geographic regions, the true prevalence of genetic DCM may even be higher, due to the age-dependent expression of the disease. An increased awareness among

G.A. Somsen (✉)
Department of Cardiology, Onze Lieve Vrouwe Gasthuis, Amsterdam, The Netherlands and
Cardiologie Centra Nederland / Cardiology Centers of The Netherlands, Amsterdam, The Netherlands
e-mail: a. somsen@cardiokliniek.nl

Table 4.1 Prevalence of DCM specified for cause[13]

Cause of DCM	# of patients	Prevalence (%)
Idiopathic	616	50.1
Peripartum	51	4.1
Myocarditis	111	9.0
Ischemic heart disease	91	7.4
Infiltrative	59	4.8
Hypertension	49	4.0
HIV	45	3.7
Connective tissue disease	39	3.2
Substance abuse	37	3.0
Doxorubicine	15	1.2
Other	117	9.5
Total	1,230	100.0

physicians of the heritability of DCM, in concert with improved imaging modalities that enable the early detection of the disease, may further increase the disease prevalence.

4.2.1 Diagnosis and Clinical Course of Familial DCM

The age of onset of symptoms differs between DCM disease genes. For example, clinical manifestations occur, in general, at young age in individuals with mutations in sarcomere protein genes or in the calcium-handling gene encoding the protein phospholamban, while in individuals with lamin A/C mutations, symptoms of the disease may become overt only after the fourth decade of life.

Presenting symptoms are often nonspecific and include exertional dyspnea, fatigue, and palpitations. Sudden death is a rare initial presentation of DCM.

Findings at physical examination are often unremarkable or subtle in DCM patients. On examination, the heart may be enlarged and the ictus cordis may be displaced. On auscultation, the heart sounds can be normal, but a S3 and S4 gallop due to reduced ventricular compliance may be present. With advanced ventricular dilation, a pansystolic blowing murmur due to mitral regurgitation may be audible at the apex. In addition, signs of heart failure, such as rales and peripheral edema, may be present. Irregular heart rhythms are common in DCM and initially include atrial or ventricular premature complexes and atrial fibrillation. In addition, ventricular arrhythmias may be present.

Diagnosis of DCM is based upon identification of ventricular dilation and impaired function. This can either be assessed by means of echocardiography, bloodpool scintigraphy, or magnetic resonance imaging (MRI). Typically, the left ventricular fractional shortening, as measured by echocardiography, is less than 25% percent. Bloodpool scintigraphy and MRI can be used to measure the end-diastolic and end-systolic volumes, from which the left ventricular ejection fraction (LVEF) can be calculated. DCM is diagnosed if LVEF is less than 40% (Table 4.2).

The end stage of the natural course of DCM is heart failure. The rate of progression toward heart failure due to DCM varies considerably – ranging from slow, indolent decline to rapid deterioration of ventricular function. An appreciation of the expected course can be gleaned from the family history. Increased morbidity and mortality in DCM is also attributed to thromboembolic events and sudden death due to arrhythmias. Although different mutations may affect myocyte function in different ways, there seems to be a common final pathway resulting in myocyte hypertrophy, premature myocyte death, and increased myocardial fibrosis. These histopathologic features are identified in virtually all postmortem and explanted cardiac tissues of genetic DCM patients.

4.2.2 Genetic Background

Familial DCM is most frequently inherited as an autosomal dominant trait, a pattern of transmission that implies considerable risk that offspring will also develop DCM. Other forms of inheritance, that is, autosomal recessive, X-linked, and matrilineal, are less common in DCM and convey different risks to offspring (Table 4.3). Mitochondrial mutations produce matrilineal inheritance, in which, all offspring from an affected mother, but no offspring from an affected father are at risk for DCM.

DCM gene mutations are known to exhibit variable penetrance, a genetic term that indicates the discrepancy between the presence of a mutation and clinical manifestation of disease. Some individuals who carry a DCM gene mutation (genotype) lack

Table 4.2 Proposed diagnostic workup of DCM patients (Modified from the 2009 Focused Update: ACCF/AHA guidelines for the diagnosis and management of heart failure in adults (http://circ.ahajournals.org/cgi/content/full/119/14/1977)

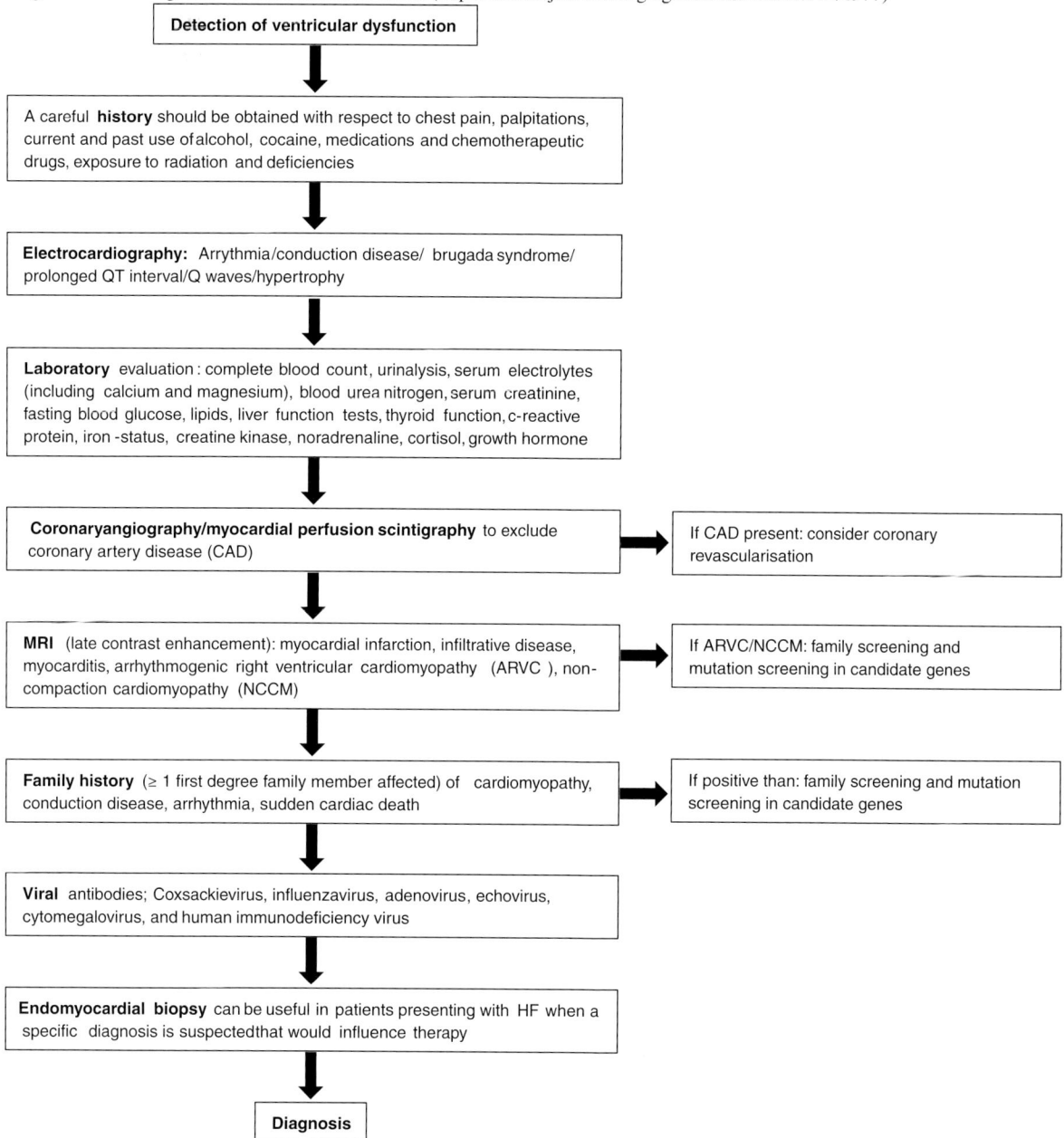

```
┌─────────────────────────────────┐
│ Detection of ventricular dysfunction │
└─────────────────────────────────┘
                │
                ▼
```

A careful **history** should be obtained with respect to chest pain, palpitations, current and past use of alcohol, cocaine, medications and chemotherapeutic drugs, exposure to radiation and deficiencies

↓

Electrocardiography: Arrythmia/conduction disease/ brugada syndrome/ prolonged QT interval/Q waves/hypertrophy

↓

Laboratory evaluation: complete blood count, urinalysis, serum electrolytes (including calcium and magnesium), blood urea nitrogen, serum creatinine, fasting blood glucose, lipids, liver function tests, thyroid function, c-reactive protein, iron-status, creatine kinase, noradrenaline, cortisol, growth hormone

↓

Coronaryangiography/myocardial perfusion scintigraphy to exclude coronary artery disease (CAD) → If CAD present: consider coronary revascularisation

↓

MRI (late contrast enhancement): myocardial infarction, infiltrative disease, myocarditis, arrhythmogenic right ventricular cardiomyopathy (ARVC), non-compaction cardiomyopathy (NCCM) → If ARVC/NCCM: family screening and mutation screening in candidate genes

↓

Family history (≥ 1 first degree family member affected) of cardiomyopathy, conduction disease, arrhythmia, sudden cardiac death → If positive than: family screening and mutation screening in candidate genes

↓

Viral antibodies; Coxsackievirus, influenzavirus, adenovirus, echovirus, cytomegalovirus, and human immunodeficiency virus

↓

Endomyocardial biopsy can be useful in patients presenting with HF when a specific diagnosis is suspected that would influence therapy

↓

Diagnosis

clinical manifestations of ventricular dysfunction (phenotype). The proportion of individuals with a gene mutation who express disease are called penetrants, whereas mutation carriers without clinical disease are called nonpenetrants. Many factors contribute to the variable penetrance of DCM mutations, including age, interacting genes, and environmental modifying factors. These parameters need to be considered when screening the relatives of individuals with DCM. The presence of a disease causing mutation can most often not be excluded based on clinical examination at a single point in time, and especially young, clinically unaffected relatives of a DCM patient may transmit a disease causing mutation to children. An appreciation of the penetrance of DCM can be obtained from large families with one specific underlying genetic defect. This information can help to define appropriate screening strategies for LV dysfunction in specific individuals at risk for developing DCM (Fig. 4.1)

Table 4.3 Known modes of inheritance of DCM[16]

Inheritance mode	% of familial DCM	Risk to offspring	Risk to sibs
Autosomal dominant	56	50%[a]	%50%[a,b]
Autosomal recessive	~16	<1%	25%
X-linked recessive[c]	~10	Dependent on sex of the index[d]	Dependent on whether or not mother is a carrier of the mutation[e]
Matrilineal (tRNA mutations in mtDNA)	?	Males and females may be equally affected. Only females will transmit mtDNA mutations to offspring. The chance of transmission may be close to 100%. Still only a proportion of offspring will be affected. Because as a result of heteroplasmy (see Chap. 2) the mutation threshold required for disease expression to occur may never be reached	

mtDNA mitochondrial DNA
[a]May be a little less in clinical practice as a result of reduced penetrance
[b]Only if one parent is carrier of the causative mutation
[c]In classical X-linked recessive disease females are asymptomatic carriers, however in the most common X-linked recessive disorders associated with DCM (Duchenne and Becker muscular dystrophy), females may develop DCM, and should remain under cardiac surveillance
[d]Affected males will transmit the mutation to none of their sons and all of their daughters
[e]Mutations in X-linked recessive disorders that are associated with DCM arise quite often de novo. If the mother is a mutation carrier, both sons and daughters have 50% chance of inheriting the mutation

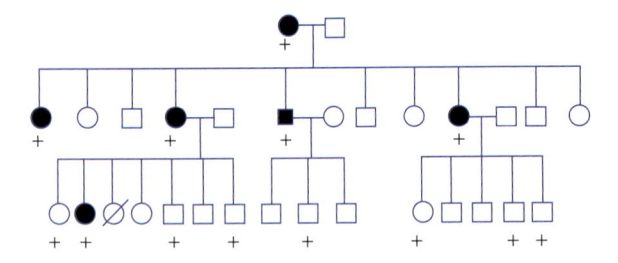

Fig. 4.1 An example of age-dependent penetrance within a dilated cardiomyopathy (*DCM*) family. Plus (+) refers to a positive genotype (carrier of a genetic defect, associated with DCM), filled symbols refer to phenotypically defined DCM

4.2.3 Genes and Mutations in DCM

Molecular studies have identified many genes and mutations that can cause DCM and more will undoubtedly be discovered. In some instances, investigations have identified only the chromosome region (or locus) that likely contains a DCM gene. Table 4.4 provides an illustration of this expanding knowledge of genes that are mutated in DCM. Many of these genes encode cytoskeletal and sarcomere proteins or proteins involved in ion homeostasis. In addition, DCM genes encode

Table 4.4 Loci and gene mutations associated with DCM as registered in the Online Mendelian Inheritance in Man (OMIM) database

Locus	Gene	Phenotype (other than DCM)	Reference
1p1–q21	LMNA	Sudden cardiac death, LBBB, SVT, sinus bradycardia, AV-conduction block. NCCM; EDMD;LGMD type 1b;partial LD; HGP;AO	12
1q31–q42	PSEN 2	Alzheimer	25
1q32	TNNT 2	NCCM; HCM	11, 32
1q42–q43	ACTN 2	HCM	33
2q31	TTN	HCM	14
2q35	DES	Progressive distal-onset skeletal myopathy, respiratory insufficiency	26
3p21	SCN5A	Sinus node dysfunction, SVT, progressive atrial and intraventricular conduction delay; lqts3;BS	30

Table 4.4 (continued)

Locus	Gene	Phenotype (other than DCM)	Reference
3p21.3–p14.3	TNNC 1	Myocyte hypertrophy, increased interstitial fibrosis, endocardial thickening, SCD	32
4q12	SGCB	LGMD type 2e	3
5q33	SGCD	LGMD type 2f (absence of skeletal muscular dystrophy)	57
6q22.1	PLN	SCD, ventricular tachycardia/fibrillation	50
6q23–q24	EYA4	Sensorineural hearing loss	51, 52
6q24	DSP	Wooly hair, keratoderma, ARVC	6
9q31	FKTN	Mild limb-girdle muscle involvement, Fukuyama muscular dystrophy, Walker–Warburg syndrome	39
10q21–23	LDB3suffices	NCCM	62
10q22.1–q23	VCL	HCM	43
11p11.2	MYBPC3	HCM	34
11p15.1	CSRP3	HCM	23
11q22.3–q23.1	CRYAB	Skeletal myopathy	49
12p12.1	ABCC9	Ventricular arrhythmia	5
12q22	LAP2	TMPO	56
13q12	SGCG	LGMD2c	42
14q12	MYH7	NCCM	22
14q24	TGFB3	Arrhythmogenic right ventricular cardiomyopathy	4
15q11–14	ACTC1	NCCM	45
15q22.1	TPM1	SCD	44
17q12	TCAP	LLGMD type 2g	23
17q21	JUP	Palmoplantar keratoderma, arrhythmogenic right ventricular cardiomyopathy (Naxos disease)	29
18q12.1	DSC2	Arrhythmogenic right ventricular cardiomyopathy	53
18q12.1	DTNA	NCCM	19
19q13.4	TNNI3	Myofibrillar loss, hyperchromatic nuclei, and myocyte hypertrophy; HCM	40
Xp21.2	DMD	Skeletal dystrophy, high serum creatine kinase, mental retardation (DMD, BMD)	38
Xq24	LAMP2	Skeletal myopathy, high serum creatine kinase, mild cognitive impairment in males, pigmentary retinopathy (Danon disease)	55
Xq28	TAZ	NCCM, death in early infancy, Barth syndrome	8

AO and others, *BMD* Becker muscular dystrophy, *BS* Brugada syndrome *DCM* dilated cardiomyopathy, *DMD* Duchenne muscular dystrophy, EDMD Emmery-Dreifus muscular dystrophy, *HCM* hypertrophic cardiomyopathy, *HGP* Hutchinson-Gilford progeria, *LBBB* left bundle *LD* lipodystrophy, *LGMD* limb girdle muscular dystrophy, *LQTS* long *QT* syndrome branch block, muscular dystrophy, *NCCM* noncompaction cardiomyopathy

proteins involved in the nuclear membrane and in mitochondrial functions.

4.3 Molecular Pathophysiology

Our current understanding of the mechanisms by which mutations result in DCM is predicted on the function of these molecules and cardiac biology. The sarcomere is the fundamental structural and functional unit of cardiac muscle that consists of an interdigitating system of thick and thin filaments (Fig. 4.2). Force is generated from the sliding movement of thick filaments relative to thin filaments, which is achieved by cyclical attachment and detachment of myosin crossbridges to actin. Hydrolysis of ATP by myosin provides energy for the detachment and subsequent reattachment of the crossbridges and results in a step-like displacement of thin filament relative to the thick filament. The troponin–tropomyosin complex provides a Ca^{2+}-sensitive switch that regulates this process. Troponin I is an inhibitory component of the troponin–tropomyosin complex which binds actin and inhibits actomyosin ATPase activity in the absence of Ca^{2+}. Ca^{2+} binding to troponin C causes the troponin–tropomyosin complex to release the myosin binding domain of actin, permitting the interaction of actin and myosin heads. The myosin light chains maintain optimal speed and efficiency of crossbridge cycling. Myosin binding protein C contributes to the organization and assembly of thick filaments and modulates crossbridge function by regulating the position of the myosin head relative to the thin filament.

DCM mutations in sarcomere proteins clearly perturb these processes. Diminished force may occur in myosin mutations that alter actin-binding residues involved in initiating the power stroke of contraction. Impaired contractile force may also occur in DCM troponin mutations that alter residues implicated in tight binary troponin interactions. Because Troponin molecules modulate calcium-stimulated actinomyosin ATPase activity, these defects may cause inefficient ATP hydrolysis and therein decrease contractile power.

Other DCM mutations may impair force transmission. Cytoskeletal proteins as desmin, dystrophin, and sarcoglycans are responsible for propagating force generated in the sarcomere to the plasma membrane and to the extracellular matrix. Mutations in cytoskeletal proteins may destabilize protein-protein interactions and compromise or cause ineffectual force propagation/force transmission throughout the myocyte.

Nuclear envelope proteins are proposed to provide structural support for the nucleus and to play an important role in chromosome organization, gene regulation, and nuclear protein handling. Mutations in these genes may cause DCM by altering the myocyte toward more susceptibility for mechanical stress.

Although mutations have long been considered to contribute specifically to the development of DCM, recent studies have shown that some of these "DCM mutations" can also cause hypertrophic cardiomyopathy (HCM) when present in a different part of the gene or in a different family.

4.3.1 Sarcomere Proteins

4.3.1.1 Actin

Actin participates in force generation by the sarcomere and also provides interaction with Z-bands and the intercalated discs (Fig. 4.2). These structures are the scaffold onto which myosin proteins generate force to support muscle contraction.

Mutations in actin and in all other sarcomere proteins can cause either hypertrophic cardiomyopathy or DCM, depending on the precise location of the mutation. DCM caused by actin mutations can be due to a failure in force generation or due to ineffectual force propagation.

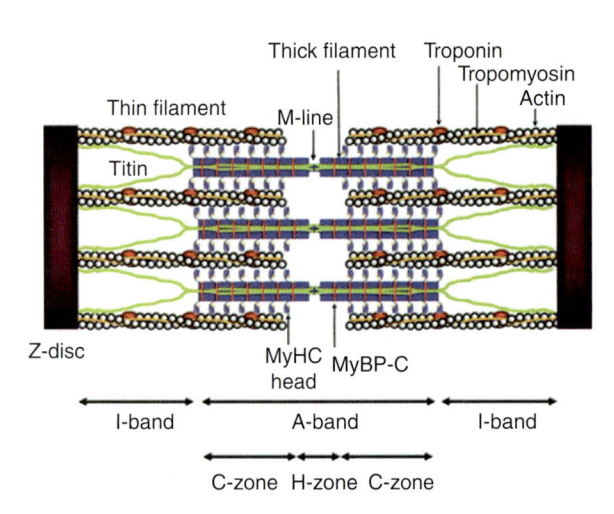

Fig. 4.2 Schematic drawing of the sarcomere structure[36]

Actin mutations are not associated with extracardiac manifestations. Although actin mutations are a rare cause of DCM, several reports indicate that these mutations can cause early onset, childhood disease.

4.3.1.2 Metavincullin

Metavincullin interacts with actin and links thin filaments to the plasma membrane. This muscle-specific isoform of vinculin is essential for force transmission during muscular contraction.[61] Metavincullin was analyzed as a candidate gene and rare mutations were identified in two patients with sporadic DCM.[43] Electron microscopy of myocytes from a DCM patient with a metavincullin mutation showed irregular and fragmented intercalated discs. Mutations in functionally distinct regions of metavincullin have also been associated with HCM.

4.3.1.3 Telethonin

Telethonin is localized in the Z disc of the skeletal and cardiac myocyte and undergoes dynamic phosphorylation following calcium/calmodulin binding. Telethonin interacts with titin, which serves as docking sites for multiple proteins and regulators of contraction and provides machinery for sensing myocyte stretch. Telethonin mutations are a rare cause of DCM[23]

4.3.1.4 Troponin-T

Cardiac troponin T participates in modulating calcium-stimulated actinomyosin ATPase activity. DCM mutations in troponin T alter residues that are involved in tight binary troponin interactions and may decrease the calcium sensitivity of force generation and therein decrease contractile power. Mutations in troponin T usually cause HCM, but DCM mutations have also been described.[37]

4.3.1.5 Troponin-I

Cardiac troponin I binds actin and inhibits actomyosin ATPase activity in the absence of calcium. The functional consequences of troponin I mutation were impairment of interactions with cardiac troponin-T but not with troponin-C. As such, this defect is presumed to function similar to troponin T mutations, and to perturb force generation. Troponin-I mutations are a rare cause of DCM – only 1 mutation was identified in 235 DCM patients. Troponin I mutations may be transmitted as recessive mutations in DCM.[40]

4.3.1.6 Troponin C

Cardiac troponin-C regulates calcium uptake in the sarcomere and promotes interaction of actin with myosin, the hydrolysis of ATP, and the generation of tension. Functional studies of mutated troponin C indicate that these disrupt normal interactions with troponin-T and I.

To date, few troponin C mutations have been defined in DCM patients,[32] but notably, the DCM phenotype was severe in these patients.

4.3.1.7 Tropomyosin

Both, alpha- and beta tropomyosin-1 mutations can cause DCM by decreasing the calcium sensitivity of the contractile apparatus or by destabilizing actin interactions. These abnormalities compromise force transmission to neighboring sarcomeres. Electron microscopy of heart tissue from a DCM patient with a tropomyosin-1 mutation revealed an abnormal sarcomere structure in which the thin filaments were irregular and fragmented, sarcomeres were also contracted with decreased distance between Z bands, and the sarcolemma had a scalloped appearance.[44] The prevalence of tropomyosin mutations is low.[63]

4.3.1.8 Beta-Myosin Heavy Chain

Beta-myosin heavy chain (beta-MHC) is the principle myosin expressed in the adult human ventricle, whereas alpha-MHC is the primary myosin isoform found in the atria. Beta-MHC is also expressed in slow skeletal muscle, type I fibers. Human myosin mutations that alter actin-binding residues cause DCM by diminishing the power stroke of contraction. Other myosin mutations that are located within the flexible fulcrum that transmits movement from myosin to the thick filament could reduce transmitted force and cause DCM.[22]

The contribution of myosin mutations to DCM has been increasingly appreciated and may account for 10% of DCM. The clinical manifestations of these mutations are restricted to cardiac disease that can present early in life or be quiescent until middle age. Symptoms and left ventricular dysfunction appear to be mild to moderate. In one report, slow progressive AV and ventricular conduction defects were found.[9]

4.3.1.9 Titin

Mutations in the giant protein titin (also known as connectin) cause DCM. Titin contributes to passive muscle stiffness and myofilament calcium sensitivity and serves as a scaffold for many molecules that regulate cardiac contraction. DCM mutations in titin decrease binding affinities of titin to Z-line proteins like T-cap/telethonin and alpha-actinin.[20] Due to its extremely large size, genetic analyses of titin have been limited. Two DCM families with titin mutations had only cardiac disease, and no skeletal muscle involvement.[14]

Other genes that encode sarcomeric proteins have been associated with familial and sporadic cases of DCM. These genes are Cypher gene (LDB3), alfa-actinin gene (ACTN2), and the CSRP3[36]

4.3.2 Nuclear Envelope Proteins

4.3.2.1 Lamin A/C

Lamin A/C encodes an inner nuclear membrane protein that interacts with emerin and other molecules. Lamin A/C is essential for maintenance of nuclear stability and human mutations in this protein may render cells more susceptible to mechanical forces. Given the constancy of myocardial contraction, lamin A/C mutations may be particularly deleterious to myocytes.

Lamin A/C mutations cause DCM that is characterized by high penetrance, adult onset, heart failure, conduction defects, arrhythmias, sudden cardiac death, and substantial premature mortality.[46] These mutations cause dominant DCM and account for at least 33% of the familial DCM and 0.5–5% of all DCM with conduction disease and bradycardia in the western populations.[2,58] Lamin A/C mutations can also cause Emery–Dreifuss muscular dystrophy, which has its onset in childhood and is characterized by joint abnormalities, conduction disease, and arrhythmias.

A distinct histopathology is found in hearts with lamin A/C mutations. There is fibrofatty degeneration of the myocardium and marked involvement of the atrioventricular node and conducting cells. A similar situation is found in skeletal muscles in patients with Emerin mutations. The marked atrioventricular node pathology in lamin A/C mutations accounts for the presence of electrophysiologic deficits (progressive atrioventricular block and atrial arrhythmias) found in these patients.

Due to the high prevalence of lamin A/C mutations in idiopathic DCM and the poor prognosis of symptomatic carriers, systematic mutation screenings have been recommended.[17,54] It has recently been confirmed that prophylactic therapy with an implantable cardioverter-defibrillator prevents sudden death in these patients more efficiently than pacemaker implantation. This might also hold for asymptomatic carriers of the mutation.

4.3.2.2 Thymopoietin

Thymopoietin is also known as lamina-associated protein (LAP2), and like lamin, contributes to the maintenance of nuclear integrity and nuclear functions. In a family with DCM, a mutation was found, which altered interaction with LAP2 and lamin A.[56] Some patients with a mutation in the thymopoietin gene have severe DCM without signs of skeletal muscle disease. The finding that thymopoietin mutation results in DCM underlines the pivotal role of nuclear functions in the pathogenesis leading to myocardial dysfunction.

4.3.3 Cytoskeletal Proteins

4.3.3.1 Desmin

Desmin is a muscle-specific intermediate filament protein that participates in maintaining the structural and mechanical integrity of the contractile apparatus in muscle tissues.[47] Mutations in the gene encoding desmin may impair normal force transmission during muscular contraction. Mutations that are found in the tail domain of the molecule may disrupt intercalated discs, whereas mutations found in the head domain may cause defects in Z-disc binding.[28]

Patients with desmin mutations have both cardiac and extracardiac manifestations. In addition to DCM, patients may develop progressive distal-onset skeletal myopathy and respiratory insufficiency.[10] The histopathology of skeletal or cardiac tissue with these mutations may show cytoplasmic aggregates of desmin, which may suggest the diagnosis.

4.3.3.2 Dystrophin

The dystrophin gene is located on chromosome X and encodes a protein that is crucial for the formation of the cytoskeleton. Dystrophin and related proteins including sarcoglycans and laminin-2 make up the dystrophin-associated glycoprotein complex, which plays a crucial role in the transmission of force from the sarcomere and the plasma membrane to the extracellular matrix.[24] Dystrophin mutations produce Duchenne muscular dystrophy (DMD) or Becker muscular dystrophy (BMD) in male subjects. In addition to these skeletal myopathies, patients also develop delayed onset of cardiomyopathy. In a minority of cases, mental retardation can be found. Severity of the disease due to these mutations is variable as DMD has a high mortality rate in young patients, whereas BMD is a milder disease. The onset of DMD usually occurs before the age of 3 years and results in physical impairment necessitating a wheelchair by the age of 10 and death in the second decade, when a DCM is present. The onset of BMD is often in the 20s and 30s and survival to a relatively advanced age is frequent. The incidence of DMD/BMD is 21.7/3.2 per 100,000 male live births.[41] In the Netherlands, the prevalence of DMD at birth was estimated to be 1:4.215 male live births.[59]

Both, DMD and BMD are X-linked inherited diseases and are characterized by DCM and progressive proximal muscular dystrophy with pseudohypertrophy of the calves. High plasma levels of creatine kinase, myopathic changes by electromyography, and myofiber degeneration with fibrosis and fatty infiltration and absence of dystrophin fibers on (heart) muscle biopsy can be found in these patients. On the electrocardiogram specific abnormalities can be found (Fig. 4.3). Even in female carriers of DMD en BMD, electrocardiographic abnormalities have been found in 41% and 27%, respectively (E.M.[18])

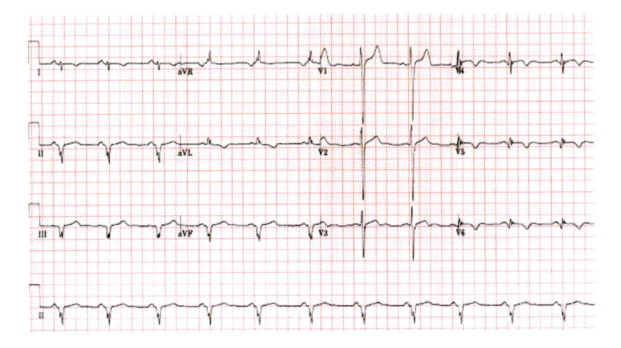

Fig. 4.3 ECG from a patient with Duchenne's muscular dystrophy (*DMD*) characterized by short PR interval, Q waves in inferior and lateral leads, and inverted T waves in the lateral leads. The extreme QRS axis is not specific for DMD (Courtesy of W.G. de Voogt, M.D., Ph.D.)

In female mutation carriers, muscle weakness and DCM can be found in 20%.[18] Left ventricular dilatation was present in 18% of these carriers.

In another study, it was shown that approximately 10% of the heterozygous females develop cardiomyopathy even in the absence of muscular weakness.[31] Therefore, screening for left ventricular dilatation and dysfunction is recommended in the female carriers of these diseases.

Therapy in DMD/BMD patients consists of corticosteroids for skeletal muscle weakness, standard pharmacologic treatment for heart failure, and noninvasive ventilation in case of respiratory failure. Currently, new therapies are studied which focus on increase of dystrophin expression, increase of muscle growth, and regeneration.

4.3.3.3 Sarcoglycans and Dystrophin-Associated Glycoproteins

The dystrophin-associated glycoprotein complex (composed of dystroglycans, delta- sarcoglycans, caveolin-3, syntrophin, and dystrobrevin) provides stability to the sarcomere and transmits force to the extracellular matrix. Recessive mutations in a subset of these genes cause DCM that is most often accompanied by skeletal myopathy.

Mutations in the delta-sarcoglycan gene (SGCD) cause mitochondrial aggregation resulting in pathological vascular smooth muscle cell proliferation and apoptosis.[27] Patients with delta-sarcoglycan mutations develop cardiomyopathy that is usually, but not invariably, accompanied by skeletal muscular dystrophy.[57] In contrast, mutations in the beta- and gamma-sarcoglycan cause both cardiomyopathy and skeletal muscular dystrophy.

4.3.4 Mitochondrial DNA Mutations

Mitochondria are crucial for the energy metabolism of the myocyte and mitochondrial mutations (which are encountered in larger quantities due to inefficient repair mechanisms in mitochondrial DNA) result in syndromes characterized by multiorgan dysfunction such as myopathy, encephalopathy, diabetes mellitus, and lactic acidosis.

4.3.4.1 Genes Involved in Electrolyte Homeostasis

SCN5A

SCN5A gene encodes the cardiac sodium channel. SCN5A mutations lead to a loss in channel function and may cause Brugada syndrome, the long QT syndrome, idiopathic ventricular fibrillation, and sick sinus syndrome. In addition, other loss-of-function mutations in SCN5A have been associated with ventricular dilatation and dysfunction accompanied by sinus bradycardia and atrial fibrillation.[1,30]

The linked molecular etiology of these different clinical disorders is presumed to reflect the essential role for ion homeostasis in ventricular function and the notion that arrhythmias can directly remodel the myocardium. The mechanisms by which similar or identical mutations in SCN5A lead to a variable expression of heart disease, even within the same family, are still unknown.

Phospholamban

Phospholamban is a critical regulator of cardiac relaxation. After contraction, calcium reuptake into the SR is mediated by the Ca^{2+}-ATPase pump (SERCA2a). Phospholamban inhibits SERCA2 activity. Phosphorylation of phospholamban relieves this inhibition and accelerates ventricular relaxation. Several human mutations in phospholamban have been defined that functionally cause constitutive inhibition of SERCA, and delayed myocardial relaxation. Phospholamban mutations cause severe DCM with rapid progression to heart failure. In addition, these patients have significant ventricular arrhythmias due to altered calcium homeostasis.[50]

4.3.5 Cardiac ATP-Sensitive Potassium Channels

Cardiac ATP-Sensitive Potassium (KATP) channels are protein complexes, which ensure the maintenance of cellular and metabolic homeostasis such that the reaction to stress is not harmful to the organism itself. During high sympathetic stimulation, KATP channels increase outward potassium current to offset the resulting calcium influx, thereby reducing the energy-demanding myocardial calcium overload and avoiding contractile dysfunction.

KATP channels contain an inwardly rectifying potassium channel pore (Kir6.2), a regulatory SUR2A subunit, and an ATPase-harboring ATP-binding cassette protein. Human mutations in the regulatory SUR2A subunit (encoded by ABCC9) cause DCM and heart failure.[5] These mutations directly result in calcium overload and contraction band necrosis. The clinical manifestations of human KATP mutations are DCM and arrhythmias such as ventricular tachycardias.

4.3.6 DCM and Ventricular Noncompaction Cardiomyopathy

Noncompaction Cardiomyopathy (NCCM) is a rare form of DCM, characterized by incomplete compaction of the left ventricle that occurs late in cardiac embryogenesis. As a result, ventricular trabeculation persists. These blood-filled deep intratrabecular recesses impair ventricular function and predispose to thrombosis. NCCM is associated with heart failure, ventricular arrhythmias, and arterial thromboembolism, and produces increased morbidity and mortality.

NCCM can be associated with other congenital heart anomalies, including ventricular septal defects, pulmonic stenosis, and atrial septal defects.[19] The clinical criteria for diagnosis in NCCM are subject to

Table 4.5 Specific morphologic criteria for noncompaction cardiomyopathy (NCCM) by echocardiography and cardiac magnetic resonance imaging (MRI)

Diagnostic modality	Diagnostic criteria
General	Absence of coexisting cardiac abnormalities
Echocardiography[21]	1. Two layer structure (compacted thin epicardial band and a much thicker noncompacted endocardial layer of trabecular meshwork)
	2. Maximum ratio of noncompacted to compacted myocardium >2:1
	3. Noncompaction predominantly localized in midlateral wall of left ventricle
	4. Deep perfused intertrabecular recesses (color Doppler and/contrast echocardiography)
	5. Reduced global left ventricular systolic and diastolic function
	6. Left ventricular thrombi
	7. Abnormal papillary muscle structure
Cardiac magnetic resonance[48]	1. Noncompaction in the apical and lateral segments
	2. Ratio of noncompacted to compacted myocardium >2.3 during diastole (sensitivity 0.86/specificity 0.99)

debate. Recently, criteria have been proposed for different diagnostic modalities as shown in Table 4.5.

NCCM can arise from defects in several different genes, including genes that encode sarcomere protein (beta-MHC, alpha cardiac actin, and cardiac troponin-T) and in genes that encode G4.5 (Tafazzin), alpha-dystrobrevin, FKBP-12-, lamin A/C- and LBD3/Cypher gene.[35] In addition, a locus (chromosome 6p24.3–21.2) has been associated with the NCCM phenotype. Due to the large number of genes that are related to NCCM, screening for mutations in NCCM families is labor intensive.

Treatment of NCCM includes management of heart failure and arrhythmias, with particular attention to the potential for thromboembolic events. Since the genetics of NCCM have been diverse, the clinical importance of the screening of first-degree relatives of affected patients is high.[7]

4.4 Therapy

At present, specific therapies for the different genetic causes of DCM have not been evaluated in clinical trials. Patients with asymptomatic and symptomatic ventricular dysfunction due to an underlying gene mutation should be treated according to the AHA/ACC/ESC guidelines for heart failure treatment. As DCM patients progress to heart failure, standard therapies are used. These include beta-receptor blockers, angiotensin-converting enzyme inhibitors, angiotensin receptor blockers, aldosterone receptor blockers, diuretics, anticoagulants when indicated, implantable cardiac defibrillator, biventricular pacing to obtain cardiac resynchronization, and surgical interventions as valvular reconstruction and heart transplantation.

Recognition of the genetic basis for DCM has led to the identification of mutation carriers who lack clinical signs of ventricular dilatation. Serial evaluations on a regular basis depending on severity and family history of these individuals are warranted to prevent sudden death and to initiate pharmacologic interventions when ventricular dilatation emerges. A more provocative issue is whether mutation carriers without clinical manifestations might benefit from prophylactic therapies to retard ventricular remodeling. While this consideration remains a research question, knowledge of genetic status poses the first clinical opportunity to intervene early in DCM and to potentially limit ventricular dysfunction and progression to heart failure.

4.5 Prognosis and Risk Stratification

The prognosis of DCM strongly depends on the precise gene mutation and interacting factors, including modifying genes and environments. Our current understanding of these interacting factors is incomplete. As such, prognosis and risk stratification are based upon

Table 4.6 Gene mutations associated with high risk of sudden cardiac death

Lamin A/C	12
Phospholamban	50
Tropomyosin-1	44
Tafazzin	8
Slow troponin-C	32

the clinical course of other affected family members and clinical parameters.

Familial history and genetic data can provide information regarding disease penetrance, age of onset, and disease progression and severity. The presence of associated phenotypes, including conduction system, atrial- and ventricular arrhythmias, and thromboembolic events will help to guide management and therapeutic strategy.

Genotype-phenotype correlations in some DCM genes have demonstrated high prevalence of sudden cardiac death and progression to heart failure (Table 4.6). However, a publication bias should always be taken into account when interpreting these data. Genetic screening and counseling of families with these genetic etiologies is particularly important. Cardiac defibrillators may improve prognosis in a subset of these patients.

4.6 Family Screening

The diagnosis of idiopathic DCM should prompt family evaluation. When one or more first-degree relatives are recognized to have DCM, a genetic etiology should be considered. Construction of pedigrees will help to delineate those at risk for disease. In addition to cardiac findings, one should search for evidence for overt or subclinical skeletal myopathy, conduction system abnormalities, and arrhythmias. Drawing a pedigree, however, is time consuming, and the accuracy of medical and family history frequently questionable. Moreover, assessment of the family history and pedigree analysis does not seem to be routine for cardiologists. Van Langen and colleagues interviewed a cohort of 643 Dutch cardiologists and it was shown that their self-reported knowledge about cardiogenetics is low[60] and not even half of them gave genetic counseling to HCM patients (let alone DCM patients). A close collaboration between clinical geneticists and cardiologist probably adds great value to patient care.

Recognition of the considerable role for genetics in DCM has prompted recommendations for screening. These are reviewed in the AHA/ACC 2005 guidelines and in more recent guidelines provided by the Heart Failure Society of America. Both documents indicate the importance of referral to a center with genetic expertise for diagnosis and counseling. [http://circ.ahajournals.org/cgi/content/full/112/12/e154] and include the performance of echocardiography and ECG in all first-degree family members.

Gene-based diagnosis of DCM is now clinically available on http://www.ncbi.nlm.nih.gov/sites/GeneTests and http://www.eddnal.com. Genetic testing requires a peripheral blood sample from one affected individual. Comprehensive screening of multiple DCM genes is simultaneously accomplished. While this is costly, once a mutation is identified, family members can be rapidly screened for the presence or absence of a mutation at very low cost. Recently, a DCM CardioChip™ test has been developed for efficient mutation screening in DCM-related genes (Fig 4.4). This commercially available chip uses microarray-based sequencing technology and screens the coding sequence and splice sites of 19 genes. In addition, screening for gene mutations related to ARVC is also available. The clinical sensitivity of this test is expected to be greater than 30%.

Gene-based diagnosis precisely defines the etiology for DCM in patients with clinically evident disease. This knowledge may provide information about clinical course and help to define the risk for associated manifestations (such as development of conduction system disease). In addition, knowledge of the causal mutation in one affected individual provides the opportunity to accurately define the risk for DCM in all relatives. When gene-based assessment reveals that a family member has *not* inherited the DCM mutation, subsequent clinical follow-up is unnecessary.

4.7 Summary

Multiple genetic defects have been shown to cause DCM. DCM mutations occur in genes that encode proteins involved in force transmission, force generation,

DCM CHIP (19 genes) (>30%)	DCM Panel A	MYBPC3	DCM	HCM	LVNC	
		MYH7	DCM	HCM	RCM	LVNC	Laing Distal Myopathy
		TNNT2	DCM	HCM	RCM	LVNC	
		TNNI3	DCM	HCM	RCM		
		TPM1	DCM	HCM			
	DCM Panel B	ACTC	DCM	HCM			
		PLN	DCM	HCM			
		LDB3	DCM	HCM	LVNC		
		LMNA	DCM		Conduction system disease	Laminopathies*
		TAZ	DCM			Barth Syndrome
		ACTN2	DCM	HCM			
		VCL	DCM	HCM			
		CSRP3	DCM	HCM			
		TCAP	DCM	HCM			
		ABCC9	DCM		Arrythmias	
		CTF1	DCM			
		SGCD	DCM			Delta-Sarcoglycanopathy
		DES	DCM		RCM	Desminopathy
		EMD	DCM			Emery-Dreifuss muscular dystrophy (EDMD)
ARVC panel		PKP2		ARVC		
		DSP	DCM	ARVC		
		DSC2		ARVC		
		DSG2	DCM?	ARVC		
		TMEM43		ARVC		

Fig. 4.4 DCM CardioChipTM test (Laboratory for Molecular Medicine Partners Center for Personalized Genetic Medicine, http://www.hpcgg.org/lmm)

nuclear function, and calcium handling in the myocyte. Disease manifestations, severity, and prognosis depend on the precise disease gene mutation. However, the clinical phenotype is also influenced by unknown genes and environmental factors.

Genetic causes for DCM are common and should be sought in patients with unexplained ventricular dysfunction. Family history is the key to defining genetic DCM and warrants prompt screening of first-degree family members. With the availability of gene-based diagnosis, molecular testing can provide precise information that defines relatives at risk who require follow-up evaluations. Equally important, gene-based diagnosis can precisely define mutation-negative family members who are not at risk for developing DCM.

New DCM genes will undoubtedly be identified. As the full repertoire of gene mutations in DCM and the associated clinical information is compiled, the value of knowing precise molecular cause will increase. As translational research in DCM harnesses gene-based diagnosis to identify and treat preclinical patients, the opportunity to improve prognosis and to prevent heart failure is expected to increase.

References

1. Adler E, Fuster V. SCN5A – a mechanistic link between inherited cardiomyopathies and a predisposition to arrhythmias? *Jama*. 2005;293(4):491-493.
2. Arbustini E, Pilotto A, Repetto A, et al. Autosomal dominant dilated cardiomyopathy with atrioventricular block: a lamin A/C defect-related disease. *J Am Coll Cardiol*. 2002;39(6):981-990.
3. Barresi R, Di Blasi C, Negri T, et al. Disruption of heart sarcoglycan complex and severe cardiomyopathy caused by beta sarcoglycan mutations. *J Med Genet*. 2000;37(2):102-107.
4. Beffagna G, Occhi G, Nava A, et al. Regulatory mutations in transforming growth factor-beta3 gene cause arrhythmogenic right ventricular cardiomyopathy type 1. *Cardiovasc Res*. 2005;65(2):366-373.
5. Bienengraeber M, Olson TM, Selivanov VA, et al. ABCC9 mutations identified in human dilated cardiomyopathy disrupt catalytic KATP channel gating. *Nat Genet*. 2004;36(4):382-387.
6. Carvajal-Huerta L. Epidermolytic palmoplantar keratoderma with woolly hair and dilated cardiomyopathy. *J Am Acad Dermatol*. 1998;39(3):418-421.
7. Chrissoheris MP, Ali R, Vivas Y, Marieb M, Protopapas Z. Isolated noncompaction of the ventricular myocardium: contemporary diagnosis and management. *Clin Cardiol*. 2007;30(4):156-160.
8. D'Adamo P, Fassone L, Gedeon A, et al. The X-linked gene G4.5 is responsible for different infantile dilated cardiomyopathies. *Am J Hum Genet*. 1997;61(4):862-867.
9. Daehmlow S, Erdmann J, Knueppel T, et al. Novel mutations in sarcomeric protein genes in dilated cardiomyopathy. *Biochem Biophys Res Commun*. 2002;298(1):116-120.
10. Dalakas MC, Park KY, Semino-Mora C, Lee HS, Sivakumar K, Goldfarb LG. Desmin myopathy, a skeletal myopathy with cardiomyopathy caused by mutations in the desmin gene. *N Engl J Med*. 2000;342(11):770-780.
11. Durand JB, Bachinski LL, Bieling LC, et al. Localization of a gene responsible for familial dilated cardiomyopathy to chromosome 1q32. *Circulation*. 1995;92(12):3387-3389.
12. Fatkin D, MacRae C, Sasaki T, et al. Missense mutations in the rod domain of the lamin A/C gene as causes of dilated cardiomyopathy and conduction-system disease. *N Engl J Med*. 1999;341(23):1715-1724.
13. Felker GM, Thompson RE, Hare JM, et al. Underlying causes and long-term survival in patients with initially unexplained cardiomyopathy. *N Engl J Med*. 2000;342(15):1077-1084.
14. Gerull B, Gramlich M, Atherton J, et al. Mutations of TTN, encoding the giant muscle filament titin, cause familial dilated cardiomyopathy. *Nat Genet*. 2002;30(2):201-204.
15. Gillum RF. Idiopathic dilated cardiomyopathy. *Epidemiology*. 1994;5(3):383-385.
16. Grunig E, Tasman JA, Kucherer H, Franz W, Kubler W, Katus HA. Frequency and phenotypes of familial dilated cardiomyopathy. *J Am Coll Cardiol*. 1998;31(1):186-194.
17. Hermida-Prieto M, Monserrat L, Castro-Beiras A, et al. Familial dilated cardiomyopathy and isolated left ventricular noncompaction associated with lamin A/C gene mutations. *Am J Cardiol*. 2004;94(1):50-54.
18. Hoogerwaard EM, Bakker E, Ippel PF, et al. Signs and symptoms of Duchenne muscular dystrophy and Becker muscular dystrophy among carriers in The Netherlands: a cohort study. *Lancet*. 1999;353(9170):2116-2119.
19. Ichida F, Tsubata S, Bowles KR, et al. Novel gene mutations in patients with left ventricular noncompaction or Barth syndrome. *Circulation*. 2001;103(9):1256-1263.
20. Itoh-Satoh M, Hayashi T, Nishi H, et al. Titin mutations as the molecular basis for dilated cardiomyopathy. *Biochem Biophys Res Commun*. 2002;291(2):385-393.
21. Jenni R, Oechslin E, Schneider J, Attenhofer Jost C, Kaufmann PA. Echocardiographic and pathoanatomical characteristics of isolated left ventricular non-compaction: a step towards classification as a distinct cardiomyopathy. *Heart*. 2001;86(6):666-671.
22. Kamisago M, Sharma SD, DePalma SR, et al. Mutations in sarcomere protein genes as a cause of dilated cardiomyopathy. *N Engl J Med*. 2000;343(23):1688-1696.
23. Knoll R, Hoshijima M, Hoffman HM, et al. The cardiac mechanical stretch sensor machinery involves a Z disc complex that is defective in a subset of human dilated cardiomyopathy. *Cell*. 2002;111(7):943-955.
24. Lapidos KA, Kakkar R, McNally EM. The dystrophin glycoprotein complex: signaling strength and integrity for the sarcolemma. *Circ Res*. 2004;94(8):1023-1031.
25. Li D, Parks SB, Kushner JD, et al. Mutations of presenilin genes in dilated cardiomyopathy and heart failure. *Am J Hum Genet*. 2006;79(6):1030-1039.
26. Li D, Tapscoft T, Gonzalez O, et al. Desmin mutation responsible for idiopathic dilated cardiomyopathy. *Circulation*. 1999;100(5):461-464.
27. Lipskaia L, Pinet C, Fromes Y, et al. Mutation of delta-sarcoglycan is associated with Ca(2+)-dependent vascular remodeling in the Syrian hamster. *Am J Pathol*. 2007;171(1):162-171.
28. Mavroidis M, Panagopoulou P, Kostavasili I, Weisleder N, Capetanaki Y. A missense mutation in desmin tail domain linked to human dilated cardiomyopathy promotes cleavage of the head domain and abolishes its Z-disc localization. *FASEB J*. 2008;22(9):3318-3327.
29. McKoy G, Protonotarios N, Crosby A, et al. Identification of a deletion in plakoglobin in arrhythmogenic right ventricular cardiomyopathy with palmoplantar keratoderma and woolly hair (Naxos disease). *Lancet*. 2000;355(9221):2119-2124.
30. McNair WP, Ku L, Taylor MR, et al. SCN5A mutation associated with dilated cardiomyopathy, conduction disorder, and arrhythmia. *Circulation*. 2004;110(15):2163-2167.
31. Mirabella M, Servidei S, Manfredi G, et al. Cardiomyopathy may be the only clinical manifestation in female carriers of Duchenne muscular dystrophy. *Neurology*. 1993;43(11):2342-2345.
32. Mogensen J, Murphy RT, Shaw T, et al. Severe disease expression of cardiac troponin C and T mutations in patients with idiopathic dilated cardiomyopathy. *J Am Coll Cardiol*. 2004;44(10):2033-2040.
33. Mohapatra B, Jimenez S, Lin JH, et al. Mutations in the muscle LIM protein and alpha-actinin-2 genes in dilated cardiomyopathy and endocardial fibroelastosis. *Mol Genet Metab*. 2003;80(1-2):207-215.
34. Moller DV, Andersen PS, Hedley P, et al. The role of sarcomere gene mutations in patients with idiopathic dilated cardiomyopathy. *Eur J Hum Genet*. 2009;17:1241-1249.

35. Moric-Janiszewska E, Markiewicz-Loskot G. Genetic heterogeneity of left-ventricular noncompaction cardiomyopathy. *Clin Cardiol.* 2008;31(5):201-204.
36. Morimoto S. Sarcomeric proteins and inherited cardiomyopathies. *Cardiovasc Res.* 2008;77(4):659-666.
37. Morimoto S, Lu QW, Harada K, et al. Ca(2+)-desensitizing effect of a deletion mutation Delta K210 in cardiac troponin T that causes familial dilated cardiomyopathy. *Proc Natl Acad Sci USA.* 2002;99(2):913-918.
38. Muntoni F, Cau M, Ganau A, et al. Brief report: deletion of the dystrophin muscle-promoter region associated with X-linked dilated cardiomyopathy. *N Engl J Med.* 1993;329(13):921-925.
39. Murakami T, Hayashi YK, Noguchi S, et al. Fukutin gene mutations cause dilated cardiomyopathy with minimal muscle weakness. *Ann Neurol.* 2006;60(5):597-602.
40. Murphy RT, Mogensen J, Shaw A, Kubo T, Hughes S, McKenna WJ. Novel mutation in cardiac troponin I in recessive idiopathic dilated cardiomyopathy. *Lancet.* 2004;363(9406):371-372.
41. Nigro G, Comi LI, Limongelli FM, et al. Prospective study of X-linked progressive muscular dystrophy in Campania. *Muscle Nerve.* 1983;6(4):253-262.
42. Noguchi S, McNally EM, Ben Othmane K, et al. Mutations in the dystrophin-associated protein gamma-sarcoglycan in chromosome 13 muscular dystrophy. *Science.* 1995;270(5237):819-822.
43. Olson TM, Illenberger S, Kishimoto NY, Huttelmaier S, Keating MT, Jockusch BM. Metavinculin mutations alter actin interaction in dilated cardiomyopathy. *Circulation.* 2002;105(4):431-437.
44. Olson TM, Kishimoto NY, Whitby FG, Michels VV. Mutations that alter the surface charge of alpha-tropomyosin are associated with dilated cardiomyopathy. *J Mol Cell Cardiol.* 2001;33(4):723-732.
45. Olson TM, Michels VV, Thibodeau SN, Tai YS, Keating MT. Actin mutations in dilated cardiomyopathy, a heritable form of heart failure. *Science.* 1998;280(5364):750-752.
46. Pasotti M, Klersy C, Pilotto A, et al. Long-term outcome and risk stratification in dilated cardiolaminopathies. *J Am Coll Cardiol.* 2008;52(15):1250-1260.
47. Paulin D, Li Z. Desmin: a major intermediate filament protein essential for the structural integrity and function of muscle. *Exp Cell Res.* 2004;301(1):1-7.
48. Petersen SE, Selvanayagam JB, Wiesmann F, et al. Left ventricular non-compaction: insights from cardiovascular magnetic resonance imaging. *J Am Coll Cardiol.* 2005;46(1):101-105.
49. Pilotto A, Marziliano N, Pasotti M, Grasso M, Costante AM, Arbustini E. alphaB-crystallin mutation in dilated cardiomyopathies: low prevalence in a consecutive series of 200 unrelated probands. *Biochem Biophys Res Commun.* 2006;346(4):1115-1117.
50. Schmitt JP, Kamisago M, Asahi M, et al. Dilated cardiomyopathy and heart failure caused by a mutation in phospholamban. *Science.* 2003;299(5611):1410-1413.
51. Schonberger J, Levy H, Grunig E, et al. Dilated cardiomyopathy and sensorineural hearing loss: a heritable syndrome that maps to 6q23-24. *Circulation.* 2000;101(15):1812-1818.
52. Schonberger J, Wang L, Shin JT, et al. Mutation in the transcriptional coactivator EYA4 causes dilated cardiomyopathy and sensorineural hearing loss. *Nat Genet.* 2005;37(4):418-422.
53. Syrris P, Ward D, Evans A, et al. Arrhythmogenic right ventricular dysplasia/cardiomyopathy associated with mutations in the desmosomal gene desmocollin-2. *Am J Hum Genet.* 2006;79(5):978-984.
54. Taylor MR, Fain PR, Sinagra G, et al. Natural history of dilated cardiomyopathy due to lamin A/C gene mutations. *J Am Coll Cardiol.* 2003;41(5):771-780.
55. Taylor MR, Ku L, Slavov D, et al. Danon disease presenting with dilated cardiomyopathy and a complex phenotype. *J Hum Genet.* 2007;52(10):830-835.
56. Taylor MR, Slavov D, Gajewski A, et al. Thymopoietin (lamina-associated polypeptide 2) gene mutation associated with dilated cardiomyopathy. *Hum Mutat.* 2005;26(6):566-574.
57. Tsubata S, Bowles KR, Vatta M, et al. Mutations in the human delta-sarcoglycan gene in familial and sporadic dilated cardiomyopathy. *J Clin Invest.* 2000;106(5):655-662.
58. van Berlo JH, Duboc D, Pinto YM. Often seen but rarely recognised: cardiac complications of lamin A/C mutations. *Eur Heart J.* 2004;25(10):812-814.
59. van Essen AJ, Busch HF, te Meerman GJ, ten Kate LP. Birth and population prevalence of Duchenne muscular dystrophy in The Netherlands. *Hum Genet.* 1992;88(3):258-266.
60. van Langen IM, Birnie E, Leschot NJ, Bonsel GJ, Wilde AA. Genetic knowledge and counselling skills of Dutch cardiologists: sufficient for the genomics era? *Eur Heart J.* 2003;24(6):560-566.
61. Vasile VC, Will ML, Ommen SR, Edwards WD, Olson TM, Ackerman MJ. Identification of a metavinculin missense mutation, R975W, associated with both hypertrophic and dilated cardiomyopathy. *Mol Genet Metab.* 2006;87(2):169-174.
62. Vatta M, Mohapatra B, Jimenez S, et al. Mutations in Cypher/ZASP in patients with dilated cardiomyopathy and left ventricular non-compaction. *J Am Coll Cardiol.* 2003;42(11):2014-2027.
63. Watkins H, McKenna WJ, Thierfelder L, et al. Mutations in the genes for cardiac troponin T and alpha-tropomyosin in hypertrophic cardiomyopathy. *N Engl J Med.* 1995;332(16):1058-1064.

Arrhythmogenic Right Ventricular Dysplasia/Cardiomyopathy

From Desmosome to Disease

Moniek G.P.J. Cox and Richard N.W. Hauer

5.1 Introduction

Arrhythmogenic Right Ventricular Dysplasia/Cardiomyopathy (ARVD/C) is a disease characterized by progressive fibrofatty replacement of primarily the *right ventricle* (RV).[1-3]

Affected individuals typically present between the second and fourth decade of life with *arrhythmias* originating from the RV. However, ARVD/C can also be the cause of *sudden death* already in adolescence, mainly in athletes.[4] From autopsy studies, it is known that already in young teenagers massive amounts of fibrofatty tissue can replace major parts of normal myocardium (Fig. 5.1).

The first series of ARVD/C patients was published in 1982, when it was called a disease in which "the right ventricular musculature is partially or totally absent and is replaced by fatty and fibrous tissue."[1] This disease was initially thought to be a defect in RV development, which is why it was first called "dysplasia." In the past 25 years, increased insight in the development of the disease as well as the discovery of pathogenic mutations involved, led to the current idea that ARVD/C is a genetically determined "cardiomyopathy."[3,5] The molecular genetic era has provided new insight in the understanding that ARVD/C is a desmosomal disease resulting from defective cell adhesion proteins. The first disease-causing gene, encoding the desmosomal protein *Plakoglobin* (*JUP*), was identified in patients with Naxos disease, an autosomal recessive variant of ARVD/C.[6] Its discovery pointed research in the direction of other desmosomal genes. Until 2004, evidence for genes underlying the autosomal dominantly inherited ARVD/C had been very limited, with three genes and six loci being identified.[7-15] The *Desmoplakin* gene (*DSP*) was the first desmosomal protein gene to be associated with the autosomal dominant form of ARVD/C.[15] It was followed by discovery of mutations in *Plakophilin-2* (*PKP2*), *Desmoglein-2* (*DSG2*), and *Desmocollin-2* (*DSC2*), also components of the cardiac desmosome.[16-18] Impaired desmosomal function may

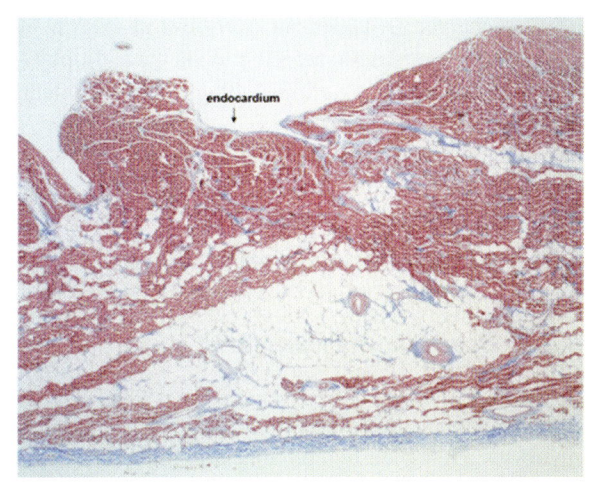

Fig. 5.1 Histology of right ventricle of a 13-year-old girl who died suddenly during exercise. AZAN staining (400×) with cardiac myocytes (red), collagen (blue), and adipocytes (*white*). Shown is the typical pattern of arrhythmogenic right ventricular dysplasia/cardiomyopathy (*ARVD/C*) with strands of fibrosis reaching all the way to the endocardium. Bundles of cardiac myocytes are embedded in between the fibrotic strands, particularly in the subendocardial layers. These interconnecting bundles of myocytes give rise to activation delay and re-entrant circuits, the typical electrophysiologic substrate for ventricular arrhythmias in ARVD/C. The large homogeneous subepicardial area of adipocytes is not arrhythmogenic, is not typical for ARVD/C, and is also observed in the cor adiposum

M.G.P.J. Cox (✉)
Department of Cardiology, University Medical Centre Utrecht, Utrecht, The Netherlands
e-mail: m.g.p.j.cox@umcutrecht.nl

result in disruption of the myocardial architecture, leading to activation delay and thereby life-threatening arrhythmias. In a few rare cases, autosomal dominant ARVD/C has been linked to other genes unrelated to the cell adhesion complex, i.e., the genes encoding the cardiac ryanodine receptor (RyR2), the transforming growth factor-β3 gene (TGFβ3), and transmembrane protein 43 (*TMEM43*).[13, 14, 19]

With mutations found in about half of the patients, mainly in desmosomal genes and *PKP2* in particular, classical ARVD/C is currently considered a genetically determined desmosomal disease.

This chapter will give an overview of ARVD/C, starting from the genetic defects and via the pathophysiologic mechanism to clinical diagnosis, treatment, and prognosis.

5.2 Molecular and Genetic Background

5.2.1 Desmosome Function

The functional and structural integrity of cardiac myocytes is enabled by cell adhesion junctions in the *intercalated disk*. Intercalated disks are located between cardiomyocytes at their longitudinal ends and contain three different kinds of intercellular connections: desmosomes, adherens junctions, and gap junctions.

Desmosomes are important for cell–cell adhesion and are predominantly found in tissues that experience mechanical stress: the heart and epidermis. They couple cytoskeletal elements to the plasma membrane at cell–cell adhesions. Desmosomes also protect the other components of the intercalated disk from mechanical stress and are involved in structural organization of the intercalated disk. Desmosomes consist of multiple proteins, which belong to three different families:

1. Transmembranous cadherins: desmogleins and desmocollins
2. Linker armadillo repeat proteins: plakoglobin and plakophilin
3. Plakins: desmoplakin and plectin

Figure 5.2 schematically represents the organization of the various proteins in the cardiac desmosome.

Within desmosomes, cadherins are connected to armadillo proteins, which for their parts interact with plakins. The plakins anchor the desmosomes to intermediate filaments, mainly desmin. Thereby, they form a three-dimensional scaffold providing mechanical support.

Adherens junctions act as bridges that link the actin filaments within sarcomeres of neighboring cells. These junctions are involved in force transmission and together with desmosomes, these mechanical junctions act as "spot welds" to create membrane domains that are protected from shear stress caused by contraction of the neighboring cells. Furthermore, they facilitate

Fig. 5.2 Schematic representation of the molecular organization of cardiac desmosomes. The plasma membrane (*PM*) spanning proteins Desmocollin-2 (*DSC2*) and Desmoglein-2 (*DSG2*) interact in the extracellular space at the dense midline (*DM*). At the cytoplasmic side, they interact with plakoglobin (*PG*) and PKP2 at the outer dense plaque (*ODP*). PKP2 and PG interact also with Desmoplakin gene (*DSP*). At the inner dense plaque (*IDP*), the C-terminus of DSP anchors the intermediate filament desmin. (Reprint with permission from Van Tintelen et al. Curr Opin Cardiol 2007[27])

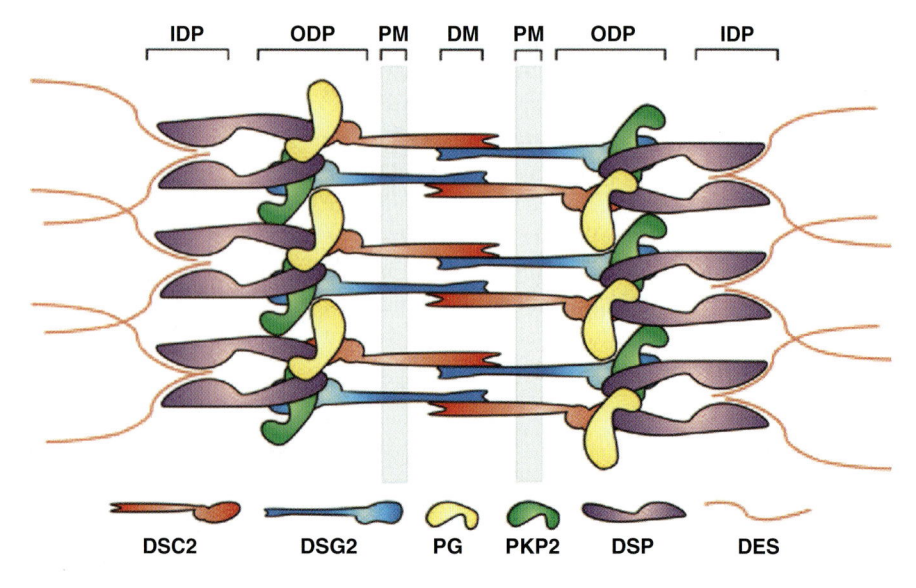

assembly and maintenance of gap junctions, securing intercellular electrical coupling.

Cardiomyocytes are individually bordered by a lipid bilayer, which gives a high degree of electrical insulation. The electrical current that forms the impulse for mechanic contraction can travel from one cell to the other via gap junctions. Gap junctions provide electrical coupling by enabling ion transfer between cells. The number, size, and distribution of gap junctions all influence impulse propagation in cardiac muscle. Consequently, alterations in function or structure of gap junctions can lead to intercellular propagation disturbances and contribute to arrhythmogenesis.[20]

Thus, the intercalated disk is an intercellular structure, where desmosomes and adherens junctions not only provide mechanical strength, but also protect the interspersed gap junctions, enabling electrical coupling between cells.

5.2.2 Desmosomal Dysfunction and ARVD/C Pathophysiology

Although the functions of different parts of the intercalated disk seem clear, the exact mechanism through which the mutations of desmosomal protein genes exactly cause disease remains to be elucidated. Various hypotheses, all based on the different functions of desmosomes, have been proposed.

First of all, genetic defects in a desmosomal protein are thought to lead to impairment in mechanical function provoking detachment of myocytes at the intercalated disks, particularly under condition of mechanical stress (like that occurring during competitive sports activity). Such defective mechanical connection followed by mechanical and electrical uncoupling of cardiomyocytes leads to cell death with fibrofatty replacement. Interconnecting bundles of surviving myocardium embedded in the fibrofatty tissue lead to lengthening of conduction pathways (Figure 5.1), and load mismatch. This results in marked *activation delay*, which is the pivotal mechanism for re-entry and thereby ventricular tachycardia (VT). Previous invasive electrophysiologic studies have, by various mapping techniques, confirmed that VT in patients with ARVD/C is due to re-entry circuits in areas of abnormal myocardium.[21] In this structural model, environmental factors such as exercise or inflammation from viral infection could aggravate impaired adhesion and accelerate disease progression. The right ventricle might be more vulnerable to disease than the left because of its thinner walls and its normal dilatory response to exercise.

Second, recent studies have shown that impairment of cell–cell adhesion due to changes in desmosomal components may affect amount and distribution of other intercalated disk proteins, including connexin43, the major protein forming gap junctions in the ventricular myocardium.[22–24] This was shown for DSP and JUP by Western blotting and confocal immunofluorescence techniques, but alterations in other desmosomal components such as PKP2, DSG2, and DSC2 are thought to have similar effects. Changes in number and function of gap junctions will diminish intercellular electrical coupling. This may contribute to intraventricular activation delay, and the substrate for re-entry.

The third hypothesis involves the canonical Wnt/β-catenin signaling pathway. Plakoglobin can localize both to the plasma membrane and the nucleus. It was demonstrated that disruption of desmoplakin frees plakoglobin from the plasma membrane allowing it to translocate to the nucleus and suppress canonical Wnt/β-catenin signaling. Wnt signaling can inhibit adipogenesis by preventing mesodermal precursors from differentiating into adipocytes.[25] Suppression of Wnt signaling by plakoglobin nuclear localization could, therefore, promote the differentiation to adipose tissue in the cardiac myocardium in patients with ARVD/C.[26]

Finally, since ion channels, like the Na$^+$ channel, are also located in the intercalated disk, they might be disrupted and contribute to arrhythmogeneity as well, although at this point this is hypothetical.

The pathophysiological mechanisms proposed above are not mutually exclusive and could occur at the same time.

5.2.3 Desmosomal Disease

Two patterns of inheritance have been described in ARVD/C. The most common or classical form of ARVD/C is inherited as an autosomal dominant trait. The rare *Naxos disease* and *Carvajal syndrome* are inherited autosomal recessively. Table 5.1 summarizes the different genes involved in ARVD/C with the corresponding phenotypes.

Table 5.1 Mutated genes with concurrent type of arrhythmogenic right ventricular dysplasia/cardiomyopathy (*ARVD/C*) (Modified from Van Tintelen et al. Curr Opin Cardiol 2007[27])

	Gene	Type of disease	Inheritance trait
Desmosomal	PKP2	Typical ARVD/C	Autosomal dominant
	DSG2	Typical ARVD/C	Autosomal dominant
	DSC2	ARVD/C	Autosomal dominant
	JUP	Naxos disease	Autosomal recessive
	DSP	Carvajal syndrome	Autosomal recessive
		ARVD/C	Autosomal dominant
		LDAC	Autosomal dominant
Nondesmosomal	RyR2	CPVT	Autosomal dominant
		ARVD/C	Autosomal dominant
	TGF-β	Typical ARVD/C	Autosomal dominant
	TMEM43	ARVD/C	Autosomal dominant

CPVT catecholaminergic polymorphic VT, *LDAC* left dominant arrhythmogenic cardiomyopathy. See text for other abbreviations

5.2.4 Autosomal Recessive Disease

In Naxos disease, the affected individuals were found to be homozygous for a 2 base pair deletion in the JUP gene.[6] All patients who are homozygous for this mutation have diffuse palmoplantar keratosis and woolly hair in infancy; children usually have no cardiac symptoms, but may have electrocardiographic abnormalities and nonsustained ventricular arrhythmias.[28] In an Arab family, an autosomal recessive mutation in the desmoplakin gene caused ARVD/C, also combined with woolly hair, and a pemphigus-like skin disorder.[29] A different autosomal recessive disease, Carvajal syndrome, is also associated with a desmoplakin gene mutation, and is manifested by woolly hair, epidermolytic palmoplantar keratoderma, and cardiomyopathy.[30] The cardiomyopathy of Carvajal syndrome was thought to have a predilection for the left ventricle, but subsequent evaluation of a deceased child revealed typical ARVD/C changes in both ventricles.[24] The cardiac phenotype in the Arab family appeared to be classic ARVD/C.

5.2.5 Autosomal Dominant Disease

Mutations in the gene encoding the intracellular desmosomal component desmoplakin lead to "classic ARVD/C" with a clinical presentation of VT, sudden death, and LV involvement as the disease progresses.[15,31,32] Desmoplakin gene mutations have also been associated with predominantly left-sided ARVD/C and, as noted above, with autosomal recessive disease.

Overall, mutations in the *PKP2* gene are the most frequently observed in ARVD/C. Figure 5.3 shows the pedigree of a family with a *PKP2* mutation. Incomplete penetrance and clinical variability is well documented.

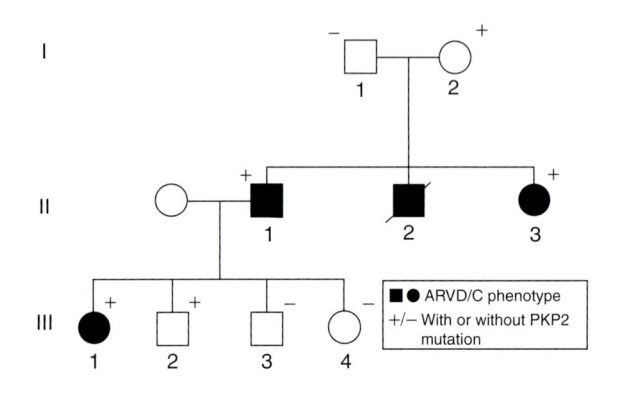

Fig. 5.3 Pedigree of family with arrhythmogenic right ventricular dysplasia/cardiomyopathy (*ARVD/C*) and plakophilin-2 (*PKP2*) mutation. This figure shows the variability in penetrance and clinical expression. Both the 72-year-old grandmother (I:2) and 20-year-old grandson (III:2) are free of any signs of disease, despite carrying the mutation. The proband (II:1) was resuscitated at age 35, his brother (II:2) died suddenly at age 18. Both the proband's sister (II:3) and daughter (III:1) were diagnosed with the disease due to a positive family history, and RV structural abnormalities. The sister (II:3) of the proband has structural and ECG abnormalities, but no arrhythmias

In four studies from different countries, analyzing 56 to 100 ARVD/C patients each, the following observations were made.[16, 33–35] *PKP2* mutations were found in 11–43% of unrelated index patients who fulfilled diagnostic Task Force criteria (TFC) for ARVD/C. In the current Dutch ARVD/C cohort, 67 of 125 (54%) probands carry a pathogenic *PKP2* mutation. In total, 21 different mutations have been observed: 10 nonsense, 4 missense, 3 frameshifts, 2 at splice sites, and 2 deletions. Haplotype analysis previously performed suggested that founder mutations were responsible for 4 of the 14 different mutations identified.[35] Among index patients with a positive family history of ARVD/C, 70% had a *PKP2* mutation.[35] No specific genotype–phenotype correlations were established, except that patients with *PKP2* mutation presented at a younger age (28 vs. 36 years), and negative T waves in V1-3 occurred more often in *PKP2* mutation carriers. Thus, *PKP2* appears to be a relatively commonly mutated gene in ARVD/C patients, particularly in cases with a documented family history. However, there does not appear to be a substantial phenotypic distinction from other mutations, and due to relatively small sample sizes and the potential for referral bias, the incidence of *PKP2* mutations in a broader population may be lower.

Because of the known association of ARVD/C with defects in other desmosomal proteins, a series of patients with ARVD/C were screened for mutations in the gene encoding the transmembranous desmosomal component DSG2 as well.[17] Among 80 unrelated probands, 26 were known to have *DSP* or *PKP2* mutations. Direct sequencing of *DSG2* in the other 54 patients revealed nine distinct mutations in eight individuals. These individuals demonstrated typical clinical characteristics of ARVD/C. An analogous study of 86 ARVD/C probands identified eight novel *DSG2* mutations in nine probands. Clinical evaluation of family members with *DSG2* mutations revealed penetrance of 58% using *Task Force criteria* and 75% using proposed modified criteria.[36] Morphological abnormalities of the right ventricle were present in 66% of gene carriers, LV involvement in 25% and classical right precordial T-wave inversion in only 26%. The authors noted that disease expression of DSG2 mutations was of variable severity, but that overall penetrance was high and LV involvement prominent.[37]

In the gene encoding DSC2, another important transmembranous desmosomal cadherin, two heterozygous mutations (a deletion and an insertion) were identified in 4 of 77 probands with ARVD/C.[18] The identification of the fifth desmosomal cell adhesion gene abnormality further supports the hypothesis that ARVC is usually a disease of cell adhesion.

5.2.6 Other, Nondesmosomal, Genes

Mutations in the gene encoding the cardiac ryanodine receptor RyR2, which is responsible for calcium release from the sarcoplasmic reticulum, have been described in only one Italian ARVD/C family.[13] Affected patients have exercise-induced polymorphic VT.[38] Mutations in *RyR2* have primarily been associated with familial catecholaminergic polymorphic VT without ARVD/C.[19, 39] RyR2 mediates the release of calcium from the sarcoplasmic reticulum that is required for myocardial contraction. The FK506 binding protein (FKBP12.6) stabilizes RyR2, preventing aberrant activation. The mutations in *RyR2* interfere with the interaction with FKBP12.6, increasing channel activity under conditions that simulate exercise.[39] Although the general opinion is that *RyR2* mutations lead to catecholaminergic polymorphic VT, without structural abnormalities, the mutations in ARVD/C have been advocated to act differently from those in familial polymorphic VT without ARVD/C.[40–42]

Transforming growth factor-β-3 (TGFβ3) regulates the production of extracellular matrix components and modulates expression of genes encoding desmosomal proteins. Its gene has been mapped to chromosome 14. Sequencing studies failed to identify any disease-causing mutations in the exonic regions of *TGFβ3*. This led to screening of the promoter and untranslated regions, where a mutation of the *TGFβ3* gene was found in all clinically affected members of a large family with ARVD/C.[14] The mutation is predicted to produce an amino acid substitution in a short peptide with an inhibitory role in TGFβ3 regulation. The implication of these observations is that regulatory mutations resulting in overexpression of TGFβ3 may contribute to the development of ARVD/C in these families. The TGFβ family of cytokines stimulates production of components of the extracellular matrix. It is therefore possible that enhanced TGFβ activity can lead to myocardial fibrosis. However, genetic analysis of two other families with ARVD/C failed to identify mutations in any of the regions of the TGFβ3 gene.

A missense mutation in the *TMEM43* gene was found in 15 unrelated ARVD/C families from a genetically isolated population in New Foundland and caused a fully penetrant, sex-influenced, high-risk form of ARVD/C.[19] The *TMEM43* gene contains the response element for PPAR gamma, an adipogenic transcription factor. The *TMEM43* gene mutation is thought to cause dysregulation of an adipogenic pathway regulated by PPAR gamma, which may explain the fibrofatty replacement of myocardium in ARVD/C patients.

5.3 Epidemiology

Estimations of the prevalence of ARVD/C in the general populations vary from 1:2,000 to 1:5,000.[43] The exact prevalence of ARVD/C, however, is unknown and is possibly higher because of the existence of many nondiagnosed or misdiagnosed cases.

The disease appears to be especially common in adolescents and young adults in northern Italy, accounting for approximately 11% of cases of sudden cardiac death overall and even 22% in athletes.[44, 45] In as many as 20% of sudden deaths occurring in people under 35 years of age, features of ARVD/C were detected at postmortem evaluation.[45] In nearly half of them, no prior symptoms had been reported. In contrast, ARVD/C has rarely been diagnosed in the United States. Founder mutations have been identified, e.g. in the Netherlands, and these could in part explain the difference in prevalence in different geographical areas.

ARVD/C has a reduced penetrance and extremely variable clinical expression. For instance, family screening has identified pathogenic mutation carriers, who had stayed free of any sign of disease up to over 70 years of age (Fig. 5.3).

Although from a genetic point of view, both men and women have to be equally affected, men are more frequently diagnosed with ARVD/C than women, with an approximate ratio of 3:1. However, as many women as men do show at least some signs of disease, but women more often do not fulfill enough criteria to meet the diagnosis. Factors explaining this difference in severity of disease expression have not yet been elucidated. It is speculated that (sports) activity or hormonal factors may play a role. A familial background has been demonstrated in >50% of ARVD/C cases.

5.4 Clinical Presentation

ARVD/C patients typically present between the second and fourth decade of life with VT originating from the right ventricle. However, in a minority of cases sudden death, possibly at a young age, or RV failure are the first signs. Based on clinicopathologic and patient follow-up studies, four different disease phases have been described for the classical form of ARVD/C, i.e., primarily affecting the RV (Table 5.2).

1. Early ARVD/C is often described as "concealed" owing to the frequent absence of clinical findings, although minor ventricular arrhythmias and subtle structural changes may be found. Although patients tend to be asymptomatic, they may nonetheless be at risk of sudden death, mainly during intense exercise.
2. The overt phase follows, in which patients suffer from palpitations, syncope, and ventricular arrhythmias of left bundle branch block morphology, ranging from isolated ventricular premature complexes to sustained VT and ventricular fibrillation (VF).
3. The third phase is characterized by RV failure due to progressive loss of myocardium with severe dilatation and systolic dysfunction, in the presence of preserved LV function.
4. Biventricular failure occurs, due to LV involvement. This phase may mimic dilated cardiomyopathy (DCM) and may require cardiac transplantation.

In the initially described classical form of ARVD/C, the RV is primarily affected with possibly (in a later stage) some LV involvement. Two additional distinct

Table 5.2 Different phases of disease severity

Phase	Characteristics
1. Concealed	Asymptomatic patients with possibly only minor ventricular arrhythmia and subtle structural changes
	However, risk of sudden death
2. Overt	Symptoms due to LBBB VT or multiple premature complexes, with more obvious structural RV abnormalities
3. RV failure	With relatively preserved LV function
4. Biventricular	Significant overt LV involvement

patterns of disease have been identified by clinicogenetic characterization of families. These are the left dominant phenotype, with early and predominant LV manifestations, and the biventricular phenotype with equal involvement of both ventricles. Recent immunohistochemical analysis of human myocardial samples demonstrated that on a desmosomal level, both ventricles are affected by the disease.[73] A marked reduction in immunoreactive signal levels for Plakoglobin was observed both in the right and left ventricle, independent of genotype. This strengthens the idea that in essence, ARVD/C is a biventricular disease. However, histologically and functionally overt manifestations of the disease usually start in the RV. The reason for this is still unclear. The most advocated idea is that the thin-walled RV is less able to withstand pressure (over)load when the mechanical junctions have an impaired function.

5.5 Clinical Diagnosis

Diagnosis of ARVD/C can be very challenging and can only be made when all other diseases causing VT episodes from the RV have been ruled out (see paragraph on differential diagnosis). Although VF and sudden death may be the first manifestations of ARVD/C, symptomatic patients typically present between age 20 and 40 years with sustained VT with left bundle branch block morphology, thus originating from the RV. The occurrence of VT episodes is usually driven by adrenergic stimulation and starts mainly during exercise, especially competitive sports. ARVD/C is a disease that shows progression over time, and may manifest differently according to the time of patient presentation.

The gold standard for ARVD/C diagnosis is the demonstration of transmural fibrofatty replacement primarily of right ventricular myocardium, determined at surgery or postmortem. Predilection sites for these structural abnormalities are the so-called triangle of dysplasia formed by the RV outflow tract (RVOT), the apex, and the subtricuspid region. In daily clinical practice, this definition of diagnosis is not usable. Even endomyocardial biopsies have major limitations. Tissue sampling from the affected often thin RV-free wall, directed by imaging techniques or voltage mapping, is rather straightforward, but is associated with a risk of perforation. Sampling from the interventricular septum is relatively safe. However, the septum is histopathologically rarely affected in ARVD/C. In addition, even in potentially affected areas, histology may be classified as normal because of the segmental nature of the lesions. Finally, since subendocardial layers are usually not affected in an early stage of the disease, histologic diagnosis may be hampered by the nontransmural nature of endomyocardial biopsies.[46,47]

Clinical diagnosis has been facilitated by a set of clinically applicable criteria for ARVD/C diagnosis defined by a Task Force in 1994.[48] Based on the evidence available at that time, the Task Force included six different groups of clinical criteria. Within these groups, diagnostic criteria were assigned major or minor according to their specificity for the disease. Every major criterion is scored as two points and every minor as one point. In total, four points have to be scored in order to fulfill the ARVD/C diagnosis, i.e., two major, one major plus two minor, or four minor criteria. From each different group, only one criterion can be counted for diagnosis, even when multiple criteria in one group are being fulfilled. Recently, these 1994 Task Force criteria have been revised. See Table in Addendum.

Specific tests are recommended in all patients suspected of ARVD/C. A 12-lead ECG, signal averaged ECG (SAECG; when available), 24 h Holter monitoring, exercise testing, and 2D-echocardiography should be performed in all. When appropriate, more detailed analysis of the RV can be done by cardiac MRI or computed tomography. Eventually, invasive tests are also available for diagnostic purposes: endomyocardial biopsy, RV cine-angiography, and electrophysiologic testing.

Table 5.3 gives an overview of the Task Force Criteria (TFC).

5.6 ECG Criteria

Criteria on ECG changes have to be obtained in normal sinus rhythm and while off anti-arrhythmic drugs. Furthermore, a complete right bundle branch block has to be absent. ECG changes are detected in up to 90% of ARVD/C patients.

5.6.1 Depolarization Abnormalities

As explained above, RV activation delay is a hallmark of ARVD/C. This delay is conveyed by the criteria of

Table 5.3 Diagnostic Task Force Criteria[48]

Factor	Major criteria	Minor criteria
Family History	Familial disease confirmed at necropsy or surgery	Family history of premature sudden death (age <35 years) due to suspected ARVD/C Family history (clinical diagnosis based on present criteria)
ECG depolarization/conduction abnormalities	Epsilon waves or localized prolongation (>110 ms) of QRS complex in leads V_1–V_3	Late potentials on signal-averaged ECG
ECG repolarization abnormalities		Inverted T waves in right precordial leads (V_2 and V_3) in people aged >12 years and in absence of right bundle branch block
Arrhythmias		VT with LBBB morphology >1,000 premature ventricular complexes in 24 h on Holter monitoring
Global or regional dysfunction and structural alterations	RV akinetic or dyskinetic areas with diastolic bulging; Severe dilatation and reduction of RV ejection fraction with no or mild LV involvement (RV > LV)	Mild global RV dilatation or ejection fraction reduction with normal LV Mild segmental dilatation of RV
Tissue characteristics of walls	Fibrofatty replacement of myocardium on endomyocardial biopsy	

epsilon waves or localized prolongation of the QRS complex in V1-3 (>110 ms) and late potentials on SAECG.

Epsilon waves are defined as low amplitude potentials after and clearly separated from the QRS complex, in at least one of V1-3 (Fig. 5.4).[49] This highly specific major criterion is unfortunately observed in only a small minority of patients.[50, 51]

Localized prolongation of the QRS complex in V1-3 >110 ms is also a major diagnostic criterion. In one study, this was observed in as many as 70% of patients. However, most reports give lower percentages.[52]

The detection of *late potentials* on SAECG is the surface counterpart of delayed activation or late potentials detected during endocardial mapping in electrophysiologic studies. They are frequently found in patients with documented VT. However, these late potentials can also be observed after myocardial infarction and other structural heart diseases, and due to this lack of specificity were considered a minor criterion.

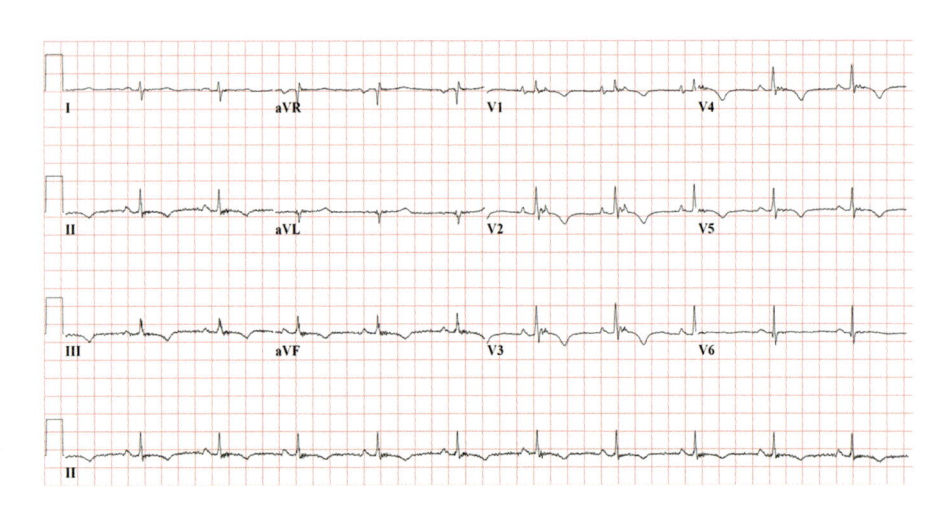

Fig. 5.4 Epsilon waves and negative T waves in V1-5

For all criteria on depolarization abnormalities, it is apparent that their finding will correlate with disease severity. For instance, a positive correlation has been found between late potentials and the extent of RV fibrosis, reduced RV systolic function, and significant morphological abnormalities on imaging.[53–55]

5.6.2 Repolarization Abnormalities

Negative T waves in leads V1 up to and including V3 form the minor ECG criterion on repolarization abnormalities (see Fig. 5.4). They are the most frequently observed criterion. In the initial series reported by Marcus et al., this was detected in over 85% of cases.[1] Subsequent studies have reported variable prevalences of right precordial T wave inversion, ranging from 19% to 94%.[48–52] The lower rates are often due to evaluation of family members, while higher rates are seen in series consisting of unrelated index patients. *T wave inversion* can be a normal feature of the ECG in children and early adolescence. Therefore, this finding is not considered pathogenic in persons aged 12 years and younger. Similarly, negative T waves are not to be judged in the presence of a right bundle branch block.

Although these negative T waves were observed consistently in a series of evaluated ARVD/C patients, T wave inversions in the right precordial leads can also be observed in 1–3% of the healthy populations aged 19–45 years, and in patients with right ventricular overload, such as major pulmonary embolism and intracardial left to right shunt, or may develop following intracranial hemorrhage as a sign of adrenergic response to the cerebral insult. Owing to this presumed lack of specificity, T wave inversions were included as a minor criterion.

5.6.3 Arrhythmias

Ventricular arrhythmias range from premature ventricular complexes to sustained VT and VF, leading to cardiac arrest.[56,57] Because of the origin in the RV, QRS complexes of ventricular arrhythmias show a left bundle branch block morphology. Moreover, the QRS axis gives an indication of the VT origin, i.e., superior axis from the RV inferior wall or apex and inferior axis from the RV outflow tract (see Fig. 5.5). Patients with extensively affected RV may show multiple VT morphologies.

VF is the mechanism of instantaneous sudden death especially occurring in young people and athletes with ARVD/C, who were often previously asymptomatic. In this subset of patients, VF may occur from deterioration of monomorphic VT, or in a phase of acute disease progression, due to acute myocyte death and reactive inflammation.[3]

5.6.4 Global and/or Regional Dysfunction and Structural Alterations

Evaluation of RV size and function can be done by various imaging modalities, including echocardiography, magnetic resonance imaging (MRI), computed

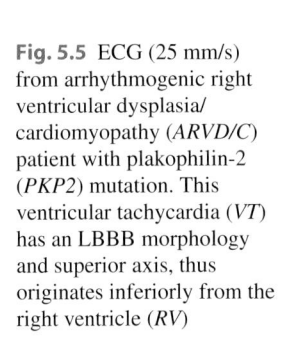

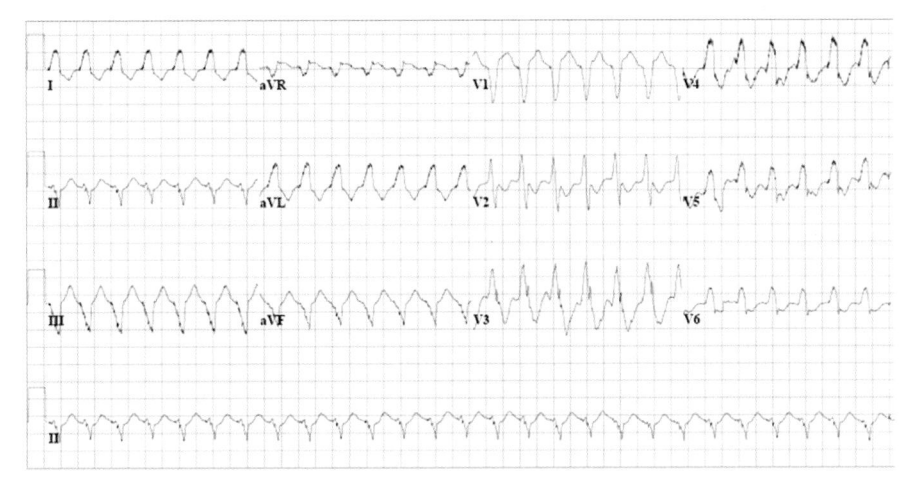

Fig. 5.5 ECG (25 mm/s) from arrhythmogenic right ventricular dysplasia/cardiomyopathy (*ARVD/C*) patient with plakophilin-2 (*PKP2*) mutation. This ventricular tachycardia (*VT*) has an LBBB morphology and superior axis, thus originates inferiorly from the right ventricle (*RV*)

tomography, and/or cine-angiography. According to the Task Force criteria, major criteria are defined as presence of large a- or dyskinetic areas in the RV or severe dilatation of the RV (RV larger than LV).[48] Less severe abnormalities are considered minor criteria. The Task Force criteria are unspecific, in the sense that no definitions of severe compared with less severe had been formulated. Although each of the different imaging modalities mentioned can detect severe structural abnormalities, the diagnostic value of each is less certain when evaluating mild cases of the disease. RV cine-angiography has historically been considered the gold standard to visualize RV structural abnormalities, with a high specificity of 90%.[58] Compared to cine-angiography, the noninvasive technique of echocardiography is widely used and serves as the first-line imaging technique in evaluating patients suspected of ARVD/C and in family screening. Especially with improvement of echocardiographic modalities, like three-dimensional echocardiography, strain and tissue Doppler, sensitivity and specificity of echocardiography have increased in the past years. Cardiac MRI is advantageous, since it has the unique possibility of visualizing the myocardium to characterize tissue composition, by differentiating fat from fibrous tissue by delayed enhancement. However, this technique is expensive and not widely available and requires great expertise to prevent mis- or overdiagnosis of ARVD/C.[59] Also, in ICD-carrying patients, this technique cannot be applied. Cardiac MRI has appeared to be the most common cause of overdiagnosis and physicians should therefore be very reluctant to diagnose ARVD/C when structural abnormalities are only present on MRI.[60,61] Furthermore, it is important to note that the presence of fat in the epi- and midmyocardial layers (without fibrosis) is known as cor adiposum and should not be considered diagnostic of ARVD/C.

5.6.5 Endomyocardial Biopsy

For reasons outlined earlier, undirected endomyocardial biopsies are rarely diagnostic. However, it had been included as a major criterion by the Task Force, since the finding of fibrofatty replacement was considered to strongly support any findings derived from other clinical investigations. The rather vague description of any "fibrofatty replacement of myocardium," had been further studied and specified by the pathology department of Padua, where histomorphometric parameters of myocytes, interstitium, fibrous tissue, and fatty tissue were evaluated. Diagnostic values indicating arrhythmogenic right ventricular cardiomyopathy were presence of <45% of cardiomyocytes, >40% fibrous tissue, and >3% of fatty tissue with 67% sensitivity and 91% specificity for at least one parameter.[62]

5.6.6 Family History

Already before the discovery of pathogenic mutations underlying the disease, it was recognized that ARVD/C often occurs in multiple members of the same family. Having a family member with proven ARVD/C was considered an increased risk for other family members to get the disease as well, and was therefore included as a minor diagnostic criterion. Also, sudden death of a family member under the age of 35 years, presumably but not proven to be due to ARVD/C-related arrhythmias, is a minor criterion. Pathologic confirmation of transmural fibrofatty replacement of the RV at autopsy or after surgical resection is the gold standard for ARVD/C diagnosis. Therefore, if this is diagnosed in a relative, it is considered a major criterion in the diagnosis of their relatives.

5.6.7 Discussion on Diagnostic Criteria

The current TFC are the essential standard for classification of individuals suspected of ARVD/C. In addition, its universal acceptance contributed importantly to unambiguous interpretation of clinical studies and facilitated comparison of their results. However, past years have shown that this set of criteria is very specific, but lacks sensitivity. From studies on larger numbers of patients and their family members, more insight has been gained into the disease. This leads to the possibility of various modifications and extensions by multiple groups.

A number of additional ECG markers have been reported, mainly reflecting right ventricular activation delay, including QRS and QT dispersion, parietal block (QRS duration in V1–V3 exceeds the QRS duration in V6 by >25 ms), localized right precordial QRS prolongation (QRS duration in leads V1+V2+V3/V4+V5+V6≥1.2). Recently, Nasir et al. demonstrated that a prolonged (≥55 ms) S-wave upstroke in the right precordial leads correlates with disease severity and induction of VT during electrophysiological studies.[63] Our group improved this criterion and introduced prolonged *terminal activation duration (TAD)*.[57] Prolonged TAD is defined as ≥55 ms, from the nadir of the S wave to the end of all depolarization deflections in V1-3, thereby covering all forms of RV activation delay, including epsilon waves (Fig. 5.6). This new criterion appeared to be more sensitive than all accepted and proposed diagnostic criteria on activation delay. Whereas other criteria were observed only in a minority, prolonged TAD was observed in as many as 71% of patients studied. Furthermore, we observed that VT with LBBB morphology and superior axis, and/or multiple different VT morphologies were recorded in 67% and 88% of ARVD/C patients, respectively. All three newly proposed additional criteria were highly specific for ARVD/C as well. However, since most markers have been deducted from ARVD/C-positive patients selected on the basis of the TFC, the additional diagnostic value still has to be evaluated in further prospective studies.

Multiple groups attempted to provide better quantification of normal and abnormal values of RV dimensions. Both for echocardiography and MRI, right ventricular dimensions of ARVD/C patients have been compared to controls.[64, 65] Normal and cut-off values have been established based on standard deviations, as was done by the initial Task Force as well. The disadvantage of this method is that the value at two standard deviations from normal do not always represent the cutoff point optimal with respect to sensitivity and specificity.

Currently, a modification of the complete set of the 1994 TFC is being created. The new TFC will differ from the current ones, by including prolonged TAD and electrical axis of VT, using quantification of RV dimensions and have a role for results of genetic screening (see Table in Addendum).

5.6.8 Modifications for Family Members

The same set of diagnostic TFC is applied whether it concerns a proband or one of his family members. Based on the experience of similar criteria in diagnosing hypertrophic and dilated cardiomyopathies showing that many relatives have phenotypic abnormalities that, although "nondiagnostic," are nevertheless indicative of disease, a new set of TFC was proposed for family members of ARVD/C patients.[36] Hamid et al. reported that by using current TFC, familial disease was observed in 28% of index patients.[36] A further 11% of their relatives had minor cardiac abnormalities, which, in the context of an autosomal dominant disease, are likely to represent early or mild disease expression. They proposed that first-degree relatives of patients with proven ARVD/C with a positive SAECG, or a minor ECG, Holter, or echocardiographic abnormality from the present diagnostic criteria should get ARVD/C diagnosis as well. Furthermore, they advocate that the frequency of ectopic activity accepted as a marker of disease expression should be reduced from 1,000 to 200 ventricular premature complexes over a 24-h period. Hereby, they expect that disease classification of family members will be improved. However, because of the positive family history, family members will already fulfill one of TFC, either major or minor,

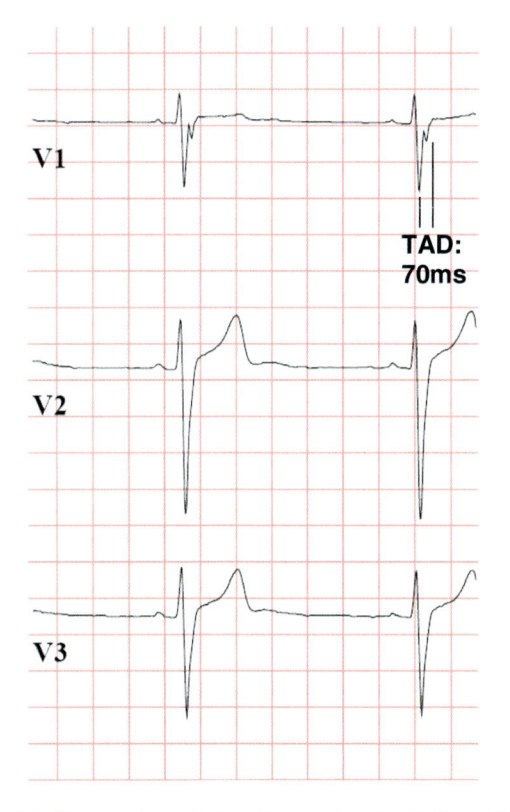

Fig. 5.6 Prolonged terminal activation duration (≥55 ms from nadir of S wave to end of depolarization)

depending on the type of history. Therefore, they have to fulfill fewer of the other criteria to get the ARVD/C diagnosis. This can possibly lead to family members being stigmatized unnecessarily, since having a positive family history indicates a higher risk to develop ARVD/C, but does not change the degree to which they are actually affected by the disease.

5.6.9 Nonclassical ARVD/C Subtypes

5.6.9.1 Naxos Disease

All patients who homozygously carry the recessive JUP mutation for Naxos disease have diffuse palmoplantar keratosis and woolly hair in infancy; children usually have no cardiac symptoms, but may have ECG abnormalities and nonsustained ventricular arrhythmias.[23, 28] The cardiac disease is 100% penetrant by adolescence, being manifested by symptomatic arrhythmias, ECG abnormalities, right ventricular structural alterations, and LV involvement. In one series of 26 patients followed for 10 years, 62% had structural progression of right ventricular abnormalities, and 27% developed heart failure due to LV involvement.[28] Almost half of the patients developed symptomatic arrhythmias and the annual cardiac and SCD mortality were 3% and 2.3%, respectively, which are slightly higher than seen in autosomal dominant forms of ARVD/C. A minority of heterozygotes have minor ECG and structural changes, but clinically significant disease is not present.

5.6.9.2 Carvajal Syndrome

Carvajal syndrome is associated with a *DSP* gene mutation, and is also a recessive disease manifested by woolly hair, epidermolytic palmoplantar keratoderma, and cardiomyopathy.[30] All patients diagnosed so far came from Ecuador. The cardiomyopathy part of Carvajal syndrome was first thought to be mainly left ventricular, with dilated left ventricular cardiomyopathy. A number of the patients with Carvajal syndrome suffered from heart failure in their teenage years, resulting in early morbidity. However, further research revealed that it is characterized mainly by ventricular hypertrophy, ventricular dilatation, and discrete focal ventricular aneurysms. In the right ventricle in particular, focal wall thinning and aneurysmal dilatation were identified in the triangle of dysplasia.

5.6.9.3 Left Dominant ARVD/C (LDAC)

As mentioned earlier, the histologic process in classic ARVD/C predominantly involves the RV and extends to the LV in more advanced stages.[52, 59, 66–68] In contrast, patients with left-dominant arrhythmogenic cardiomyopathy (LDAC, also known as left-sided ARVD/C or arrhythmogenic left ventricular cardiomyopathy) have fibrofatty changes that predominantly involve the LV.[68] Clinically, this disease entity is characterized by (infero) lateral T-wave inversion, arrhythmia of LV origin, and/or proven LDAC.

Patients presented with arrhythmia or chest pain at ages ranging from adolescence to over 80 years. By cardiovascular MRI, about one third of patients show an LV ejection fraction <50%. Furthermore, MRI with late gadolinium enhancement (LGE) of the LV demonstrated late enhancement in a subepicardial/midwall distribution. Similar to ARVD/C, some patients with LDAC have desmosomal gene mutations (see later).

5.7 Differential Diagnosis

Although diagnosis in an overt case of ARVD/C is often not difficult, early and occasionally late stages of the disease may show similarities with a few other diseases. Especially differentiation from idiopathic VT originating from the RV outflow tract (RVOT) can be challenging. However, *idiopathic RVOT VT* is a benign nonfamilial condition, in which the ECG shows no depolarization or repolarization abnormalities and no RV structural changes can be detected. Furthermore, VT episodes have a single morphology (LBBB morphology with inferior axis) and are usually not reproducibly inducible by premature extrastimuli at programmed stimulation during EP studies.[70, 71] In contrast, idiopathic RVOT VT may be inducible by regular burst pacing and isoproterenol infusion. It is important to differentiate idiopathic RVOT VT from ARVD/C for several reasons. The first is the known genetic etiology in ARVD/C whereas RVOT tachycardia has not. Therefore, it has implications with regard to screening of family members. The prognosis of RVOT tachycardia is uniformly excellent with sudden death occurring extremely rarely. Finally, in contrast to ARVD/C, catheter ablation is usually curative in idiopathic RVOT tachycardia.

Another disease mimicking ARVD/C is cardiac *sarcoidosis*. Sarcoidosis is a disease with unknown etiology, characterized by the presence of noncaseating granulomas in affected tissues; mainly lungs, but heart, skin, eyes, reticuloendothelial system, kidneys, and central nervous system can also be affected. The prevalence of this condition varies between geographical regions, and the disease may also be familial and occurring in specific racial subgroups.[72] Clinical symptoms of cardiac involvement are present in about 5% of all patients with sarcoidosis. The clinical manifestations of cardiac sarcoidosis depend on the location and extent of granulomatous inflammation and include conduction abnormalities, ventricular arrhythmias, valvular dysfunction, and congestive heart failure. Myocardial sarcoid granulomas or areas of myocardial scarring are typically present in the left ventricle and septum of patients with this condition, yet the right ventricle can be predominantly affected. A VT associated with right ventricular abnormalities can, therefore, result in diagnostic confusion, especially if there is no systemic evidence of sarcoidosis. Patients can present with clinical features similar to those of ARVD/C including arrhythmias and sudden cardiac death.[73] Cardiac sarcoidosis can only be diagnosed definitively by endomyocardial biopsy, when granulomas are visualized. To strengthen differentiation from ARVD/C, gadolinium-enhanced MRI may be beneficial by detecting located abnormalities in the septum, which is typical for sarcoidosis but hardly ever seen in ARVD/C. Active foci of sarcoidosis can be visualized by positron emission (PET) scan. Therapy with corticosteroids is recommended for patients with a clear diagnosis of cardiac sarcoidosis. Treatment aims to control inflammation and fibrosis in order to maintain cardiac structure and function.

Also, any other form of *myocarditis* has to be excluded before diagnosis of ARVD/C can be made. Myocarditis may arise from viral or other pathogen exposure as well as toxic or immunologic insult. In general, endomyocardial biopsy is required to distinguish it from ARVD/C.

Especially in more advanced stages of the disease, when LV ejection fraction drops below 50%, ARVD/C may mimic dilated cardiomyopathy (DCM). Patients with DCM usually present with heart failure or thromboembolic disease, including stroke. It is uncommon to have sustained ventricular tachycardia or sudden death as the initial presenting symptom of DCM. Therefore, in that case patients should be first suspected to have ARVD/C.

5.8 Molecular Genetic Analysis

It is important to realize that ARVD/C diagnosis is based on the clinical diagnostic TFC. To date, molecular information has not been incorporated into the diagnostic criteria. Mutations underlying the disease show incomplete penetrance and variable clinical expression. Some genetically affected patients can have no signs or symptoms whatsoever, whereas no mutations can be identified in a large minority of clinically diagnosed patients. Therefore, DNA analyses will not be of any consequence for the index patient, but can be used to identify whether family members are predisposed to disease development.

The strategy for genetic testing in ARVD/C is as follows.

Individuals with clinical diagnosis of ARVD/C are the first to be tested. The detection of a pathogenic mutation does not make a diagnosis of ARVD/C. In contrast, if no mutation can be identified in a patient diagnosed with ARVD/C, the clinical diagnosis of ARVD/C is still applicable. In case a pathogenic mutation is identified in the proband, parents, siblings, and possibly children of this patient can be subsequently tested for the mutation concerned, via the cascade method. When an (asymptomatic) relative is found to carry a pathogenic mutation, cardiologic screening is required at least biannually.

Table 5.1 shows the different genes related to ARVD/C. Although numbers vary per country, *PKP2* accounts for the large majority of mutations found. Currently, DNA analysis on *PKP2*, *DSG2*, *DSC2*, *DSP*, and *JUP* is recommended in all ARVD/C patients.

5.9 Prognosis and Therapy

Although the prognosis of ARVD/C is considerably better than that of sustained ventricular tachycardia with left ventricular structural heart disease, ARVD/C is a progressive disease and will probably lead to right ventricular failure in the long term unless sudden cardiac death occurs. The death rate for patients with ARVD/C has been estimated at 2.5% per year.[74] Retrospective analysis of clinical and pathologic studies identified several risk factors for sudden death, such as previously aborted sudden death, syncope, young age, malignant family history, severe RV dysfunction, and LV involvement.[75,76]

Electrophysiologic testing by programmed ventricular stimulation can be useful for diagnostic purposes by induction of multiple VT morphologies and VT with LBBB morphologies and superior axis.[57] However, EP studies have proven not to be useful in risk stratifying patients with ARVD/C. This was illustrated in the multicenter study of 132 patients with ARVD/C in whom electrophysiologic study was performed prior to ICD implantation.[77] The positive and negative predictive values of VT inducibility for subsequent appropriate device therapy were only 49% and 54%, respectively.

In addition to symptomatic treatment, prevention of sudden death is the most important therapeutic goal in ARVD/C. Most data available on effective treatment strategies refer to retrospective analyses in single centers with only limited number of patients, of which results are difficult to compare due to different patient selections and treatment strategies. There is limited data on long-term outcomes and no controlled randomized trials have been performed. International registries are established, but have not yet reported results on treatment.

Evidence suggests that asymptomatic patients and healthy mutation carriers do not require any prophylactic treatment. They should instead undergo regular cardiologic check-ups including 12-lead ECG, 24 h Holter monitoring, echocardiography, and exercise testing for early identification of unfavorable signs. To all patients diagnosed with or showing multiple signs of ARVD/C as well as all mutation carriers, specific life style advises have to be given, indifferent from which additional therapeutic measures are taken. Sports participation has been shown to increase the risk of sudden death in ARVD/C patients fivefold.[78] Furthermore, excessive mechanical stress, such as during competitive sports activity and training, may aggravate the underlying myocardial lesion and accelerate disease progression. Therefore, patients with ARVD/C have to be advised against practicing highly competitive sports and sports with long endurance, like running marathons.

Therapeutic options in patients with ARVD/C include anti-arrhythmic drugs, catheter ablation, and implantation of cardioverter defibrillators (*ICDs*).

Patients with recurrences of ventricular tachycardia have a favorable outcome when they are treated medically and therefore pharmacologic treatment is the first choice. This concerns not only patients who have presented with sustained VT, but also patients and family members with nonsustained VT or >500 ventricular extrasystoles on 24-h Holter monitoring. Since ventricular arrhythmias and cardiac arrest occur frequently during or after physical exercise or may be triggered by catecholamines, anti-adrenergic β-blockers are recommended. Sotalol is the drug of first choice. Alternatively, other β-receptor blocking agents, amiodarone and flecainide, have all been reported as useful. Efficacy of drug treatment has to be evaluated by serial Holter monitoring and/or exercise testing. This strategy has proven to have better long-term outcome than standard empirical treatment (Fig. 5.7).[79]

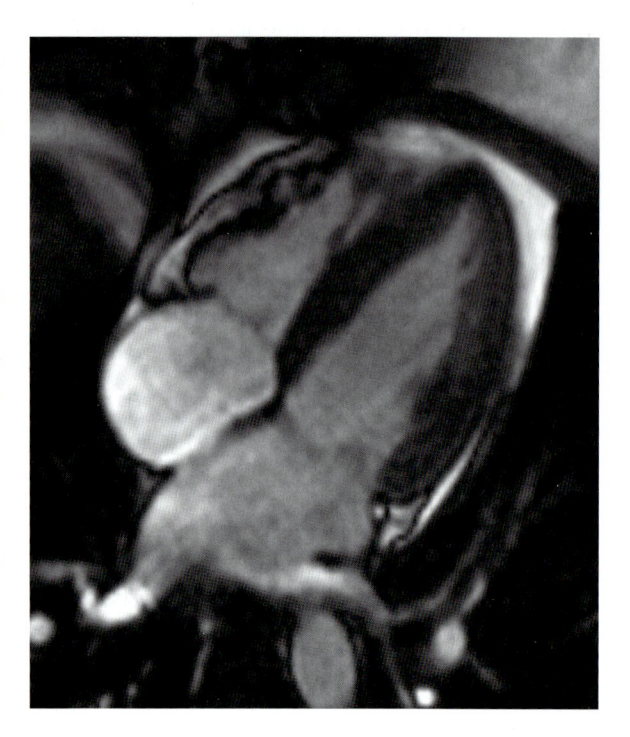

Fig. 5.7 Magnetic resonance imaging (*MRI*) image of arrhythmogenic right ventricular dysplasia/cardiomyopathy (*ARVD/C*) patient at the end of systole. Dyskinetic bulgings are clearly visible in the right ventricle (*RV*) free wall

Catheter ablation is an alternative in patients who are refractory to drug treatment and have frequently recurring VT episodes with predominantly a single morphology. In addition, catheter ablation has been shown to improve the effectiveness of pharmacological treatment: 70% of patients may respond to antiarrhythmic agents to whom they were unresponsive prior to ablation therapy.[80] Marchlinski et al. performed VT ablation in 19 ARVD/C patients by use of focal and/or linear lesions, in 17 of whom no VT recurred during the subsequent 7 ± 22 months.[81] In a series of 50 consecutive patients studied during 16 years, Fontaine et al. reached a 40% success rate by radiofrequency ablation after multiple ablation sessions, which increased to 81% when fulguration was used additionally.[82] However, these successes have been reported by single centers with highly experienced electrophysiologists, and may not hold true in general practice. Catheter ablation is usually considered as only palliative and not curative. In general, long-term success rates are poor. Owing to disease progression, new VTs with different morphologies will occur after a certain period of time.[83]

Although *anti-arrhythmic drugs* and catheter ablation may reduce VT burden, there is no proof from prospective trials that these therapies will also prevent sudden death. ICD implantation is indicated in patients who are intolerant to anti-arrhythmic drug therapy and who are at serious risk for sudden death. Implantation of an ICD has to be considered in ARVD/C patients with aborted cardiac arrest, intolerable fast VT, and those with risk factors as mentioned above.

5.10 Summary

ARVD/C is most often a genetically determined disease characterized by fibrofatty replacement of myocardial tissue. Clinically, it affects primarily the right ventricle, but extension to the left ventricle occurs, especially in more advanced stages of the disease. Patients typically present between the second and fourth decade of life with exercise-induced tachycardias originating from the right ventricle. However, it is also a major cause of sudden death in the young and athletes.

Its prevalence has been estimated to vary from 1:2,000 to 1:5,000.

The causative genes encode proteins of mechanical cell junctions (plakoglobin, plakophilin2, desmoglein, desmocollin, desmoplakin) and account for intercalated disk remodeling. The classical form of ARVD/C is inherited in an autosomal dominant trait, but shows reduced penetrance and variable expression. The more rare recessively inherited variants are often associated with palmoplantar keratoderma and woolly hair. Clinical diagnosis is made according to a set of Task Force criteria, based on family history, depolarization and repolarization abnormalities, arrhythmias with a left bundle branch block morphology, functional and structural alterations of the right ventricle, and fibrofatty replacement through endomyocardial biopsy. Two-dimensional echocardiography, cine-angiography, and magnetic resonance are the imaging tools for visualizing structural/functional abnormalities. The main differential diagnoses are idiopathic right ventricular outflow tract tachycardia, myocarditis, and sarcoidosis. Only palliative therapy is available and consists of antiarrhythmic drugs, catheter ablation, and implantable cardioverter defibrillator. Young age, family history of juvenile sudden death, overt left ventricular involvement, ventricular tachycardia, syncope, and previous cardiac arrest are the major risk factors for adverse prognosis.

References

1. Marcus FI, Fontaine GH, Guiraudon G, et al. Right ventricular dysplasia: a report of 24 adult cases. *Circulation*. 1982;65(2):384-398.
2. Corrado D, Basso C, Thiene G, et al. Spectrum of clinicopathologic manifestations of arrhythmogenic right ventricular cardiomyopathy/dysplasia: a multicenter study. *J Am Coll Cardiol*. 1997;30:1512-1520.
3. Basso C, Thiene G, Corrado D, Angelini A, Nava A, Valente M. Arrhythmogenic right ventricular cardiomyopathy: dysplasia, dystrophy, or myocarditis? *Circulation*. 1996;94:983-991.
4. Thiene G, Nava A, Corrado D, Rossi L, Pennelli N. Right ventricular cardiomyopathy and sudden death in young people. *N Engl J Med*. 1988 Jan 21;318(3):129-133.
5. Richardson P, McKenna W, Bristow M, et al. Report of the 1995 World Health Organization/International Society and Federation of Cardiology Task Force on the Definition and Classification of cardiomyopathies. *Circulation*. 1996 Mar 1;93(5):841-842.
6. McKoy G, Protonotarios N, Crosby A, et al. Identification of a deletion in plakoglobin in arrhythmogenic right ventricular cardiomyopathy with palmoplantar keratoderma and woolly hair (Naxos disease). *Lancet*. 2000;355:2119-2124.

7. Rampazzo A, Nava A, Miorin M, et al. ARVD4, a new locus for arrhythmogenic right ventricular cardiomyopathy, maps to chromosome 2 long arm. *Genomics*. 1997;45:259-263.
8. Ahmad F, Li D, Karibe A, et al. Localization of a gene responsible for arrhythmogenic right ventricular dysplasia to chromosome 3p23. *Circulation*. 1998;98:2791-2795.
9. Li D, Ahmad F, Gardner MJ, et al. The locus of a novel gene responsible for arrhythmogenic right-ventricular dysplasia characterized by early onset and high penetrance maps to chromosome 10p12–p14. *Am J Hum Genet*. 2000;66:148-156.
10. Melberg A, Oldfors A, Blomstrom-Lundqvist C, et al. Autosomal dominant myofibrillar myopathy with arrhythmogenic right ventricular cardiomyopathy linked to chromosome 10q. *Ann Neurol*. 1999;46:684-692.
11. Rampazzo A, Nava A, Danieli GA, et al. The gene for arrhythmogenic right ventricular cardiomyopathy maps to chromosome 14q23–q24. *Hum Mol Genet*. 1994;3:959-962.
12. Severini GM, Krajinovic M, Pinamonti B, et al. A new locus for arrhythmogenic right ventricular dysplasia on the long arm of chromosome 14. *Genomics*. 1996;31:193-200.
13. Tiso N, Stephan DA, Nava A, et al. Identification of mutations in the cardiac ryanodine receptor gene in families affected with arrhythmogenic right ventricular cardiomyopathy type 2 (ARVD2). *Hum Mol Genet*. 2001 Feb 1; 10(3):189-194.
14. Beffagna G, Occhi G, Nava A, et al. Regulatory mutations in transforming growth factor-beta3 gene cause arrhythmogenic right ventricular cardiomyopathy type 1. *Cardiovasc Res*. 2005;65:366-373.
15. Rampazzo A, Nava A, Malacrida S, et al. Mutation in human desmoplakin domain binding to plakoglobin causes a dominant form of arrhythmogenic right ventricular cardiomyopathy. *Am J Hum Genet*. 2002;71:1200-1206.
16. Gerull B, Heuser A, Wichter T, et al. Mutations in the desmosomal protein plakophilin-2 are common in arrhythmogenic right ventricular cardiomyopathy. *Nat Genet*. 2004;36:1162-1164.
17. Pilichou K, Nava A, Basso C, et al. Mutations in desmoglein-2 gene are associated to arrhythmogenic right ventricular cardiomyopathy. *Circulation*. 2006;113:1171-1179.
18. Syrris P, Ward D, Evans A, et al. Arrhythmogenic right ventricular dysplasia/cardiomyopathy associated with mutations in the desmosomal gene desmocollin-2. *Am J Hum Genet*. 2006;79:978.
19. Merner ND, Hodgkinson KA, Haywood AF, et al. Arrhythmogenic right ventricular cardiomyopathy type 5 is a fully penetrant, lethal arrhythmic disorder caused by a missense mutation in the TMEM43 gene. *Am J Hum Genet*. 2008 Apr;82(4):809-821.
20. Bernstein SA, Morley GE. Gap junctions and propagation of the cardiac action potential. *Adv Cardiol*. 2006;42:71-85.
21. Ellison KE, Friedman PL, Ganz LI, Stevenson WG. Entrainment mapping and radiofrequency catheter ablation of ventricular tachycardia in right ventricular dysplasia. *J Am Coll Cardiol*. 1998 Sept;32(3):724-728.
22. Saffitz JE. Dependence of electrical coupling on mechanical coupling in cardiac myocytes: insights gained from cardiomyopathies caused by defects in cell-cell connections. *Ann N Y Acad Sci*. 2005 Jun;1047:336-344.
23. Kaplan SR, Gard JJ, Protonotarios N, et al. Remodeling of myocyte gap junctions in arrhythmogenic right ventricular cardiomyopathy due to a deletion in plakoglobin (Naxos disease). *Heart Rhythm*. 2004 May;1(1):3-11.
24. Kaplan SR, Gard JJ, Carvajal-Huerta L, Ruiz-Cabezas JC, Thiene G, Saffitz JE. Structural and molecular pathology of the heart in Carvajal syndrome. *Cardiovasc Pathol*. 2004 Jan-Feb;13(1):26-32.
25. Ross SE, Hemati N, Longof KA, et al. Inhibition of adipogenesis by Wnt signaling. *Science*. 2000;289(5481):950-953.
26. Garcia-Gras E, Lombardi R, Giocondo MJ, et al. Suppression of canonical Wnt/beta-catenin signaling by nuclear plakoglobin recapitulates phenotype of arrhythmogenic right ventricular cardiomyopathy. *J Clin Invest*. 2006;116:2012-2021.
27. Van tentelen et al. *Curr Opin Cardiol*. 2007 May;22: 185-192.
28. Protonotarios N, Tsatsopoulou A, Anastasakis A, et al. Genotype-phenotype assessment in autosomal recessive arrhythmogenic right ventricular cardiomyopathy (Naxos disease) caused by a deletion in plakoglobin. *J Am Coll Cardiol*. 2001 Nov 1;38(5):1477-1484.
29. Alcalai R, Metzger S, Rosenheck S, Meiner V, Chajek-Shaul T. A recessive mutation in desmoplakin causes arrhythmogenic right ventricular dysplasia, skin disorder, and woolly hair. *J Am Coll Cardiol*. 2003;42:319.
30. Norgett EE, Hatsell SJ, Carvajal-Huerta L, et al. Recessive mutation in desmoplakin disrupts desmoplakin-intermediate filament interactions and causes dilated cardiomyopathy, woolly hair and keratoderma. *Hum Mol Genet*. 2000 Nov 1;9(18):2761-2766.
31. Bauce B, Basso C, Rampazzo A, et al. Clinical profile of four families with arrhythmogenic right ventricular cardiomyopathy caused by dominant desmoplakin mutations. *Eur Heart J*. 2005;26:1666.
32. Sen-Chowdhry S, Syrris P, McKenna WJ. Desmoplakin disease in arrhythmogenic right ventricular cardiomyopathy: early genotype-phenotype studies. *Eur Heart J*. 2005;26: 1582.
33. Syrris P, Ward D, Asimaki A, et al. Clinical expression of plakophilin-2 mutations in familial arrhythmogenic right ventricular cardiomyopathy. *Circulation*. 2006;113:356.
34. Dalal D, Molin LH, Piccini J, et al. Clinical features of arrhythmogenic right ventricular dysplasia/cardiomyopathy associated with mutations in plakophilin-2. *Circulation*. 2006;113:1641.
35. Van Tintelen JP, Entius MM, Bhuiyan ZA, et al. Plakophilin-2 mutations are the major determinant of familial arrhythmogenic right ventricular dysplasia/cardiomyopathy. *Circulation*. 2006;113:1650.
36. Hamid MS, Norman M, Quraishi A, et al. Prospective evaluation of relatives for familial arrhythmogenic right ventricular cardiomyopathy/dysplasia reveals a need to broaden diagnostic criteria. *J Am Coll Cardiol*. 2002 Oct 16;40(8): 1445-1450.
37. Syrris P, Ward D, Asimaki A, et al. Desmoglein-2 mutations in arrhythmogenic right ventricular cardiomyopathy: a genotype-phenotype characterization of familial disease. *Eur Heart J*. 2007;28:581.
38. Rampazzo A, Beffagna G, Nava A, et al. Arrhythmogenic right ventricular cardiomyopathy type 1 (ARVD1): confirmation of locus assignment and mutation screening of four candidate genes. *Eur J Hum Genet*. 2003;11:69.

39. Wehrens XH, Lehnart SE, Huang F, et al. FKBP12.6 deficiency and defective calcium release channel (ryanodine receptor) function linked to exercise-induced sudden cardiac death. *Cell*. 2003;113:829.
40. Bauce B, Nava A, Rampazzo A, et al. Familial effort polymorphic ventricular arrhythmias in arrhythmogenic right ventricular cardiomyopathy map to chromosome 1q42–43. *Am J Cardiol*. 2000;85:573.
41. Priori SG, Napolitano C, Memmi M, et al. Clinical and molecular characterization of patients with catecholaminergic polymorphic ventricular tachycardia. *Circulation*. 2002;106:69.
42. Tiso N, Salamon M, Bagattin A, Danieli GA, Argenton F, Bortolussi M. The binding of the RyR2 calcium channel to its gating protein FKBP12.6 is oppositely affected by ARVD2 and VTSIP mutations. *Biochem Biophys Res Commun*. 2002;299:594.
43. Gemayel C, Pelliccia A, Thompson PD. Arrhythmogenic right ventricular cardiomyopathy. *J Am Coll Cardiol*. 2001 Dec;38(7):1773-1781.
44. Corrado D, Pelliccia A, Bjørnstad HH, et al. Cardiovascular pre-participation screening of young competitive athletes for prevention of sudden death: proposal for a common European protocol. Consensus Statement of the Study Group of Sport Cardiology of the Working Group of Cardiac Rehabilitation and Exercise Physiology and the Working Group of Myocardial and Pericardial Diseases of the European Society of Cardiology. *Eur Heart J*. 2005 Mar;26(5):516-524.
45. Basso C, Corrado D, Thiene G. Cardiovascular causes of sudden death in young individuals including athletes. *Cardiol Rev*. 1999 May–June;7(3):127-135.
46. Corrado D, Basso C, Thiene G. Arrhythmogenic right ventricular cardiomyopathy: diagnosis, prognosis, and treatment. *Heart*. 2000;83:588-595.
47. Wiesfeld AC, Crijns HJ, van Dijk RB, et al. Potential role of endomyocardial biopsy in the clinical characterization of patients with idiopathic ventricular fibrillation: arrhythmogenic right ventricular dysplasia – an undervalued cause. *Am Heart J*. 1994;127:1421-1424.
48. McKenna WJ, Thiene G, Nava A, et al. Diagnosis of arrhythmogenic right ventricular dysplasia/cardiomyopathy. Task Force of the Working Group Myocardial and Pericardial Disease of the European Society of Cardiology and of the Scientific Council on Cardiomyopathies of the International Society and Federation of Cardiology. *Br Heart J*. 1994 Mar;71(3):215-218.
49. Fontaine G, Umemura J, Di Donna P, Tsezana R, Cannat JJ, Frank R. Duration of QRS complexes in arrhythmogenic right ventricular dysplasia. A new non-invasive diagnostic marker. *Ann Cardiol Angeiol (Paris)*. 1993;42:399-405.
50. Peters S, Trümmel M. Diagnosis of arrhythmogenic right ventricular dysplasia-cardiomyopathy: value of standard ECG revisited. *Ann Noninvasive Electrocardiol*. 2003 July;8(3):238-245.
51. Pinamonti B, Sinagra G, Salvi A, et al. Left ventricular involvement in right ventricular dysplasia. *Am Heart J*. 1992 Mar;123(3):711-724.
52. Nava A, Bauce B, Basso C, et al. Clinical profile and long-term follow-up of 37 families with arrhythmogenic right ventricular cardiomyopathy. *J Am Coll Cardiol*. 2000 Dec;36(7):2226-2233.
53. Nasir K, Rutberg J, Tandri H, Berger R, Tomaselli G, Calkins H. Utility of SAECG in arrhythmogenic right ventricle dysplasia. *Ann Noninvasive Electrocardiol*. 2003 Apr;8(2):112-120.
54. Oselladore L, Nava A, Buja G, et al. Signal-averaged electrocardiography in familial form of arrhythmogenic right ventricular cardiomyopathy. *Am J Cardiol*. 1995 May 15;75(15):1038-1041.
55. Turrini P, Angelini A, Thiene G, et al. Late potentials and ventricular arrhythmias in arrhythmogenic right ventricular cardiomyopathy. *Am J Cardiol*. 1999 Apr 15;83(8):1214-1219.
56. Zareba W, Piotrowicz K, Turrini P. Electrocardiographic manifestations. In: Marcus FI, Nava A, Thiene G, eds. *Arrhythmogenic right ventricular dysplasia/cardiomyopathy, recent advances*. Milano: Springer Verlag; 2007:121-128.
57. Cox MG, Nelen MR, Wilde AA, et al. Activation delay and VT parameters in arrhythmogenic right ventricular dysplasia/cardiomyopathy: toward improvement of diagnostic ECG criteria. *J Cardiovasc Electrophysiol*. 2008 Aug;19(8):775-781.
58. White JB, Razmi R, Nath H, Kay GN, Plumb VJ, Epstein AE. Relative utility of magnetic resonance imaging and right ventricular angiography to diagnose arrhythmogenic right ventricular cardiomyopathy. *J Interv Card Electrophysiol*. 2004 Feb;10(1):19-26.
59. Bluemke DA, Krupinski EA, Ovitt T, et al. MR Imaging of arrhythmogenic right ventricular cardiomyopathy: morphologic findings and interobserver reliability. *Cardiology*. 2003;99(3):153-162.
60. Nasir K, Bomma C, Khan FA, et al. Utility of a combined signal-averaged electrocardiogram and QT dispersion algorithm in identifying arrhythmogenic right ventricular dysplasia in patients with tachycardia of right ventricular origin. *Am J Cardiol*. 2003 July 1;92(1):105-109.
61. Tandri H, Calkins H, Nasir K, et al. Magnetic resonance imaging findings in patients meeting task force criteria for arrhythmogenic right ventricular dysplasia. *J Cardiovasc Electrophysiol*. 2003 May;14(5):476-482.
62. Angelini A, Thiene G, Boffa GM, et al. Calliaris. Endomyocardial biopsy in right ventricular cardiomyopathy. *I Int J Cardiol*. 1993 July 15;40(3):273-282.
63. Nasir K, Bomma C, Tandri H, et al. Electrocardiographic features of arrhythmogenic right ventricular dysplasia/cardiomyopathy according to disease severity: a need to broaden diagnostic criteria. *Circulation*. 2004 Sept 21;110(12):1527-1534.
64. Yoerger DM, Marcus F, Sherrill D, et al. Multidisciplinary Study of Right Ventricular Dysplasia Investigators. Echocardiographic findings in patients meeting task force criteria for arrhythmogenic right ventricular dysplasia: new insights from the multidisciplinary study of right ventricular dysplasia. *J Am Coll Cardiol*. 2005 Mar 15;45(6):860-865.
65. Tandri H, Macedo R, Calkins H, et al. Multidisciplinary Study of Right Ventricular Dysplasia Investigators. Role of magnetic resonance imaging in arrhythmogenic right ventricular dysplasia: insights from the North American arrhythmogenic right ventricular dysplasia (ARVD/C) study. *Am Heart J*. 2008 Jan;155(1):147-153.
66. Tandri H, Saranathan M, Rodriguez ER, et al. Noninvasive detection of myocardial fibrosis in arrhythmogenic right ventricular cardiomyopathy using delayed-enhancement magnetic resonance imaging. *J Am Coll Cardiol*. 2005;45:98.
67. Sen-Chowdhry S, Prasad SK, Syrris P, et al. Cardiovascular magnetic resonance in arrhythmogenic right ventricular car-

diomyopathy revisited: comparison with task force criteria and genotype. *J Am Coll Cardiol*. 2006;48:2132.
68. Corrado D, Basso C, Thiene G, et al. Spectrum of clinicopathologic manifestations of arrhythmogenic right ventricular cardiomyopathy/dysplasia: a multicenter study. *J Am Coll Cardiol*. 1997;30:1512.
69. Sen-Chowdhry S, Syrris P, Prasad SK, et al. Left-dominant arrhythmogenic cardiomyopathy: an under-recognized clinical entity. *J Am Coll Cardiol*. 2008;52:2175.
70. Lerman BB, Stein KM, Markowitz SM. Idiopathic right ventricular outflow tract tachycardia: a clinical approach. *PACE*. 1996;19:2120-2137.
71. Markowitz SM, Litvak BL, Ramirez de Arellano EA, Markisz JA, Stein KM, Lerman BB. Adenosine-sensitive ventricular tachycardia, right ventricular abnormalities delineated by magnetic resonance imaging. *Circulation*. 1997;96:1192-1200.
72. Thomas KW, Hunninghake GW. Sarcoidosis. *JAMA*. 2003; 289:3300-3303.
73. Chapelon C, Piette JC, Uzzan B, et al. The advantages of histological samples in sarcoidosis. Retrospective multicenter analysis of 618 biopsies performed on 416 patients. *Rev Med Intern*. 1987 Mar–Apr;8(2):181-185.
74. Fontaine G, Fontaliran F, Hebert J, et al. Arrhythmogenic right ventricular dysplasia. *Annu Rev Med*. 1999;50:17-35.
75. Hulot JS, Jouven X, Empana JP, Frank R, Fontaine G. Natural history and risk stratification of arrhythmogenic right ventricular dysplasia/cardiomyopathy. *Circulation*. 2004 Oct 5;110(14):1879-1884.
76. Peters S. Long-term follow-up and risk assessment of arrhythmogenic right ventricular dysplasia/cardiomyopathy: personal experience from different primary and tertiary centres. *J Cardiovasc Med*. 2007 July;8(7):521-526.
77. Corrado D, Leoni L, Link MS, et al. Implantable cardioverter-defibrillator therapy for prevention of sudden death in patients with arrhythmogenic right ventricular cardiomyopathy/dysplasia. *Circulation*. 2003 Dec 23;108(25):3084-3091.
78. Corrado D, Basso C, Rizzoli G, Schiavon M, Thiene G. Does sports activity enhance the risk of sudden death in adolescents and young adults? *J Am Coll Cardiol*. 2003 Dec 3;42(11):1959-1963.
79. Wichter T, Paul TM, Eckardt L, et al. Arrhythmogenic right ventricular cardiomyopathy. Antiarrhythmic drugs, catheter ablation, or ICD? *Herz*. 2005;30(2):91-101.
80. Movsowitz C, Schwartzman D, Callans DJ, et al. Idiopathic right ventricular outflow tract tachycardia: narrowing the anatomic location for successful ablation. *Am Heart J*. 1996;131:930-936.
81. Marchlinski FE, Zado E, Dixit S, et al. Electroanatomic substrate and outcome of catheter ablative therapy for ventricular tachycardia in setting of right ventricular cardiomyopathy. *Circulation*. 2004 Oct 19;110(16):2293-2298.
82. Fontaine G, Tonet J, Gallais Y, et al. Ventricular tachycardia catheter ablation in arrhythmogenic right ventricular dysplasia: a 16-year experience. *Curr Cardiol Rep*. 2000 Nov; 2(6):498-506.
83. Dalal D, Jain R, Tandri H, et al. Long-term efficacy of catheter ablation of ventricular tachycardia in patients with arrhythmogenic right ventricular dysplasia/cardiomyopathy. *J Am Coll Cardiol*. 2007 July 31;50(5):432-440.

Noncompaction Cardiomyopathy

6

Yvonne M. Hoedemaekers, Kadir Caliskan, and Danielle F. Majoor-Krakauer

6.1 Introduction

Noncompaction of the left ventricle or *noncompaction cardiomyopathy* (*NCCM*) is a relatively new clinicopathologic entity, first described by Feldt et al. in 1969.[1] NCCM is characterized by a prominent trabecular meshwork and deep intertrabecular recesses communicating with the left ventricular (LV) cavity, morphologically reminiscent of early cardiac development, and is therefore thought to be caused by an arrest of normal embryogenesis of the myocardium.[2,3] Initial presentation includes congestive heart failure, thrombo-embolic events, and (potentially lethal) arrhythmias, including sudden cardiac death. NCCM may be a part of a more generalized cardiomyopathy, involving both the morphologically normal and the predominantly apical, abnormal LV segments. The cardiologic features of NCCM range from asymptomatic in adults to severe congenital forms.[4–7] Recently, NCCM was classified by the American Heart Association (AHA) as a separate primary, genetic cardiomyopathy, based on the predominant myocardial involvement and genetic etiology.[8] The European Society of Cardiology (ESC) considers NCCM as unclassified, due to the lack of consensus whether NCCM is a separate individual cardiomyopathy or a nonspecific morphological trait that can be found solitary or in combination with other forms of cardiomyopathy like hypertrophic cardiomyopathy (HCM), dilated cardiomyopathy (DCM), or with congenital heart disease.[9] The majority of NCCM diagnosed in adults is isolated. Nonisolated forms of NCCM are more frequent in childhood and may co-occur with congenital heart malformations, or may be part of a malformation or chromosomal syndrome.[7] The combination of NCCM and neuromuscular disorders is observed in adults as well as in children.

The majority of NCCM, isolated and nonisolated, is hereditary and NCCM appears to be genetically heterogenous.[10,11] An important proportion of isolated NCCM in children and adults has been associated with mutations in the same *sarcomere* genes that are involved in HCM, DCM, and restrictive cardiomyopathy (RCM).[10] Absence of a genetic defect does not preclude a genetic cause of NCCM. In approximately half of the familial NCCM, the genetic defect remains unkown.[11] Shared sarcomere defects and the occurrence of HCM and DCM in families with NCCM patients indicate that at least some forms of NCCM are part of a broader cardiomyopathy spectrum.

The literature differentially refers to this form of cardiomyopathy as left ventricular noncompaction (LVNC), noncompaction cardiomyopathy (NCCM), noncompaction of the left ventricular myocardium (NCLVM), left ventricular hypertrabeculation (LVHT), spongiform cardiomyopathy, embryonic myocardium, honeycombed myocardium, persisting myocardial sinusoids, myocardial dysgenesis, ventricular dysplasia, or spongy myocardium. In analogy with the nomenclature of hypertrophic (HCM) and dilated cardiomyopathy (DCM), the term noncompaction cardiomyopathy is preferable. Therefore, noncompaction cardiomyopathy, abbreviated as NCCM, will be used in this chapter to denote this entity.

Y.M. Hoedemaekers (✉)
Department of Clinical Genetics, Erasmus Medical Center, Rotterdam, The Netherlands
e-mail: y.hoedemaekers@erasmusmc.nl

6.1.1 Definition

NCCM is defined by prominent *trabeculations* on the luminal surface of the left ventricular apex, the lateral wall, and rarely the septum in association with deep recesses that extend into the ventricular wall, which do not communicate with the coronary circulation. It is associated with a clinical triad of heart failure, arrhythmias, and/or thrombo-embolic events.[12,13] Diagnosis of NCCM relies on two-dimensional transthoracic echocardiography and/or cardiac magnetic resonance imaging (MRI) (Table 6.1). Improvements in cardiac imaging techniques have led to increased recognition and diagnosis of NCCM. Figure 6.1 displays echocardiographic and cardiac MRI images of two NCCM patients, showing the abnormal segmental trabeculations as the hallmark of this new entity.

Features of noncompaction observed in cardiologic patients and normal controls illustrate the necessity of defining criteria in order to differentiate accurately normal physiological trabecularization from NCCM.[14]

Table 6.1 Echocardiographic diagnostic criteria for NCCM

I. Chin et al.[2]
Focusing on trabeculae localized at the LV apex on the parasternal short axis and apical views and on LV free-wall thickness at end-diastole NCCM is defined by a ratio of X/Y ≤ 0.5 with
X = distance from the epicardial surface to the trough of the trabecular recess
Y = distance from the epicardial surface to the peak of the trabeculation

II. Jenni et al.[12]
1. An excessively thickened left ventricular myocardial wall with a two-layered structure consisting of a compact epicardial layer (C) and a noncompacted endocardial layer (NC) of prominent trabeculations and deep intertrabecular recesses
2. A maximal end-systolic NC/C ratio > 2, measured at the parasternal short axis
3. Color Doppler evidence of deep perfused intertrabecular recesses
4. Absence of coexisting cardiac anomalies

III. Stollberger et al.[15]
1. More than three trabeculations protruding from the left ventricular wall, apical to the papillary muscles and visible in a single image
2. Perfusion of the intertrabecular spaces from the ventricular cavity visualized on color Doppler imaging

In 1990, the first diagnostic criteria for NCCM by Chin et al. were derived from the observations made in eight NCCM patients.[2] These diagnostic criteria defined NCCM by the ratio of the distance from the epicardial surface to the trough of the trabecular recess (X) to the distance from the epicardial surface to the peak of the trabeculations (Y), with ratio X/Y ≤ 0.5.

More than a decade later, Jenni et al. proposed new diagnostic criteria for isolated NCCM, consisting of four echocardiographic features: (1) an excessively thickened left ventricular myocardial wall with a two-layered structure consisting of a compact epicardial layer (C) and a noncompacted endocardial layer (NC) of prominent trabeculations and deep intertrabecular recesses; (2) a maximal end-systolic *NC/C ratio > 2*, measured at the parasternal short axis; (3) color-Doppler evidence of deeply perfused intertrabecular recesses; (4) absence of coexisting cardiac anomalies.[12]

In 2002, Stollberger et al. proposed other diagnostic criteria for NCCM, wherein the diagnosis was a function of the number of trabeculations (>3) protruding from the left ventricular wall, apically to the papillary muscles and visible in a single image plane with obligatory perfusion of the intertrabecular spaces from the ventricular cavity visualized on color-Doppler imaging.[15]

More recently, MRI criteria for NCCM introduced by Petersen et al. indicated that a noncompacted/compacted ratio (NC/C) of >2.3, measured in end-diastole, can differentiate with sufficient sensitivity between the normal variation of noncompaction of the LV in the population, noncompaction in other cardiovascular disorders, and NCCM.[16]

The most recent classification system of NCCM as proposed by Belanger et al. (2008) included dividing noncompaction into four categories (none, mild, moderate, and severe) according to noncompaction to compaction ratio and the size of the noncompaction area.[13] This new classification scheme used the following criteria: (1) absence of congenital heart disease, hypertrophic or infiltrative cardiomyopathy, and coronary artery disease; (2) evidence of prominent trabeculations in the apex in any view (noncompacted to compacted ratio does not require to be >2); (3) concentration of the noncompacted area in the apex; (4) blood flow through the area of noncompaction.

The Jenni echo criteria have been the most convenient to work with in daily clinical practice and have

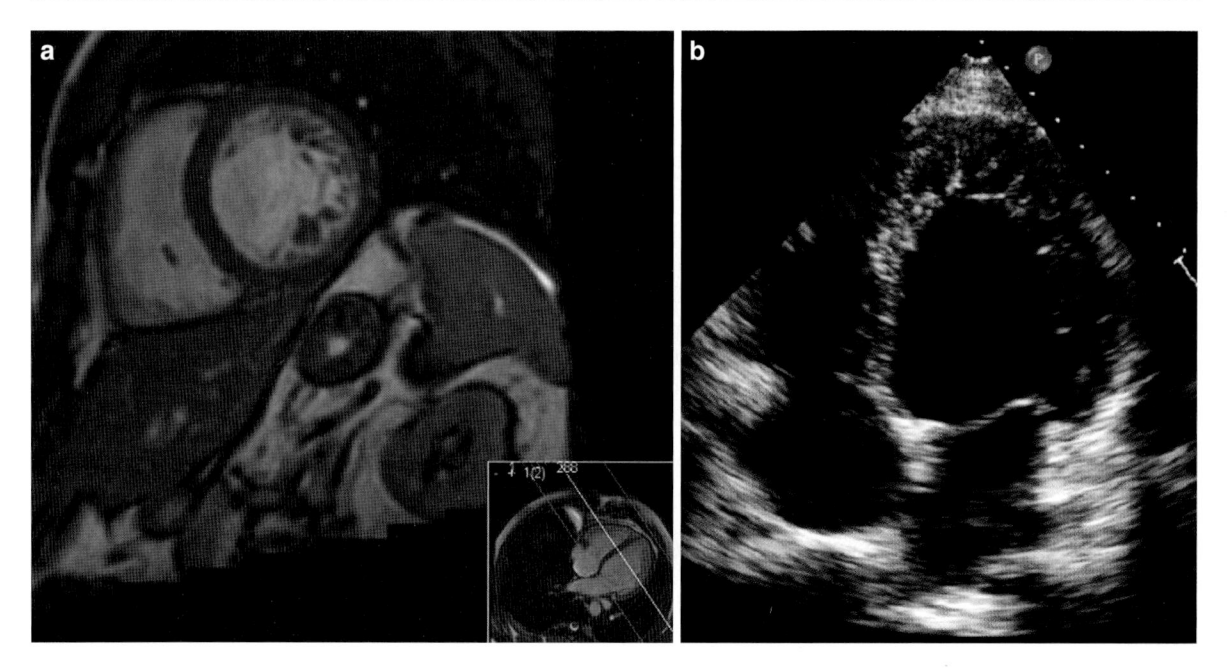

Fig. 6.1 (**a**, **b**) Cardiac MRI and echocardiography of a 43-year-old patient illustrating a two-layered myocardium with prominent intertrabecular recesses

been most widely applied in studies. However, further efforts to reach universal consensus with respect to the diagnosis of NCCM are clearly needed. A disparity in diagnosis has been observed when comparing the application of three different sets of NCCM criteria (Chin, Jenni, and Stollberger) in a cohort of 199 heart failure patients; 79% fulfilled the Chin criteria, 64% fulfilled the Jenni criteria, and 53% the criteria proposed by Stollberger. In only 30% of patients, there was consensus among the three criteria on the diagnosis. Moreover, 8.3% of normal controls fulfilled one or more criteria with a higher prevalence in black controls.[14]

For now, it is disputable whether any of these diagnostic criteria are sufficiently sensitive to diagnose patients with mild noncompaction, and identify patients who may benefit from careful surveillance. For instance, in NCCM family studies, a substantial proportion of (mostly asymptomatic) relatives showed mild to moderate features of NCCM.[11] Longitudinal studies of mild forms of NCCM will be needed to determine whether the current diagnostic criteria are suitable for diagnosis of family members in familial NCCM, or should be adapted in analogy to the criteria proposed for diagnosis of attenuated forms of familial HCM in relatives.

6.1.2 Pathology

6.1.2.1 Macroscopy

The noncompacted endocardial layer of the myocardium comprises excessively numerous and prominent trabeculations with deep intertrabecular recesses that extend into the compacted myocardial layer. The apical and mid ventricular segments of the left ventricular inferior and lateral wall are predominantly affected.[17, 18] In a pathoanatomical study of NCCM, Burke et al. described the morphology and microscopy of 14 pediatric NCCM cases.[18] The macroscopic appearance varied from anastomosing trabeculae to a relatively smooth endocardial surface, with narrow openings of the recesses to the ventricular cavity. Three types of recess patterns were distinguished: (1) anastomosing broad trabeculae; (2) coarse trabeculae resembling multiple papillary muscles; (3) interlacing smaller muscle bundles or relatively smooth endocardial surface with compressed invaginations, identified primarily microscopically (Fig. 6.2). In this study, no morphological differences were found between isolated and nonisolated NCCM.[18]

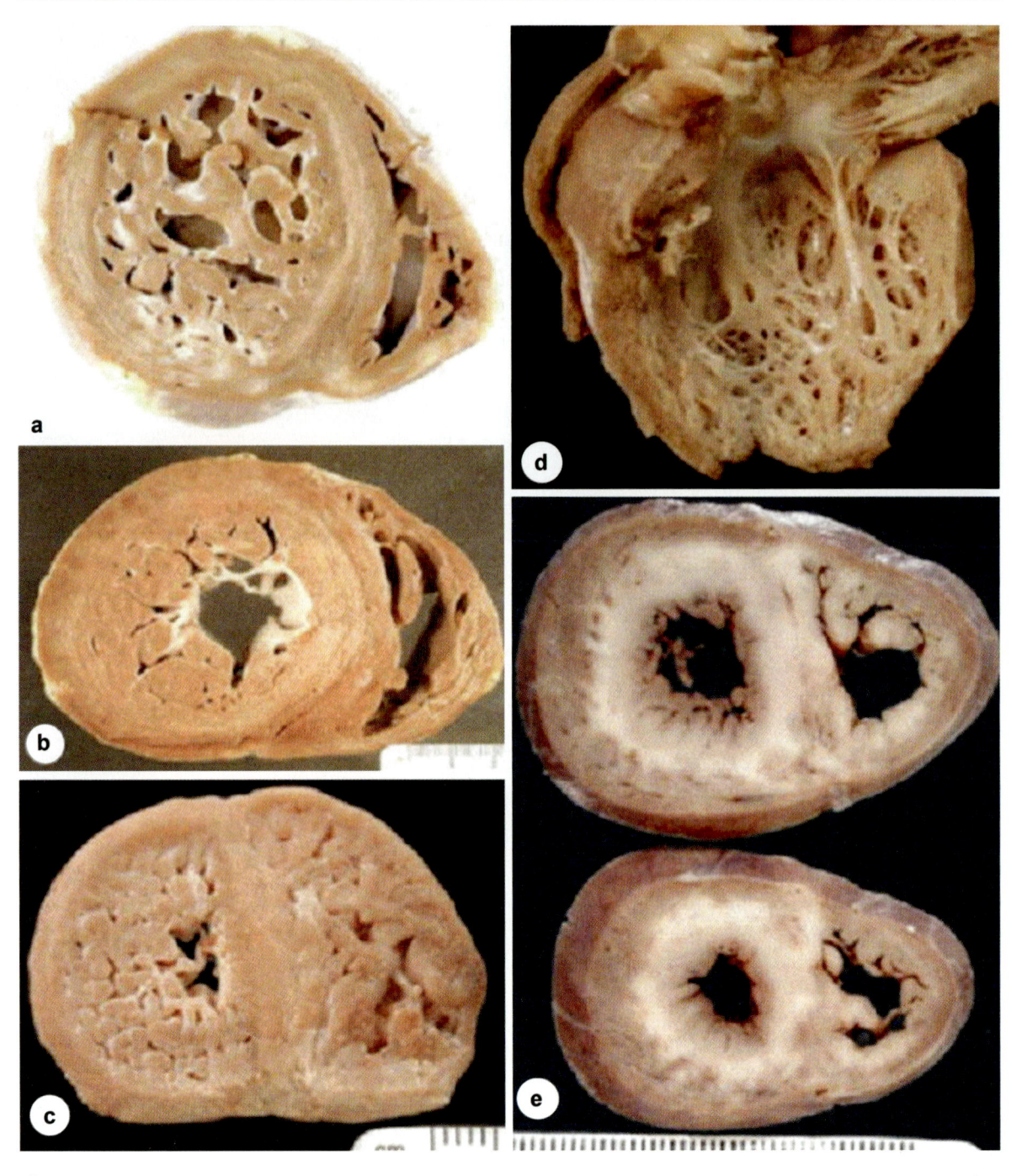

Fig. 6.2 NCCM gross pathology with a variety of NCCM patterns: (**a**) Anastomosing broad trabeculae. (**b**) Coarse trabeculae resembling multiple papillary muscles. (**c**) Interlacing smaller muscle bundles resembling a sponge. (**d**) Trabeculae viewed en face. (**e**) Subtle NCCM on gross section, requires histological confirmation (Reproduced from Burke et al.[18] With permission)

Jenni et al. described pathology of seven adult NCCM cases.[12] The pathoanatomical localization of the noncompacted myocardium corresponded to the echocardiographic findings. Two patients also showed involvement of the right ventricular apex.[12]

In a review of published pathology of NCCM, Stollberger et al. distinguished three particular morphologic features of NCCM in adults and children: (1) Extensive spongiform transformation of the LV. (2) Prominent coarse trabeculations and deep recesses, covered with endocardial tissue and not communicating with coronary arteries. (3) Dysplastic thinned myocardium with excessive trabeculations.[19] The first morphology was frequently associated with other cardiac malformations, compared to the second and third.

In 1987, in an autopsy study of 474 normal hearts of all ages, it was found that prominent trabeculations may be observed in as many as 68% of the hearts, although more than three trabeculations were only identified in 3.4%.[20]

6.1.2.2 Microscopy

Two patterns of myocardial structure in the superficial noncompacted layer in NCCM have been described by Burke et al.: (1) anastomosing muscle bundles forming irregularly branching endocardial recesses with a staghorn-like appearance; (2) multiple small papillary muscles, resulting in an irregular surface appearance (Fig. 6.3).[18] In most patients, these patterns overlapped. Endocardial fibrosis with prominent elastin deposition was found in all 14 cases and subendocardial replacement fibrosis, consistent with microscopic ischemic infarcts, was present in 10.[18] Right ventricular involvement was identified in six cases.[18]

Histological examination in another study showed that ventricular endocardium covered the recesses in continuity with the LV cavity and identified ischemic lesions in the thickened endocardium and the prominent trabeculae.[12] Interstitial fibrosis ranged from absence to severe. No fiber disarray was identified in any of these cases. Signs of chronic inflammation and abnormalities of intramyocardial blood vessels were present in some patients.[12]

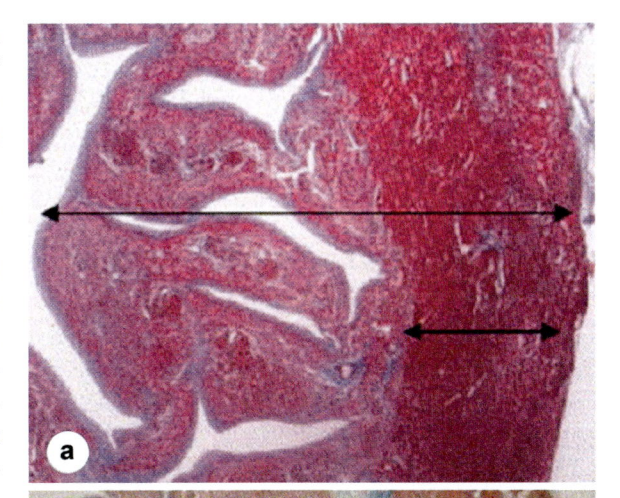

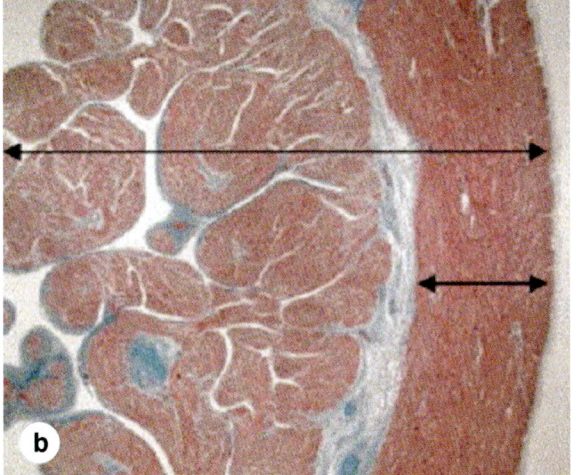

Fig. 6.3 Histological features in NCCM The ratio of noncompact versus compact myocardium is larger than 2. (**a**) Relatively smooth endocardial surface (*left*) with anastomosing broad trabeculae. (**b**) Polypoid pattern of trabeculae; prominent fibrous band separating the noncompact from the compact myocardium (Reproduced from Burke et al.[18] With permission)

In one adult case report, abundant extracellular matrix and myocardial fiber disarray were reported.[21]

Freedom et al. proposed two criteria for the pathological diagnosis of NCCM: (1) absence of well-formed LV papillary muscles and (2) histological verification of more than 50% penetration of invaginated endocardial recesses toward the epicardial surface. The endothelium that covers the recesses extends close to the surface of the compact layer. The recesses neither communicate nor connect with the coronary circulation.[22]

6.2 Epidemiology

Estimates of prevalence of NCCM were derived from large retrospective studies of patients referred for echocardiography. Population studies for NCCM have not been performed. In 1997 Ritter et al. identified NCCM in 17 of 37,555 (0.045%) patients who had an echocardiographic exam.[23] Similarly, in 2006 Aras et al. reported a prevalence of 0.14% in over 42,000 patients and in 2008 Sandhu identified definite or possible NCCM in 13/4,929 (0.26%) patients referred for echocardiography.[24,25] Prevalence was much higher (3.7%) in patients selected for a LV ejection fraction ≤45%.[25] Depending on the diagnostic criteria applied, even higher prevalence of NCCM (15.8% by Belanger; 23.6% by Kohli) were reported recently, indicating that NCCM may be more prevalent than previously indicated.[13,14] A substantial proportion of individuals is asymptomatic, suggesting that true prevalence of NCCM may be higher, because asymptomatic individuals may go unnoticed in the studies of cardiologic patients.[11,13] In a large study on childhood cardiomyopathies, NCCM was the most frequent cardiomyopathy after DCM and HCM, with an estimated prevalence of 9% in pediatric cardiomyopathies.[26]

6.3 Etiology and Molecular Genetics

The etiology of NCCM is rapidly being unravelled as more and more genetic defects in different genes are found, indicating that NCCM is genetically heterogeneous. Causes for acquired NCCM are scarce. One report suggested that candida sepsis was associated with cardiologic features mimicking NCCM.[27] Currently, genetic defects are identified in 42% of NCCM patients (35% of adults and 78% of children).[10] Most genetic defects are inherited as autosomal dominant trait (Table 6.2), with exception of rare genetic causes of syndromal NCCM, predominantly diagnosed in children. A small proportion of patients have a de novo mutation.

However, absence of a genetic defect does not exclude a genetic etiology. By performing systematic cardiologic family studies, it was shown that no genetic defect could be found in approximately half of the familial forms of NCCM, indicating that further studies are needed to find additional genetic causes for NCCM.[11]

There is evidence that some forms of NCCM are part of a spectrum of cardiomyopathies, including hypertrophic, dilated, and restrictive cardiomyopathy. A shared etiology consisting of genetic defects in the same sarcomere genes, sometimes even with identical mutations, has been found in these types of cardiomyopathy. Co-occurrence of NCCM, HCM, and DCM within families endorses a shared genetic susceptibility to these different forms of cardiomyopathy.[10,11] The phenotypic variability of cardiomyopathies within families, including variability in age at onset and severity of clinical features, might be explained by additional modifying factors, additional genetic variants or defects, or may depend on yet unidentified exogenous or systemic factors.

6.3.1 Molecular Defects in NCCM

Isolated NCCM has been associated with mutations in 14 different genes (Table 6.2). Defects in sarcomere genes have been identified to be the most prevalent genetic cause occurring in 33% of all patients with isolated NCCM.[11] In two DNA studies in cohorts of approximately 60 isolated NCCM patients, mutations were identified in 17–41% of the patients depending on the number and choice of analyzed genes.[10,28] In the study by Dooijes et al. of 56 patients, the yield was slightly higher 41% in all and 50% in case of confirmed familial disease.[11] In children with isolated NCCM, the yield of testing for sarcomere genes was as high as 75%.[10,11]

Over 40 different mutations in sarcomere genes encoding thick (*MYH7*), intermediate (*MYBPC3*), and thin filaments (*TNNT2, TNNI3, TPM1, ACTC*) have been described. In particular in *MYH7,* the most frequent NCCM-associated gene, accounting for up to 21% of isolated NCCM (19% in adults and 25% in children).[10,28] Fifty percent of the *MYH7* mutations currently associated with NCCM cluster in the ATPase active site of the head-region in the N-terminal part of *MYH7*.[10] This is an evolutionary well-conserved region of MYH7. As the ATP-ase active site is required

Table 6.2 Genes associates with noncompaction cardiomyopathy (NCCM)

Gene	Locus	Protein	Other associated disorders	Reference
ACTC1	15q14	α-Cardiac actin	Hypertrophic and dilated cardiomyopathy Congenital myopathy with fiber-type disproportion	10, 11, 28, 50
CASQ2	1p13.3-p11	Calsequestrin	Catecholaminergic polymorphic ventricular tachycardia Hypertrophic cardiomyopathy	10, 11
DTNA	18q12.1-q12.2	α-Dystrobrevin		35, 134
KCNH2	7q35-q36	Potassium voltage-gated channel, subfamily H, member 2	Long QT syndrome 2 Short QT syndrome	135
LDB3[a]	10q22.2-q23.3	LIM-Domain binding protein	Dilated cardiomyopathy Late onset distal myopathy Myofibrillar myopathy	10, 11, 36, 134, 136
LMNA	1q21.2	Lamin A/C	Dilated cardiomyopathy Emery–Dreifuss muscular dystrophy Lipodystrophy Restrictive dermopathy Werner syndrome Hutchinson–Gilford Progeria Limb girdle muscular dystrophy 1B Charcot–Marie–Tooth 2B1	10, 11, 61, 62
MYBPC3	11p11.2	Cardiac myosin-binding protein C	Hypertrophic and dilated cardiomyopathy	10, 11
MYH7	14q12	β-Myosin heavy chain	Hypertrophic, dilated, and restrictive cardiomyopathy Myosin storage myopathy Distal myopathy Scapuloperoneal myopathy	10, 11, 28, 29
PLN	6q22.1	Phospholamban	Hypertrophic and dilated cardiomyopathy	10, 11
SCN5A	3p21	Sodium channel type 5 α-subunit	Long QT syndrome 3 Brugada syndrome Sick sinus syndrome Familial heart block Paroxysmal ventricular fibrillation Cardiac conduction defect Dilated cardiomyopathy	137
TAZ[b]	Xq28	Taffazin	Barth syndrome Dilated cardiomyopathy	10, 11, 35, 134, 136, 138–145
TNNI3	19p13.4	Cardiac troponin I	Hypertrophic, dilated, and restrictive cardiomyopathy	10, 11
TNNT2	1q32	Cardiac troponin T	Hypertrophic, dilated, and restrictive cardiomyopathy	10, 11, 28
TPM1	15q22.1	A-tropomyosin	Hypertrophic and dilated cardiomyopathy	10, 11

Except *TAZ* related disorders, all are autosomal dominantly inherited
[a]Cypher/ZASP
[b]G4.5

for normal force production, impaired force generation might play a role in the etiology of NCCM. Mutations in this region have been associated with NCCM with or without Ebstein anomaly.[28,29] Other *MYH7* mutations (30%) were found in the C-terminal rod-region of the MYH7 protein that plays an important role in the formation of the core of the thick filament. Mutations in this region of the gene are more commonly associated with skeletal myopathies. Relatively few cardiomyopathy mutations are situated in this region.

Sarcomere mutations were common causes for NCCM in adults as well as in children.[10,11] Multiple or compound/double heterozygous mutations were identified in 25% of the children and in 10% of the adult NCCM patients.[10] HCM complex genotypes have been described in 7%.[30] In HCM, double heterozygosity for truncating sarcomere mutations have been previously associated with severe congenital forms mostly inherited in an autosomal recessive mode.[31-33] In NCCM, double mutations were associated with severe disease in two children and were also observed in adults.[10] Nonsarcomere genetic causes for isolated NCCM include mutations in the calcium-handling genes calsequestrin (*CASQ2*) and phospholamban (*PLN*), in taffazin (*TAZ*), α-dystrobrevin (*DTNA*), lamin A/C (*LMNA*) and LIM domain binding 3 (*LDB3*), potassium voltage-gated channel (*KCNH2*), and sodium channel type 5 (*SCN5A*) genes.[34-36] However, mutations in these genes were only rare causes of NCCM in single families.[37]

The absence of a mutation in approximately half of familial NCCM could be explained by phenotype assignment errors, the involvement of other yet unidentified genes, the presence of mutations in non-analyzed gene sequences, and incomplete sensitivity of the methods used.[10]

6.4 Pathogenesis

Mutations in different genes associated with NCCM affect different mechanisms in the cardiomyocyte leading to changes that may individually cause NCCM or lead to a common cellular disturbance resulting in NCCM.

Mutations in sarcomere genes may have their effect through defective force generation (either by a dominant negative mechanism where the mutant protein acts as a "poison polypeptide" or by haploinsufficiency resulting in less protein); mutated cytoskeletal proteins may lead to a defective force transmission; myocardial energy deficits may be the result of mutations in ATP-regulatory genes and a fourth possible mechanism is abnormal calcium homeostasis either due to changes in calcium availability or myofibrillar sensitivity for calcium.[38]

The development of NCCM features might be a compensatory response to dysfunction in one of these mechanisms.

The variable phenotypic expression of (sarcomere) gene mutations leading to different types of cardiomyopathy has not been explained. The localization of the mutations may partly explain phenotypic diversity. Another theory is "dose-effect"; the extent of the defective mechanism may determine which phenotype develops. Third, there might be independent pathways leading to the different cardiomyopathies. Finding identical mutations in different phenotypes suggests a role for additional factors, either environmental or molecular.

6.4.1 Isolated NCCM

The first hypothesis on the pathogenesis of NCCM stemmed from observations that the morphology of NCCM was reminiscent of the embryonic stages of cardiac development. Consequently, it was postulated that NCCM could be the result from an arrest of compaction of myocardial fibers.[39] Figure 6.4 illustrates the striking resemblance between NCCM and the physiological embryonic noncompaction in the 8th–10th embryonic week. However, the possible mechanisms causing the arrest remain unclear. Epicardium derived cells are thought to play an important role in myocardial architecture and in the development of noncompaction.[40,41] Mutations in genes involved in myocardial genesis like peroxisome proliferator activator receptor binding protein (*PBP*), jumonji (*JMJ*), FK506 binding protein (*FKBP12*), transcription factor specificity protein (*Sp3*), homeobox factor *NKX2.5*, bone morphogenetic protein 10 (*BMP10*) lead to congenital NCCM in knock out mice.[42-46] However, in human NCCM, no mutations in these genes have been described.

6 Noncompaction Cardiomyopathy

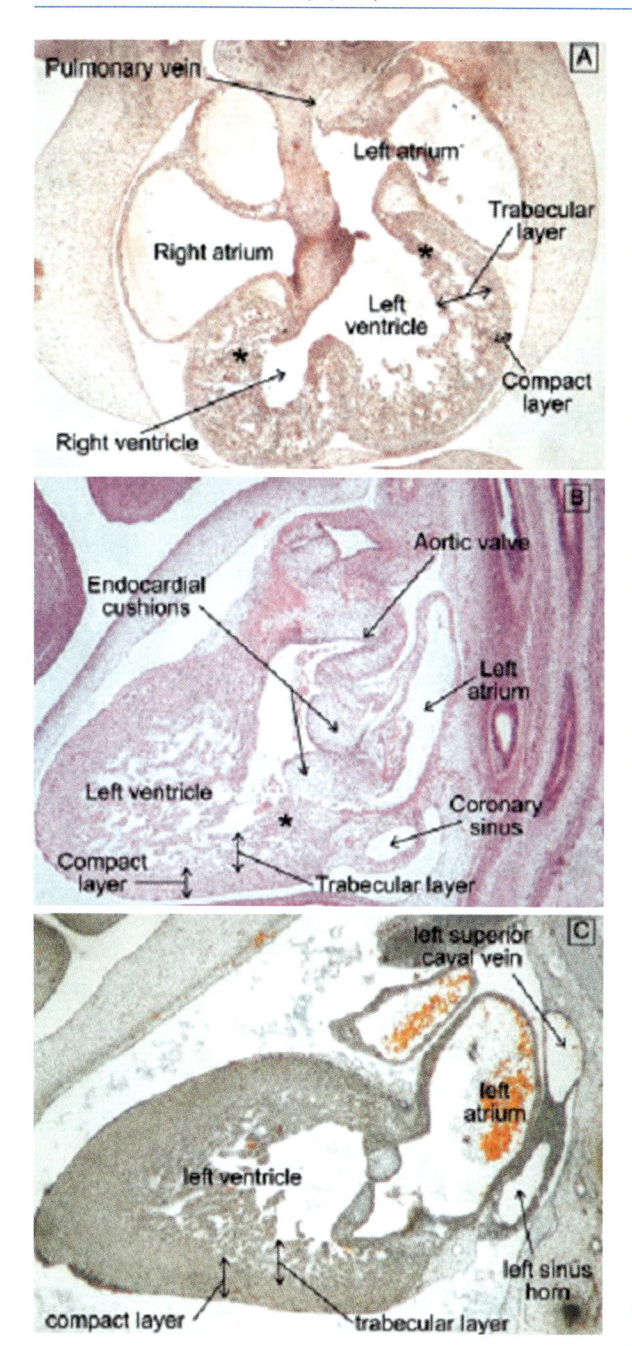

Fig. 6.4 Human embryos at Carnegie stage 16 (**a**), stage 18 (**b**) and after closing of the embryonic interventricular foramen (**c**). During development, there is an extensive trabecular layer forming the greater part of the ventricular wall thickness compared to the extent of the compact layer. The trabecular layer becomes compacted and forms the papillary muscles of the atrioventricular valves (*asterisks*) (Reproduced from Freedom et al.[22] With permission)

Until now, there is very little insight into factors that influence the variability in age at onset and severity of symptoms of NCCM, or any other familial form of cardiomyopathy.

In the majority of patients, NCCM is diagnosed in adulthood, similar to HCM and DCM, which are rarely congenital.[47,48] Of course, it could be that in NCCM the lesions detected in adult patients were present from birth on, but remained unnoticed until symptoms developed and high-resolution cardiac imaging techniques were applied. However, the detection of sarcomere defects in NCCM patients may suggest otherwise, since mutations in sarcomere genes are known to cause late-onset HCM and DCM. Similarly, sarcomere mutations might lead to late onset NCCM. Longitudinal cardiologic studies of unaffected carriers of pathogenic mutations are necessary to provide insight whether noncompaction may develop later in life. The pathogenetic mechanism(s) of sarcomere defects in cardiomyopathies are not fully understood. It is possible that the pathological myocardial changes in the adult onset sarcomere related cardiomyopathies are caused by a compensatory response to impaired myocyte function resulting from mutations in the sarcomere genes.[38,49]

6.4.2 Nonisolated NCCM

NCCM has been observed in a number of neuromuscular disorders, metabolic and mitochondrial disease, congenital malformations, and chromosomal syndromes.

Some of these disorders may share pathogenetic mechanisms with NCCM. Alternatively, NCCM might be secondary to other cardiac malformations or other malformations or even vice versa. Another possibility is that the co-occurrence is coincidental. Congenital heart malformations for instance are relatively frequent (birth prevalence 0.008) and may therefore occasionally coincide with NCCM without a mutual etiology.

6.4.3 Congenital Heart Disease

The co-occurrence of congenital heart disease and noncompaction is predominantly observed in children.

Tsai et al. showed that 78% of 46 children with NCCM had a congenital heart defect.[7] The large number of structural heart malformations reported in association with noncompaction are presented in Table 6.3, indicating that septal defects, patent ductus arteriosus, and Ebstein's anomaly are the most prevalent congenital heart defects in NCCM.

Increasingly, *congenital cardiac malformations* (septal defects, Ebstein anomaly, patent ductus arteriosus, Fallot's tetralogy, aortic coarctation, and aortic aneurysms) are being reported in familial cardiomyopathies (HCM, DCM, and NCCM) linked to sarcomere mutations, suggesting that these specific sarcomere defects may have been involved in cardiac morphogenesis.[11,29,50-54] But

Table 6.3 Congenital heart disease associated with noncompaction cardiomyopathy (NCCM)

Congenital heart disease in NCCM	Proportion of CHD In NCCM studies[a]	Case reports	References
Aberrant origin of right/left subclavian artery	1/12 (8%)	1	146, 147
Absent aortic valve		1	148
Anomalous pulmonary venous return	2/26 (8%)		18, 146
Aortic coarctation	6/204 (3%)		7, 11, 113, 146, 149
Aortico-left ventricular tunnel		1	150
Aortic stenosis	2/46 (4%)	2	7, 22, 151
Aortopulmonary window	1/21 (5%)		113
Atrial septal defect	22/135 (16%)	3	7, 11, 29, 113, 152, 153
Atrio-ventricular diverticulum		1	154
Bicuspid aortic valves	3/64 (5%)	3	7, 113, 119, 155
Bicuspid pulmonary valve	1/14 (7%)		18
Cardiac aneurysms		4	81, 156–158
Coronary ostial stenosis	1/14 (7%)		18
Cor triatriatum	1/46 (2%)		7
Dextrocardia	2/58 (3%)	1	1, 7, 146
Dextro malposed great arteries	1/12 (8%)		146
Dextroversion		1	159
Double inlet left ventricle	1/46 (2%)		7
Double orifice mitral valve		4	160–162
Double outlet right ventricle	1/54 (2%)		149
Ebstein's anomaly	6/117 (5%)	10	7, 11, 153, 163–168
Fallot's tetralogy	1/71 (1%)	1	11, 147
Hypoplastic left heart syndrome	3/54 (6%)		149
Hypoplastic right ventricle	3/58 (5%)		7, 146
Isomerism of the left atrial appendage	4/66 (6%)	8	22, 146, 149, 169
Left-sided superior vena cava	1/46 (2%)		7
Mitral valve atresia		1	148
Mitral valve cleft	2/54 (4%)	1	149, 158

Table 6.3 (continued)

Congenital heart disease in NCCM	Proportion of CHD In NCCM studies[a]	Case reports	References
Mitral valve dysplasia	2/14 (14%)		18
Mitral valve prolaps	1/46 (2%)		7
Patent ductus arteriosus	16/182 (9%)	1	7, 11, 149, 153
Persistent left superior vena cava	1/14 (7%)	1	18, 157
Pulmonary atresia	6/125 (5%)	1	11, 149, 153
Pulmonary valve dysplasia	2/14 (14%)		18
Pulmonary stenosis	4/97 (4%)	1	11, 18, 146, 153
Single ventricle	1/12 (8%)	1	146, 170
Subaortic membrane	2/55 (4%)		149
Transposition of the great arteries	1/46 (2%)	1	7, 171
Tricuspid atresia	2/54 (4%)		149
Tricuspid valve dysplasia	1/14 (7%)		18
Ventricular septal defect	23/218 (11 %)	3	1, 7, 11, 18, 113, 146, 149, 151, 157

[a]Cumulative number of NCCM patients with congenital heart defect (CHD) described in one or more NCCM studies

since there is rarely more than one patient with a congenital heart defect, even in families with multiple cardiomyopathy patients, the association of sarcomere defects and heart defects still demands further exploration.

6.4.4 Neuromuscular Disease

Similar to HCM and DCM, NCCM has been associated with neuromuscular disorders. Stollberger and Finsterer identified NCCM-like morphological features in Duchenne and Becker muscular dystrophy and in myotonic dystrophy (see chapter *neuromuscular disorders*).[55–58] The gene mutated in Duchenne and Becker muscular dystrophy is a part of the dystrophine complex, a complex of muscle membrane associated proteins, connecting the cytoskeleton to the surrounding extracellular matrix and may also play a role in cell signaling. The dystrophine gene is expressed in skeletal and cardiac myocytes. A large proportion of patients and also female carriers have cardiac symptoms, including DCM.[59,60] Other genes previously associated with neuromuscular disorders, like adult onset myofibrillar myopathy (*LDB3* or *Cypher/ZASP*), limb girdle muscular dystrophy (LGMD) (*LMNA*), scapuloperoneal myopathy (*MYH7*), myosin storage distal myopathy (*MYH7*), and Barth syndrome (*TAZ*) have recently been associated with isolated NCCM (Tables 6.1 and 6.4). ZASP, lamin A and C, β-myosin heavy chain, and taffazin are all expressed in cardiac and skeletal muscle tissue. ZASP has a function in cytoskeletal assembly. Mutations in ZASP can lead to DCM and to skeletal myopathy. Lamin A and C, proteins situated in the nuclear membrane, play an important role in maintaining nuclear architecture. *LMNA* mutations have been described in three NCCM patients.[11,61,62] In one of them, there was familial limb girdle muscular dystrophy (LGMD) as well as DCM.[11] Over 200 mutations have been described in *LMNA*, causing over 20 different phenotypes, including isolated DCM, LGMD, Emery–Dreifuss muscular dystrophy, Hutchinson–Gilford progeria, partial lipodystrophy, and peripheral neuropathy. For many of the phenotypes, there is no clear genotype--phenotype correlation, phenotypes may overlap, and different phenotypes are associated with single mutations.[62] Up to 25% of patients with an LMNA mutation may remain cardiologically asymptomatic.[63] The β-myosin heavy chain is

Table 6.4 Neuromuscular disorders associated with noncompaction cardiomyopathy (NCCM)/hypertrabeculation

Neuromuscular disorders	Gene	Inheritance	Features	Reference
Adenosine Monophosphate Deaminase 1 (MADA deficiency)	AMPD1	AD	Exercise-induced myopathy, muscle weakness, cramps; prolonged fatigue after exertion; benign congenital hypotonia	172
Becker and Duchenne muscular dystrophy	DMD	XR	Muscle weakness and wasting; hypotonia; waddling gait; pseudohypertrophy; cognitive impairment; cardiomyopathy; respiratory failure	55–57, 132, 173, 174
Charcot-Marie-Tooth 1A (HMSN IA)	PMP22	AD	Distal limb muscle weakness and atrophy; distal sensory impairment	175
Myotonic dystrophy I	DMPK	AD	Myotonia; weakness; muscle wasting; adult cognitive deterioration; cataract; arrhythmia	58, 176, 177
Myotonic dystrophy II	ZNF9	AD	Muscle pain; myotonia; weakness (proximal/deep finger/neck flexor); cataract; cardiac conduction abnormalities; palpitations; tachycardia; hypogonadism; frontal balding	178
Infantile epilepsy-encephalopathy syndrome (Ohtahara syndrome)	ARX	XR	Age-dependent epileptic encephalopathy with "burst-suppression" on EEG; physical and mental retardation	179
Limb girdle muscular dystrophy 1B	LMNA	AD	Muscle weakness and wasting restricted to the limb musculature, proximal greater than distal	11, 132
Succinate dehydrogenase deficiency		AR	Encephalomyopathy; cardiomyopathy; generalized muscle weakness; cerebellar ataxia; optic atrophy; tumor formation in adulthood	180

AD autosomal dominant, *XR* X-linked recessive, *AR* autosomal recessive

part of type II myosin that generates the mechanical force needed for muscle contraction.

Tafazzins have no known similarities to other proteins. Two regions of the protein may be functionally significant, one serving as a membrane anchor and soluble cytoplasmic protein and the other may serve as an exposed loop, interacting with other proteins.

Table 6.4 presents a list of neuromuscular disorders in which NCCM has been identified. In addition, one case of noncompaction in a patient with Friedreich ataxia has been reported.[64] Friedreich ataxia is associated with symmetric, concentric, hypertrophic cardiomyopathy.

6.4.5 Syndromes

NCCM can occur as part of a *syndrome* in combination with dysmorphic features and other congenital malformations. When there are other congenital defects or when there are dysmorphic features in a patient, one of the listed syndromes in Table 6.5 or one of the *chromosomal defects* in Table 6.6 could be considered in the differential diagnosis.

6.4.6 Mitochondrial

Mitochondrial disorders often lead to multi-organ disease, including central and peripheral nervous system, eyes, heart, kidney, and endocrine organs. One of the cardiac features observed in mitochondrial disease is noncompaction cardiomyopathy. Cardiac features may be the first or only feature in patients suffering from a mitochondrial disorder. In a study of 113 pediatric patients with mitochondrial disease, NCCM was

Table 6.5 Syndromes associated with noncompaction cardiomyopathy (NCCM)/hypertrabeculation

Syndrome	Gene	Inheritance	Features	Reference
Barth syndrome/3-methylglutaconic aciduria	TAZ	XR	Growth retardation, dilated cardiomyopathy, skeletal myopathy, intermittent lactic acidemia, granulocytopenia, recurrent infections	35, 37, 134, 136, 138–145
Branchio-oto-renal syndrome I/Melnick Fraser syndrome	EYA1	AD	Long narrow face; hearing loss (sensory/conductive/mixed); preauricular pits; microtia; cup-shaped ears; lacrimal duct stenosis; cleft palate; bifid uvula; branchial cleft fistulas/cysts; renal dysplasia/aplasia; polycystic kidneys; vesico-ureteric reflux	181
Congenital adrenal hypoplasia	NR0B1	XR	Failure to thrive; hypogonadotropic hypogonadism; cryptorchidism; hyperpigmentation; primary adrenocortical failure; adrenal insufficiency; glucomineralocorticoid insufficiency; salt-wasting; delayed puberty	66
Contractural arachnodactyly/Beals syndrome	FBN2	AD	Marfanoid habitus; micrognathia; frontal bossing; crumpled ear helices; ectopia lentis; high-arched palate; septal defects; bicuspid aortic valve; mitral valve prolapse; patent ductus arteriosus; aortic root dilatation; pectus carinatum; kypkoscoliosis; hip/knee/elbow contractures; arachnodactyly; ulnar deviation of fingers; talipes equinovarus; hypoplastic calf muscles; motor development delay	182
Cornelia de Lange Syndrome I	NIPBL	AD	Short stature; microcephaly; long philtrum; micrognathia; low-set ears; sensorineural hearing loss; synophrys; myopia; long curly eyelashes; ptosis; anteverted nostrils; depressed nasal bridge; cleft lip/palate; thin upper lip; widely spaced teeth; congenital heart defect; pyloric stenosis; hypoplastic male genitalia; structural renal anomalies; phocomelia; oligodactyly; syndactyly of 2nd and 3^{d} toes; single transverse palmar crease; cutis marmorata; hirsutism; low posterior hair line; mental retardation; language delay; automutilation	113
Leopard syndrome	PTPN11, RAF1	AD	Short stature; triangular face; low-set ears; sensorineural hearing loss; hypertelorism; ptosis; epicanthal folds; broad flat nose; cleft palate; short neck; pulmonic stenosis; HCM; subaortic stenosis; complete heart block; bundle branch block; winged scapulae; hypospadia; absent/hypoplastic ovary; unilateral renal agenesis; spina bifida occulta; dark lentigines (mostly neck and trunk); café-au-lait spots	183
Melnick Needles osteodysplasty	FLNA	XD	Short stature; micrognathia; large ears; hypertelorism; exophthalmos; cleft palate; misaligned teeth; long neck; mitral/tricuspid valve prolapse; NCCM; pulmonary hypertension; pectus excavatum; omphalocele; hydronephrosis; tall vertebrae; bowing of humerus/radius/ulna/tibia; short distal phalanges of the fingers; pes planus; coarse hair; delayed motor development; hoarse voice	184
Nail Patella Syndrome	LMX1B	AD	Short stature; sensorineural hearing loss; ptosis; cataract; cleft lip/palate; malformed sternum; hypoplasia of first ribs; glomerulanephritis; renal failure; scoliosis; elbow deformities; hypoplastic or absent patella; clinodactyly; talipes equinovarus; longitudinal ridging nails; slow nail growth; koilonychias; anonychia; aplasia pectaralis minor/biceps/triceps/quadriceps	185, 186

(continued)

Table 6.5 (continued)

Syndrome	Gene	Inheritance	Features	Reference
Noonan syndrome	PTPN11 KRAS SOS1 RAF1	AD	Short stature; triangular face; low-set ears; hypertelorism; downslanting palpebral fissures; epicanthal folds; myopia; micrognathia; high arched palate; low posterior hairline; webbed neck; septal defects; pulmonic stenosis; patent ductus arteriosus; pectus carinatum superiorly/pectus excavatum inferiorly; cryptorchidism; clinodactyly; woolly hair; mental retardation (mild); bleeding tendency; malignant schwannoma	187
Roifman syndrome		XR	Short-trunk dwarfism; long philtrum; strabismus; narrow and downslanting palpebral fissures; long eyelashes; retinal dystrophy; narrow upturned nose; NCCM; hepato-splenomegaly; spondylo-epiphyseal dysplasia; eczema; hyperconvex nails; hypotonia; (mild) mental retardation; hypogonadotropic hypogonadism; recurrent infections; antibody deficiency	188
Syndromic microphtalmia/MIDAS syndrome (MIcrophtalmia, Dermal Aplasia, Sclerocornea)	HCCS	XD	Short stature; microcephaly; hearing loss; microphtalmia; sclerocornea; cataract; iris coloboma; retinopathy; septal defects; cardiac conduction defects; cardiomyopathy; overriding aorta; anteriorly placed anus; hypospadia; linear skin defects; corpus callosum agenesis; hydrocephalus; mental retardation; seizures	189, 190

AD autosomal dominant, XD X-linked dominant, XR X-linked recessive

Table 6.6 Chromosomal defects associated with noncompaction cardiomyopathy (NCCM)

Chromosomal defects	Features	Reference
Deletion		
1p36	Microcephaly; sensorineural hearing loss; deep-set eyes; flat nose; cleft lip/palate; cardiomyopathy; septal defects; patent ductus arteriosus; dilated aortic root; feeding problems; gastro-oesophageal reflux; short fifth finger and clinodactyly; mental retardation (severe); seizures; hypotonia	191–194
1q43-q43	Microcephaly; upslanting palpebral fissures; epicanthus, broad nasal bridge, micrognathia; low set ears; bow-shaped upper lip; widely spaced teeth; short webbed neck; congenital heart defects; mental retardation (severe); speech impairment; seizures; corpus callosum agenesis	195
5q35.1q35.3	Facial hirsutism; synophrys; downslanting palpebral fissures; atrial septal defect and patent ductus arteriosus; NCCM with sick sinus syndrome and second degree heart block; feeding problems; gastro-oesophageal reflux; joint hypermobility	196
22q11.2	Velo-cardio-facial syndrome: short stature; microcephaly; retrognathia; narrow palpebral fissures; square nasal root; prominent tubular nose; cleft palate; velopharyngeal insufficiency; congenital heart defect (85%): ventricular septal defect; Fallot's tetralogy; inguinal/umbilical hernia; slender hands and digits; learning disability; mental retardation; schizophrenia; bipolar disorder	66
Numeric		
4q trisomy/1q monosomy	Senile-like appearance; narrow palpebral fissures; telecanthus; epicanthus; broad nasal bridge; low-set ears; long philtrum; dimple below lower lip; anteriorly displaced anus; rocker-bottom feet; mental retardation; hypotonia, hypoplastic corpus callosum	197
Trisomy 13	Microcephaly; hypotelorism; cleft lip/palate; coloboma; low-set ears; septal defects; patent ductus arteriosus Polydactyly; overlapping fingers; mental retardation (severe); hypotonia; seizures	198

Table 6.6 (continued)

Chromosomal defects	Features	Reference
Trisomy 21	Short stature; bachycephaly; flat facial profile; conductive hearing loss; epicanthal folds; upslant; iris brushfield spots; protruding tongue; congenital heart malformation; duodenal atresia; Hirschsprung disease; joint laxity; single transverse palmar crease; excess nuchal skin; mental retardation; hypothyroidism; leukemia	11, 149
Mosaic trisomy 22	Microcephaly; hypertelorism; preauricular pits/tags; low-set ears; micrognathia, long philtrum; septal defects; double aortic arch; clinodactyly; hypoplastic nails; hemiatrophy; mental retardation	199
45,X0	Turner syndrome: short stature; short webbed neck; low hair line; broad nasal bridge; low-set ears; congenital heart defects: aortic coarctation; bicuspid aortic valves; aortic dilatation; lymph-edema of hands and feet; renal abnormalities: single horseshoe kidney; renal vascular abnormalities; delayed puberty; amenorrhea; infertility; hypothyroidism	200, 201
Loci		
6p24.3-21.1	NCCM; bradycardia; pulmonary valve stenosis; atrial septal defect; left bronchial isomerism; azygous continuation of the inferior vena cava; polysplenia; intestinal malrotation	153
11p15	NCCM; mild pulmonary stenosis; mild mitral valve prolapse; atrial septal defect	202

identified in 13%.[65] Pignatelli et al. showed that 5 of the 36 pediatric NCCM patients who underwent a skeletal muscular biopsy, had morphologic and biochemical evidence for a mitochondrial defect, including a partial deficiency of complex I-III of the mitochondrial respiratory chain.[66] Mutations in mitochondrial DNA (mtDNA) and in nuclear DNA have been identified in the mitochondrial disorders associated with NCCM.[67–69]

6.4.7 Miscellaneous

NCCM has been described in patients with heterotaxy with polysplenia, polycystic kidney disease, congenital adrenal hyperplasia, nephropathic cystinosis, and myelofibrosis.[11,66,70–74] Whether these co-occurrences are coincidental or represent shared etiologies with NCCM is unknown.

Among the possibly acquired forms of NCCM, there are reports about an infectious cause.[27] Recently, an etiologic role for macro- and microvascular abnormalities was suggested.[75–84] NCCM has also been described in patients with coronary heart disease.[85–87] Since coronary artery disease is a frequent disorder, this association may well be coincidental. Aortic elasticity was significantly altered in a group of 20 NCCM patients (aortic stiffness index of 8.3 ± 5.2).[78]

Microvascular abnormalities in NCCM including decreased coronary flow reserve with wall motion abnormalities in more extended regions of the myocardium than the noncompacted area have been observed.[77] In addition, several case studies reported hypoperfusion of the noncompacted region in NCCM patients using myocardial perfusion SPECT, positron emission tomography, Thallium myocardial imaging, or MRI.[75,76,79,80,82,83] It is thought that fibrosis, thrombus formation, hypokinesis, and necrosis may be the underlying mechanisms of hypoperfusion.[76,79,80]

Other pathogenic hypotheses for NCCM include adaptation to changes in the cardiovascular and/or hemodynamic climate; myocardial dissection or tearing of the inner layer of the cardiac muscle due to dilatation.[19,22]

6.5 Clinical Aspects

Heart failure is among the most frequent presentations of NCCM, followed by supraventricular and ventricular arrhythmias, including sudden cardiac death, and thrombo-embolic events. However, as in other cardiomyopathies, there is a great variability in presentation, even within families, ranging from a fully asymptomatic course to severe heart failure necessitating cardiac transplantation. The age of presentation is also highly variable varying from prenatal and neonatal diagnosis to diagnosis at the age of 94 years.[6, 11, 88–93] Prenatal diagnostic imaging detects more often bilateral ventricular hypertrophy/*hypertrabeculation*s than the typical left ventricular morphologic changes observed postnatally and in adults (unpublished observation). The fourth to fifth decade is the median age for diagnosis in adult isolated NCCM, constituting a relatively young population in adult cardiologic practice. Many patients remain asymptomatic and may be detected due to an asymptomatic heart murmur, or by chance by preoperative cardiac evaluation or medical assessment for insurance or jobs or because they participated in cardiologic family screening, after a relative had been diagnosed with NCCM.[11, 13] Symptomatic patients may present clinical symptoms of dyspnea, fatigue (atypical) chest pain, and/or (pre) syncope. NCCM may also present as a peripartum cardiomyopathy.[11, 94–96] Review of the literature revealed a male to female ratio of almost 2:1.[19] This gender difference cannot be fully explained by the occurrence of X-linked forms of NCCM.

Different *arrhythmias* and *conduction disorders* may occur in NCCM patients (Table 6.7). None of these arrhythmias is characteristic or pathognomonic for NCCM. Thrombo-embolic events may include stroke (cerebrovascular event or transient ischemic attack), peripheral embolism, and mesenterial thrombosis.

6.6 Differential Diagnosis

The definitive diagnosis of NCCM relies on the morphological features of the LV myocardium, as defined by an imaging modality, like echocardiography, MRI, CT, or LV angiography. The variability in the extent of physiological trabecularization may complicate distinction of NCCM from normal physiological left ventricular trabeculations. Especially in the area around the base of the papillary muscles of the mitral valve, more trabeculations may be present. However, in the normal heart, there is no excessive segmental thickening (due to hypertrabeculation) like in NCCM and the thickness of these physiological trabeculations does not exceed the thickness of the compact layer. Also, the area of noncompaction is larger in NCCM than in physiological trabeculations.[13]

Secondary forms of (acquired) NCCM may be the result of hypertension, chronic volume or pressure overload,[97] ischemic heart disease or extreme physical activity (i.e., athletes), leading to NCCM-like abnormalities. These are referred to as pseudo-noncompaction cardiomyopathy or an NCCM look-alike. Hypertensive patients are diagnostically challenging, because of the occurrence of LV hypertrophy due to hypertension. Further studies are needed to confirm whether excessive trabeculation is more prevalent in specific ethnic groups, as suggested by one study.[14]

Furthermore, dilated, hypertrophic, and ischemic cardiomyopathy may be mistaken for NCCM or vice versa, due to prominent trabeculations or abnormal myocardial thickening. Candida sepsis with intramyocardial abscesses and intramyocardial hematoma may mimic NCCM.[27, 98, 99]

The neuromuscular disorders, syndromes, and chromosomal abnormalities mentioned earlier (Tables 6.3–6.5) should be considered in the differential diagnosis of nonisolated NCCM, especially when NCCM occurs in patients with dysmorphism, growth retardation, or skeletal muscle weakness.

Table 6.7 Arrhythmia and conduction disorders associated with noncompaction cardiomyopathy (NCCM)

Arrhythmia/conduction disorders associated with NCCM	Reference
Atrial fibrillation	15, 113, 203
Atrioventricular nodal re-entrant tachycardia	204
Bigemini ventricular extra systole	146
Complete atrioventricular block	1, 158, 205, 206
Complete left bundle branch block	109, 146
Giant P-waves and focal atrial tachycardia	207
Long QT syndrome 2	135
Narrow QRS complex	106, 107, 110
Persistent atrial standstill	208
Sick sinus syndrome	209
Sinus bradycardia	153, 210
Supraventricular tachyarrhythmia	7, 113, 130, 146, 211
Ventricular fibrillation	106, 205, 212
Ventricular tachycardia	7, 79, 106, 109, 210
Wolff–Parkinson–White syndrome	2, 7, 146, 210, 213

6.7 Therapy, Follow-up, and Prognosis

6.7.1 Therapy and Follow-up

Current guidelines for heart failure, arrhythmias, cardiac resynchronization therapy, and ICD implantation for primary and secondary prevention are applied for NCCM.[100–102] *β-Blockers* and *Angiotensin-converting-enzyme (ACE) – inhibitors* are the cornerstones of the treatment in the presence of LV dysfunction and/or arrhythmias. Establishing an expert consensus rapport, similar to HCM, based on case reports, small cohorts and clinical registries would be recommended since no randomized trials or studies on management of NCCM have been conducted, and clear-cut evidence-based clinical guidelines for this disorder are therefore missing.[103] An important issue is the use of prophylactic anticoagulants, in view of frequent thrombo-embolic events. The early case reports and case series emphasized the high risk of thrombo-embolism and advised routine anticoagulation therapy. However, a review of 22 publications addressing the issue concluded that thromboembolic events are rare in NCCM.[104] Fazio et al. came to the same conclusion.[105] Currently, in our hospital, anticoagulation therapy is advised only in patients with an ejection fraction less than 40% (cut off arbitrary), paroxysmal or persistent atrial fibrillation and/or previous thrombo-embolic events.

Successful cardiac resynchronization therapy has been described in several NCCM patients, leading to left ventricular reverse remodeling and an increase in left ventricular function.[106–110]

Heart transplantation has been performed in some NCCM patients with severe heart failure.[3, 11, 23, 111–117] Left ventricular restoration surgery has been reported successful in a single patient.[118] Treatment with an *implantable cardioverter defibrillator* (ICD) will be discussed further on.

The indication for cardiologic follow-up depends on individual symptoms and cardiac abnormalities. In asymptomatic patients with preserved LV function, annual or biannual cardiologic follow-up is recommended, including ECG and echocardiography. If necessary, these could be extended with 24-h-Holter monitoring and exercise-testing. When EF is below 50%, β-blocker therapy and ACE-inhibitors should be prescribed, especially when NCCM is accompanied by hypertension or arrhythmias.

6.7.2 Prognosis

Initially, NCCM was reported to have a grave prognosis.[2, 3, 12, 19, 23, 119–125] However, the application of new imaging techniques allowing diagnosing NCCM in asymptomatic individuals suggests that the first observations were influenced by selection of the most severely affected individuals. It has become clear that prognosis of NCCM is as variable as the prognosis in other cardiomyopathies. Even in those with presentation in early childhood, gradual improvement in cardiac function may be observed, although in others evolvement to severe heart failure requiring heart transplantation does occur.[6, 88, 90, 92, 126, 127] Similarly, in some adult patients a rapid deterioration of heart function occurs, whereas in others the disease remains stable up to old age.[89] Malignant arrhythmias leading to sudden cardiac death and heart failure are the main indicators of poor prognosis. The establishment of appropriate risk stratification will be an important issue in the near future in order to identify patients at risk and to help prevent sudden cardiac death.

6.8 Risk Stratification and Indication for ICD

Patients at the highest risk for sudden death are patients who previously experienced (aborted) cardiac arrest, ventricular fibrillation, and sustained VF. Family history of sudden death, unexplained syncope (especially during exercise), abnormal blood pressure response during exercise tests, frequent premature ventricular beats on the resting ECG, and /or nonsustained ventricular tachycardia on Holter monitoring and significantly impaired left ventricular function may be considered risk factors. The results from longitudinal studies and the understanding of underlying disease mechanisms will hopefully help to gain more insight into the risk factors and allow more appropriate risk stratification.[128]

Consensus and guidelines for prophylactic ICD treatment in NCCM patients are also needed. Regular ICD indications include primary and secondary prevention. For secondary prevention, i.e., after a previous episode of aborted cardiac death or collapse due to sustained VT or VF, current ICD guidelines advise

ICD implantation. In the Rotterdam NCCM cohort of 67 patients, an ICD was indicated in 42% according to the current ICD guidelines ($n=28$:21 primary and 7 for secondary prevention). After long-term follow-up, appropriate ICD therapy occurred only in patients with secondary prevention ($n=3$). Inappropriate ICD therapy occurred in 33% of the patients with primary prevention and in 29% of the patients with secondary prevention.[129] In another study, follow-up of 12 patients who received an ICD showed overall appropriate therapy in 42% in primary and secondary prevention combined.[130] In primary prevention, 25% of ICD therapy was appropriate opposed to 50% in secondary prevention.[130] This accentuates the need for further research of appropriate risk stratification of sudden cardiac death in patients with NCCM.

6.9 Cardiogenetic Aspects

6.9.1 Molecular and Cardiologic Family Screening

Familial NCCM has been estimated to occur in 18–71% of adults with isolated NCCM, mostly consistent with an autosomal dominant mode of inheritance, indicating the importance of informing and examining relatives of patients with isolated NCCM.[2, 11, 24, 66, 120, 131–133] Since extensive family studies showed that the majority of affected relatives are asymptomatic, cardiologic evaluation should include all adult relatives irrespective of medical history. Obviously, taking a family history is by itself insufficient to identify familial disease, given the high frequency of asymptomatic disease in families.[11] In families where a pathogenic mutation has been identified, relatives can be offered predictive DNA analysis. In families without a pathogenic mutation, cardiac family screening remains the method of choice to identify relatives at risk of developing symptomatic cardiomyopathy, who may benefit from early treatment.

Apart from NCCM, other cardiomyopathies may co-occur within families, like hypertrophic and dilated cardiomyopathy, so cardiac screening should aim at identifying all cardiomyopathies. Cardiac screening of relatives may show minor abnormalities not fulfilling NCCM criteria, which may be difficult to differentiate from normal physiologic trabecularization. Hypothetically, these minor abnormalities might develop into NCCM eventually. Longitudinal studies of patients with mild NCCM features are needed to investigate the natural history of these forms of noncompaction.

6.9.2 Genotype–Phenotype Correlations

Molecular studies of NCCM have thus far shown that there are few recurrent mutations.[10] Therefore, it is difficult to establish genotype-phenotype correlations. Additionally, intrafamilial phenotypic variability complicates predictions based on an identified mutation. Multiple (truncating) sarcomere mutations appear to result in a more severe phenotype with childhood onset.[10, 11] Multiple mutations identified in adults mostly also comprise involvement of a nonsarcomere gene. Adult patients with multiple mutations seem to have more symptoms than adults with a single mutation.[10, 11] These observations may indicate that the combination of a sarcomere mutation and a nonsarcomere mutation causes a less severe phenotype than when a patient has two sarcomere mutations. Mutations in *DTNA* and *TAZ* seem to transfer the strongest predisposition to childhood onset NCCM.

6.9.3 Molecular Strategies

The proposed strategies for the molecular and cardiologic evaluation of NCCM are depicted in the flowchart in Fig. 6.5. Extensive genetic screening may lead to the identification of a molecular defect in over 40% of isolated NCCM patients and in half of these patients an *MYH7* mutation is found.[10] *MYH7* gene sequencing should be considered as an initial approach, being the most prevalent cause for NCCM in adults and children. Further molecular analyses of the other genes within the NCCM spectrum, which quantitatively have a relatively modest contribution to NCCM morbidity, may be considered when no mutation in *MYH7* can be identified. Sarcomere gene analysis is also warranted in pediatric patients, given the high percentage of sarcomere mutations in this group.[10] When an adult or pediatric patient is severely affected, screening for a

Fig. 6.5 Flowchart for family screening in NCCM * Including unclassified variants; ** if clinically indicated; *** earlier when symptomatic

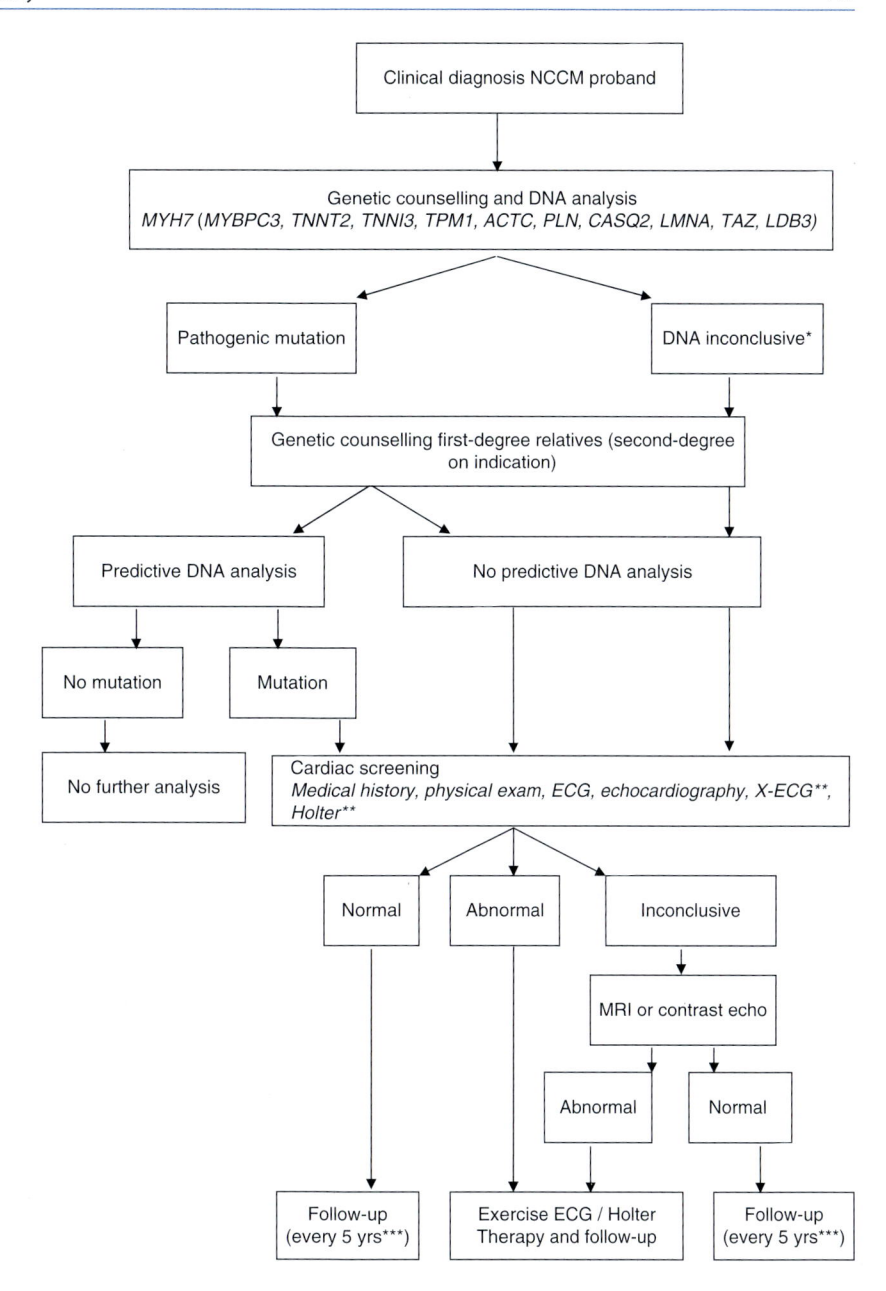

second molecular defect is advised, given the high frequency of multiple mutations in NCCM.

6.10 Summary

NCCM is a relatively new, genetically heterogeneous, cardiomyopathy. Clinical presentation and prognosis range from asymptomatic disease with no or slow progression, to severe disabling, rapidly progressive cardiac failure. Initial presentation includes the triad of heart failure (potentially lethal) arrhythmias and/or thrombo-embolism. In adults, the majority of NCCM is isolated.

The first clinical presentation of NCCM may occur at all ages, even prenatally. In childhood, clinical features are often more severe and NCCM is frequently associated with congenital heart defects. The echocardiographic diagnostic criteria as proposed by Jenni et al. are convenient in daily practice and currently the most widely applied. The general cardiac guidelines for chronic heart failure and ICDs are suitable and applicable to the NCCM population.

In as much as 41% of isolated NCCM, molecular testing may yield a genetic defect, mostly in sarcomere genes. The *MYH7* gene is the most prevalent disease gene. The nonisolated forms of NCCM are caused by a range of different (rare) genetic defects. Until now, in half of familial isolated NCCM, the genetic defect remains unknown. Genetic defects in a large number of sarcomere and other cardiomyopathy genes and in genes primarily associated with skeletal myopathies indicate that NCCM may result from a wide range of pathophysiologic mechanisms.

Shared genetic defects and familial aggregation of NCCM, HCM, and DCM indicates that NCCM may be part of a broad spectrum of cardiomyopathies.

The genetic etiology of NCCM requires that patients and their relatives are offered genetic testing and counseling. This may include (predictive) molecular analysis of relatives, when applicable, and/or cardiac evaluation of at-risk relatives, even when they are as yet asymptomatic.

Acknowledgments The authors thank Michelle Michels and Dennis Dooijes for their scientific contribution and the Dutch Heart Foundation for their financial support.

References

1. Feldt RH, Rahimtoola SH, Davis GD, Swan HJ, Titus JL. Anomalous ventricular myocardial patterns in a child with complex congenital heart disease. *Am J Cardiol*. 1969; 23(5):732-734.
2. Chin TK, Perloff JK, Williams RG, Jue K, Mohrmann R. Isolated noncompaction of left ventricular myocardium. A study of eight cases. *Circulation*. 1990;82(2):507-513.
3. Oechslin EN, Attenhofer Jost CH, Rojas JR, Kaufmann PA, Jenni R. Long-term follow-up of 34 adults with isolated left ventricular noncompaction: a distinct cardiomyopathy with poor prognosis. *J Am Coll Cardiol*. 2000;36(2):493-500.
4. Hoedemaekers YM, Caliskan K, Majoor-Krakauer D, et al. Cardiac {beta}-myosin heavy chain defects in two families with non-compaction cardiomyopathy: linking non-compaction to hypertrophic, restrictive, and dilated cardiomyopathies. *Eur Heart J*. 2007;28(22):2732-2737.
5. Nemes A, Caliskan K, Geleijnse ML, Soliman OI, Vletter WB, ten Cate FJ. Reduced regional systolic function is not confined to the noncompacted segments in noncompaction cardiomyopathy. *Int J Cardiol*. 2009;134(3):366-370.
6. Moura C, Hillion Y, Daikha-Dahmane F, et al. Isolated noncompaction of the myocardium diagnosed in the fetus: two sporadic and two familial cases. *Cardiol Young*. 2002; 12(3):278-283.
7. Tsai SF, Ebenroth ES, Hurwitz RA, Cordes TM, Schamberger MS, Batra AS. Is left ventricular noncompaction in children truly an isolated lesion? *Pediatr Cardiol*. 2009;30(5): 597-602.
8. Maron BJ, Towbin JA, Thiene G, et al. Contemporary definitions and classification of the cardiomyopathies: an American Heart Association Scientific Statement from the Council on Clinical Cardiology, Heart Failure and Transplantation Committee; Quality of Care and Outcomes Research and Functional Genomics and Translational Biology Interdisciplinary Working Groups; and Council on Epidemiology and Prevention. *Circulation*. 2006;113(14): 1807-1816.
9. Elliott P, Andersson B, Arbustini E, et al. Classification of the cardiomyopathies: a position statement from the European Society Of Cardiology Working Group on Myocardial and Pericardial Diseases. *Eur Heart J*. 2008; 29(2):270-276.
10. Dooijes D, Hoedemaekers YM, Michels M, van der Smagt J, van de Graaf R, van Tienhoven M, ten Cate FJ, Caliskan K, Majoor-Krakauer DF. Left ventricular noncompaction cardiomyopathy: disease genes, mutation spectrum and diagnostic implications. *Submitted*. 2009.
11. Hoedemaekers YM, Caliskan K, Michels M, Frohn-Mulder I, van der Smagt JJ, Phefferkorn JJ, Wessels MW, ten Cate FJ, Sijbrands EJG, Dooijes D, Majoor-Krakauer DF. The importance of genetic counseling, DNA diagnostics and cardiologic family screening in left ventricular noncompaction cardiomyopathy. *Circ Cardiovasc Genet*. 2010;3(3):232-239.
12. Jenni R, Oechslin E, Schneider J, Attenhofer Jost C, Kaufmann PA. Echocardiographic and pathoanatomical characteristics of isolated left ventricular non-compaction: a step towards classification as a distinct cardiomyopathy. *Heart*. 2001;86(6):666-671.
13. Belanger AR, Miller MA, Donthireddi UR, Najovits AJ, Goldman ME. New classification scheme of left ventricular noncompaction and correlation with ventricular performance. *Am J Cardiol*. 2008;102(1):92-96.
14. Kohli SK, Pantazis AA, Shah JS, et al. Diagnosis of left-ventricular non-compaction in patients with left-ventricular systolic dysfunction: time for a reappraisal of diagnostic criteria? *Eur Heart J*. 2008;29(1):89-95.
15. Stollberger C, Finsterer J, Blazek G. Left ventricular hypertrabeculation/noncompaction and association with additional cardiac abnormalities and neuromuscular disorders. *Am J Cardiol*. 2002;90(8):899-902.
16. Petersen SE, Selvanayagam JB, Wiesmann F, et al. Left ventricular non-compaction: insights from cardiovascular magnetic resonance imaging. *J Am Coll Cardiol*. 2005;46(1): 101-105.
17. Hughes SE, McKenna WJ. New insights into the pathology of inherited cardiomyopathy. *Heart*. 2005;91(2):257-264.
18. Burke A, Mont E, Kutys R, Virmani R. Left ventricular noncompaction: a pathological study of 14 cases. *Hum Pathol*. 2005;36(4):403-411.
19. Stollberger C, Finsterer J. Left ventricular hypertrabeculation/ noncompaction. *J Am Soc Echocardiogr*. 2004;17(1):91-100.

20. Boyd MT, Seward JB, Tajik AJ, Edwards WD. Frequency and location of prominent left ventricular trabeculations at autopsy in 474 normal human hearts: implications for evaluation of mural thrombi by two-dimensional echocardiography. *J Am Coll Cardiol*. 1987;9(2):323-326.
21. Pujadas S, Bordes R, Bayes-Genis A. Ventricular non-compaction cardiomyopathy: CMR and pathology findings. *Heart*. 2005;91(5):582.
22. Freedom RM, Yoo SJ, Perrin D, Taylor G, Petersen S, Anderson RH. The morphological spectrum of ventricular noncompaction. *Cardiol Young*. 2005;15(4):345-364.
23. Ritter M, Oechslin E, Sutsch G, Attenhofer C, Schneider J, Jenni R. Isolated noncompaction of the myocardium in adults. *Mayo Clin Proc*. 1997;72(1):26-31.
24. Aras D, Tufekcioglu O, Ergun K, et al. Clinical features of isolated ventricular noncompaction in adults long-term clinical course, echocardiographic properties, and predictors of left ventricular failure. *J Card Fail*. 2006;12(9):726-733.
25. Sandhu R, Finkelhor RS, Gunawardena DR, Bahler RC. Prevalence and characteristics of left ventricular noncompaction in a community hospital cohort of patients with systolic dysfunction. *Echocardiography*. 2008;25(1):8-12.
26. Nugent AW, Daubeney PE, Chondros P, et al. The epidemiology of childhood cardiomyopathy in Australia. *N Engl J Med*. 2003;348(17):1639-1646.
27. Stollberger C, Preiser J, Finsterer J. Candida sepsis with intramyocardial abscesses mimicking left ventricular noncompaction. *Eur J Echocardiogr*. 2004;5(1):76-78.
28. Klaassen S, Probst S, Oechslin E, et al. Mutations in sarcomere protein genes in left ventricular noncompaction. *Circulation*. 2008;117(22):2893-2901.
29. Budde BS, Binner P, Waldmuller S, et al. Noncompaction of the ventricular myocardium is associated with a de novo mutation in the beta-myosin heavy chain gene. *PLoS ONE*. 2007;2(12):e1362.
30. Morita H, Rehm HL, Menesses A, et al. Shared genetic causes of cardiac hypertrophy in children and adults. *N Engl J Med*. 2008;358(18):1899-1908.
31. Lekanne Deprez RH, Muurling-Vlietman JJ, Hruda J, et al. Two cases of severe neonatal hypertrophic cardiomyopathy caused by compound heterozygous mutations in the MYBPC3 gene. *J Med Genet*. 2006;43(10):829-832.
32. Van Driest SL, Vasile VC, Ommen SR, et al. Myosin binding protein C mutations and compound heterozygosity in hypertrophic cardiomyopathy. *J Am Coll Cardiol*. 2004;44(9):1903-1910.
33. Zahka K, Kalidas K, Simpson MA, et al. Homozygous mutation of MYBPC3 associated with severe infantile hypertrophic cardiomyopathy at high frequency among the Amish. *Heart*. 2008;94(10):1326-1330.
34. Hermida-Prieto M, Monserrat L, Castro-Beiras A, et al. Familial dilated cardiomyopathy and isolated left ventricular noncompaction associated with lamin A/C gene mutations. *Am J Cardiol*. 2004;94(1):50-54.
35. Ichida F, Tsubata S, Bowles KR, et al. Novel gene mutations in patients with left ventricular noncompaction or Barth syndrome. *Circulation*. 2001;103(9):1256-1263.
36. Vatta M, Mohapatra B, Jimenez S, et al. Mutations in Cypher/ZASP in patients with dilated cardiomyopathy and left ventricular non-compaction. *J Am Coll Cardiol*. 2003;42(11):2014-2027.
37. Kenton AB, Sanchez X, Coveler KJ, et al. Isolated left ventricular noncompaction is rarely caused by mutations in G4.5, alpha-dystrobrevin and FK Binding Protein-12. *Mol Genet Metab*. 2004;82(2):162-166.
38. Fatkin D, Graham RM. Molecular mechanisms of inherited cardiomyopathies. *Physiol Rev*. 2002;82(4):945-980.
39. Sedmera D, Pexieder T, Vuillemin M, Thompson RP, Anderson RH. Developmental patterning of the myocardium. *Anat Rec*. 2000;258(4):319-337.
40. Lie-Venema H. The role of epicardium-derived cells (EPDCs) in the development of non-compaction cardiomyopathy. *Florence International Course on Advances in Cardiomyopathies – 5th meeting of the European Myocardial and Pericardial Disease WG of the ESC*. Florence, Italy; 2008.
41. Lie-Venema H, van den Akker NM, Bax NA, et al. Origin, fate, and function of epicardium-derived cells (EPDCs) in normal and abnormal cardiac development. *ScientificWorldJournal*. 2007;7:1777-1798.
42. Crawford SE, Qi C, Misra P, et al. Defects of the heart, eye, and megakaryocytes in peroxisome proliferator activator receptor-binding protein (PBP) null embryos implicate GATA family of transcription factors. *J Biol Chem*. 2002;277(5):3585-3592.
43. Lee Y, Song AJ, Baker R, Micales B, Conway SJ, Lyons GE. Jumonji, a nuclear protein that is necessary for normal heart development. *Circ Res*. 2000;86(9):932-938.
44. Shou W, Aghdasi B, Armstrong DL, et al. Cardiac defects and altered ryanodine receptor function in mice lacking FKBP12. *Nature*. 1998;391(6666):489-492.
45. van Loo PF, Mahtab EA, Wisse LJ, et al. Transcription factor Sp3 knockout mice display serious cardiac malformations. *Molec Cell Biol*. 2007;27(24):8571-8582.
46. Breckenridge RA, Anderson RH, Elliott PM. Isolated left ventricular non-compaction: the case for abnormal myocardial development. *Cardiol Young*. 2007;17(2):124-129.
47. Palloshi A, Puccetti P, Fragasso G, et al. Elderly manifestation of non-compaction of the ventricular myocardium. *J Cardiovasc Med*. 2006;7(9):714-716.
48. Finsterer J, Stollberger C. Acquired noncompaction. *Int J Cardiol*. 2006;110(3):288-300.
49. Lombardi R, Betocchi S. Aetiology and pathogenesis of hypertrophic cardiomyopathy. *Acta Paediatr Suppl*. 2002;91(439):10-14.
50. Monserrat L, Hermida-Prieto M, Fernandez X, et al. Mutation in the alpha-cardiac actin gene associated with apical hypertrophic cardiomyopathy, left ventricular non-compaction, and septal defects. *Eur Heart J*. 2007;28(16):1953-1961.
51. Zhu L, Vranckx R, Khau Van Kien P, et al. Mutations in myosin heavy chain 11 cause a syndrome associating thoracic aortic aneurysm/aortic dissection and patent ductus arteriosus. *Nat Genet*. 2006;38(3):343-349.
52. Guo DC, Pannu H, Tran-Fadulu V, et al. Mutations in smooth muscle alpha-actin (ACTA2) lead to thoracic aortic aneurysms and dissections. *Nat Genet*. 2007;39(12):1488-1493.
53. Xin B, Puffenberger E, Tumbush J, Bockoven JR, Wang H. Homozygosity for a novel splice site mutation in the cardiac myosin-binding protein C gene causes severe neonatal hypertrophic cardiomyopathy. *Am J Med Genet*. 2007;143A(22):2662-2667.

54. Wessels MW, Willems PJ. Mutations in sarcomeric protein genes not only lead to cardiomyopathy but also to congenital cardiovascular malformations. *Clin Genet*. 2008;74(1): 16-19.
55. Finsterer J, Gelpi E, Stollberger C. Left ventricular hypertrabeculation/noncompaction as a cardiac manifestation of Duchenne muscular dystrophy under non-invasive positive-pressure ventilation. *Acta Cardiol*. 2005;60(4):445-448.
56. Finsterer J, Stollberger C. Spontaneous left ventricular hypertrabeculation in dystrophin duplication based Becker's muscular dystrophy. *Herz*. 2001;26(7):477-481.
57. Finsterer J, Stollberger C, Feichtinger H. Noncompaction in Duchenne muscular dystrophy: frustrated attempt to create a compensatory left ventricle? *Cardiology*. 2006;105(4): 223-225.
58. Finsterer J, Stollberger C, Wegmann R, Jarius C, Janssen B. Left ventricular hypertrabeculation in myotonic dystrophy type 1. *Herz*. 2001;26(4):287-290.
59. Hoogerwaard EM, Bakker E, Ippel PF, et al. Signs and symptoms of Duchenne muscular dystrophy and Becker muscular dystrophy among carriers in The Netherlands: a cohort study. *Lancet*. 1999;353(9170):2116-2119.
60. Hoogerwaard EM, van der Wouw PA, Wilde AA, et al. Cardiac involvement in carriers of Duchenne and Becker muscular dystrophy. *Neuromuscul Disord*. 1999;9(5):347-351.
61. Hermida-Prieto MML, Castro-Beiras A, Castro-Beiras A, et al. Familial dilated cardiomyopathy and isolated left ventricular noncompaction associated with Lamin A/C gene mutations. *Am J Cardiol*. 2004;94:50-54.
62. Rankin J, Auer-Grumbach M, Bagg W, et al. Extreme phenotypic diversity and nonpenetrance in families with the LMNA gene mutation R644C. *Am J Med Genet*. 2008; 146A(12):1530-1542.
63. Malhotra R, Mason PK. Lamin A/C deficiency as a cause of familial dilated cardiomyopathy. *Curr Opin Cardiol*. 2009;24(3):203-208.
64. Finsterer J, Stollberger C, Blazek G. Neuromuscular implications in left ventricular hypertrabeculation/noncompaction. *Int J Cardiol*. 2006;110(3):288-300.
65. Scaglia F, Towbin JA, Craigen WJ, et al. Clinical spectrum, morbidity, and mortality in 113 pediatric patients with mitochondrial disease. *Pediatrics*. 2004;114(4):925-931.
66. Pignatelli RH, McMahon CJ, Dreyer WJ, et al. characterization of left ventricular noncompaction in children: a relatively common form of cardiomyopathy. *Circulation*. 2003;108(21):2672-2678.
67. Finsterer J, Stollberger C, Schubert B. Acquired left ventricular hypertrabeculation/noncompaction in mitochondriopathy. *Cardiology*. 2004;102(4):228-230.
68. Finsterer J, Stollberger C, Wanschitz J, Jaksch M, Budka H. Nail-patella syndrome associated with respiratory chain disorder. *Eur Neurol*. 2001;46(2):92-95.
69. Finsterer J, Bittner R, Bodingbauer M, Eichberger H, Stollberger C, Blazek G. Complex mitochondriopathy associated with 4 mtDNA transitions. *Eur Neurol*. 2000;44(1):37-41.
70. Cho YH, Jin SJ, Je HC, et al. A case of noncompaction of the ventricular myocardium combined with situs ambiguous with polysplenia. *Yonsei Med J*. 2007;48(6):1052-1055.
71. Finsterer J, Stollberger C, Michaela J. Familial left ventricular hypertrabeculation in two blind brothers. *Cardiovasc Pathol*. 2002;11(3):146-148.
72. Finsterer J, Stollberger C, Prainer C, Hochwarter A. Lone noncompaction in Leber's hereditary optic neuropathy. *Acta Cardiol*. 2004;59(2):187-190.
73. Moon JY, Chung N, Seo HS, Choi EY, Ha JW, Rim SJ. Noncompaction of the ventricular myocardium combined with polycystic kidney disease. *Heart Vessels*. 2006;21(3): 195-198.
74. Ahmed I, Phan TT, Lipkin GW, Frenneaux M. Ventricular noncompaction in a female patient with nephropathic cystinosis: a case report. *J Med Case Rep*. 2009;3:31.
75. Borges AC, Kivelitz D, Baumann G. Isolated left ventricular non-compaction: cardiomyopathy with homogeneous transmural and heterogeneous segmental perfusion. *Heart*. 2003;89(8):e21.
76. Hamamichi Y, Ichida F, Hashimoto I, et al. Isolated noncompaction of the ventricular myocardium: ultrafast computed tomography and magnetic resonance imaging. *Int J Cardiovasc Imaging*. 2001;17(4):305-314.
77. Jenni R, Wyss CA, Oechslin EN, Kaufmann PA. Isolated ventricular noncompaction is associated with coronary microcirculatory dysfunction. *J Am Coll Cardiol*. 2002; 39(3):450-454.
78. Nemes A, Caliskan K, Geleijnse ML, Soliman OI, Anwar AM, Ten Cate FJ. Alterations in aortic elasticity in noncompaction cardiomyopathy. *Int J Cardiovasc Imaging*. 2008; 24(1):7-13.
79. Sato Y, Matsumoto N, Matsuo S, et al. Subendomyocardial perfusion abnormality and necrosis detected by magnetic resonance imaging in a patient with isolated noncompaction of the ventricular myocardium associated with ventricular tachycardia. *Cardiovasc Revasc Med*. 2009;10(1):66-68.
80. Sato Y, Matsumoto N, Matsuo S, et al. Myocardial perfusion abnormality and necrosis in a patient with isolated noncompaction of the ventricular myocardium: evaluation by myocardial perfusion SPECT and magnetic resonance imaging. *Int J Cardiol*. 2007;120(2):e24-26.
81. Sato Y, Matsumoto N, Yoda S, et al. Left ventricular aneurysm associated with isolated noncompaction of the ventricular myocardium. *Heart Vessels*. 2006;21(3):192-194.
82. Soler R, Rodriguez E, Monserrat L, Alvarez N. MRI of subendocardial perfusion deficits in isolated left ventricular noncompaction. *J Comput Assist Tomogr*. 2002;26(3):373-375.
83. Junga G, Kneifel S, Von Smekal A, Steinert H, Bauersfeld U. Myocardial ischaemia in children with isolated ventricular non-compaction. *Eur Heart J*. 1999;20(12):910-916.
84. Caliskan K, Balk AHMM, Wykrzykowska JJ, van Geuns RJ, Serruys PW. How Should I Treat An unusual referral for heart transplantation? *EuroIntervention*. 2010;5(7):861-5.
85. Gabrielli FA, Lombardo A, Natale L, et al. Myocardial infarction in isolated ventricular non-compaction: contrast echo and MRI. *Int J Cardiol*. 2006;111(2):315-317.
86. Ito H, Dajani KA. A case with noncompaction of the left ventricular myocardium detected by 64-slice multidetector computed tomography. *J Thorac Imaging*. 2009;24(1):38-40.
87. Martini DB, Sperotto C, Zhang L. "Cardiac incidentaloma": Left ventricular non-compaction in a kindred with familial coronary artery disease. *Cardiol J*. 2007;14(4):407-410.
88. Ozkutlu S, Bostan O, Karagoz T, Deren O, Tekinalp G. Prenatal diagnosis of isolated non-compaction of the ventricular myocardium: study of six cases. *Pediatr Int*. 2007; 49(2):172-176.

89. Sato Y, Matsumoto N, Matsuo S, et al. Isolated noncompaction of the ventricular myocardium in a 94-year-old patient: depiction at echocardiography and magnetic resonance imaging. *Int J Cardiol*. 2007;119(1):e32-34.
90. Guntheroth W, Komarniski C, Atkinson W, Fligner CL. Criterion for fetal primary spongiform cardiomyopathy: restrictive pathophysiology. *Obstetr Gynecol*. 2002;99 (5 Pt 2):882-885.
91. Halbertsma FJ, Van't Hek LG, Daniels O. Spongy cardiomyopathy in a neonate. *Cardiol Young*. 2001;11(4): 458-460.
92. Winer N, Lefevre M, Nomballais MF, et al. Persisting spongy myocardium. A case indicating the difficulty of antenatal diagnosis. *Fetal Diagn Ther*. 1998;13(4):227-232.
93. Whitham JK, Hasan BS, Schamberger MS, Johnson TR. Use of cardiac magnetic resonance imaging to determine myocardial viability in an infant with in utero septal myocardial infarction and ventricular noncompaction. *Pediatr Cardiol*. 2008;29(5):950-953.
94. Bahl A, Swamy A, Sharma Y, Kumar N. Isolated noncompaction of left ventricle presenting as peripartum cardiomyopathy. *Int J Cardiol*. 2006;109(3):422-423.
95. Patel C, Shirali G, Pereira N. Left ventricular noncompaction mimicking peripartum cardiomyopathy. *J Am Soc Echocardiogr*. 2007;20(8):e1009-1012.
96. Rehfeldt KH, Pulido JN, Mauermann WJ, Click RL. Left ventricular hypertrabeculation/noncompaction in a patient with peripartum cardiomyopathy. *Int J Cardiol*. 2008. Epub ahead of print.
97. Caliskan K, Balk AHMM, Jordaens L, Szili-Torok T. Bradycardiomyopathy: the case for a causative relationship between severe sinus bradycardia and heart failure. *J Cardiovasc Electrophysiol*. 2010;21(7):822-824.
98. Stollberger C, Preiser J, Finsterer J. Histological detection of intramyocardial abscesses in Candida sepsis mimicking left ventricular non-compaction/hypertrabeculation on echocardiography. *Mycoses*. 2004;47(1–2):72-75.
99. Stollberger C, Finsterer J, Waldenberger FR, Hainfellner JA, Ullrich R. Intramyocardial hematoma mimicking abnormal left ventricular trabeculation. *J Am Soc Echocardiogr*. 2001;14(10):1030-1032.
100. Hunt SA. ACC/AHA 2005 guideline update for the diagnosis and management of chronic heart failure in the adult: a report of the American College of Cardiology/American Heart Association Task Force on Practice Guidelines (Writing Committee to Update the 2001 Guidelines for the Evaluation and Management of Heart Failure). *J Am Coll Cardiol*. 2005;46(6):e1-82.
101. Hunt SA, Baker DW, Chin MH, et al. ACC/AHA Guidelines for the Evaluation and Management of Chronic Heart Failure in the Adult: Executive Summary A Report of the American College of Cardiology/American Heart Association Task Force on Practice Guidelines (Committee to Revise the 1995 Guidelines for the Evaluation and Management of Heart Failure): Developed in Collaboration With the International Society for Heart and Lung Transplantation; Endorsed by the Heart Failure Society of America. *Circulation*. 2001; 104(24):2996-3007.
102. Nieminen MS, Bohm M, Cowie MR, et al. Executive summary of the guidelines on the diagnosis and treatment of acute heart failure: the Task Force on Acute Heart Failure of the European Society of Cardiology. *Eur Heart J*. 2005;26(4):384-416.
103. Maron BJ, McKenna WJ, Danielson GK, et al. American College of Cardiology/European Society of Cardiology clinical expert consensus document on hypertrophic cardiomyopathy. A report of the American College of Cardiology Foundation Task Force on Clinical Expert Consensus Documents and the European Society of Cardiology Committee for Practice Guidelines. *J Am Coll Cardiol*. 2003; 42(9):1687-1713.
104. Stollberger C, Finsterer J. Thrombi in left ventricular hypertrabeculation/noncompaction – review of the literature. *Acta Cardiol*. 2004;59(3):341-344.
105. Fazio G, Corrado G, Zachara E, et al. Anticoagulant drugs in noncompaction: a mandatory therapy? *J Cardiovasc Med*. 2008;9(11):1095-1097.
106. Oginosawa Y, Nogami A, Soejima K, et al. Effect of cardiac resynchronization therapy in isolated ventricular noncompaction in adults: follow-up of four cases. *J Cardiovasc Electrophysiol*. 2008;19(9):935-938.
107. Saito K, Ibuki K, Yoshimura N, et al. Successful cardiac resynchronization therapy in a 3-year-old girl with isolated left ventricular non-compaction and narrow QRS complex. *Circ J*. 2009;73(11):2173-2177.
108. Stollberger C, Blazek G, Bucher E, Finsterer J. Cardiac resynchronization therapy in left ventricular hypertrabeculation/non-compaction and myopathy. *Europace*. 2008; 10(1):59-62.
109. Okubo K, Sato Y, Matsumoto N, et al. Cardiac resynchronization and cardioverter defibrillation therapy in a patient with isolated noncompaction of the ventricular myocardium. *Int J Cardiol*. 2008;136(3):e66-68.
110. Kubota S, Nogami A, Sugiyasu A, Kasuya K. Cardiac resynchronization therapy in a patient with isolated noncompaction of the left ventricle and a narrow QRS complex. *Heart Rhythm*. 2006;3(5):619-620.
111. Andrews RE, Fenton MJ, Ridout DA, Burch M. New-onset heart failure due to heart muscle disease in childhood: a prospective study in the United kingdom and Ireland. *Circulation*. 2008;117(1):79-84.
112. Barbosa ND, Azeka E, Aiello VD, et al. Isolated left ventricular noncompaction: unusual cause of decompensated heart failure and indication of heart transplantation in the early infancy–case report and literature review. *Clinics*. 2008;63(1):136-139.
113. Fazio G, Pipitone S, Iacona MA, et al. The noncompaction of the left ventricular myocardium: our paediatric experience. *J Cardiovasc Med*. 2007;8(11):904-908.
114. Kovacevic-Preradovic T, Jenni R, Oechslin EN, Noll G, Seifert B, Attenhofer Jost CH. Isolated left ventricular noncompaction as a cause for heart failure and heart transplantation: a single center experience. *Cardiology*. 2009;112(2): 158-164.
115. Spieker T, Krasemann T, Hoffmeier A, et al. Heart transplantation for isolated noncompaction of the left ventricle in an infant. *Thorac Cardiovasc Surg*. 2007;55(2): 127-129.
116. Stamou SC, Lefrak EA, Athari FC, Burton NA, Massimiano PS. Heart transplantation in a patient with isolated noncompaction of the left ventricular myocardium. *Annals Thora Surg*. 2004;77(5):1806-1808.

117. Tigen K, Karaahmet T, Kahveci G, Mutlu B, Basaran Y. Left ventricular noncompaction: case of a heart transplant. *Eur J Echocardiogr.* 2008;9(1):126-129.
118. Shimamoto T, Marui A, Yamanaka K, et al. Left ventricular restoration surgery for isolated left ventricular noncompaction: report of the first successful case. *J Thorac Cardiovasc Surg.* 2007;134(1):246-247.
119. Cavusoglu Y, Ata N, Timuralp B, et al. Noncompaction of the ventricular myocardium: report of two cases with bicuspid aortic valve demonstrating poor prognosis and with prominent right ventricular involvement. *Echocardiography.* 2003;20(4):379-383.
120. Ichida F, Hamamichi Y, Miyawaki T, et al. Clinical features of isolated noncompaction of the ventricular myocardium: long-term clinical course, hemodynamic properties, and genetic background. *J Am Coll Cardiol.* 1999;34(1):233-240.
121. Neudorf UE, Hussein A, Trowitzsch E, Schmaltz AA. Clinical features of isolated noncompaction of the myocardium in children. *Cardiol Young.* Jul 2001;11(4):439-442.
122. Oechslin E, Jenni R. Isolated left ventricular non-compaction: increasing recognition of this distinct, yet "unclassified" cardiomyopathy. *Eur J Echocardiogr.* 2002;3(4):250-251.
123. Rigopoulos A, Rizos IK, Aggeli C, et al. Isolated left ventricular noncompaction: an unclassified cardiomyopathy with severe prognosis in adults. *Cardiology.* 2002;98(1-2):25-32.
124. Weiford BC, Subbarao VD, Mulhern KM. Noncompaction of the ventricular myocardium. *Circulation.* 2004;109(24):2965-2971.
125. Zambrano E, Marshalko SJ, Jaffe CC, Hui P. Isolated non-compaction of the ventricular myocardium: clinical and molecular aspects of a rare cardiomyopathy. *Lab Invest.* 2002;82(2):117-122.
126. Kitao K, Ohara N, Funakoshi T, et al. Noncompaction of the left ventricular myocardium diagnosed in pregnant woman and neonate. *J Perinat Med.* 2004;32(6):527-531.
127. Menon SC, O'Leary PW, Wright GB, Rios R, MacLellan-Tobert SG, Cabalka AK. Fetal and neonatal presentation of noncompacted ventricular myocardium: expanding the clinical spectrum. *J Am Soc Echocardiogr.* 2007;20(12):1344-1350.
128. Caliskan K, Hoedemaekers YM, ten Cate FJ, Theuns DJAM, Majoor-Krakauer DF, Balk AHMM, Jordaens L, Szili-Torok T. Sudden cardiac death as a first sign of noncompaction cardiomyopathy. *ESC Scientific Sessions.* Barcelona, Spain; 2009.
129. Caliskan K, Theuns DJAM, Hoedemaekers YM, ten Cate FJ, Jordaens L, Szili Torok T. Implantable cardioverter-defibrillators for primary and secondary prevention in patients with noncompaction cardiomyopathy. *J Am Coll Card.* 2009;53(10 (suppl 1)):A136.
130. Kobza R, Jenni R, Erne P, Oechslin E, Duru F. Implantable cardioverter-defibrillators in patients with left ventricular noncompaction. *Pacing Clin Electrophysiol.* 2008;31(4):461-467.
131. Espinola-Zavaleta N, Soto ME, Castellanos LM, Jativa-Chavez S, Keirns C. Non-compacted cardiomyopathy: clinical-echocardiographic study. *Cardiovasc Ultrasound.* 2006;4:35.
132. Lofiego C, Biagini E, Pasquale F, et al. Wide spectrum of presentation and variable outcomes of isolated left ventricular non-compaction. *Heart.* 2007;93(1):65-71.
133. Murphy RT, Thaman R, Blanes JG, et al. Natural history and familial characteristics of isolated left ventricular non-compaction. *Eur Heart J.* 2005;26(2):187-192.
134. Xing Y, Ichida F, Matsuoka T, et al. Genetic analysis in patients with left ventricular noncompaction and evidence for genetic heterogeneity. *Mol Genet Metab.* 2006;88(1):71-77.
135. Ogawa K, Nakamura Y, Terano K, Ando T, Hishitani T, Hoshino K. Isolated non-compaction of the ventricular myocardium associated with long QT syndrome. *Circ J.* 2009;73(11):2169-2172.
136. Marziliano N, Mannarino S, Nespoli L, et al. Barth syndrome associated with compound hemizygosity and heterozygosity of the TAZ and LDB3 genes. *Am J Med Genet.* 2007;143(9):907-915.
137. Shan L, Makita N, Makita N, et al. SCN5A variants in Japanese patients with left ventricular noncompaction and arrhythmia. *Mol Genet Metab.* 2008;93(4):468-474.
138. Bione S, D'Adamo P, Maestrini E, Gedeon AK, Bolhuis PA, Toniolo D. A novel X-linked gene, G4.5. is responsible for Barth syndrome. *Nat Genet.* 1996;12(4):385-389.
139. Bleyl SB, Mumford BR, Brown-Harrison MC, et al. Xq28-linked noncompaction of the left ventricular myocardium: prenatal diagnosis and pathologic analysis of affected individuals. *Am J Med Genet.* 1997;72(3):257-265.
140. Bleyl SB, Mumford BR, Thompson V, et al. Neonatal, lethal noncompaction of the left ventricular myocardium is allelic with Barth syndrome. *Am J Hum Genet.* 1997;61(4):868-872.
141. Brady AN, Shehata BM, Fernhoff PM. X-linked fetal cardiomyopathy caused by a novel mutation in the TAZ gene. *Prenat Diagn.* 2006;26(5):462-465.
142. Chen R, Tsuji T, Ichida F, et al. Mutation analysis of the G4.5 gene in patients with isolated left ventricular noncompaction. *Mol Genet Metab.* 2002;77(4):319-325.
143. Cortez-Dias N, Varela MG, Sargento L, et al. Left ventricular non-compaction: a new mutation predisposing to reverse remodeling? *Rev Port Cardiol.* 2009;28(2):185-194.
144. D'Adamo P, Fassone L, Gedeon A, et al. The X-linked gene G4.5 is responsible for different infantile dilated cardiomyopathies. *Am J Hum Genet.* 1997;61(4):862-867.
145. Yen TY, Hwu WL, Chien YH, et al. Acute metabolic decompensation and sudden death in Barth syndrome: report of a family and a literature review. *Eur J Pediatr.* 2008;167(8):941-944.
146. Ozkutlu S, Ayabakan C, Celiker A, Elshershari H. Noncompaction of ventricular myocardium: a study of twelve patients. *J Am Soc Echocardiogr.* 2002;15(12):1523-1528.
147. Tunaoglu FS, Kula S, Olgunturk R, Ozturk G. Noncompaction with arcus aorta anomalies. *Turk J Pediatr.* 2003;45(4):363-366.
148. Niwa K, Ikeda F, Miyamoto H, Nakajima H, Ando M. Absent aortic valve with normally related great arteries. *Heart Vessels.* 1987;3(2):104-107.
149. Ali SK. Unique features of non-compaction of the ventricular myocardium in Arab and African patients. *Cardiovas J Afr.* 2008;19(5):241-245.

150. Vijayalakshmi IB, Chitra N, Prabhu Deva AN. Use of an Amplatzer duct occluder for closing an aortico-left ventricular tunnel in a case of noncompaction of the left ventricle. *Pediatr Cardiol*. 2004;25(1):77-79.
151. Tatu-Chitoiu A, Bradisteanu S. A rare case of biventricular non-compaction associated with ventricular septal defect and descendent aortic stenosis in an young man. *Eur J Echocardiogr*. 2006;9(2):306-308.
152. Song ZZ. A combination of right ventricular hypertrabeculation/noncompaction and atrial septal defect. *Int J Cardiol*. 2009. Epub ahead of print.
153. Wessels MW, De Graaf BM, Cohen-Overbeek TE, et al. A new syndrome with noncompaction cardiomyopathy, bradycardia, pulmonary stenosis, atrial septal defect and heterotaxy with suggestive linkage to chromosome 6p. *Hum Genet*. 2008;122(6):595-603.
154. Salazar Gonzalez JJ, Rite Montanes S, Asso Abadia A, Pueo Crespo E, Salazar Gonzalez E, Placer Peralta LJ. Isolated non-compaction of the ventricular myocardium. *An Esp Pediatr*. 2002;57(6):570-573.
155. Cavusoglu Y, Aslan R, Birdane A, Ozbabalik D, Ata N. Noncompaction of the ventricular myocardium with bicuspid aortic valve. *Anadolu Kardiyol Derg*. 2007;7(1):88-90.
156. Cavusoglu Y, Tunerir B, Birdane A, et al. Transesophageal echocardiographic diagnosis of ventricular noncompaction associated with an atrial septal aneurysm in a patient with dilated cardiomyopathy of unknown etiology. *Can J Cardiol*. 2005;21(8):705-707.
157. Unlu M, Ozeke O, Kara M, Yesillik S. Ruptured sinus of Valsalva aneurysm associated with noncompaction of the ventricular myocardium. *Eur J Echocardiogr*. 2008; 9(2):311-313.
158. Dagdeviren B, Eren M, Oguz E. Noncompaction of ventricular myocardium, complete atrioventricular block and minor congenital heart abnormalities: case report of an unusual coexistence. *Acta Cardiol*. 2002;57(3):221-224.
159. Friedman MA, Wiseman S, Haramati L, Gordon GM, Spevack DM. Noncompaction of the left ventricle in a patient with dextroversion. *Eur J Echocardiogr*. 2007; 8(1):70-73.
160. Gorgulu S, Celik S, Eksik A, Tezel T. Double-orifice mitral valve associated with nonisolated left ventricular noncompaction – a case report. *Angiology*. 2004;55(6):707-710.
161. Sugiyama H, Hoshiai M, Toda T, Nakazawa S. Double-orifice mitral valve associated with noncompaction of left ventricular myocardium. *Pediatr Cardiol*. 2006;27(6):746-749.
162. Wang XX, Song ZZ. A combination of left ventricular noncompaction and double orifice mitral valve. *Cardiovasc Ultrasound*. 2009;7:11.
163. Arslan S, Gurlertop HY, Gundogdu F, Senocak H. Left ventricular noncompaction and mid-caviter narrowing associated with Ebstein's anomaly: three-dimensional transthoracic echocardiographic image. *Int J Cardiol*. 2007; 115(1):e52-55.
164. Attenhofer Jost CH, Connolly HM, Warnes CA, et al. Noncompacted myocardium in Ebstein's anomaly: initial description in three patients. *J Am Soc Echocardiogr*. 2004;17(6):677-680.
165. Bagur RH, Lederlin M, Montaudon M, et al. Images in cardiovascular medicine Ebstein anomaly associated with left ventricular noncompaction. *Circulation*. 2008;118(16): e662-664.
166. Betrian Blasco P, Gallardo Agromayor E. Ebstein's anomaly and left ventricular noncompaction association. *Int J Cardiol*. 2007;119(2):264-265.
167. Ilercil A, Barack J, Malone MA, Barold SS, Herweg B. Association of noncompaction of left ventricular myocardium with Ebstein's anomaly. *Echocardiography*. 2006; 23(5):432-433.
168. Sinkovec M, Kozelj M, Podnar T. Familial biventricular myocardial noncompaction associated with Ebstein's malformation. *Int J Cardiol*. 2005;102(2):297-302.
169. Friedberg MK, Ursell PC, Silverman NH. Isomerism of the left atrial appendage associated with ventricular noncompaction. *Am J Cardiol*. 2005;96(7):985-990.
170. Vanpraagh R, Ongley PA, Swan HJ. Anatomic types of single or common ventricle in man. Morphologic and geometric aspects of 60 necropsied cases. *Am J Cardiol*. 1964; 13:367-386.
171. Dogan R, Dogan OF, Oc M, Duman U, Ozkutlu S, Celiker A. Noncompaction of ventricular myocardium in a patient with congenitally corrected transposition of the great arteries treated surgically: case report. *Heart Surg Forum*. 2005;8(2):E110-113.
172. Finsterer J, Schoser B, Stollberger C. Myoadenylatedeaminase gene mutation associated with left ventricular hypertrabeculation/non-compaction. *Acta Cardiol*. 2004; 59(4):453-456.
173. Stollberger C, Finsterer J, Blazek G, Bittner RE. Left ventricular non-compaction in a patient with becker's muscular dystrophy. *Heart*. 1996;76(4):380.
174. Finsterer J, Stollberger C, Gaismayer K, Janssen B. Acquired noncompaction in Duchenne muscular dystrophy. *Int J Cardiol*. 2006;106(3):420-421.
175. Corrado G, Checcarelli N, Santarone M, Stollberger C, Finsterer J. Left ventricular hypertrabeculation/noncompaction with PMP22 duplication-based charcot-marie-tooth disease type 1A. *Cardiology*. 2006;105(3):142-145.
176. Finsterer J, Stolberger C, Kopsa W. Noncompaction in myotonic dystrophy type 1 on cardiac MRI. *Cardiology*. 2005;103(3):167-168.
177. Sa MI, Cabral S, Costa PD, et al. Cardiac involveent in type 1 myotonic dystrophy. *Rev Port Cardiol*. 2007;26(9):829-840.
178. Wahbi K, Meune C, Bassez G, et al. Left ventricular noncompaction in a patient with myotonic dystrophy type 2. *Neuromuscul Disord*. 2008;18(4):331-333.
179. Hussein A, Schmaltz AA, Trowitzsch E. Isolated abnormality ("noncompaction") of the myocardium in 3 children. *Klin Padiatr*. 1999;211(3):175-178.
180. Davili Z, Johar S, Hughes C, Kveselis D, Hoo J. Succinate dehydrogenase deficiency associated with dilated cardiomyopathy and ventricular noncompaction. *Eur J Pediatr*. 2006;166(8):867-870.
181. Stollberger C, Finsterer J. Noncompaction in Melnick Fraser syndrome. *Pacing Clin Electrophysiol*. 2007; 30(8):1047. author reply 1048.
182. Matsumoto T, Watanabe A, Migita M, et al. Transient cardiomyopathy in a patient with congenital contractural arachnodactyly (Beals syndrome). *J Nippon Med Sch*. 2006;73(5):285-288.

183. Limongelli G, Pacileo G, Marino B, et al. Prevalence and clinical significance of cardiovascular abnormalities in patients with the LEOPARD syndrome. *Am J Cardiol.* 2007;100(4):736-741.
184. Wong JA, Bofinger MK. Noncompaction of the ventricular myocardium in Melnick-Needles syndrome. *Am J Med Genet.* 1997;71(1):72-75.
185. Finsterer J, Stollberger C, Kopsa W. Noncompaction on cardiac MRI in a patient with nail-patella syndrome and mitochondriopathy. *Cardiology.* 2003;100(1):48-49.
186. Finsterer J, Stollberger C, Steger C, Cozzarini W. Complete heart block associated with noncompaction, nail-patella syndrome, and mitochondrial myopathy. *J Electrocardiol.* 2007;40(4):352-354.
187. Amann G, Sherman FS. Myocardial dysgenesis with persistent sinusoids in a neonate with Noonan's phenotype. *Pediatr Pathol.* 1992;12(1):83-92.
188. Mandel K, Grunebaum E, Benson L. Noncompaction of the myocardium associated with Roifman syndrome. *Cardiol Young.* 2001;11(2):240-243.
189. Happle R, Daniels O, Koopman RJ. MIDAS syndrome (microphthalmia, dermal aplasia, and sclerocornea): an X-linked phenotype distinct from Goltz syndrome. *Am J Med Genet.* 1993;47(5):710-713.
190. Kherbaoui-Redouani L, Eschard C, Bednarek N, Morville P. Cutaneous aplasia, non compaction of the left ventricle and severe cardiac arrhythmia: a new case of MLS syndrome (microphtalmia with linear skin defects). *Arch Pediatr.* 2003;10(3):224-226.
191. Battaglia A, Hoyme HE, Dallapiccola B, et al. Further delineation of deletion 1p36 syndrome in 60 patients: a recognizable phenotype and common cause of developmental delay and mental retardation. *Pediatrics.* 2008;121(2):404-410.
192. Cremer K, Ludecke HJ, Ruhr F, Wieczorek D. Left-ventricular non-compaction (LVNC): a clinical feature more often observed in terminal deletion 1p36 than previously expected. *Eur J Med Genet.* 2008;51(6):685-688.
193. Saito S, Kawamura R, Kosho T, et al. Bilateral perisylvian polymicrogyria, periventricular nodular heterotopia, and left ventricular noncompaction in a girl with 10.5–11.1 Mb terminal deletion of 1p36. *Am J Med Genet.* 2008; 146A(22):2891-2897.
194. Thienpont B, Mertens L, Buyse G, Vermeesch JR, Devriendt K. Left-ventricular non-compaction in a patient with monosomy 1p36. *Eur J Med Genet.* 2007;50(3):233-236.
195. Kanemoto N, Horigome H, Nakayama J, et al. Interstitial 1q43-q43 deletion with left ventricular noncompaction myocardium. *Eur J Med Genet.* 2006;49(3):247-253.
196. Pauli RM, Scheib-Wixted S, Cripe L, Izumo S, Sekhon GS. Ventricular noncompaction and distal chromosome 5q deletion. *Am J Med Genet.* 1999;85(4):419-423.
197. De Rosa G, Pardeo M, Bria S, et al. Isolated myocardial non-compaction in an infant with distal 4q trisomy and distal 1q monosomy. *Eur J Pediatr.* 2005;164(4):255-256.
198. McMahon CJ, Chang AC, Pignatelli RH, et al. Left ventricular noncompaction cardiomyopathy in association with trisomy 13. *Pediatr Cardiol.* 2005;26(4):477-479.
199. Wang JC, Dang L, Mondal TK, Khan A. Prenatally diagnosed mosaic trisomy 22 in a fetus with left ventricular non-compaction cardiomyopathy. *Am J Med Genet.* 2007; 143A(22):2744-2746.
200. Altenberger H, Stollberger C, Finsterer J. Isolated left ventricular hypertrabeculation/noncompaction in a Turner mosaic with male phenotype. *Acta Cardiol.* 2009;64(1):99-103.
201. van Heerde M, Hruda J, Hazekamp MG. Severe pulmonary hypertension secondary to a parachute-like mitral valve, with the left superior caval vein draining into the coronary sinus, in a girl with Turner's syndrome. *Cardiol Young.* 2003;13(4):364-366.
202. Sasse-Klaassen S, Probst S, Gerull B, et al. Novel gene locus for autosomal dominant left ventricular noncompaction maps to chromosome 11p15. *Circulation.* 2004; 109(22):2720-2723.
203. Sajeev CG, Francis J, Shanker V, Vasudev B, Abdul Khader S, Venugopal K. Young male with isolated noncompaction of the ventricular myocardium presenting with atrial fibrillation and complete heart block. *Int J Cardiol.* 2006; 107(1):142-143.
204. Enriquez SG, Entem FR, Cobo M, Olalla JJ. Uncommon etiology of syncope in a patient with isolated ventricular noncompaction. *Pacing Clin Electrophysiol.* 2007;30(4): 577-579.
205. Celiker A, Kafali G, Dogan R. Cardioverter defibrillator implantation in a child with isolated noncompaction of the ventricular myocardium and ventricular fibrillation. *Pacing Clin Electrophysiol.* 2004;27(1):104-108.
206. Taniguchi M, Hioka T, Maekawa K, Takagagi K, Shoji K, Yoshida K. Adult case of isolated ventricular noncompaction discovered by complete atrioventricular block. *Circ J.* 2004;68(9):873-875.
207. Zhou Y, Zhang P, Zhou Q, Guo J, Xu Y, Li X. Giant P waves and focal atrial tachycardia in a patient with ventricular noncompaction. *Int J Cardiol.* 2008;123(2):210-212.
208. El Menyar AA, Gendi SM. Persistent atrial standstill in noncompaction cardiomyopathy. *Pediatr Cardiol.* 2006; 27(3):364-366.
209. Ozkutlu S, Onderoglu L, Karagoz T, Celiker A, Sahiner UM. Isolated noncompaction of left ventricular myocardium with fetal sustained bradycardia due to sick sinus syndrome. *Turk J Pediatr.* 2006;48(4):383-386.
210. Celiker A, Ozkutlu S, Dilber E, Karagoz T. Rhythm abnormalities in children with isolated ventricular noncompaction. *Pacing Clin Electrophysiol.* 2005;28(11):1198-1202.
211. Fazio G, Corrado G, Pizzuto C, et al. Supraventricular arrhythmias in noncompaction of left ventricle: is this a frequent complication? *Int J Cardiol.* 2008;127(2):255-256.
212. Sato Y, Matsumoto N, Takahashi H, et al. Cardioverter defibrillator implantation in an adult with isolated noncompaction of the ventricular myocardium. *Int J Cardiol.* 2006; 110(3):417-419.
213. Fichet J, Legras A, Bernard A, Babuty D. Aborted sudden cardiac death revealing isolated noncompaction of the left ventricle in a patient with wolff-Parkinson-white syndrome. *Pacing Clin Electrophysiol.* 2007;30(3):444-447.

Mitochondrial Cardiomyopathy

N. de Jonge and J.H. Kirkels

7.1 Introduction

Mitochondria are the major sites of energy production in the cell as they harbor the process of *oxidative phosphorylation* (OXPHOS). OXPHOS is performed by proteins at the mitochondrial respiratory chain, comprising complexes I–IV and adenosine triphosphate (ATP) synthase (complex V). As the heart is an energy-dependent tissue, mitochondria constitute 20–40% of the cellular volume of cardiomyocytes. The mitochondrial energy production is under the genetic control of both nuclear and mitochondrial genes. Mutations within these genes may cause defects in oxidative phosphorylation and have severe consequences for those organs which are heavily dependent on energy production like the heart, the brain, and skeletal muscle. Because myopathy is often one of the main presenting symptoms, patients with mitochondrial diseases tend to be seen primarily by neurologists and pediatricians. However, the importance of mitochondrial disease in cardiology is being more and more recognized, not only because cardiomyopathy may be the only manifestation of mitochondrial disease (Fig. 7.1).

Mitochondrial DNA (mtDNA) is a circular double-stranded genome of 16.5 kilobases, encoding 13 polypeptides of the respiratory chain subunits, 28 ribosomal RNAs, and 22 transfer RNAs (tRNAs). All these mitochondrial gene products are used in the mitochondrion for energy production, but apart from these, many other components of the respiratory chain and regulatory mitochondrial proteins are coded by nuclear genes.

Mitochondrial myopathies can thus be caused both by mutations in mtDNA as well as nuclear DNA. Therefore, different modes of inheritance may be observed in mitochondrial disease, as mtDNA is exclusively maternally inherited, while nuclear DNA follows Mendelian inheritance. The maternal inheritance of mtDNA is due to the fact that the mammalian egg contains about 100,000 mitochondria and mtDNA, whereas the sperm contains only in the order of 100 mtDNA.[1]

Mammalian mtDNA has a very high mutation rate in comparison to nuclear DNA. Each cell contains hundreds to thousands of mitochondria and each mitochondrion contains several copies of mtDNA. Mutations in mtDNA therefore result in *heteroplasmy*: the presence of two or more different genomes (with and without a mutation) in one cell, the proportion of which may change over time as the mitochondria multiply and are randomly distributed over daughter cells during cell division. Due to this process the proportion of mutant mtDNA varies considerably between organ systems and even within a specific tissue, resulting in different phenotypes and marked variability in severity and symptom patterns. The heart, central nervous system, and the skeletal muscles are particularly vulnerable to defects in energy metabolism, and therefore are often involved in mitochondrial disease.

Phenotype–genotype correlation in mitochondrial disease is complex: patients with the same clinical syndrome do not always show the same mutation in the mtDNA and, conversely, a single mutation can be associated with different clinical syndromes.[2]

Many mutations in mtDNA may lead to *cardiomyopathy,* mostly *hypertrophic*, but *dilating cardiomyopathy* and *left ventricular noncompaction* are also possible.[3] A list of known mutations leading to cardiomyopathy is reviewed elsewhere and shown in Table 7.1.[4] Besides mutations in the mtDNA, many

---

N. de Jonge (✉)
Department of cardiology, University Medical Centre, Utrecht, The Netherlands
e-mail: n.dejonge@umcutrecht.nl

Fig. 7.1 A cardiomyocyte demonstrating the high numbers of mitochondria in between the contractile filaments

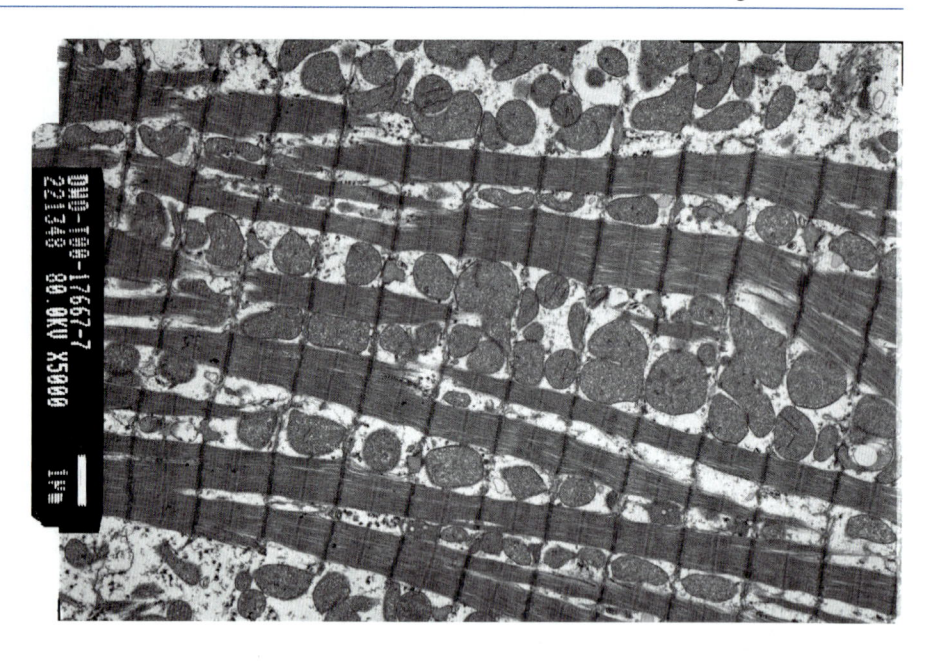

Table 7.1 Specific mitochondrial DNA (mtDNA) point mutations in cardiac disease (Adapted from Marin-Garcia[4])

Gene	Site	Cardiac phenotype
tRNA mutations		
Leu	3243 A->G	DCM
Leu	3260 A->G	Tachycardia, adult onset
Leu	3303 C->T	Fatal infantile CM
Leu	3254 C->G	HCM
Leu	12997 T->C	DCM
Ile	4300 A->G	HCM, adult onset
Ile	4317 A->G	Fatal infantile CM
Ile	4320 C->T	Fatal infantile CM
Ile	4269 A->G	CF at 18 year, adult onset
Ile	4295 A->G	HCM
Ile	4284 G->A	CM
Lys	8363 G->A	HCM
Lys	8334 A->G	HCM
Lys	8269 A->G	HCM
Lys	8348 A->G	HCM
Gly	9997 T->C	Ventricular arrhythmia, HCM
Cys	5814 A->G	HCM
Ala	5587 T->C	DCM
Arg	10415 T->C	DCM
Arg	10424 T->C	Fatal DCM
rRNA mutations		
12s	1555 A->G	CM
16s	3093 C->G	CM

Table 7.1 (continued)

Gene	Site	Cardiac phenotype
Structural gene mutations		
Cytb	14927 A->G	HCM
Cytb	15236 A->G	DCM
Cytb	15508 C->G	DCM
Cytb	15509 A->C	Fatal postpartum CM
Cytb	15498 G->A	Histiocytoid CM
COI	6860 A->C	DCM
COII	7923 A->G	DCM
COIII	9216 A->G	DCM
ND5	14069 C->T	DCM
ATPase6	8993 T->G	Leigh syndrome/HCM

DCM dilated cardiomyopathy, *HCM* hypertrophic cardiomyopathy, *CF* cardiac failure

mutations in nuclear genes encoding mitochondrial proteins may also cause cardiomyopathy. Some examples include mutations in the mitochondrial transport protein *frataxin* leading to *Friedreich's ataxia* with hypertrophic cardiomyopathy and mutations in the gene encoding the protein *tafazzin*, resulting in *Barth syndrome*, an X-linked neonatal disorder characterized by dilating cardiomyopathy, cyclic neutropenia, and skeletal myopathy.

The best known cardiac manifestations of mitochondrial disease are mentioned in the Table 7.2.

Table 7.2 Cardiac manifestations of mitochondrial disease

Hypertrophic (non-obstructive) cardiomyopathy
Dilated cardiomyopathy
Left ventricular noncompaction
Left ventricular hypertrophy
WPW-syndrome
Long QT-syndrome
Ventricular tachycardia
Left anterior hemiblock
Right bundle branch block
Total AV block
Mitral valve prolapse

Many mitochondrial disorders become manifest in the first years of life. The frequency of cardiomyopathy in mitochondrial disease has been reported to be from 17% to 40% and the incidence of mitochondrial cardiomyopathy in children and young adults is estimated to be at least 1/50,000.[3,5] Children with mitochondrial cardiomyopathy generally have an earlier onset, more severe morbidity, and increased mortality compared with children who have mitochondrial disorders without cardiac involvement.[6] One study showed that of the patients with cardiomyopathy 71% died or underwent heart transplantation, in contrast to 26% in patients with mitochondrial disease without cardiomyopathy.[5]

As mentioned before, cardiac involvement in mitochondrial disease is usually part of multisystem manifestations of the disorders in oxidative phosphorylation. It is important to realize, however, that isolated cardiac pathology may be the presenting symptom in mitochondrial disease. In one study, approximately 10% of patients presented with symptoms of cardiac involvement, defined as the presence of symptoms of heart failure, or abnormalities on echo, ECG, or chest x-ray.[7] Recently, a novel mutation in mtDNA *(m.8528 T>C)* was described in four young patients presenting with an isolated hypertrophic cardiomyopathy, further underlining OXPHOS defects as a potential cause of isolated cardiomyopathy.[6]

In this Chapter, two syndromes will be described in more detail: *MELAS* and *Kearns–Sayre syndrome*.

7.2 MELAS Syndrome

This is a multisystem clinical syndrome manifested by mitochondrial *encephalomyopathy, lactic acidosis,* and *recurrent stroke-like episodes.*[8] The most commonly described gene mutation causing MELAS syndrome is a mitochondrial adenine-to-guanine transition at nucleotide pair 3243 *(m.3243A>G)* encoding the mitochondrial tRNA$^{(Leu)}$.[9] At least 29 other specific point mutations have been associated with the MELAS syndrome.[10] These mutations lead to impaired oxidative phosphorylation, resulting in the inability of the mitochondria to produce sufficient ATP to meet the energy needs of the cell. This causes a shift to lactate production, which can be systemically noticed as lactate acidosis.

Due to the variability in severity and symptoms and the problems confirming the diagnosis, the incidence of MELAS syndrome is difficult to assess. It is estimated to be as common as neuromuscular diseases like Duchenne muscular dystrophy (frequency 18 per 100,000).[10]

The clinical features of MELAS syndrome vary widely, but almost all include stroke-like episodes before 40 years of age, encephalopathy characterized by seizures, dementia, or both, and lactic acidosis. Although age at onset may be high in some patients, most patients, however, present with initial symptoms between 2 and 20 years of age.[8] Other symptoms related to MELAS syndrome are hearing loss, migraine headaches, peripheral neuropathy, depression, learning disabilities, growth failure, diabetes mellitus, gastrointestinal symptoms, renal involvement, and myopathy.

Cardiac involvement in MELAS syndrome is reported to be as high as 18–100%.[11-13] The most common pathology is *non-obstructive concentric hypertrophy*, although dilatation is also reported and might be seen as progression of the initial hypertrophic cardiomyopathy.[14] In children, cardiomyopathy may actually be the first manifestation of MELAS syndrome. Wolff—Parkinson–White (WPW) syndrome has also been reported in MELAS syndrome in up to 17% of patients.[11,15]

The clinical suspicion for mitochondrial disease is based on the combination of symptoms related to different organ systems. On the other hand, especially in young children, the presence of a cardiomyopathy may be the only manifestation of a mitochondrial disorder.

Laboratory examination will show lactic acidosis in almost all patients. MRI of the brain in MELAS syndrome will typically show asymmetric lesions of the occipital and parietal lobes, mimicking ischemia, although not restricted to one specific vascular region.

ECGs may demonstrate aspecific abnormalities suggestive of cardiomyopathy, like left ventricular hypertrophy, negative T-waves in the precordial leads, a left-oriented electrical axis, and prolonged QT_c.[5] Echocardiographic examination is mandatory in demonstrating cardiac involvement in mitochondrial disease. Next to left ventricular hypertrophy, diastolic and systolic dysfunction may be present.

Muscular biopsy in most patients will show *ragged red fibers*: deposits of mitochondrial material beneath the sarcolemma, visualized by Gomori trichrome staining or succinate dehydrogenase.[10,12]

Ultra structural analysis of the heart demonstrates abnormal and markedly enlarged mitochondria.

Molecular diagnosis of mtDNA mutations is complicated by the variability in heteroplasmy depending on the specific tissue sampled. A detectable mutation in muscle cells is not necessarily detectable in leucocytes, cells regularly used for DNA analysis. Urine sediment cells and cheek mucosa appear to be a better alternative for DNA analysis.[16]

No specific treatments are available for mitochondrial cardiomyopathies, although there are some suggestions that the use of l-*arginine* and *coenzyme Q10* in addition to vitamin supplementation might be advantageous.[10] As in other cardiomyopathies, regular heart failure therapy is indicated, consisting of diuretics, ACE-inhibitors, and Beta-blockers. In case of refractory heart failure, despite optimal medical therapy, heart transplantation can be considered in selected patients.[17] This requires extensive evaluation of extracardiac involvement, especially with regard to potential contraindications such as recurrent strokes, dementia, and muscle waisting.

Furthermore, heart transplantation and other operations are generally accompanied by a significantly increased perioperative risk, in particular due to stroke, coma, seizures, respiratory failure, and cardiac arrhythmias.[18] *Perioperative management* includes generous hydration, loading with intravenous glucose, and careful control of body temperature and pH. Ringer's solution should be avoided because of the lactate load. *Anesthetic agents* in these patients may increase the susceptibility to reactive oxygen species (ROS) and apoptosis, resulting in neurotoxicity. In general, therefore, an increased sensitivity to anesthetics is noted, requiring adjustment of dosing and careful management, including optimal oxygenation.[18]

In summary, given the high incidence of cardiac involvement, all patients with MELAS syndrome should undergo cardiac examination because this may have therapeutic and prognostic consequences. On the other hand, patients with hypertrophic cardiomyopathy at a younger age should be considered having mitochondrial disease, especially when they also suffer from short stature, seizures, hemiparesis, hemianopsia or cortical blindness. MELAS syndrome is maternally inherited, but genotype–phenotype correlation is complex, which hampers the role of genetic counseling in this syndrome.

7.3 Kearns–Sayre Syndrome

Clinically, this mitochondrial disease is characterized by progressive *external ophthalmoplegia* resulting in ptosis, and *pigmentary retinopathy*. Other manifestations of KSS are short stature, cerebellar signs, hearing loss, mental retardation, vestibular system dysfunction, delayed puberty, and high cerebrospinal fluid protein content. Typical onset is before the age of 20. Progression of the disease can be accompanied by proximal myopathy.[19]

Cardiac pathology consists of conduction defects caused by *infra-His block*, resulting in *total AV-block*, *right bundle branch* block, or *left anterior hemiblock*.[20,21] These conduction defects may be rapidly progressive and result in *acute cardiac death*. Transition of a normal electrocardiogram into total AV-block has been reported within the course of 10 months.[22] Complete heart block may also be the presenting symptom of Kearns–Sayre syndrome in some patients.[23] It seems plausible that early pacemaker implantation improves survival, but criteria for prophylactic implantation are not yet clear. The ACC guidelines state that third degree and advanced second degree AV block associated with neuromuscular disease like Kearns–Sayre, with or without symptoms constitute a class I indication for permanent pacemaker implantation.[24] Some authors suggest that, given the rapid progression to potential fatal complete AV block, the presence of fascicular block in Kearns–Sayre syndrome warrants prophylactic implantation of a pacemaker.[25] In patients with a normal electrocardiogram, regular ECG follow-ups at least every 6 months are advisable.

Pathologic examination of cardiac biopsies especially shows enlarged and abnormally structured mitochondria, but loss of myofibrils in skeletal and cardiac muscle may also be seen.[25]

Macroscopically, mitral valve prolapse may be noted, as well as cardiomyopathy in some cases.[5,26] The incidence of cardiomyopathy in KSS, in future may increase due to the prolonged longevity in patients treated by early pacemaker implantation.[25] Clinical manifestations of cardiac disease in KSS have been reported to occur in 57% of patients affected and include syncopal attacks, cardiac arrest, and congestive heart failure.

In contrast to the MELAS syndrome, which is mostly caused by a point mutation in the mtDNA and maternally inherited, genetic analysis in KSS mainly shows a large *deletion of mtDNA* (nucleotide positions 8483 to 13483).[21] Most cases are sporadic with heteroplasmy of 34–60% and a severely compromised life expectancy: patients rarely survive beyond the age of 30.[19] This high mortality is partly related to sudden cardiac death due to AV-block, which can be prevented in some patients by timely pacemaker implantation.

7.4 Conclusion

The mitochondrial diseases are a heterogeneous group of disorders which can affect virtually all organ systems, not only in infancy, but also during the early-to-mid adult years. Most of the mitochondrial diseases are caused by mutations in the nuclear DNA, of which only a few have been identified thus far. A small percentage is caused by mutations in the mitochondrial DNA.

Mitochondrial diseases should be included in the differential diagnosis whenever a patient presents with progressive multisystem involvement that does not clearly fit with an established pattern of disease. The combination of cardiomyopathy, deafness, diabetes, together with encephalopathy and myopathy are highly susceptible of mitochondrial disease.[19] Especially in very young children hypertrophic cardiomyopathy can be the predominant symptom of mitochondrial disease.

Cardiomyopathy may be the presenting and predominant clinical expression of MELAS syndrome and is one of the causes of death in this disease, underlining the importance of this condition to the cardiologist. The same holds for the progressive conduction disorders in Kearns–Sayre syndrome, which may require pacemaker implantation to prevent sudden death.

Apart from these rather well-delineated disorders, many others exist and the phenotypes frequently overlap complicating things even more.

Given all these facts, genetic counseling in mitochondrial disease is difficult. There is only a very small chance that males with mtDNA mutations will transmit the disease. The risk in females is depending on the level of heteroplasmy, but it remains difficult to give advice in the clinical routine.

Diagnosis of mitochondrial diseases is notoriously difficult and relies on a high level of suspicion, but is important, given the potential management implications, not only with respect to cardiac disease, but also more in general like decreased anesthetic requirement during surgical procedures.

References

1. Wallace DC. Mitochondrial defects in cardiomyopathy and neuromuscular disease. *Am Heart J*. 2000;139:S70-S85.
2. Ashizawa T. What is Kearns-Sayre syndrome after all? *Arch Neurol*. 2001;58:1053-1054.
3. Scaglia F, Towbin JA, Craigen WJ, et al. Clinical spectrum, morbidity, and mortality in 113 pediatric patients with mitochondrial disease. *Pediatrics*. 2004;114:925-931.
4. Garcia J-M, Goldenthal MJ. Understanding the impact of mitochondrial defects in cardiovascular disease: a review. *J Cardiac Failure*. 2002;8:347-361.
5. Holmgren D, Wahlander H, Eriksson BO, Oldfors A, Holme E, Tulinius M. Cardiomyopathy in children with mitochondrial disease. *Eur Heart J*. 2003;24:280-288.
6. Ware SM, El-Hassan N, Kahler SG, et al. Infantile cardiomyopathy caused by a mutation in the overlapping region of mitochondrial ATPase 6 and 8 genes. *J Med Genet*. 2009; 46:308-314.
7. Yaplito-Lee J, Weintraub R, Jamsen K, et al. Cardiac manifestations in oxidative phosphorylation disorders of childhood. *J Pediatr*. 2007;150:407-411.
8. Pavlakis SG, Phillips PC, DiMauro S, De Vivo DC, Rowland LP. Mitochondrial myopathy, encephalopathy, lactic acidosis, and strokelike episodes: a distinctive clinical syndrome. *Ann Neurol*. 1984;16:481-488.
9. Goto Y, Nonaka I, Horai S. A mutation in the tRNA (Leu) (UUR) gene associated with the MELAS subgroup of mitochondrial encephalomyopathies. *Nature*. 1990;348:651-653.
10. Sproule D, Kaufman P. Mitochondrial encephalopathy, lactic acidosis and strokelike episodes. Basic concepts, clinical phenotype, and therapeutic management of MELAS syndrome. *Ann NY Acad Sci*. 2008;1142:133-158.

11. Hirano M, Pavlakis SG. Mitochondrial myopathy, encephalopathy, lactic acidosis, and strokelike episodes (MELAS) current concepts. *J Child Neurol*. 1994;9:4-13.
12. Vydt TCG, de Coo RFM, Soliman OII, ten Cate FJ, van Geuns RJM, et al. Cardiac involvement in adults with m.3242A>G MELAS gene mutation. *Am J Card*. 2007;99: 264-269.
13. Wortmann SB, Rodenburg RJ, Backx AP, Schmitt E, Smeitink JAM, Morava E. Early cardiac involvement in children carrying the A3243G mtDNA mutation. *Acta Paediatrica*. 2007;96:450-451.
14. Okajima Y, Tanabe Y, Takayanagi M, Aotsuka H. A follow up study of myocardial involvement in patients with mitochondrial encephalomyopathy, lactic acidosis and strokelike episodes (MELAS). *Heart*. 1998;80:292-295.
15. Sproule DM, Kaufmann P, Engelstad K, et al. Wolff-Parkinson-White syndrome in patients with MELAS. *Arch Neurol*. 2007;64:1625-1627.
16. Shanske S, Pancrudo J, Kaufmann P, et al. Varying loads of mitochondrial DNA A3243G mutation in different tissues: implications for diagnosis. *Am J Med Genet*. 2004;130A: 134-137.
17. Bhati RS, Sheridan BC, Mill MR, Selzman CH. Heart transplantation for progressive cardiomyopathy as a manifestation of MELAS syndrome. *J Heart Lung Transplant*. 2005; 24:2286-2289.
18. Muravchick S. Clinical implications of mitochondrial disease. *Adv Drug Deliv Rev*. 2008;60:1553-1560.
19. Finsterer J. Mitochondriopathies. *Eur J Neurol*. 2004;11: 163-186.
20. Roberts NK, Perloff JK, Kark RAP. Cardiac conduction in the Kearns-Sayre Syndrome (a neuromuscular disorder associated with progressive external ophthalmoplegia and pigmentary retinopathy). *Am J Cardiol*. 1979;44:1396-1400.
21. Anan R, Nakagawa M, Miyata M, Higuchi I, Nakao S, et al. Cardiac involvement in mitochondrial diseases. A study on 17 patients with documented mitochondrial DNA defects. *Circulation*. 1995;91:955-961.
22. Welzing L, von Kleist-Retzow JC, Kribs A, et al. Rapid development of life-threatening complete atrioventricular block in Kearns-Sayre syndrome. *Eur J Pediatr*. 2009; 168:757-759.
23. Chawla S, Coku J, Forbes T, Kannan S. Kearns-Sayre syndrome presenting as complete heart block. *Pediatr Cardiol*. 2008;29:659-662.
24. Epstein AE, DiMarco JP, Ellenbogen KA, et al. ACC/AHA/HRS 2008 guidelines for device based therapy of cardiac rhythm abnormalities. *J Am Coll Cardiol*. 2008;51:e1-e62.
25. Charles R, Holt S, Kay JM, et al. Myocardial ultrastucture and the development of atrioventricular block in Kearns-Sayre syndrome. *Circulation*. 1981;63:214-219.
26. Channer KS, Channer JL, Campbell MJ, Russel RJ. Cardiomyopaty in the Kearns-Sayre syndrome. *Br Heart J*. 1988;59:486-490.

Restrictive Cardiomyopathy

J.H. Kirkels and N. de Jonge

8.1 Introduction

Restrictive cardiomyopathy (RCM) is a rare disease, characterized by increased stiffness of the ventricular walls, which causes heart failure because of impaired diastolic filling. In the early stages, systolic function may be normal, but when the disease progresses, systolic function usually declines as well. There is an overlap with other types of cardiomyopathy, such as hypertrophic cardiomyopathy (HCM), dilated cardiomyopathy (DCM), and left ventricular noncompaction. Indeed, an autosomal dominantly segregating cardiomyopathy has been described where a single sarcomere gene mutation caused idiopathic RCM in some and HCM in other family members.[1]

According to the most recent AHA classification of cardiomyopathies[2,] RCM is defined by restrictive ventricular physiology associated with normal or reduced diastolic volumes (of one or both ventricles), normal or near-normal systolic function, and normal or only mildly increased ventricular wall thickness. Several studies indicate that RCM is not a single entity; it is a heterogeneous group of disorders that can present with a spectrum of cardiac phenotypes.[3]

Classification of RCM is based on the underlying pathophysiological process: non-infiltrative, infiltrative, storage diseases, and endomyocardial (Table 8.1). Approximately 50% of cases are caused by a specific clinical disorder, the majority in western countries being amyloidosis, whereas the remainder represents an "idiopatic" or "primary" process. Restrictive cardiomyopathy may also be associated with neuromuscular disorders, both congenital and acquired forms.[4] Hypertrophic cardiomyopathy may be particularly difficult to distinguish, since HCM in a late phase may start to dilate and wall thickness may appear normal or even reduced. Conversely, thickening of ventricular walls in cardiac infiltration or storage disease may resemble HCM. RCM must also be clinically distinguished from constrictive pericarditis, which is also characterized by abnormal ventricular filling with (near) normal systolic function.

Hereditary forms of RCM can be found in all subgroups, with both autosomal dominant and recessive genetic properties. Family history and investigation of first-degree relatives may therefore be important.

8.2 Molecular Background

Several inherited and acquired disorders may cause RCM, but many cases remain idiopathic. Familial RCM has been reported but it remains uncertain whether this is a distinct genetic entity.

Restrictive cardiomyopathy may be due to myocardial fibrosis, hypertrophy, or infiltration of varying compounds, like amyloid or storage of, for example, glycogen. The terms hypertrophic and restrictive cardiomyopathy do not refer to specific diseases, but are instead purely descriptive terms used to characterize myocardial disease associated with a broad spectrum of genetic syndromes or systemic diseases.

Cardiomyocyte contraction is dependent on intracellular calcium concentration and regulated by the troponin complex. In vitro studies have shown that RCM-causing mutations in TNNI3 show a greater

J.H. Kirkels (✉)
Department of cardiology, University Medical Centre,
Utrecht, The Netherlands
e-mail: j.h.kirkels@umcutrecht.nl

Table 8.1 Classification of restrictive cardiomyopathy

			Genetic	Primay cardiac presentation	Common primary site or presentation
Myocardial	*Non-infiltrative*				
		Idiopathic restrictive cardiomyopathy	+	+	n.a.
		Scleroderma	±	–	Skin, joints, Raynaud, GI-tract, lungs
		Pseudoxanthoma elasticum	+		Skin, vascular wall (GI-tract)
		Diabetic cardiomyopathy		–	
	Infiltrative				
		Amyloidosis	±(AL)/+(AA)	+	AL: bone marrow, kidneys AA: peripheral neuropathy
		Sarcoidosis		±	Lungs
		Gaucher disease	+	–	Spleen, liver, bone marrow, bone
		Hurler disease	+	–	Bone, liver, spleen, brain
	Storage disease				
		Hemochromatosis	+		Liver, skin pigmentation, diabetes mellitus, arthropathy, impotence in male
		Fabry disease	+	+	Neuropathy, skin, kidney, stroke
		Glycogen storage disease	+	(+)/–	Hypoglycemia, muscle weakness, fatigability
Endomyocardial					
		Endomyocardial fibrosis	?	+	n.a.
		Hypereosinophilic syndrome	?	+	Systemic thromboemboli, neuropathy, GI-tract inflammation, lungs, bone marrow
		Carcinoid heart disease		–	Flushing, diarrhea, bronchospasm
		Metastatic cancers		–	n.a.
		Radiation		(+)[a]	n.a.
		Anthracyclin toxicity		(+)[a]	n.a.
		Fibrous endocarditis caused by drugs (serotonin, methysergide, ergotamine, mercurial agents, busulfan)		+	n.a.

[a]Primary cardiac presentation after treatment of previous malignancy

increase in Ca^{2+} sensitivity than HCM-causing mutations, resulting in more severe diastolic impairment and potentially accounting for the RCM phenotype in humans.[5]

The molecular background of different forms of RCM is highly variable, depending on the underlying cause, and will be discussed in more detail in specific clinical entities.

8.3 Clinical Aspects

Inability of the ventricles to fill limits cardiac output and raises filling pressures, leading to exercise intolerance and dyspnea. In most patients, venous pressure is elevated, which may lead to edema, ascites, and liver enlargement. Palpitations are often seen, with a relatively high occurrence of atrial fibrillation, which in turn may lead to rapid clinical deterioration due to high ventricular rates with short diastolic filling times. Third and fourth heart sounds may be present on physical examination.

8.4 Diagnosis

In typical cases, echocardiography, cardiac CT, or MRI will reveal normal or concentric thickened ventricles with normal or reduced intraventricular volumes. In contrast to hypertrophic cardiomyopathy, macroscopic hypertrophy and reduction of intraventricular volume are not pronounced. The atria are usually enlarged, sometimes exceeding ventricular volume. Systolic function may be normal or slightly reduced; diastolic function is reduced, with high E-wave, shortened deceleration time (<150 ms), and an E/A ratio of >2 on transmitral Doppler echocardiography (Fig. 8.1). Especially in infiltrative cardiomyopathies, the ECG may show low-voltage and nonspecific ST segment or T-wave abnormalities (Fig. 8.2). Cardiac catheterization shows a reduced cardiac output and elevation of left and right ventricular end-diastolic pressures with a dip-plateau representing an abrupt termination of filling in the first one-third to one-half of diastole. This configuration may resemble constrictive pericarditis; however, in constrictive pericarditis there usually is a thickened pericardium, best seen on CT or MRI. In addition, interventricular dependence and respiratory variation of transmitral inflow on Doppler examination will be more pronounced in constrictive pericarditis; in difficult cases, volume challenge and simultaneous LV and RV pressure recording in relation to respiratory activity may be of help. Recently, tissue Doppler imaging was

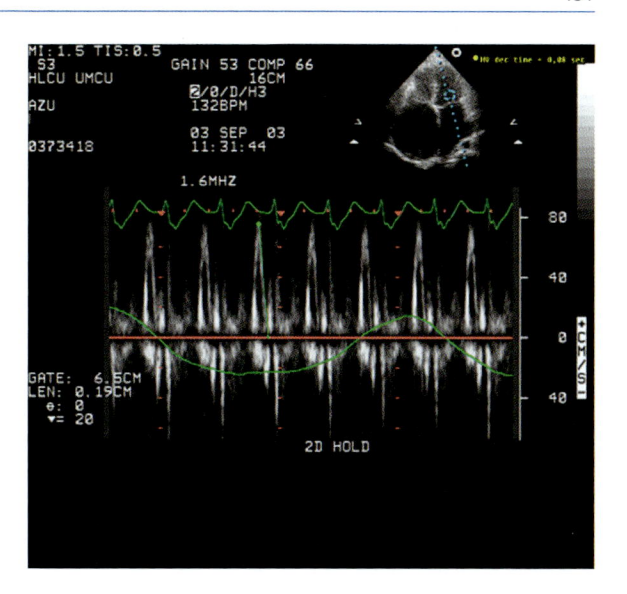

Fig. 8.1 2D-echocardiogram and transmitral Doppler signals in restrictive cardiomyopathy. In the upper part the 2D echocardiogram (apical four-chamber view) is shown, with normal sized ventricles (*top*) and huge atria below. In the main panel the transmitral Doppler recording is shown in relation to the ECG, indicating E-waves with short deceleration time (*green line*) and almost absent A-waves (high E/A ratio)

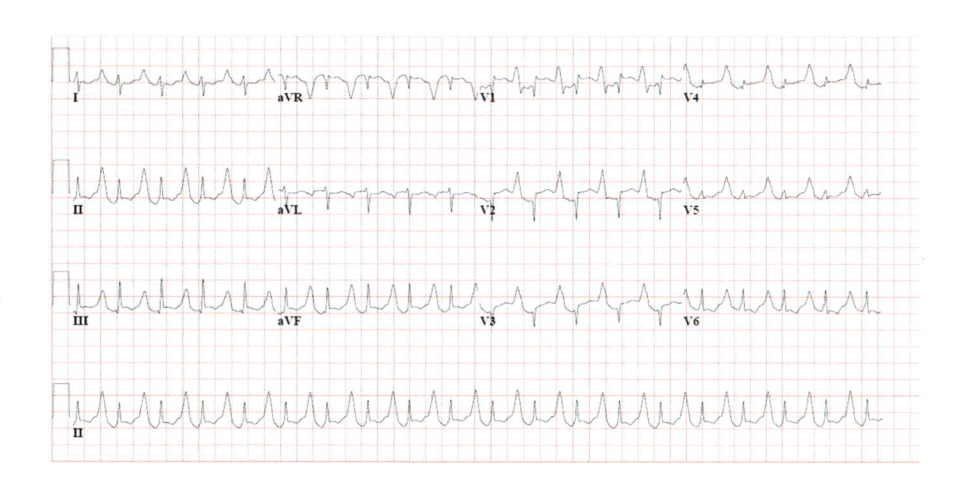

Fig. 8.2 ECG in restrictive cardiomyopathy. Low voltage abnormal QRS-complexes, preceded by huge P-waves in a 16-year-old girl with restrictive cardiomyopathy

shown to reliably discriminate between the two conditions, with a cut-off value of >5 cm/s mean annular velocity (averaged from four walls) ruling out RCM.[6] Surprisingly, BNP values showed a large overlap between the two conditions.

The mainstay of diagnosis is endomyocardial biopsy, revealing fibrosis or the underlying specific infiltration or storage.

8.5 Clinical Approach and Differential Diagnosis

Since RCM often occurs in the setting of a systemic disease, in many cases the primary underlying disease is already known, like in Gaucher disease (GD), where noncardiac manifestations usually precede cardiac involvement. In these cases the clinical question may not be making the right diagnosis, but proving or excluding cardiac involvement. This may have consequences for the work-up; for instance, in case of a patient with known hemochromatosis, it may be best to start with cardiac MRI in order to find cardiac iron overload, whereas in suspected amyloidosis it may be best to start with endomyocardial biopsy.

Clinical history taking and clinical examination should be directed at symptoms indicative of underlying disease.[4] Ophthalmologic, otologic, dermatologic, gastroenterologic, nephrologic, hematologic, and neurologic examination may be necessary to help establishing a possibly treatable cause of restrictive cardiomyopathy before the disease becomes intractable.

In apparently idiopathic restrictive cardiomyopathy, it may be necessary to clinically exclude other causes of restriction, like hypertension, and to exclude the presence of specific infiltration or storage in an endomyocardial biopsy. In addition, taking an extensive family history including other phenotypes of cardiomyopathy and performing a genetic evaluation may be of help.[7]

8.6 Treatment

In many cases, treatment is disappointing since myocardial damage is progressive and irreversible, with a possible exception for hemochromatosis and Fabry's disease (see below). In case of amyloidosis, aggressive anticancer treatment and/or bone marrow transplantation may slow progression of the disease, but this does not remove the already existing deposits of amyloid. In general, there is no specific medication for diastolic heart failure, other than diuretics to treat pulmonary or systemic congestion. The balance between pulmonary congestion due to fluid overload on the one hand, and forward failure due to too low filling pressures on the other hand, often is very delicate. Controlling heart rate with betablockers to allow adequate filling time is important; however, when restriction progresses, ventricular filling may no longer improve with longer diastole. In end-stage disease, a higher heart rate may even be the only way to compensate for a very low stroke volume. Atrial fibrillation occurs very often in RCM as a result of chronically elevated filling pressures and dilated atria, warranting oral anticoagulation to prevent stroke or embolism and adequate rate control when rhythm control is not possible anymore. Like in mitral stenosis and sinus rhythm there is no consensus on the preventive use of anticoagulants in RCM and sinus rhythm. The only exception may be endomyocardial fibrosis (EMB) and hypereosinophylic syndrome, where endocavitary thrombosis and fibrosis with apical filling are thought to occur.

Heart transplantation may be an option in carefully selected cases, but due to the malignant nature or the multiorgan involvement of many underlying diseases, heart transplantation is often contraindicated. In idiopathic RCM heart transplantation may offer good survival.

8.7 Prognosis

Prognosis is very much dependent on the underlying disease. In a study of 94 patients with idiopathic RCM after 68 months 50% had died.[8] The causes of death were: heart failure (47%), sudden death (17%), cancer (13%), infection (13%), and arrhythmias (11%).

8.8 Idiopatic and Familial Restrictive Cardiomyopathy

Idiopatic RCM is characterized by myocyte hypertrophy and interstitial fibrosis, with a restrictive hemodynamic pattern of the ventricles with reduced diastolic volumes, in the presence of normal or near-normal wall thickness and systolic function. By definition, there is no known underlying or related disease to explain cardiac involvement. In childhood, RCM is very rare, accounting for 2–5% of pediatric cardiomyopathies.[3] About 30% of children with RCM have a family history of cardiomyopathy and prognosis is poor (2-year mortality >50%).[3] In their study of 12 children, in one third a mutation in genes coding for sarcomeric proteins were found; in the other two-thirds it was speculated that some might have been caused by mutations in genes encoding cytoskeletal or nuclear envelope proteins, more commonly associated with DCM. Others might have been associated to – as yet unknown – inborn errors of metabolism or storage disorders with predominant cardiac involvement.[3]

Familial RCM is an autosomal dominant cardiomyopathy with incomplete penetrance,[10] generally considered in the absence of specific genetic abnormalities known to cause hypertrophic cardiomyopathy. However, some have suggested that RCM is part of the clinical expression of cardiac troponin I mutations.[3] A bundle branch block leading to complete heart block usually develops in the third or fourth decade.[11] Those who survive the fifth decade may develop a progressive myopathy,[3,9] although there are also reports of families with multiple affected individuals without skeletal myopathy.[11] Mogensen et al.[1] described a large family in which individuals were affected by either idiopathic RCM or hypertrophic cardiomyopathy (HCM). Linkage analysis to selected sarcomeric contractile protein genes identified cardiac troponin I (TNNI3) as the likely disease gene. Several mutations were found, which also appeared to be present in six of nine unrelated RCM patients. They conclude that the restrictive phenotype is part of the spectrum of hereditary sarcomeric contractile protein disease. Changes in actin-binding affinity, affinity to troponin C, and the ability to inhibit thin filaments during diastole, caused by certain TNNI3 mutations, may lead to an altered interaction within the actin–troponin–tropomyosin complex, and thus may cause either severe diastolic dysfunction and RCM, or myocardial hypertrophy.[12]

Genetic engineering of adult cardiac myocytes[13] was used to identify effects of mutant cardiac troponin I (cTnI). The R193H mutant cTnI was associated with incomplete relaxation and acute remodeling to a contracted state as a direct correlate of the stiff heart characteristic of RCM in vivo. This occurred independently of Ca^{2+} concentration or sensitivity. Transgenic mice, expressing R193H cTnI in the heart, showed gradual changes in 12 months from impaired relaxation to diastolic dysfunction and eventually a phenotype similar to human RCM.[14] These results demonstrate a critical role of the COOH-terminal domain of cTnI in the development of RCM. On the other hand, Cubero et al.[11] present a family of RCM patients with autosomal dominant heredity, without signs of skeletal myopathy and no troponin I mutations.

To make it even more complex, a recent study[15] described a unique family with autosomal dominant heart disease variably expressed as RCM, HCM, and dilated cardiomyopathy. They showed that a cardiac troponin T (TNNT2) marker cosegregated with the disease phenotype. A missense mutation resulting in an I79N substitution was found in all nine affected family members, but none of the six unaffected relatives. Segregation analyses excluded a primary pathogenic role for eight other sarcomeric protein genes; however, this does not exclude a potential modifying effect of variants within these or other genes on cardiac phenotype.[15]

8.9 Restrictive Cardiomyopathy as Part of Specific Clinical Conditions with Known or Suspected Genetic Background (Selected Subjects)

8.9.1 Non-infiltrative Restrictive Cardiomyopathy

8.9.1.1 Scleroderma/Systemic sclerosis (SSc)

Apart from cardiac complications due to systemic or pulmonary hypertension, primary cardiac involvement can also occur in SSc. Patchy myocardial fibrosis as a result of recurrent vasospasm of small vessels may lead to the clinical picture of restrictive cardiomyopathy. Extensive fibrosis may be seen in patients with a long history of Raynaud phenomenon.[16] Familial clustering and ethnic influence have been demonstrated. Polymorphisms in genes coding for extracellular matrix proteins and cell-signaling molecules implicate non-MHC areas in SSc pathogenesis.[17] There are associations of polymorphisms in several genes with susceptibility and severity of systemic sclerosis. All patients showed genetically predisposed high TGFbeta1 production, with polymorphisms at codons 10 and 25 of the TGFbeta1 gene.[18] Current data suggest that SSc is a multigenic complex disorder.

8.9.1.2 Pseudoxanthoma Elasticum

Pseudoxanthoma elasticum (PXE) is an inherited disorder that is associated with accumulation of mineralized and fragmented elastic fibers in the skin, vascular walls, and Bruch's membrane in the eye. It may lead to peripheral and coronary arterial occlusive disease as well as gastrointestinal bleedings. There is yet no definitive therapy. Recent studies suggest that PXE is inherited almost exclusively as an autosomal recessive trait. Its prevalence has been estimated to be 1:25,000–100,000. Very recently, the ABCC6 gene on chromosome 16p13.1 was found to be associated with the disease. Mutations within ABCC6 cause reduced or absent transmembraneous transport that leads to accumulation of extracellular material. Presumably, this mechanism causes calcification of elastic fibers

In a study of 19 patients it was found that systolic function was normal, but diastolic parameters were abnormal in seven patients. Explanations for these abnormalities could be silent myocardial ischemia due to early coronary involvement and/or the direct consequences of ultrastructural defects of the elastic tissue of the heart.[19]

8.9.1.3 Diabetic Cardiomyopathy

In diabetes mellitus, alterations in cardiac structure or function in the absence of ischemic heart disease, hypertension, or other cardiac pathologies is termed diabetic cardiomyopathy.

Structural changes include myocardial hypertrophy, fibrosis, and fat droplet deposition, initially leading to abnormal diastolic function. Advanced glycation endproducts (AGEs) are thought to be important in the pathophysiology of diabetic cardiomyopathy. Irreversible modification of proteins by glucose results in the formation of AGEs, a heterogeneous family of biologically and chemically reactive compounds with cross-linking properties. This process of protein modification is magnified by the high ambient glucose concentration present in diabetes.[20]

The genetic background of diabetes is beyond the scope of this chapter. However, there are some very interesting studies pointing to a genetic link between diabetes and cardiac damage. Oxidative stress is known to be enhanced with diabetes and oxygen toxicity may alter cardiac progenitor cell (CPC) function resulting in defects in CPC growth and myocyte formation, which may favor premature myocardial aging and heart failure. Ablation of the p66shc gene in a mouse model[21] prevented these negative effects, pointing at a possible genetic link between diabetes, reactive oxygen species, and the development of heart failure.

8.9.2 Infiltrative

8.9.2.1 Cardiac Amyloidosis

Amyloidosis comprises a group of diseases characterized by extracellular deposition of insoluble fibrillar proteins with concomitant destruction of normal tissue structure and function. This results in stiffening and thickening of the myocardial walls, which can be easily demonstrated by echocardiography and often has a granular sparkling appearance. Absence of high ECG voltages further strengthens the suspicion of amyloidosis. Cardiac clinical manifestations include diastolic and systolic dysfunction, arrhythmias and conduction disturbances, orthostatic hypotension, coronary insufficiency, valvular dysfunction, and pericardial effusion.

Endomyocardial biopsy is the method of choice to diagnose cardiac amyloidosis and also allows characterization of the amyloid protein.[22]

About 20 different proteins are known to form amyloid fibrils in vivo. The nomenclature is based on these proteins.[23] In clinical practice, however, amyloidosis is often classified as primary, secondary, hereditary, and age related.

Primary amyloidosis or *systemic AL amyloidosis* is the result of monoclonal immunoglobulin light chains secreted by clonal plasma cells (multiple myeloma) and predominantly deposited in the heart, kidney, and nerves. Congestive heart failure and conduction disturbances are frequent cardiovascular complications and often result in early death of the patients.

Although multiple myeloma is not considered a genetic disease, there are reports of around 130 families with two or more cases of multiple myeloma, MGUS or Waldenström's macroglobulinemia.[24]

Secondary or systemic AA amyloidosis is associated with chronic diseases and manifested mainly in the kidney, liver and spleen, and, only rarely, in the heart. Proteinuria and renal failure are paramount.

Hereditary systemic amyloidosis is predominantly caused by deposition of amyloid fibrils derived from genetic variants of transthyretin (TTR), a transport protein synthesized mainly by the liver. More than 100 mutations are known already, of which the Val 122 Ile variant is the most common.[25] Inheritance is often autosomal dominant with varying degree of penetrance.[26,27] Clinical syndromes include cardiomyopathy, nephropathy, and neuropathy. The presenting symptom often is the peripheral ascending neuropathy; cardiac involvement often is the final cause of death.

Senile systemic amyloidosis (SSA) is caused by the deposition of amyloid fibrils from normal nonmutant TTR, especially in the heart. It is age related, with male predominance and rare in patients younger than 60 years of age. Clinically it manifests as congestive heart failure, relatively frequently accompanied by carpal tunnel syndrome.[25] Progression of this disease is much slower than in AL amyloidosis, despite the more severe hypertrophy present in the senile form.[25] Autopsy studies suggest that in up to 25% of individuals over the age of 80 years this type of TTR-derived amyloid can be found in the heart.[27]

8.9.2.2 Sarcoidosis

Myocardial sarcoidosis generally occurs in association with other manifestations of the systemic disease, but primary cardiac symptomatology does occur. Cardiac infiltration by sarcoid granulomas may result in increased stiffness of the heart, with overt features of restrictive cardiomyopathy. In addition, systolic dysfunction, conduction abnormalities, and arrhythmias may be seen. Treatment is empirically with glucocorticoids.

A genetic predisposition is likely, based on increased familial occurrences and different disease modes in different ethnic groups.[28] The strongest genetic associations are found within the human leucocyte antigen (HLA) antigens and functional polymorphisms within the butyrophilin-like 2 (BTNL2) gene.[29]

8.9.2.3 Gaucher Disease

Although Gaucher disease (GD) is the most common lysosomal storage disease, it very rarely affects the heart (only subtype 3, occurring 1 in 200,000). It is caused by deficiency of glucocerebrosidase, which results in abnormal accumulation of glycolipids within cellular lysosomes. GD is one of the few inherited metabolic disorders that can be treated by enzyme replacement therapy with recombinant enzyme; early identification is crucial to improving ultimate outcome.

Gaucher disease is inherited as an autosomal recessive disorder. The glucocerebrosidase gene is located on chromosome 1q21, and more than 180 distinct mutations are known. However, three mutant alleles account for most cases: N370S, L444P, and 84GG. The prevalence of these alleles varies with ethnicity. N370S is present exclusively in Ashkenazi Jews and non-Jewish Europeans, whereas L444P is common in northern Sweden. The diagnosis of GD is confirmed by the finding of reduced glucocerebrosidase activity in peripheral leukocytes. Diagnosis also can be confirmed by mutation analysis, which is an effective method for patient classification and carrier diagnosis.

8.9.2.4 Mucopolysaccharidoses: Hurler Disease, Hunter Disease

The mucopolysaccharidoses (MPS) are lysosomal storage disorders caused by the deficiency of enzymes required for the stepwise breakdown of glycosaminoglycans (GAGs), previously known as mucopolysaccharides. Fragments of partially degraded GAGs accumulate in the lysosomes, resulting in cellular dysfunction and clinical abnormalities. These are rare conditions, with an estimated total incidence of all types of MPS of approximately one in 20,000 live births. Hurler syndrome is the severe form of MPS I and is characterized by a broad spectrum of clinical problems including skeletal abnormalities, hepatosplenomegaly, and severe mental retardation. The incidence is approximately one in 100,000 births.

Cardiac abnormalities become apparent between birth and 5 years of age. These include cardiomyopathy, endocardial fibroelastosis, and valvular regurgitation, which alone or in combination can lead to heart failure. GAG storage within blood vessels causes irregular and diffuse narrowing of the coronary arteries and irregular lesions of the aorta. Coronary artery disease is often unrecognized until autopsy examination; it should be considered in affected patients with cardiac problems.

Mucopolysaccharidosis II (Hunter syndrome) is caused by a deficiency of iduronate 2-sulfatase (IDS), which results in storage of heparan and dermatan sulfate. MPS II is caused by mutations in the gene encoding for IDS, which is located on chromosome Xq28. Although the disorder is X-linked, cases in females have been reported on rare occasions.

8.9.3 Storage Diseases

8.9.3.1 Hemochromatosis

Iron-overload cardiomyopathy is often the result of multiple transfusions or a hemoglobinopathy, most frequently B-thalassemia. If cardiomyopathy occurs in the presence of diabetes, hepatic cirrhosis, and increased skin pigmentation, it may also result from familial hemochromatosis, an autosomal recessive disorder that arises from a mutation in the HFE gene, which codes for a transmembrane protein that is responsible for regulating iron uptake in the intestine and liver. The HFE gene is tightly linked to the HLA-A locus on chromosome 6p. The most common mutation is C282Y, identified in 85–90% of hemochromatosis patients in Northern Europe.[30] A second, relatively common HFE mutation (H63D) is not associated with clinically relevant iron overload but in case of compound heterozygosity with C282Y, iron overload can occur.

Cardiac involvement causes a mixture of systolic and diastolic dysfunction, often with arrhythmias. Cardiac dysfunction is due to direct toxicity from free iron and to adverse effects caused by myocardial cell infiltration, preferentially in the sarcoplasmic reticulum. The ventricles are more affected than the atria and the conduction system is often involved. Loss of myocytes occurs with replacement fibrosis. Macroscopically, the heart may be dilated or non-dilated with thickened ventricular walls.

On cardiac MRI, a reduced T2* signal will be seen with increasing cardiac iron storage.

Phlebotomy and iron chelators like desferoxamine may reduce cardiac and other iron stores and result in clinical improvement.

8.9.3.2 Fabry Disease

This is an X-linked lysosomal storage disorder, caused by deficiency of lysosomal α-galactosidase A (GLA), leading to the accumulation of glycosphingolipids in tissues like the heart. The ensuing ventricular hypertrophy is often classified as a restrictive cardiomyopathy, although it may also resemble hypertrophic cardiomyopathy. It is the second most prevalent lysosomal storage disease after Gaucher disease. The gene is located on the long arm (Xq22.1 region) of the X chromosome. Several hundred mutations in the gene have been identified.

The prevalence of Fabry disease is estimated to range from 1:17,000 to 1:117,000 males in Caucasians. Clinical manifestations are usually evident by the age of 10, often starting with neuropathy (burning pains of the palms and soles) and skin lesions (angiokeratomas). At higher age cardiac and renal disease and stroke become more important. Cardiac involvement may lead to (symmetrical) ventricular hypertrophy, conduction defects, coronary artery disease, valve insufficiencies, and aortic root dilatation.[31] In general, cardiac involvement will be accompanied by other signs of Fabry disease, although these may be missed. Sometimes, the disease is limited to the myocardium. Therefore, screening for Fabry disease is advised in patients with otherwise unexplained LVH. Tissue Doppler may provide a preclinical diagnosis of cardiac involvement, even in patients without LVH. Echocardiographic appearance of Fabry disease may be distinguished from other forms of LVH based on a thickened, hyperechogenic layer in the endocardium and subendocardial myocardium, caused by local intracellular glycolipid deposition. This is paralleled by a hypoechogenic layer, representing the mildly affected midwall myocardium. A definitive diagnosis can be made based on a low plasma α-galactosidase A level in males or by endomyocardial biopsy, showing concentric lamellar bodies in the sarcoplasm of heart cells on electron microscopy. In females the diagnosis can be made by analysis of the GLA gene.

Although the disease is generally considered X-linked recessive, a better name would be X-linked semidominant. LVH may occur in heterozygous females in up to 64%; end-stage renal disease and stroke may also develop and the overall negative effect on life span may be as much as 15 years.

Enzyme replacement therapy is available, albeit very expensive. Recombinant Agalsidase-β may partly clear microvascular endothelial deposits in the heart and kidneys. Therapy can reduce LVH and enhance myocardial function.

8.9.3.3 Glycogen Storage Disease

Disorders of glycogen metabolism most often affect the liver and skeletal muscle, where glycogen is most abundant. To date, 12 subforms of glycogen storage disease (GSD) have been identified. The physiologic importance of a given enzyme determines the clinical manifestations of the disease. In general, hypoglycemia, hepatomegaly, and skeletal muscle weakness and easy fatigability are the predominant clinical features. In GSD type II (Pompe disease) and IIa (Danon disease), cardiac involvement may occur. The classic infantile form is characterized by cardiomyopathy and severe generalized muscular hypotonia.[32] The tongue may be enlarged. Hepatomegaly also may be present and is usually due to heart failure. Pompe disease is an autosomal recessive disorder with considerable allelic heterogeneity. It is caused by mutations in the gene encoding lysosomal alpha-1,4-glucosidase (GAA) located at 17q25.2-q25. More than 200 mutations have been reported.[33]

8.9.4 Endomyocardial Causes of Restrictive Cardiomyopathy

8.9.4.1 Endomyocardial Fibrosis

Endomyocardial fibrosis (EMB) is an obliterative cardiomyopathy characterized by fibrotic thickening and obliteration of left, right, or both ventricles, with a predilection to selectively involve the apices and inflow region and spare the outflow tract. The fibrotic process does not affect the valve leaflets, the atria, or the great vessels and extracardiac involvement is not known. There is a peculiar distribution of the disease in very specific areas within some countries around the equator.[34] In an epidemiological study of 214 families in Mozambique, 99 had no cases of EMB, 63 had one case, and 52 had more than one case.[35] The familial occurrence could be caused by genetic factors or susceptibility; however, this has not been elucidated. It may also rely on environmental factors, like the abundance of thorium and cerium in the soil, accompanied by magnesium deficiency. It has also been related to filariasis and altered immunological response to streptococcal infection in individuals whose immune status had been altered by parasitic infections

8.9.4.2 Hypereosinophylic Syndrome

Hypereosinophilic syndrome (HES) is a heterogeneous group of disorders characterized by unexplained persistent primary eosinophilia causing end-organ damage.

In the acute necrotic stage, there is endocardial damage, myocardial infiltration with eosinophils and lymphocytes, eosinophil degranulation, and myocardial necrosis. This phase may be clinically silent without abnormalities on echocardiography. However, serum troponin levels may be raised and contrast-enhanced MRI may detect myocardial inflammation. In the second stage, thrombus formation occurs along areas of damaged endocardium. This may lead to systemic embolization. In the third phase, progressive scar formation produces endomyocardial fibrosis and finally a restrictive cardiomyopathy.

Apart from cardiac manifestations and thromboembolic (cerebral) complications, encephalopathy and peripheral neuropathy may occur.

One HES variant, myeloproliferative, is actually chronic eosinophilic leukemia, which has a unique genetic marker, FIP1L1-PDGFRA, with consequences for the treatment.[36] Loeffler endocarditis, eosinophilic endomyocardial disease, or fibroplastic endocarditis appears to be a subcategory of the hypereosinophilic syndrome in which the heart is predominantly involved.

Autosomal dominant transmission of marked eosinophilia has been reported. In one family, the gene has been mapped to chromosome 5q31–33.[37]

MRI may be helpful in cases of restrictive cardiomyopathy with luminal obliteration to differentiate perfused and enhancing myocardium from poorly vascularized and hypoenhancing thrombus or eosinophilic infiltrate.[38]

8.10 Summary

Restrictive cardiomyopathy is a rare disease, often presenting with fatigue, exercise intolerance, or dyspnea. In many cases, RCM occurs as part of a multiorgan disease or malignancy, where cardiac involvement may occur early or late in time. Therefore, depending on clinical suspicion or other, noncardiac symptoms and findings, additional investigations are necessary before a definitive diagnosis can be made. The diagnosis of idiopathic RCM can only be made by exclusion. Idiopatic RCM sometimes presents as a familial or genetic form, related to mutations in the cardiac troponin I genes. There may also be overlap, both clinically and genetically, with family members with hypertrophic or dilated cardiomyopathy. Prognosis is often poor and treatment options scarce: symptomatic therapy with diuretics and/or betablockers and occasionally specific therapy for the underlying disease.

References

1. Mogensen J, Kubo T, Duque M, Uribe W, Shaw A, Murphy R, Gimeno JR, Elliott P, McKenna WJ (2003) Idiopathic restrictive cardiomyopathy is part of the clinical expression of cardiac troponin I mutations. J Clin Invest 111:209–216
2. Maron BJ, Towbin JA, Thiene G, Antzelevitch C, Corrado D, Arnett D, Moss AJ, Seidman CE, Young JB (2006) Contemporary definitions and classification of the cardiomyopathies: an American Heart Association Scientific Statement from the Council on Clinical Cardiology, Heart Failure and Transplantation Committee; Quality of Care and Outcomes Research and Functional Genomics and Translational Biology Interdisciplinary Working Groups; and Council on Epidemiology and Prevention. Circulation 113:1807–1816
3. Kaski JP, Syrris P, Burch M, Tomé-Esteban MT, Fenton M, Christiansen M, Andersen PS, Sebire N, Ashworth M, Deanfield JE, McKenna WJ, Elliott PM (2008) Idiopathic restrictive cardiomyopathy in children is caused by mutations in cardiac sarcomere protein genes. Heart 94:1478–1484
4. Stöllberger C, Finsterer J (2007) Extracardiac medical and neuromuscular implications in restrictive cardiomyopathy. Clin Cardiol 30:375–380
5. Gomes AV, Liang J, Potter JD (2005) Mutations in human cardiac troponin I that are associated with restrictive cardiomyopathy affect basal ATPase activity and the calcium sensitivity of force development. J Biol Chem 280:30909–30915
6. Sengupta PP, Krishnamoorthy VK, Abhayaratna WP, Korinek J, Belohlavek M, Sundt TM 3rd, Chandrasekaran K, Seward JB, Tajik AJ, Khandheria BK (2008) Comparison of usefulness of tissue Doppler imaging versus brain natriuretic peptide for differentiation of constrictive pericardial disease from restrictive cardiomyopathy. Am J Cardiol 102:357–362
7. Hershberger RE, Lindenfeld J, Mestroni L, Seidman CE, Taylor MR, Towbin JA (2009) Genetic evaluation of cardiomyopathy – a Heart Failure Society of America practice guideline. J Card Fail 15:83–97
8. Ammash NM, Seward JB, Bailey KR, Edwards WD, Tajik AJ (2000) Clinical profile and outcome of idiopathic restrictive cardiomyopathy. Circulation 101:2490–2496
9. Fitzpatrick AP, Shapiro LM, Rickards AF, Poole-Wilson PA (1990) Familial restrictive cardiomyopathy wieth atrioventricular block and skeletal myopathy. Br Heart J 63:114–118
10. Katritsis D, Wilmshurst PT, Wendon JA, Davies MJ, Webb-Peploe MM (1991) Primary restrictive cardiomyopathy: clinical and pathologic characteristics. J Am Coll Cardiol 18:1230–1235
11. Cubero GI, Larraya GL, Reguero JR (2007) Familial restrictive cardiomyopathy with atrioventricular block without skeletal myopathy. Exp Clin Cardiol 12:54–55
12. Kostareva A, Gudkova A, Sjöberg G, Mörner S, Semernin E, Krutikov A, Shlyakhto E, Sejersen T (2009) Deletion in TNNI3 gene is associated with restrictive cardiomyopathy. Int J Cardiol 131:410–412
13. Davis J, Wen H, Edwards T, Metzger JM (2007 May 25) Thin filament disinhibition by restrictive cardiomyopathy mutant R193H troponin I induces Ca2+-independent mechanical tone and acute myocyte remodeling. Circ Res 100(10):1494–1502
14. Du J, Liu J, Feng HZ, Hossain MM, Gobara N, Zhang C, Li Y, Jean-Charles PY, Jin JP, Huang XP (2008) Impaired relaxation is the main manifestation in transgenic mice expressing a restrictive cardiomyopathy mutation, R193H, in cardiac TnI. Am J Physiol Heart Circ Physiol 294:H2604–H2613
15. Menon SC, Michels VV, Pellikka PA, Ballew JD, Karst ML, Herron KJ, Nelson SM, Rodeheffer RJ, Olson TM (2008) Cardiac troponin T mutation in familial cardiomyopathy with variable remodeling and restrictive physiology. Clin Genet 74:445–454
16. Tzelepis GE, Kelekis NL, Plastiras SC, Mitseas P, Economopoulos N, Kampolis C, Gialafos EJ, Moyssakis I, Moutsopoulos HM (2007) Pattern and distribution of myocardial fibrosis in systemic sclerosis: a delayed enhanced magnetic resonance imaging study. Arthritis Rheum 56:3827–3836
17. Johnson RW, Tew MB, Arnett FC (2002) The genetics of systemic sclerosis. Curr Rheumatol Rep 4:99–107
18. Crilly A, Hamilton J, Clark CJ, Jardine A, Madhok R (2002) Analysis of transforming growth factor beta1 gene polymorphisms in patients with systemic sclerosis. Ann Rheum Dis 61:678–681
19. Nguyen LD, Terbah M, Daudon P, Martin L (2006) Left ventricular systolic and diastolic function by echocardiogram in pseudoxanthoma elasticum. Am J Cardiol 97:1535–1537

20. Brownlee M (1994 Jun) Lilly Lecture 1993. Glycation and diabetic complications. Diabetes 43(6):836–841
21. Rota M, LeCapitaine N, Hosoda T, Boni A, De Angelis A, Padin-Iruegas ME, Esposito G, Vitale S, Urbanek K, Casarsa C, Giorgio M, Lüscher TF, Pelicci PG, Anversa P, Leri A, Kajstura J (2006) Diabetes promotes cardiac stem cell aging and heart failure, which are prevented by deletion of the p66shc gene. Circ Res 99:42–52
22. Kholova I, Niessen HWM (2005) Amyloid in the cardiovascular system: a review. J Clin Pathol 58:125–133
23. Selvanayagam JB, Hawkins PN, Paul B, Myerson G, Neubauer S (2007) Evaluation and management of cardiac amyloidosis. J Am Coll Cardiol 50:2101–2110
24. Ogmundsdóttir HM, Einarsdóttir HK, Steingrímsdóttir H, Haraldsdóttir V (2009) Familial predisposition to monoclonal gammopathy of unknown significance, Waldenström's macroglobulinemia, and multiple myeloma. Clin Lymphoma Myeloma 9:27–29
25. Rapezzi C, Merlini G, Quarta CC, Riva L, Longhi S, Leone O et al (2009) Systemic cardiac amyloidosis. Disease profiles and clinical courses of the 3 main types. Circulation 120:1203–1212
26. Hesse A, Altland K, Linke RP, Almeida MR, Saraiva MJ, Steinmetz A, Maisch B (1993) Cardiac amyloidosis: a review and report of a new transthyretin (prealbumin) variant. Br Heart J 70:111–115
27. Ruberg FL, Judge DP, Maurer MS (2009) Familial amyloid cardiomyopathy due to TTR mutations: an underground cause of restrictive cardiomyopathy. Letter to the editor. J Card Fail 15:464
28. Grunewald J (2008) Genetics of sarcoidosis. Curr Opin Pulm Med 14:434–439
29. Spagnolo P, Sato H, Grutters JC, Renzoni EA, Marshall SE, Ruven HJ, Wells AU, Tzouvelekis A, van Moorsel CH, van den Bosch JM, du Bois RM, Welsh KI (2007) Analysis of BTNL2 genetic polymorphisms in British and Dutch patients with sarcoidosis. Tissue Antigens 70:219–227
30. Pietrangelo A (2004) Hereditary hemochromatosis – a new look at an old disease. N Engl J Med 350:2383–2397
31. Mehta A, Ricci R, Widmer U, Dehout F, Garcia de Lorenzo A, Kampmann C, Linhart A, Sunder-Plassmann G, Ries M, Beck M (2004) Fabry disease defined: baseline clinical manifestations of 366 patients in the Fabry Outcome Survey. Eur J Clin Invest 34:236–242
32. Howell RR, Byrne B, Darras BT, Kishnani P, Nicolino M, van der Ploeg A (2006) Diagnostic challenges for Pompe disease: an under-recognized cause of floppy baby syndrome. Genet Med 8:289–296
33. Kroos M, Pomponio RJ, van Vliet L, Palmer RE, Phipps M, Van der Helm R, Halley D, Reuser A (2008) Update of the Pompe disease mutation database with 107 sequence variants and a format for severity rating. Hum Mutat 29:E13–E26
34. Sivasankaran S (2009) Restrictive cardiomyopathy in India: the story of a vanishing mystery. Heart 95:9–14
35. Mocumbi AO, Ferreira MB, Sidi D, Yacoub MH (2008) A population study of endomyocardial fibrosis in a rural area of Mozambique. N Engl J Med 359:43–49
36. Gleich GJ, Leiferman KM (2009) Hypereosinophilic syndromes: current concepts and treatments. Br J Haematol 145:271–285
37. Rioux JD, Stone VA, Daly MJ, Cargill M, Green T, Nguyen H, Nutman T, Zimmerman PA, Tucker MA, Hudson T, Goldstein AM, Lander E, Lin AY (1998) Familial eosinophilia maps to the cytokine gene cluster on human chromosomal region 5q31-q33. Am J Hum Genet 63:1086–1094
38. Kleinfeldt T, Ince H, Nienaber CA. Hypereosinophilic syndrome: a rare case of Loeffler's endocarditis documented in cardiac MRI. *Int J Cardiol.* 2009 Apr 15

Part III
Primary Electrical Heart Diseases

Congenital Long QT-Syndrome

Hubert F. Baars and Jeroen F. van der Heijden

9.1 Introduction

The *congenital long QT syndrome* (LQTS) is considered to be one of the hereditary cardiac arrhythmia syndromes, nowadays also known as cardiac *channelopathies*. The syndrome is characterized by prolongation of the heart rate *corrected QT-interval* (QTc) on the 12-lead electrocardiogram (ECG). In affected family members, it is associated with recurrent syncope, seizures, and *sudden cardiac death* due to *ventricular arrhythmias (Torsade des Pointes (TdP)* and *ventricular fibrillation)*, which typically follow a precipitating event such as exertion, extreme emotion, swimming and diving, or auditory stimulation.

The congenital LQTS was first described in 1957 by Jervell and Lange-Nielsen in a family consisting of four children with a prolonged QT-interval on the electrocardiogram, congenital sensorineural hearing loss, recurrent syncope, and sudden cardiac death.[1] The inheritance pattern appeared to be autosomal recessive. A few years later, in the early 1960s *Romano*[2] and *Ward*[3] described independently of each other, families in which affected members had recurrent syncope, sudden cardiac death and QT prolongation on the electrocardiogram, however, without the associated deafness, and with an autosomal dominant mode of inheritance.

Breakthroughs in molecular genetics in the mid-1990s revealed the fundamental molecular basis of the LQTS, which in general relates to a defect in specific cardiac ion channels, caused by specific gene mutations.[4,5] These defects result in either a decrease in repolarizing potassium currents or an inappropriate late entry of sodium in the myocyte eventually resulting in a prolonged QT-interval on the electrocardiogram.[6]

The diagnosis of LQTS is made by a careful evaluation of the personal and family history of the patient and a detailed examination of the electrocardiogram in order to detect either prolongation of the QT-interval or specific changes in the morphology of the ST-T segment. In addition to the 12-lead electrocardiogram, exercise testing and Holter monitoring may be useful.

In the majority of patients, the diagnosis can be confirmed by molecular genetic testing, which will reveal mutations in specific genes, mostly encoding for cardiac K^+ or Na^+ channels. Once the diagnosis has been made and therapy has been started, it is essential to identify the patient at risk for malignant ventricular arrhythmias because in that case a prophylactic ICD might be pertinent. In addition, screening of the first degree family members should be initiated, preferably in close collaboration with a clinical geneticist, in order to detect the asymptomatic or at the moment not yet diagnosed patient at risk.

This chapter will review the normal electrophysiological mechanisms involved in the cardiac action potential as well as the structure and function of the cardiac ion channels. Subsequently, a description of the congenital LQTS will be given.

9.2 Electrophysiological and Molecular Mechanisms

9.2.1 Cardiac Action Potential

The cardiac action potential (AP) reflects the integrated electrical activity of many ionic currents across the cell

H.F. Baars (✉)
Department of Cardiology, TweeSteden Hospital, Tilburg and Department of Cardiology, Heart and Lung Center Utrecht, University Medical Centre Utrecht, The Netherlands
e-mail: bbaars@tsz.nl

membrane through voltage-gated ion channels, ionic pumps, and ionic exchangers (Fig. 9.1). Depolarizing currents convey positively charged ions (Na+, Ca++) into the cell, while the repolarizing K+ current flows to the outside of the cell (extensively reviewed in, e.g., Tan et al.[7] and Grunnet et al.[8])

The AP can be divided into five phases:

- Phase 4: the resting membrane potential is maintained by a balance between Na+ and Ca++ leak currents and the inward rectifier current (I_{K1}).
- Phase 0: The cardiomyocyte membrane is depolarized by a rapid, transient influx of Na+ ions through voltage-gated sodium channels (I_{Na}). This is reflected as the upstroke of the AP.
- Phase 1: during early repolarization, the transient efflux of K+ through transient outward channels (I_{to}), which inactivate rapidly, terminates the AP upstroke.
- Phase 2: the plateau phase of the AP is maintained by a balance between an inward Ca++ current through L-type Ca++ channels and a K+ efflux (I_{Kr} and I_{Ks}).
- Phase 3: during late repolarization the Ca++ channels have inactivated. The outward K+ currents continue mainly through the slowly activating delayed rectifier K+ channel (I_{Ks}). At the end a large outward K+ current (I_{K1}) brings the membrane potential to its resting level.

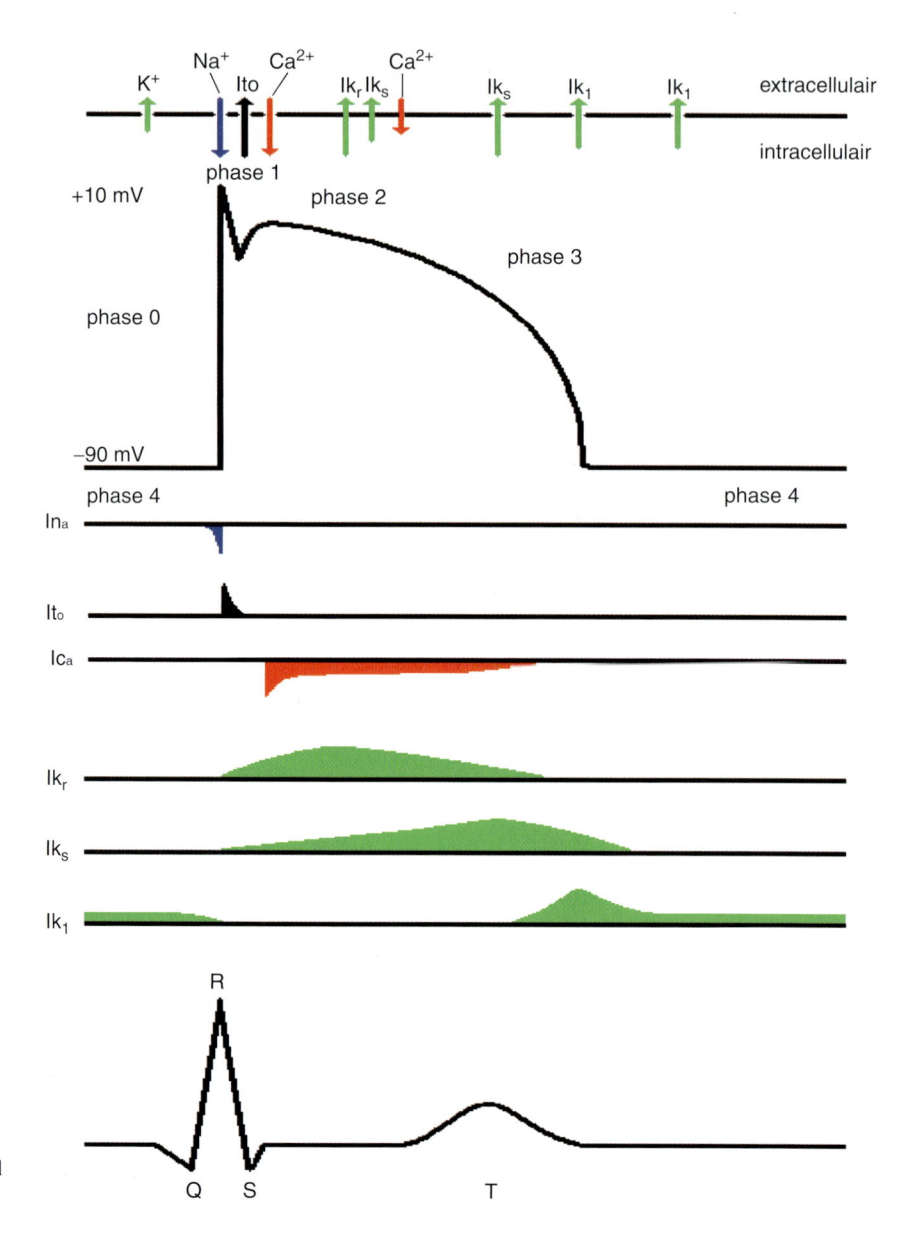

Fig. 9.1 Cardiac action potential (see text for explanation)

On the surface ECG, the QRS complex reflects the depolarization, and the T-wave reflects the repolarization phase of the ventricle. The T-wave is much longer than the QRS complex simply because repolarization takes longer than cardiac excitation.

The repolarization phase is not homogeneous in the ventricle, because of the fact that the ventricular wall consists of three different layers (epicardium, midmyocardium (M cell layer) and endocardium) with different action potential duration and configuration.[9,10] Voltage gradients between these layers during repolarization determine the height and the shape of the T-wave.

9.2.2 Cardiac Ion Channels

The structure of the voltage-gated potassium channel involved in most cases of LQTS includes a monomer of six membrane-spanning regions (S1–S6) with connecting intracellular cytoplasmatic loops, a voltage-sensing domain (S4), and a pore region located between S5 and S6 (Fig. 9.2). A tetrameric structure of four individual monomers together will eventually form a functional hydrophobic envelope, which surrounds a central cavity or pore. These four pore forming α-subunits are known to be modulated in their biophysical properties, pharmacologic responses, tissue distribution, and intracellular trafficking by smaller accessory β-subunits. These β-subunits often have a single membrane-spanning domain (Fig. 29.2).

The structure of the cardiac sodium (SCN5A) and L-type calcium (CACNA1C) channels is more complex. They consist of a single protein with four homologous domains (the alpha-subunits), each of which contain six transmembrane domains, similar to the four linked potassium channel modules (for figure see chapter on Brugada syndrome).

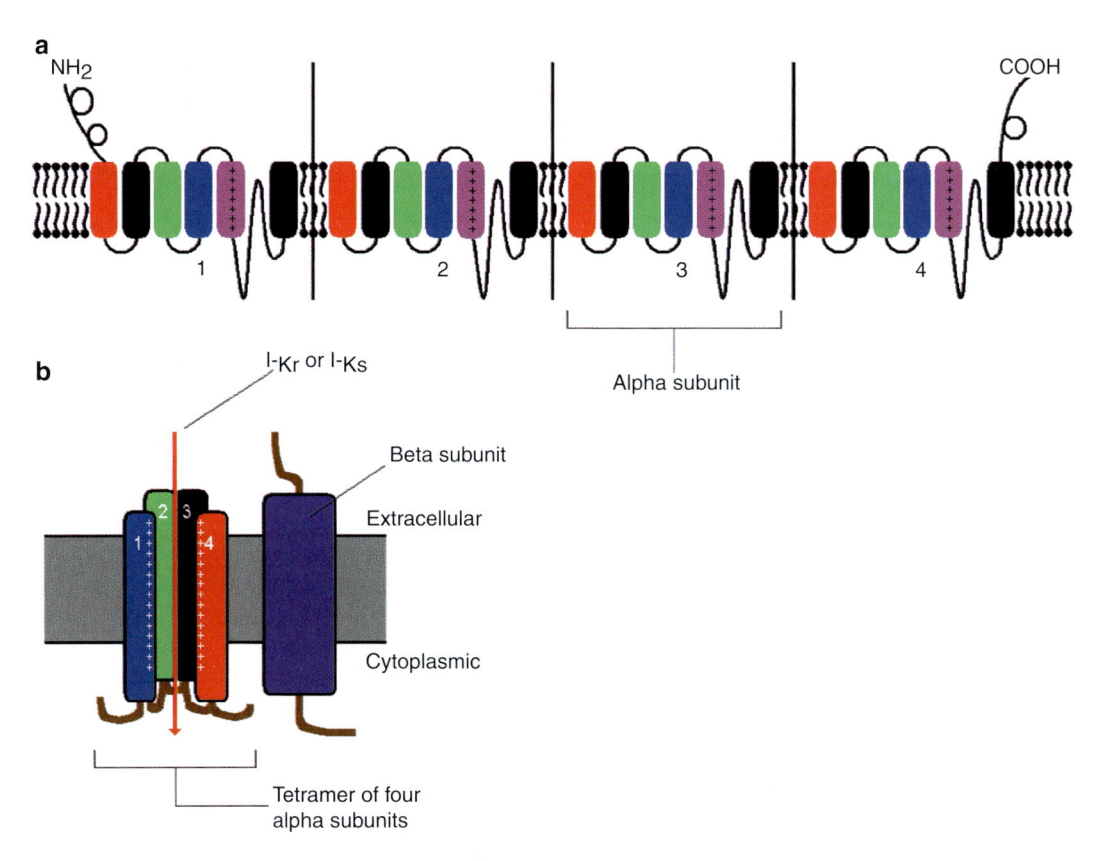

Fig. 9.2 (**a**) Schematic topology of the four α-subunits (K_vLQT1 or hERG) of a voltage-gated potassium channel. Characteristic are the six membrane-spanning segments of the α-subunit. Four of those α-subunits (number 1–4) form a tetrameric potassium channel, which is responsible for the I_{Ks} or I_{Kr}. This is showed in (**b**), which includes the β-subunit (respectively MinK or MiRP1) that can modulate the function of the α-subunits

9.3 Epidemiology and Prevalence

The prevalence of congenital LQTS in the USA has been estimated to be approximately 1 in 5,000 individuals, causing hundreds to thousands of sudden cardiac deaths in children, adolescents, and young adults each year.[11] However, the prevalence of congenital LQTS may be as high as 1 in 2,000–3,000 live births. This figure is derived from the largest prospective study of neonatal electrocardiography ever performed in 44,596 infants at age 3–4 weeks.[12] The investigators concluded that the prevalence in their population must be close to 1:2,500 at least. Since most mutation carriers remain asymptomatic throughout life, clinical disease is less common. Nevertheless, LQTS is one of the most common causes of autopsy-negative sudden unexplained death.[13]

The congenital long QT syndrome can be divided in two distinguishable forms: the Romano-Ward syndrome (RWS) and the Jervell and Lange-Nielsen syndrome (JLNS). The RWS is the most common form of the LQTS and accounts for over 99% of cases. It is transmitted as an autosomal dominant trait and it can be divided into 12 subtypes (LQT1 to LQT12). The extremely rare, autosomal recessive, JLNS is associated with congenital deafness.

9.4 Molecular Genetics and Pathogenesis

9.4.1 Molecular Genetics

Advantages of molecular genetics in the last decade of the twentieth century have provided insights into the mechanisms underlying the LQTS. The genetic defects causing the LQTS are for the majority of cases located in genes encoding for the α- and β-subunits of the voltage-gated potassium channels (e.g., KCNQ1, KCNH2, KCNE1 and KCNE2). On the other hand, mutations in the genes encoding for the cardiac sodium (SCN5A) and calcium channel (CACNA1C) can also result in prolongation of the QT-interval. Finally, genes encoding for several structural membrane scaffolding proteins or proteins interacting with these cardiac channels are responsible for some exceptional types of LQTS. Since the first description of LQTS, over 300 mutations have been described in 12 distinct LQTS-related genes[14–17] (Table 9.1).

The dysfunction of the ion channels is caused by two distinct biophysical mechanisms, consisting of coassembly or trafficking defects and of the formation of channels with aberrant function.[15] In the case of coassembly

Table 9.1 LQTS subtypes and corresponding determinants

	Chromosome	Gene	Protein	Function	Current
LQT 1	11p15.5	KCNQ1	KvLQT1	α-Subunit	I_{Ks}
LQT 2	7q35-36	KCNH2	hERG	α-Subunit	I_{Kr}
LQT 3	3p21-24	SCN5A	Nav1.5	α-Subunit	I_{Na}
LQT 4	4q25-27	Ankyrin-B	Ankyrin-B	Scaffolding	I_{Na}
LQT 5	21q22	KCNE1	MinK	β-Subunit	I_{Ks}
LQT 6	21q22	KCNE2	MiRP1	β-Subunit	I_{Kr}
LQT 7	17q23	KCNJ2	Kir2.1	α-Subunit	I_{K1}
LQT 8	1q42	CACNA1C	Ca(v)1.2	α-Subunit	I_{Ca}
LQT 9	3p25	CAV3	Caveoline-3	Scaffolding	I_{Na}
LQT 10	11q23.3	SCN4B	Nav1.5	β4-Subunit	I_{Na}
LQT 11	7q21-22	AKAP9	Yotiao	ChIP	I_{Ks}
LQT 12	20q11.2	SNTA1	α1-syntrophin	ChIP	I_{Na}

ChIP channel interacting protein; *Scaffolding* structural membrane scaffolding proteins

defects, the mutant subunits are not transported properly to the cell membrane and fail to incorporate into the tetrameric channel. The net effect will be a 50% or less reduction in channel function (haploinsufficiency). The second biophysical mechanism consist of the formation of defective channels in which the product of the mutated gene, being an abnormal protein, is transported to the cell membrane and incorporated in the channel subunit. This will lead to a dysfunctional channel, resulting in a more than 50% reduction in channel current (dominant-negative effect).

The LQTS type 1 (50% of cases) and type 2 (35–40% of cases) are caused by pathogenic mutations in genes encoding the α-subunits of two specific voltage-gated potassium channels resulting in impairment of the delayed rectifier current. The delayed rectifier current comprises of two independent components: one slowly activating (I_{Ks}, KCNQ1, LQT1) and one rapidly activating (I_{Kr}, KCNH2, LQT2) component. The pathogenic mutations will lead to a "loss of function" of the relevant ion channel and as a consequence of that to a reduction of repolarizing potassium current.[18] This reduction in the repolarizing potassium current results in a prolongation of the action potential duration and thereby to the creation of the arrhythmogenic substrate.

"Gain of function" mutations in the SCN5A-encoded sodium channel protein, which are associated with prolonged depolarization due to a small persistent inward current (I_{Na}), are the cause of the LQTS type 3 (LQT3, 10–15%).

Rare subtypes of the LQTS stem from mutations in genes encoding for the β-subunits of several cardiac ion channels, which causes a disturbance of their function. They include KCNE1 encoding the β-subunit of the voltage-gated potassium channel responsible of the slowly activating potassium current I_{Ks} (minK, LQT5, 2–3%), KCNE2 encoding the β-subunit of the voltage-gated potassium channel responsible for the rapidly activating potassium current I_{Kr} (MiRP1, LQT6, <1%) and the SCN4B-gene encoding the β4-subunit of the I_{Na} sodium channel (LQT10, <1%).

The LQT1, LQT2, LQT3, LQT5, LQT6, and LQT10 make up the classic forms of the congenital LQTS and it is this group of ion channel genes that has characterized this primary electrical heart disease as a pure cardiac channelopathy.

During the past years, mutations in other genes have been identified resulting in QT prolongation in single individuals or just a few families. These LQTS-related disorders involve mutations in the ankyrin-B gene (LQT4), which product function as a cytoskeletal membrane adapter as ankyrin-B is involved in anchoring of ion channels to the cellular membrane and mutations in the caveolin-3 gene with increase in late sodium current (LQT9).[19, 20] Other rare LQTS-variants are syndromes associated with abnormal cardiac repolarization. One of these syndromes is the Andersen–Tawil syndrome (LQT7 or ATS1), due to mutations in the KCNJ2-encoded Kir2.1 potassium channel (reduced I_{K1} current), with a phenotype dominated by minor skeletal (facial) abnormalities and periodic hypokalemic paralysis.[21,22] Another syndrome is the Timothy syndrome (LQT8 or TS1), which is caused by mutations in the a-subunit of the L-type calcium channel, which leads to an increase in Ca(v) 1.2 current. The Timothy syndrome involves, among others, a typical syndactyly in hands and feet, mental retardation/autism, and many more features.[23]

The autosomal recessive variant of the LQTS (JLN syndrome) arises in patients who inherit abnormal KCNQ1 or KCNE1 alleles from both parents, who are usually without symptoms themselves. The abnormal alleles can bear the same mutation as is usually the case in consanguineous families or be different (compound heterozygosity).[24]

In this chapter, we will focus on the clinically most relevant forms of the LQTS, which are the LQT1, LQT2, and LQT3 subtypes.

9.4.2 Mechanisms Involved in Arrhythmia

Torsade de Pointes is the classical ventricular arrhythmia associated with LQTS. These TdP are either pause dependent (LQT2) or may be induced during higher heart rates without a preceding pause (LQT1).[25] Pause-dependent TdP, common in LQT2, is triggered by early afterdepolarizations (EAD), caused by Ca^{2+} release from intracellular Ca^{2+} stores.[26] Subsequently, Ca^{2+}-dependent transmembrane currents are altered in such a way as to allow L-type Ca^+ channels to recover more readily from inactivation and to reactivate before repolarization is complete, thus generating EADs. On the other hand, in LQT1 the induction of TdP is not pause dependent, suggesting a role for delayed afterdepolarizations (DAD), secondary to intracellular Ca^{2+} overload.[27] Experimental studies have shown that blockade of the I_{Ks} causes DAD.[28]

Prolonged repolarization may result in reentry through dispersion of repolarization. Augmented spatial dispersion of repolarization within the ventricular myocardium can lead to a unidirectional block in conduction and set the stage for reentry, which is the principal arrhythmogenic substrate.[29,30] The reentry mechanism is initiated by an extraventricular beat (EAD or DAD).

Especially, in the LQT1 patients, events will occur during higher heart rates. Stimulation of the β-adrenergic receptor by epinephrine will lead to an increased activity of the intracellular adenylyl cyclase activity, which increases cAMP.[31] cAMP will activate protein kinase A, which phosphorylates among others, the K$^+$ channels. Phosphorylation of K$^+$ channels will enhance I_{Ks}, resulting in a reduction of the action potential duration.[32] The shorter duration of the action potential is reflected in a shorter QT-interval during an increase in heart rate.[33] In the case of LQT1 and LQT5, the I_{Ks} channels are not able to adjust to an increase in heart rate, which will lead to an abnormal response to adrenergic stimulation with impaired shortening of the action potential and subsequent progressively prolonged QTc during exercise and early recovery. Beta-adrenergic regulation of I_{Kr} (LQT2) is more complex (reviewed[34]). Differential results were found in studies indicating either a reduction or an increase of I_{Kr} in response to β-adrenergic stimulation. The exact mechanisms of action are beyond the scope of this chapter.

The rationale for treatment with β-blockers in patients with LQTS (especially LQT1 and LQT2) is the prevention of calcium overload in both types of LQTS by preventing the loading of intracellular Ca^{2+} stores by cyclic adenosine monophosphate level (cAMP)-dependent processes, notably Ca^{2+} influx through L-type Ca$^+$ channels.[35] Furthermore, direct effect on the β-adrenergic modulation of I_{Ks} and I_{Kr} currents are responsible for the therapeutic effect. The exact mechanisms are again beyond the scope of this chapter.

The "gain of function" mutations in the sodium channel in the case of LQT3 results in a persisting depolarizing inward sodium current during the plateau phase of the action potential and as a consequence of that to a prolongation of the repolarization as well. The increase in the I_{Na} is especially important during slow heart rates. During fast heart rates, the effect of the pathogenic mutation is less because Na$^+$ will accumulate in the cell.

This phenomenon will lower the Na$^+$ gradient across the membrane and thereby the magnitude of I_{Na},[36] resulting in a shorter QT-interval during exercise, even more than normal.[37]

9.5 Clinical Aspects

9.5.1 Clinical Presentation in General

The clinical picture of the LQTS is diverse. It can vary from a lifelong asymptomatic course in approximately half of the patients with genetically proven LQTS to recurrent syncopal attacks, seizure like episodes, and premature sudden cardiac death in others. The considerable variation in the phenotype of LQTS (incomplete penetrance) is presumably caused by several modifying factors, including genetic polymorphism at the mutated locus and environmental factors. Underlying the presenting symptoms, syncope, ACA and SCD are polymorphic ventricular tachyarrhythmias in the form of TdP, and ventricular fibrillation. These TdP are most often self-limiting but in some cases they degenerate into ventricular fibrillation leading to sudden death. Syncope is the most frequent symptom in LQTS. Among symptomatic probands, approximately 50% will experience their first cardiac event (syncope or death) before the age of 12 and by the age of 40 this is increased to approximately 90%.[38] Less than 5% of the LQTS patients will present with sudden cardiac death (SCD) or aborted cardiac arrest (ACA) as first symptom. Conversely, however, less than half of the sudden death victims with LQTS experienced a prior warning episode of syncope, which means that the occurrence of ventricular fibrillation is unpredictable in many cases. In the majority of LQTS patients, cardiac events will be triggered by physical activity, emotional stress, or rest.

9.5.2 Genotype–Phenotype Specific Correlations

The clinical course of LQTS is in particular dependent on the genotype concerned (Table 9.2). By the age of 15, more than 60% of LQT1 patients have had a cardiac

Table 9.2 Clinical characteristics in the common forms of LQTS

	LQT1	LQT2	LQT3
Prevalence (%)[45]	50	35–40	10–15
Events occurring with exercise or emotion (%)[40]	88	56	32
Events occurring at rest or sleep (%)[40]	3	29	39
Lethal events occurring with exercise or emotion (%)[40]	82	29	16
Lethal events occurring at rest or sleep (%)[40]	9	49	64
Specific triggers	Swimming/diving	Loud noise	Sleep/rest
Pause dependency in TdP onset[25]	–	++	+/-?
Augmented risk postpartum[94]	+	+++	+
Median age first event[39]	9	12	16
Events < 10 year (%)[39]	40	16	2
Events < 40 year (%)[39]	63	46	18
Death during event (%)[39]	4	4	20
Sudden death as presenting symptom (%)[39]	2	1	3
Efficacy of β-blocker	+++	++	+?

TdP torsades de pointes

event, compared to less than 10% in LQT3 patients.[39] The number of events that occurred till the age of 40 was also higher in LQT1 (63%) or LQT2 (46%) than LQT3 (18%). In contrast, the likelihood of death per cardiac event is much higher in LQT3 (20%) than in LQT1 and LQT2 (4%).[39]

The triggers for arrhythmic episodes and cardiac events also depend on the genotype.

LQT1 patients experience the majority of their events during intensive physical exercise, emotional stress, or other conditions associated with elevated sympathetic activity such as anger and fright.[40] Very specific and particularly frequently occurring triggers for malignant ventricular arrhythmias in LQT1 patients are swimming and diving.[41,42] The symptoms in LQT2 patients occur both during exercise and in rest. In fact, 15% of events occurred in rest and/or sleep. Auditory stimuli such as telephone ringing, doorbell ringing, or the sound of an alarm clock are very specific triggers for LQT2-related life-threatening arrhythmias.[43] Patients with LQT3 are at particularly high risk for cardiac events at rest or during sleep, because their QT-interval is excessively prolonged at slow heart rates.[40]

9.5.3 Clinical Diagnosis

The clinical diagnosis of LQTS is made by a solid evaluation of the history of the patient. In addition, a careful inquiry about family history of unexplained sudden death or recurrent syncope at young age is of great importance. Take notice of unexpected drowning in a good swimmer, road traffic accidents without obvious cause, familial epilepsy, and sudden infant death, which are all suspicions for possible malignant ventricular arrhythmias. Finally, sudden death with negative postmortem examination should trigger a family investigation for several cardio-genetic diseases including LQTS.

In addition to clinical history, it is important to conduct a precise examination of the electrocardiogram (ECG) at rest and/or during exercise (Table 9.3).

Table 9.3 Electrocardiographic characteristics in LQTS type 1-3

	LQT 1	LQT 2	LQT 3
Normal QTc (<440 ms) male (%)[109]	56	33	29
Normal QTc (<460 ms) female (%)[109]	64	55	61
QTc shortening with exercise	<Normal	Normal	>Normal
Specific ST-T wave abnormalities	Broad-based and prolonged T waves	Widened, bifid T; low amplitude T in lead II, III, and avF	Long isoelectric segment with late appearing T with sharp deflection
Signs of sinus node dysfunction at rest[38,63]	+	–	++
Signs of sinus node dysfunction during exercise[38,63]	+	–	–
QT prolongation in response to epinephrine (steady state)[67]	+	–	–

Screening ECGs from family members may be informative as well. Structural heart diseases should be excluded by echocardiography or MRI-scanning.

9.5.4 Clinical Diagnostic Criteria

The typical cases of LQTS patients present no diagnostic difficulty for most of the physicians aware of the disease. In other cases it can be troublesome and more variables are necessary than clinical history and ECG only. To overcome these problems, diagnostic criteria were first proposed in 1985 by Schwartz and colleagues.[44] The latest diagnostic criteria are listed in Table 9.4.[45] The score ranges from 0 to 9 points and contains three diagnostic probabilities: low probability (0–1 point), intermediate probability (2–3 points), and high probability of LQTS (3½ points or more).

One should consider that these criteria have been defined in the pre-molecular era and should be used with common sense. They should be used only for the diagnosis of LQTS for a patient who is suspected to have the disease on clinical ground.

9.5.5 QT-Interval

The QT-interval on the ECG is defined as the time interval between the onset of QRS complex and the end of the T-wave. The end of the T-wave is defined as the intersection of a tangent[46] to the steepest slope of the last limb of the T-wave and the baseline, in lead II or V5 and V6. This value has to be corrected for the heart rate according to the Bazett's formula by dividing it by the square root of the preceding RR interval in seconds. (QTc = QT-interval/square root of RR interval).[47] However, at slow heart rates there is an overcorrection and at fast heart rates there is an undercorrection of the QTc. Calculations should be based on the longest QT-interval measured. When sinus arrhythmia exists, an average QTc of at least three consecutive QRST complexes in lead II has to be calculated.[48] It is of great importance that the QTc is calculated manually by a physician with expertise in LQTS because it can be quite difficult in some cases.[49,50]

Before the beginning of the molecular era in LQTS, QTc in excess of 440 ms in males and 460 ms in females were considered prolonged.[51,52] The longer QTc in women become evident after puberty, suggesting a role for hormonal changes. Nowadays we know that there is a considerable overlap in QTc between "normals" and LQTS patients. Healthy individuals can have QTc up to 480 ms and many genetically proven LQTS patients have QTc within normal limits (<440–460 ms). This means that a normal QTc does not rule out a genetic predisposition to LQTS,[53] a consequence of the reduced penetrance of the LQTS. In fact, 10–50% (depending on genotype) of genetically proven LQTS patients do not show overt QT prolongation on a baseline ECG.[11,54,55] However, data

Table 9.4 1993–2006 Long QT syndrome (LQTS) diagnostic criteria (From Crotti et al.[45])

			Points
Electrocardiographic findings [a]			
A	QTc[b]	> 480 ms	3
		460 – 470 ms	2
		450 – 459 (male) ms	1
B	Torsade de pointes [c]		2
C	T wave alternans		1
D	Notched T wave in 3 leads		1
E	Low heart rate for age [d]		0.5
Clinical history			
A	Syncope [c]	with stress	2
		without stress	1
B	Congenital deafness		0.5
Family history [e]			
A	Family members with definite lqts		1
B	Unexplained sudden cardiac death below age 30 among immediate family members		0.5

[a]In the absence of medications or disorders known to affect these electrocardiographic features
[b]QTc calculated by Bazett's formula where QTc = QT/√RR
[c]Mutually exclusive
[d]Resting heart rate below the 2nd percentile for age
[e]The same family member cannot be counted in A and B
Score: ≤1 point = low probability of LQTS
>1 to 3 points = intermediate probability of LQTS
≥3.5 points = high probability of LQTS

from the Mayo LQTS clinic show that a QTc value < 400 ms has a virtually 100% negative predictive value and that a QTc of 480 ms or more almost always indicates acquired or congenital LQTS. For screening purposes, in the general population, we advise to consider QTc of 470 ms or more as prolonged to maximize both the positive and negative predictive values.[12] In general practice, one can use the three-level ECG classification, which is described by Goldenberg et al.[56] (Table 9.5).

9.5.6 T Wave Morphology

LQTS patients often have an abnormal, genotype specific morphology of the ST-T segments, independently of the duration of the QT-interval (Fig. 9.3a–c). These T-wave abnormalities may be evident particularly in the lateral precordial leads. LQT1 patients typically have broad-based and prolonged T waves.[57] In contrast, LQT2 patients typically have characteristic widened bifid T-waves, mostly low in amplitude in the extremity leads.[58] A notch above the apex of the T-wave (G2 notch) often indicate LQT2, but is not commonly found

Table 9.5 Practical three-level ECG classification (Adapted from Goldenberg et al.[56])

Rating	1–15 years	Adult male	Adult female
Normal	<440 ms	<430 ms	<450 ms
Borderline	440–460	430–450	450–470
Prolonged	>460 ms	>450 ms	>470 ms

Fig. 9.3 (a) LQT1; (b) LQT2; (c) LQT3

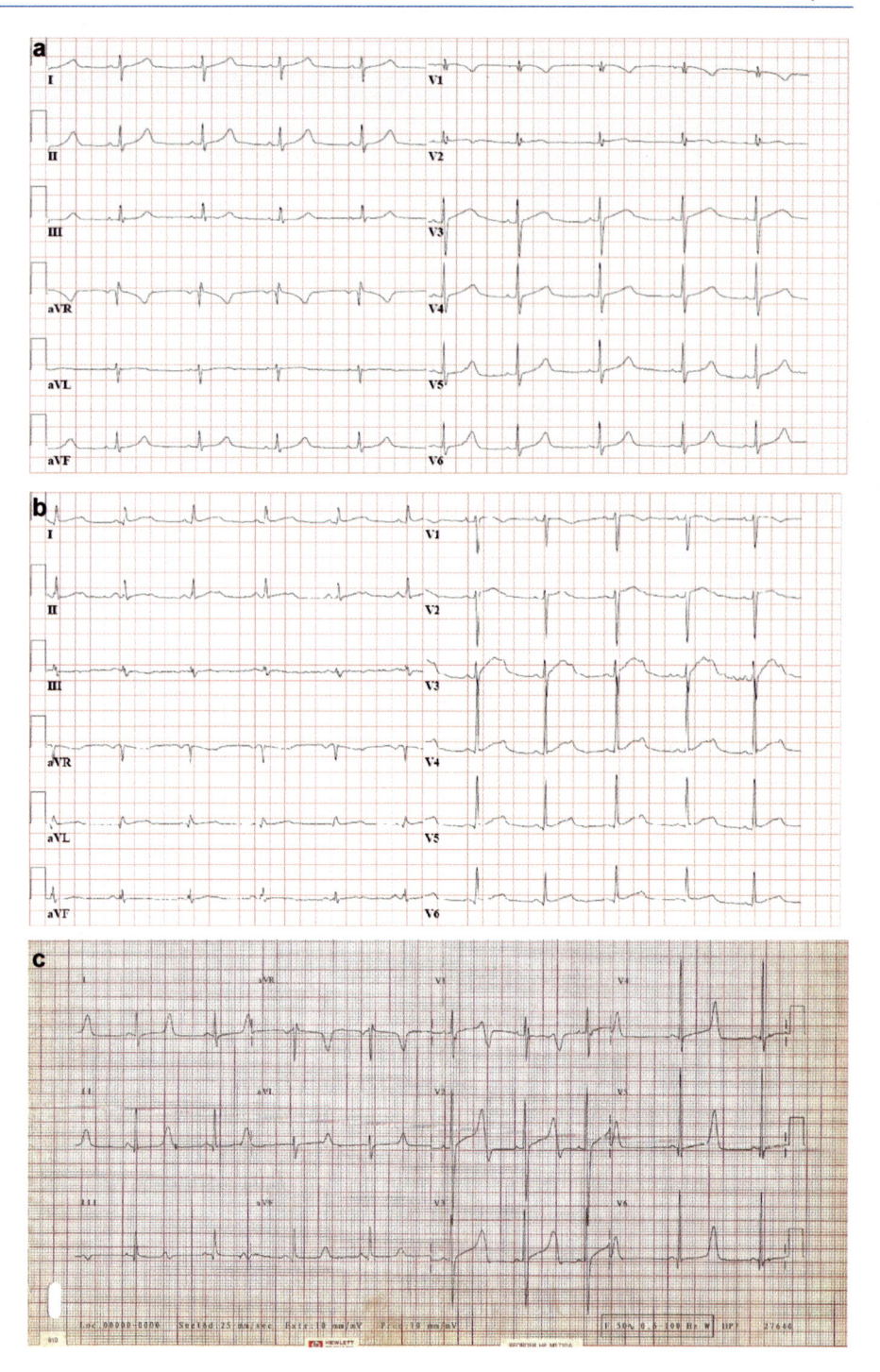

on the baseline ECG.[59] They can appear, however, under low-dose epinephrine testing and so unmask a concealed LQT2.[60] Finally, LQT3 patients have a long isoelectric segment with a late-appearing T-wave with a relatively sharp deflection and a normal duration; however, this can also be seen in LQT1 patients.

9.5.7 T Wave Alternans

Beat-to-beat alternation in the T wave morphology, in polarity or amplitude, may be present in rest for brief moments but most commonly appears during emotional or physical stress and may precede TdP. It is a marker of major electrical instability and regional heterogeneity of repolarization and it identifies patients at high risk for malignant arrhythmias.[61]

9.5.8 QT-Dispersion

QT-dispersion is defined as the difference between the maximal and minimal QT-intervals in the 12 standard leads. It is increased in LQTS patients and it has been described as an arrhythmic marker; however, normal values are not available.[62]

9.5.9 Sinus Node Dysfunction

Signs of sinus node dysfunction, consisting of sinus bradycardia, sinus pauses, and a lower than expected heart rate during exercise have been reported in LQTS patients.[38] Slow heart rates can be particularly striking in younger children. In addition, LQT3 carriers regularly present with significant sinusbradycardia and sinus pauses at rest.[63]

9.5.10 Holter Monitoring

Holter monitoring during 24 or 48 h may aid in the evaluation of LQTS, but caution must be taken because normal values of maximal QTc are lacking at the moment. A holter-recorded maximum QTc greater than 500 ms does not equal LQTS! The value in holter monitoring lies in detecting T wave abnormalities suggestive of LQTS in patients with borderline QT prolongation and uncertain clinical diagnosis.[59] Sometimes 24-h ECG-monitoring can be useful in detecting bradycardia-induced QT prolongation in LQT3 patients or pause-dependent QT prolongation in LQT2 patients. In fact, in many cases of LQT2 patients, the onset of TdP is typically pause dependent, being initiated by a short-long-short sequence of preceding RR intervals.[25] This phenomenon is absent or rare in LQT1 patients. In addition, one can use holter monitoring for detection of ECG-signs of high electrical instability, for example, T wave alternans and QT-dispersion for risk stratification in known LQTS patients.

9.5.11 Exercise Testing

Exercise testing can be performed to identify a concealed LQTS patient (especially LQT1) or can be done in patients having already a diagnostic QTc at rest. LQT1 patients have an inadequate shortening of QTc with increasing heart rates and an exaggerated prolongation of the QT-interval after exercise.[64] In addition, they may have a diminished chronotropic response.[65] In LQT2 patients, the chronotropic response to exercise usually is normal as is the shortening of the QT-interval. In LQT3, the QT-interval shortening is slightly more than normal.[37] Exercise-induced ventricular ectopy is uncommon in LQTS and should prompt suspicion for catecholamine-induced polymorphic ventricular tachycardia.[66]

9.5.12 Epinephrine Stress Test

A substantial part of patients with LQTS have a normal QTc at baseline (10–50%), especially LQT1 patients. To unmask these concealed mutation carriers or low penetrant patients, a provocation test with epinephrine can be useful, again especially in LQT1.[67] The two proposed protocols that exists for epinephrine QT stress testing include the escalating-dose protocol by Ackerman's group ("the Mayo protocol")[67] and the bolus injection followed by brief continuous infusion by Shimizu's group ("the Shimizu protocol").[68]

Both protocols are safe and well tolerated. Induction of TdP or ventricular fibrillation is uncommon. However, their use in clinical practice is debated because normal subjects also showed QT- and QTc prolongation, in varying degree, in response to epinephrine infusion, and normal values are missing at the moment. The specificity of the proposed criteria for epinephrine provocation in diagnosis of the LQTS is variable; however, paradoxical QT prolongation at low-dose epinephrine (Mayo protocol) or a QTc > 600 ms at any dose is highly specific.[69]

9.5.13 Differential Diagnosis

In typical cases of syncope and clear prolonged QT-interval on the ECG the diagnosis of congenital LQTS can be quite simple. In borderline cases, however, the following conditions should be considered: acquired long QT syndrome, vasovagale syncope, orthostatic hypotension, hypertrophic cardiomyopathy, arrhythmogenic right ventricular cardiomyopathy, catecholamine-induced polymorphic ventricular tachycardia, and epilepsy. Most of these disease-entities can be demonstrated or ruled out with additional cardiological evaluation.

The acquired long QT syndrome can be secondary to or associated with drug-therapy, myocardial ischemia, several cardiomyopathies, heart failure, left ventricle hypertrophy, hypokalemia, hypocalcemia, hypomagnesemia, autonomic influences, hypothyroidism, and hypothermia.

Drugs that prolong the QT-interval include antiarrhythmic agents such as sotalol and amiodarone and many non-cardiovascular drugs such as haloperidol. Most drugs that cause TdP block the rapid component of the delayed rectifier current, and previously unrecognized LQTS, of any sub-type, can be identified in 5–20% of patients with drug-induced TdP.[70] In addition, several genetic alterations have been identified that predispose people to drug induced QT-interval prolongation.[70,71]

9.6 Molecular Genetic Diagnosis

Until recently the LQTS was a pure clinical diagnosis, but nowadays commercially available molecular screening is becoming more and more part of the routine diagnostic process. The genetic testing may also provide additional information, which can be used for the risk stratification process and the way of treatment. Detailed genetic counseling by a clinical geneticist and a cardiologist is warranted before testing, particularly for asymptomatic persons for whom the option of not testing must also be discussed.

Genetic testing for the common subtypes of LQTS can identify a mutation in 50–75% of probands in whom the diagnosis appears to be certain on clinical grounds.[72] It is important to notice that a negative genetic test in a subject with clinical LQTS does not rule out the diagnosis! The false negative ones are probably due to technical difficulties with genotyping, noncoding variants, or as yet unidentified disease-associated genes.

The specific clinical picture and/or the typical abnormalities on the ECG can guide the mutation analysis as they can suggest the affected gene in 70–90% of patients.[73,74]

The identification of the pathogenic mutation in the proband is also of great importance for family screening, because it allows for the identification of all affected family members potentially at risk for sudden cardiac death, even those who are asymptomatic and have a normal QTc. These concealed LQTS patients constitute a significant minority (25–50%) of the total LQTS population.[75] Identification of asymptomatic mutation carriers is important because preventive measures as avoidance of QT-interval prolonging drugs can be taken and screening of their children, who may have or develop more severe disease, is indicated as well.

Hundreds of mutations in the 12 genes linked to the LQTS have been identified thus far. Most reported mutations are in the coding regions of the gene, although noncoding mutations, which result in the loss of allele expression, have also been described (in the potassium channel genes). Also larger genomic rearrangements, for example, deleting a large part or the entire KCNQ1 gene, can occur. Most families have their own mutations, which are often termed "private" mutations. In most cases of LQTS, the mutation involved is a missense mutation (>70%). In other cases, it can be a frameshift mutation (10%), in-frame deletion, nonsense mutation, or splice-site mutation.[14]

One should be aware of the fact that genetic testing for LQTS-genes also has the potential for false positive results, since detection of a previously undescribed mutation with unknown significance does not establish the diagnosis. Further analyses by linkage within a family or in vitro studies may be necessary to establish the functional significance of the specific mutation found.

Molecular genetic testing may also be of help in avoiding misdiagnoses. This was demonstrated by the recent data from Taggart and colleagues, who showed that genetic testing of patients in combination with a superb clinical evaluation by a specialized cardiologist may improve the diagnostic process because a large minority of individuals referred as having LQTS were found to be unaffected based on frequently occurring miscalculation of QTc or misinterpretation of symptoms.[76]

In addition, testing of non-LQTS-related genes can lead to the correct diagnosis in specific cases. Tester and colleagues found ryanodine receptor gene mutations responsible for the primary arrhythmia syndrome "catecholaminergic polymorphic ventricular tachycardia" in 17 of 269 (6.3%) patients with negative LQTS genetic testing.[77]

Genetic testing has not been evaluated so far in patients presenting with syncope, borderline QTc interval, and a negative family history. For them the incidence of false positive and false negative test results and their implications for therapy currently remain unknown.

9.7 Risk Stratification

The clinical course of LQTS is variable even within families because of incomplete penetrance. It is influenced by age, gender, genotype, environmental factors, therapy, and probably modifier genes. Continuous *risk assessment* for life-threatening cardiac arrhythmias is warranted in LQTS patients, because the risk for the individual patient may vary during life.

The main clinical risk factors consist of gender, QTc duration, and a history of syncope.

The genetic risk factors for malignant ventricular arrhythmias are mainly determined by the biophysical function of the mutations and the location of the mutation in the gene, rather than the specific type of LQTS.

9.7.1 Gender

Data from the international LQTS Registry demonstrate that the phenotype shows major time-dependent gender differences in the risk of cardiac events (syncope, ACA, SCD).[78] The rate of fatal or near-fatal events in children is significantly higher among boys than among girls throughout childhood, despite the fact that girls have longer QTc intervals. Male gender is independently associated with a significant increase in the risk of these events before age 15, whereas a gender risk reversal was shown to occur after age 14.[79]

During childhood, LQT1 boys have significantly more risk of syncope, ACA or SCD, than LQT1 girls.[80] There was no difference between boys and girls with LQT2 and LQT3. More recent studies analyzed only risk factors for life-threatening cardiac events and did not investigate syncopal events.[81–83] These studies demonstrated that the cumulative risk for ACA and SCD in children (age 1–12) was 5% in boys and 1% in girls ($p<0.001$).[83] After 16 years, females, both in LQT1 and 2, have a higher risk of cardiac events than males. They maintain higher risk than male patients throughout late adolescence and during adulthood.[81–83] In adulthood (age 40–60) the LQTS-related risk in women continues to be high, whereas event rates among affected men are not significantly different from those observed in unaffected men.[84] However, in affected men over age 40 the risk of ACA or death is similar to the risk in affected women. This means that the higher arrhythmic risk for LQTS women over 40 counterbalances the increased male risk due to acquired cardiovascular diseases. The mechanisms behind these age-specific gender differences are unknown. In theory, they are caused by environmental (epi), genetic, or hormonal factors.

9.7.2 QTc Duration

A baseline QTc of >500 ms in LQTS patients is associated with a high risk of syncope, ACA or SCD. Recent data showed that the baseline QTc in LQTS patients is also a major risk factor for life-threatening cardiac events (ACA or SCD) only.[81–83] In adolescents a QTc > 530 ms was associated with a significant increase in the risk of ACA or SCD compared with shorter values.[81] Adults with a QTc > 500 ms carry a significantly increased risk.[82] There appears to be considerable variability in the QTc interval when serial ECGs of one patient are recorded The maximum QTc interval measured at any time seems to be the strongest predictor.[85]

9.7.3 Time-Dependent Syncope

Recent data from the international LQTS Registry demonstrated that a history of syncope, assessed as a time-dependent factor, is the most powerful predictor of life-threatening cardiac events.[81–83] The time of occurrence and frequency of the syncopal events affect outcome. In adolescents (age 10–20) the risk is increased 18-fold when there have been two or more syncopal events in the last 2 years. The risk is increased 12-fold when there was only one syncopal event during the last 2 years. Any syncope 2–10 years ago increased the risk by a factor 3.[81] In adults (age 18–40), time-dependent syncope after age 18 gives a >fivefold increase of the risk, while a syncopal event before age 18 was not a significant factor.[82]

9.7.4 Stratification of Risk

LQTS risk groups may, in general, be categorized as high, intermediate, and low.[86] The high risk group consists of patients with a history of aborted cardiac arrest and/or documented torsades des pointes. The intermediate risk group consists of subjects with time-dependent syncope and/or QTc >500 ms. Patients in the low risk group did not experience any syncope and have a QTc <500 ms.

9.8 Specific Risk Factors

9.8.1 Biophysical Function and Location of the Mutation

Recent genotype–phenotype studies from the international LQTS Registry have provided important information about the effect of location, coding type, and biophysical function of the channel mutations on the manifestation and clinical course of LQTS patients.[87,88] The biophysical function appeared to be an important determinant of outcome. In a study of 600 patients with 77 different KCNQ1 mutations, the dominant-negative ion channel dysfunction had a >twofold increase in the risk of cardiac events compared with those patients who had mutations with haploinsufficiency effects.[88] The same study showed that patients with transmembrane mutations had a significantly higher risk of cardiac events compared with C-terminus mutations. In addition, several recent studies showed that the dominant-negative KCNQ1-A341V mutation is associated with a particularly high clinical severity independently of the ethnic origin of the families.[89–91] The location of the mutation was also shown to be an important factor in LQT2 patients.[87] Patients with pore mutations in the KCNH2 gene were shown to have more severe ECG-signs and clinical manifestations occurring at an earlier age compared with subjects with non-pore mutations.[87] Additional genetic variants, however, may be present and may modify clinical severity of otherwise less severe mutations.[92]

9.8.2 LQTS Genotypes

Recent reports only assessing the life-threatening events ACA and SCD suggest that a specific genotype (LQT1-2-3) does not contribute significantly to the outcome after correcting for the clinical risk factors.[81–83]

9.8.3 Postpartum Period

The risk of syncope and sudden death is decreased during pregnancy. The postpartum period, however, is associated with a significant augmented risk of cardiac events in all types of LQTS, especially in LQT2.[93,94] The increased risk clusters in the 9-month period after delivery. In this period, cardiac events are more common in LQT2 (16%) than in LQT1 (<1%).[95,96] Nearly 10% of female probands experience their first event during this period.[93] The mechanisms by which the arrhythmias are generated are not clear. Many cofactors potentially play a role. Among these, changes in hormone balance (high levels of estrogens and progesterone), fatigue and sleep deprivation, stress, noise (crying of the baby), and anemia might play a role.[97] It is not known whether there is also an association with breastfeeding or the number of previous pregnancies.

Treatment with β-blockers should be continued during and directly after pregnancy. A close cardiac follow-up with serial ECGs and monitoring in a clinical setting after delivery is recommended when QT duration is significantly prolonged in comparison with prepregnancy values or when QTc exceeds 500 ms.[97]

9.8.4 Family History

The severity of symptoms in the proband does not predict the severity in affected family members.[98] A family history of SCD in a first-degree relative is not a significant predictor of outcome during childhood.[83]

9.9 Therapy and Prognosis

Data to guide the management of patients with LQTS have become available generally from large registries and referral centers and therefore may not reflect every patient since more severe cases may be overrepresented. Randomized trials of therapy are lacking because of the low prevalence of the LQTS and the variable penetrance of the disease.

General and genotype-specific lifestyle measures as well as the avoidance of QT-prolonging drugs are applicable to every LQTS patient. In addition to these obligate measures one often has to initiate β-blocker therapy, as most of the episodes of malignant ventricular arrhythmias are due to a sudden increase in sympathetic activity.[44] The efficacy of β-blockers has been recognized since the first description of the syndrome. However, one should remember that β-blocker therapy is not entirely effective in preventing malignant ventricular arrhythmias. Beta-blocker failure can be due to inadequate dosage, noncompliance, concomitant use of QT-prolonging medication, and/or incomplete effectiveness of β-blockers themselves. In addition, one has to know that there might be a period of extra risk after stopping β-blockers due to the up-regulation of β-receptors while on treatment. After initiating β-blocker therapy one has to decide whether this treatment will be safe enough for that patient in particular, because in some cases additional non-pharmacological interventions are necessary.[86]

The overall mortality in untreated, symptomatic LQTS patients is high and approached 50% over 10 years in early series.[99] Especially, remarkable is the mortality of 20% in the first year after the initial syncope.

9.9.1 Symptomatic Patients

9.9.1.1 General Lifestyle Measures

As stated LQTS-related syncope and death are most often adrenergically mediated (Table 9.6). Therefore, restriction of participation in competitive sports and/or athletic activities is generally recommended. However, it is not known whether this restriction should apply to patients, in which adrenergic stressors are not that prominent.

Table 9.6 Therapy in symptomatic LQTS patients

	LQT 1	LQT 2	LQT3
Lifestyle measurements, contraindicated medication, and gene-specific measurements	+	+	+
β-blocker therapy[100,101]	+	+	?
Pacemaker therapy (always in combination with β-blocker therapy)[104]		Pause-dependent TdP	Pronounced sinus bradycardia
Left cardiac sympathetic denervation (LCSD)[103]	Recurrent syncope despite β-blocker therapy Absolute contraindication for β-blocker therapy Electrical storm with ICD shocks		
ICD therapy[106]	Aborted cardiac arrest with or without therapy[a] Recurrent syncope despite β-blocker therapy and/or LCSD Absolute contraindication for β-blocker therapy in some cases Most LQT three patients (QTc > 500 ms) Special request patient or his/her parents, after thorough evaluation of the pros and cons		

TdP torsade de pointes
[a]Debatable in LQT 1 without β-blocker therapy (see Vincent et al.[101])

Cardiac and noncardiac drugs that block the I_{Kr} current and thereby prolong the QT-interval are contraindicated. A drug list of contraindicated medication can be found on www.azcert.org/medical-pros/druglists/drug-lists.cfm or on www.qtdrugs.org or on www.cardiogenetica.nl. Such a list should be given to all LQTS patients.

9.9.1.2 Beta-Adrenergic Blockade

Beta-adrenergic blocking agents represent the therapy of choice in symptomatic LQTS patients, unless specific and valid contraindications exist. Long-acting preparations such as propranolol retard, metoprolol retard, or nadolol are usually used in a maximal tolerated dose. Beta-blockers seldom result in excessive bradycardia, especially if the dosage is increased over several weeks very gradually.

Beta-blockers are extremely effective in LQT1 patients because the impairment of the I_{Ks} current makes them particularly sensitive to cathecholamines and very responsive to β-blockade. Priori and her colleagues demonstrated that cardiac events among genotyped patients receiving β-blockers occurred in 10% of LQT1 patients versus 23% in LQT2 patients and 32% in LQT3 patients during a mean follow-up of 5.2 years. Cardiac arrest occurred in 1.1%, 6.6%, and 14% for LQT1, LQT2, and LQT3, respectively.[100] These data and others showed that therapy with β-blockers alone is most often sufficient for LQT1 patients, also in case of aborted cardiac arrest as initial symptom.[101] Beta-blocker noncompliance and use of QT-prolonging drugs appeared responsible for almost all life-threatening "β-blocker failures" in LQT1 patients.[101] LQT2 and LQT3 patients are less well protected by β-blockers and for them (LQT3 in particular) additional therapies seem necessary in many cases.[100,102]

9.9.1.3 Left Cardiac Sympathetic Denervation

Left cardiac sympathetic denervation (LCSD) consists of the removal of the first four thoracic ganglia and can be performed quite safe without the necessity of thoracotomy. The major complication of the operation, the Horner's syndrome, can almost always be avoided. In approximately 30% of patients, a very modest ptosis will remain after the operation.

Peter Schwartz and his group reported their results on 147 LQTS patients who underwent LCSD in the last 35 years.[103] LCSD is associated with a significant reduction in the incidence of aborted cardiac arrest and syncope in high-risk LQTS patients. However, LCSD is not entirely effective in preventing cardiac events including sudden cardiac death during long-term follow-up. LCSD should be considered in patients with recurrent syncope despite beta-blockade and in patients who experience arrhythmia storms with an implanted defibrillator. Potentially LQT3 patients with major QTc prolongation and a personal or family history of events during rest or sleep should be considered for LCSD.

9.9.1.4 Pacemaker Therapy

Cardiac pacing as a therapy to prevent malignant ventricular arrhythmias in LQTS patients is seldom indicated. However, in some cases it can be highly effective.[104] For patients whose onset of TdP is preceded by a pause, as can be the case in LQT2, a pacemaker as adjunct to other therapy programmed with pause-preventing algorithms might be useful.[25] LQT3 patients with pronounced sinus bradycardia, which can be more prominent on β-blocker therapy, may benefit as well because the QTc lengthened disproportional during slow heart rhythms. Also in infants or young children with 2:1 AV-block pacemaker, implantation remains a reasonable choice as a bridge to the ICD.

However, if a pacemaker is considered, it is probably more logical to implant an ICD with adequate pacing modes. In addition, one should remember that cardiac pacing should always be combined with β-blocker therapy.

9.9.1.5 Implantable Cardioverter Defibrillator

Implantable cardioverter defibrillator (ICD) therapy is widely considered in LQTS patients at high risk for sudden death, but the clinical problem is how to select the appropriate patient for this therapy. It is commonly agreed that in the case of a documented cardiac arrest in a LQTS patient, either on or off therapy, an ICD should be implanted along with the use of β-blockers (Class I indication, level of evidence: A).[105] Recently, however, Vincent et al. showed that resuscitated LQT1 patients (in the absence of β-blocker) may do well on β-blockade alone.[101] Furthermore, implantation of an ICD with continued use of β-blockers can be effective to reduce sudden cardiac death in LQTS patients experiencing syncope and/or VT while receiving β-blockers (Class IIa, level of evidence: B).[105]

The indications for prophylactic ICD implantation in asymptomatic patients are less clear and less uniform. The guidelines indicate that implantation of an ICD with the use of β-blockers may be considered for prophylaxis of sudden cardiac death for patients in categories possibly associated with higher risk of cardiac arrest such as LQT2 and LQT3 (Class IIb, level of evidence: B).[105] Long-term follow-up studies are needed to help decision making. Nevertheless, there has been a major increase in the number of ICDs implanted in LQTS patients, probably in many cases not justified.

The available data from both the USA[106] ICD-LQTS Registry shows that the majority of the implanted patients has not suffered a cardiac arrest and, moreover, many had not even failed β-blocker therapy. One should not forget that the ICD does not prevent occurrence of malignant ventricular arrhythmias and that TdP are frequently self-terminating in these patients. In addition, ICD therapy is not without complications, such as infection and lead complications. The need for several battery and leads replacements, especially when implanted at young age, also remains a major clinical problem.

In our practice, the practical way to interpret the guidelines are as follows: (1) after a cardiac arrest (with or without other therapy) with possible exception for LQT1 patients not on β-blocker, (2) recurrent syncope despite full dose of β-blockade and possibly LCSD, (3) definite contraindication for β-blockers (exceptional), (4) specific patients with high risk characteristics based on age, sex, previous history, genetic subgroup (including sometime mutation-specific features), or presence of typical ECG signs indicating high electrical instability, (5) when requested by the patient or his/her parents after thoroughly explaining the risk/benefit ratio of an ICD and of alternative treatment modalities (β-blocker and LCSD).

When the indication for an ICD is established a decision needs to be made on the type of ICD. A factor to consider is the massive release of catecholamine, triggered by pain and fear that follows an ICD discharge in a conscious (young) patient, which may give rise to further arrhythmias and to further discharges.

At the end of this dramatic vicious circle, the ICD can stop while the ventricular arrhythmia is not terminated. Specifically, to mitigate this effect, one can choose an ICD with special features like a long time prior to discharge to allow a spontaneous return to sinus rhythm after onset of TdP/VF. In addition, a period of relatively rapid pacing in the atrium after an appropriate shock could be of significance in preventing new ventricular arrhythmias and new shocks.[107] However, this latter algorithm is not standard in all ICDs available at this moment.

9.9.1.6 Genotype Specific Measures and Therapies

LQT1 patients should not be allowed to participate in competitive sports. Swimming is particularly dangerous, as 99% of the arrhythmic episodes associated with swimming were shown to occur in LQT1 patients.[40] Hence, swimming should be allowed only under guidance.

As LQT2 patients are at higher risk while awaking from sleep or rest by a sudden noise, it is recommended to remove telephones and alarm clocks from the bedroom, which can cause a startle reaction and initiate a polymorphic ventricular arrhythmia. In addition, it is sensible to wake children in the morning with caution. LQT2 patients are especially vulnerable when their potassium level is low and efforts should be made to maintain a serum potassium level>4 mEq/L. In case of hypokalemia, oral K$^+$-supplements should be given.

Specific data regarding management of LQT3 patients are more limited because of the lower prevalence. There seems no benefit in restricting normal physical activity. An intercom system in the bedroom can be advised to detect a noisy gasping preceding death because of a progressive but slow fall in blood pressure as a consequence of the TdP.

The response to β-blocker therapy seems relatively poor in contrast to LQT1 and LQT2. Because LQT3 patients were shown to have excessive further prolongation of the QT-interval at slow heart rates, β-blocker therapy can even be harmful, although this has not been demonstrated convincingly.[102] Therefore, an early prophylactic ICD implantation might be considered in high-risk LQT3 patients. Sodium channel blockade represents a rational approach for a gene-specific therapy in LQT3, since the causative mutation precipitate an increase in late sodium current via the Nav 1.5 sodium channel. Therefore, in some cases mexiletine can be given in addition to β-blocker therapy after tested efficacious in shortening QTc significantly by more than 40 ms.[102] However, long-term studies are not available and these drugs can therefore not replace the regular treatment options. Ranolazine, a novel drug that reduces late sodium channel current, might be of great potential in the future.[108]

9.9.2 Asymptomatic Patients

Asymptomatic LQTS patients should get the same general and specific lifestyle measures as their symptomatic counterparts. They should also have a list of drugs to be avoided. Given the fact that sudden cardiac death can be the first manifestation of LQTS patients it is considered necessary to prescribe a β-blocker to nearly all asymptomatic patients. However, there are some reasonable exceptions to make. Patients with a normal QTc (<440 ms), for example, have a very low risk and their treatment seems optional; the specific type of mutation they have could possibly influence the choice. In addition, LQT1 males older than 20–25 years, with a QTc < 500 ms, also carry a very low risk. In a minority of cases of asymptomatic patients a prophylactic ICD is indicated or wanted. These are patients with earlier mentioned high risk characteristics or with a special request.

9.10 Family Screening

Since the LQTS is a familial, monogenetic condition, all first degree family members should be screened for this potentially dangerous disease. First-degree family members are those who are sibling (brother or sister), parent, or child of the index patient. Given the autosomal dominant mode of inheritance, all of them have a 50% change of being an asymptomatic or symptomatic gene carrier.

Predictive genetic testing in first degree family members of a LQTS patient is a very important part of prevention, because it enables timely prophylactic measures and therapies in mutation carriers and therefore helps to reduce sudden death at young age. In addition, asymptomatic family members who are carriers may still pass on the mutation to 50% of their children, who in turn may have a more malignant phenotype due to the difference in penetrance.

In the evaluation of first degree relatives of a definitely affected LQTS proband (the index patient) with a documented pathogenic mutation in one of the LQTS-related genes, it is not acceptable to exclude LQTS based upon a normal QTc or a low "Schwartz score." One third of the asymptomatic gene mutation carriers have QTc values within the normal range. Using the "Schwartz criteria," a very low sensitivity of only 19% and a specificity of 99% for diagnosing LQTS in family members of proven LQTS patients were found.[109] Alternatively, these investigators found a cut-off value of the QTc of 430 ms a better way of screening of the family members with a sensitivity of 72% and a specificity of 86%. Therefore, molecular genetic testing (genotyping) is the only definitive diagnostic test for them.

Careful counseling prior to genetic testing is time-consuming but essential and therefore should preferably be done in a specialized multidisciplinary cardio genetic center, in close collaboration between a clinical geneticist and cardiologist. In case of children, a pediatric cardiologist together with psychosocial workers should work in a team to support the children and also the parents. Testing of asymptomatic persons or not yet diagnosed patients can have great influence on very different matters like psychosocial well-being, employment, or insurance issues. It has been documented that the testing procedure in LQTS patients leads to distress, which decreases over time. However, disease-related anxiety persists in the subgroup of carriers, which indicates the need for ongoing psychosocial care.[110]

Although there are no international guidelines, screening of children and prophylactic treatment when indicated has to start before life-threatening symptoms can be expected. This will depend on age and genotype. In the Netherlands we will screen children with LQTS type 1 long before the age of 5 and in particular before they start to swim. In families with LQTS type 2, children should be screened at age 8 and in the case of LQTS type 3 at the beginning of puberty.

As mentioned earlier one cannot exclude the diagnosis of LQTS in a family member on the basis of a normal ECG. If thereforethe person to be screened does not want a DNA-test, one has to deal with him or her as a potential patient with subsequent consequences

9.11 Summary

The congenital LQTS constitute a family of clinically heterogeneous entities, characterized by prolonged QT-intervals on the ECG and often abnormally appearing ST-T segments. The clinical course of the disease is time-dependent and age-specific with respect to gender differences. It predisposes especially young individuals without structural heart disease to malignant polymorphic ventricular arrhythmias, which in turn can lead to (recurrent) syncope, ACA, or even SCD. As such, the LQTS is one of the important causes of sudden cardiac death at age <45 year.

The molecular background of this hereditary disorder became apparent in 1995 and since then 12 different types of LQTS have been described. Hence, the congenital LQTS appears to be a genetically heterogeneous primary electrical heart disease with a monogenetic origin. It is part of the greater family of the genetic cardiac channelopathies.

The diagnosis of the LQTS relies mainly on the personal and family history of the patient in combination with more or less specific abnormalities on the electrocardiogram. In most cases (50–75%), the clinical diagnosis can be confirmed by DNA testing.

All symptomatic and asymptomatic patients with (suspected) LQTS have to receive certain lifestyle measures (general and genotype-specific) and must have cardiac follow-up visits on a structural basis in a specialized center in order to control therapy compliance. In addition, they should get a regularly updated list of prohibited QT-prolonging medications. The pharmacological treatment of patients with the LQTS consists principally of β-blockers, whether the patient is symptomatic or not. LQT1 patients can be treated with β-blockers as sole therapy with reasonable safety as long as they are fully compliant with their medications in a sufficiently high dose and avoid QT-prolonging drugs. It is important to realize that this can be a problem in some adolescents! In LQT2 and LQT3 patients β-blocker failures in arrhythmia prevention are more common and additional non-pharmacological treatment modalities (LCSD, pacemaker, ICD) are needed more often. and permanent cardiological follow-up is indicated. The same applies for all family members of genotype-negative LQTS patients.

In addition to treating the index patient, one has to examine his or her first-degree relatives, who have 50% chance of being a mutation carrier or having the disease, in close collaboration with a clinical geneticist, psychosocial workers, and a pediatric cardiologist.

References

1. Jervell A, Lange-Nielsen F. Congenital deaf-mutism, functional heart disease with prolongation of the Q-T interval and sudden death. *Am Heart J*. 1957 July;54(1):59-68.
2. Romano C, Gemme G, Pongiglione R. Rare cardiac arrythmias of the pediatric age. II. Syncopal attacks due to paroxysmal ventricular fibrillation (presentation of 1st case in Italian pediatric literature. *Clin Pediatr (Bologna)*. 1963 September;45:656-683.
3. Ward OC. A new familial cardiac syndrome in children. *J Irish Med Assoc*. 1964;54:103-106. Ref Type: Generic.
4. Curran ME, Splawski I, Timothy KW, Vincent GM, Green ED, Keating MT. A molecular basis for cardiac arrhythmia: HERG mutations cause long QT syndrome. *Cell*. 1995 March 10;80(5):795-803.
5. Wang Q, Shen J, Splawski I, et al. SCN5A mutations associated with an inherited cardiac arrhythmia, long QT syndrome. *Cell*. 1995 March 10;80(5):805-811.
6. Moss AJ, Kass RS. Long QT syndrome: from channels to cardiac arrhythmias. *J Clin Invest*. 2005 August;115(8):2018-2024.
7. Tan HL, Hou CJ, Lauer MR, Sung RJ. Electrophysiologic mechanisms of the long QT interval syndromes and torsade de pointes. *Ann Intern Med*. 1995 May 1;122(9):701-714.
8. Grunnet M, Hansen RS, Olesen SP. hERG1 channel activators: a new anti-arrhythmic principle. *Prog Biophys Mol Biol*. 2008 October;98(2–3):347-362.
9. el-Sherif N, Caref EB, Yin H, Restivo M. The electrophysiological mechanism of ventricular arrhythmias in the long QT syndrome. Tridimensional mapping of activation and recovery patterns. *Circ Res*. 1996 September;79(3):474-492.
10. Murakawa Y, Sezaki K, Yamashita T, Kanese Y, Omata M. Three-dimensional activation sequence of cesium-induced ventricular arrhythmias. *Am J Physiol*. 1997 September;273(3 Pt 2):H1377-H1385.
11. Ackerman MJ. The long QT syndrome: ion channel diseases of the heart. *Mayo Clin Proc*. 1998 March;73(3):250-269.
12. Schwartz PJ, Stramba-Badiale M, Crotti L, et al. Prevalence of the congenital long-QT syndrome. *Circulation*. 2009 November 3;120(18):1761-1767.
13. Tester DJ, Ackerman MJ. Postmortem long QT syndrome genetic testing for sudden unexplained death in the young. *J Am Coll Cardiol*. 2007 January 16;49(2):240-246.
14. Splawski I, Shen J, Timothy KW, et al. Spectrum of mutations in long-QT syndrome genes. KVLQT1, HERG, SCN5A, KCNE1, and KCNE2. *Circulation*. 2000 September 5;102(10):1178-1185.
15. Goldenberg I, Moss AJ. Long QT syndrome. *J Am Coll Cardiol*. 2008 June 17;51(24):2291-2300.
16. Chen L, Marquardt ML, Tester DJ, Sampson KJ, Ackerman MJ, Kass RS. Mutation of an A-kinase-anchoring protein

causes long-QT syndrome. *Proc Natl Acad Sci USA*. 2007 December 26;104(52):20990-20995.
17. Ueda K, Valdivia C, Medeiros-Domingo A, et al. Syntrophin mutation associated with long QT syndrome through activation of the nNOS-SCN5A macromolecular complex. *Proc Natl Acad Sci USA*. 2008 July 8;105(27):9355-9360.
18. Sanguinetti MC, Curran ME, Spector PS, Keating MT. Spectrum of HERG K+-channel dysfunction in an inherited cardiac arrhythmia. *Proc Natl Acad Sci USA*. 1996 March 5;93(5):2208-2212.
19. Mohler PJ, Schott JJ, Gramolini AO, et al. Ankyrin-B mutation causes type 4 long-QT cardiac arrhythmia and sudden cardiac death. *Nature*. 2003 February 6;421(6923): 634-639.
20. Vatta M, Ackerman MJ, Ye B, et al. Mutant caveolin-3 induces persistent late sodium current and is associated with long-QT syndrome. *Circulation*. 2006 November 14; 114(20):2104-2112.
21. Andersen ED, Krasilnikoff PA, Overvad H. Intermittent muscular weakness, extrasystoles, and multiple developmental anomalies. A new syndrome? *Acta Paediatr Scand*. 1971 September;60(5):559-564.
22. Tawil R, Ptacek LJ, Pavlakis SG, et al. Andersen's syndrome: potassium-sensitive periodic paralysis, ventricular ectopy, and dysmorphic features. *Ann Neurol*. 1994 March;35(3):326-330.
23. Reichenbach H, Meister EM, Theile H. The heart-hand syndrome. A new variant of disorders of heart conduction and syndactylia including osseous changes in hands and feet. *Kinderarztl Prax*. 1992 April;60(2):54-56.
24. Schulze-Bahr E, Wang Q, Wedekind H, et al. KCNE1 mutations cause jervell and Lange-Nielsen syndrome. *Nat Genet*. 1997 November;17(3):267-268.
25. Tan HL, Bardai A, Shimizu W, et al. Genotype-specific onset of arrhythmias in congenital long-QT syndrome: possible therapy implications. *Circulation*. 2006 November 14; 114(20):2096-2103.
26. Viswanathan PC, Rudy Y. Pause induced early afterdepolarizations in the long QT syndrome: a simulation study. *Cardiovasc Res*. 1999 May;42(2):530-542.
27. Marban E, Robinson SW, Wier WG. Mechanisms of arrhythmogenic delayed and early afterdepolarizations in ferret ventricular muscle. *J Clin Invest* 1986 November;78(5): 1185-1192.
28. Burashnikov A, Antzelevitch C. Acceleration-induced action potential prolongation and early afterdepolarizations. *J Cardiovasc Electrophysiol*. 1998 September;9(9):934-948.
29. Antzelevitch C, Pollevick GD, Cordeiro JM, et al. Loss-of-function mutations in the cardiac calcium channel underlie a new clinical entity characterized by ST-segment elevation, short QT intervals, and sudden cardiac death. *Circulation*. 2007 January 30;115(4):442-449.
30. Conrath CE, Opthof T. Ventricular repolarization: an overview of (patho)physiology, sympathetic effects and genetic aspects. *Prog Biophys Mol Biol*. 2006 November;92(3): 269-307.
31. Tsien RW, Giles W, Greengard P. Cyclic AMP mediates the effects of adrenaline on cardiac purkinje fibres. *Nat New Biol*. 1972 December 6;240(101):181-183.
32. Priori SG, Corr PB. Mechanisms underlying early and delayed afterdepolarizations induced by catecholamines. *Am J Physiol*. 1990 June;258(6 Pt 2):H1796-H1805.
33. Yang T, Kanki H, Roden DM. Phosphorylation of the IKs channel complex inhibits drug block: novel mechanism underlying variable antiarrhythmic drug actions. *Circulation*. 2003 July 15;108(2):132-134.
34. Thomas D, Kiehn J, Katus HA, Karle CA. Adrenergic regulation of the rapid component of the cardiac delayed rectifier potassium current, I(Kr), and the underlying hERG ion channel. *Basic Res Cardiol*. 2004 July;99(4):279-287.
35. Veldkamp MW, Verkerk AO, van Ginneken AC, et al. Norepinephrine induces action potential prolongation and early afterdepolarizations in ventricular myocytes isolated from human end-stage failing hearts. *Eur Heart J*. 2001 June;22(11):955-963.
36. Roden DM, Lazzara R, Rosen M, Schwartz PJ, Towbin J, Vincent GM. Multiple mechanisms in the long-QT syndrome. Current knowledge, gaps, and future directions. The SADS Foundation Task Force on LQTS. *Circulation*. 1996 October 15;94(8):1996-2012.
37. Schwartz PJ, Priori SG, Locati EH, et al. Long QT syndrome patients with mutations of the SCN5A and HERG genes have differential responses to Na+channel blockade and to increases in heart rate. Implications for gene-specific therapy. *Circulation*. 1995 December 15;92(12):3381-3386.
38. Moss AJ, Schwartz PJ, Crampton RS, et al. The long QT syndrome. Prospective longitudinal study of 328 families. *Circulation*. 1991 September;84(3):1136-1144.
39. Zareba W, Moss AJ, Schwartz PJ, et al. Influence of genotype on the clinical course of the long-QT syndrome. International Long-QT Syndrome Registry Research Group. *N Engl J Med*. 1998 October 1;339(14):960-965.
40. Schwartz PJ, Priori SG, Spazzolini C, et al. Genotype-phenotype correlation in the long-QT syndrome: gene-specific triggers for life-threatening arrhythmias. *Circulation*. 2001 January 2;103(1):89-95.
41. Moss AJ, Robinson JL, Gessman L, et al. Comparison of clinical and genetic variables of cardiac events associated with loud noise versus swimming among subjects with the long QT syndrome. *Am J Cardiol*. 1999 October 15; 84(8):876-879.
42. Ackerman MJ, Tester DJ, Porter CJ. Swimming, a gene-specific arrhythmogenic trigger for inherited long QT syndrome. *Mayo Clin Proc*. 1999 November;74(11):1088-1094.
43. Wilde AA, Jongbloed RJ, Doevendans PA, et al. Auditory stimuli as a trigger for arrhythmic events differentiate HERG-related (LQTS2) patients from KVLQT1 related patients (LQTS1). *J Am Coll Cardiol*. 1999 February; 33(2):327-332.
44. Schwartz PJ. Idiopathic long QT syndrome: progress and questions. *Am Heart J*. 1985 February;109(2):399-411.
45. Crotti L, Celano G, Dagradi F, Schwartz PJ. Congenital long QT syndrome. *Orphanet J Rare Dis*. 2008;3:18.
46. Lepeschkin E, Surawicz B. The measurement of the Q-T interval of the electrocardiogram. *Circulation*. 1952 September;6(3):378-388.
47. Bazett HC. An analysis of the time-relations of electrocardiograms. *Heart*. 1920;7:353-370.
48. Allan WC, Timothy K, Vincent GM, Palomaki GE, Neveux LM, Haddow JE. Long QT syndrome in children: the value of the rate corrected QT interval in children who present with fainting. *J Med Screen*. 2001;8(4):178-182.
49. Viskin S, Rosovski U, Sands AJ, et al. Inaccurate electrocardiographic interpretation of long QT: the majority of physi-

cians cannot recognize a long QT when they see one. *Heart Rhythm.* 2005 June;2(6):569-574.
50. Postema PG, De Jong JS, van der Bilt IA, Wilde AA. Accurate electrocardiographic assessment of the QT interval: teach the tangent. *Heart Rhythm.* 2008 July;5(7):1015-1018.
51. Moss AJ, Schwartz PJ, Crampton RS, Locati E, Carleen E. The long QT syndrome: a prospective international study. *Circulation.* 1985 January;71(1):17-21.
52. Merri M, Benhorin J, Alberti M, Locati E, Moss AJ. Electrocardiographic quantitation of ventricular repolarization. *Circulation.* 1989 November;80(5):1301-1308.
53. Vincent GM, Timothy KW, Leppert M, Keating M. The spectrum of symptoms and QT intervals in carriers of the gene for the long-QT syndrome. *N Engl J Med.* 1992 September 17;327(12):846-852.
54. Moss AJ. Long QT syndromes. *Curr Treat Options Cardiovasc Med.* 2000 August;2(4):317-322.
55. Schwartz PJ. Clinical applicability of molecular biology: the case of the long QT syndrome. *Curr Control Trials Cardiovasc Med.* 2000;1(2):88-91.
56. Goldenberg I, Moss AJ, Zareba W. QT interval: how to measure it and what is "normal". *J Cardiovasc Electrophysiol.* 2006 March;17(3):333-336.
57. Moss AJ, Zareba W, Benhorin J, et al. ECG T-wave patterns in genetically distinct forms of the hereditary long QT syndrome. *Circulation.* 1995 November 15;92(10):2929-2934.
58. Lehmann MH, Suzuki F, Fromm BS, et al. T wave "humps" as a potential electrocardiographic marker of the long QT syndrome. *J Am Coll Cardiol.* 1994 September;24(3):746-754.
59. Lupoglazoff JM, Denjoy I, Berthet M, et al. Notched T waves on Holter recordings enhance detection of patients with LQt2 (HERG) mutations. *Circulation.* 2001 February 27;103(8):1095-1101.
60. Khositseth A, Hejlik J, Shen WK, Ackerman MJ. Epinephrine-induced T-wave notching in congenital long QT syndrome. *Heart Rhythm.* 2005 February;2(2):141-146.
61. Zareba W, Moss AJ, le CS, Hall WJ. T wave alternans in idiopathic long QT syndrome. *J Am Coll Cardiol.* 1994 June;23(7):1541-1546.
62. Napolitano C, Priori SG, Schwartz PJ. Significance of QT dispersion in the long QT syndrome. *Prog Cardiovasc Dis.* 2000 March;42(5):345-350.
63. Veldkamp MW, Wilders R, Baartscheer A, Zegers JG, Bezzina CR, Wilde AA. Contribution of sodium channel mutations to bradycardia and sinus node dysfunction in LQT3 families. *Circ Res.* 2003 May 16;92(9):976-983.
64. Vincent GM, Jaiswal D, Timothy KW. Effects of exercise on heart rate, QT, QTc and QT/QS2 in the Romano-Ward inherited long QT syndrome. *Am J Cardiol.* 1991 August 15;68(5):498-503.
65. Swan H, Viitasalo M, Piippo K, Laitinen P, Kontula K, Toivonen L. Sinus node function and ventricular repolarization during exercise stress test in long QT syndrome patients with KvLQT1 and HERG potassium channel defects. *J Am Coll Cardiol.* 1999 September;34(3):823-829.
66. Horner JM, Ackerman MJ. Ventricular ectopy during treadmill exercise stress testing in the evaluation of long QT syndrome. *Heart Rhythm.* 2008 December;5(12):1690-1694.
67. Ackerman MJ, Khositseth A, Tester DJ, Hejlik JB, Shen WK, Porter CB. Epinephrine-induced QT interval prolongation: a gene-specific paradoxical response in congenital long QT syndrome. *Mayo Clin Proc.* 2002 May;77(5):413-421.
68. Shimizu W, Noda T, Takaki H, et al. Epinephrine unmasks latent mutation carriers with LQT1 form of congenital long-QT syndrome. *J Am Coll Cardiol.* 2003 February 19;41(4):633-642.
69. Magnano AR, Talathoti N, Hallur R, Bloomfield DM, Garan H. Sympathomimetic infusion and cardiac repolarization: the normative effects of epinephrine and isoproterenol in healthy subjects. *J Cardiovasc Electrophysiol.* 2006 September;17(9):983-989.
70. Paulussen AD, Gilissen RA, Armstrong M, et al. Genetic variations of KCNQ1, KCNH2, SCN5A, KCNE1, and KCNE2 in drug-induced long QT syndrome patients. *J Mol Med.* 2004 March;82(3):182-188.
71. Yang P, Kanki H, Drolet B, et al. Allelic variants in long-QT disease genes in patients with drug-associated torsades de pointes. *Circulation.* 2002 April 23;105(16):1943-1948.
72. Roden DM. Clinical practice. Long-QT syndrome. *N Engl J Med.* 2008 January 10;358(2):169-176.
73. Van LI, Birnie E, Alders M, Jongbloed RJ, Le MH, Wilde AA. The use of genotype-phenotype correlations in mutation analysis for the long QT syndrome. *J Med Genet.* 2003 February;40(2):141-145.
74. Donger C, Denjoy I, Berthet M, et al. KVLQT1 C-terminal missense mutation causes a forme fruste long-QT syndrome. *Circulation.* 1997 November 4;96(9):2778-2781.
75. Priori SG, Napolitano C, Schwartz PJ. Low penetrance in the long-QT syndrome: clinical impact. *Circulation.* 1999 February 2;99(4):529-533.
76. Taggart NW, Haglund CM, Tester DJ, Ackerman MJ. Diagnostic miscues in congenital long-QT syndrome. *Circulation.* 2007 May 22;115(20):2613-2620.
77. Tester DJ, Kopplin LJ, Will ML, Ackerman MJ. Spectrum and prevalence of cardiac ryanodine receptor (RyR2) mutations in a cohort of unrelated patients referred explicitly for long QT syndrome genetic testing. *Heart Rhythm.* 2005 October;2(10):1099-1105.
78. Goldenberg I, Moss A, Zabera W. Time-dependent gender differences in the clinical course of patients with the congenital long-QT syndrome. In: Wang P, Hsia H, Al-Ahmad A, Zei P, eds. *Ventricular Arrhythmias and Sudden Cardiac Death Mechanism.* Malden, MA: Blackwell Publishing; 2008:28-36.
79. Locati EH, Zareba W, Moss AJ, et al. Age- and sex-related differences in clinical manifestations in patients with congenital long-QT syndrome: findings from the International LQTS Registry. *Circulation.* 1998 June 9;97(22):2237-2244.
80. Zareba W, Moss AJ, Locati EH, et al. Modulating effects of age and gender on the clinical course of long QT syndrome by genotype. *J Am Coll Cardiol.* 2003 July 2;42(1):103-109.
81. Hobbs JB, Peterson DR, Moss AJ, et al. Risk of aborted cardiac arrest or sudden cardiac death during adolescence in the long-QT syndrome. *JAMA.* 2006 September 13;296(10):1249-1254.
82. Sauer AJ, Moss AJ, McNitt S, et al. Long QT syndrome in adults. *J Am Coll Cardiol.* 2007 January 23;49(3):329-337.
83. Goldenberg I, Moss AJ, Peterson DR, et al. Risk factors for aborted cardiac arrest and sudden cardiac death in children with the congenital long-QT syndrome. *Circulation.* 2008 April 29;117(17):2184-2191.

84. Goldenberg I, Moss AJ, Bradley J, et al. Long-QT syndrome after age 40. *Circulation*. 2008 April 29;117(17):2192-2201.
85. Goldenberg I, Mathew J, Moss AJ, et al. Corrected QT variability in serial electrocardiograms in long QT syndrome: the importance of the maximum corrected QT for risk stratification. *J Am Coll Cardiol*. 2006 September 5;48(5):1047-1052.
86. Moss AJ, Zareba W, Hall WJ, et al. Effectiveness and limitations of beta-blocker therapy in congenital long-QT syndrome. *Circulation*. 2000 February 15;101(6):616-623.
87. Moss AJ, Zareba W, Kaufman ES, et al. Increased risk of arrhythmic events in long-QT syndrome with mutations in the pore region of the human ether-a-go-go-related gene potassium channel. *Circulation*. 2002 February 19;105(7):794-799.
88. Moss AJ, Shimizu W, Wilde AA, et al. Clinical aspects of type-1 long-QT syndrome by location, coding type, and biophysical function of mutations involving the KCNQ1 gene. *Circulation*. 2007 May 15;115(19):2481-2489.
89. Crotti L, Spazzolini C, Schwartz PJ, et al. The common long-QT syndrome mutation KCNQ1/A341V causes unusually severe clinical manifestations in patients with different ethnic backgrounds: toward a mutation-specific risk stratification. *Circulation*. 2007 November 20;116(21):2366-2375.
90. Brink PA, Crotti L, Corfield V, et al. Phenotypic variability and unusual clinical severity of congenital long-QT syndrome in a founder population. *Circulation*. 2005 October 25;112(17):2602-2610.
91. Liu JF, Goldenberg I, Moss AJ, et al. Phenotypic variability in Caucasian and Japanese patients with matched LQT1 mutations. *Ann Noninvasive Electrocardiol*. 2008 July;13(3):234-241.
92. Crotti L, Lundquist AL, Insolia R, et al. KCNH2-K897T is a genetic modifier of latent congenital long-QT syndrome. *Circulation*. 2005 August 30;112(9):1251-1258.
93. Rashba EJ, Zareba W, Moss AJ, et al. Influence of pregnancy on the risk for cardiac events in patients with hereditary long QT syndrome. LQTS Investigators. *Circulation*. 1998 February 10;97(5):451-456.
94. Seth R, Moss AJ, McNitt S, et al. Long QT syndrome and pregnancy. *J Am Coll Cardiol*. 2007 March 13;49(10):1092-1098.
95. Khositseth A, Tester DJ, Will ML, Bell CM, Ackerman MJ. Identification of a common genetic substrate underlying postpartum cardiac events in congenital long QT syndrome. *Heart Rhythm*. 2004 May;1(1):60-64.
96. Heradien MJ, Goosen A, Crotti L, et al. Does pregnancy increase cardiac risk for LQT1 patients with the KCNQ1-A341V mutation? *J Am Coll Cardiol*. 2006 October 3;48(7):1410-1415.
97. Meregalli PG, Westendorp IC, Tan HL, Elsman P, Kok WE, Wilde AA. Pregnancy and the risk of torsades de pointes in congenital long-QT syndrome. *Neth Heart J*. 2008 December;16(12):422-425.
98. Kimbrough J, Moss AJ, Zareba W, et al. Clinical implications for affected parents and siblings of probands with long-QT syndrome. *Circulation*. 2001 July 31;104(5):557-562.
99. Chiang CE. Congenital and acquired long QT syndrome. Current concepts and management. *Cardiol Rev*. 2004 July;12(4):222-234.
100. Priori SG, Napolitano C, Schwartz PJ, et al. Association of long QT syndrome loci and cardiac events among patients treated with beta-blockers. *JAMA*. 2004 September 15;292(11):1341-1344.
101. Vincent GM, Schwartz PJ, Denjoy I, et al. High efficacy of beta-blockers in long-QT syndrome type 1: contribution of noncompliance and QT-prolonging drugs to the occurrence of beta-blocker treatment "failures". *Circulation*. 2009 January 20;119(2):215-221.
102. Schwartz PJ, Spazzolini C, Crotti L. All LQT3 patients need an ICD: true or false? *Heart Rhythm*. 2009 January;6(1):113-120.
103. Schwartz PJ, Priori SG, Cerrone M, et al. Left cardiac sympathetic denervation in the management of high-risk patients affected by the long-QT syndrome. *Circulation*. 2004 April 20;109(15):1826-1833.
104. Viskin S. Cardiac pacing in the long QT syndrome: review of available data and practical recommendations. *J Cardiovasc Electrophysiol*. 2000 May;11(5):593-600.
105. Zipes DP, Camm AJ, Borggrefe M, et al. ACC/AHA/ESC 2006 guidelines for management of patients with ventricular arrhythmias and the prevention of sudden cardiac death: a report of the American College of Cardiology/American Heart Association Task Force and the European Society of Cardiology Committee for Practice Guidelines (Writing Committee to Develop guidelines for management of patients with ventricular arrhythmias and the prevention of sudden cardiac death) developed in collaboration with the European Heart Rhythm Association and the Heart Rhythm Society. *Europace*. 2006 September;8(9):746-837.
106. Zareba W, Moss AJ, Daubert JP, Hall WJ, Robinson JL, Andrews M. Implantable cardioverter defibrillator in high-risk long QT syndrome patients. *J Cardiovasc Electrophysiol*. 2003 April;14(4):337-341.
107. Udo EO, Baars HF, Winter JB, Wilde AA. Not just any ICD device in patients with long-QT syndrome. *Neth Heart J*. 2007 December;15(12):418-421.
108. Moss AJ, Zareba W, Schwarz KQ, Rosero S, McNitt S, Robinson JL. Ranolazine shortens repolarization in patients with sustained inward sodium current due to type-3 long-QT syndrome. *J Cardiovasc Electrophysiol*. 2008 December;19(12):1289-1293.
109. Hofman N, Wilde AA, Tan HL. Diagnostic criteria for congenital long QT syndrome in the era of molecular genetics: do we need a scoring system? *Eur Heart J*. 2007 June;28(11):1399.
110. Hendriks KS, Grosfeld FJ, van Tintelen JP, et al. Can parents adjust to the idea that their child is at risk for a sudden death?: psychological impact of risk for long QT syndrome. *Am J Med Genet A*. 2005 October 1;138A(2):107-112.

The Brugada Syndrome

Begoña Benito, Josep Brugada, Ramon Brugada, and Pedro Brugada

10.1 Introduction

The syndrome of right bundle branch block, persistent ST-segment elevation, and sudden cardiac death, today better known as the *Brugada syndrome*, was described in 1992 as a new clinical entity characterized by a typical electrocardiographic pattern and a susceptibility to develop polymorphic ventricular arrhythmias in the absence of structural heart disease.[1] The description of the first eight patients was followed by other case reports[2,3] and subsequently numerous works appeared either focusing on clinical characteristics of larger populations of patients[4-7] or defining the genetic, molecular, and cellular aspects of the disease.[8-10] In recent years, major advances in clinical and mechanistic knowledge have provided very valuable information about the disease, but remaining questions still propel today a large research activity on the subject. This chapter reviews the current knowledge on clinical, genetic, and molecular features of the Brugada syndrome, and provides updated information supplied by recent clinical and basic studies.

10.2 Definition and Epidemiology

Certain ambiguities appeared in the years following the initial description of the syndrome, basically concerning the characteristic electrocardiographic hallmark and the specific diagnostic criteria. Three repolarization patterns were soon identified (Fig. 10.1)[11]: (a) *type 1* electrocardiogram (ECG) pattern, the one described in the initial report in 1992, in which a coved ST-segment elevation greater than or equal to 2 mm is followed by a negative T-wave, with little or no isoelectric separation, this feature being present in more than one right precordial lead (from V1 to V3); (b) *type 2* ECG pattern, also characterized by an ST-segment elevation but followed by a positive or biphasic T-wave, which results in a saddle-back configuration; (c) *type 3* ECG pattern, a right precordial ST-segment elevation less than or equal to 1 mm either with a coved-type or a saddle-back morphology.

Although all three patterns can be present in Brugada syndrome patients, only the type 1 ECG diagnostic of the syndrome was stated in the first consensus report of the Arrhythmia Working Group of the European Society of Cardiology[11] and subsequently confirmed in the II Consensus Conference published in 2005.[12] Both documents also held that, in order to establish the definite diagnosis of the Brugada syndrome, the type 1 ECG pattern should be documented in combination with one of the following clinical criteria: documented ventricular fibrillation (VF), polymorphic ventricular tachycardia (VT), a family history of sudden death (SD) at an age younger than 45 years, the presence of coved-type ECG in family members, inducibility of ventricular arrhythmias with programmed electrical stimulation, syncope, or nocturnal agonal respiration.[11,12] However, this definition should be applied with caution, especially when causative mutations have been identified and the disorder can be understood as a disease rather than a syndrome.[13,14] In this regard, our data confirm that the only presence of the characteristic type 1 ECG pattern, even with no further clinical criteria, may be associated with SD in the follow-up.[14] This attests for

B. Benito (✉)
Electrophysiology Research Program, Research Center, Montreal Heart Institute, 5000 Rue Belanger, Montreal (QC) H1T 1C8, Canada
e-mail: bbenito@clinic.ub.es

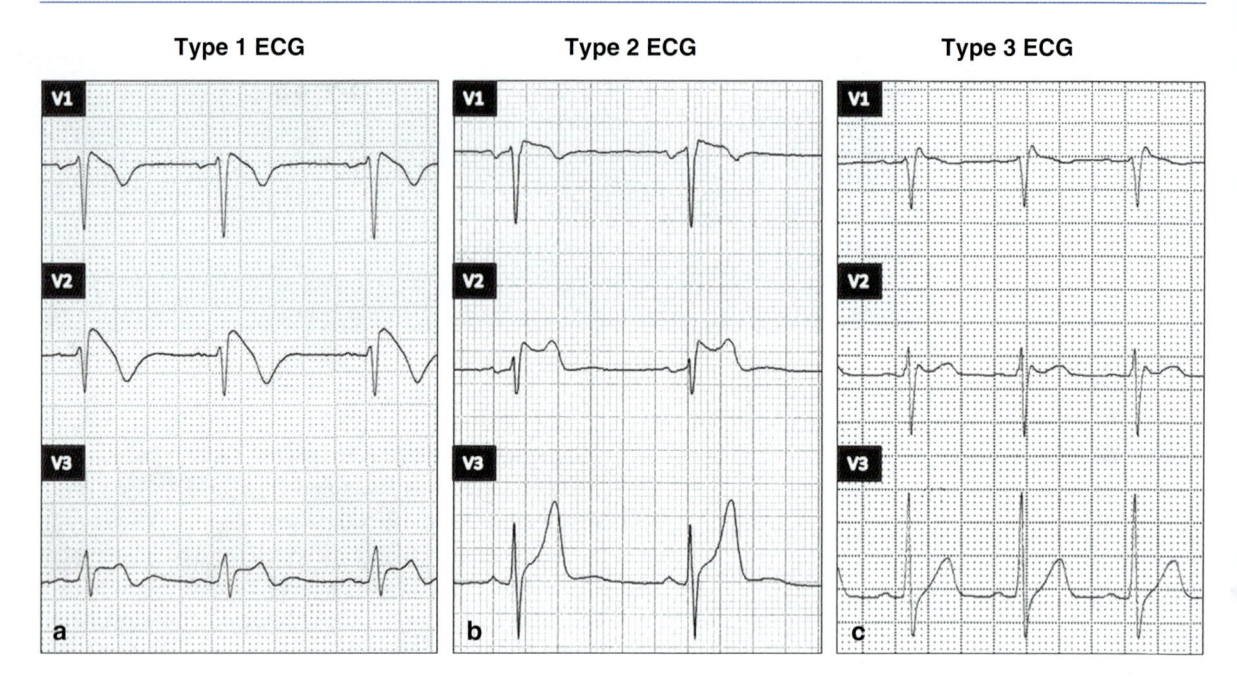

Fig. 10.1 Three different electrocardiogram (*ECG*) patterns in right precordial leads frequently observed in patients with Brugada syndrome: (**a**) type 1, also called coved-type ECG pattern, in which a descendent ST-segment elevation is followed by negative T-waves; (**b**) type 2 or saddle-back pattern, a ST-segment elevation followed by positive or biphasic T-waves; (**c**) type 3, either a coved-type or a saddle-back morphology with ST-segment elevation < 1 mm (see text for more detailed description). A type 1 ECG pattern is required to establish the definite diagnosis of Brugada syndrome

the need of following all patients even when a type 1 ECG pattern is found isolated. Moreover, these patients should be instructed not to use sodium channel blocking agents or other contraindicated medications and (high) fever should be treated promptly. First-degree relatives should be screened and have an ECG.

The Brugada syndrome is included among the so-called *channelopathies*, that is, primary electrical disorders produced by the dysfunction of a cardiac channel participating in the action potential, the electrical change favoring the development of arrhythmias. Characteristically, no underlying structural heart disease exists concomitantly. In fact, the Brugada syndrome is thought to be responsible for 4–12% of all SD and for up to 20% of SD in subjects with structurally normal heart.[12]

The *prevalence* of the Brugada syndrome has been estimated in 5/10,000 inhabitants, although this rate should be understood cautiously, first, because many patients present concealed forms of the disease, thus making it likely that the real prevalence is higher, and second, because important ethnic and geographic differences have been described.[12] For example, a type 1 ECG pattern was observed in 12/10,000 Japanese inhabitants,[15] whereas the few available data on North American and European populations point to a much lower prevalence.[16,17] In fact, the syndrome is considered to be endemic in certain Southeast Asian areas, where it has been long recognized as the sudden unexplained death syndrome (SUDS), also named *bangungot* (in the Philippines), *pokkuri* (in Japan), or *lai tai* (in Thailand), all of them known to be phenotypically, genetically, and functionally identical disorders as the Brugada syndrome.[18]

10.3 Genetics of the Brugada Syndrome

Inheritance in the Brugada syndrome occurs via an autosomal dominant mode of transmission,[12] although in some patients the disease can be sporadic, that is, absent in parents and other relatives.[19] The first *mutations* related to the syndrome were described in 1998 by Chen and coworkers, and were identified in *SCN5A*, the gene encoding the α- subunit of the cardiac sodium channel (locus 3p21, 28 exons).[8] To date, more than 100 other different mutations associated with the syndrome have been found in the same gene.[9,18,20–23] Functional studies performed with expression systems have demonstrated, for the majority of them, a loss of function of the sodium channel that translates into a decrease in sodium current (INa). This can be achieved either through a quantitative decrease (failure in expression) or through a qualitative dysfunction (impaired kinetics) of the sodium channels (Fig. 10.2).[9,18,20–23]

Although *SCN5A* has been the only gene linked to the Brugada syndrome for almost a decade, mutations

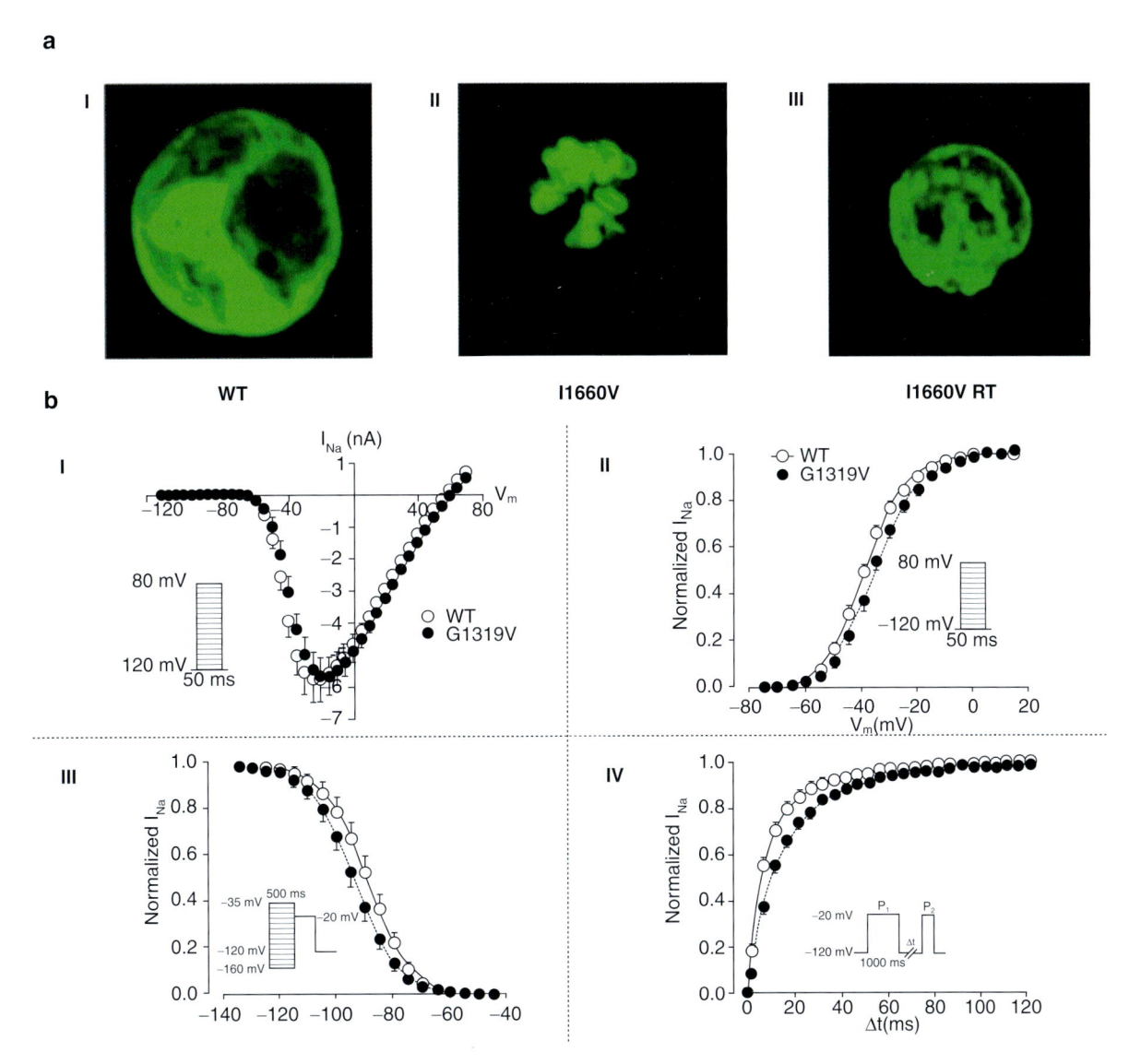

Fig. 10.2 Examples of two different mutations in *SCN5A* leading to a loss of function of the sodium (Na) channel: (**a**) Mutation I1660V, producing a trafficking defect of the Na channel, and thus a decrease of Na channels present in the sarcolemma. Mutant and WT Na channels have been expressed in TSA201 cells and tagged with green fluorescent protein. (A-I) WT channels are present both in the center and the periphery of the cell, suggesting that WT channels are manufactured in the cell center and trafficked to the cell membrane. (A-II) The fluorescence distribution of I1660V channels is essentially localized in intracellular organelles, which suggests that mutant channels are manufactured but remain trapped within the cell. (A-III) Rescue of the mutant channels by incubation at room temperature (Modified from Cordeiro et al.[21] With permission). (**b**) Mutation G1319V, which modifies the kinetics of the sodium channel. Functional studies performed in HEK-293 cells. (B-I) Maximal peak current amplitudes are similar in WT and mutant cells, indicating that the number of functional channels is similar for WT and mutants. (B-II) Voltage dependence of activation, showing a small depolarizing shift in mutant channels compared with WT channels, with no change in slopes. (B-III) Voltage dependence of steady-state inactivation, reflecting enhanced inactivation in mutant channels compared with WT. (B-IV) Recovery from inactivation, which is markedly slowed in G1319V channels (Modified from Casini et al.[22] With permission). *WT* wild type

in *SCN5A* are generally found in only 18–30% of patients,[12] this rate suggesting a genetic heterogeneity of the disease. Accordingly, in the last 2 years four other genes have been found to be linked to the Brugada syndrome. The first of them, the glycerol-3-phosphate dehydrogenase 1-like (*GPD-1 L*), was described in 2007[24] after previous identification of the locus on chromosome 3 (3p22–p24) in 2002.[25] The A280V mutation in *GPD1-L* was shown to induce a sodium loss-of-function effect by affecting the trafficking of the cardiac sodium channel to the cell surface.[24] Very interestingly, two recent reports demonstrate that mutations in genes other than those involved with sodium channel function can be responsible for some cases of Brugada syndrome. Loss-of-function mutations in the genes encoding the cardiac calcium channel Cav1.2 (*CACNA1c*) and its β subunit *CACNB2b* have been linked to a syndrome overlapping short QT and the Brugada ECG pattern.[26] On the other hand, Delpon et al. have described the first family with Brugada syndrome carrying a mutation (R99H) in the *KCNE3* gene, which encodes a beta-subunit that is thought to modulate Kv4.3 channels and be responsible for an increase in transient potassium Ito currents.[27] Together, these findings open up new lines of research, where the concept of Brugada syndrome as a pure sodium channelopathy gives way to the concept of the syndrome as an ionic imbalance between the inward and outward currents during the phase 1 of the action potential.

10.4 Pathophysiology: Cellular and Ionic Mechanisms

Experimental studies have elucidated the cellular and molecular basis for the two main characteristic features of the Brugada syndrome: the specific ECG morphology (ST-segment elevation in right precordial leads) and the susceptibility for VF and SD. Figure 10.3 represents the normal ventricular myocyte action potential and the major ionic currents involved in each one of the phases. Sodium loss-of function conditions, the most encountered disorder in *SCN5A* mutations related to Brugada syndrome,[9,18,20–23] create an *ionic imbalance* between outward and inward positive currents during phase 1. The imbalance favors cell repolarization and the appearance of a particular notch in the action potential (dashed line), which is mediated by a relative increase in the outward transient potassium currents (Ito). From Fig. 10.3, it is easy to conclude that comparable imbalances may appear either by decrease in ICaL (in calcium channel loss-of-function mutations[26]) or absolute increase in Ito (in the recently described *KCNE3* mutation[27]).

Because the Ito density is constitutionally greater in epicardium than in endocardium, the ionic imbalance underlying the Brugada syndrome is heterogeneous through the myocardial wall. This creates a *transmural voltage gradient* between epicardium and endocardium responsible for the characteristic *ST-segment elevation* on the ECG (Fig. 10.4).[28] The imbalance between outward and inward positive currents during the phase 1 sets also the basis for the development of ventricular arrhythmias in the Brugada syndrome. The proposed mechanism would be a *phase-2 reentry*, which is represented in Fig. 10.5. When the notch is such that phase 1 reaches approximately −30 mV, all-or-none repolarization can lead to a complete loss of the action potential dome. The heterogeneity of the loss of the dome among different sites within the epicardium and between the epicardium and the endocardium results in *epicardial and transmural dispersion of repolarization*, respectively (Fig. 10.5a). This substrate may facilitate the development of premature beats, by means of conduction of the action potential dome from the sites where it is maintained to the sites where it is lost (Fig. 10.5b).[10,28] Studies with high-resolution optical mapping in arterially perfused canine right ventricular preparations confirm the presence of a gradient between dome-loss regions and dome-restoration regions in the epicardium, and a subsequent development of a reentrant pathway that rotates in the epicardium and gradually involves the transmural myocardium (Fig. 10.5c).[29]

The understanding that the imbalance between inward and outward ionic currents at phase 1 defines the pathologic substrate of the Brugada syndrome entails multiple applications. First, it provides the basis for the development of experimental models, which have been successfully created by administration of potassium openers (pinacidil), the combination of acetylcholine and a sodium channel blocker (flecainide), or the administration of drugs with combined sodium channel and calcium channel blocker effect (terfenadine).[10,29,30] These interventions create a relative predominance of outward positive currents at the end of phase 1, and thus accentuate the notch. The ionic imbalance hypothesis

Fig. 10.3 Ventricular myocyte action potential and main underlying ionic currents. The shaded area highlights phase 1, mostly determined by the balance between INa, ICa, and Ito. When positive inward currents are impaired with respect to positive outward currents (*), the cell achieves greater degree of repolarization, and the normal dome of the action potential is lost, leading to the development of a particular notch at the end of phase 1 (dashed line). This is the mechanism that is thought to underlie the Brugada syndrome. *ICa* inward calcium current, *INa* inward sodium current, *Ito* transient outward potassium current

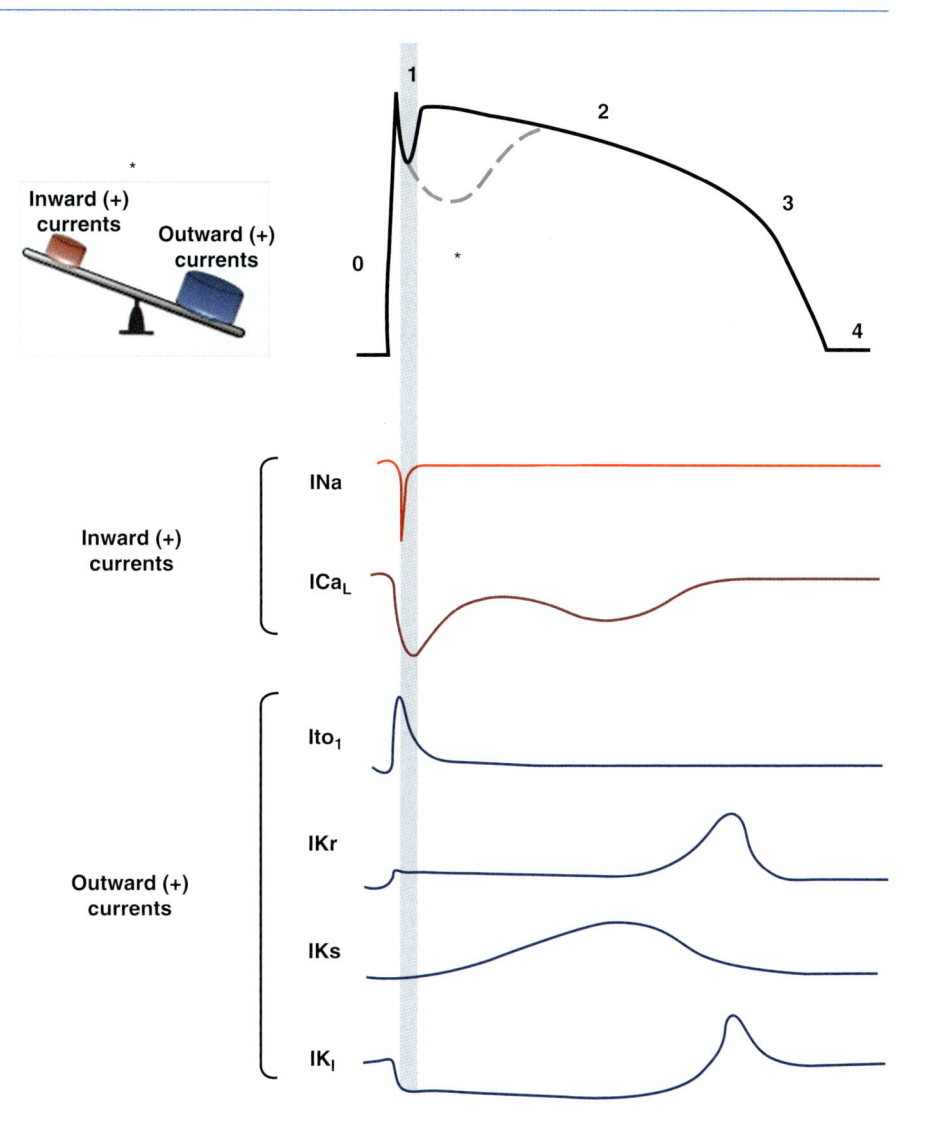

also explains the effects of certain modulators and certain particularities of the syndrome, such as the enhanced phenotypic expression (accompanied by an increased risk of arrhythmias) during vagal situations[31–34] (acetylcholine inhibits calcium currents whereas beta-adrenergic drugs increase them[10,35]) or the worse prognosis in men than in women affected with the disease[36] (men could have constitutionally greater Ito density than women).[37] Likewise, it appears that interventions that decrease inward positive currents (as do sodium channel blockers) could be potentially harmful in patients with Brugada syndrome by increasing ST-segment elevation and the risk of arrhythmic events, although they have been proven on the other hand notwithstanding useful for unmasking concealed forms of the disease.[38]

In contrast, Ito blockers such as quinidine, which reduce the notch at the end of the phase 1, could represent a good therapeutic option for Brugada syndrome patients (see Treatment).

10.5 Clinical Manifestations of the Brugada Syndrome

Patients with Brugada syndrome usually remain asymptomatic. However, *syncope* or *cardiac arrest*, a consequence of an arrhythmic complication such as polymorphic VT or VF, has been described in up to 17–42% of diagnosed individuals.[39–42] This rate probably

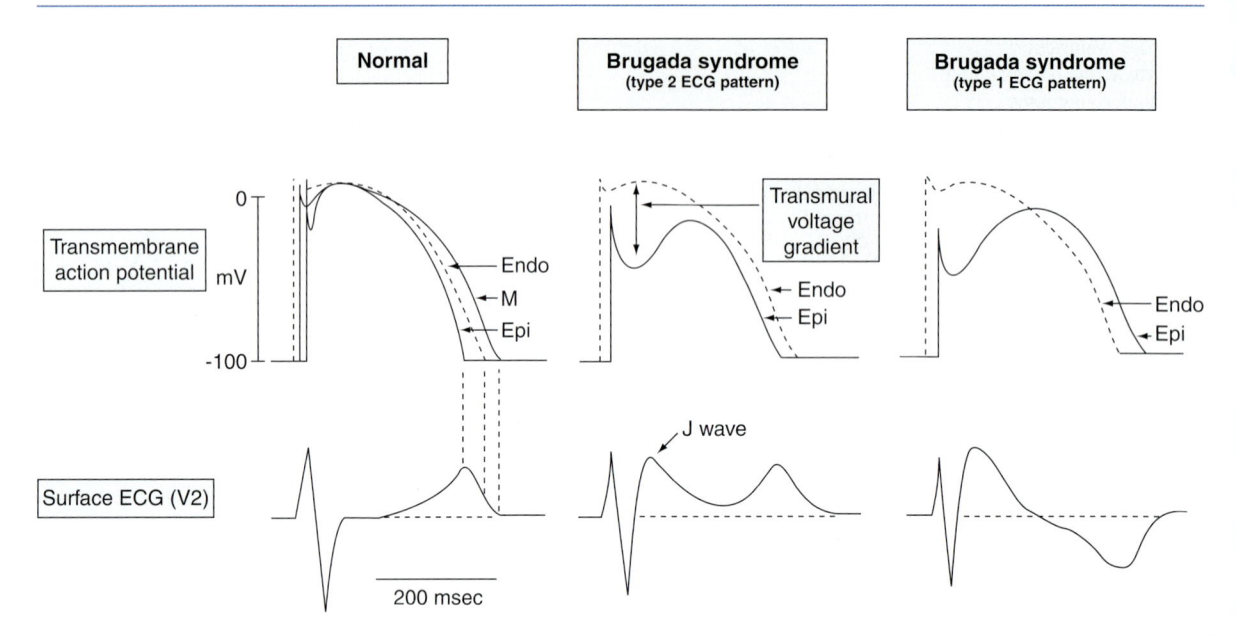

Fig. 10.4 Proposed mechanism that underlies ST-segment elevation in Brugada syndrome. The accentuated notch present in epicardium but not in endocardium gives rise to transmural voltage gradient and J-point elevation (Brugada saddle-back). Further accentuation of the notch may be accompanied by a prolongation of the action potential in epicardium, which becomes longer than in endocardium, thus leading to the development of negative T-waves in addition to the ST-segment elevation (Brugada coved-type) (Modified from Antzelevitch.[28] With permission)

overestimates the real prevalence of symptoms among Brugada syndrome patients, given that most asymptomatic patients remain underdiagnosed. The age of symptom occurrence (especially cardiac arrest) is consistently around the fourth decade of life in all the series (Fig. 10.6),[19] with no definite explanation for this observation thus far. Previous syncope may be present in up to 23% of patients who present with cardiac arrest.[40]

Up to 20% of patients with Brugada syndrome may present *supraventricular arrhythmias*[43] and thus complain of palpitations and/or dizziness. An increased atrial vulnerability to both spontaneous and induced *atrial fibrillation* (AF) has been reported in patients with Brugada syndrome.[44] Other symptoms, such as neurally mediated syncope have been also recently associated with the Brugada syndrome, but their implications on prognosis have not yet been established.[45,46]

As in the case of other sodium channel-related disorders as type-3 long QT syndrome, ventricular arrhythmias in the Brugada syndrome typically occur at rest, especially during nighttime or sleep. In a study by Matsuo et al.,[32] 26 of 30 episodes of VF documented in implantable cardioverter defibrillator (ICD) recordings of Brugada syndrome patients appeared during sleep. This finding has been confirmed in more recent series.[34] As mentioned before, the increase in *vagal tone* mediated by acetyl-choline decreases calcium currents,[10] which could favor arrhythmogenesis through a phase-2 reentry mechanism.

It is currently accepted that the clinical phenotype of the Brugada syndrome is eight to ten times more prevalent in male than in female patients.[12] Consequently, the main clinical studies published thus far include a 71–77% of men, which generally are more symptomatic as compared to women.[39–42] In fact, the observation that SD mainly occur in young men at the time of sleep has long been recognized in South-East Asia for the SUDS, where males from certain small villages used to dress in women's bedclothes since the syndrome is understood as a female spirit searching for young males at night. We recently conducted a study aimed to analyze *gender differences* in a large population of patients with Brugada syndrome.[36] The study population ($n=384$) included 272 men (70.8%) and 112 women (29.2%). General demographic characteristics were similar between male and female patients (mean age 45.8), but, at diagnosis, men presented more frequently with symptoms (syncope in 18%, previous aborted SD in 6%) than women (14% and 1% respectively, $p=0.04$). Male patients also had higher rates of

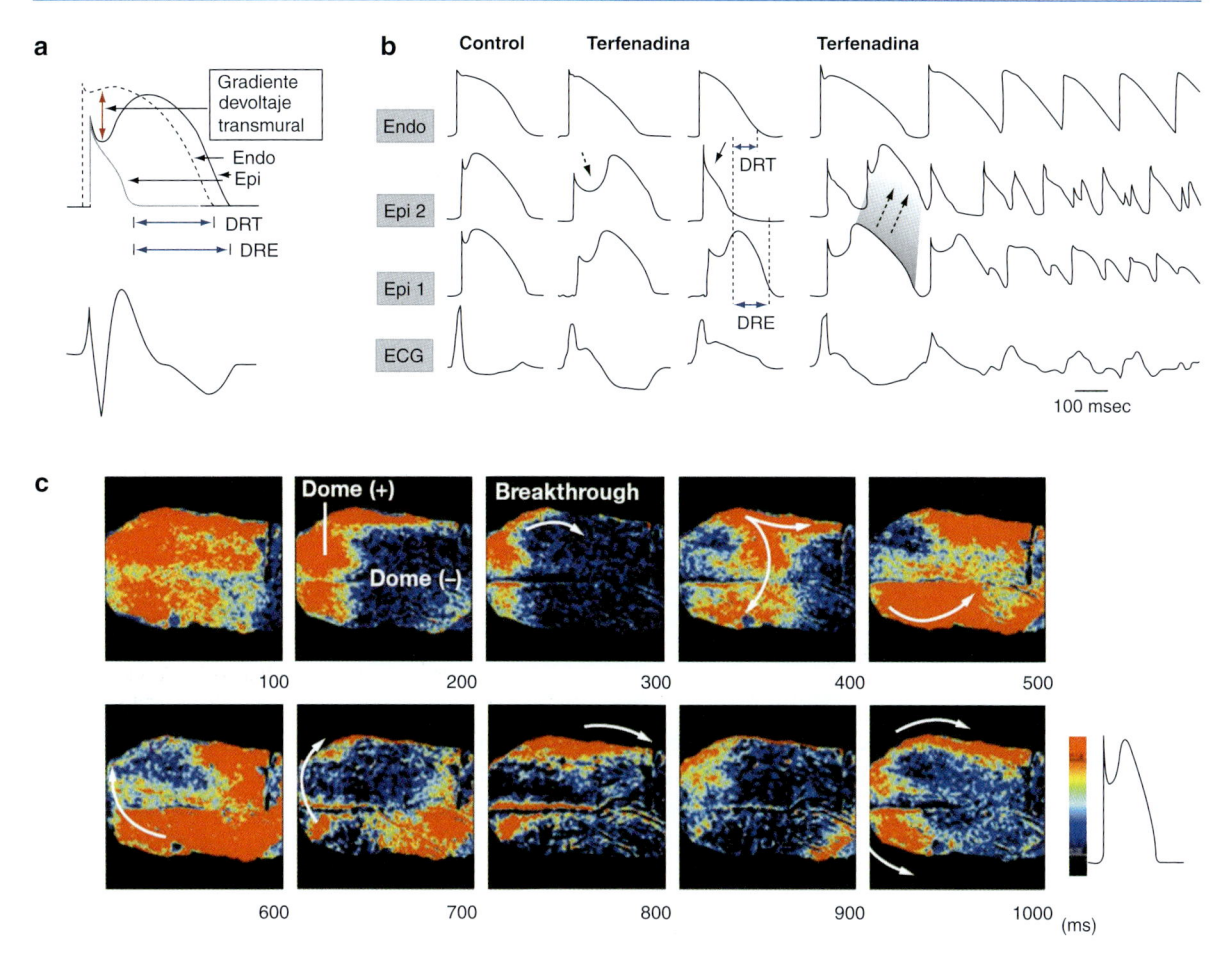

Fig. 10.5 Proposed mechanism that underlies ventricular arrhythmias in Brugada syndrome. (**a**) With a further shift in the balance of currents at the end of phase 1, all-or-none repolarization occurs and leads to a complete loss of the action potential dome (silver drawing). The arrythmogenic substrate is thought to develop when the loss of dome appears at some epicardial sites but not at others, creating both transmural dispersion of repolarization and epicardial dispersion of repolarization (*blue arrows*). At this point, a premature impulse or extrasystole can induce a reentrant arrhythmia (Modified from Antzelevitch.[28] With permission). (**b**) Simultaneous transmembrane action potentials at two epicardial sites and one endocardial site together with a transmural electrocardiogram recorded from a canine arterially perfused right ventricular wedge preparation. The administration of terfenadine (5 μM), a potent sodium and calcium channel blocker, accentuates the epicardial action potential notch (*dashed arrow*), induces all-or-none repolarization heterogeneously at the end of phase 1 (*solid arrow*), and creates epicardial transmural of repolarization (*EDR*) and transmural dispersion of repolarization (*TDR*). Propagation from the site where the dome is maintained (epicardial site 1) to the site where it is lost (epicardial site 2) results in the development of a premature beat that leads to polymorphic ventricular tachycardia by phase 2 reentry (*dashed double arrow*) (Modified from Antzelevitch.[28] With permission). (**c**) High-resolution optical mapping system with transmembrane action potentials from 256 sites simultaneously (epicardial and endocardial surface) of an arterially perfused canine right ventricular wedge preparation. Recording at the beginning of a polymorphic ventricular tachycardia. Propagation by phase-2 reentry occurs from red areas (where the dome is maintained) toward blue areas (where the dome is lost) (Modified from Shimizu et al.[29] With permission)

spontaneous type 1 ECG (47% versus 23%, $p=0.0001$) and inducibility of VF during the electrophysiologic study (32% versus 12% $p = 0.0001$) (Fig. 10.7a).[36] Prognosis also differed between men and women. Cardiac events (defined as SD or documented VF) appeared in 31 males (11.6%) but only in three females (2.8%) during a mean follow-up period 58 ± 48 months (log-rank test $p=0.007$). Kaplan-Meier estimate of cardiac event-free survival according to gender is represented in Fig. 10.7b.

Two main hypotheses have been proposed to explain the gender distinction, perhaps interacting with each other: the sex-related intrinsic differences in ionic currents and the hormonal influence. Di Diego and

Fig. 10.6 Incidence of spontaneous ventricular fibrillation (*VF*) or sudden death (*SD*) according to age in patients with Brugada syndrome. Data from 370 updated patients of the international registry. SD or VF occurred in 120 (32.4%) patients

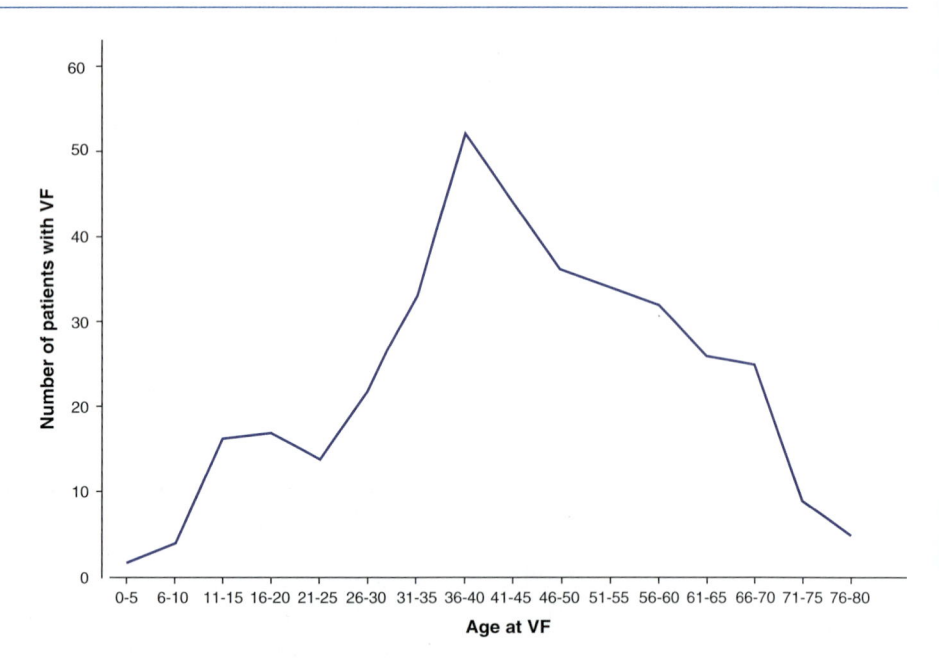

coworkers showed by whole-cell patch-clamp techniques that Ito density was significantly greater in male than female right ventricle epicardia of arterially perfused canine heart preparations, thus providing the basis for the greater ST-segment elevation and the higher occurrence of arrhythmias in men as compared to women.[37] Sex hormones seem also to play a role in the final phenotype. Regression of the typical ECG features has been reported in castrated men,[47] and levels of testosterone seem to be higher in Brugada male patients as compared to controls.[48] In accordance with the hormonal hypothesis, the few available data existing thus far of Brugada syndrome in childhood have not shown a difference in clinical presentation between boys and girls before the age of 16.[49]

Although three of the eight patients reported in the first description of the disease were within pediatric ages,[1] to date little information has been available on the behavior of the Brugada syndrome during *childhood*. Probst et al. recently provided data from a multicenter study including 30 Brugada patients aging less than 16 years (mean age 8 ± 5).[49] More than half (*n* = 17) had been diagnosed during family screening, but symptoms were present in 11 patients (one aborted SD and ten syncope). Interestingly, 10 of the 11 symptomatic patients displayed spontaneous type 1 ECG, and in five of them, symptoms were precipitated by *fever* illnesses. Five patients received an ICD and four were treated with hydroquinidine.[49] During a mean follow-up period of 37 ± 23 months, three patients (10% of the population) experienced SD (*n* = 1) or appropriate shock by ICD (*n* = 2). Importantly, all the three patients had presented with syncope at the time of diagnosis and displayed spontaneous type 1 ECG. The four patients under quinidine remained asymptomatic during 28 ± 24 months of follow-up.[49] Our results on 58 pediatric patients younger than 18 are in line with the ones by Probst et al.[50] In our population, six patients experienced cardiac events during a mean follow-up of 48.8 ± 48 months. Cardiac events occurred more frequently among patients with spontaneous type 1 ECG and among those with inducible VF at the EPS; but, in our series, symptoms at diagnosis were the strongest variable to predict events during follow-up.[50] Though small, these two studies suggest that:

- The Brugada syndrome can manifest during childhood.
- Symptoms in pediatric patients may appear particularly during febrile episodes.
- Symptomatic patients, especially if they present spontaneous type 1 ECG, may be at a high risk of cardiac events in a relatively short period of follow-up.
- Individuals at risk can be protected with an ICD, although *quinidine* could be a good option in certain patients, particularly the youngest.

Fig. 10.7 Gender differences in clinical manifestations of Brugada syndrome. Data from Benito et al.[36] (**a**) Differences on clinical characteristics at the time of first evaluation. Males are more symptomatic, display more pathological electrocardiogram (*ECG*) at baseline and present more inducibility of ventricular fibrillation than females. (**b**) Kaplan-Meier analysis of cardiac events defined as sudden death or documented ventricular fibrillation during follow-up. A total of 31/272 males (11.6%) and 3/112 females (2.8%) experienced cardiac events during a mean follow-up period 58 ± 48 months (log-rank test $p=0.007$). *VF* ventricular fibrillation, *SD* sudden death

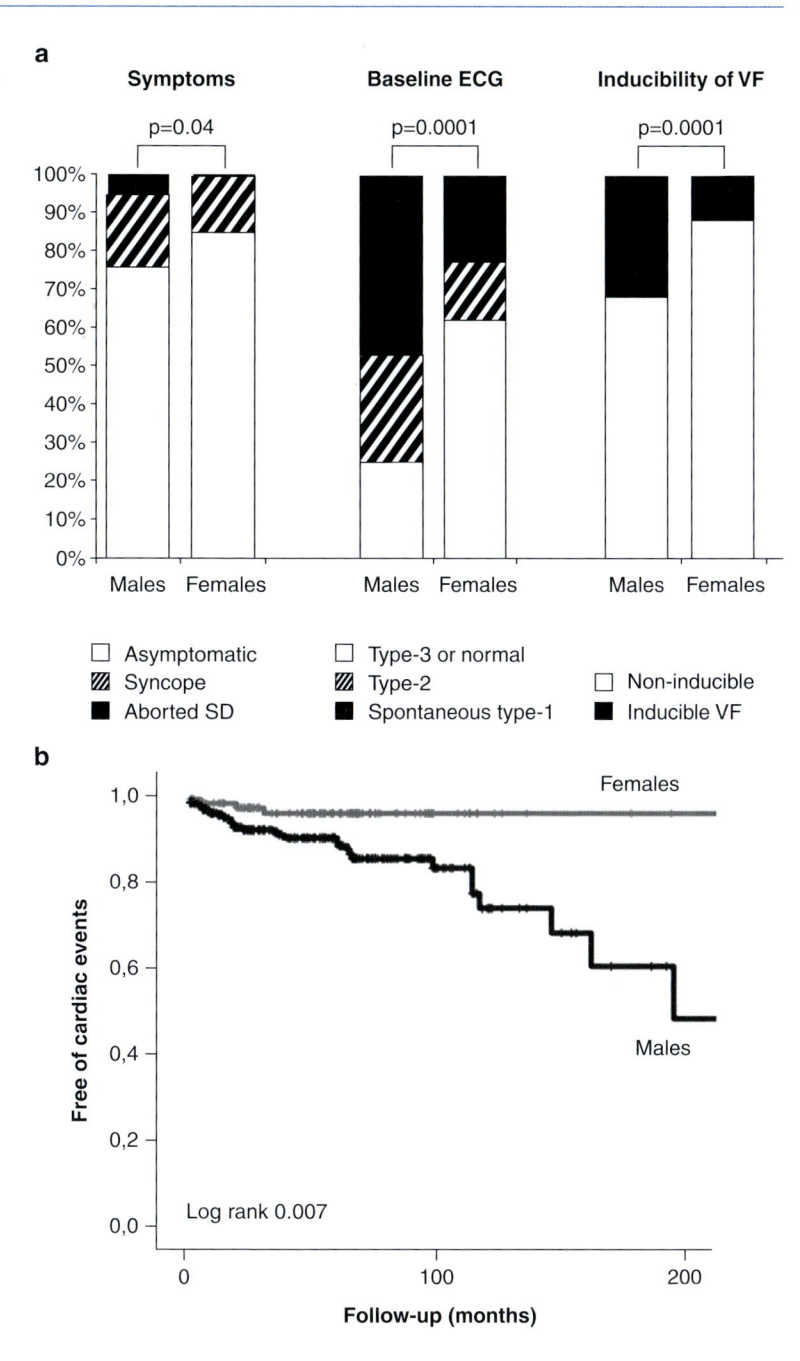

10.6 Diagnosis

10.6.1 ECG Findings

As mentioned above, three types of repolarization have been described (Fig. 10.1), but only is the coved-type ST-segment elevation (type 1 ECG pattern) diagnostic of the syndrome. However, it is important to underline that the ECG typically fluctuates over time in Brugada patients, and thus can change from type 1 to type 2 or type 3 within the same individual or even be transiently normal. The prevalence of spontaneous *ECG fluctuations* was assessed in a work by Veltmann et al., including 310 ECGs on 43 patients followed during 17.7 months.[51] Among 15 patients with an initial diagnostic ECG, 14 revealed at least one nondiagnostic ECG in a median time of 12 days, while 8 out of 28 patients with nondiagnostic ECG developed a type 1 ECG pattern in a median time of 16 days. On the basis of these results, it seems that repetitive ECG recordings may be mandatory in patients with the syndrome.[51,52]

Numerous studies have analyzed the ECG of the Brugada syndrome aiming to identify new electrocardiographic hallmarks and/or risk markers. Pitzalis and coworkers described a prolongation of the corrected QT interval (QTc) in the right but not the left precordial leads after administration of flecainide to patients with Brugada syndrome and nondiagnostic basal ECG.[53] Subsequently, other groups have correlated a QTc ≥ 460 ms in V2 to the occurrence of life-threatening arrhythmias.[54] More recently, the *aVR sign* (an R wave ≥ 3 mm or an R/q ratio ≥ 0.75 in lead aVR) has also been defined as a risk marker of cardiac events in Brugada syndrome, the prominent R-wave possibly reflecting some degree of right ventricular conduction delay and consequently more electrical heterogeneity (Fig. 10.8a).[55] In addition, *T-wave alternans*, also indicative of transmural dispersion of repolarization, has been reported in some cases after administration of sodium blockers, and associated with an increased risk for development of VF (52.9% vs 8.3%, $p<0.001$) (Fig. 10.8b).[56] Finally, a very recent study points out that up to 11% of patients with Brugada syndrome may have a spontaneous *inferior-lateral repolarization pattern*, which also has been linked to a greater rate of symptoms (Fig. 10.8c).[57]

Cardiac conduction disturbances may be present in patients with Brugada syndrome. Both phenotypes (Brugada syndrome and cardiac conduction disorders) can be explained by a reduction in the sodium current, and have been described within the same family carrying a mutation on the *SCN5A* gene.[58] Consequently, conduction parameters (specifically PQ interval, QRS duration, and HV interval) seem to be longer among those patients with Brugada syndrome who are *SCN5A* genetic carriers (and do have a mutation in the sodium channel) as compared to non-*SCN5A* genetic carriers, in which the underlying mechanism or mutation is not identified.[59] These differences have been recently shown to accentuate progressively during follow-up.[60] In a recent study by our group, we observed that some conduction parameters such as QRS duration are increased among symptomatic patients. Indeed, in a population of 200 Brugada patients, of whom 66 (33%) presented symptoms, the optimized cut-off point of QRS in lead V2 ≥ 120 ms gave an odds ratio (OR) of 2.5 (95% CI: 1.4–4.6, $p=0.003$) for being symptomatic.[61]

Although sinus rhythm is the most common, supraventricular arrhythmias, and especially AF, can be found in up to one third of patients with Brugada syndrome.[43,44,62] Other rhythm disorders, as bradycardia secondary to sick sinus syndrome or atrial standstill, have also been reported in association to Brugada syndrome.[28]

10.6.2 Differential Diagnosis and ECG Modulators

It is worth to note that some factors can account for an ECG abnormality that can closely resemble the Brugada ECG (Table 10.1). Importantly, some of them are conditions different than the syndrome and should be carefully excluded during the differential diagnosis, while others may induce ST-segment elevation probably when an underlying genetic predisposition is present.[19]

Modulating factors play a major role in the dynamic nature of the ECG and also may be responsible for the ST-segment elevation in genetically predisposed patients (Table 10.1). As mentioned before, sympathovagal balance, hormones, metabolic factors, and pharmacological agents, by means of specific effects on transmembrane ionic currents, are thought to modulate not only the ECG morphology but also explain the development of ventricular arrhythmias under certain conditions (see Pathophysiology: cellular and ionic mechanisms).[10,31,32,34,38,47,48,63] Temperature may be an important modulator in some patients with Brugada syndrome. Premature inactivation of the sodium channel has been shown to be accentuated at higher temperatures in some *SCN5A* mutations, suggesting that febrile states could unmask certain Brugada patients or temporarily increase the risk of arrhythmias.[64] In fact, several case reports in which fever precipitate the syndrome or an arrhythmic complication have been published in the last years.[65,66] It seems that fever would be a particularly important trigger factor among the pediatric population.[49]

Fig. 10.8 Other electrocardiogram (*ECG*) findings that have been associated to higher risk of ventricular arrhythmias in patients with Brugada syndrome. For detailed explanation, refer to the text. (**a**) aVR sign (Reproduced from Babai et al.[55] With permission); (**b**) T-wave alternans (Adapted from Tada et al.[56] With permission); (**c**) Inferior-lateral early repolarization pattern (Reproduced from Sarzoky et al.[57] With permission)

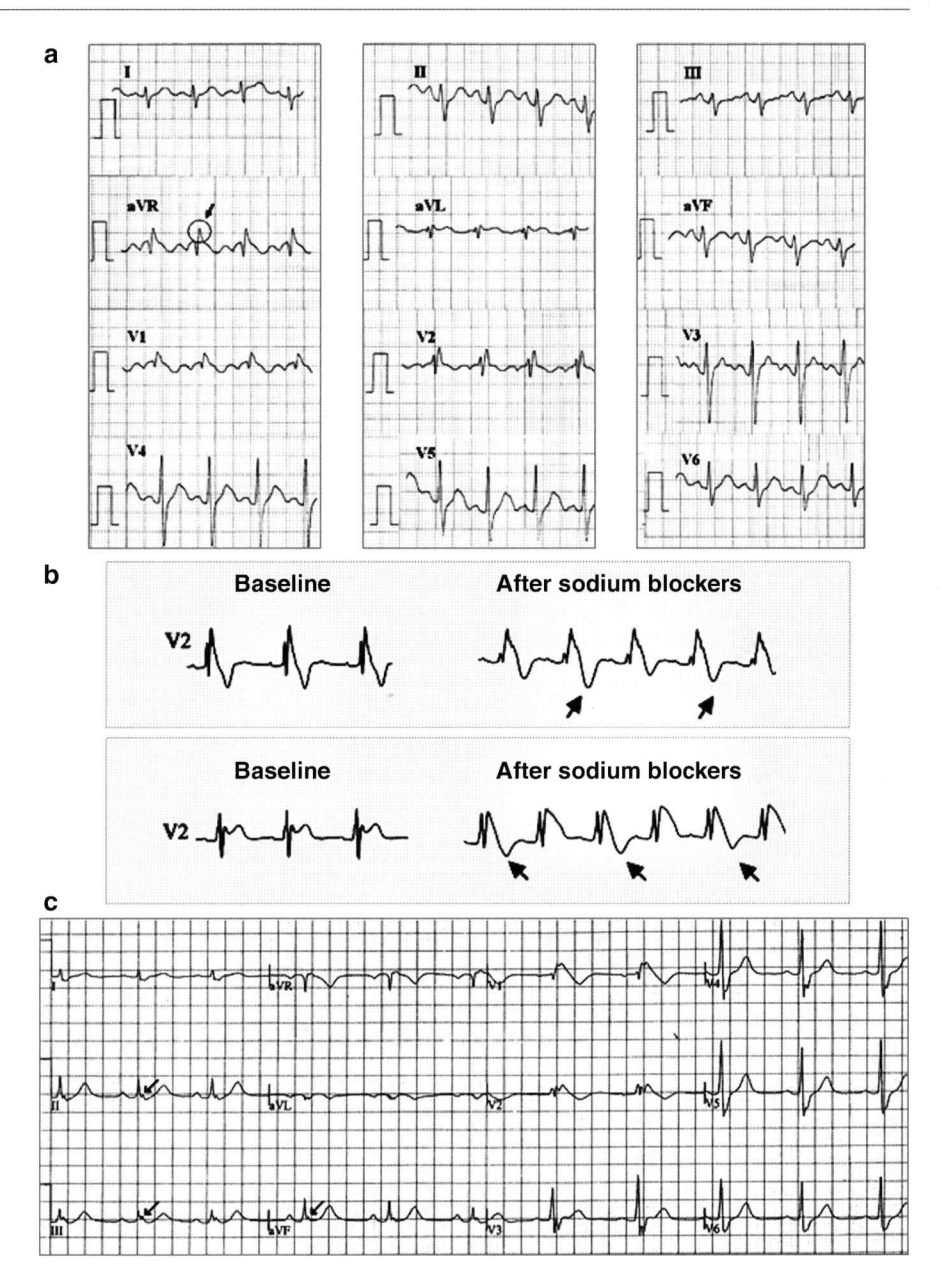

10.6.3 Diagnostic Tools: Pharmacological Tests and Upper Right Precordial Leads

Because the ECG in the Brugada syndrome is dynamic in nature and even can be transiently normal in affected patients, pharmacological provocative tests have been used in an attempt to unmask concealed forms of the disease. Sodium channel blockers, which increase the ionic imbalance at the end of the phase 1 of the action potential by decreasing sodium currents, appear as the most attractive option.[38] *Ajmaline*, *flecainide*, procainamide, pilsicainide, disopyramide, and propafenone have been used,[12] although the specific diagnostic value for all of them has not yet been systematically studied. The recommended dose regimens for the most commonly used drugs are listed in Table 10.2. The

Table 10.1 ECG abnormalities that can lead to ST-segment elevation in V1-V3

Differential diagnosis	Genetic predisposition?
- Atypical right bundle branch block	- Hyperkalemia
- Acute myocardial infarction, especially of RV	- Hypercalcemia
- Acute pericarditis/miopericarditis	- Cocaine intoxication /Alcohol intoxication
- Hemopericardium	- Treatment with:
- Pulmonary embolism	I. Antiarrhythmic drugs:
- Dissecting aortic aneurysm	- Na channel blockers (class IC, class IA)
- Central and autonomic nervous system disorders	- Ca channel blockers
- Duchenne muscular dystrophy	- β-blockers
- Friedreich Ataxia	II. Antianginal drugs:
- LV hypertrophy	- Ca channel blockers
- Arrhythmogenic RV cardiomyopathy	- Nitrates
- Mechanical compression of RV outflow tract	III. Psychotropic drugs:
• Mediastinal tumor	- Tricyclic antidepressants
• Pectus excavatum	- Tetracyclic antidepressants
- After electrical cardioversion	- Phenothiazines
- Early repolarization, especially in athletes	- Selective serotonin reuptake inhibitors
- Hypothermia	- Lithium

Reproduced with permission from Benito et al.[19]

Table 10.2 Drugs used to unmask Brugada syndrome

Drug	Dosage	Administration
Ajmaline	1 mg/kg over 5 min	IV
Flecainide	2 mg/kg over 10 min	IV
	400 mg	PO
Procainamide	10 mg/kg over 10 min	IV
Pilsicainide	1 mg/kg over 10 min	IV

Reproduced with permission from Antzelevtich et al.[12]

diagnosis of Brugada syndrome can be established if a coved-type (type 1) ECG pattern appears or accentuates after the sodium blocker administration. The pharmacological test should be monitored under continuous ECG recording and should be terminated when: (1) the diagnostic test is positive, (2) premature ventricular beats or other arrhythmias develop, and/or (3) QRS widens to greater than or equal to 130% of baseline.[12]

Current data point out that ajmaline is probably the most useful drug in the diagnosis of Brugada syndrome. In a study with 147 individuals from four large families with identified *SCN5A* mutations, ajmaline provided a sensitivity of 80%, a specificity of 94.4%, a positive predictive value of 93.3%, and a negative predictive value of 82.9% for the diagnosis of Brugada syndrome.[67] These results are considerably higher than those obtained for flecainide in another study with 110 genotyped patients, in which the sensitivity, the specificity, and the positive and negative predictive values for the diagnosis were 77%, 80%, 96%, and 36%, respectively.[68] It is important to note the low negative predictive value, which should be taken into account when using flecainide, especially during genetic screening. Ajmaline and flecainide were directly compared in a study with 22 patients with confirmed Brugada syndrome who were subjected to both pharmacological tests. Although the test was positive in 22 of 22 patients after ajmaline administration, only 15 patients showed a positive response to flecainide.[69] Whole-cell patch-clamp experiments revealed that, although both of them have a sodium blocker effect, flecainide reduced Ito to a greater extent than ajmaline, thus explaining its lesser effectiveness.[69]

Given the limitations of the standard ECG, even after administration of sodium blockers, new strategies have been proposed to help in the clinical diagnosis of the Brugada syndrome. Placement of the right precordial leads in an upper position (*second or third intercostal spaces*) can increase the sensitivity of the ECG to detect the Brugada phenotype, both in the presence or absence of a drug challenge,[70,71] although whether the greater sensitivity is at the expense of a lower specificity is still uncertain. Recent data demonstrate that the presence of a type 1 ECG pattern recorded at higher intercostal spaces, even when the standard ECG is normal, can identify a subgroup of patients that behaves similarly in terms of prognosis to those with spontaneous type 1 ECG pattern at standard leads (Fig. 10.9).[72] Therefore, this strategy seems to allow the identification of a subset of patients at risk that would otherwise be underdiagnosed.

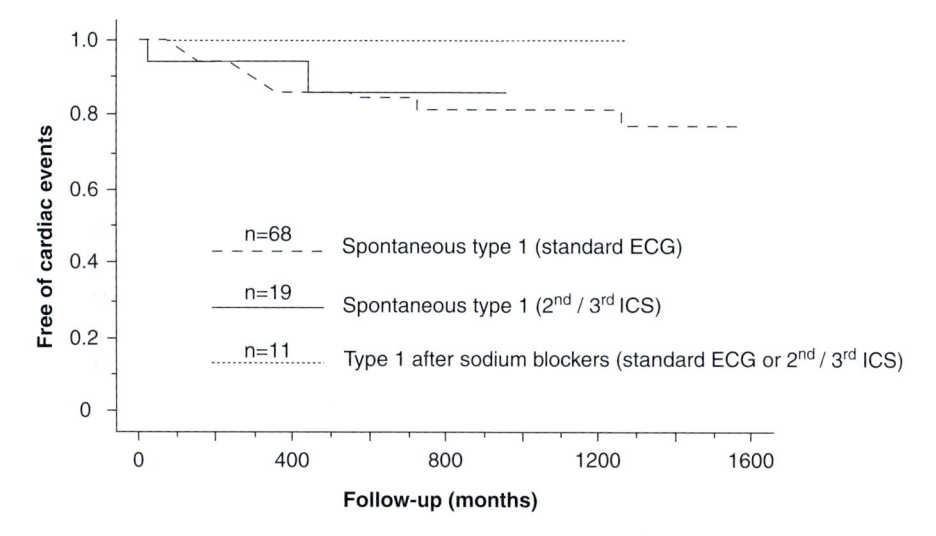

Fig. 10.9 Kaplan-Meier analysis of cardiac events (documented VF or SD) during follow-up in patients with spontaneous type 1 ECG pattern at standard leads (*dashed line*), patients with spontaneous type 1 ECG recorded only at a higher intercostal space (*solid line*), and patients with type 1 ECG pattern at standard and/or higher intercostals spaces only after receiving a sodium channel blocker (*dotted line*). No significant difference was observed in the frequency of cardiac events between the first two groups (Modified from Miyamoto et al.[72] With permission). *ECG* electrocardiogram, *SD* sudden death, *STD* standard, *VF* ventricular fibrillation

10.7 Prognosis and Risk Stratification

Prognosis and risk stratification are probably the most controversial issues in Brugada syndrome. The main clinical studies arising from the largest databases differ on the risk of SD or VF in the population with Brugada syndrome, and particularly on defining the specific risk markers with regard to prognosis.

In our most updated population with Brugada syndrome coming from the international registry, the percentage of patients who experienced *SD or VF* throughout lifetime was 25% (177 out of 724 patients).[73] The mean age at cardiac events was 42±15 years. Of course, such a high rate might have been influenced by a baseline high-risk population referred for the international registry and included in this analysis. In fact, our reported annual rate of events has decreased from the first patients included in the registry[4] to the most recent published series,[39,41,73] the change probably reflecting the inherent bias during the first years following the description of a novel disease, in which particularly severe forms are most likely to be diagnosed. It is important to note that, in the global series, the probability of having a cardiac event during lifetime varied widely (from 3% to 45%) depending on the baseline characteristics of the individuals. Thus, a careful risk stratification of every individual seems mandatory.

Several clinical variables have been demonstrated to predict a worse outcome in patients with Brugada syndrome. In all the analysis of our series over time, the presence of symptoms before diagnosis, a spontaneous type 1 ECG at baseline, the inducibility of ventricular arrhythmias at the electrophysiological study (EPS) and male gender have consistently shown to be related to the occurrence of cardiac events in follow-up.[4,36,39,41,73]

Little controversy exists on the value of a *previous cardiac arrest* as a risk marker for future events. Our data state that up to 62% of patients recovered from an aborted SD are at risk of a new arrhythmic event within the following 54 months.[39] Thus, these patients should be protected with an ICD irrespective of the presence of other risk factors (indication class I).[12] Because there is not such a general agreement on the best approach toward patients who have never developed VF, we conducted a prospective study including 547 individuals with Brugada syndrome and no previous cardiac arrest.[41] Of them (mean age 41±45 years, 408 males), 124 had presented previous syncope (22.7%) and 423 (77.3%) were asymptomatic and had been diagnosed during routine ECG or family screening. The baseline ECG showed a type 1 ECG pattern spontaneously in 391 patients (71.5%). During a mean follow-up of 24±32 months, 45 individuals (8.2%) developed a first major cardiac event (documented VF or SD).[41] By univariable analysis, a history of *previous syncope* (HR 2.79 [1.5–5.1] 95% CI, $p=0.002$), a *spontaneous type 1 ECG* (HR 7.69 [1.9–33.3] 95% CI, $p = 0.0001$), *male*

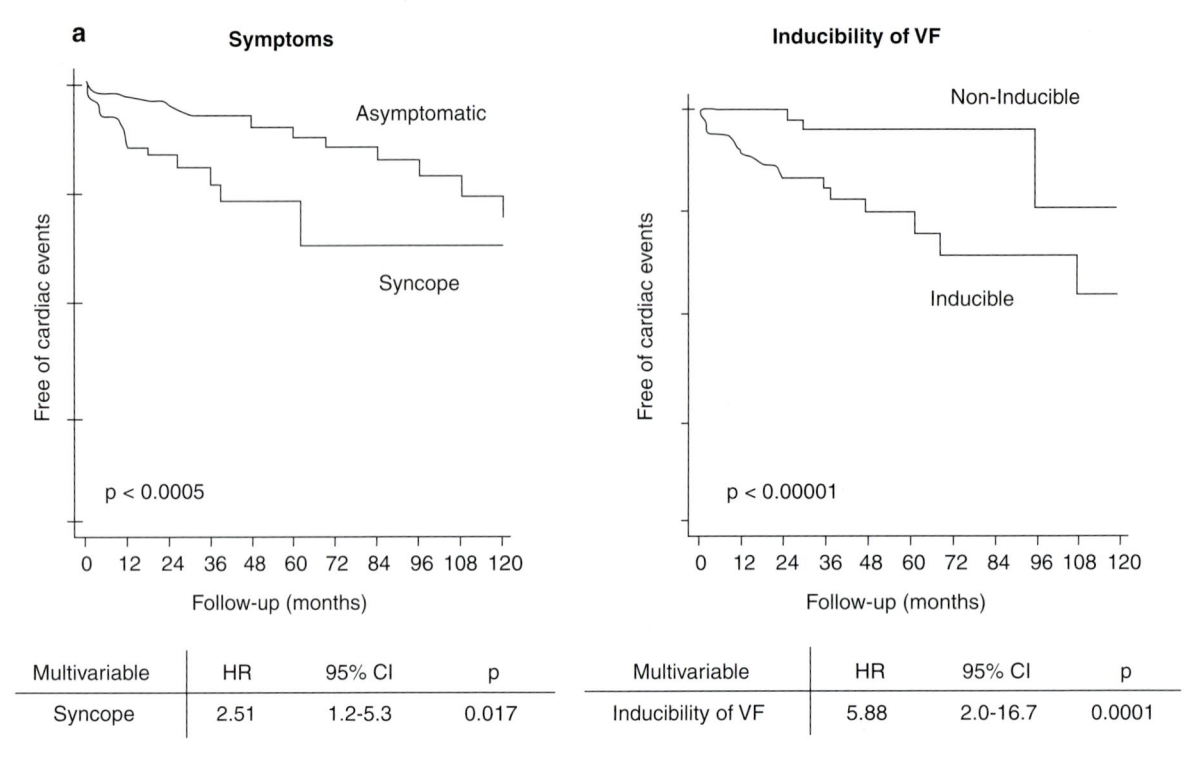

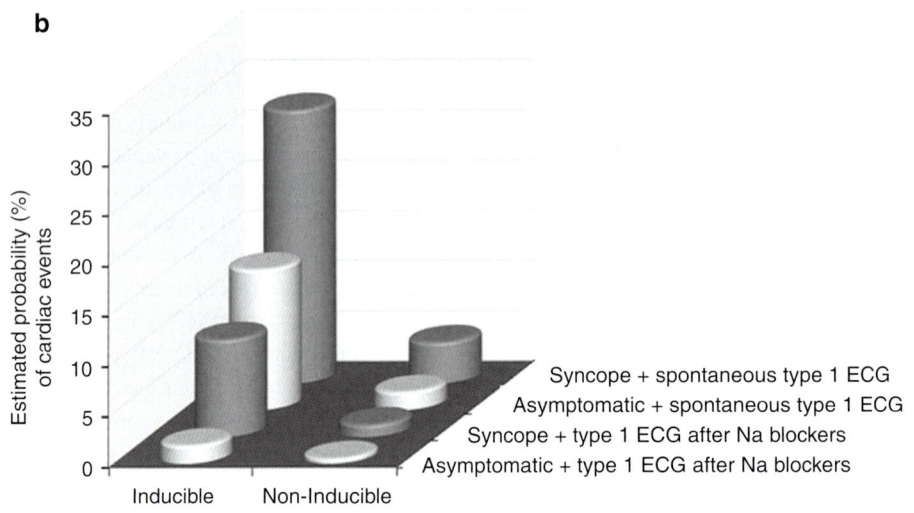

Fig. 10.10 Cardiac events (sudden death or documented ventricular fibrillation) during follow-up. (**a**) Kaplan Meier analysis according to previous symptoms and inducibility of ventricular fibrillation in the electrophysiological study, both independent predictors in the series by Brugada et al.[41] (**b**) Estimated probability of events in follow-up by logistic regression according to symptoms, inducibility of ventricular arrhythmias at the electrophysiological study and baseline ECG (Data from Brugada et al.[41]) *SD* sudden death, *VF* ventricular fibrillation, *EPS* electrophysiological study. *Induced type 1 ECG after administration of sodium channel blockers

gender (HR 5.26 [1.6–16.6] 95% CI, $p = 0.001$) and inducibility of ventricular arrhythmias at the *EPS* (HR 8.33 [2.8–25] 95% CI, $p = 0.0001$) were significantly related to VF or SD in follow-up. Multivariable analysis identified previous syncope and inducibility of VF as the only independent risk factors for the occurrence of events in follow-up (Fig. 10.10a).[41] Logistic regression analysis allowed the definition of eight categories of risk, of which asymptomatic patients with normal ECG at baseline and noninducible VF at the EPS would represent the lowest-risk population, and patients with syncope, spontaneous type 1 ECG, and inducibility at EPS would have the worst outcome (Fig. 10.10b). Further analysis indicated that EPS was particularly useful in predicting cardiac events among asymptomatic patients with no family history of SD (named fortuitous cases, n = 167).[73] Indeed, 11 out of 167 patients (6%) presented documented VF during follow-up, and the only independent predictor in this subgroup was the inducibility at EPS. In contrast, not performing an EPS in this subgroup of patients with the aim of identifying those at risk was shown to be predictive of effective SD (p = 0.002).[73]

Other groups agree that previous symptoms and a spontaneous type 1 ECG are risk factors, although they have found a much lower incidence of arrhythmic events for the whole population (6.5% in 34 ± 44 months of follow-up in the work of Priori et al.[40] and 4.2% in 40 ± 50 months of follow-up in that of Eckardt et al.[42]). The worse outcome in our series may probably reflect a more severely ill baseline population.[42] The other large registries also agree that EPS inducibility is greatest among patients with previous SD or syncope,[40,42] but failed to demonstrate a value of the EPS in predicting outcome. Several reasons could explain this discrepancy[73]: (1) the use of multiple testing centers with non-standardized stimulation protocols; (2) the inclusion of patients with type 2 and type 3 ST-segment elevation (and not type 1) in some series, suggesting that they may contain individuals who do not have the syndrome; (3) the lack of events during follow-up in the other registries. The latter might change when longer follow-ups are available because events can only increase in follow-up and so does the positive predictive value.[73]

Male gender has consistently shown a trend to present more arrhythmic events in all the studies.[39–42] In a series of 384 patients, we recently observed that men and women differed in baseline characteristics at time of first evaluation (Fig. 10.7a) and also in follow-up (Fig. 10.7b).[36] Importantly, our data also confirmed that the classical risk markers defined for mixed populations (symptoms, spontaneous type 1 ECG, and inducibility of VF) were all useful for identifying male patients at risk. However, in the presence of a very low rate of events, none of these markers showed power enough to predict outcome in the female population. In contrast, conduction disturbances were found to correlate with prognosis in women, and specifically the PR interval was the only independent predictor of outcome in the female population.[36]

Spontaneous *AF*, which can appear in 10–53% of cases,[44,62] has been recently shown to have prognostic significance. Kusano et al., in a series of 73 patients with Brugada syndrome, observed that spontaneous AF was associated with higher incidence of syncopal episodes (60.0% vs 22.2%, $p < 0.03$) and documented VF (40.0% vs 14.3%, $p < 0.05$).[62] Multiple ECG parameters have been assessed in the search for new risk markers, of which a prolonged QTc in V2, the aVR sign, the presence of T wave alternans, the early repolarization pattern in inferior or lateral leads, and probably a wide QRS complex seem to be the most important (see Diagnosis: ECG findings).[54–57,61]

Interestingly, a positive family history of SD or the presence of a *SCN5A* mutation have not been proven to be risk markers in any of the large studies conducted thus far.[40,42,74] However, recent data suggest that other genetic findings might correlate with phenotype and even have prognostic implications (see Cardiogenetics aspects).

10.8 Treatment

10.8.1 Implantable Cardioverter Defibrillator

The ICD is the only proven effective treatment of the Brugada syndrome thus far. On the basis of available clinical and basic science data, a II consensus conference was held in September 2003, focused on risk stratification schemes and approaches to therapy.[12] The *recommendations* for ICD implantation stated by this consensus are summarized in Fig. 10.11. Briefly, symptomatic patients should always receive an ICD. Asymptomatic patients may benefit from EPS for risk stratification: ICD should be implanted in those with inducible VF having a spontaneous type 1 ECG at baseline or a sodium channel blocker–induced ECG with a positive family history of SD. Finally, asymptomatic patients who have no family history of SD and who develop a type 1 ECG only after sodium channel blockade should be closely followed up, without enough evidence existing for the usefulness of EPS or a direct indication for ICD.[12] It is important to note that a type 1 ECG can also be induced by certain circumstances other than the deliberate administration of sodium blockers during a pharmacological test. This is the case of patients who "spontaneously"

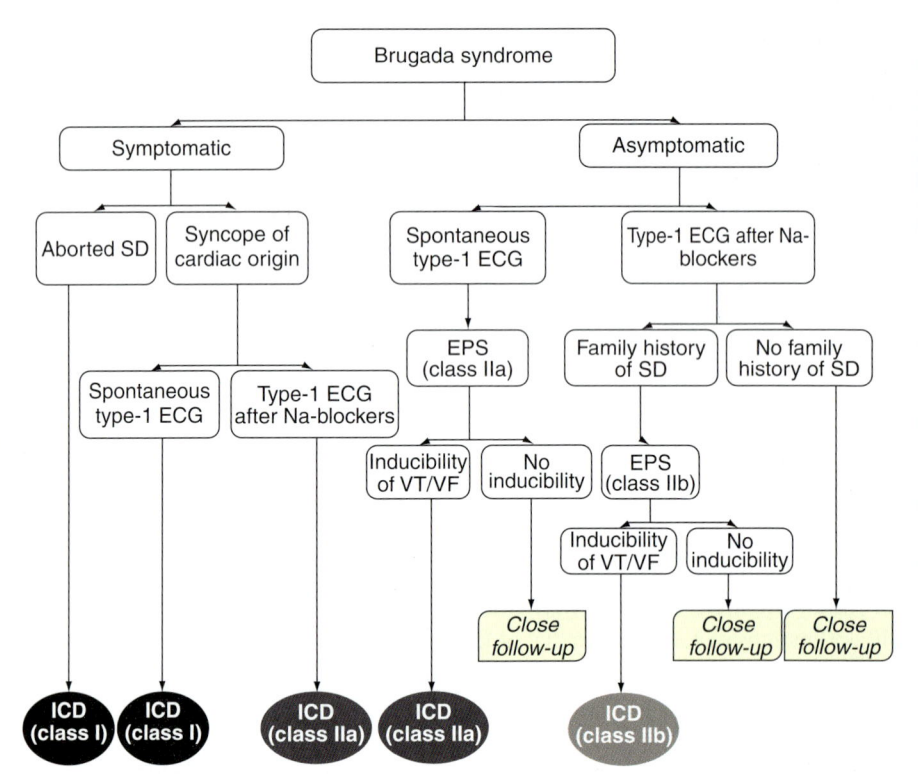

Fig. 10.11 Indications for implantable cardioverter defibrillator (*ICD*) implantation in patients with Brugada syndrome. Class I designation indicates clear evidence that the procedure or treatment is useful or effective; Class II, conflicting evidence about usefulness or efficacy; Class IIa, weight of evidence in favor of usefulness or efficacy; Class IIb, usefulness or efficacy less well established (Modified from Antzelevitch et al.[12] With permission)

develop a type 1 ECG during electrolyte imbalances, administration of certain drugs (particularly anesthetics) and, more importantly, febrile episodes.[75] Although these should probably be considered cases of induced rather than spontaneous type 1 ECG with regard to therapy (Fig. 10.11), to date there is no definite information concerning the long-term prognosis (and thus, indications of ICD implantation) in this particular subgroup of patients. Our data, however, do show that up to 50% of these patients are at risk of developing arrhythmic complications, possibly fatal, during the acute event that unmasks the characteristic type 1 ECG.[75] Consequently, close monitoring and therapeutic intervention with antipyretics or by termination of the culprit medication is mandatory in these circumstances.[75]

From the 2 main retrospective studies conducted on patients with Brugada syndrome who have received primary prophylactic ICD, it can be concluded that ICD is an effective therapy for patients at risk,[76,77] which can have an annual rate of *appropriate shocks* of up to 3.7%.[77] It is important to note that this rate is not only comparable to those reported in other ICD trials dealing with other cardiac diseases,[78,79] but also is affecting young and otherwise healthy people, whose life expectancy could be more than 30 years. Therefore, should this rate remain constant in time, it seems that most patients would likely experience an appropriate shock in a lifetime. However, perhaps just because of being young, a noteworthy rate of *inappropriate shocks* by the device has also been reported. In the study by Sacher et al.,[76] 45 (20%) out of 220 patients had inappropriate shocks in follow-up. In our series, the rate was even higher (36%).[77] The reasons for inappropriate therapies were mainly sinus tachycardia, supraventricular arrhythmias, T-wave oversensing, and lead failure in both studies.[76,77] On the basis of these results and because ICD is not affordable worldwide, there is growing effort to find pharmacological approaches to help treat the disease.

10.8.2 Pharmacological Options

With the aim of rebalancing the ion channel currents active during phase 1 of the action potential, so as to reduce the magnitude of the notch (Fig. 10.3), two main pharmacologic approaches have been assessed

Table 10.3 Pharmacological approach to therapy in the Brugada syndrome

Action	Proved on
Ito blockers:	
4-aminopyridine	Effective in experimental models (suppression of phase-2 reentry)[10] Probable neurotoxicity in humans
Quinidine	Effective in experimental models (suppression of phase-2 reentry)[10] Initial results showing effectiveness in clinical practice: ↓ inducibility of VF[63] ↓ spontaneous VF in follow-up[63, 80] Adjunctive therapy in patients with ICD and multiple shocks[80] Effective in electrical storm[81] A possible option in children[49]
Tedisamil	Effective in experimental models (suppression of phase-2 reentry)
AVE0118	Effective in experimental models (suppression of phase-2 reentry)
ICa activators:	
Isoproterenol	Effective in experimental models (suppression of phase-2 reentry)[10] Effective in electrical storm[82]
Cilostazol	Controversial preliminary results in preventing VF[83,84]
INa openers:	
Dimethyl Lithospermate B (dmLSB)	Effective in experimental models (suppression of phase-2 reentry)

Reproduced with permission from Benito et al.[19]

(Table 10.3): (1) drugs that decrease outward positive currents, like Ito inhibitors; and (2) drugs that increase inward positive currents (ICa, INa).

Quinidine, a drug with Ito- and IKr-blocking properties, has been the most assayed drug in clinical studies. In a work by Belhassen et al.,[63] 25 patients with inducible VF were treated with quinidine (1,483 ± 240 mg orally). After treatment, 22 (88%) of 25 patients were no longer inducible at the EPS, and none of the 19 patients with ongoing medical therapy with oral quinidine developed arrhythmias during follow-up (56 ± 67 months).[63] However, 36% of the patients had transient side effects that led to drug discontinuation. Preliminary data have also proven quinidine as a good adjunctive therapy in patients with ICD and multiple shocks[80] and as an effective treatment of electrical storms associated to Brugada syndrome.[81] More recently, quinidine has been proposed as a good alternative to ICD implantation in children with the syndrome and at high risk for malignant arrhythmias.[49]

β-Adrenergic agents, through an increase in ICa currents, decrease transmural dispersion of repolarization and epicardial dispersion of repolarization in experimental models.[10] Clinically, they have proved to be effective in the treatment of electrical storms associated to Brugada syndrome.[82] Recently, phosphodiesterase III inhibitors have appeared as a new appealing option because they would increase ICa and decrease Ito. Indeed, *cilostazol* was effective in preventing ICD shocks in a patient with recurrent episodes of VF.[83] However, a recent publication reports the failure of such drug in another patient with multiple ICD discharges despite sustained therapy.[84]

10.9 Cardiogenetics Aspects

10.9.1 Genetic Diagnosis and Genotype–Phenotype Correlation

Since the first description in 1998,[8] *SCN5A* encoding the cardiac sodium channel has been the gene most frequently linked to the Brugada syndrome. Although mutations in new genes have been described in the last 2 years (*GPD1-L, CACNA1c, CACNB2b, KCNE3*),[24,26,27] their contribution to the overall population with the disease in still unknown. However, initial studies performed in negative-*SCN5A* carriers suggest that these and other candidate genes (*Caveolin-3, Irx-3, Irx-4, Irx-5, Irx-6, Plakoglobin, Plakophilin-2, SCN1B, SCN2B, SCN3B,* and *SCN4B*) are unlikely to be major causal genes of the Brugada syndrome.[85,86] Therefore, *SCN5A* remains the first candidate gene to analyze during genetic testing, although screening of other genes involved in the action potential could be useful when SCN5A analysis is found to be negative. It is important to note that *SCN5A* mutations are only found in around 18–30% of cases of Brugada syndrome,[12] leaving a great majority of patients with a negative genetic test. This implies that, whereas the finding of a *SCN5A* mutation provides the diagnosis in patients with phenotype of Brugada syndrome, a negative result does not rule out the disease. In these cases, especially if the test is also negative for mutations in other genes, the diagnosis must be based on the clinical phenotype, mainly through the basal or induced ECG findings.

Because so far mutations in the *SCN5A* gene are responsible for the greatest proportion of cases with Brugada syndrome, the first studies on genotype–phenotype correlation have been focused only in *SCN5A* mutation carriers. Very recently, Meregalli et al. presented their results on 147 patients with Brugada syndrome or progressive cardiac conduction disease carrying 32 different mutations in the *SCN5A* gene. The authors found that those patients carrying a mutation leading to a truncated protein (nonsense or frameshift with premature stop codon) were more likely to present syncope (25% of cases) than patients with a missense mutation functionally known to be inactive (and thus generating a >90% of peak INa reduction, syncope rate of 11%) or missense mutations known to be functionally active (≤ 90% of peak INa reduction, syncope rate of 5%, $p = 0.03$).[87] However, no differences in the rate of major arrhythmic complications (SD or VF) were found according to the type of mutation.[87] Our data on 188 patients (all with Brugada syndrome) carrying 69 different mutations in *SCN5A* demonstrate that the presence of a mutation leading to a premature stop codon is related to a greater rate of SD and/or documented VF.[88] Indeed, the incidence of major cardiac events during a lifetime was 23.9% in patients with truncated protein, and 7.7% in patients with other types of mutation ($p = 0.01$).[88] Moreover, in our series, the presence of a mutation leading to a truncated protein was confirmed as an independent predictor of cardiac events (HR 2.9, 95% CI 1.2–7.2, $p = 0.02$), together with the classical clinical risk factors reported in previous series.[88]

In recent years, *polymorphisms* have been acquiring greater importance to explain certain phenotypes of genetic diseases. In the *SCN5A* locus, the common H558R polymorphism has been shown to restore (at least partially) the sodium current impaired by other simultaneous mutations causing either cardiac conduction disturbances (T512I)[89] or Brugada syndrome (R282H).[90] Thus, this polymorphism seems to give rise to less severe phenotypes by mitigating the effect of nearby mutations. According to this, our data on 75 genotyped Brugada patients with SCN5A mutation demonstrate that those carrying the common AA genotype (H558H) have longer QRS duration in lead II ($p = 0.017$), higher J-point elevation in lead V2 ($p = 0.013$), higher rate of "aVR sign" ($p = 0.005$) and a trend toward more symptoms ($p = 0.06$) than carriers of the polymorphisms AG (H558R) or GG (R558R).[91]

From these data, one can conclude that genetic testing may be useful in the risk stratification of patients with Brugada syndrome who are carriers of a *SCN5A* mutation. This concept is particularly important because, in contrast to previously defined clinical variables, genetic information is constitutional and thus invariable over time within the same individual.

10.9.2 Family Screening

The first step after identification of a proband with Brugada syndrome is to evaluate his or her individual risk of SD and indicate ICD if necessary (see Prognosis and risk stratification). Genetic testing is recommended because a positive result helps confirm the disease in patients with borderline phenotype, may provide information on arrhythmic risk during lifetime (see above), and allows the identification of family members who are also carriers of the disease.

Given that the Brugada syndrome is commonly an inherited disorder, family screening should always be performed to identify possible relatives who are unknowingly at risk of SD. The first step would be to construct a family pedigree as the one shown in Fig. 10.12. It is important to remind that hereditary forms of Brugada syndrome are autosomal dominant not linked to sex, and thus each affected patient has a 50% probability of transmitting the disease to his/her offspring.

If genetic testing is available and the responsible mutation has been identified in the proband, genetic testing of all other family members (starting by first degree-related) is the best approach for family screening because it allows both to establish and rule out the disease with the maximum sensitivity and specificity. If genetic testing is not available or the responsible mutation has not been identified, all direct relatives should be studied first with basic anamnesis and a basal ECG. For those with diagnostic type 1 ECG, and thus, carriers of the disease, conventional risk stratification should be performed in order to estimate their probability of cardiac events in follow-up. In those family members with normal ECG at baseline, as long as they are asymptomatic, our data indicate that the probability of cardiac events in

10 The Brugada Syndrome

Fig. 10.12 Example of a family pedigree. Squares indicate males, circles indicate females. Affected patients are represented in black solid figures, while open figures indicate non-affected individuals. Sloping lines indicate deceased individuals

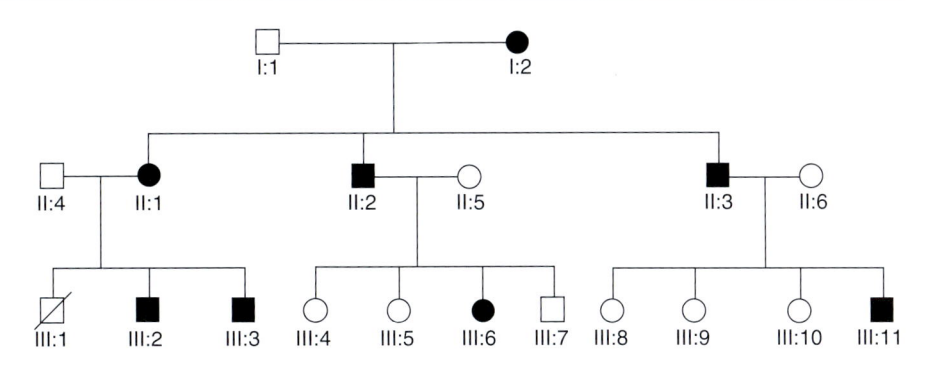

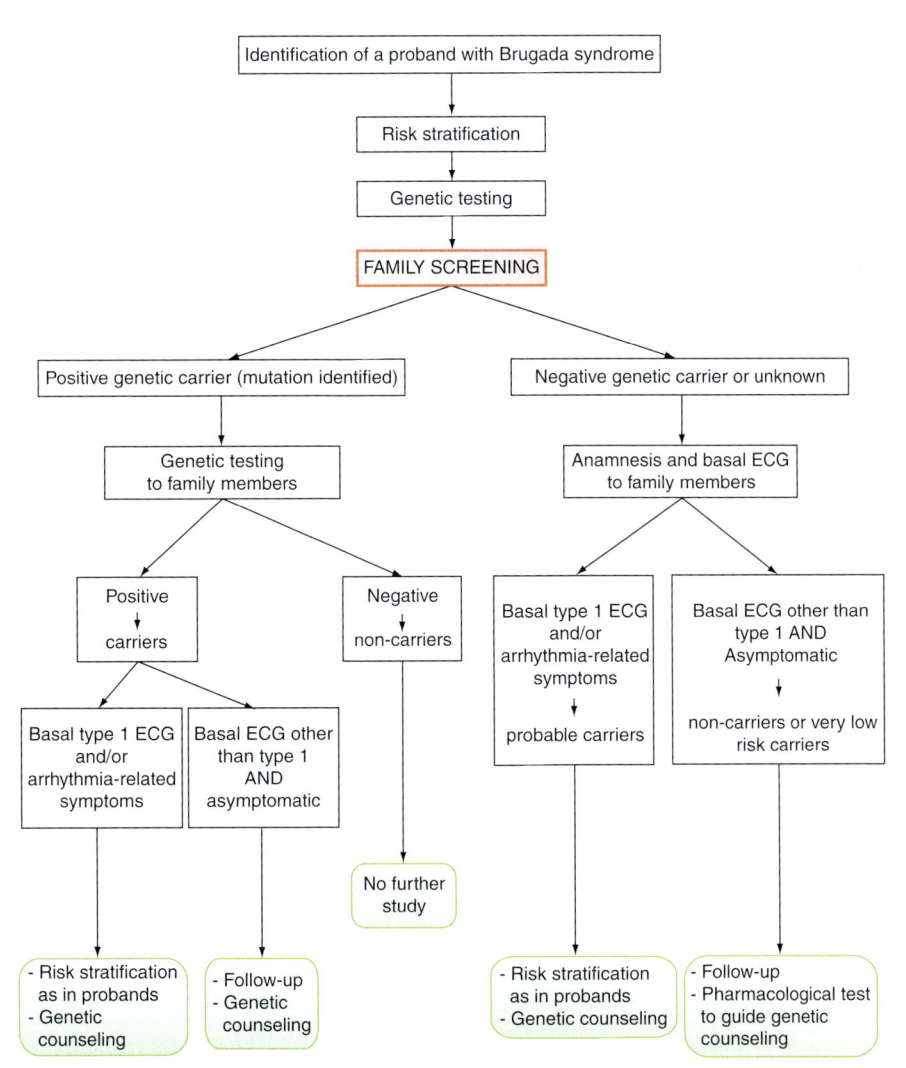

Fig. 10.13 Proposed algorithm for family screening after identification of a proband with confirmed Brugada syndrome

follow-up is extremely low (1/131 at 100 months, unpublished data), meaning that either they are not carriers of the disease, or, if they are, have little expressivity and thus are at a very low risk of arrhythmias. Pharmacological provocative test could be performed in these patients to increase the penetrance and better identify possible disease carriers, which would be useful for family planning and genetic counseling. The proposed algorithm for family screening is presented in Fig. 10.13.

10.10 Summary

The Brugada syndrome is characterized by a typical ECG morphology and an increased susceptibility to present ventricular arrhythmias and sudden death in the absence of structural heart disease. The characteristic ECG pattern, known as coved-type or type 1, consists of a persistent ST-segment elevation in right precordial leads followed by negative T-waves, and must be distinguished from other conditions that also present with right ST-segment elevation. A number of mutations in several genes encoding cardiac transmembrane channels have been linked to the syndrome, but loss-of-function mutations in *SCN5A* encoding the sodium channel remain the most common genetic finding in patients with the syndrome. In recent years, experimental studies have confirmed that it is an imbalance between inward and outward positive currents during the phase 1 of the action potential the main mechanism that underlies both the ST-segment elevation and the susceptibility to develop ventricular arrhythmias. This hypothesis also explains the effect of certain modulators and certain drugs with clinical applications, for example, sodium-channel blockers for unmasking concealed forms of the disease or Ito blockers such as quinidine for pharmacological therapy, usually coadjunctive to ICD. Because the clinical phenotype can be widely variable, risk stratification is mandatory in all diagnosed patients, and should be based on the presence of previous symptoms, basal ECG findings, results of the EPS, gender, and probably also on genetic findings. ICD is the only proven effective therapy for patients at high risk. Genetic testing, although only positive in a minority of patients, can help confirm the disease in patients with borderline phenotype, may provide prognostic information, and can be extremely valuable for family screening.

References

1. Brugada P, Brugada J. Right bundle branch block, persistent ST segment elevation and sudden cardiac death: a distinct clinical and electrocardiographic syndrome. A multicenter report. *J Am Coll Cardiol*. 1992 November 15;20(6):1391-1396.
2. Proclemer A, Facchin D, Feruglio GA, Nucifora R. Recurrent ventricular fibrillation, right bundle-branch block and persistent ST segment elevation in V1-V3: a new arrhythmia syndrome? A clinical case report. *G Ital Cardiol*. 1993 December;23(12):1211-1218.
3. Ferracci A, Fromer M, Schlapfer J, Pruvot E, Kappenberger L. Primary ventricular fibrillation and early recurrence: apropos of a case of association of right bundle branch block and persistent ST segment elevation. *Arch Mal Coeur Vaiss*. 1994 October;87(10):1359-1362.
4. Brugada J, Brugada R, Brugada P. Right bundle-branch block and ST-segment elevation in leads V1 through V3: a marker for sudden death in patients without demonstrable structural heart disease. *Circulation*. 1998 February 10;97(5):457-460.
5. Alings M, Wilde A. "Brugada" syndrome: clinical data and suggested pathophysiological mechanism. *Circulation*. 1999 February 9;99(5):666-673.
6. Brugada P, Brugada R, Brugada J. Sudden death in patients and relatives with the syndrome of right bundle branch block, ST segment elevation in the precordial leads V1to V3and sudden death. *Eur Heart J*. 2000 February 2;21(4): 321-326.
7. Priori SG, Napolitano C, Gasparini M, et al. Clinical and genetic heterogeneity of right bundle branch block and ST-segment elevation syndrome: a prospective evaluation of 52 families. *Circulation*. 2000 November 14;102(20):2509-2515.
8. Chen Q, Kirsch GE, Zhang D, et al. Genetic basis and molecular mechanism for idiopathic ventricular fibrillation. *Nature*. 1998 March 19;392(6673):293-296.
9. Rook MB, Bezzina AC, Groenewegen WA, et al. Human SCN5A gene mutations alter cardiac sodium channel kinetics and are associated with the Brugada syndrome. *Cardiovasc Res*. 1999;44(3):507-517.
10. Yan GX, Antzelevitch C. Cellular basis for the Brugada syndrome and other mechanisms of arrhythmogenesis associated with ST-segment elevation. *Circulation*. 1999 October 12;100(15):1660-1666.
11. Wilde AAM, Antzelevitch C, Borggrefe M, et al. Proposed diagnostic criteria for the Brugada syndrome. *Eur Heart J*. 2002 November 1;23(21):1648-1654.
12. Antzelevitch C, Brugada P, Borggrefe M, et al. Brugada Syndrome: Report of the Second Consensus Conference: Endorsed by the Heart Rhythm Society and the European Heart Rhythm Association. *Circulation*. 2005 February 8;111(5):659-670.
13. van den Berg MP, de Boer RA, van Tintelen JP. Brugada syndrome or Brugada electrocardiogram? *J Am Coll Cardiol*. 2009 April 28;53(17):1569.
14. Benito B, Brugada J, Brugada R, Brugada P. Brugada Syndrome or Brugada Electrocardiogram? Authors' Reply. *J Am Coll Cardiol*. 2009 April 28;53(17):1569-1570.
15. Miyasaka Y, Tsuji H, Yamada K, et al. Prevalence and mortality of the Brugada-type electrocardiogram in one city in Japan. *J Am Coll Cardiol*. 2001 September;38(3):771-774.
16. Hermida JS, Lemoine JL, Aoun FB, Jarry G, Rey JL, Quiret JC. Prevalence of the Brugada syndrome in an apparently healthy population. *Am J Cardiol*. 2000 July 1;86(1):91-94.
17. Donohue D, Tehrani F, Jamehdor R, Lam C, Movahed MR. The prevalence of Brugada ECG in adult patients in a Large University Hospital in the Western United States. *Am Heart Hosp J*. 2008;6(1):48-50.
18. Vatta M, Dumaine R, Varghese G, et al. Genetic and biophysical basis of sudden unexplained nocturnal death syndrome (SUNDS), a disease allelic to Brugada syndrome. *Hum Mol Genet*. 2002 February 1;11(3):337-345.

19. Benito B, Brugada R, Brugada J, Brugada P. Brugada syndrome. *Prog Cardiovasc Dis*. 2008 July;51(1):1-22.
20. Deschênes I, Baroudi G, Berthet M, et al. Electrophysiological characterization of SCN5A mutations causing long QT (E1784K) and Brugada (R1512W and R1432G) syndromes. *Cardiovasc Res*. 2000;46(1):55-65.
21. Cordeiro JM, Barajas-Martinez H, Hong K, et al. Compound heterozygous mutations P336L and I1660V in the human cardiac sodium channel associated with the Brugada syndrome. *Circulation*. 2006 November 7;114(19):2026-2033.
22. Casini S, Tan HL, Bhuiyan ZA, et al. Characterization of a novel SCN5A mutation associated with Brugada syndrome reveals involvement of DIIIS4-S5 linker in slow inactivation. *Cardiovasc Res*. 2007 December 1;76(3):418-429.
23. Pfahnl AE, Viswanathan PC, Weiss R, et al. A sodium channel pore mutation causing Brugada syndrome. *Heart Rhythm*. 2007 January;4(1):46-53.
24. London B, Michalec M, Mehdi H, et al. Mutation in glycerol-3-phosphate dehydrogenase 1 like gene (GPD1-L) decreases cardiac Na+ current and causes inherited arrhythmias. *Circulation*. 2007 November 13;116(20):2260-2268.
25. Weiss R, Barmada MM, Nguyen T, et al. Clinical and molecular heterogeneity in the Brugada syndrome: a novel gene locus on chromosome 3. *Circulation*. 2002 February 12;105(6):707-713.
26. Antzelevitch C, Pollevick GD, Cordeiro JM, et al. Loss-of-function mutations in the cardiac calcium channel underlie a new clinical entity characterized by ST-segment elevation, short QT intervals, and sudden cardiac death. *Circulation*. 2007 January 30;115(4):442-449.
27. Delpon E, Cordeiro JM, Nunez L, et al. Functional effects of KCNE3 mutation and its role in the development of Brugada syndrome. *Circ Arrhythm Electrophysiol*. 2008;1(3):209-218.
28. Antzelevitch C. Brugada syndrome. *Pacing Clin Electrophysiol*. 2006;29(10):1130-1159.
29. Shimizu W, Aiba T, Kamakura S. Mechanisms of disease: current understanding and future challenges in Brugada syndrome. *Nat Clin Pract Cardiovasc Med*. 2005 August;2(8):408-414.
30. Fish J, Antzelevitch C. Role of sodium and calcium channel block in unmasking the Brugada syndrome. *Heart Rhythm*. 2004;1(2):210-217.
31. Miyazaki T, Mitamura H, Miyoshi S, Soejima K, Aizawa Y, Ogawa S. Autonomic and antiarrhythmic drug modulation of ST segment elevation in patients with Brugada syndrome. *J Am Coll Cardiol*. 1996 April;27(5):1061-1070.
32. Matsuo K, Kurita T, Inagaki M, et al. The circadian pattern of the development of ventricular fibrillation in patients with Brugada syndrome. *Eur Heart J*. 1999 March 2;20(6):465-470.
33. Tatsumi H, Takagi M, Nakagawa E, Yamashita H, Yoshiyama M. Risk stratification in patients with Brugada syndrome: analysis of daily fluctuations in 12-lead electrocardiogram (ECG) and signal-averaged electrocardiogram (SAECG). *J Cardiovasc Electrophysiol*. 2006 July;17(7):705-711.
34. Takigawa M, Noda T, Shimizu W, et al. Seasonal and circadian distributions of ventricular fibrillation in patients with Brugada syndrome. *Heart Rhythm*. 2008 November;5(11):1523-1527.
35. Litovsky SH, Antzelevitch C. Differences in the electrophysiological response of canine ventricular subendocardium and subepicardium to acetylcholine and isoproterenol. A direct effect of acetylcholine in ventricular myocardium. *Circ Res*. 1990 September;67(3):615-627.
36. Benito B, Sarkozy A, Mont L, et al. Gender differences in clinical manifestations of Brugada syndrome. *J Am Coll Cardiol*. 2008;52:1567-1573.
37. Di Diego JM, Cordeiro JM, Goodrow RJ, et al. Ionic and cellular basis for the predominance of the Brugada syndrome phenotype in males. *Circulation*. 2002 October 8;106(15):2004-2011.
38. Brugada R, Brugada J, Antzelevitch C, et al. Sodium channel blockers identify risk for sudden death in patients with ST-segment elevation and right bundle branch block but structurally normal hearts. *Circulation*. 2000 February 8;101(5):510-515.
39. Brugada J, Brugada R, Antzelevitch C, Towbin J, Nademanee K, Brugada P. Long-term follow-up of individuals with the electrocardiographic pattern of right bundle-branch block and ST-segment elevation in precordial leads V1 to V3. *Circulation*. 2002 January 1;105(1):73-78.
40. Priori SG, Napolitano C, Gasparini M, et al. Natural history of Brugada syndrome: insights for risk stratification and management. *Circulation*. 2002 March 19;105(11):1342-1347.
41. Brugada J, Brugada R, Brugada P. Determinants of sudden cardiac death in individuals with the electrocardiographic pattern of Brugada syndrome and no previous cardiac arrest. *Circulation*. 2003 December 23;108(25):3092-3096.
42. Eckardt L, Probst V, Smits JPP, et al. Long-term prognosis of individuals with right precordial st-segment-elevation Brugada syndrome. *Circulation*. 2005 January 25;111(3):257-263.
43. Eckardt L, Kirchhof P, Loh P, et al. Brugada syndrome and supraventricular tachyarrhythmias: a novel association? *J Cardiovasc Electrophysiol*. 2001 June;12(6):680-685.
44. Morita H, Kusano-Fukushima K, Nagase S, et al. Atrial fibrillation and atrial vulnerability in patients with Brugada syndrome. *J Am Coll Cardiol*. 2002 October 16;40(8):1437-1444.
45. Benito B, Brugada J. Recurrent syncope: an unusual presentation of Brugada syndrome. *Nat Clin Pract Cardiovasc Med*. 2006 October;3(10):573-577.
46. Makita N, Sumitomo N, Watanabe I, Tsutsui H. Novel SCN5A mutation (Q55X) associated with age-dependent expression of Brugada syndrome presenting as neurally mediated syncope. *Heart Rhythm*. 2007 April;4(4):516-519.
47. Matsuo K, Akahoshi M, Seto S, Yan K. Disappearance of the Brugada-type electrocardiogram after surgical castration: a role for testosterone and an explanation for the male preponderance. *Pacing Clin Electrophysiol*. 2003;26:1551-1553.
48. Shimizu W, Matsuo K, Kokubo Y, et al. Sex hormone and gender difference-role of testosterone on male predominance in Brugada syndrome. *J Cardiovasc Electrophysiol*. 2007 April 9;18(4):415-421.
49. Probst V, Denjoy I, Meregalli PG, et al. Clinical aspects and prognosis of Brugada syndrome in children. *Circulation*. 2007 April 17;115(15):2042-2048.
50. Benito B, Sarkozy A, Berne P, et al. Características clínicas y pronóstico del síndrome de Brugada en la población pediátrica. *Rev Esp Cardiol*. 2007;60(Suppl 2):70. Ref Type: Abstract.
51. Veltmann C, Schimpf R, Echternach C, et al. A prospective study on spontaneous fluctuations between diagnostic and non-diagnostic ECGs in Brugada syndrome: implications for correct phenotyping and risk stratification. *Eur Heart J*. 2006 November 1;27(21):2544-2552.

52. Wilde AA. Spontaneous electrocardiographic fluctuations in Brugada syndrome: does it matter? *Eur Heart J.* 2006 November;27(21):2493-2494.

53. Pitzalis MV, Anaclerio M, Iacoviello M, et al. QT-interval prolongation in right precordial leads: an additional electrocardiographic hallmark of Brugada syndrome. *J Am Coll Cardiol.* 2003 November 5;42(9):1632-1637.

54. Castro Hevia J, Antzelevitch C, Tornés Bárzaga F, et al. Tpeak-Tend and Tpeak-Tend dispersion as risk factors for ventricular tachycardia/ventricular fibrillation in patients with the Brugada syndrome. *J Am Coll Cardiol.* 2006;47(9):1828-1834.

55. Babai Bigi MA, Aslani A, Shahrzad S. aVR sign as a risk factor for life-threatening arrhythmic events in patients with Brugada syndrome. *Heart Rhythm.* 2007 August;4(8): 1009-1012.

56. Tada T, Kusano KF, Nagase S, et al. The relationship between the magnitude of T wave alternans and amplitude of the corresponding T Wave in patients with Brugada syndrome. *J Cardiovasc Electrophysiol.* 2008 February 20;19(1):56-61.

57. Sarkozy A, Chierchia GB, Paparella G, et al. Inferior and lateral electrocardiographic repolarization abnormalities in Brugada syndrome. *Circ Arrhythmia Electrophysiol.* 2009 April 1;2(2):154-161.

58. Kyndt F, Probst V, Potet F, et al. Novel SCN5A mutation leading either to isolated cardiac conduction defect or Brugada syndrome in a large French family. *Circulation.* 2001 December 18;104(25):3081-3086.

59. Smits JP, Eckardt L, Probst V, et al. Genotype-phenotype relationship in Brugada syndrome: electrocardiographic features differentiate SCN5A-related patients from non-SCN5A-related patients. *J Am Coll Cardiol.* 2002 July 17;40(2):350-356.

60. Yokokawa M, Noda T, Okamura H, et al. Comparison of long-term follow-up of electrocardiographic features in Brugada syndrome between the SCN5A-positive probands and the SCN5A-negative probands. *Am J Cardiol.* 2007 August 15;100(4):649-655.

61. Junttila MJ, Brugada P, Hong K, et al. Differences in 12-Lead electrocardiogram between symptomatic and asymptomatic Brugada syndrome patients. *J Cardiovasc Electrophysiol.* 2007 December 12;19(4):380-383.

62. Kusano KF, Taniyama M, Nakamura K, et al. Atrial fibrillation in patients with Brugada syndrome relationships of gene mutation, electrophysiology, and clinical backgrounds. *J Am Coll Cardiol.* 2008 March 25;51(12):1169-1175.

63. Belhassen B, Glick A, Viskin S. Efficacy of quinidine in high-risk patients with Brugada syndrome. *Circulation.* 2004 September 28;110(13):1731-1737.

64. Dumaine R, Towbin JA, Brugada P, et al. Ionic mechanisms responsible for the electrocardiographic phenotype of the Brugada syndrome are temperature dependent. *Circ Res.* 1999 October 29;85(9):803-809.

65. Gonzalez Rebollo JM, Hernandez MA, Garcia A, de CA Garcia, Mejias A, Moro C. Recurrent ventricular fibrillation during a febrile illness in a patient with the Brugada syndrome. *Rev Esp Cardiol.* 2000 May;53(5):755-757.

66. Porres JM, Brugada J, Urbistondo V, Garcia F, Reviejo K, Marco P. Fever unmasking the Brugada syndrome. *Pacing Clin Electrophysiol.* 2002 November;25(11):1646-1648.

67. Hong K, Brugada J, Oliva A, et al. Value of electrocardiographic parameters and Ajmaline test in the diagnosis of Brugada syndrome caused by SCN5A mutations. *Circulation.* 2004 November 9;110(19):3023-3027.

68. Meregalli PG, Ruijter JM, Hofman N, Bezzina CR, Wilde AA, Tan HL. Diagnostic value of flecainide testing in unmasking SCN5A-related Brugada syndrome. *J Cardiovasc Electrophysiol.* 2006 August;17(8):857-864.

69. Wolpert C, Echternach C, Veltmann C, et al. Intravenous drug challenge using flecainide and ajmaline in patients with Brugada syndrome. *Heart Rhythm.* 2005 March;2(3):254-260.

70. Shimizu W, Matsuo K, Takagi M, et al. Body surface distribution and response to drugs of ST segment elevation in Brugada syndrome: clinical implication of eighty-seven-lead body surface potential mapping and its application to twelve-lead electrocardiograms. *J Cardiovasc Electrophysiol.* 2000 April;11(4):396-404.

71. Sangwatanaroj S, Prechawat S, Sunsaneewitayakul B, Sitthisook S, Tosukhowong P, Tungsanga K. New electrocardiographic leads and the procainamide test for the detection of the Brugada sign in sudden unexplained death syndrome survivors and their relatives. *Eur Heart J.* 2001 December;22(24): 2290-2296.

72. Miyamoto K, Yokokawa M, Tanaka K, et al. Diagnostic and prognostic value of a type 1 Brugada electrocardiogram at higher (third or second) V1 to V2 recording in men with Brugada syndrome. *Am J Cardiol.* 2007 January 1;99(1):53-57.

73. Brugada P, Brugada R, Brugada J, et al. Should patients with an asymptomatic Brugada electrocardiogram undergo pharmacological and electrophysiological testing? *Circulation.* 2005 July 12;112(2):279-292.

74. Gehi A, Duong T, Metz L, Gomes J, Mehta D. Risk stratification of individuals with the Brugada electrocardiogram: a meta-analysis. *J Cardiovasc Electrophysiol.* 2006;17(6):577-583.

75. Junttila MJ, Gonzalez M, Lizotte E, et al. Induced Brugada-type electrocardiogram, a sign for imminent malignant arrhythmias. *Circulation.* 2008 April 8;117(14):1890-1893.

76. Sacher F, Probst V, Iesaka Y, et al. Outcome after implantation of a cardioverter-defibrillator in patients with Brugada syndrome: a multicenter study. *Circulation.* 2006 November 28;114(22):2317-2324.

77. Sarkozy A, Boussy T, Kourgiannides G, et al. Long-term follow-up of primary prophylactic implantable cardioverter-defibrillator therapy in Brugada syndrome. *Eur Heart J.* 2007 February;28(3):334-344.

78. Maron BJ, Shen WK, Link MS, et al. Efficacy of implantable cardioverter-defibrillators for the prevention of sudden death in patients with hypertrophic cardiomyopathy. *N Engl J Med.* 2000 February 10;342(6):365-373.

79. Bardy GH, Lee KL, Mark DB, et al. Amiodarone or an implantable cardioverter-defibrillator for congestive heart failure. *N Engl J Med.* 2005 January 20;352(3):225-237.

80. Hermida JS, Denjoy I, Clerc J, et al. Hydroquinidine therapy in Brugada syndrome. *J Am Coll Cardiol.* 2004 May 19;43(10):1853-1860.

81. Mok NS, Chan NY, Chiu AC. Successful use of quinidine in treatment of electrical storm in Brugada syndrome. *Pacing Clin Electrophysiol.* 2004 June;27(6 Pt 1):821-823.

82. Ohgo T, Okamura H, Noda T, et al. Acute and chronic management in patients with Brugada syndrome associated with electrical storm of ventricular fibrillation. *Heart Rhythm.* 2007 June;4(6):695-700.

83. Tsuchiya T, Ashikaga K, Honda T, Arita M. Prevention of ventricular fibrillation by cilostazol, an oral phosphodiesterase inhibitor, in a patient with Brugada syndrome. *J Cardiovasc Electrophysiol*. 2002 July;13(7):698-701.
84. Abud A, Bagattin D, Goyeneche R, Becker C. Failure of cilostazol in the prevention of ventricular fibrillation in a patient with Brugada syndrome. *J Cardiovasc Electrophysiol*. 2006 February;17(2):210-212.
85. Koopmann TT, Beekman L, Alders M, et al. Exclusion of multiple candidate genes and large genomic rearrangements in SCN5A in a Dutch Brugada syndrome cohort. *Heart Rhythm*. 2007 June;4(6):752-755.
86. Makiyama T, Akao M, Haruna Y, et al. Mutation analysis of the glycerol-3 phosphate dehydrogenase-1 like (GPD1L) gene in Japanese patients with Brugada syndrome. *Circ J*. 2008 October;72(10):1705-1706.
87. Meregalli PG, Tan HL, Probst V, et al. Type of SCN5A mutation determines clinical severity and degree of conduction slowing in loss-of-function sodium channelopathies. *Heart Rhythm*. 2009 March;6(3):341-348.
88. Benito B, Campuzano O, Ishac R, et al. Role of genetic testing in risk stratification of Brugada syndrome. *Heart Rhythm*. 2009;6:S102. Ref Type: Abstract.
89. Viswanathan PC, Benson DW, Balser JR. A common SCN5A polymorphism modulates the biophysical effects of an SCN5A mutation. *J Clin Invest*. 2003 February;111(3):341-346.
90. Poelzing S, Forleo C, Samodell M, et al. SCN5A polymorphism restores trafficking of a Brugada syndrome mutation on a separate gene. *Circulation*. 2006 August 1;114(5):368-376.
91. Lizotte E, Junttila MJ, Dube MP, et al. Genetic modulation of Brugada syndrome by a common polymorphism. *J Cardiovasc Electrophysiol*. 2009;20:1137-1141.

Short QT Syndrome

Christian Wolpert, Christian Veltmann, Rainer Schimpf, and Martin Borggrefe

11.1 Introduction

In 2000, Gussak et al. described an idiopathic short QT interval associated with atrial fibrillation in a family and sudden death in an unrelated individual.[1] Three years later, in 2003, Gaita et al. reported the association of a short QT interval and sudden cardiac death in two unrelated European families.[2] Within the following years, mutations in five different genes causative for the short QT interval were identified. The mutations either cause a gain of function of cardiac potassium channels I_{Kr}, I_{Ks}, and I_{K1}, or a loss of function in the cardiac L-type calcium channel (I_{Ca}).[3-6]

The scope of this chapter is to provide a comprehensive description of the short QT syndrome (SQTS) including the clinical, genetic, and pathophysiologic aspects, as well as therapeutic consequences and treatment options.

11.2 Molecular and Genetic Background

The SQTS is a genetically heterogeneous disease just like the congenital long QT syndrome. By now, mutations in five different genes have been identified.[3-6] The mutations are located on chromosomes 7, 10, 11, 12, and 17 and encode for different cardiac ion channels. According to the chronology of their first description, the mutations are termed SQT1 to SQT5 (Table 11.1).

The first mutation identified to be causative for the short QT syndrome (SQT1) was a gain of function mutation leading to an increase of the rapid component of the delayed rectifier potassium current (I_{Kr}).[7] Two different missense mutations were identified resulting in the same amino acid change in HERG (KCNH2). These mutations at nucleotide 1764 in the KCNH2 gene substitute the asparagine at codon 588 for a positively charged lysine (N588K). The residue is located in the S5-P loop region of HERG at the outer mouth of the channel. The N588K mutation causes a loss of the normal rectification of the current at plateau voltages, which results in a significant increase of I_{Kr} during phase 2 and 3 of the action potential leading to abbreviation of the action potential and both, atrial and ventricular refractoriness.

Genetic heterogeneity in SQTS was stressed by findings of Bellocq et al. in 2004 who identified a mutation in a single sporadic case of a 70-year-old patient with SQTS (QTc 302 ms) and sudden cardiac arrest. They identified a gain-of-function mutation (V307L) in the KCNQ1 gene, which encodes the slow component of the delayed rectifier potassium channel (I_{Ks}) (SQT2). The mutation causes a −20 mV shift of the half-activation potential and acceleration of the activation kinetics and activation of the mutant channels at more negative potentials. This results in a gain of function of I_{Ks} and abbreviation of the action potential. A further missense mutation in the same gene (V141M) was identified in a baby with bradycardia and atrial fibrillation in utero.[3] The ECG revealed a shortened QT interval and episodes of atrial fibrillation.

The third mutation responsible for SQTS was identified in 2005 by Priori and coworkers in two relatives without sudden cardiac arrest (SQT-3).[4] A gain of function in KCNJ2, encoding the inward rectifier potassium channel (I_{K1}), causes abbreviation of the QT

C. Veltmann (✉)
1st Department of Medicine – Cardiology, University Medical Centre Mannheim, Theodor-Kutzer-Ufer 1-3, 68167 Mannheim, Germany
e-mail: christian.veltmann@umm.de

11.6 Diagnosis of Short QT Syndrome

The electrocardiogram of the first patients identified with a SQTS (SQT1) showed very short QT intervals and in addition short QT intervals corrected for heart rate (QTc <300 ms). The patients identified as SQT2 – SQT5 exhibited QTc of up to 360 ms. The ECG in SQT1-3 reveals tall, symmetrical, and asymmetrical peaked T wave especially in the precordial leads (Fig. 11.2). In SQT3, the T wave has a less steep ascending part and a steep downslope.[4] In most cases, a ST segment is absent with the T wave originating directly from the S wave. Another finding in SQTS is a prolonged $T_{peak} - T_{end}$ interval. Recently, Anttonen et al. compared the $J_{point} - T_{peak}$ interval in symptomatic patients with SQTS, probands with a short QT interval, and a control group of subjects with normal QT interval.[17] Symptomatic patients with SQTS had significantly shorter $J_{point} - T_{peak}$ intervals and higher corrected $T_{peak} - T_{end}/QTc$ ratio compared to asymptomatic probands with a short QT interval and subjects with a normal QT interval. Patients diagnosed with SQT4 and SQT5 on the basis of a mutation in the cardiac calcium channel exhibit shorter than normal QT intervals of 330–360 ms, which is relatively longer than in SQT1–SQT3. These patients additionally show ST segment elevation diagnostic of Brugada syndrome either spontaneously or after the administration of intravenous ajmaline.[5]

Another important finding in SQT1 is the inappropriate adaptation of the QT interval to heart rate. In patients with SQTS the QT interval does not shorten adequately compared to normal controls.[18]

Quinidine was able to restore the QTc/heart rate ratio toward the normal range. A lack of adaptation of QT interval with heart rate may be one additional criterion for the diagnosis of SQTS.

A further diagnostic tool in SQTS is the electrophysiological study. Atrial and ventricular effective refractory periods are significantly shortened especially in SQT1. Atrial refractory period of 140 ms and ventricular effective refractory period of 150 ms or less are highly suspicious criteria of the SQTS. Another finding is the high inducibility of ventricular fibrillation during programmed ventricular stimulation in patients with SQTS (Fig. 11.3).[16]

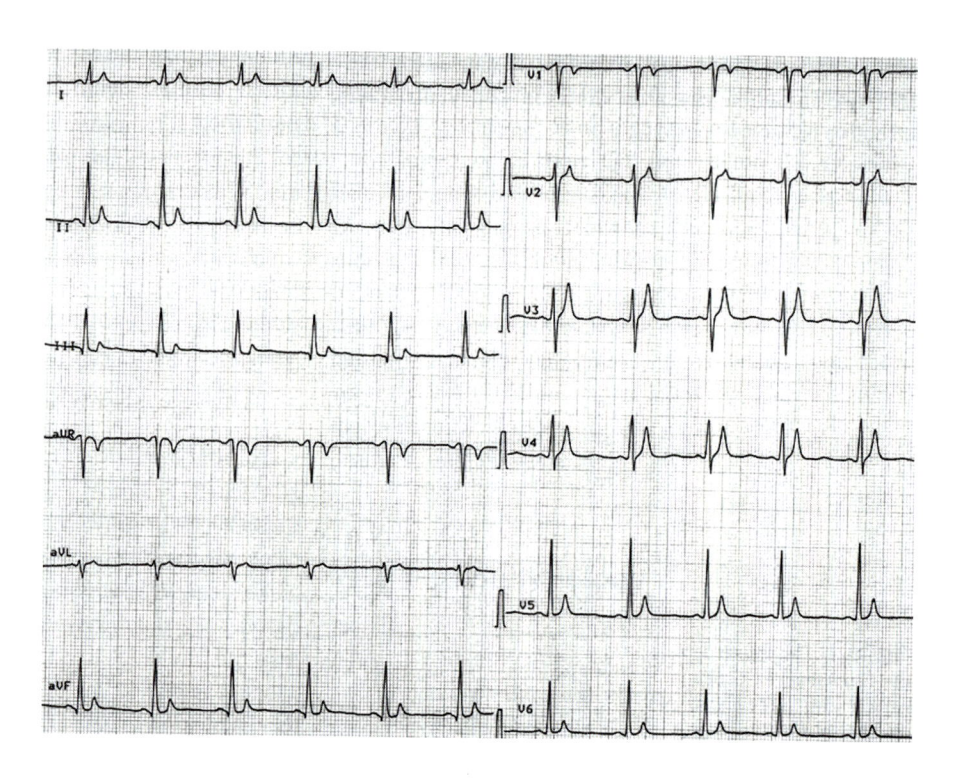

Fig. 11.2 In this ECG one can appreciate the typical T wave morphology of a patient with SQT1

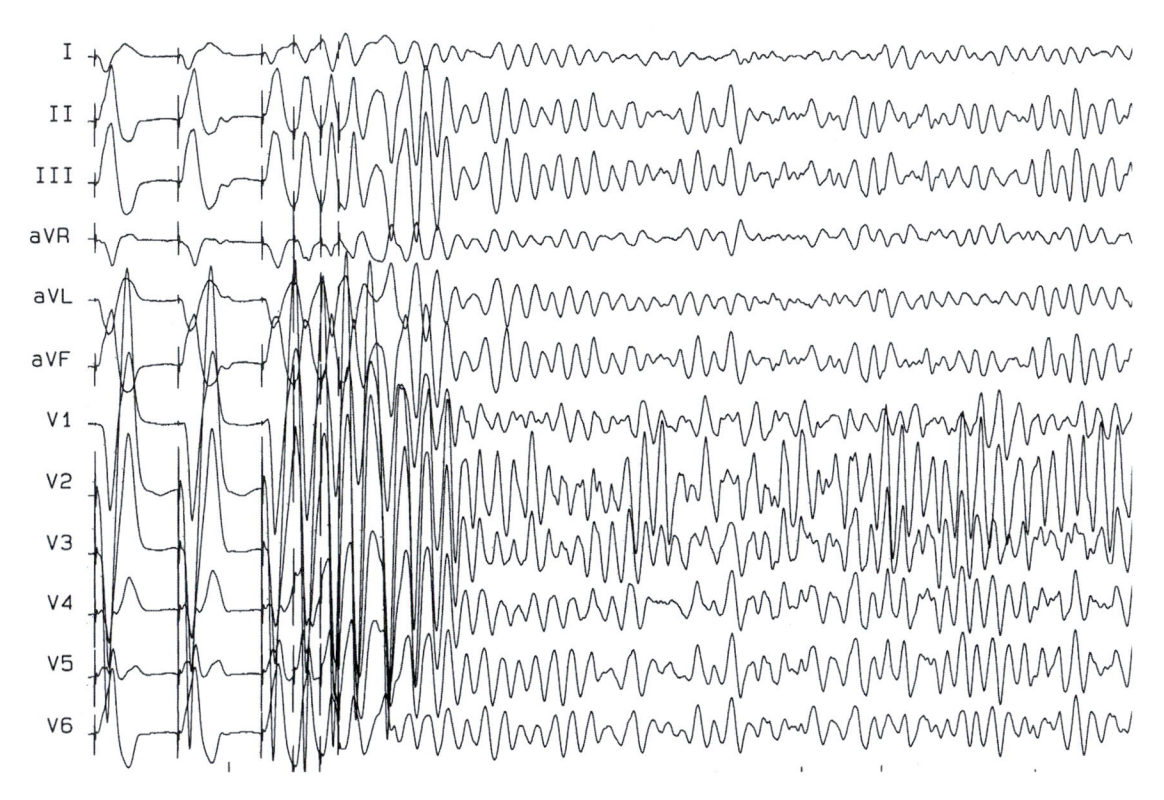

Fig. 11.3 This figure depicts the induction of ventricular fibrillation in a patient with SQT1 using triple extrastimuli with very short coupling intervals

11.7 Differential Diagnosis

The heritable SQTS should be differentiated from the acquired or secondary forms of QT shortening. Documentation of a short QT interval on the ECG should therefore lead to the exclusion of structural heart diseases and underlying conditions such as hyperkalemia, hypercalcemia, hyperthermia, the period immediately following VF, acidosis, and/or digitalis overdose.[19] Furthermore, structural heart disease, especially dilated cardiomyopathy, should be ruled out.[20]

11.8 Therapy, Follow-up, and Prognosis

11.8.1 Pharmacologic Therapy of Short QT Syndrome

After the identification of the genetic background and the cellular mechanisms of the SQTS clinical and experiment studies have been conducted with respect to the pharmacologic treatment. However, data on the pharmacology treatment of patients with SQTS are still limited with respect to the clinical use and long-term outcome because of the low number of patients diagnosed with SQTS at the moment.

Most of the experiences in vitro and in vivo are available for SQT1. Heterogeneous expression studies exhibited that the N588K mutation increased the density of I_{Kr} and reduced the affinity of I_{Kr} blockers like D-sotalol 20-fold.[7] Thus, in vitro experiments could proof the failure of D-sotalol restoring QT interval in vivo. McPate et al. could demonstrate that the effect of E-4031, a specific I_{Kr} blocker, was also significantly attenuated by the N588K mutation, whereas quinidine was less and disopyramide the least affected by N588K-HERG.[21] Cordeiro et al. could nicely show that these findings are based on the +90mV shift in the voltage-dependence of inactivation of the HERG channels. Most I_{Kr}-blockers interact with the HERG channels in the inactivated state. Thus, a failure of inactivation of the HERG channel leads to the inefficacy of the specific

I_{Kr} blockers.[22] Recently, McPate et al. could demonstrate that besides disopyramide and quinidine, propafenone and amiodarone also were only slightly inhibited by the mutant N588K.[21] Thus, these drugs may serve as an additional option in the pharamacologic treatmet of SQT1. For SQT3 El Harchi et al. could identify in in vitro experiments that chloroquine is an effective pharamcologic inhibitor of the SQT3 D172N mutant Kir2.1.[23]

In the clinical setting, several class I and III antiarrhythmic drugs have been tested in patients with the gain of function mutation in HERG (SQT1).[7,11,24] For class III antiarrhythmics, neither D-sotalol nor ibutilide were able to prolong QT interval. Flecainide, a Na+ channel blocker, which has in addition a blocking effect on I_{Kr} and on the transient outward potassium current (I_{to}), led to an increase in ventricular effective refractory periods. However, acute administration of flecainide did cause prolongation of refractoriness and only slight prolongation of the QT interval.[24] In contrast, the class I antiarrhythmic agent quinidine, was able to normalize the QT interval and to prolong the ventricular effective refractory period in patients with a SQT1.[18] Additionally, quinidine restored the heart rate dependence of the QT interval toward the normal range and rendered ventricular tachyarrhythmias non-inducible in patients in whom baseline electrophysiological studies demonstrated reproducible inducibility of ventricular fibrillation.[18,24] Following the positive effects of disopyramide in in vitro experiments, disopyramide has also been shown to be effective in a pilot study in patients with a SQT1.[21,25] No patient on quinidine therapy suffered from ventricular fibrillation or a recurrence of atrial fibrillation during mid-term follow-up.[16,26] A subset of patients treated with propafenone is free of recurrences of atrial fibrillation without prolongation of the QT interval. Whether quinidine, propafenone, or disopyramide represent an alternative to ICD therapy in prevention of sudden cardiac death cannot be finally answered. Drugs may be an alternative in patients refusing ICD implantation or for those who are not eligible for ICD therapy. In addition, drugs can be given to ICD-bearing patients who experience recurrent electrical shocks.

Whether the effects of the investigated class I and III drugs can be translated to SQT2–SQT5 is not clear. However, in a patient with SQT4 quinidine was able to prolong QT interval and suppress paroxysms of atrial fibrillation.

Due to the electrophysiological and genetic heterogeneity of the SQTS, therapy may have very different effects depending on the type of mutation and the affected channel. Further studies of pharmacologic therapy are warranted to elucidate the potential long-term benefit of pharmacologic treatment. However, such studies are complicated by the low number of SQTS patients identified so far, especially in SQT2 and SQT3.

11.8.2 ICD Therapy

By now the only reliable treatment to prevent patients from sudden cardiac death is the implantation of an implantable cardioverter-defibrillator (ICD). In symptomatic patients with SQTS, the ICD is the therapy of choice, while antiarrhythmic drug therapy may represent an adjunct or an alternative therapy in children or in newborns where ICD implantation is technically challenging and often associated with high morbidity. The risk for inappropriate ICD discharges due to T wave oversensing is increased in patients with SQTS compared to other conditions with ICD implanted, since intracardiac T waves are high and closely coupled to the preceding R wave. This issue can be solved by individual ICD programming of the sensing parameters and selection of specific devices. Additionally, quinidine therapy helped to avoid T wave oversensing by increasing the QT interval.[27]

11.9 Risk Stratification and Indication for ICD

Since the number of patients with SQTS is still low, we are lacking reliable data and relevant follow-up duration for a conclusive statement on risk stratification. In patients with a short QT interval in the presence of syncope of unknown origin or aborted sudden cardiac death an implantation of an ICD is indicated. In asymptomatic patients the indication for primary prophylactic ICD implantation is not clear. In the families with SQT1 most of the asymptomatic relatives with SQTS underwent primary prophylactic ICD due to the high incidence of familiar sudden death and syncope. Whether the family history of sudden cardiac death associated with a short QT interval serves as a risk factor is unknown. As stated above inducibility at programmed stimulation is high, which may be explained by the extremely short refractory periods. However, whether the inducibility of ventricular arrhythmias is predicting future cardiac events is doubtful. In the only patient with SQTS receiving an appropriate ICD shock so far, ventricular tachyarrhythmias were not inducible.[26] Whether the degree of QT interval shortening can be used as a marker of risk is still unknown.

11.10 Cardiogenetics Aspects

Because of the high genetic heterogeneity in SQTS and the low number of patients, a genotype–phenotype correlation cannot be established. Furthermore, only approximately 25% of patients diagnosed with SQTS were carriers of an underlying potassium or calcium channel mutation. This suggests that other, unknown, genetic defects may be involved. After the identification of a patient with SQTS, genetic analysis should be attempted and family screening initiated, irrespective of having a causal mutation found.

11.11 Summary

The SQTS is one of the primary electrical diseases of the heart with a high incidence of syncope and sudden cardiac death. The hallmark for the diagnosis is a short QT interval on the baseline ECG with a QTc <350 ms for males and a QTc<360 ms for females, respectively-. Until now, approximately 50 patients have been identified worldwide. Because of the limited number of patients and the genetic heterogeneity of the disease, strong genotype–phenotype correlation and a conclusive risk stratification are not yet available. The class I antiarrhythmic drug quinidine was shown to prolong QT interval and refractoriness and rendered previously inducible ventricular tachyarrhythmias non-inducible. The only reliable treatment so far in the prevention of sudden cardiac death is the implantation of an ICD and anyone with the diagnosis of symptomatic SQTS should have an ICD implanted. In addition, patients with SQTS should be referred for genetic counseling, molecular genetic analysis, and initiation of family screening.

> **Take Home Message**
> - SQTS is a very rare but potentially highly malignant disease.
> - SQTS should be considered in anyone with a QT<350 ms without potential other causes.
> - One always must think about SQTS in the following special cases:
>   — Aborted cardiac arrest or sudden cardiac death of unknown origin
>   — Atrial fibrillation at young age

References

1. Gussak I, Brugada P, Brugada J, et al. Idiopathic short QT interval: a new clinical syndrome? *Cardiology*. 2000;94:99-102.
2. Gaita F, Giustetto C, Bianchi F, et al. Short QT Syndrome: a familial cause of sudden death. *Circulation*. 2003;108:965-970.
3. Bellocq C, van Ginneken AC, Bezzina CR, et al. Mutation in the KCNQ1 gene leading to the short QT-interval syndrome. *Circulation*. 2004;109:2394-2397.
4. Priori SG, Pandit SV, Rivolta I, et al. A novel form of short QT syndrome (SQT3) is caused by a mutation in the KCNJ2 gene. *Circ Res*. 2005;96:800-807.
5. Antzelevitch C, Pollevick GD, Cordeiro JM, et al. Loss-of-function mutations in the cardiac calcium channel underlie a new clinical entity characterized by ST-segment elevation, short QT intervals, and sudden cardiac death. *Circulation*. 2007;115:442-449.
6. Hong K, Piper DR, Diaz-Valdecantos A, et al. De novo KCNQ1 mutation responsible for atrial fibrillation and short QT syndrome in utero. *Cardiovasc Res*. 2005;68:433-440.
7. Brugada R, Hong K, Dumaine R, et al. Sudden death associated with short-QT syndrome linked to mutations in HERG. *Circulation*. 2004;109:30-35.
8. Anttonen O, Junttila MJ, Rissanen H, Reunanen A, Viitasalo M, Huikuri HV. Prevalence and prognostic significance of short QT interval in a middle-aged Finnish population. *Circulation*. 2007;116:714-720.
9. Funada A, Hayashi K, Ino H, et al. Assessment of QT intervals and prevalence of short QT syndrome in Japan. *Clin Cardiol*. 2008;31:270-274.
10. Gallagher MM, Magliano G, Yap YG, et al. Distribution and prognostic significance of QT intervals in the lowest half centile in 12, 012 apparently healthy persons. *Am J Cardiol*. 2006;98:933-935.
11. Kobza R, Roos M, Niggli B, et al. Prevalence of long and short QT in a young population of 41, 767 predominantly male Swiss conscripts. *Heart Rhythm*. 2009;6:652-657.

12. Viskin S. The QT interval: too long, too short or just right. *Heart Rhythm.* 2009;6:711-715.
13. Moriya M, Seto S, Yano K, Akahoshi M. Two cases of short QT interval. *Pacing Clin Electrophysiol.* 2007;30: 1522-1526.
14. Extramiana F, Antzelevitch C. Amplified transmural dispersion of repolarization as the basis for arrhythmogenesis in a canine ventricular-wedge model of short-QT syndrome. *Circulation.* 2004;110:3661-3666.
15. Milberg P, Tegelkamp R, Osada N, et al. Reduction of dispersion of repolarization and prolongation of postrepolarization refractoriness explain the antiarrhythmic effects of quinidine in a model of short QT syndrome. *J Cardiovasc Electrophysiol.* 2007;18:658-664.
16. Giustetto C, Di Monte F, Wolpert C, et al. Short QT syndrome: clinical findings and diagnostic-therapeutic implications. *Eur Heart J.* 2006;27:2440-2447.
17. Anttonen O, Junttila MJ, Maury P, et al. Differences in twelve-lead electrocardiogram between symptomatic and asymptomatic subjects with short QT interval. *Heart Rhythm.* 2009;6:267-271.
18. Wolpert C, Schimpf R, Giustetto C, et al. Further insights into the effect of quinidine in short QT syndrome caused by a mutation in HERG. *J Cardiovasc Electrophysiol.* 2005; 16:54-58.
19. Cheng TO. Digitalis administration: an underappreciated but common cause of short QT interval. *Circulation.* 2004;109:e152. author reply e152.
20. Bohora S, Namboodiri N, Tharakan J, Vk AK, Nayyar S. Dilated cardiomyopathy with short QT interval: is it a new clinical entity? *Pacing Clin Electrophysiol.* 2009;32:688-690.
21. McPate MJ, Zhang H, Adeniran I, Cordeiro JM, Witchel HJ, Hancox JC. Comparative effects of the short QT N588K mutation at 37 degrees C on hERG K+channel current during ventricular, Purkinje fibre and atrial action potentials: an action potential clamp study. *J Physiol Pharmacol.* 2009; 60:23-41.
22. Cordeiro JM, Brugada R, Wu YS, Hong K, Dumaine R. Modulation of I(Kr) inactivation by mutation N588K in KCNH2: a link to arrhythmogenesis in short QT syndrome. *Cardiovasc Res.* 2005;67:498-509.
23. El Harchi A, McPate MJ, Zhang YH, Zhang H, Hancox JC. Action potential clamp and chloroquine sensitivity of mutant Kir2.1 channels responsible for variant 3 short QT syndrome. *J Mol Cell Cardiol.* 2009;47(5):743-747.
24. Gaita F, Giustetto C, Bianchi F, et al. Short QT syndrome: pharmacological treatment. *J Am Coll Cardiol.* 2004;43: 1494-1499.
25. Schimpf R, Veltmann C, Giustetto C, Gaita F, Borggrefe M, Wolpert C. In vivo effects of mutant HERG K+channel inhibition by disopyramide in patients with a short QT-1 syndrome: a pilot study. *J Cardiovasc Electrophysiol.* 2007;18:1157-1160.
26. Schimpf R, Bauersfeld U, Gaita F, Wolpert C. Short QT syndrome: successful prevention of sudden cardiac death in an adolescent by implantable cardioverter-defibrillator treatment for primary prophylaxis. *Heart Rhythm.* 2005;2: 416-417.
27. Schimpf R, Wolpert C, Bianchi F, et al. Congenital short QT syndrome and implantable cardioverter defibrillator treatment: inherent risk for inappropriate shock delivery. *J Cardiovasc Electrophysiol.* 2003;14:1273-1277.

Catecholaminergic Polymorphic Ventricular Tachycardia

12

Christian van der Werf and Arthur A.M. Wilde

12.1 Introduction

Catecholaminergic polymorphic ventricular tachycardia (CPVT) is a highly malignant inherited arrhythmia syndrome, characterized by bidirectional or polymorphic ventricular tachycardia (VT) and ventricular fibrillation (VF) during physical or emotional stress in structurally normal hearts.[1] This disease is illustrative of the progress that has been made in the field of cardiogenetics in the last decennia. The first case reports on CPVT date from the 1960s and 1970s, among others a case of a 6-year-old girl with bidirectional tachycardia triggered by exercise.[2] An important series describing the phenotype of 21 children with CPVT was published in 1995,[1] and the underlying molecular genetic basis was discovered in 2001.[3–5] In recent years significant progress has been made in the treatment of this potentially lethal condition.[6–9]

12.2 Etiology

12.2.1 Molecular Background

CPVT is caused by modifications in the normal cardiac **calcium** homeostasis (Fig. 12.1).[10–13] In the normal heart, a small quantity of calcium enters the cell via voltage-dependent L-type calcium channels during the plateau phase of the action potential. This triggers calcium release from the sarcoplasmic reticulum (SR) into the cytosol through the cardiac ryanodine type 2 receptor (RyR2) channel, which is located in the membrane of the SR. This entire process is termed calcium-induced calcium release (CICR). As a consequence, the cytosolic calcium concentration increases

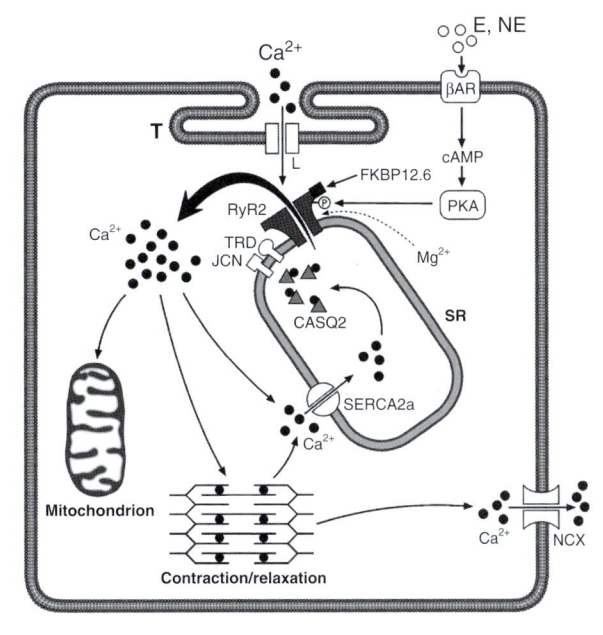

Fig. 12.1 Regulation of the RyR2 channel function in cardiac myocytes. *T* T-tubule; *L* L-type of Ca2+ channel; *E*, *NE* epinephrine, norepinephrine; *hAR* beta-adrenergic receptor; *cAMP* cyclic AMP; *PKA* protein kinase A; *FKBP12.6* calstabin2; *TRD* triadin 1; *JCN* junctin; *CASQ2* calsequestrin 2; *SR* sarcoplastic reticulum; *SERCA2a* sarcoplastic reticulum Ca2+-ATPase; *NCX* Na+/Ca+exchanger (Adapted from [14]. With permission)

C. van der Werf (✉)
Department of Cardiology, Heart Failure Research Centre,
Academic Medical Centre, Amsterdam, The Netherlands
e-mail: c.vanderwerf@amc.uva.nl

significantly, which causes activation of the actin-myosin filaments with subsequent myocyte shortening and systole. Stimulation of the sympathetic nervous system is associated with activation of intracellular cyclic AMP, which results in phosphorylation of RyR2 through protein kinase A (PKA) and activation of CICR.[14]

During diastole, most of the reuptake of cytosolic calcium into the SR is performed via SR calcium adenosine triphosphatase (SERCA).[11] In addition, calcium is removed from the cell by sodium/calcium exchanger (NCX).

12.2.2 Pathophysiology

Genetic defects in the genes encoding RyR2 and cardiac calsequestrin (CASQ2) underlie CPVT.[3,5] To date, over 70 mutations have been identified in the *RyR2* gene, which causes an autosomal dominant CPVT phenotype and is termed CPVT type 1. The rare CPVT type 2 phenotype is caused by mutations in the gene encoding CASQ2 and shows an autosomal recessive inheritance pattern. A third and highly malignant type of CPVT maps to a yet unknown gene that resides on chromosome 7p14–p22 and also segregates as an autosomal recessive trait.[15]

Both RyR2 and CASQ2 are calcium regulatory proteins.[12] Out of the three isoforms of ryanodine receptor (RyR1 to RyR3), RyR2 is considered to be the cardiac isoform.[11] RyR2 is one of the largest cardiac ion channel proteins, comprising 4,967 amino acids and the gene encoding RyR2 is located on chromosome 1 (1q42–q43).[16] CASQ2 is a 399 amino acid protein and is located in the cardiac SR as well and modulates the activity of the RyR2 channels.[11] When the calcium concentration in the SR is low, CASQ2 inhibits RyR2 activity and this inhibition is relieved when the calcium concentration increases.[17] The gene encoding CASQ2 also resides on chromosome 1 (1p11–p13.3).[5]

Mutations in the genes encoding RyR2 or CASQ2 modify their function, which results in calcium leakage from the SR into the cytosol.[18,19] This intracellular calcium overload results in activation of NCX. NCX exchanges one calcium ion for three sodium ions, which produces a transient inward current (I_{Ti}). I_{Ti} generates delayed afterdepolarizations (DADs), which can lead to triggered activity in myocytes and ventricular arrhythmias, in particular under conditions of β-adrenergic stimulation.[20–22]

12.3 Clinical Aspects

12.3.1 Epidemiology

There are relatively few epidemiological data on CPVT. The prevalence in the general population is unknown, but is estimated to be approximately 1 in 10 000.[13]

However, this rare disease seems to be much more common in young victims of sudden unexplained death (SUD), that is, sudden death victims with no significant anatomical abnormalities at autopsy, or in which no autopsy is performed. This was, for example, shown by Tester et al., who performed nontargeted screening of the cardiac ion channel genes in 49 SUD victims aged 1–43 years.[23] In seven patients (14%), a putative causal mutation was detected in the *RyR2* gene. In similar subsequent series, Creighton et al. identified three *RyR2* mutations (21%) in 14 exercise-triggered SUD victims aged 1–43 years,[24] and Nishio et al. found three *RyR2* mutations (18%) out of 17 cases of SUD at the age of 12–42 years.[25]

Another clue of the significant contribution of CPVT to SUD in the young was derived from studies by Tan et al. and Hofman et al.[26,27] In these series, the yield of cardiological and genetic examination in surviving relatives of the SUD victims was assessed. In the series by Tan et al., comprising 43 SUD victims aged 1–40 years, CPVT was diagnosed in five cases (12%).[26] Hofman et al. included 25 pediatric SUD victims (partly overlapping with the series decribed by Tan et al.) and found CPVT in the relatives of two of them (8%).[27] In contrast, in two comparable series by Behr et al., comprising 32 and 57 SUD victims aged 4–64 years in which autopsy was negative in all cases, CPVT was diagnosed in none of the families.[28,29]

In sudden infant death syndrome (SIDS), which is defined as the sudden death of an infant under 1 year of age that remains unexplained after thorough investigation including autopsy, CPVT might also play a role. Tester et al. demonstrated that 1–2% of SIDS may be due to CPVT.[30]

12.3.2 Clinical Diagnosis

The first key element in diagnosing CPVT is the patient's clinical history.[1,31] CPVT patients experience complaints, which are typically induced by exercise or emotional stress, in particular dizziness or syncope. Most symptomatic patients are young children. The first syncope usually occurs at the age of 7–9 years and not before the age of 3.[1] Sometimes these children know exactly which activities induce their complaints, and most often they try to avoid these triggers. Family history frequently contains syncope or sudden death in young relatives related to similar triggers.

Many of the patients who in the end are diagnosed with CPVT, have been misdiagnosed previously, predominantly with epilepsy. This is probably because syncope in CPVT can resemble epilepsy. It can include a hypertonic phase, convulsive movements, and loss of urine or feces.

The second key element, the typical bidirectional or **polymorphic VT**, is sometimes registered with Holter monitoring during syncope. However, most often VT is provoked and registered during exercise testing. The typical exercise test in an untreated CPVT patient is as follows. Ventricular premature beats (VPBs) appear during sinustachycardia, first isolated and monomorphic. The number of VPBs increases with sinus rate, for example, to bigeminy. Finally, the VPBs form rarely monomorphic and mostly polymorphic salvos. These rapid, irregular, and polymorphic VTs may persist when exercise is continued. When the exercise test is ceased, ventricular ectopy frequently disappears immediately. The typical **bidirectional VT**, which is infrequently observed, refers to ventricular complexes with beat-to-beat alternating right and left QRS axis (Fig. 12.2). Sinus rate at the start of ventricular ectopy is often well reproducible within a patient. The onset focus of ventricular ectopy has been observed to be located frequently in the right ventricular outflow tract.[7]

Supraventricular arrhythmias, mainly atrial ectopic beats, nonsustained supraventricular tachycardia, and short runs of atrial fibrillation, can also appear in CPVT patients during exercise.[31]

When exercise test is not possible or not preferred, adrenaline (0.1–0.3 µg/kg/min) or isoproterenol infusion can be used for simulation.[1,17,32] These tests can also be considered in a patient with a normal resting ECG and a negative exercise test in whom syncope or aborted cardiac arrest (ACA) occurred during exertion or emotion.[17] Programmed electrical stimulation has not been proven valuable to induce VT in CPVT.[1]

Resting ECG has inconsistently been described to show several abnormalities, in particular bradycardia and prominent U-waves.[1,31,33] Physical examination, echocardiography, and other cardiological examinations are typically normal in CPVT patients.

12.4 Genetic Diagnosis

The yield of genotyping in CPVT depends on the certainty of the clinical diagnosis.[34,35] In patients with a conclusive diagnosis, that is, documented exercise or emotion-induced bidirectional or polymorphic VT, in the absence of structural cardiac disease, a *RyR2* mutation is found in 50–70%.[34–36] CPVT patients with a *RyR2* mutation are more often male, and syncope occurs at significantly younger age as compared to nongenotyped patients.[36] The yield of *RyR2* genotyping is 38% in patients with exertion-induced syncope and exercise test-induced ventricular ectopy, but no bidirectional or polymorphic VT,[35] and 5% in patients with exercise or emotion-induced syncope without documented bidirectional or polymorphic VT,[34] both a possible diagnosis of CPVT. Notably, in 45 cases with exercise-induced syncope, corrected QT interval values <480 ms, and no mutations in the 12 long QT syndrome (LQTS)-associated genes, *RyR2* missense mutations were found in 31%.[35] Mutations in *RyR2* are clustered in certain regions.[35] This facilitates a tiered genotyping strategy, which was recently proposed and seems attractive and more cost-effective because of the large size of this gene.[35] The *RyR2* genotyping strategy starts with screening the 16 exons known to host three of more unique CPVT-associated mutations, which (in case of negative results) can be followed by screening the 13 exons with two mutations reported and 16 exons with one mutation reported. Finally, when no mutation is detected yet, the normal pseudo-comprehensive *RyR2* scan can be performed.

Fig. 12.2 Exercise test recordings in a female CPVT patient. (**a**) The resting ECG shows sinusbradycardia with no ventricular ectopy or other abnormalities. (**b**) At a sinus rate of 80/min the first ventricular ectopy appears, that is, VPBs in bigeminy (indicated with an asterisk). (**c**) During maximum exercise the last four VPBs of a bidirectional NSVT are shown on the left. After one regular sinus beat a NSVT of 15 beats starts. In the extremity leads the NSVT is bidirectional and at the switch to the precordial leads it becomes polymorphic. At the right, after one regular sinus beat a NSVT initiates again. *VPB* ventricular premature beat; *NSVT* nonsustained ventricular tachycardia

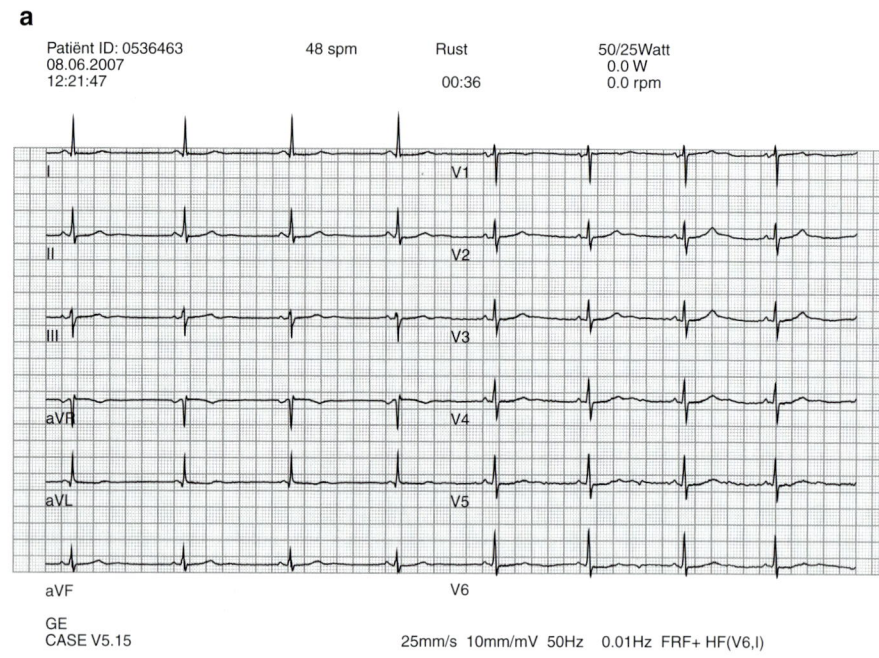

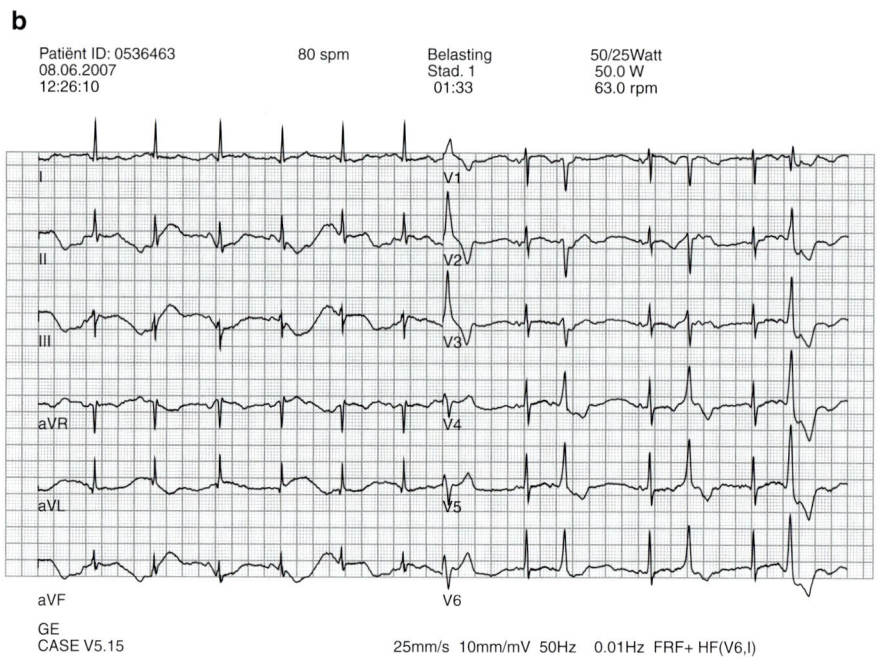

Fig. 12.2 (continued)

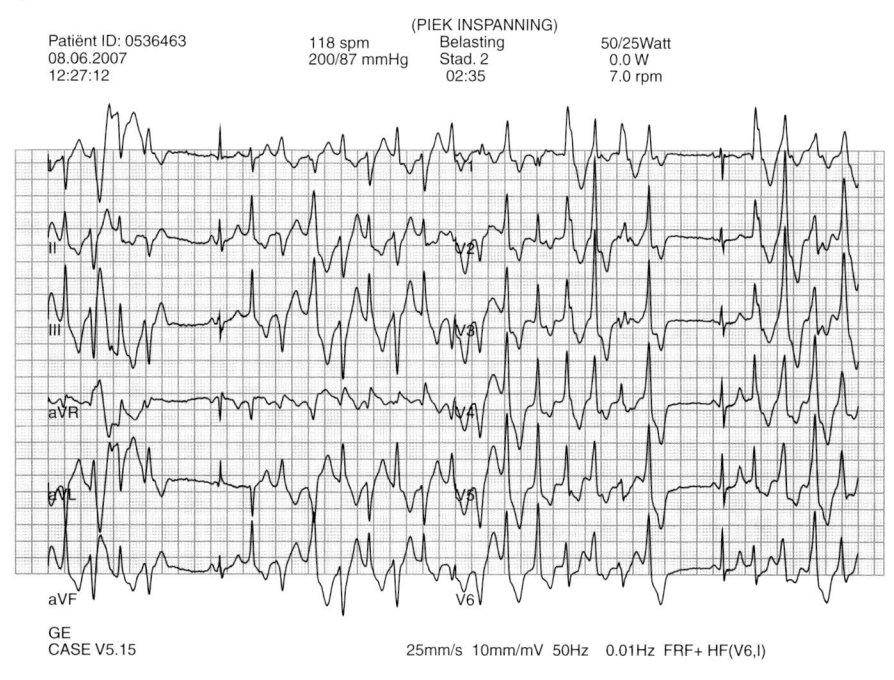

Identifying a putative pathogenic mutation is important, because it provides the opportunity to genotype relatives of the index patient. The phenotype is similar in genotype-positive and nongenotyped CPVT patients, thus the identification of a mutation does not influence treatment or risk stratification in the index patient.

When relatives of an index patient appear to be carrier of the same mutation, sometimes complaints and exercise-induced VT can be completely absent. In one series of 43 *RyR2* mutation carriers, 35% was completely asymptomatic.[37] However, these silent mutation carriers are at risk of cardiac events as well.[37,38]

12.4.1 Differential Diagnosis

When a patient presents with exercise or emotional stress-induced syncope or ACA, LQTS, particularly type 1, is another possible diagnosis. In most cases, resting 12-lead ECG can immediately differentiate between LQTS and CPVT. In CPVT, usually no abnormalities are seen, whereas LQTS can be revealed most often by a prolonged QTc interval and/or an abnormal T wave morphology. When resting ECG is normal; exercise testing can help to discriminate between LQTS and CPVT. Ventricular ectopy is rarely observed in LQTS and ventricular ectopy beyond a single intermittent VPB is much more common in CPVT.[39] In a cohort of 381 patients who were referred because of a suspicion of LQTS, VPBs in bigeminy had a sensitivity of 81%, specificity of 96%, positive predictive value of 45%, and negative predictive value of 99% for CPVT.[39]

When bidirectional VT is observed during exercise, Andersen–Tawil syndrome (LQTS type 7) should also be considered.[13] This condition is caused by loss of function mutations in the *KCNJ2* gene, and is usually associated with extracardiac manifestations, that is, periodic paralysis and facial dysmorphisms.

RyR2 mutations have been reported in few families with arrhythmogenic right ventricular dysplasia/cardiomyopathy as well.[40] However, as in CPVT no structural cardiac abnormalities are present, the differences between these two phenotypes are usually very clear.

12.4.2 Therapy and Prognosis

12.4.2.1 Lifestyle Modifications

CPVT patients are advised not to perform any (competitive) sports, especially when VT is observed during exercise testing. In pediatric CPVT patients, the parents are recommended to observe their children while playing and swimming. Although cardiac events during swimming are more typical for LQTS type 1, they also occur in CPVT patients.[41]

12.4.2.2 β-Blocker Therapy

As the trigger for arrhythmias in CPVT is an adrenergic state, β-blocker therapy is the cornerstone of therapy. β-blockers have been empirically used with success for the treatment of CPVT for decades, but not until in 2009 Hayashi et al. provided statistical evidence that β-blockers protect for cardiac events and fatal or near fatal events.[38] In this series of 101 CPVT patients, 20 patients were not treated with β-blockers by their physician, most frequently because cardiac symptoms and stress-induced VT were absent. The 8-year cardiac event rate, which was defined as stress-induced syncope, ACA (including appropriate implantable cardioverter-defibrillator (ICD) discharge) or SCD, was 27% in the patients treated with β-blockers, as compared to 58% in the untreated patients. In 11% of the patients treated with β-blockers a fatal or near-fatal event (SCD or ACA) occurred during 8 year of follow-up, in comparison to a 25% fatal or near-fatal event rate in the untreated patients. In multivariate analysis, the absence of β-blocker treatment was an independent predictor for any cardiac event (hazard ratio (HR), 5.48; 95% CI, 1.80–16.68; $P=0.003$) and fatal or near-fatal events (HR, 5.54; 95% CI, 1.17–26.15; $P=0.03$).

An other important finding in this study was that the event rates within patients with a syncopal history before diagnosis and patients with no history of syncope do not differ.[38] Moreover, cardiac events also occurred in silent mutation carriers. This signifies that every CPVT patient and/or *RyR2* or *CASQ2* mutation carrier should be treated aggressively, regardless of any prior syncopal events or documented ventricular arrhythmias.

Furthermore, a younger age at the time of the diagnosis independently predicted cardiac events, both in the entire series (HR, 0.54 per decade; 95% CI, 0.33–0.89, $P=0.02$), and in the subgroup treated with β-blockers (HR, 0.31 per decade, 95% CI, 0.14–0.69, $P=0.004$).[38] Children are most often treated with a β-blocker dosage based on bodyweight, but these results suggest that this might be inadequate.

Although the authors state that in their series, the cardiac events in patients on β-blocker therapy might be due to an insufficient dosage or noncompliance,[38] there have been more observations of suboptimal treatment with the maximum tolerable dosage of β-blockers in some patients. For example, in a series by Sumitomo et al., seven of the 29 CPVT patients died suddenly during a follow-up of 6.8 years, including four patients on β-blocker therapy.[7] and Priori et al. even reported sustained VT or VF in 18 out of 39 (46%) on β-blocker therapy during 40–52 months of follow-up.[36]

The effectiveness of the different β-blockers in CPVT have never been compared with one another. In most case series nadolol is used.[1] A dosage of >1.5 mg/kg is advisable.[38] Atenolol, bisoprolol, metoprolol, propanolol, and sotalol are also used. β-blockers with intrinsic sympathomimetic activity are not recommended. Whatever β-blocker prescribed, the optimal dosage to suppress VT should be titrated by performing exercise tests on a regular base.

12.4.2.3 Other Pharmacologic Therapy

As β-blocker therapy is not sufficient in all CPVT patients, calcium channel blocker therapy has been suggested as an adjuvant.[6,42] Rosso et al. showed that oral verapamil (2–3 mg per kg bodyweight per day in children, and 240 mg per day in adults) in addition to β-blocker therapy decreased VT during exercise test in five CPVT patients when compared to monotherapy with β-blockers.[6] Swan et al. reported similar findings in six CPVT patients who were on β-blocker therapy and underwent an exercise test after intravenous infusion of verapamil (0.2 mg per kg bodyweight).[42]

Watanabe et al. recently provided evidence for a beneficial effect of flecainide in two highly symptomatic CPVT patients despite conventional drug therapy.[9] In a CPVT mouse model, flecainide proved to inhibit

RyR2-mediated calcium release into the cytoplasm, which prevented ventricular arrhythmias. Subsequently, this mechanism-based approach suppressed VT in two patients, who had remained highly symptomatic despite β-blocker and calcium channel blocker therapy. These promising results will need to be confirmed in a larger series of conventional therapy-resistant CPVT patients.

Other pharmacologic agents, such as amiodarone and magnesium, have been tested in CPVT patients, but lacked efficacy.[1,7] In the acute setting, adenosine and propanolol have been reported in case reports to terminate VT in CPVT patients.[43,44]

12.4.2.4 Nonpharmacologic Therapy

Surgical left cardiac sympathetic denervation (LCSD) has been proven to be a highly effective therapy in CPVT patients whose symptoms are not controlled by pharmacologic therapy.[8,45–47] During this procedure, the lower part of the stellate ganglion and the second and third thoracic ganglia are ablated. As a result, the release of norepinephrine in the heart is highly diminished, which has an antiarrhythmic effect. A few cases with excellent results after this procedure have been published,[8] including one case with 10 year follow-up,[45] and several patients in whom video-assisted thoracoscopic surgery was used.[46,47]

12.4.3 Follow-up

The basis of follow-up of CPVT patients is performance of exercise testing and Holter monitoring on a regular basis. With these examinations, the efficacy of treatment can be monitored and, if necessary, adjustments can be made. It is important to emphasize that in a part of CPVT patients VPBs remain present during exercise testing despite optimal therapy. In clinical practice this is considered safe and acceptable, though its prognostic relevance is actually unknown. However, the presence of couplets or more successive VPBs should not be accepted, as these are significantly associated with future cardiac events (sensitivity, 0.62; specificity, 0.67; $P=0.03$).[38]

12.4.4 Risk Stratification/ Indication for ICD

The 2006 ACC/AHA/ESC Guidelines for Management of Patients With Ventricular Arrhythmias and the Prevention for Sudden Cardiac Death give a class I recommendation for implantation of an ICD in addition to β-blocker therapy for CPVT patients who are survivors of ACA.[48] A class IIa recommendation is given for ICD implantation in CPVT patients with syncope and/or documented sustained VT despite β-blocker therapy.

However, as the malignancy of this disease has been well recognized, in the recent years ICD implantation has been performed more liberal than the guidelines recommend. This is an unfortunate development, because ICD implantation has some serious disadvantages in the setting of this specific condition and can even be harmful. Both appropriate and inappropriate shocks cause pain and fear, which raises the release of catecholamines and can result in VT, electrical storm or VF. This mechanism was illustrated in two case reports on death of a CPVT patient with an ICD.[49,50] In one case, a young male CPVT patient received inappropriate ICD therapy as a result of paroxysmal atrial fibrillation, which triggered the fatal ventricular arrhythmias.[49] In the second case, a young woman with incessant rapid polymorphic VT could not be saved despite appropriate ICD therapy.[50] Hence, lessons to be learned from these cases are that one should be very reluctant to implant an ICD in CPVT patients. Indeed, ICDs can be proarrhythmic in this setting.

12.4.5 Cardiogenetics Aspects

The odds of positive genotyping are highest in patients with a conclusive diagnosis of CPVT.[14,34] Bai et al. found a *RyR2* mutation in 62% of 81 patients with a conclusive diagnosis of CPVT, which resulted in an estimated cost of US $5263 per one positive *RyR2* genotyping. In patients with exercise or emotion-induced syncope without documented bidirectional or polymorphic VT, ACA survivors, and relatives of SCD victims the yield of genotyping was much lower, resulting in considerably higher costs. Thus, it is highly recommendable to reserve genotyping for index patients with a certain diagnosis of CPVT. When no mutation is found in *RyR2*, *CASQ2* genotyping can be considered. Apart from families with consanguinity, this is also recommended in cases with a clear CPVT phenotype, as heterozygous carriership of *CASQ2* mutations can cause CPVT.[51]

When a particular CPVT genotype is detected, first-degree relatives should be tested to identify those relatives at risk of cardiac events. This provides the unique opportunity to take preventive measure in mutation carriers, as the first manifestation of CPVT can be cardiac arrest. In contrast to many other inherited diseases, genotyping in CPVT is strongly recommended in children from the age of 4 years, because the disease can manifest itself at young age. In case no mutation is detected in the index patient, thorough clinical evaluation of first-degree relatives is recommended.

12.5 Summary

CPVT is a rare, but very lethal primary arrhythmia syndrome, most often caused by gain-of-function mutations in the gene encoding RyR2, which result in an intracellular calcium overload. Increased adrenergic stimulation induces the classic bidirectional or polymorphic ventricular VT or even ventricular fibrillation. The typical pediatric patients often experience syncopes during stress and have structurally normal hearts. The primary treatment consists of cessation of physical exertion and β-blocker therapy, with regular follow-up including exercise, testing, and Holter monitoring. This treatment should be offered to all CPVT patients, that is, clinically diagnosed patients with or without syncope and silent mutation carriers. In case of insufficient suppression of ventricular arrhythmias, addition of a calcium channel blocker or perhaps flecainide are alternative options. LCSD can be performed in the most severe cases. Implantation of an ICD should be avoided as much as possible because of its potential proarrhythmic effect in this specific condition. In all first-degree relatives above the age of 4 of an index patient, cardiological and genetic evaluation is recommended to provide the opportunity of taking preventive measures in the affected individuals.

References

1. Leenhardt A, Lucet V, Denjoy I, Grau F, Ngoc DD, Coumel P. Catecholaminergic polymorphic ventricular tachycardia in children. A 7-year follow-up of 21 patients. *Circulation*. 1995;91:151-159.
2. Reid DS, Tynan M, Braidwood L, Fitzgerald GR. Bidirectional tachycardia in a child. A study using His bundle electrography. *Br Heart J*. 1975;37:339-344.
3. Priori SG, Napolitano C, Tiso N, et al. Mutations in the cardiac ryanodine receptor gene (hRyR2) underlie catecholaminergic polymorphic ventricular tachycardia. *Circulation*. 2001;103:196-200.
4. Laitinen PJ, Swan H, Kontula K. Molecular genetics of exercise-induced polymorphic ventricular tachycardia: identification of three novel cardiac ryanodine receptor mutations and two common calsequestrin 2 amino-acid polymorphisms. *Eur J Hum Genet*. 2003;11:888-891.
5. Lahat H, Pras E, Olender T, et al. A missense mutation in a highly conserved region of CASQ2 is associated with autosomal recessive catecholamine-induced polymorphic ventricular tachycardia in Bedouin families from Israel. *Am J Hum Genet*. 2001;69:1378-1384.
6. Rosso R, Kalman JM, Rogowski O, et al. Calcium channel blockers and beta-blockers versus beta-blockers alone for preventing exercise-induced arrhythmias in catecholaminergic polymorphic ventricular tachycardia. *Heart Rhythm*. 2007;4:1149-1154.
7. Sumitomo N, Harada K, Nagashima M, et al. Catecholaminergic polymorphic ventricular tachycardia: electrocardiographic characteristics and optimal therapeutic strategies to prevent sudden death. *Heart*. 2003;89:66-70.
8. Wilde AA, Bhuiyan ZA, Crotti L, et al. Left cardiac sympathetic denervation for catecholaminergic polymorphic ventricular tachycardia. *N Engl J Med*. 2008;358:2024-2029.
9. Watanabe H, Chopra N, Laver D, et al. Flecainide prevents catecholaminergic polymorphic ventricular tachycardia in mice and humans. *Nat Med*. 2009;15:380-383.
10. Liu N, Priori SG. Disruption of calcium homeostasis and arrhythmogenesis induced by mutations in the cardiac ryanodine receptor and calsequestrin. *Cardiovasc Res*. 2008;77:293-301.
11. Katz G, Arad M, Eldar M. Catecholaminergic polymorphic ventricular tachycardia from bedside to bench and beyond. *Curr Probl Cardiol*. 2009;34:9-43.
12. Gyorke S. Molecular basis of catecholaminergic polymorphic ventricular tachycardia. *Heart Rhythm*. 2009;6:123-129.
13. Liu N, Ruan Y, Priori SG. Catecholaminergic polymorphic ventricular tachycardia. *Prog Cardiovasc Dis*. 2008;51:23-30.
14. Kontula K, Laitinen PJ, Lehtonen A, Toivonen L, Viitasalo M, Swan H. Catecholaminergic polymorphic ventricular tachycardia: recent mechanistic insights. *Cardiovasc Res*. 2005;67:379-387.
15. Bhuiyan ZA, Hamdan MA, Shamsi ET, et al. A novel early onset lethal form of catecholaminergic polymorphic ventricular tachycardia maps to chromosome 7p14-p22. *J Cardiovasc Electrophysiol*. 2007;18:1060-1066.
16. Swan H, Piippo K, Viitasalo M, et al. Arrhythmic disorder mapped to chromosome 1q42-q43 causes malignant

polymorphic ventricular tachycardia in structurally normal hearts. *J Am Coll Cardiol*. 1999;34:2035-2042.
17. Cerrone M, Napolitano C, Priori SG. Catecholaminergic polymorphic ventricular tachycardia: A paradigm to understand mechanisms of arrhythmias associated to impaired Ca(2+) regulation. *Heart Rhythm*. 2009;6:1652-1659.
18. Jiang D, Xiao B, Yang D, et al. RyR2 mutations linked to ventricular tachycardia and sudden death reduce the threshold for store-overload-induced Ca2+ release (SOICR). *Proc Natl Acad Sci U S A*. 2004;101:13062-13067.
19. di Barletta MR, Viatchenko-Karpinski S, Nori A, et al. Clinical phenotype and functional characterization of CASQ2 mutations associated with catecholaminergic polymorphic ventricular tachycardia. *Circulation*. 2006;114: 1012-1019.
20. Schlotthauer K, Bers DM. Sarcoplasmic reticulum Ca(2+) release causes myocyte depolarization. Underlying mechanism and threshold for triggered action potentials. *Circ Res*. 2000;87:774-780.
21. Cerrone M, Noujaim SF, Tolkacheva EG, et al. Arrhythmogenic mechanisms in a mouse model of catecholaminergic polymorphic ventricular tachycardia. *Circ Res*. 2007; 101:1039-1048.
22. Knollmann BC, Chopra N, Hlaing T, et al. Casq2 deletion causes sarcoplasmic reticulum volume increase, premature Ca2+ release, and catecholaminergic polymorphic ventricular tachycardia. *J Clin Invest*. 2006;116:2510-2520.
23. Tester DJ, Spoon DB, Valdivia HH, Makielski JC, Ackerman MJ. Targeted mutational analysis of the RyR2-encoded cardiac ryanodine receptor in sudden unexplained death: a molecular autopsy of 49 medical examiner/coroner's cases. *Mayo Clin Proc*. 2004;79:1380-1384.
24. Creighton W, Virmani R, Kutys R, Burke A. Identification of novel missense mutations of cardiac ryanodine receptor gene in exercise-induced sudden death at autopsy. *J Mol Diagn*. 2006;8:62-67.
25. Nishio H, Suzuki K. Postmortem molecular analysis for fatal arrhythmogenic disease in sudden unexplained death. Leg Med (Tokyo) 2009.
26. Tan HL, Hofman N, van Langen IM, van der Wal AC, Wilde AA. Sudden unexplained death: heritability and diagnostic yield of cardiological and genetic examination in surviving relatives. *Circulation*. 2005;112:207-213.
27. Hofman N, Tan HL, Clur SA, Alders M, van Langen I, Wilde AA. Contribution of inherited heart disease to sudden cardiac death in childhood. *Pediatrics*. 2007;120:e967-e973.
28. Behr E, Wood DA, Wright M, et al. Cardiological assessment of first-degree relatives in sudden arrhythmic death syndrome. *Lancet*. 2003;362:1457-1459.
29. Behr ER, Dalageorgou C, Christiansen M, et al. Sudden arrhythmic death syndrome: familial evaluation identifies inheritable heart disease in the majority of families. *Eur Heart J*. 2008;29:1670-1680.
30. Tester DJ, Dura M, Carturan E, et al. A mechanism for sudden infant death syndrome (SIDS): stress-induced leak via ryanodine receptors. *Heart Rhythm*. 2007;4:733-739.
31. Napolitano C, Priori SG. Diagnosis and treatment of catecholaminergic polymorphic ventricular tachycardia. *Heart Rhythm*. 2007;4:675-678.
32. Krahn AD, Gollob M, Yee R, et al. Diagnosis of unexplained cardiac arrest: role of adrenaline and procainamide infusion. *Circulation*. 2005;112:2228-2234.
33. Postma AV, Denjoy I, Kamblock J, et al. Catecholaminergic polymorphic ventricular tachycardia: RYR2 mutations, bradycardia, and follow up of the patients. *J Med Genet*. 2005;42:863-870.
34. Bai R, Napolitano C, Bloise R, Monteforte N, Priori SG. Yield of genetic screening in inherited cardiac channelopathies: how to prioritize access to genetic testing. *Circ Arrhythmia Electrophysiol*. 2009;2:6-15.
35. Medeiros-Domingo A, Bhuiyan ZA, Tester DJ, et al. The RYR2-encoded ryanodine receptor/calcium release channel in patients diagnosed previously with either catecholaminergic polymorphic ventricular tachycardia or genotype negative, exercise-induced long QT syndrome: a comprehensive open reading frame mutational analysis. *J Am Coll Cardiol*. 2009;54:2065-2074.
36. Priori SG, Napolitano C, Memmi M, et al. Clinical and molecular characterization of patients with catecholaminergic polymorphic ventricular tachycardia. *Circulation*. 2002; 106:69-74.
37. Bauce B, Rampazzo A, Basso C, et al. Screening for ryanodine receptor type 2 mutations in families with effort-induced polymorphic ventricular arrhythmias and sudden death: early diagnosis of asymptomatic carriers. *J Am Coll Cardiol*. 2002;40:341-349.
38. Hayashi M, Denjoy I, Extramiana F, et al. Incidence and risk factors of arrhythmic events in catecholaminergic polymorphic ventricular tachycardia. *Circulation*. 2009;119:2426-2434.
39. Horner JM, Ackerman MJ. Ventricular ectopy during treadmill exercise stress testing in the evaluation of long QT syndrome. *Heart Rhythm*. 2008;5:1690-1694.
40. Tiso N, Stephan DA, Nava A, et al. Identification of mutations in the cardiac ryanodine receptor gene in families affected with arrhythmogenic right ventricular cardiomyopathy type 2 (ARVD2). *Hum Mol Genet*. 2001;10:189-194.
41. Choi G, Kopplin LJ, Tester DJ, Will ML, Haglund CM, Ackerman MJ. Spectrum and frequency of cardiac channel defects in swimming-triggered arrhythmia syndromes. *Circulation*. 2004;110:2119-2124.
42. Swan H, Laitinen P, Kontula K, Toivonen L. Calcium channel antagonism reduces exercise-induced ventricular arrhythmias in catecholaminergic polymorphic ventricular tachycardia patients with RyR2 mutations. *J Cardiovasc Electrophysiol*. 2005;16:162-166.
43. De Rosa G, Delogu AB, Piastra M, Chiaretti A, Bloise R, Priori SG. Catecholaminergic polymorphic ventricular tachycardia: successful emergency treatment with intravenous propranolol. *Pediatr Emerg Care*. 2004;20:175-177.
44. Sumitomo N, Sakurada H, Mugishima H, Hiraoka M. Adenosine triphosphate terminates bidirectional ventricular tachycardia in a patient with catecholaminergic polymorphic ventricular tachycardia. *Heart Rhythm*. 2008;5:496-497.
45. Makanjee B, Gollob MH, Klein GJ, Krahn AD. Ten-year follow-up of cardiac sympathectomy in a young woman with catecholaminergic polymorphic ventricular tachycardia and an implantable cardioverter defibrillator. *J Cardiovasc Electrophysiol*. 2009;20(10):1167-1169.

46. Collura CA, Johnson JN, Moir C, Ackerman MJ. Left cardiac sympathetic denervation for the treatment of long QT syndrome and catecholaminergic polymorphic ventricular tachycardia using video-assisted thoracic surgery. *Heart Rhythm*. 2009;6:752-759.
47. Atallah J, Fynn-Thompson F, Cecchin F, DiBardino DJ, Walsh EP, Berul CI. Video-assisted thoracoscopic cardiac denervation: a potential novel therapeutic option for children with intractable ventricular arrhythmias. *Ann Thorac Surg*. 2008;86:1620-1625.
48. Zipes DP, Camm AJ, Borggrefe M, et al. ACC/AHA/ESC 2006 Guidelines for Management of Patients With Ventricular Arrhythmias and the Prevention of Sudden Cardiac Death: a report of the American College of Cardiology/American Heart Association Task Force and the European Society of Cardiology Committee for Practice Guidelines (writing committee to develop Guidelines for Management of Patients With Ventricular Arrhythmias and the Prevention of Sudden Cardiac Death): developed in collaboration with the European Heart Rhythm Association and the Heart Rhythm Society. *Circulation*. 2006;114:e385-e484.
49. Pizzale S, Gollob MH, Gow R, Birnie DH. Sudden death in a young man with catecholaminergic polymorphic ventricular tachycardia and paroxysmal atrial fibrillation. *J Cardiovasc Electrophysiol*. 2008;19:1319-1321.
50. Mohamed U, Gollob MH, Gow RM, Krahn AD. Sudden cardiac death despite an implantable cardioverter-defibrillator in a young female with catecholaminergic ventricular tachycardia. *Heart Rhythm*. 2006;3:1486-1489.
51. Postma AV, Denjoy I, Hoorntje TM, et al. Absence of calsequestrin 2 causes severe forms of catecholaminergic polymorphic ventricular tachycardia. *Circ Res*. 2002;91:e21-e26.

13. A Molecular Genetic Perspective on Atrial Fibrillation

Jason D. Roberts and Michael H. Gollob

13.1 Introduction

Atrial fibrillation (AF), the most common sustained cardiac arrhythmia, is a source of significant morbidity and mortality predominantly through its sequelae of heart failure and stroke. Despite its prevalence and clinical significance, its pathophysiology remains incompletely understood and treatment strategies remain relatively ineffective. In recent years, the importance of *genetics* in predisposing to AF has been clearly recognized, and in a remarkably short time period there have been a flurry of landmark discoveries. Insight into the molecular genetics of this condition promises to lead to more effective forms of therapy that will help reduce the burden currently carried by patients and health care systems.

13.2 Epidemiology

AF represents the most common cardiac arrhythmia and affects well over two million Americans.[1] Its prevalence increases with age and ranges from less than 0.5% in those aged less than 55–9% in octogenarians.[2] Because of the advancing age of Western populations, its incidence is expected to increase over the coming years and has been projected to affect as many as 16 million Americans by 2050.[3] AF independently increases the risk of mortality with an age-adjusted odds ratio for death of 1.9 in females and 1.5 in males; however, its sequelae of stroke and heart failure represent the greatest sources of morbidity and mortality.[4] Accounting for a large proportion of strokes in the elderly, AF has been estimated to cost the American health care system over US $1 billion annually.[1,5] Its burden on the global population, in terms of both health and health care dollars, will likely only get worse in the coming years as populations age. This is further exacerbated by the current lack of highly effective therapies for this exceedingly common condition.

In addition to advancing age, structural heart disease also represents a major risk factor for AF.[6] However, approximately 10–20% of cases of AF occurs in the absence of known risk factors and have been termed *lone AF*.[7] Without obvious contributing factors, genetics have been hypothesized to play an important role in the development of this form of the arrhythmia. Indeed, a family with lone AF transmitted with an autosomal dominant pattern of inheritance was first documented by Wolff in 1943.[8] A recent study of siblings has found that brothers and sisters of patients with lone AF have a 70-fold and 34-fold increased risk of developing the arrhythmia relative to the general population, respectively.[9] Previously felt to be rare, contemporary work has begun to suggest that familial AF may be much more common than previously expected. Defining familial lone AF as being the presence of the arrhythmia in both the proband and at least one first degree relative, it has been shown that it accounts for 15% of all cases of lone AF.[10]

Other studies have found evidence for a genetic component in the more common form of AF associated with structural heart disease. A prospective cohort analysis from the Framingham Heart Study involving 2,243 subjects found that parental AF conferred an increased risk for development of the arrhythmia in

M.H. Gollob (✉)
Inherited Arrhythmia Clinic and Research Laboratory, University of Ottawa Heart Institute, Rm H350, 40 Ruskin Street, Ottawa, ON, K1Y 4W7, Canada
e-mail: mgollob@ottawaheart.ca

their offspring (odds ratio of 1.85).[11] A similar study from Iceland involving 5,269 patients identified a 1.77-fold increased risk of developing the arrhythmia in first degree relatives.[12] These data emphasize that a genetic predisposition is likely important in the development of all forms of AF.

13.3 Molecular Background

AF reflects a disturbance of the electrical activity within the top chambers of the heart. Transmission of electrical impulses within the heart occurs through *ion channels*, pore forming proteins present within the plasma membranes of cardiomyocytes.[13] There are a variety of different types of cardiac ion channels, each of which contributes to the cardiac action potential. The two major types of cardiomyocytes within the heart include pacemaker cells and cardiac muscle cells, each with its own distinct *action potential*.[14] The action potential for the cardiac muscle cell is designed to allow for rapid conduction of electrical impulses. In contrast, cardiac pacemaker cells are endowed with a property termed intrinsic automaticity, which allows these cells to function as pacemakers within the heart. The action potential of the cardiac muscle cell will be the main focus of this discussion.

Figure 13.1 depicts the action potential of the cardiac muscle cell, which is divided into five phases. Phase 0 reflects the rapid depolarizing upstroke that occurs secondary to rapid sodium influx into cells and is referred to as I_{Na}. It is mediated by a *voltage-gated sodium channel*, termed $Na_V 1.5$, which is the protein

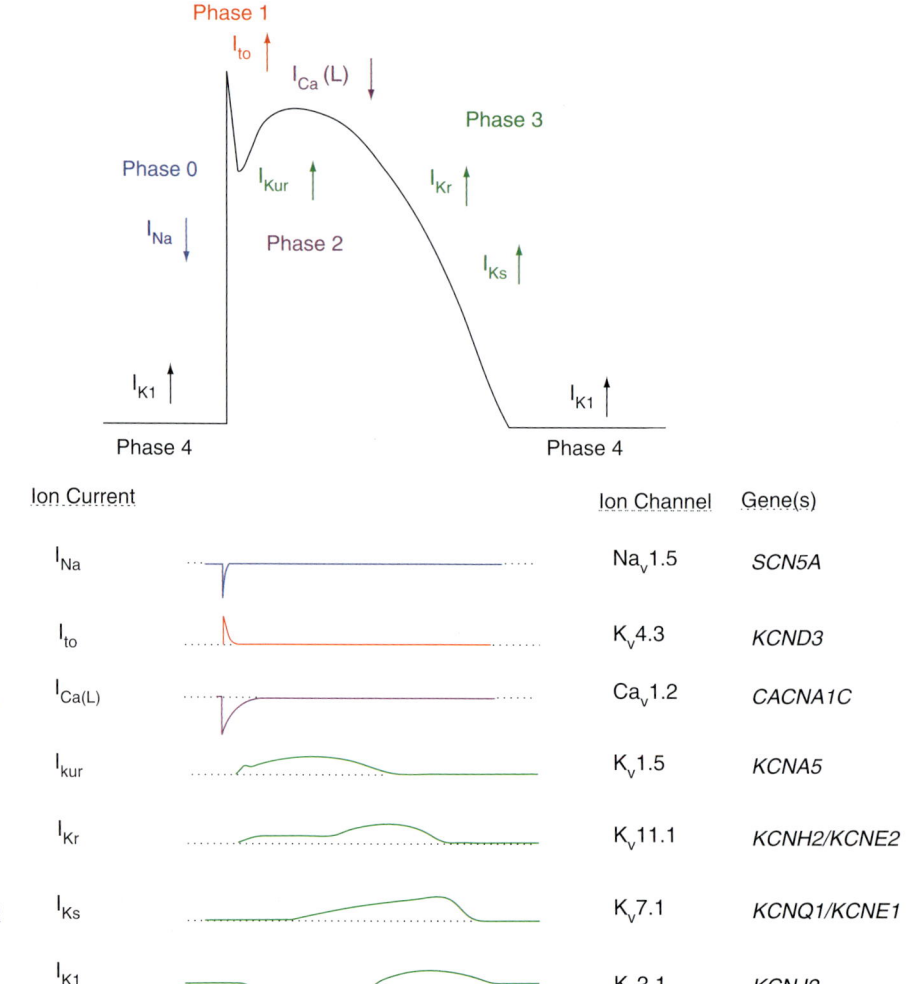

Fig. 13.1 The atrial action potential. The action potential is divided into four phases mediated by distinct ionic currents. These currents are driven by voltage-gated ion channels that allow specific ions to pass across the cardiac sarcolemma. The identity of the currents, the voltage-gated ion channels, and their encoding genes is provided (Adapted from Marban)

product of the *SCN5A* gene.[15] The rapid upstroke, reflective of the rapid flow of current, endows these cells with an ability to transmit electrical impulses with a high conduction velocity. Phase 1 involves a transient current of *repolarization*, termed I_{to} that occurs secondary to an efflux of potassium ions from the cell through the $K_v4.3$ *voltage-gated potassium channel* encoded by *KCND3*.[16] Phase 2, also referred to as the plateau or dome phase, reflects a balance of inward calcium current and outward potassium current.[14] Flow of calcium occurs through voltage-gated L-type calcium channels. The calcium influx during phase 2 not only plays a critical role in the kinetics of the cardiac action potential, but is also important in excitation-contraction coupling.[17]

Repolarization is mediated by a current that arises secondary to the flow of potassium ions out of the cell. Referred to as the delayed rectifier potassium current, it is divided into 3 different components on the basis of timing. The first, termed the *ultrarapid component of the delayed rectifier potassium current* (I_{Kur}), is unique to the atria and is felt to be mediated by Kv1.5, a voltage-gated potassium channel encoded by *KCNA5*.[18,19] Following I_{Kur} is the *rapid component of the delayed rectifier potassium current* (I_{Kr}) and involves the gene products of both *KCNH2* (HERG) and *KCNE2*.[20] Lastly, there is the *slow component of the delayed rectifier potassium current* (I_{Ks}), which involves the gene products of both *KCNQ1* and *KCNE1*.[21] Although I_{Kur} only occurs within the atria, I_{Kr} and I_{Ks} occur in both the atria and the ventricles.[22]

Phase 4 reflects a resting phase whose properties are in part modulated by the cardiac *inward rectifier potassium current*, or I_{K1}, mediated by Kir2.1 and encoded by the *KCNJ2* gene.[23] Although I_{K1} is voltage-gated, its activity differs markedly relative to the previously mentioned voltage-gated potassium channels involved in Phase 3. While the delayed rectifier potassium current is triggered by cellular repolarization, I_{K1} activity predominates when the cell is hyperpolarized or near resting potential.[22] The efflux of potassium ions mediated by I_{K1} during Phase 4 serves as an important contributor to the resting membrane potential of the cell and in this context has the potential to influence cellular excitability.[24] Upon depolarization, the magnitude of I_{K1} is dramatically reduced; a property that may be mediated by intracellular magnesium ions and polyamines interfering with the flow of potassium ions through the channel.[25] This reduced current persists until the terminal portion of Phase 3 when I_{K1} increases and exerts influence on cellular repolarization and action potential duration.[22]

Another important current within cardiac atria is the *muscarinic receptor activated potassium current*, I_{KACh}, which mediates the flow of potassium ions across the membrane in response to a vagal stimulus.[26] The cardiac muscarinic receptor, M2, is a G protein-coupled receptor that, upon binding of a cholinergic agonist (acetylcholine), permits the $G_{\beta\gamma}$ subunit to dissociate and subsequently bind and activate I_{KACh}.[27] The constituents of I_{KACh} include Kir3.1 and Kir3.4 encoded by *KCNJ3* and *KCNJ5*, respectively.[28] I_{KACh}, similar to I_{Kur}, is considered to be exclusive to the atria, although mRNA of both subunits has been detected in the ventricles.[29] Activation of I_{KACh} while the cell is depolarized results in a further efflux of potassium ions, which has the potential to shorten the action potential duration and, as will be discussed in subsequent sections, may influence the development of AF.[30]

Two major concepts that can be derived from knowledge of the cardiac action potential and its associated currents include the electrical properties of *conduction velocity* and *refractory period*. Conduction velocity reflects the velocity at which an electrical impulse is transmitted through myocardial tissue.[14] Two of the major determinants of conduction velocity include sodium channels and gap junctions. As discussed above, the kinetics of the $Na_v1.5$ channel allow for rapid conduction within the heart.[15] Gap junctions represent intercellular pores that allow electrical current to flow between cells.[31] It is this intercellular coupling, along with the rapid conduction along the cell membrane mediated by $Na_v1.5$ that results in coordinated activity between individual cells of the heart and an ensuing functional electrical syncitium.

The second concept involves the refractory period and refers to the length of time following excitation that a cell requires before it can be re-excited.[13] An electrical impulse that encounters refractory tissue dies out. The length of the refractory period is dependent upon the rate at which a cell is able to repolarize to a potential compatible with re-excitation and therefore phase 3 of the action potential plays a critical role. The mediators of phase 3, namely, the potassium channels, are therefore important contributors to the refractory period of the cell. Inhibition of potassium channels, as achieved with potassium channel blockers, results in a prolonged repolarization time manifested on the electrocardiogram as a prolonged QT interval in the case of ventricular repolarization.

Heterogeneity of refractory periods and conduction velocities within the heart, also referred to as *dispersion*, results in a substrate that is capable of sustaining *reentry*.[32,33] Reentrant circuits represent a major mechanism for tachyarrhythmias and are particularly important in the pathophysiology of AF, as will be discussed. Dispersion arises secondary to heterogeneous distributions of ion channels within the heart and occurs in normal individuals due to the different current magnitudes intrinsic to specific cardiac layers such as the endocardium and epicardium.[34] However, the degree of dispersion can be exacerbated from birth secondary to genetic variations altering the function of key protein mediators, or can occur over time as a result of asymmetric cardiac electrical and structural remodeling processes. An example of dispersion on the 12-lead ECG is that of QT dispersion reflecting regional heterogeneity of ventricular repolarization.

An understanding of these concepts is necessary in order to properly address the pathophysiology of AF and its associated molecular genetics.

13.4 Pathophysiology

AF reflects disorganized and chaotic activity of the atria with impulses firing at a rate of approximately 400–600 times per minute. The mechanisms underlying the development and maintenance of AF remain incompletely understood and there continues to be a variety of competing theories.[35] The dominant conceptual model of AF over the past 50 years, the *multiple wavelet hypothesis*, is derived from the work of Gordon Moe and involves multiple circuit reentry excitation.[36] In this model, which has been confirmed by high resolution electrical mapping during AF, irregular atrial activity arises from multiple self-perpetuating micro-reentrant circuits that exhibit spatial and temporal variability.[37,38] The second model implicates rapidly discharging atrial ectopic foci, a concept that has been strengthened following the recognition that ectopic beats originating from pulmonary veins frequently initiate AF.[39-41] This had led to the use of radiofrequency catheter ablation techniques, in which the pulmonary veins are electrically isolated from the surrounding atria, in order to treat AF.[42]

The multiple wavelet hypothesis suggests that increasing numbers of reentrant wavelets within the atria favor the maintenance of AF. A wavelet is a small wave of depolarizing current that may circle back upon itself to form a micro-reentrant circuit. In order to appreciate the conditions governing the number of wavelets that can be established, it is important to have an understanding of the *wavelength of reentry* concept. The wavelength of a circulating electrical impulse is defined as the distance traveled within one refractory period and can be calculated as the product of conduction velocity and refractory period.[43] In contrast, the pathlength represents the distance traveled by an electrical impulse during one complete circuit. As denoted in Fig. 13.2b, a wavelength that is greater in size than its pathlength will result in the circulating impulse encountering refractory tissue and the circuit will be terminated.[43] However, a pathlength traveled, that is greater in size than the circulating wavelength, will introduce an excitable gap that will permit ensuing circus movement allowing the reentrant circuit to perpetuate (Fig. 13.2c).[44] In accordance with the leading circle hypothesis, a circulating wavelet automatically establishes itself within a pathlength equivalent to its wavelength (Fig. 13.2a).[43] On this basis,

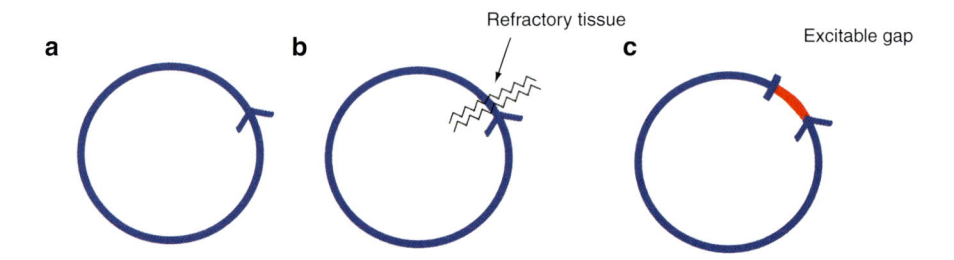

Fig. 13.2 Micro reentrant circuits in atrial fibrillation. (**a**) A circulating wavelet, whose wavelength is equivalent to its pathlength, exhibiting circus activity. (**b**) An increase in the refractory period of the micro reentrant circuit has resulted in a wavelength that exceeds the pathlength. The depolarizing current encounters refractory tissue and the circuit is terminated. (**c**) A reduction in the refractory period of the circulating wavelet from A generates a wavelength that is shorter than the corresponding pathlength resulting in the introduction of an excitable gap (Adapted from Ref.35)

coupled with wavelength being the product of conduction velocity and refractory period, the number of wavelets that can be supported by atria of a given size is inversely proportional to both conduction velocity and refractory period.

The notion of an increased number of wavelets promoting the maintenance of AF is supported by increased atrial size serving as a risk factor for the arrhythmia.[45] In a similar fashion, an increase in circulating wavelets through a reduction in refractory period and an ensuing reduction in wavelength theoretically promotes the maintenance of AF. The theory that a shorter wavelength, through a reduced refractory period, predisposes to AF serves as the rationale for using potassium channel blockers to terminate AF. Potassium channel blockers prolong atrial repolarization and result in an increased refractory period thereby reducing the potential number of circulating wavelets that can be supported by atria of a given size. Of note, the use of sodium channel blockers in AF is not supported by the wavelength of reentry concept. These medications decrease conduction velocity and, given the associated reduction in wavelength, would increase the number of circulating atrial wavelets. Their efficacy provides support for mechanistic heterogeneity within AF, a concept that will become evident as the genetic heterogeneity of the arrhythmia is further explored. Sodium channel blockers may be effective in treating a form of AF characterized by focal firing, a concept that will be addressed in the discussion surrounding *gain-of-function* sodium channel mutations.

The aforementioned atrial electrical properties, conduction velocity, and refractory period, are often viewed as contributing to the substrate for arrhythmogenesis. The development of a reentry circuit is dependent upon both substrate and trigger, examples being an early afterdepolarization or enhanced automaticity resulting in a premature beat. It is important to note that the physiology responsible for both trigger and substrate are not static but dynamic secondary to modulation by the autonomic nervous system. As such, the *autonomic nervous system* has been recognized as a critical component of arrhythmogenesis. In the setting of lone AF, the sentinel observations of the eminent electrophysiologist Phillipe Coumel implicated the parasympathetic nervous system as the major culprit.[46] He noted that the arrhythmia tended to be triggered during periods of high vagal tone such as sleep and postprandially. The mechanism through which the parasympathetic nervous system mediates lone AF appears to be in part dependent upon I_{KAch}.[22] As previously discussed, activation of I_{KAch} when the cell is depolarized triggers an efflux of potassium ions shortening the atrial action potential duration and the corresponding refractory period. The heterogeneous vagal innervation of the atria has the potential to result in regional variation of refractory periods.[30] The resultant dispersion in cellular refractoriness throughout the atria has the potential to serve as an ideal substrate for reentry and arrhythmogenesis.

As evidenced by this discussion, there are multiple variables within atrial electrical physiology that can help contribute to the development of AF. The pathophysiological heterogeneity of this disorder is further supported by the varied genetics that characterize the condition. Effective treatment is likely dependent upon targeted therapy that addresses the specific aberrant pathway, which triggers the arrhythmia development in an individual. As a result, a detailed understanding of the molecular mechanisms underlying AF is warranted.

13.5 Molecular Genetics

A genetic contribution to AF was supported by informal observations of familial clustering of the arrhythmia and previous reports of an inheritance pattern consistent with autosomal dominant transmission.[8] However, the first form of definitive evidence did not come until 1997 when a genetic locus was found to segregate with the arrhythmia in a Spanish family that suffered from an autosomal dominant form of lone AF.[47] Linkage analysis localized the culprit locus to the long arm of chromosome 10 (10q22–24). Despite isolation of the locus to a relatively small genomic region, a culprit gene could not be identified. Candidate genes in that region included the β-adrenergic receptor (*ADRB1*), the α-adrenergic receptor (*ADRA2*), and a G-protein-coupled receptor kinase (*GPRK5*). Sequencing of these genes, however, did not reveal a mutation that segregated with disease. Over 10 years later, the culprit gene within this locus remains unknown.

13.6 Potassium Channels: Gain-Of-Function Mutations

13.6.1 KCNQ1

The first causative gene for familial AF was not found until 6 years following identification of the 10q22–24 locus. The discovery came from the study of a four generation Chinese family that also exhibited an autosomal dominant pattern of inheritance for lone AF.[48] Linkage analysis mapped the culprit locus to the short arm of chromosome 11 (11p15.5), a region distinct from that found in the Spanish family. This on its own was a significant finding given that it demonstrated that AF was a genetically heterogeneous disorder that could be caused by more than one gene, a finding corroborated by a separate group in the same year.[10] Review of the genetic contents of the 11p15.5 region found that it contained the *KCNQ1* gene whose protein product encodes the pore forming α-subunit of I_{Ks} (*KCNQ1/KCNE1*). *Loss-of-function* mutations within *KCNQ1* had been previously implicated with congenital Long QT Syndrome type 1 and therefore its previous association with arrhythmia made it an ideal candidate gene.[49] Sequencing of *KCNQ1* found a Ser140Gly mutation that was present in all affected family members and absent from all but one of the unaffected members. The finding that the Ser140Gly mutation appeared to segregate with disease was further strengthened by its absence in 188 healthy control individuals coupled with it being a highly conserved residue across different species.

Following identification of the putative culprit mutation, functional studies were undertaken in an effort to elucidate the mechanism through which it resulted in a phenotype of AF. Coexpression of mutant *KCNQ1* with *KCNE1* in COS-7 cells resulted in a markedly increased current density relative to the wild-type gene at all voltages, consistent with a gain-of-function mutation. Given that *KCNQ1* contributes to the slow component of the delayed rectifier potassium current (I_{Ks}) and is responsible for repolarization of cardiomyocytes, a gain-of-function mutation could result in more rapid repolarization and reduce the effective refractory period of cells (Fig. 13.3). As discussed previously, this would create a substrate ideal for multiple circuit reentry and promote maintenance of the arrhythmia in a manner consistent with the multiple wavelet hypothesis. This notion is supported by the observation that up to 30% of patients with short QT syndrome, a condition characterized by enhanced ventricular repolarization and malignant ventricular arrhythmias, suffer from AF.[50]

Although this theory fits nicely, it is worth noting that nine of the 16 patients with the Ser140Gly mutation were actually found to have a prolonged QT interval on 12-lead electrocardiography. A prolonged QT interval, being consistent with a slower rate of repolarization within the ventricles, is in contrast with the in vitro functional data. It is conceivable that the

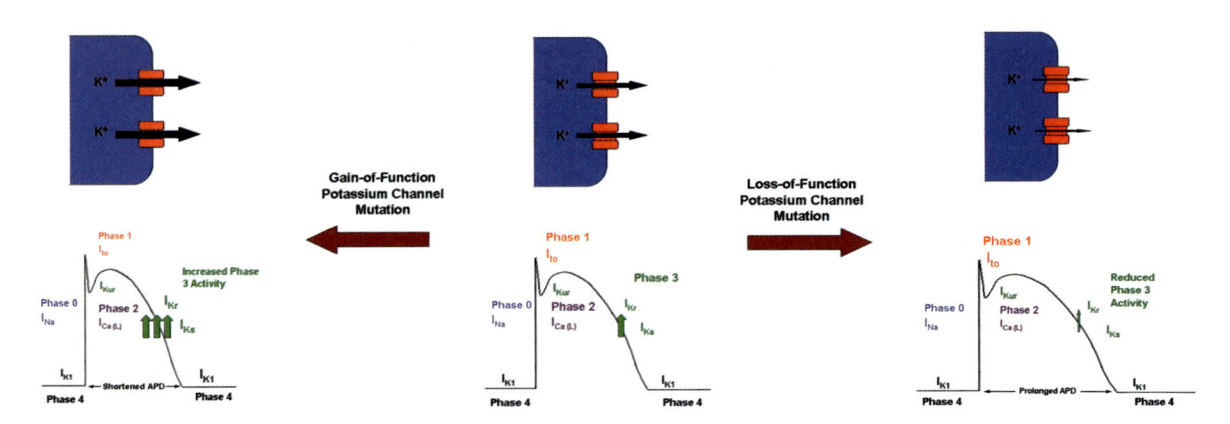

Fig. 13.3 The effect of potassium channel gene mutations on atrial action potential duration. A gain-of-function mutation in a voltage-gated potassium channel increases the efflux of potassium ions during Phase 3 resulting in a shortening of the action potential duration. Conversely, the decrease in Phase 3 current secondary to a loss-of-function mutation in a voltage-gated potassium channel causes prolongation of the atrial action potential

mutation may have different effects on the atria and ventricles as a result of the different electrical and structural properties of these chambers. This theory was recently hypothesized by Lundby, et al. who identified a Gln147Arg *KCNQ1* substitution in a patient with lone AF and a prolonged QT interval.[51] When mutant *KCNQ1* was coexpressed with *KCNE1* in *Xenopus laevis* oocytes, a loss-of-function was observed; however, coexpression with *KCNE2* resulted in a gain-of-function. Although these findings are intriguing, it is difficult to arrive at firm conclusions as the relative distributions of *KCNE1* and *KCNE2* within the atria and ventricles are largely unknown.[51] An alternative explanation for the seemingly discordant in vitro and electrocardiographic findings may be secondary to single cell in vitro studies not accurately recapitulating the complex physiology of the heart. It is conceivable that the Ser140Gly substitution within *KCNQ1* actually results in a prolonged action potential duration and effective refractory period within the atria of the intact heart. This could result in AF through an alternative mechanism, which will be further addressed when the *KCNA5* gene is discussed.

13.6.2 KCNE2 and KCNJ2

The discovery that a mutation within a potassium channel gene caused an autosomal dominant form of AF alluded to the possibility that other potassium channel genes may contribute to the arrhythmia. Given that linkage analysis studies are frequently limited by small pedigree size, subsequent studies employed a candidate gene approach in which multiple potassium channel genes were screened for mutations in families with AF. This approach led to the identification of two additional potassium channel genes, namely *KCNE2* and *KCNJ2*, as being causal for familial forms of lone AF in two separate studies performed by the same group. A mutation within *KCNE2*, commonly known to encode the β-subunit of I_{Kr}, was discovered following screening of 28 unrelated Chinese kindreds with familial AF for mutations within eight different potassium channels genes (*KCNQ1, KCNH2, KCNE1-5*, and *KCNJ2*).[52] Two of the 28 probands were found to carry an Arg27Cys mutation that was subsequently found in affected members of the two kindreds and was absent in 462 healthy controls. It is important to note that each family only had two affected members, while multiple unaffected members carried the Arg27Cys mutation. This may potentially be accounted for by the mutation carrying a low degree of penetrance and necessitating additional genetic and environmental factors in order for the phenotype of AF to be expressed. Additionally, although *KCNE2* is generally considered to serve as the β-subunit of I_{Kr}, coexpression of Arg27Cys *KCNE2* with *KCNH2* did not result in a change in current relative to wild-type. However, there was an increase in current noted when it was coexpressed with *KCNQ1*. Previous work with COS cells has demonstrated that the protein products of *KCNE2* and *KCNQ1* are capable of interacting to generate a background current that is not voltage dependent.[53] It is conceivable that the mutant *KCNE2* may predispose to AF through a background current that may affect cellular repolarization.

KCNJ2 was identified using a similar approach in which 30 Chinese AF kindreds were screened for mutations in ten ion channel or transporter-related genes (*KCNQ1, KCNH2, SCN5A, ANK-B, KCNJ2*, and *KCNE1-5*).[54] *KCNJ2* encodes Kir2.1, which is responsible for the cardiac inward rectifier potassium current I_{K1}. As discussed, this channel mediates a background potassium current that contributes to the resting membrane potential of the cell and influences cellular excitability and repolarization within the heart. It is also the causative gene for congenital long QT syndrome Type 7, also referred to as Andersen–Tawil Syndrome.[55] The proband and the other four affected family members were all found to carry a Val93Ile mutation within *KCNJ2*, a mutation not found in 420 healthy individuals. In this instance, two unaffected family members were found to carry the mutation; however, their unaffected status may have been secondary to their relatively young ages (33 and 42 years old). Functional analysis of the mutant protein revealed increased current density at potentials ranging from −140 to −80 mV and from −60 to −40 mV consistent with a gain-of-function effect. The putative predisposing mechanism of Val93Ile *KCNJ2* for AF involves enhanced repolarization and a reduction in refractory period, as hypothesized with *KCNQ1*.

13.7 Potassium Channels: Loss-of-Function Mutations

13.7.1 KCNA5

Up until this point, all of the potassium channel genes implicated in the development of lone AF had been shown to exhibit gain-of-function effects on in vitro functional analysis. As discussed, the purported mechanism involved a reduction in effective refractory period and reentrant wavelength, which in accordance with the multiple wavelet hypothesis would promote and maintain AF. However, was it conceivable that a loss-of-function effect in a potassium channel could result in AF?

Using the candidate gene approach, 154 patients with lone AF were screened for mutations within the *KCNA5* gene, encoding the atrial specific voltage-gated potassium channel $K_v1.5$ responsible for I_{Kur}.[56] A unique sequence variant was identified in a patient with a family history of the arrhythmia. The patient carried a nonsense mutation (E375X) that resulted in the production of a truncated protein that lacked the S4–S6 voltage sensor, the pore region, and the C-terminus. Because of a lack of available DNA, stringent genetic support for the mutation segregating with the arrhythmia was not possible. Subsequent functional studies revealed that expression of mutant E375X *KCNA5* within HEK293 cells failed to generate current. This was consistent with a loss-of-function effect, and not unexpected given the drastic effect of the nonsense mutation on the mature protein. In addition, when coexpressed with wild-type *KCNA5*, cells exhibited a significant reduction in current density compatible with a dominant negative effect, which accounted for the autosomal dominant pattern of inheritance in the setting of a loss-of-function mutation.

Given that loss-of-function within a voltage-gated potassium channel involved in repolarization presumably results in a prolonged refractory period, the mechanism through which mutant E375X *KCNA5* predisposes to AF would have to be different from the previously described gain-of-function potassium channel mutations (Fig. 13.3). In vitro studies using human atrial myocytes and in vivo studies with a murine model found that administration of 4-aminopyridine, a known blocker of I_{Kur}, dramatically increased the incidence of early afterdepolarizations. The authors hypothesized that increased early afterdepolarizations, in combination with a prolonged atrial action potential duration, could result in disorganized atrial activity akin to that seen in torsade de pointes within the ventricles. Early afterdepolarizations serving as a trigger for AF had previously been suggested in an animal model whereby injection of cesium chloride, a potassium channel blocker, into the sinus node artery of dogs resulted in a polymorphic atrial tachycardia that subsequently degenerated into AF, leading the investigators to coin the term "atrial torsade."[57]

It is important to note that the mechanisms underlying a form of AF caused by a gain-of-function voltage-gated potassium channel are dramatically different, and essentially opposing to a form of the arrhythmia driven by a loss-of-function in a similar channel. The distinct triggers and substrates in these phenotypically identical forms of the arrhythmia serve to emphasize the marked heterogeneity that likely underlies the pathophysiology of AF and provides insight into the variable efficacies of many therapies.

13.8 A Potassium Channel Variant and "Secondary Hit" Hypothesis

Although genetics play an important role in the development of AF, the critical role of the environment is emphasized by the notion that the arrhythmia rarely develops in childhood and becomes increasingly common with advancing age. The interaction of genetics with environmental influences was eloquently illustrated in a family with autosomal dominant AF who also suffered from a high incidence of hypertension.[58] Mutation screening in four genes (*KCNQ1*, *KCNE1-3*) identified a novel missense mutation within *KCNQ1*, namely Arg14Cys. Analysis of the family pedigree suggested that the mutation segregated with the arrhythmia, while it was absent in 100 control patients. The interesting findings came following in vitro functional studies in which the mutant Arg14Cys *KCNQ1* was coexpressed with *KCNE1* in CHO cells. The mutant potassium channel initially behaved identically to wild-type; however, following treatment with a hypotonic solution to simulate the cell swelling and stretching consistent with the atrial milieu in a hypertensive patient, the mutant channels exhibited a marked increase in current and a leftward shift in the voltage-dependence of activation consistent with a gain-of-function effect. Wild-type channel properties were unaffected by exposure to the hypotonic media. The authors hypothesized that the inherited ion channel defect represented the "first hit," however, a "second hit" mediated by environmental factors such as hypertension was necessary for development of the arrhythmia. This phenomenon would help to account for the increasing prevalence of AF with aging.

13.9 Are Potassium Channel Mutations Common?

Although many of the genes responsible for familial AF encode potassium channel proteins, an obvious question is whether potassium channel mutations are a common cause of AF. Furthermore, most of the genetic studies discussed involved Chinese kindreds and it was therefore uncertain if these results were applicable to other ethnicities. In order to address this question, two separate studies from the same group screened over 200 patients with lone AF and AF with hypertension for mutations within potassium channel genes (*KCNQ1*, *KCNJ2*, and *KCNE1-5*).[59,60] The studies involved predominantly patients of Western European ancestry. Although a number of common polymorphisms were detected, no disease-causing mutations were discovered. It was, therefore, concluded that potassium channel gene mutations represent a rare cause of AF in patients of Western European descent.

13.10 Connexins

Gap junctions are specialized channels that directly connect cytoplasmic compartments of adjacent cells, allowing for passage of charged ions and coordinated propagation of cardiac action potentials.[31] The molecular constituents of gap junction channels are connexin proteins, which oligomerize into hexameric structures known as connexons or hemichannels (Fig. 13.4). Adjacent cells each contribute a hemichannel to form a functional gap junction channel (Fig. 13.4).[61] There are multiple different isoforms of connexin proteins; however, the two most highly expressed isoforms within

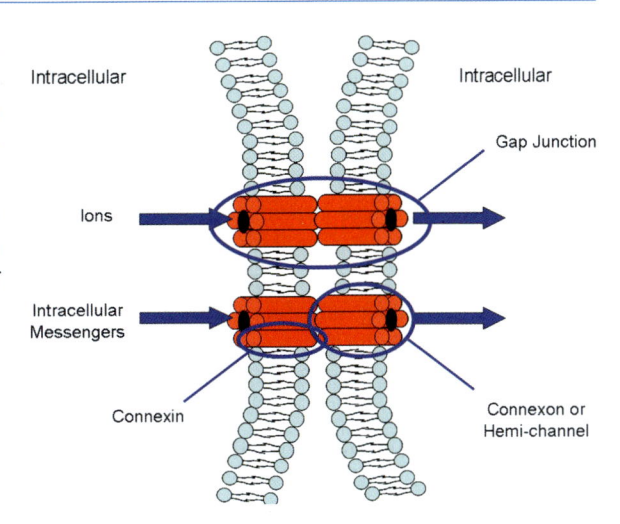

Fig. 13.4 Micro circuit reentry, secondary to conduction velocity heterogeneity. An ectopic impulse within the atria initiates depolarizing currents in different directions. Depolarizing wavefront A encounters refractory tissue and terminates. Wavefront B travels a different course and encounters a region of slow conduction velocity. The resultant delay allows time for the refractory tissue originally encountered by A to repolarize and conduct allowing a reentry circuit to be established. The region of slow conduction velocity encountered by B may be secondary to reduced connexin activity

the heart are *connexin 40* and *43*.[62] Connexin 40 is of particular interest in AF, since it is expressed in atrial myocytes and is absent from ventricular cells.[62] The importance of connexins to AF has been well established in animal studies. In knockout mice lacking the connexin 40 gene, atrial tachyarrhythmias could be induced by burst atrial pacing whereas this was not possible in wild-type mice.[63] A goat model of persistent AF revealed that connexin 40 distribution within the atria was markedly heterogeneous, a phenomenon also seen in humans with AF.[64,65] Although the latter finding does not establish causality, it does suggest that heterogeneous distribution of connexins within the heart may form an ideal substrate for AF.

In the light of the atrial specific expression of Cx40, and the vulnerability of Cx40 knockout mice to AF, our group subsequently screened a group of 15 patients with sporadic, lone AF for somatic mutations within connexin 40 and 43.[66] Somatic mutations, as opposed to germ-line, could potentially account for the more common scenario of nonfamilial AF in otherwise healthy individuals. DNA was obtained from both peripheral blood lymphocytes and resected atrial tissue of patients who had undergone an open-heart pulmonary vein isolation procedure. In four of the 15 patients, genetic mutations within the connexin 40 gene (*GJA5*) were identified, while no mutations were found within the connexin 43 gene (*GJA1*). Findings consistent with a tissue-specific or somatic basis of the mutations were found in three of the four patients, as evidenced by the mutations being present only in the resected atrial tissue of patients and not in their peripheral blood lymphocytes. This finding introduced the paradigm that this common arrhythmia may occur secondary to tissue specific mutations in a fashion similar to that seen in more rare somatic genetic diseases, most commonly tumor-prone syndromes.[67] Since myocardial cells do not divide, somatic *mosaicism* within the cardiac tissue must have resulted from a *somatic mutation* in an early myocardial progenitor cell during embryogenesis.

All of the mutations occurred within the highly conserved transmembrane-spanning domains of the connexin 40 protein. The three somatic mutations included Gly38Asp, Pro88Ser, and Met163Val, while the only germ-line mutation was Ala96Ser. Functional studies of the mutant Cx40 proteins were performed in a gap junction-deficient cell line, N2A cells. Cells expressing the Pro88Ser mutation showed a profound trafficking defect, showing intracellular retention of the mutant protein. This cell model was consistent with the finding of immunostaining from the atrial tissue of an affected patient harboring this mutation, which exhibited a mosaic pattern of abnormal gap junction formation and intracellular retention of connexin 40. In contrast, cells expressing the Ala96Ser mutation displayed appropriate trafficking; however, functional electrical cell-to-cell coupling through these channels was significantly reduced. Lastly, these mutant connexins demonstrated a dominant negative effect on wild-type Cx40, as well as a trans-dominant negative effect on wild-type Cx43. This latter finding provides strong support for the concept of heteromeric interaction of Cx40 and Cx43 in hemichannel formation.

Thus, somatic or atrial-specific genetic defects of Cx40 were associated with sporadic, lone AF, and may promote an arrhythmic substrate by a loss-of-function mechanism due to either impaired trafficking or loss of functional cell-to-cell coupling. The somatic and mosaic pattern of these mutant connexins further promote heterogeneous conduction velocity, creating an ideal substrate for electrical reentry.

13.11 Sodium Channels: Loss-of-Function Mutations

SCN5A encodes the sodium channel, $Na_v1.5$, responsible for the rapid depolarization upstroke of the cardiac action potential. It has been associated with numerous arrhythmic disorders including the Brugada Syndrome, congenital long QT syndrome type 3, and sick sinus syndrome.[68–70] Given its obvious importance with the electrical properties of the heart, multiple groups employed a candidate gene approach screening patients with AF for mutations within *SCN5A*. The first study involved 157 patients with lone AF; screening did not identify any novel mutations felt to be causative for AF.[71] The H558R single nucleotide polymorphism (SNP), of which approximately one third of the population are heterozygous, was also examined in these patients along with 314 matched controls. The R558 allele, which had previously been shown to alter $Na_v1.5$ function by reducing depolarizing sodium current, was found to confer an increased risk of developing AF (odds ratio: 1.6).[71,72] However, the sample size in this study was rather small and the data has yet to be duplicated in larger cohorts.

A second study involving 375 patients with AF (118 had lone AF while 257 had AF associated with heart disease) and 360 well-matched controls, identified eight novel mutations in ten separate AF patients.[73] None of the variants were found in controls and all involved highly conserved residues within $Na_v1.5$. Six of the patients appeared to represent familial cases and in each case the variant appeared to segregate with the disease. Functional studies were not performed and therefore the mechanism through which these variants cause AF is unknown. These findings suggest, in contrast to the previous study, that mutations within *SCN5A* represent a relatively common cause of AF in patients with and without heart disease.

A third group screened *SCN5A* in 57 patients with lone AF or AF with hypertension and a confirmed family history of the arrhythmia.[74] A single novel mutation was found, Asn1986Lys, which was not found in 300 ethnically matched controls. The father of the proband, who also suffered from AF, was found to be a carrier of the mutation, while the unaffected mother did not have any sequence variants within *SCN5A*. Unfortunately, further genetic profiling of the family was not possible due to unwillingness to participate in the study. Expression of the mutant gene within *Xenopus laevis* oocytes suggested a loss-of-function effect as evidenced by a significant hyperpolarizing shift in the midpoint of steady-state inactivation. This alteration was predicted to prolong the atrial action potential duration and, therefore, Asn1986Lys-SCN5A presumably triggers AF through a manner akin to the aforementioned atrial torsade. These findings, which confirmed the association of *SCN5A* with AF, however, suggested that it is not a frequent cause of the condition.

13.12 Sodium Channels: Gain-of-Function Mutations

Although previous studies had implicated loss-of-function *SCN5A* mutations in association with AF, subsequent work indicated that gain-of-function mutations within *SCN5A* are also part of the genetic spectrum responsible for AF. Prior to these studies, the only disease related to an *SCN5A* gain-of-function effect was long QT syndrome type 3, which is mediated by a persistent late sodium current.[69]

In a four generation Japanese family with an autosomal dominant form of AF, a novel Met1875Thr mutation was identified within *SCN5A*.[75] The proband was reported to have increased right atrial excitability during radiofrequency catheter ablation for AF. All affected family members were found to carry the mutation, while the mutation was absent in all unaffected family members and 210 ethnically matched controls. Functional analysis of Met1875Thr revealed a pronounced depolarizing shift in the midpoint of steady-state inactivation consistent with a gain-of-function effect. No persistent sodium current was observed, consistent with the observation that affected individuals had QT intervals within the normal range.

A second study from our group involving a mother and son with lone AF identified a Lys1493Arg mutation involving a highly conserved residue within the DIII-IV linker located six amino acids downstream from the fast inactivation motif of sodium channels.[76] Biophysical studies demonstrated a significant positive shift in the voltage-dependence of inactivation and a large ramp current near resting membrane potential, consistent with a gain-of-function. When expressed in HL-1 atrial cardiomyocytes, enhanced cellular excitability was observed in the form of spontaneous action potential depolarizations and a lower threshold for action potential firing as compared to wild-type cells. Collectively, these studies suggest that both gain- and loss-of-function mutations within *SCN5A* are associated with AF.

The existing evidence suggests that *SCN5A* gain-of-function mutations predispose to AF by enhancing cellular hyperexcitability. The depolarizing shift in steady-state inactivation increases the probability that the channel will be in the open conformation and capable of conducting current.[76] This alteration in the gating of the Na_v1-5 mediated current will presumably result in a predisposition for cells to reach threshold potential and fire, consistent with enhanced automaticity. This increase in focal discharges has the potential to serve as the trigger for AF (Fig. 13.5). In addition, $Na_v1.5$ channels have

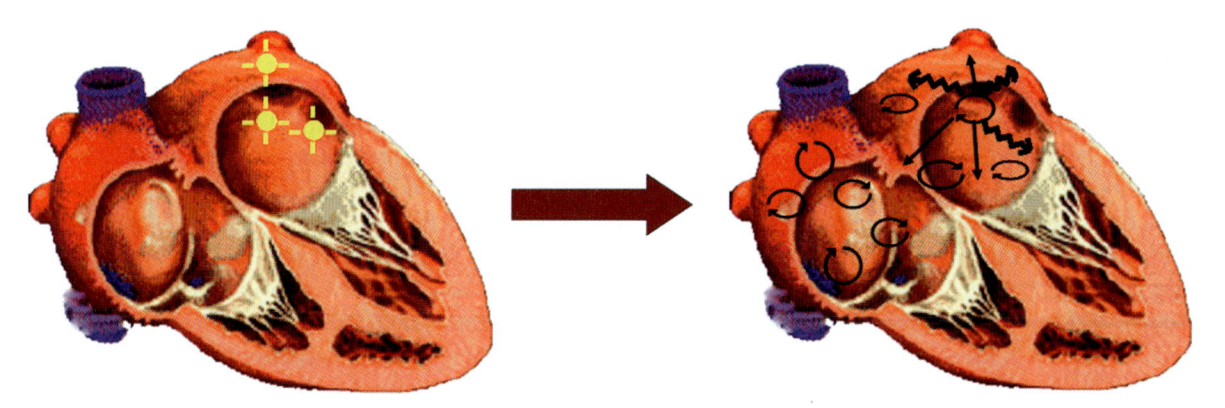

Fig. 13.5 Cellular hyperexcitability triggering atrial fibrillation. Ectopic foci originating from the pulmonary veins contribute to the development of a self-perpetuating micro reentrant circuit. Rapid, heterogeneous conduction from the reentrant circuit to the surrounding atria results in electrical activity consistent with atrial fibrillation

recently been identified in the autonomic ganglia that surround the pulmonary veins.[77] Mutations within *SCN5A* may therefore result in neuronal hyperexcitability that may trigger AF through a parasympathetic pathway and contribute to the rapidly firing ectopic foci observed in the region of the pulmonary veins in some patients with the arrhythmia.

13.13 Atrial Natriuretic Peptide

The most recent gene to be associated with AF does not implicate an ion channel, but instead involves a circulating hormone, the *atrial natriuretic peptide* (ANP). Although known to be important in cardiac physiology, ANP had been largely viewed as cardioprotective in the setting of heart failure.[78] It was known, however, to be capable of modulating the electrical activities of the heart and there were reports of its effects on specific ion channels.[79–81] However, little work had been done on ANP in the context of AF and previous studies that had examined for a possible relationship had been negative.[82]

Linkage analysis of a Caucasian family of northern European ancestry with autosomal dominant AF mapped the causative locus to the small arm of chromosome 1 (1p36–35).[83] Review of the genes within this region revealed the presence of *NPPA*, the gene encoding ANP, and subsequent sequencing revealed a two base pair deletion in exon 3 that resulted in a frameshift associated with loss of the stop codon. Extension of the reading frame results in an elongated peptide that is 40 amino acids in length relative to the 28 amino acid length of the wild-type. The deletion was present in all of the affected family members and absent in unaffected family members and 560 control patients. Functional studies involving an isolated rat whole-heart model suggested that the mutant ANP resulted in a reduced effective refractory period; however, the mechanism was not entirely clear. ANP mediates its effects on cells through binding to natriuretic peptide receptors that possess intracellular guanylate cyclase activity.[84] Previous work has suggested that ANP molecules with an elongated C-terminus may be more resistant to degradation and therefore may circulate at higher levels.[85] Therefore, the authors hypothesized that increased circulating ANP may produce increased intracellular levels of cGMP that may in turn, through an unknown mechanism, reduce the effective refractory period.

13.14 Unknown Loci

Finally, in addition to 10q22–24, there are multiple other loci that have been identified through linkage analysis in kindreds with autosomal dominant forms of lone AF. These loci include 6q14–16, 10p11–q21, and 5p15; the specific causative genes remain unknown.[86–88]

13.15 Genome Wide Association Studies

The availability of DNA microarrays containing hundreds of thousands of *single nucleotide polymorphisms* (SNPs) has resulted in the opportunity to screen the entire genome for regions that may confer an increased risk for disease. This robust mechanism is proving to be an invaluable tool in unraveling the complex genetics underlying many common diseases including coronary artery disease and AF.

A *genome wide association study* using a DNA microarray containing 316,515 SNPs was performed on 550 patients with AF or flutter in combination with 4,476 control patients from Iceland.[89] They discovered a strong association with SNPs on chromosome 4q25, the most significant being rs2200733 with an odds ratio of 1.84 (95% confidence intervals: 1.54–2.21). Replication studies using additional samples from Iceland (2,251 cases and 13,238 controls), Sweden (143 cases and 738 controls), the USA (636 cases and 804 controls), and China (333 cases and 2,836 controls) further reinforced the association with rs2200733. The odds ratio for the combined European population was 1.72 (95% confidence interval: 1.59–1.86), while that for the Chinese cohort was 1.42 (95% confidence interval: 1.16–1.73) The haplotype block corresponding to the associated SNPs does not contain a known gene and, therefore, the mechanism for this association is currently unknown. Candidate genes present within the adjacent haplotype block include *PITX2*, which encodes a protein involved in cardiac development, and *ENPEP*, whose protein product is involved in angiotensin II breakdown.[90,91] Research to delineate the genetic factors at the 4q25 locus responsible for the increased risk for the development of AF is ongoing.

Following identification of the 4q25 locus, two subsequent genome wide association studies were performed with larger numbers of cases and controls in order to improve power and identify previously undetected loci associated with AF. Both groups independently identified two separate SNPs, rs7193343 and rs2106261, which localized to an intronic region within the *ZFHX3* gene on chromosome 16q22.[92,93] *ZFHX3* encodes a transcription factor, AT motif-binding factor 1, whose function in the heart is currently unclear. The *ZFHX3* gene has recently been implicated in a vasculitis involving the coronary arteries (Kawasaki disease).[94] The association of 16q22 with AF was not as strong as for 4q25, with an odds ratio in the range of 1.2 in most European populations. Furthermore, it was not significantly associated with AF in the Chinese population.[93] Lastly, it is important to note that although the 16q22 SNPs did localize to a gene, it does not necessarily implicate *ZFHX3* in the pathogenesis of AF. These SNPs may appear to associate with AF due to linkage disequilibrium with the true causal variants in surrounding regions. As with the 4q25 locus, further work is necessary in order to better appreciate the apparent relationship between 16q22 and AF.

13.16 Other Gene Associations

13.16.1 ACE

The renin-angiotensin-system, a pathway of importance in cardiovascular disease, was first implicated with AF following evidence that it was activated in the atria of humans and dogs with the arrhythmia.[95,96] Multiple studies have since suggested that treatment with ACE inhibitors may be protective against the development of AF, and has triggered multicenter randomized controlled trials.[97,98] The possibility that common polymorphisms within genes of the RAS pathway may contribute to AF was examined in a case control study involving 250 patients with documented nonfamilial structural AF and 250 well-matched controls.[99] Polymorphisms within the genes for ACE, angiotensinogen, and the angiotensin II type 1 receptors were compared between cases and controls. Single locus analysis found that SNPs from the angiotensinogen gene, namely, M235T, G-6A, and G-217A, were significantly associated with AF. However, as with any small study, validation in a much larger cohort is required prior to reaching any definitive conclusions of such an association. Further, recent data indicates that pharmacologic targeting of the renin-–angiotensin–system pathway had no effect on minimizing recurrence rates of AF.[100]

13.16.2 GNB3, Enos, MMP-2, IL-10

The *GNB3* gene encodes the β_3 subunit of a heterotrimeric G protein, which, in broad terms, is important in coordinating the cellular response to extracellular receptor stimulation via signal transduction.[101] A C825T polymorphism within exon 10 of *GNB3* results in an alternative splicing pattern, such that the 825T allele generates a modified β_3-subunit that is 41 amino acids shorter and more active than the wild-type form.[102] Previous work in humans had demonstrated that the TT genotype was associated with an increase in I_{K1}, but a decrease in I_{Kur}.[103] This triggered an association study involving 291 patients with AF and possible structural heart disease along with 292 controls patients without the arrhythmia.[104] The prevalence of the *GNB3* TT genotype was found to be significantly lower in patients with AF relative to the control group suggesting that the TT genotype is protective against the development of the arrhythmia. Functional studies were not performed.

SNPs within a variety of other genes including endothelial nitric oxide synthase (eNOS), matrix metalloprotease-2 (MMP-2), and interleukin-10 (IL-10) have also been implicated with the development of AF.[105,106] These studies are intriguing; however, similar to many of the previously described association studies they are limited by relatively small numbers and a frequent lack of functional data corroborating the apparent associations.

13.17 The Autonomic Nervous System

Clinical observations suggest that the autonomic nervous system plays a critical role in the pathogenesis of AF and this has been supported in different animal models. AF can be readily triggered in structurally normal hearts through exposure to a cholinergic agonist such as carbachol followed by burst pacing. In a canine model, atrial vagal denervation through radiofrequency catheter ablation prevented subsequent induction of the arrhythmia through burst pacing and vagal stimulation.[107] In an effort to investigate the molecular mechanisms underlying this phenomenon, knockout mice that lacked Kir3.4 (previously referred to as GIRK4) were developed.[29] As discussed previously, Kir3.1 and Kir3.4 encode the protein products responsible for I_{KACh} and the absence of either results in the complete loss of I_{KACh}. Unlike wild-type mice, burst pacing in the presence of carbachol was unable to induce AF in the I_{KACh} deficient knockout mice. This data serves to implicate I_{KACh} in the pathogenesis of AF and suggests that blockers of I_{KACh} may potentially serve as an effective treatment for the arrhythmia. This is an especially attractive treatment option given that I_{KACh}, like I_{Kur}, appears to be localized predominantly to the atria.

Although there is relatively robust data supporting the involvement of the cholinergic system in the pathogenesis of AF, genetic mutations in genes encoding the molecular mediators of the cholinergic response in the heart have not been reported.

13.18 Clinical Aspects – Genetic Diagnosis and Targeted Therapy

The contribution of genetics to AF has only recently begun to be appreciated and has not yet been incorporated into routine clinical practice. At the present time, two unresolved issues preclude the routine use of clinical genetic testing for AF cases. First, although a number of genetic etiologies have now been identified, each currently known gene accounts for a very small percentage of cases. Secondly, specific types of AF (vagal, postoperative, for example) do not reliably predict genotype, and thus phenotype–genotype correlations are not yet apparent. The goal of further research is to identify more common genetic etiologies of AF, and to establish phenotype–genotype correlations, with the ultimate achievement of targeted therapy based on established genotype.

The current status of clinical genetics in AF is in contrast to other known inherited arrhythmia diseases, such as long QT syndrome and Brugada Syndrome. In these conditions, common genetic causes are known, and the yield of clinical genetic testing in excess of 50% and 20%, respectively. The use of genetic testing in these scenarios is most useful for screening of asymptomatic, phenotype negative family members. In families where a preponderance of AF exists, clinical screening for AF in symptomatic individuals remains the only practical tool available. Routine clinical screening for asymptomatic individuals is not warranted beyond annual routine primary care physical examinations.

A detailed understanding of the genetics contributing to the pathophysiology of AF will likely allow for the development of a pharmacogenomic strategy that improves treatment efficacy and reduces adverse events. AF is a heterogeneous disorder from the perspective of both genetics and pathophysiology. In one instance, it may arise secondary to a gain-of-function effect within a potassium channel, whereas in another case, it may result from a loss-of-function effect in the same channel. Although phenotypically they may be indistinguishable on electrocardiography, the most efficacious treatment choice is likely to be markedly different in the light of differing electrophysiologic triggers. Given their identical phenotype, genetic characterization will likely be necessary in order to identify the particular AF subtype. In the first example given above, the arrhythmia has likely developed secondary to a shortened atrial refractory period, which has resulted in an ideal substrate for multiple reentrant wavelets within the atria. Effective treatment with a potassium channel blocker that restores the atrial effective refractory period to its normal length, thereby disrupting the reentrant wavelets, may be the optimal agent to restore and maintain sinus rhythm. However, the same treatment in the second case would likely exacerbate the arrhythmia, given that it is secondary to a prolonged atrial refractory period, which may have resulted in "atrial torsade." The genetic and pathophysiologic heterogeneity underlying the arrhythmia is likely responsible for the variable treatment response observed in cases of lone AF.

In addition to the above example with potassium channel mutations, a similar approach can be extended to the other genes that have been implicated in AF. For example, forms of AF that develop secondary to cellular hyperexcitability as a result of gain-of-function mutations within sodium channels may benefit from

Table 13.1 The genes associated with atrial fibrillation and the putative mechanisms leading to the arrhythmia

Gene	Study method	Mode of inheritance	Protein and function	Functional Effect of Mutation	Mechanism for AF
Potassium Channels					
KCNQ1	Linkage Analysis	Autosomal Dominant	α-subunit of I_{Ks}	Gain-of-Function	Reduced atrial ERP
KCNE2	Candidate Gene Approach	Autosomal Dominant	β-subunit of background potassium current	Gain-of-Function	Reduced atrial ERP
KCNJ2	Candidate Gene Approach	Autosomal Dominant	$K_{ir}2.1$ responsible for I_{K1}	Gain-of-Function	Reduced atrial ERP
KCNA5	Candidate Gene Approach	Autosomal Dominant	$K_v1.5$ responsible for I_{Kur}	Loss-of-Function	Prolonged atrial APD
Connexins					
GJA5	Candidate Gene Approach	Sporadic	Connexin 40 responsible for cell coupling	Loss-of-Function	Conduction velocity dispersion
Sodium Channels					
SCN5A	Candidate Gene Approach	Autosomal Dominant	$Na_v1.5$ responsible for I_{Na}	Loss-of-Function	Prolonged atrial APD
		Autosomal Dominant		Gain-of-Function	Cellular hyperexcitability
Circulating Hormones					
NPPA	Linkage Analysis	Autosomal Dominant	Atrial Natriuretic Peptide	Unknown	Unknown
Unknown Loci					
10q22–24	Linkage Analysis	Autosomal Dominant	Unknown	Unknown	Unknown
6q14–16	Linkage Analysis	Autosomal Dominant	Unknown	Unknown	Unknown
10p11–q21	Linkage analysis	Autosomal Dominant	Unknown	Unknown	Unknown
5p15	Linkage analysis	Autosomal Dominant	Unknown	Unknown	Unknown
4q25	Genome wide association scan	Sporadic	Unknown	Unknown	Unknown
16q22	Genome wide association scan	Sporadic	Unknown	Unknown	Unknown

sodium channel blockers. In the context of an AF subtype characterized by conduction velocity heterogeneity that arose secondary to a loss-of-function connexin mutation, such a patient may benefit from a form of therapy that keeps gap junctions in their open state. Although not currently available, there are emerging gap junction pharmacophores that may serves this purpose.[108] Targeted therapy, consistent with a pharmacogenomic approach, should be a goal that is strived for in the coming years. Along with being more efficacious, this strategy should also reduce the unwanted proarrhythmic effects seen with antiarrhythmic drugs.

13.19 Summary

An understanding of the genetic factors that lead to the development of AF holds great promise for the development of effective therapies against this exceedingly common arrhythmia. The ability to identify the specific electrophysiologic mechanisms, on the basis of genetic discovery, should lead to more effective forms of targeted therapy that carry minimal risk. This era of pharmacogenomics has yet to arrive for AF, however it is slowly becoming within reach.

References

1. Feinberg WM, Blackshear JL, Laupacis A, Kronmal R, Hart RG. Prevalence, age distribution, and gender of patients with atrial fibrillation. Analysis and implications. *Arch Intern Med*. 1995;155:469-473.
2. Go AS, Hylek EM, Phillips KA, et al. Prevalence of diagnosed atrial fibrillation in adults: national implications for rhythm management and stroke prevention: the AnTicoagulation and Risk Factors In Atrial Fibrillation (ATRIA) Study. *JAMA*. 2001;285:2370-2375.
3. Miyasaka Y, Barnes ME, Gersh BJ, et al. Secular trends in incidence of atrial fibrillation in Olmsted County, Minnesota, 1980 to 2000, and Implications on the Projections for Future Prevalence. *Circulation*. 2006;114:119-125.
4. Benjamin EJ, Wolf PA, D'Agostino RB, Silbershatz H, Kannel WB, Levy D. Impact of atrial fibrillation on the risk of death: the Framingham Heart Study. *Circulation*. 1998; 98:946-952.
5. Wolf PA, Abbott RD, Kannel WB. Atrial fibrillation: a major contributor to stroke in the elderly. The Framingham Study. *Arch Intern Med*. 1987;147:1561-1564.
6. Benjamin EJ, Levy D, Vaziri SM, D'Agostino RB, Belanger AJ, Wolf PA. Independent risk factors for atrial fibrillation in a population-based cohort. *JAMA*. 1994;271: 840-844.
7. Kopecky SL, Gersh BJ, McGoon MD, et al. The natural history of lone atrial fibrillation. A population-based study over three decades. *N Engl J Med*. 1987;317:669-674.
8. Wolff L. Familial auricular fibrillation. *N Engl J Med*. 1943;229:396-397.
9. Ellinor PT, Yoerger DM, Ruskin JN, MacRae CA. Familial aggregation in lone atrial fibrillation. *Hum Genet*. 2005; 188:179-184.
10. Darbar D, Herron KJ, Ballew JD, et al. Familial atrial fibrillation is a genetically heterogeneous disorder. *J Am Coll Cardiol*. 2003;41:2185-2192.
11. Fox CS, Parise H, D'Agostino RB Sr, et al. Parental atrial fibrillation as a risk factor for atrial fibrillation in offspring. *JAMA*. 2004;291:2851-2855.
12. Arnar DO, Thorvaldsson S, Manolio TA, et al. Familial aggregation of atrial fibrillation in Iceland. *Eur Heart J*. 2006;27:708-712.
13. Katz AM. Cardiac ion channels. *N Engl J Med*. 1993; 328:1244-1251.
14. Katz AM. *Physiology of the Heart*. 2nd ed. New York: Raven Press; 1992.
15. Abriel H. Cardiac sodium channel $Na_v1.5$ and its associated proteins. *Arch Mal Coeur Vaiss*. 2007;100:787-793.
16. Kaab S, Dixon J, Duc J, et al. Molecular basis of transient outward potassium current downregulation in human heart failure: a decrease in $K_v4.3$ mRNA correlates with a reduction in current density. *Circulation*. 1998;98:1383-1393.
17. Bers DM. Cardiac excitation-contraction coupling. *Nature*. 2002;415:198-205.
18. Tamkun MM, Knoth KM, Walbridge JA, Kroemer H, Roden DM, Glover DM. Molecular cloning and characterization of two voltage-gated K+ channel cDNAs from human ventricle. *FASEB J*. 1991;5:331-337.
19. Wang Z, Fermini B, Nattel S. Sustained depolarization-induced outward current in human atrial myocytes. Evidence for a novel delayed rectifier K+ current similar to $K_v1.5$ cloned channel currents. *Cir Res*. 1993;73:1061-1076.
20. Abbott GW, Sesti F, Splawski I, et al. MiRP1 forms I_{Kr} potassium channels with HERG and is associated with cardiac arrhythmia. *Cell*. 1999;97:175-187.
21. Sanguinetti MC, Curran ME, Zou A, et al. Coassembly of K(v)LQT1 and MinK (IsK) proteins to form cardiac I_{Ks} potassium channel. *Nature*. 1996;384:80-83.
22. Tamargo J, Caballero R, Gomez R, Valenzuela C, Delpon E. Pharmacology of cardiac potassium channels. *Cardiovasc Res*. 2004;62:9-33.
23. Kubo Y, Baldwin TJ, Jan YN, Jan LY. Primary structure and functional expression of a mouse inward rectifier potassium channel. *Nature*. 1993;362:127-133.
24. Dhamoon AS, Jalife J. The inward rectifier current (I_{K1}) controls cardiac excitability and is involved in arrhythmogenesis. *Heart Rhythm*. 2005;2:316-324.
25. Lu Z. Mechanism of rectification in inward-rectifier K+ channels. *Annu Rev Physiol*. 2004;66:103-129.
26. Sakmann B, Noma A, Trautwein W. Acetylcholine activation of single muscarinic K+ channels in isolated pacemaker cells of the mammalian heart. *Nature*. 1983;303: 250-253.

27. Pfaffinger PJ, Martin JM, Hunter DD, Nathanson NM, Hille B. GTP-binding proteins couple cardiac muscarinic receptors to a K⁺ channel. *Nature*. 1985;317:536-538.
28. Corey S, Krapivinsky G, Krapivinsky L, Krapivinsky L, Clapham DE. Number and stoichiometry of subunits in the native atrial G-protein-gated K⁺ channel, I$_{KACh}$. *J Biol Chem*. 1998;273:5217-5218.
29. Kovoor P, Wickman K, Maguire CT, et al. Evaluation of the role of I$_{KACh}$ in atrial fibrillation using a mouse knockout model. *J Am Coll Cardiol*. 2001;37:2136-2143.
30. Liu L, Nattel S. Differing sympathetic and vagal effects on atrial fibrillation in dogs: role of refractoriness heterogeneity. *Am J Physiol*. 1997;273:H805-H816.
31. Rohr S. Role of gap junctions in the propagation of the cardiac action potential. *Cardiovasc Res*. 2004;62:309-322.
32. Brachmann J, Karolyi L, Kubler W. Atrial dispersion of refractoriness. *J Cardiovasc Electrophysiol*. 1998;9(8 Suppl): S35-S39.
33. Weiss JN, Qu Z, Chen PS, et al. The dynamics of cardiac fibrillation. *Circulation*. 2005;112:1232-1240.
34. Antzelevitch C. Modulation of transmural repolarization. *Ann N Y Acad Sci*. 2005;1047:314-323.
35. Nattel S. New ideas about atrial fibrillation 50 years on. *Nature*. 2002;415:219-226.
36. Moe GK, Rheinboldt WC, Abildskov JA. A computer model of atrial fibrillation. *Am Heart J*. 1964;67:200-220.
37. Konings KT, Kirchhof CJ, Smeets JR, Wellens HJ, Penn OC, Allessie MA. High-density mapping of electrically induced atrial fibrillation in humans. *Circulation*. 1994;89:1665-1680.
38. Cox JL, Canavan TE, Schuessler RB, et al. The surgical treatment of atrial fibrillation II. Intraoperative electrophysiologic mapping and description of the electrophysiologic basis of atrial flutter and fibrillation. *J Thorac Cardiovasc Surg*. 1991;101:406-426.
39. Mandapati R, Skanes A, Chen J, Berenfeld O, Jalife J. Stable microreentrant sources as a mechanism of atrial fibrillation in the isolated sheep heart. *Circulation*. 2000;101:194-199.
40. Jalife J, Berenfeld O, Mansour M. Mother rotors and fibrillatory conduction: a mechanism of atrial fibrillation. *Cardiovasc Res*. 2002;54:204-216.
41. Haissaguerre M, Jais P, Shah DC, et al. Spontaneous initiation of atrial fibrillation by ectopic beats originating in the pulmonary veins. *N Engl J Med*. 1998;339:659-666.
42. Pappone C, Rosanio S, Oreto G, et al. Circumferential radiofrequency ablation of pulmonary vein ostia: A new anatomic approach for curing atrial fibrillation. *Circulation*. 2000;102:2619-2628.
43. Allessie MA, Bonke FI, Schopman FJ. Circus movement in rabbit atrial muscle as a mechanism of tachycardia. III. The "leading circle" concept: a new model of circus movement in cardiac tissue without the involvement of an anatomical obstacle. *Circ Res*. 1977;41:9-18.
44. Rensma PL, Allessie MA, Lammers WJ, Bonke FI, Schalij MJ. Length of excitation wave and susceptibility to reentrant atrial arrhythmias in normal conscious dogs. *Circ Res*. 1988;62:395-410.
45. Psaty BM, Manolio TA, Kuller LH, et al. Incidence of and risk factors for atrial fibrillation in older adults. *Circulation*. 1997;96:2455-2461.
46. Coumel P, Attuel P, Lavallee J, Flammang D, Leclercq JF, Slama R. The atrial arrhythmia syndrome of vagal origin. *Arch Mal Coeur Vaiss*. 1978;71:645-656.
47. Brugada R, Tapscott T, Czernuszewicz GZ, et al. Identification of a genetic locus for familial atrial fibrillation. *N Engl J Med*. 1997;336:905-911.
48. Chen YH, Xu WJ, Bendahhou S, et al. *KCNQ1* gain-of-function mutation in familial atrial fibrillation. *Science*. 2003;299:251-254.
49. Wang Q, Curran ME, Splawski I, et al. Positional cloning of a novel potassium channel gene: *KvLQT1* mutations cause cardiac arrhythmias. *Nat Genet*. 1996;12:17-23.
50. Giustetto C, Di Monte F, Wolpert C, et al. Short QT syndrome: clinical findings and diagnostic-therapeutic implications. *Eur Heart J*. 2006;27:2440-2447.
51. Lundby A, Ravn LS, Svendsen JH, Olesen SP, Schmitt N. *KCNQ1* mutation Q147R is associated with atrial fibrillation and prolonged QT interval. *Heart Rhythm*. 2007;4:1532-1541.
52. Yang Y, Xia M, Jin Q, et al. Identification of a *KCNE2* gain-of-function mutation in patients with familial atrial fibrillation. *Am J Hum Genet*. 2004;75:899-905.
53. Tinel N, Diochot S, Borsotto M, Lazdunski M, Barhanin J. KCNE2 confers background current characteristics to the cardiac *KCNQ1* potassium channel. *EMBO J*. 2000;19: 6326-6330.
54. Xia M, Jin Q, Bendahhaou S, et al. A Kir2.1 gain-of-function mutation underlies familial atrial fibrillation. *Biochem Biophys Res Commun*. 2005;332:1012-1019.
55. Plaster NM, Tawil R, Tristani-Firouzi M, et al. Mutations in Kir2.1 cause the developmental and episodic electrical phenotypes of Andersen's syndrome. *Cell*. 2001;105:511-519.
56. Olson TM, Alekseev AE, Liu XK, et al. K$_v$1.5 channelopathy due to KCNA5 loss-of-function mutation causes human atrial fibrillation. *Hum Mol Genet*. 2006;15:2185-2191.
57. Satoh T, Zipes DP. Cesium-induced atrial tachycardia degenerating into atrial fibrillation in dogs: Atrial torsades de pointes? *J Cardiovasc Electrophysiol*. 1998;9:970-975.
58. Otway R, Vandenberg JI, Guo G, et al. Stretch-sensitive *KCNQ1* mutation. A link between genetic and environmental factors in the pathogenesis of atrial fibrillation? *J Am Coll Cardiol*. 2007;49:578-586.
59. Ellinor PT, Moore RK, Patton KK, Ruskin JN, Pollak MR, MacRae CA. Mutations in the long QT gene, *KCNQ1*, are an uncommon cause of atrial fibrillation. *Heart*. 2004;90: 1487-1488.
60. Ellinor PT, Petrov-Kondratov VI, Zakharova E, Nam EG, MacRae CA. Potassium channel gene mutations rarely cause atrial fibrillation. *BMC Med Genet*. 2006;7:70.
61. Herve JC, Bourmeyster N, Sarrouilhe D, Duffy HS. Gap junctional complexes: from partners to functions. *Prog Biophys Mol Biol*. 2007;94:29-65.
62. Vozzi C, Dupont E, Coppen SR, Yeh HI, Severs NJ. Chamber-related differences in connexin expression in the human heart. *J Mol Cell Cardiol*. 1999;31:991-1003.
63. Hagendorff A, Schumacher B, Kirchhoff S, Luderitz B, Willecke K. Conduction disturbances and increased atrial vulnerability in Connexin40-deficient mice analyzed by transesophageal stimulation. *Circulation*. 1999;99:1508-1515.
64. van der Velden HM, Ausma J, Rook MB, et al. Gap junctional remodeling in relation to stabilization of atrial fibrillation in the goat. *Cardiovasc Res*. 2000;46:476-486.
65. Wilhelm M, Kirste W, Kuly S, et al. Atrial distribution of connexin 40 and 43 in patients with intermittent, persistent, and postoperative atrial fibrillation. *Heart Lung Circ*. 2006;15:30-37.

66. Gollob MH, Jones DL, Krahn AD, et al. Somatic mutations in the connexin 40 gene (*GJA5*) in atrial fibrillation. *N Engl J Med*. 2006;354:2677-2688.
67. Erickson RP. Somatic gene mutation and human disease other than cancer. *Mutat Res*. 2003;543:125-136.
68. Chen Q, Kirsch GE, Zhang D, et al. Genetic basis and molecular mechanisms for idiopathic ventricular fibrillation. *Nature*. 1998;392:293-296.
69. Wang Q, Shen J, Splawski I, et al. SCN5A mutations associated with an inherited cardiac arrhythmia, long QT syndrome. *Cell*. 1995;80:805-811.
70. Benson DW, Wang DW, Dyment M, et al. Congenital sick sinus syndrome caused by recessive mutations in the cardiac sodium channel gene (*SCN5A*). *J Clin Invest*. 2003;112:1019-1028.
71. Chen LY, Ballew JD, Herron KJ, Rodeheffer RJ, Olson TM. A common polymorphism in *SCN5A* is associated with lone atrial fibrillation. *Clin Pharmacol Ther*. 2007;81:35-41.
72. Makielski JC, Ye B, Valdivia CR, et al. A ubiquitous splice variant and a common polymorphism affect heterologous expression of recombinant human *SCN5A* heart sodium channels. *Circ Res*. 2003;93:821-828.
73. Darbar D, Kannankeril PJ, Donahue BS, et al. Cardiac sodium channel (*SCN5A*) variants associated with atrial fibrillation. *Circulation*. 2008;117:1927-1935.
74. Ellinor PT, Nam EG, Shea MA, Milan DJ, Ruskin JN, MacRae CA. Cardiac sodium channel mutation in atrial fibrillation. *Heart Rhythm*. 2008;5:99-105.
75. Makiyama T, Akao M, Shizuta S, et al. A novel *SCN5A* gain-of-function mutation M1875T associated with familial atrial fibrillation. *J Am Coll Cardiol*. 2008;52:1326-1334.
76. Li Q, Huang H, Liu G, et al. Gain-of-function mutation of $Na_v1.5$ in atrial fibrillation enhances cellular excitability and lowers the threshold for action potential firing. *Biochem Biophys Res Commun*. 2009;380:132-137.
77. Scornik FS, Desai M, Brugada R, et al. Functional expression of "cardiac-type" $Na_v1.5$ sodium channel in canine intracardiac ganglia. *Heart Rhythm*. 2006;3:842-850.
78. Rubattu S, Sciarretta S, Valenti V, Stanzione R, Volpe M. Natriuretic peptides: an update on bioactivity, potential therapeutic use, and implication in cardiovascular diseases. *Am J Hypertens*. 2008;21:733-741.
79. Crozier I, Richards AM, Foy SG, Ikram H. Electrophysiological effects of atrial natriuretic peptide on the cardiac conduction system in man. *Pacin Clin Electrophysiol*. 1993;16:738-742.
80. Sorbera LA, Morad M. Atrionatriuretic peptide transforms cardiac sodium channels into calcium-conducting channels. *Science*. 1990;247:969-973.
81. Le Grand B, Deroubaix E, Coueil JP, Coraboeuf E. Effects of atrionatriuretic factor on Ca^{2+} current and Cai-independent transient outaward K^+ current in human atrial cells. *Pflugers Arch*. 1992;421:486-491.
82. Ellinor PT, Low AF, Patton KK, Shea MA, Macrae CA. Discordant atrial natriuretic peptide and brain natriuretic peptide levels in lone atrial fibrillation. *J Am Coll Cardiol*. 2005;45:82-86.
83. Hodgson-Zingman DM, Karst ML, Zingman LV, et al. Atrial natriuretic peptide frameshift mutation in familial atrial fibrillation. *N Engl J Med*. 2008;359:158-165.
84. Levin ER, Gardner DG, Samson WK. Natriuretic peptides. *N Engl J Med*. 1998;339:321-328.
85. Chen HH, Lainchbury JG, Burnett JC Jr. Natriuretic peptide receptors and neutral endopeptidase in mediating the renal actions of a new therapeutic synthetic natriuretic peptide *Dendroaspis* natriuretic peptide. *J Am Coll Cardiol*. 2002;40:1186-1191.
86. Ellinor PT, Shin JT, Moore RK, Yoerger DM, MacRae CA. Locus for atrial fibrillation maps to chromosome 6q14–16. *Circulation*. 2003;107:2880-2883.
87. Volders PG, Zhu Q, Timmermans C, et al. Mapping a novel locus for familial atrial fibrillation on chromosome 10p11–q21. *Heart Rhythm*. 2007;4:469-475.
88. Darbar D, Hardy A, Haines JL, Roden DM. Prolonged signal-averaged P-wave duration as an intermediate phenotype for familial atrial fibrillation. *J Am Coll Cardiol*. 2008;51:1083-1089.
89. Gudbjartsson DF, Arnar DO, Helgadottir A, et al. Variants conferring risk of atrial fibrillation on chromosome 4q25. *Nature*. 2007;448:353-357.
90. Franco D, Campione M. The role of PITX2 during cardiac development. Linking left-right signaling and congenital heart diseases. *Trends Cardiovasc Med*. 2003;13:157-163.
91. Zini S, Fournie-Zaluski MC, Chauvel E, Roques BP, Corvol P, Llorens-Cortes C. Identification of metabolic pathways of brain angiotensin II and III using specific aminopeptidase inhibitors: predominant role of angiotensin III in the control of vasopressin release. *Proc Natl Acad Sci USA*. 1996;93:11968-11973.
92. Benjamin EJ, Rice KM, Arking DE, et al. Variants in *ZFHX3* are associated with atrial fibrillation in individuals of European ancestry. *Nat Genet*. 2009;41:879-881.
93. Gudbjartsson DF, Holm H, Gretarsdottir S, et al. A sequence variant in *ZFHX3* on 16q22 associates with atrial fibrillation and ischemic stroke. *Nat Genet*. 2009;41:876-878.
94. Burgner D, Davila S, Breunis WB, et al. A genome-wide association study identifies novel and functionally related susceptibility Loci for Kawasaki disease. *PLoS Genet*. 2009;5:e1000319.
95. Li D, Shinagawa K, Pang L, et al. Effects of angiotensin-converting enzyme inhibition on the development of the atrial fibrillation substrate in dogs with ventricular tachypacing-induced congestive heart failure. *Circulation*. 2001;104:2608-2614.
96. Goette A, Staack T, Rocken C, et al. Increased expression of extracellular signal-regulated kinase and angiotensin-converting enzyme in human atria during atrial fibrillation. *J Am Coll Cardiol*. 2000;35:1669-1677.
97. Vermes E, Tardif JC, Bourassa MG, et al. Enalapril decreases the incidence of atrial fibrillation in patients with left ventricular dysfunction: insight from the Studies of Left Ventricular Dysfunction (SOLVD) trials. *Circulation*. 2003;107:2926-2931.
98. Pedersen OD, Bagger H, Kober L, Torp-Pedersen C. Trandolapril reduces the incidence of atrial fibrillation after acute myocardial infarction in patients with left ventricular dysfunction. *Circulation*. 1999;100:376-380.
99. Tsai CT, Lai LP, Lin JL, et al. Renin-angiotensin system gene polymorphisms and atrial fibrillation. *Circulation*. 2004;109:1640-1646.
100. GISSI-AF Investigators, Disertori M, Latini R, et al. Valsartan for prevention of recurrent atrial fibrillation. *N Engl J Med*. 2009;360:1606-1617.
101. Siffert W. G-protein beta3 subunit 825T allele and hypertension. *Curr Hypertens Rep*. 2003;5:47-53.

102. Siffert W, Rosskopf D, Siffert G, et al. Association of a human G-protein beta3 subunit variant with hypertension. *Nat Genet*. 1998;18:45-48.
103. Dobrev D, Wettwer E, Himmel HM, et al. G-Protein beta(3)-subunit 825T allele is associated with enhanced human atrial inward rectifier potassium currents. *Circulation*. 2000;102:692-697.
104. Schreieck J, Dostal S, von Beckerath N, et al. C825T polymorphism of the G-protein beta3 subunit gene and atrial fibrillation: association of the TT genotype with a reduced risk for atrial fibrillation. *Am Heart J*. 2004;148:545-550.
105. Bedi M, McNamara D, London B, Schwartzman D. Genetic susceptibility to atrial fibrillation in patients with congestive heart failure. *Heart Rhythm*. 2006;3:808-812.
106. Kato K, Oguri M, Hibino T, et al. Genetics for lone atrial fibrillation. *Int J Mol Med*. 2007;19:933-939.
107. Chiou CW, Eble JN, Zipes DP. Efferent vagal innervation of the canine atria and sinus and atrioventricular nodes: the third fat pad. *Circulation*. 1997;95:2573-2584.
108. Verma V, Larsen BD, Coombs W, et al. Novel pharmacophores of connexin 43 based on the "RXP" series of Cx43-binding peptides. *Circ Res*. 2009;105:176-184.

Part IV

Other Hereditary Arrythmias

Idiopathic Ventricular Fibrillation

14

Pieter G. Postema, Christian van der Werf, and Arthur A.M. Wilde

Funding: Netherlands Heart Foundation (grant 2005T024 to P.G.P); Fondation Leducq Trans-Atlantic Network of Excellence, Preventing Sudden Death (grant 05-CVD-01 to A.A.M.W.); ZorgOnderzoek Nederland Medische Wetenschappen (ZonMW, grant 120610013 to C.W.)

14.1 Introduction

Cardiac arrest (CA) mostly follows from cardiac disorders that elicit lethal ventricular tachyarrhythmias, primarily *ventricular fibrillation* (VF).[1,2] The predominant cause of VF in the general population is considered to be previously silent coronary artery disease resulting in myocardial infarction with VF and CA as its first symptoms.[3] However, there are many other cardiac as well as non-cardiac causes that can result in VF, such as intracranial haemorrhage, pulmonary embolism, myocarditis, cardiomyopathies, valvular heart disease, congenital cardiac anomalies and accessory pathways.[4–8] Despite the advances in our medical emergency systems, only 2–30% of patients survive VF.[9–11] The remaining suffer a *sudden (cardiac) death* (SD/SCD).

In approximately 5% of CAs at all ages, there is no explanation for the event, not even after extensive evaluation (Fig. 14.1).[12] Within this group, several categories can be distinguished based on the initial rhythm and whether or not the patient survived the event. In case of unexplained aborted CA with documented VF, *idiopathic VF* (IVF) is the terminology that best acknowledges our current inability to identify a plausible cause for the occurrence of VF in these patients who were previously considered healthy.[12] In cases of unexplained aborted CA with another initial rhythm or unexplained SD, unexplained cardiac arrest or sudden unexplained death are the terms most commonly used. It is important to recognise that the diagnostic workup for these patients and/or their family members is similar and aimed at diagnosing or excluding specific causes of VF and CA.

As the total burden of CA and SD is enormous, the burden of IVF is considerable. In the United States, the estimated *incidence of SD* is 180,000 to 250,000 cases per year, which results in an estimated incidence of IVF of up to 9,000 or 12,500 cases per year.[1,3,13,14] Although IVF constitutes a minority in aborted CA and SD, it is an intriguing and notoriously difficult condition that still affects many. Its cause is unknown, there are no clinical signs that identify individuals at risk and its first symptom may be CA. Besides the possibility of recurrent VF episodes in those individuals who survived IVF, it is clear that there are several hereditary forms of IVF. The latter implies that whole families may be predisposed to IVF and SCD.[5] As coronary artery disease becomes more prevalent with increasing age, it is comprehensible that the majority of IVF will occur in young patients, i.e. in those less than 40–45 years old.[5] Further, many cases that have been described as IVF over the last decades can now be categorised into distinct clinical entities. As a result, the contribution of IVF to CA has declined.

Clinical decision making in these patients and/or in their families is seriously hampered by the absence of risk markers for the onset of VF, also after extensive evaluation. Clinical decision making is further complicated by the lack of treatment options in order to make

P.G. Postema (✉)
Department of Cardiology, Academic Medical Centre, Amsterdam, The Netherlands
e-mail: p.g.postema@amc.uva.nl

Fig. 14.1 Diagnostic flow chart in cardiac arrest based on the initial rhythm and whether or not the patient survived the event. Note that if the patient did not survive the event, the presence or absence of ventricular fibrillation is of no value

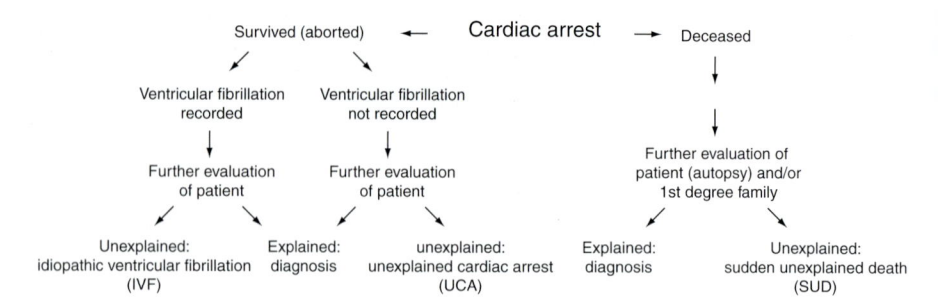

VF in (yet) asymptomatic family members less likely to occur.[15,16] In the current era of implantable cardioverter defibrillators (ICDs), survivors of a CA due to IVF will often be pragmatically treated with an ICD. This device can restore normal cardiac rhythm in case of recurrent ventricular arrhythmias, but is not without complications and does not prevent VF recurrence.[17]

14.2 Clinical Diagnosis

The difficulties associated with the diagnosis of IVF may be classified into two components: (1) IVF is a diagnosis *per exclusionem*, and (2) the paucity of IVF survivors. Diagnoses *per exclusionem* are hard to establish with our current diagnostic tools, and even more so if the patient did not survive the event. The diagnosis IVF is made in those who survived VF (Fig. 14.1). In patients with other initial rhythms or in patients who did not survive the event, when further evaluations of the patients or first-degree relatives do not result in a diagnosis, we diagnose *unexplained cardiac arrest* (UCA) or *sudden unexplained death* (SUD).

In the *survivors of CA,* the first effort in establishing a diagnosis is of course a detailed documentation of the event, medical history, family history, physical examination and blood chemistry.[12] The second effort is non-invasive cardiological evaluation, including resting ECG, exercise ECG and an echocardiogram. Registration of arrhythmias can be extremely helpful, if not the key to a diagnosis. If a diagnosis is still not (fully) established, additional evaluations should be performed, which include Holter-monitoring, coronary angiography (consider intra-coronary ergonovine or equivalents for evaluation of coronary spasm), toxicology screening, cardiac biopsies, cardiac magnetic resonance imaging, and drug provocation testing with sodium channel blockers (using, for example, ajmaline or flecainide[18,19]) and adrenaline.[20] Clearly, there is no established chronological order in which these evaluations should be performed, for example, coronary angiography will often be one of the first diagnostic (and possibly therapeutic) investigations in the workup of a resuscitated patient.

In case of SD, cardiac and genetic examination of first-degree relatives is recommended when autopsy did not reveal a cause of death or was not performed. The first step in this approach is to collect detailed information on the event and on the medical history of the SD victim and his or her first-degree relatives. A resting ECG in the relatives is made and if *autopsy* was performed, the autopsy report is reviewed. Ideally, the autopsy should have been performed according to the histopathological guideline that Basso et al. established for post-mortem evaluation of SD victims.[21] Subsequently, depending on the information available, further steps can be taken. These may consist of revision of cardiac autopsy by a specialised pathologist and/or further cardiac examinations in the attending relatives (Fig. 14.2). Tan et al.[22] and Behr et al.[23,24] established the contributions of these modalities in relatives of the SD victim. In these studies, it appeared that the rest and exercise ECGs were most valuable for establishing a diagnosis.

Despite the *per exclusionem* character of the IVF diagnosis, Haissaguerre et al. recently recognised a clinical sign that was positively associated with IVF: the electrocardiographic finding of "*early repolarization*" was found in a third of IVF victims as opposed to 5% in a control population.[25] Furthermore, those IVF victims with early repolarization had more recurrent episodes of VF than IVF victims without early repolarization. However, as early repolarization is indeed a rather common electrocardiographic finding (present in about 1–5% of the population[26,27]), it cannot be used

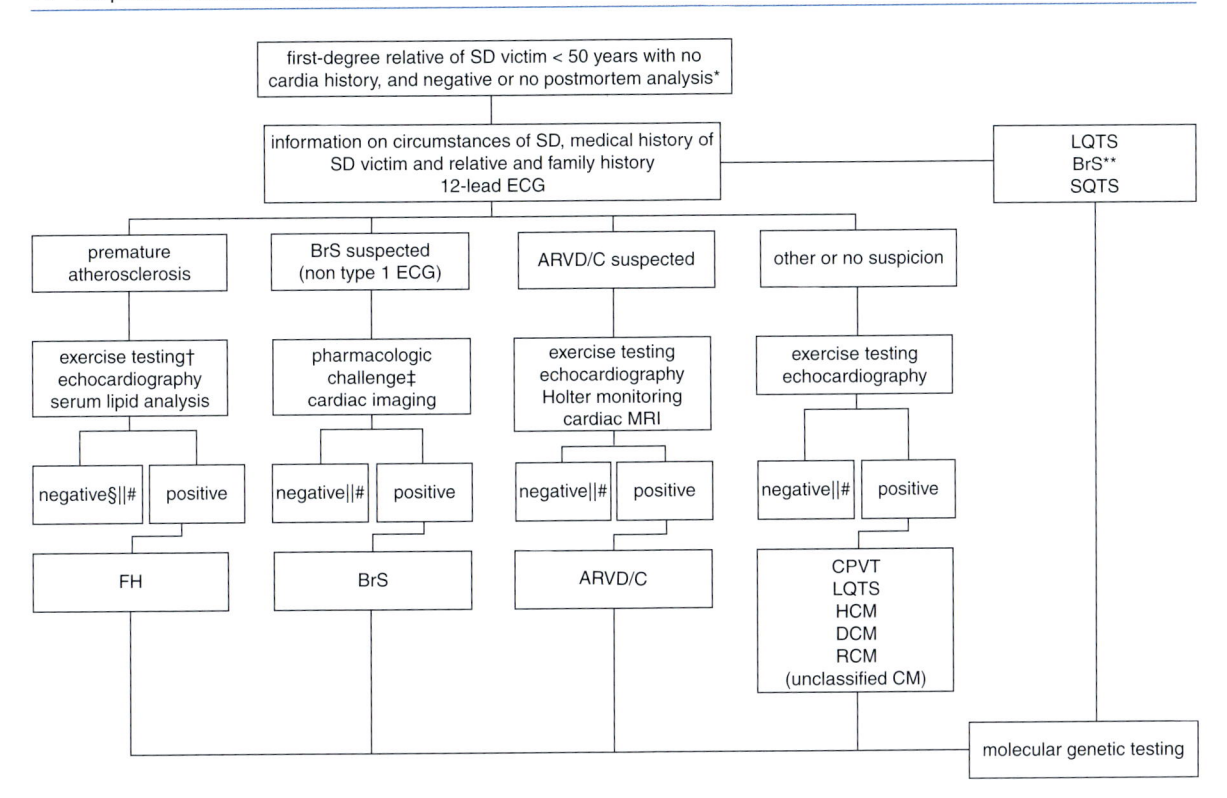

Fig. 14.2 Algorithm for cardiological and genetic examination in relatives of SD victims. *Including revision of autopsy by cardiac pathologists, if possible; †Consider adrenaline provocation; ‡Class I drugs (ajmaline, flecainide or pilsicainide; preferably not procainamide); §Advice to monitor traditional cardiovascular risk factors (e.g. hypertension, diabetes mellitus, hypercholesterolemia, overweight); ||Advice repetition of cardiological examination in 3–5 years; #When not performed yet, consider examinations in one or more of the other pathways; *SD* indicates sudden death, *LQTS* long QT syndrome, *BrS* Brugada syndrome, *SQTS* short QT syndrome, *ARVD/C* arrhythmogenic right ventricular dysplasia/cardiomyopathy, *FH* familial hypercholesterolemia, *CPVT* catecholaminergic polymorphic ventricular tachycardia, *HCM* hypertrophic cardiomyopathy; *DCM* dilated cardiomyopathy, *RCM* restrictive cardiomyopathy, *CM* cardiomyopathy. Please note that this algorithm is not exhaustive

as a risk marker in patients who are not suspected of an increased risk of CA. Further, its value in patients suspected of a CA risk still needs further study.

14.3 Historical Diagnoses

Medicine has dealt with many cases of unexplained CA in the past centuries. An informative and early case of familial SUD was, for example, composed by the German physician Meissner in 1856.[28] He described a young girl living in an institute for deaf children, who had stolen something from one of her peers. When the offence was discovered, she was summoned to explain her action, but the moment she arrived in front of the director, she sank to the floor and died. Meissner could not explain this acute and premature death. He did not entirely exclude a punishment from God, but he considered sudden cardiac failure, possibly with some form of preexisting cardiac pathology to be more likely. This notwithstanding, he had to inform the parents of the girl about the tragedy. Unexpectedly, the SUD of their daughter did not surprise the parents. They had already lost two of their children in similar conditions: one, immediately following a vigorous fright and the other, following a fit of intense anger.

At present, many SUD cases, like the one described by Meissner, can be clarified with our immensely increased knowledge and diagnostic repertoire over the past decades. It is tempting to speculate that Meissner's case is one of the earliest descriptions of *Jervell Lange-Nielsen syndrome* (autosomal recessive form of *Long-QT syndrome* with deafness). However, it may be that even at present we would not be able to explain these SUDs and would be obliged to diagnose familial SUD.

During the course of the twentieth century, many more UCA/SUD cases have been published, but now as distinct clinical diagnostic entities, which could be recognised and sometimes also effectively treated and/or explained. Primarily, this development has become possible with the application of *electrocardiography* to study abnormal cardiac behaviour.[29,30] The progress in our understanding of CA and cases previously considered to be IVF, further expanded in the last decades with the development of cardiac imaging techniques such as *echocardiography*,[31] *electrophysiological* and *molecular research* and with the elucidation of *DNA*[32] and the *human genome*.[33] The evolution of genetics has enabled a search for genes underpinning the risk for familial IVF and CA even when phenotypic markers are lacking or difficult to interpret. The increasing insights in cardiac pathophysiology and its relation with ventricular arrhythmias and CA have increased the contribution of known causes of aborted CA and, consistently, decreased the contribution of unknown causes of CA (and thus IVF). The differential diagnosis of CA currently includes multiple different pathologies[6,34,35] and, as IVF is a diagnosis *per exclusionem*, our diagnostic repertoire has increased simultaneously. In Table 14.1, we describe a limited number of milestone publications in the history of IVF. These publications cover several (familial) causes of VF and SD that previously would probably have been considered IVF. Importantly, this list excludes many causes which may also have (in part) a familial background, such as coronary artery disease (e.g. familial hypercholesterolemia) or muscle disorders (e.g. Duchenne's muscular dystrophy), and more clear non-familial causes such as myocarditis, cardiomyopathies caused by nutritional deficiencies or valvular disease. Furthermore, the authors of most of these publications were not the first to associate a certain observation with VF or SD; rather, they were recognised for their collection of several cases into a separate clinical entity.

14.4 Genetic Diagnosis

As there are *familial forms of IVF* (in 5–20% of cases[5,36,37]), there must be underlying genetic defects that are transmitted through these IVF families. In solitary cases, similar genetic defects or genetic variations vulnerable to interaction with, for example, drugs can be expected to be important. The identification of a genetic defect that predisposes an individual to IVF may certainly save lives as it offers the unique possibility of assessing the risk status of, and to treat accordingly, pre-symptomatic individuals with a potentially fatal disease, who do not manifest the disease otherwise. However, the identification of genetic defects involved in IVF is very difficult for the same two reasons that complicate its diagnosis. First, unlike the other arrhythmia syndromes, there is no clinical phenotype, except aborted CA or SD, that reveals an individual's risk. For example, in Long-QT syndrome or Brugada syndrome, the typical ECG characteristics

Table 14.1 Several milestone publications explaining ventricular fibrillation/sudden (cardiac) death (*VF/SCD*) previously considered idiopathic

Year	Syndrome	Authors	Diagnosis by:
1951	Long-QT syndrome	Jervell and Lange-Nielsen[66]	ECG, exercise-ECG, Holter-monitoring
1958	Hypertrophic cardiomyopathy	Teare[67]	Post-mortem, echocardiography, ECG, biopsy
1978	Catecholaminergic polymorphic ventricular tachycardia	Coumel et al.[68]	Exercise-ECG, Holter-monitoring, adrenaline provocation
1982	Arrhythmogenic right ventricular cardiomyopathy/dysplasia	Marcus et al.[69]	Post-mortem, ECG, echocardiography, MRI, biopsy, Holter-monitoring
1992	Brugada syndrome	Brugada et al.[70]	ECG, sodium channel blocker provocation
2000	Short-QT syndrome	Gussak et al.[71]	ECG
2008	IVF associated with early repolarization	Haissaguerre et al.[25]	ECG
2009	IVF associated with DPP6	Alders et al.[38]	Genetic analyses

can be used to classify family members as affected or unaffected, and this is subsequently used to correlate with the genetic data. Second, as many IVF patients die young, this leaves little patients and material available for analysis.

Recently, two breakthroughs in a genetic diagnosis of IVF were established. Alders et al. uncovered a *haplotype* (a combination of alleles transmitted together) on chromosome 7q36 that harbours the *DPP6* gene in 10 distantly related IVF families using a *genome-wide haplotype-sharing analysis*.[38] That familial IVF can indeed be an extremely malignant disease was apparent from the poor event-free survival in carriers of this risk-haplotype: before the age of 58 years, 50% of the risk-haplotype carriers had died or had been resuscitated. From expression analyses in heart biopsies, it became clear that the risk-haplotype carriers had on average a 22-fold higher expression of *DPP6* than controls. This makes overexpression of *DPP6* the likely pathogenetic mechanism underlying IVF in these families. Furthermore, *DPP6* is probably involved in the transient outward current (I_{to}) in the heart, which is responsible for phase 1 of the cardiac action potential by the Kv4.2 and Kv4.3 subunits.[39] Because the *DPP6* gene had previously not been related to CA or IVF, and at present, all detected families constitute a founder population, the importance of this gene in other IVF patients and families still requires further study. Interestingly, affected patients in the *DPP6* families present with very short-coupled monomorphic extrasystoles (predominantly from the right ventricular apex/free wall) initiating VF. This finding suggests possible overlap with similar descriptions of IVF[36] and *short-coupled Torsade de pointes*.[40]

A second study into IVF genetics was performed by Haissaguerre et al. concerning IVF associated with *early repolarization*.[41] In 157 IVF patients with early repolarization, mutation analysis was conducted by direct sequencing of candidate genes for cardiac arrhythmias. This included genes encoding potassium channels, their subunits and transcriptional regulators (*KCNQ1, KCNE1, KCNH2, KCNE2, KCNJ2, KCNJ8, KCNJ11, ABCC9, KCNJ5, KCNJ3, KCND3, IRX3, IRX5*), sodium channels (*SCN5A, SCN1B*), Na+/Ca2+ exchanger (*NCX1*), calcium channels (*CACNA1C, CACNB2*), Ca2+-binding proteins (*CALR, CASQ2*), and cytoskeletal proteins interacting with ion channels (*ANK2*). In just one patient, a missense mutation in *KCNJ8* encoding the Kir6.1 subunit of the K_{ATP} channel was identified. Although the function of the K_{ATP} channel and its Kir6.1 subunit is still not fully elucidated, it could be that the channels act in synergy with the Kv4.2 and Kv4.3 subunits responsible for I_{to}.[41] Clearly, the importance of *KCNJ8* mutations in IVF also demands further study, but judging from this report, the quantitative contribution of *KCNJ8* to IVF morbidity will be limited. Moreover, it is apparent that a candidate gene approach, even using a population of definitely affected patients, has a low yield of mutations, further establishing the current idiopathic nature of the condition and the difficulties associated with genetic research in this field.

Earlier, a *SCN5a* mutation had been associated with IVF by Akai et al.[42] However, these patients had (as expected with loss-of-function *SCN5a* mutations) severe conduction disease. Therefore, they did not meet the strict criteria for IVF used here, as (progressive) conduction disease is already established as an entity associated with VF.

14.5 Therapy, Follow-up and Prognosis

As mentioned, therapeutic decision making in IVF is complicated. First, there needs to be a therapeutic strategy for patients who survived IVF. Second, regardless of survival, there needs to be a diagnostic and preventive strategy for their first-degree relatives. Unfortunately, *risk-stratification in IVF* is hardly possible because of its idiopathic nature. Obviously, it is extremely important that any cause of VF that would require specific treatment (such as Long-QT syndromes type-1 and type-2, which are primarily treated with beta-blockers), is excluded in these patients and in their families.

In aborted CA patients, a first logical therapeutic option is an ICD for secondary prevention of SD (restoration of normal cardiac rhythm in case of VF recurrence, Fig. 14.3) resulting in a near normal life expectancy. One of the first three ICDs implanted worldwide was actually implanted in an IVF patient (the two other implanted patients had recurrent VF after a myocardial infarction and recurrent VF with hypertrophic cardiomyopathy respectively).[43] Importantly, an ICD will not lower the chance of VF recurrence, which is reported in 25–43% of IVF patients.[12,44,45] Further, from other arrhythmia syndromes, we know that ICDs will

Fig. 14.3 Ventricular fibrillation (*VF*) is treated using implantable cardioverter defibrillators (*ICD*) in a patient with idiopathic VF (*IVF*) associated with DPP6[38]

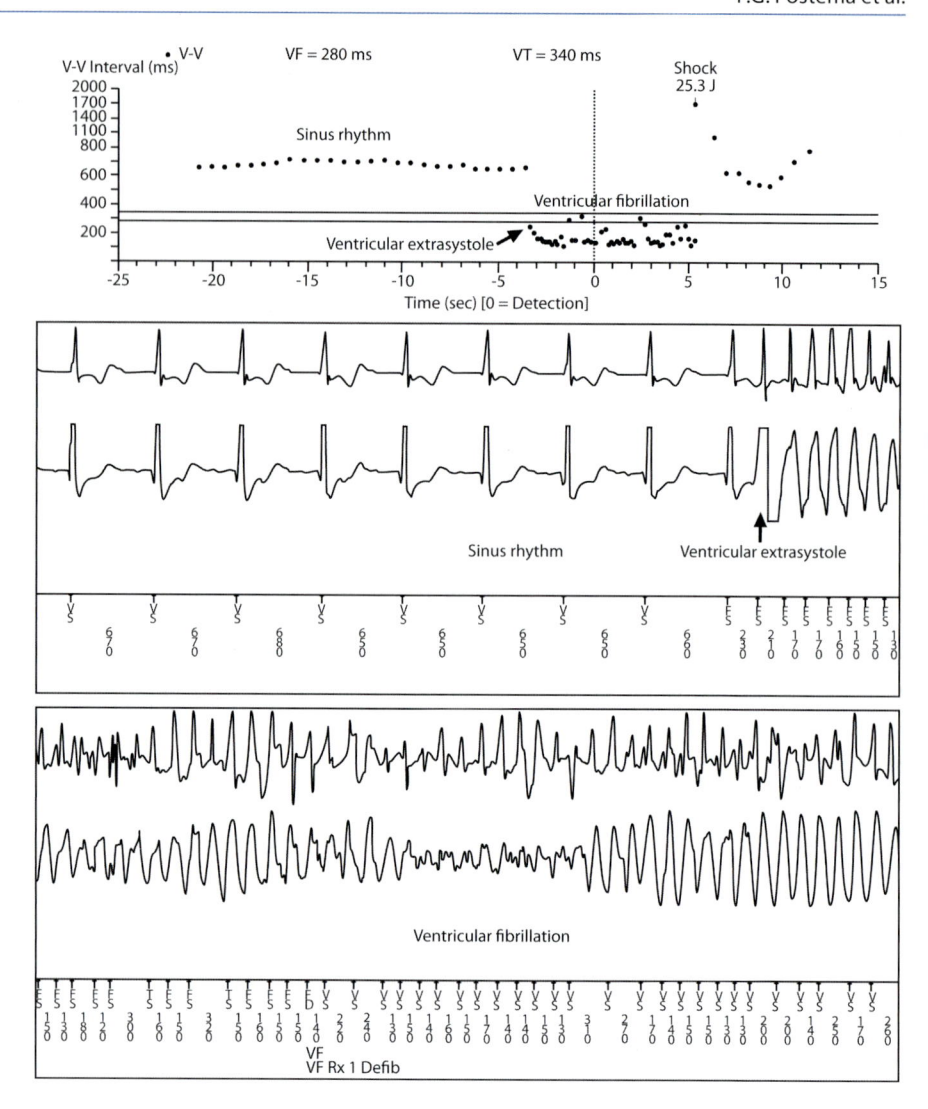

not always prove to be life saving,[46,47] and that complications as a result of the implantation or the device can be severe, perhaps even life-threatening.[48–51] Especially when primary prevention of SCD with an ICD is considered, it is noteworthy that *complications of ICD therapy* in arrhythmia syndromes such as Long-QT syndrome, Brugada syndrome, and IVF seem to be higher than the ICD complication rates in the large primary and secondary prevention trials of structural heart disease.[52] For example, this becomes apparent when comparing a study in Brugada syndrome patients with a study in patients with structural heart disease.[52] In an European Brugada syndrome multicenter study with 38 ± 27 months follow-up and the inclusion of both primary (92%) and secondary prevention (8%) ICDs, the potential benefit for the 8% of patients who received appropriate ICD therapy during follow-up was accompanied by 20% of patients with inappropriate ICD shocks.[49] In contrast, in the AVID trial (Antiarrhythmics Versus Implantable Defibrillators) with 31 ± 14 months follow-up and the inclusion of patients with structural heart disease for secondary prevention ICDs, a 12% inappropriate shock rate was documented against 65% of appropriate therapy.[53,54] Furthermore, as the patients in AVID were, on average, 65-years old and already had structural heart disease, 16% of them died during follow-up.[55] Patients with arrhythmia syndromes are much younger, more active, have no gross structural cardiac abnormalities and (thus) have a much longer life expectancy than the

AVID patients. It is easy to see that the exposure time to ICD complications is thus much longer, inappropriate shocks on supraventricular tachycardias (e.g. during exercise) will be more prevalent, lead-systems are much longer and heavier exposed to physical activity, and the many ICD replacements will expose these patients to repeated implantation-related complications (e.g. infections, bleeding and pneumothorax).

ICD implantation as primary prevention of SD in family members of IVF victims should therefore not be a decision taken lightly. First, it should be confirmed that there is indeed a familial form of IVF present in these patients. Second, socioeconomic results of ICD implantation (e.g. difficulties in obtaining mortgages) and possible psychological effects should be discussed. Third, the possible beneficial and possible harmful effects of an ICD on the short and long term should be carefully weighted. Naturally, it would be extremely helpful if a form of risk-stratification could be applied, for example, if age-categories that seem to be at highest risk could be determined In the rare occasion that a genetic cause for IVF can be identified, this can be used in asymptomatic family members for risk-stratification for CA in a disease that does not express otherwise.[38]

Obviously, it would be preferable to lower or eradicate the chance of VF recurrences, rather than treating them when they arose. This can be potentially achieved in two ways: (1) erasure of the arrhythmic substrate by ablation therapy and/or (2) antiarrhythmic drug therapy. Because of the invasive character of option 1, this will be attempted only in (severely) symptomatic patients. But when effective, *ablation therapy,* in particular, is an elegant and rewarding treatment for patients who suffer from ICD shock after ICD shock for repeated VF episodes. Electrocardiographic documentation of at least one arrhythmia episode (and preferably its start) will be essential for ablation therapy to be considered. Haissaguerre et al. performed ablations in 27 IVF patients who had experienced 10 ± 12 episodes of recurrent VF; 24 of these patients (89%) had no recurrence of VF after 24 ± 28 months of follow-up, and 4 of the patients were even primarily treated with ablation therapy and did not receive an ICD.[36] Although very promising, ablation therapy for IVF (and other arrhythmia syndromes) is only incidentally performed in experienced centres and currently lacks long-term follow-up. Nevertheless, despite possible serious complications, lowering the number of IVF episodes in addition to ICD therapy will obviously be of great benefit to patients. Cardiac transplantation has incidentally been used in other arrhythmia syndromes as a last resort in the treatment of an intolerable high incidence of recurrent VF,[56] and might be of similar value in otherwise uncontrollable IVF patients.

The only antiarrhythmic drug with potential in IVF is *quinidine*.[57-59] Other antiarrhythmic drugs have not proven to be beneficial,[12,44] although some promote the use of beta-blockers anyway.[45] Already in 1929, Dock described a patient with recurrent VF without any evidence of heart disease in whom VF recurrences could be prevented by using quinidine.[60] A comparable IVF case was presented in 1949 by Moe.[61] This patient was recurrence free on quinidine for his remaining 40 years.[62] The mechanism of the favourable effect of quinidine on IVF remains uncertain, but it may be that its I_{to} blocking properties underlie its success in IVF (and other arrhythmia syndromes) during the past 80 years.[57] It is therefore injudicious that the large pharmaceutical companies have ceased or lowered its production, making it increasingly difficult to obtain quinidine supplies.[52,63-65] However, regardless of the proven value of quinidine in the treatment of several IVF victims, its value for preventive treatment in as yet asymptomatic family members in designated IVF families is certainly not clear.

14.6 Conclusions

IVF is a rather rare but notoriously difficult and malignant condition that affects young patients and sometimes a large number of their family members. IVF is a diagnosis *per exclusionem,* which necessitates detailed and thorough assessments of IVF victims and their first-degree relatives in the search for better-defined conditions that may require specific treatment. IVF recurrence is present in 25–43% of IVF patients, and in some IVF families 50% mortality before the age of 60 may exist in predisposed individuals. Stratification of IVF risk or risk of IVF recurrence is hardly possible, but recently uncovered genetic associations with IVF may prove to be of use in the future. Secondary prevention of SD in IVF may be achieved with ICDs, ablation therapy and/or quinidine treatment. Primary prevention of SD in relatives is seriously hampered by its idiopathic nature, but may still be indicated in selected cases. Importantly, complication rates from ICD therapy in IVF and other arrhythmia syndromes, involving many young patients, will be high.

References

1. Zipes DP, Wellens HJ. Sudden cardiac death. *Circulation*. 1998;98:2334-2351.
2. Myerburg RJ, Castellanos A. Cardiac Arrest and Sudden Cardiac Death. In: Libby P, Bonow RO, Mann DL, Zipes DP, eds. *Braunwald's Heart Disease: A Textbook of Cardiovascular Medicine*. 8th ed. Philadelphia, PA: Saunders Elsevier; 2008:933-973.
3. Chugh SS, Reinier K, Teodorescu C, et al. Epidemiology of sudden cardiac death: clinical and research implications. *Prog Cardiovasc Dis*. 2008;51:213-228.
4. Puranik R, Chow CK, Duflou JA, Kilborn MJ, McGuire MA. Sudden death in the young. *Heart Rhythm*. 2005;2:1277-1282.
5. Viskin S, Belhassen B. Idiopathic ventricular fibrillation. *Am Heart J*. 1990;120:661-671.
6. Elliott P, Andersson B, Arbustini E, et al. Classification of the cardiomyopathies: a position statement from the European Society Of Cardiology Working Group on Myocardial and Pericardial Diseases. *Eur Heart J*. 2008; 29:270-276.
7. Corrado D, Basso C, Rizzoli G, Schiavon M, Thiene G. Does sports activity enhance the risk of sudden death in adolescents and young adults? *J Am Coll Cardiol*. 2003;42:1959-1963.
8. Maron BJ, Doerer JJ, Haas TS, Tierney DM, Mueller FO. Sudden deaths in young competitive athletes: analysis of 1866 deaths in the United States, 1980-2006. *Circulation*. 2009;119:1085-1092.
9. Hallstrom AP, Ornato JP, Weisfeldt M, et al. Public-access defibrillation and survival after out-of-hospital cardiac arrest. *N Engl J Med*. 2004;351:637-646.
10. Steill IG, Wells GA, Field BJ, et al. Improved out-of-hospital cardiac arrest survival through the inexpensive optimization of an existing defibrillation program: OPALS study phase II. Ontario Prehospital Advanced Life Support. *JAMA*. 1999;281:1175-1181.
11. Hollenberg J, Herlitz J, Lindqvist J, et al. Improved survival after out-of-hospital cardiac arrest is associated with an increase in proportion of emergency crew-witnessed cases and bystander cardiopulmonary resuscitation. *Circulation*. 2008;118:389-396.
12. Survivors of out-of-hospital cardiac arrest with apparently normal heart. Need for definition and standardized clinical evaluation. Consensus Statement of the Joint Steering Committees of the Unexplained Cardiac Arrest Registry of Europe and of the Idiopathic Ventricular Fibrillation Registry of the United States. Circulation 1997;95:265–72.
13. Huikuri HV, Castellanos A, Myerburg RJ. Sudden death due to cardiac arrhythmias. *N Engl J Med*. 2001;345:1473-1482.
14. Chugh SS, Jui J, Gunson K, et al. Current burden of sudden cardiac death: multiple source surveillance versus retrospective death certificate-based review in a large U.S. community. *J Am Coll Cardiol*. 2004;44:1268-1275.
15. Viskin S, Belhassen B. Clinical problem-solving. When you only live twice. *N Engl J Med*. 1995;332:1221-1225.
16. Viskin S. Brugada syndrome in children: don't ask, don't tell? *Circulation*. 2007;115:1970-1972.
17. Tung R, Zimetbaum P, Josephson ME. A critical appraisal of implantable cardioverter-defibrillator therapy for the prevention of sudden cardiac death. *J Am Coll Cardiol*. 2008;52: 1111-1121.
18. Wolpert C, Echternach C, Veltmann C, et al. Intravenous drug challenge using flecainide and ajmaline in patients with Brugada syndrome. *Heart Rhythm*. 2005;2:254-260.
19. Postema PG, Wolpert C, Amin AS et al. Drugs and Brugada syndrome patients: review of the literature, recommendations and an up-to-date website (www.brugadadrugs.org). Heart Rhythm 2009;9:1335-1341.
20. Krahn AD, Gollob M, Yee R, et al. Diagnosis of unexplained cardiac arrest: role of adrenaline and procainamide infusion. *Circulation*. 2005;112:2228-2234.
21. Basso C, Calabrese F, Corrado D, Thiene G. Postmortem diagnosis in sudden cardiac death victims: macroscopic, microscopic and molecular findings. *Cardiovasc Res*. 2001; 50:290-300.
22. Tan HL, Hofman N, van Langen I, van der Wal AC, Wilde AA. Sudden unexplained death: heritability and diagnostic yield of cardiological and genetic examination in surviving relatives. *Circulation*. 2005;112:207-213.
23. Behr E, Wood DA, Wright M, et al. Cardiological assessment of first-degree relatives in sudden arrhythmic death syndrome. *Lancet*. 2003;362:1457-1459.
24. Behr ER, Dalageorgou C, Christiansen M, et al. Sudden arrhythmic death syndrome: familial evaluation identifies inheritable heart disease in the majority of families. *Eur Heart J*. 2008;29:1670-1680.
25. Haissaguerre M, Derval N, Sacher F, et al. Sudden cardiac arrest associated with early repolarization. *N Engl J Med*. 2008;358:2016-2023.
26. Klatsky AL, Oehm R, Cooper RA, Udaltsova N, Armstrong MA. The early repolarization normal variant electrocardiogram: correlates and consequences. *Am J Med*. 2003;115: 171-177.
27. Mehta M, Jain AC, Mehta A. Early repolarization. *Clin Cardiol*. 1999;22:59-65.
28. Meissner FL. *Taubstummheit und Taubstummenbildung: Beobachtungen und Erfahrungen*. Leipzig/Heidelberg: C.F. Winter'sche Verslagehandlung; 1856:119-120.
29. Einthoven W. Ueber die Form des menschlichen Electrocardiogramms. *Pflugers Arch Gesamte Physiol*. 1895;60: 101-123.
30. Einthoven W. The different forms of the human electrocardiogram and their signification. *Lancet*. 1912;179:853-861.
31. Edler I, Hertz CH. The use of the ultrasonic reflectoscope for the continuous recording of the movements of heart walls. *Kungl Fysiogr Sällsk Lund Förhandl*. 1954;24:1-19.
32. Watson JD, Crick FH. Molecular structure of nucleic acids; a structure for deoxyribose nucleic acid. *Nature*. 1953; 171:737-738.
33. Lander ES, Linton LM, Birren B, et al. Initial sequencing and analysis of the human genome. *Nature*. 2001;409:860-921.

34. Maron BJ, Towbin JA, Thiene G et al. Contemporary definitions and classification of the cardiomyopathies: an American Heart Association Scientific Statement from the Council on Clinical Cardiology, Heart Failure and Transplantation Committee; Quality of Care and Outcomes Research and Functional Genomics and Translational Biology Interdisciplinary Working Groups; and Council on Epidemiology and Prevention. Circulation 2006;113:1807–1816.
35. Lambiase PD, Elliott PM. Genetic aspects and investigation of sudden death in young people. *Clin Med.* 2008;8: 607-610.
36. Haissaguerre M, Shoda M, Jais P, et al. Mapping and ablation of idiopathic ventricular fibrillation. *Circulation.* 2002; 106:962-967.
37. Noda T, Shimizu W, Taguchi A, et al. Malignant entity of idiopathic ventricular fibrillation and polymorphic ventricular tachycardia initiated by premature extrasystoles originating from the right ventricular outflow tract. *J Am Coll Cardiol.* 2005;46:1288-1294.
38. Alders M, Koopmann TT, Christiaans I, et al. Haplotype-sharing analysis implicates chromosome 7q36 harboring DPP6 in familial idiopathic ventricular fibrillation. *Am J Hum Genet.* 2009;84:468-476.
39. Radicke S, Cotella D, Graf EM, Ravens U, Wettwer E. Expression and function of dipeptidyl-aminopeptidase-like protein 6 as a putative beta-subunit of human cardiac transient outward current encoded by Kv4.3. *J Physiol.* 2005; 565:751-756.
40. Leenhardt A, Glaser E, Burguera M, Nurnberg M, Maison-Blanche P, Coumel P. Short-coupled variant of torsade de pointes. A new electrocardiographic entity in the spectrum of idiopathic ventricular tachyarrhythmias. *Circulation.* 1994;89:206-215.
41. Haissaguerre M, Chatel S, Sacher F, et al. Ventricular Fibrillation with Prominent Early Repolarization Associated with a Rare Variant of KCNJ8/K Channel. *J Cardiovasc Electrophysiol.* 2009;20:93-98.
42. Akai J, Makita N, Sakurada H, et al. A novel SCN5A mutation associated with idiopathic ventricular fibrillation without typical ECG findings of Brugada syndrome. *FEBS Lett.* 2000;479:29-34.
43. Mirowski M, Reid PR, Mower MM, et al. Termination of malignant ventricular arrhythmias with an implanted automatic defibrillator in human beings. *N Engl J Med.* 1980; 303:322-324.
44. Wever EF, Hauer RN, Oomen A, Peters RH, Bakker PF, de Medina EO Robles. Unfavorable outcome in patients with primary electrical disease who survived an episode of ventricular fibrillation. *Circulation.* 1993;88:1021-1029.
45. Wever EF, de Medina EO Robles. Sudden death in patients without structural heart disease. *J Am Coll Cardiol.* 2004;43:1137-1144.
46. Coronel R, Casini S, Koopmann TT, et al. Right ventricular fibrosis and conduction delay in a patient with clinical signs of Brugada syndrome: a combined electrophysiological, genetic, histopathologic, and computational study. *Circulation.* 2005;112:2769-2777.
47. Brugada P, Brugada J, Brugada R. When our best is not enough: the death of a teenager with brugada syndrome. *J Cardiovasc Electrophysiol.* 2009;20:108-109.
48. Sherrid MV, Daubert JP. Risks and challenges of implantable cardioverter-defibrillators in young adults. *Prog Cardiovasc Dis.* 2008;51:237-263.
49. Sacher F, Probst V, Iesaka Y, et al. Outcome after implantation of a cardioverter-defibrillator in patients with brugada syndrome. A multicenter study. *Circulation.* 2006;114:2317-2324.
50. Viskin S, Halkin A. Treating the long-QT syndrome in the era of implantable defibrillators. *Circulation.* 2009;119:204-206.
51. Messali A, Thomas O, Chauvin M, Coumel P, Leenhardt A. Death due to an implantable cardioverter defibrillator. *J Cardiovasc Electrophysiol.* 2004;15:953-956.
52. Viskin S, Wilde AA, Tan HL, Antzelevitch C, Shimizu W, Belhassen B. Empiric quinidine therapy for asymptomatic Brugada syndrome: time for a prospective registry. *Heart Rhythm.* 2009;6:401-404.
53. Kron J, Herre J, Renfroe EG, et al. Lead- and device-related complications in the antiarrhythmics versus implantable defibrillators trial. *Am Heart J.* 2001;141:92-98.
54. Raitt MH, Klein RC, Wyse DG, et al. Comparison of arrhythmia recurrence in patients presenting with ventricular fibrillation versus ventricular tachycardia in the Antiarrhythmics Versus Implantable Defibrillators (AVID) trial. *Am J Cardiol.* 2003;91:812-816.
55. A comparison of antiarrhythmic-drug therapy with implantable defibrillators in patients resuscitated from near-fatal ventricular arrhythmias. The Antiarrhythmics versus Implantable Defibrillators (AVID) Investigators. N Engl J Med 1997;337:1576–1583.
56. Ayerza MR, de Zutter M, Goethals M, Wellens F, Geelen P, Brugada P. Heart transplantation as last resort against Brugada syndrome. *J Cardiovasc Electrophysiol.* 2002; 13:943-944.
57. Belhassen B, Glick A, Viskin S. Excellent long-term reproducibility of the electrophysiologic efficacy of quinidine in patients with idiopathic ventricular fibrillation and Brugada syndrome. *Pacing Clin Electrophysiol.* 2009; 32:294-301.
58. Belhassen B, Viskin S, Fish R, Glick A, Setbon I, Eldar M. Effects of electrophysiologic-guided therapy with Class IA antiarrhythmic drugs on the long-term outcome of patients with idiopathic ventricular fibrillation with or without the Brugada syndrome. *J Cardiovasc Electrophysiol.* 1999; 10:1301-1312.
59. Haissaguerre M, Sacher F, Nogami A, et al. Characteristics of recurrent ventricular fibrillation associated with inferolateral early repolarization role of drug therapy. *J Am Coll Cardiol.* 2009;53:612-619.
60. Dock W. Transitory ventricular fibrillation as a cause of syncope and its prevention by quinidine sulphate: With case report and discussion of diagnostic criteria for ventricular fibrillation. *Am Heart J.* 1929;4:709-714.
61. Moe T. Morgagni-Adams-Stokes attacks caused by transient recurrent ventricular fibrillation in a patient without apparent organic heart disease; a case report. *Am Heart J.* 1949; 37:811-819.
62. Kontny F, Dale J. Self-terminating idiopathic ventricular fibrillation presenting as syncope: a 40-year follow-up report. *J Intern Med.* 1990;227:211-213.

63. Viskin S, Antzelevitch C, Marquez MF, Belhassen B. Quinidine: a valuable medication joins the list of 'endangered species'. *Europace*. 2007;9:1105-1106.
64. Viskin S. Idiopathic ventricular fibrillation "Le Syndrome d'Haissaguerre" and the fear of J waves. *J Am Coll Cardiol*. 2009;53:620-622.
65. Wilde AA, Langendijk PN. Antiarrhythmic drugs, patients, and the pharmaceutical industry: value for patients, physicians, pharmacists or shareholders? *Neth Heart J*. 2007;15:127-128.
66. Jervell A, Lange-Nielsen F. Congenital deaf-mutism, functional heart disease with prolongation of the Q-T interval and sudden death. *Am Heart J*. 1957;54:59-68.
67. Teare D. Asymmetrical hypertrophy of the heart in young adults. *Br Heart J*. 1958;20:1-8.
68. Coumel P, Fidelle J, Lucet V, Attuel P, Bouvrain Y. Catecholamine-induced severe ventricular arrhythmias with Adams-Stokes syndrome in children: report of four cases. *Br Heart J*. 1978;40:28-37.
69. Marcus FI, Fontaine GH, Guiraudon G, et al. Right ventricular dysplasia: a report of 24 adult cases. *Circulation*. 1982;65:384-398.
70. Brugada P, Brugada J. Right bundle branch block, persistent ST segment elevation and sudden cardiac death: a distinct clinical and electrocardiographic syndrome. A multicenter report. *J Am Coll Cardiol*. 1992;20:1391-1396.
71. Gussak I, Brugada P, Brugada J, et al. Idiopathic short QT interval: a new clinical syndrome? *Cardiology*. 2000;94:99-102.

The Genetics of Mitral Valve Prolapse

Paul L. van Haelst, Toon Oomen, and J. Peter van Tintelen

15.1 Introduction

Mitral valve prolapse (MVP) is one of the most common forms of valvular heart disease. The clinical presentation can be very diverse, ranging from an incidental finding within asymptomatic patients to dramatic cases with severe mitral regurgitation, heart failure, bacterial endocarditis, or even sudden cardiac death. Mitral valve prolapse appears to be one of the most common Mendelian cardiovascular abnormalities in humans. This chapter briefly discusses the epidemiologic aspects of mitral valve prolapse, its pathophysiology, and the current status of genetic knowledge of this intriguing valvular disorder.

15.2 Epidemiology and Definitions

Mitral valve prolapse has long been recognized as an auscultatory phenomenon. It was not until 1966 that Barlow discovered the reason for the often-heard mid-systolic click.[1] Shortly afterwards, the introduction of echocardiography led to a tsunami of patients diagnosed with MVP. Due to incorrect echocardiographic definitions and selection bias of studied populations, prevalences of up to 35% were reported in the 1970s and early 1980s.[2] A redefinition of echocardiographic criteria due to improved knowledge of mitral valve architecture provided a more accurate insight into the extent of the problem. Currently, the prevalence of MVP is known to range from 0.6[3] to 2.5%[4] in the general population. It is equally distributed between men and women, but patients with MVP tend to have a leaner stature.[4]

Mitral valve prolapse (MVP) is defined as the billowing of one or both mitral valve leaflets across the plane of the mitral valve annulus into the left atrium during systole. By definition, the leaflets should reach more than 2 mm above the annular plane on the parasternal long axis view with echocardiography (Fig. 15.1).

MVP may present as part of a systemic disorder or as a solitary disease. It may occur more frequently in connective tissue disorders such as Marfan syndrome, Ehlers Danlos syndrome, and osteogenesis imperfecta. However, in most cases, it presents as a solitary entity; only 1–2% of all patients with MVP do have a connective tissue disorder. This chapter focuses on the solitary forms.

15.3 Pathophysiology and Clinical Aspects of Mitral Valve Disease

The solitary forms of mitral valve disease are referred to as the classical prolapse, with the valve leaflet thickness exceeding 5 mm on echocardiography, and the nonclassical form, with leaflet thickness of less than 5 mm (both in the presence of a systolic upward displacement of 2 mm). Leaflet thickening or myxomatous degeneration is characterized by expansion of the spongiosa layer due to accumulation of proteoglycans. Also, structural alterations of collagen in all components of the valvular system and chordae are found. It is thought that the mechanism underlying the expansion of the spongiosa layer is the result of a dysregulation of the balance between matrix protein synthesis

P.L. van Haelst (✉)
Department of Cardiology, Antonius Hospital Sneek,
P.O. Box 20000, BA 8600 Sneek, The Netherlands
e-mail: p.vhaelst@antonius-sneek.nl

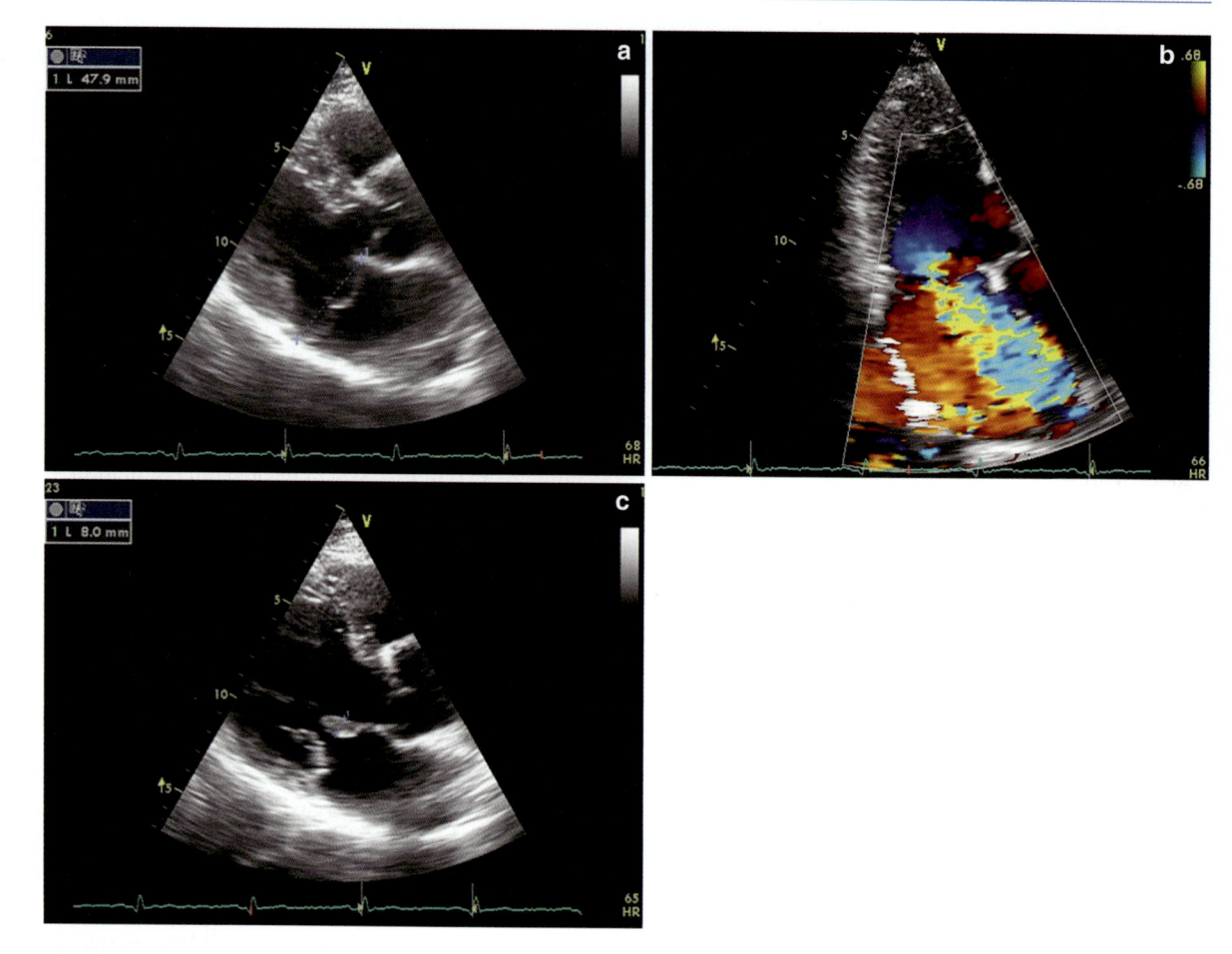

Fig. 15.1 (**a**) Parasternal long axis echocardiography showing classical mitral valve prolapse. Both posterior and anterior myxomatous mitral valve leaflets are billowing up to 6.5 mm in the left atrium during systole. (**b**) Apical three-chamber echocardiography with color Doppler flow measurement showing moderate to severe mitral regurgitation in the same patient. The left atrium is enlarged. (**c**) Parasternal long axis demonstrating myxomatous tips of both anterior and posterior mitral valves measuring 8 mm

and degradation.[5] Most cases come along with a family history of valvular disease. MVP was found in 46% of first-degree relatives older than 20 years, whereas only 16% of patients below that age were affected, suggesting progressive disease with age-dependent penetrance.[6]

The presentation of MVP is extremely heterogeneous, and to date no specific set of predictors for disease progression has been identified.

The diagnosis is made by physical examination. Typically, a midsystolic click is heard, often followed by a late systolic murmur.[2] The diagnosis is confirmed by two-dimensional echocardiography. MVP generally has a good prognosis. Complications such as severe mitral insufficiency, heart failure, thromboembolic complications, and sudden cardiac death are rare, especially in patients with nonclassical prolapse. Freed and coworkers, from the Framingham study group, found that complications affected only 3% of all patients with MVP.[4] However, patients with classical prolapse carry a 14-fold risk for complications.[7] It is not surprising that patients above 50 years who have a decreased left ventricular function, moderate to severe mitral regurgitation, and atrial fibrillation exhibit more complications. The risk for infective endocarditis is raised three to eight times. However, current guidelines for infectious endocarditis no longer advocate the use of prophylactic antibiotics in patients with MVP. Only patients who are known to have had endocarditis should receive infective endocarditis prophylaxis when appropriate.[8]

Mitral regurgitation may lead to atrial fibrillation and therefore to thromboembolic complications. Antithrombotic medication, on the other hand, should be given only if classical risk factors unrelated to MVP are present. Nowadays, mitral valve reconstruction or replacement is advocated even in asymptomatic patients with moderate to severe mitral regurgitation.[9]

Sudden cardiac death occurs twice as often in patients with myxomatous valve disease as compared to the general population. SCD is found more often in patients with impaired left ventricular function, moderate to severe mitral regurgitation, and redundant chordate.[10] Interestingly, in a series of 200 victims of sudden cardiac death younger than 35 years, mitral valve prolapse was the only cardiac abnormality that could be found in as many as 10% of cases.[11] Although patients with MVP more often exhibit atrial and ventricular arrhythmias during Holter monitoring, the exact mechanism of SCD is yet to be elucidated.[5] Cardiac screening of probands and siblings may be considered in patients with classical MVP. When conducting family studies, echocardiography should be performed in all cases. Holter monitoring can be performed in the presence of complaints or a family history of sudden cardiac death.

15.4 Genetic Aspects of Mitral Valve Disease

Classical mitral valve prolapse has long been recognized as an autosomal dominant inherited condition with reduced penetrance, influenced by age and sex.[12–14] Until today, three loci have been identified, on chromosomes 16p11.2-p12.1, 11p15.4, and 13q31.3-q32.1 for myxomatous mitral valve prolapse (MMVP1),[15] MMVP2,[16] and MMPV3,[17] respectively. All these studies were performed using linkage analysis; however, no underlying genes have been identified yet.

Recently, an important observation was made by Kyndt et al. who identified a mutation of the filamin A gene to be responsible for X-linked myxomatous mitral valves in four unrelated families.[18] Interestingly, both males and females were affected in these families although females showed milder manifestations of the disorder. Filamin A is an actin binding protein that plays a pivotal role in cell motility and membrane stability. Filamin A may contribute to the development of myxomatous changes of the cardiac valves by regulation of transforming growth factor-β (TGF-β) signaling through its interaction with Smads activated by TGF-β receptors.[19,20] Defective signaling cascades that involve members of the TGF-β superfamily have been described in impaired remodeling of cardiac valves during development.

15.5 Summary and Conclusions

Mitral valve prolapse is the most common valvular disorder with a strong genetic contribution. The course of disease is benign in most cases, but serious complications such as heart failure, severe mitral regurgitation, bacterial endocarditis, and sudden cardiac death occur, especially in patients with myxomatous degenerated valves. Until now, three chromosomal loci have been identified in autosomal dominant MVP without a specific gene identified yet. One gene, encoding filamin A, has been identified in X-linked myxomatous MVP, suggesting an underlying mechanism in the regulation of the valvular cytoskeleton. More genetic as well as clinical research is warranted to define patients at risk for this potentially lethal condition. First-degree relatives of patients with classical MVP should undergo cardiac screening and, if sudden death has occurred, genetic screening should be part of the workup.

References

1. Barlow JB, Bosman CK. Aneurysmal protrusion of the posterior leaflet of the mitral valve. An auscultatory-electrocardiographic syndrome. *Am Heart J*. 1966;71:166-178.
2. Weisse AB. Mitral valve prolapse: now you see it; now you don't: recalling the discovery, rise and decline of a diagnosis. *Am J Cardiol*. 2007;99:129-133.
3. Hepner AD, Ahmadi-Kashani M, Movahed MR. The prevalence of mitral valve prolapse in patients undergoing echocardiography for clinical reason. *Int J Cardiol*. 2007; 123:55-57.
4. Freed LA, Levy D, Levine RA, et al. Prevalence and clinical outcome of mitral-valve prolapse. *N Engl J Med*. 1999;341:1-7.
5. Hayek E, Gring CN, Griffin BP. Mitral valve prolapse. *Lancet*. 2005;365:507-518.
6. Grau JB, Pirelli L, Yu PJ, et al. The genetics of mitral valve prolapse. *Clin Genet*. 2007;72:288-295.
7. Nishimura RA, McGoon MD, Shub C, et al. Echocardiographically documented mitral-valve prolapse. Long-term

follow-up of 237 patients. *N Engl J Med.* 1985;313: 1305-1309.
8. Wilson W, Taubert KA, Gewitz M, et al. Prevention of infective endocarditis: guidelines from the American Heart Association: a guideline from the American Heart Association Rheumatic Fever, Endocarditis, and Kawasaki Disease Committee, Council on Cardiovascular Disease in the Young, and the Council on Clinical Cardiology, Council on Cardiovascular Surgery and Anesthesia, and the Quality of Care and Outcomes Research Interdisciplinary Working Group. *Circulation.* 2007;116:1736-1754.
9. Bonow RO, Carabello BA, Chatterjee K, et al. Focused update incorporated into the ACC/AHA 2006 guidelines for the management of patients with valvular heart disease: a report of the American College of Cardiology/American Heart Association Task Force on Practice Guidelines (Writing Committee to Revise the 1998 Guidelines for the Management of Patients With Valvular Heart Disease): endorsed by the Society of Cardiovascular Anesthesiologists, Society for Cardiovascular Angiography and Interventions, and Society of Thoracic Surgeons. *Circulation.* 2008;118:e523-e661.
10. Kligfield P, Levy D, Devereux RB, et al. Arrhythmias and sudden death in mitral valve prolapse. *Am Heart J.* 1987;113:1298-1307.
11. Basso C, Corrado D, Thiene G. Cardiovascular causes of sudden death in young individuals including athletes. *Cardiol Rev.* 1999;7:127-135.
12. Devereux RB, Brown WT, Kramer-Fox R, et al. Inheritance of mitral valve prolapse: effect of age and sex on gene expression. *Ann Intern Med.* 1982;97:826-832.
13. Strahan NV, Murphy EA, Fortuin NJ, et al. Inheritance of the mitral valve prolapse syndrome. Discussion of a three-dimensional penetrance model. *Am J Med.* 1983;74:967-972.
14. Weiss AN, Mimbs JW, Ludbrook PA, et al. Echocardiographic detection of mitral valve prolapse. Exclusion of false positive diagnosis and determination of inheritance. *Circulation.* 1975;52:1091-1096.
15. Disse S, Abergel E, Berrebi A, et al. Mapping of a first locus for autosomal dominant myxomatous mitral-valve prolapse to chromosome 16p11.2-p12.1. *Am J Hum Genet.* 1999;65: 1242-1251.
16. Freed LA, Acierno JS Jr, Dai D, et al. A locus for autosomal dominant mitral valve prolapse on chromosome 11p15.4. *Am J Hum Genet.* 2003;72:1551-1559.
17. Nesta F, Leyne M, Yosefy C, et al. New locus for autosomal dominant mitral valve prolapse on chromosome 13: clinical insights from genetic studies. *Circulation.* 2005;112:2022-2030.
18. Kyndt F, Gueffet JP, Probst V, et al. Mutations in the gene encoding filamin A as a cause for familial cardiac valvular dystrophy. *Circulation.* 2007;115:40-49.
19. Sasaki A, Masuda Y, Ohta Y, et al. Filamin associates with Smads and regulates transforming growth factor-beta signaling. *J Biol Chem.* 2001;276:17871-17877.
20. Derynck R, Zhang YE. Smad-dependent and Smad-independent pathways in TGF-beta family signalling. *Nature.* 2003;425:577-584.

Atrioventricular (AV) Reentry Tachycardia

Denise P. Kolditz-Muijs, Monique R.M. Jongbloed, and Martin J. Schalij

16.1 Introduction

Supraventricular tachycardia (SVT) is the most common and best known cardiac arrhythmia in both children and adults, with an estimated prevalence of 2.25 per 1,000 in the normal population worldwide.[1-3] These supraventricular arrhythmias are often repetitive, occasionally persistent, and rarely life threatening.[4] Macro-reentry through an anomalous atrioventricular (AV) pathway is by far the most frequent mechanism of SVT in children (80%) and accounts for ~30% of SVT cases in adults.[1,5] In adults, AV Nodal Reentrant Tachycardia (AVNRT) accounts for most SVTs (approximately 80%); yet, this type of SVT accounts for only 5% of SVT cases in infants and toddlers and comprises only 13–16% of SVTs in children and adolescents.[6] Much less prevalent forms of SVT in both children and adults include Permanent-Junctional-Ectopic-Tachycardia (PJET) and Ectopic-Atrial-Tachycardia (EAT).

Knowledge about cardiac arrhythmias has significantly improved in the last decades. New forms of arrhythmias have been identified, electrophysiological mechanisms have been investigated, and genetic studies have identified gene alterations to be the basis of a few types of arrhythmias. While most studies have focused on arrhythmia mechanisms in ventricular tachycardia, the genetic and molecular bases of some forms of SVT are being recognized with increased frequency. In this chapter, the epidemiology, pathogenesis, clinical aspects, electrocardiographic diagnosis, therapy, and genetic aspects of AVRT, and in particular, the Wolff-Parkinson-White (WPW) syndrome, concealed bypasses, and Mahaim tachycardia are discussed.

16.2 Epidemiology in AVRT

Atrioventricular reentrant tachycardia (AVRT) accounts for ~30% of SVT cases in adults and is the most frequent mechanism of SVT in children (80%).[1,5] AV Reentrant Tachycardia involves the presence of an *accessory myocardial AV pathway (AP)* that bypasses the isolating *annulus fibrosus cordis*. The overall incidence of APs in the general population is 0.1–0.3%.[7] Most APs are manifest and exhibit bidirectional conduction (~60–75%) as in the WPW syndrome, while ~5–27% conduct only anterogradely as in Mahaim tachycardia, and ~17–37% are concealed (conduct only retrogradely).[8]

16.2.1 Wolff–Parkinson–White Syndrome

The prevalence of WPW syndrome is 0.1–3.1 per 1,000 persons, while the overall incidence of APs in first-degree relatives of patients with ventricular preexcitation on the ECG is 3.4%.[7,9]

Because of its intermittent pattern, the precise incidence of WPW is unknown.[10] The incidence of WPW in males is more than twice that in females.[9,10] Additionally, in female patients, an incidence peak around the age of 7 years has been reported, while men are significantly younger at first presentation than women.[10] In approximately 60% of pediatric patients, the first episode of AVRT occurs before birth or in infancy and appears to spontaneously resolve completely in two thirds of cases

---

D.P. Kolditz-Muijs (✉)
Leiden University Medical Center, Leiden, The Netherlands
e-mail: d.p.kolditz@lumc.nl

before the age of 1 year, while more than 80–90% of patients become asymptomatic after the first year of life.[2, 11] Relapse during follow-up is, however, observed in 20–30% of these cases.[12] A second new onset AVRT incidence peak is seen around the age of 8–12 years, in which case, spontaneous regression is observed only in ~20% of patients.[13]

Approximately 10% of patients may have two or more AV bypass tracts. Most patients with WPW syndrome demonstrate isolated ventricular preexcitation in a structurally normal heart.[14] In a small percentage of patients with WPW, however, APs occur in association with other cardiac abnormalities or congenital heart disease.[15] WPW syndrome is more prevalent in children with Ebstein's anomaly of the tricuspid valve, AV septal defects, and ventricular septal defects, while occasionally, the WPW syndrome may be inherited.[16–18] Moreover, coronary artery disease was found to be associated with WPW in 6% of patients.[10]

In the vast majority of cases, WPW syndrome has no clear familial involvement, but a significant minority of cases result from a clearly inherited disease or occurs as part of a syndrome with a strong genetic basis. Identification of WPW in more than one family member should prompt clinical evaluation of relatives for additional findings of ventricular hypertrophy or conduction abnormalities as preexcitation and cardiomyopathy (e.g. PRKAG2, Danon disease, Pompe disease) and preexcitation and progressive AV conduction block (e.g. PRKAG2) are a rare combination and might imply an underlying mutation and/or autosomal dominant trait.[19] Familial hypertrophic cardiomyopathy (HCM) is an autosomal dominant disorder that is characterized by unexplained ventricular hypertrophy with histological evidence of myocyte and myofibrillar disarray and is associated with ventricular preexcitation in 5–10% of patients and was first described in 1960 by Braunwald et al.[20, 21]

16.2.2 Mahaim Tachycardia

Mahaim fibers were first identified in 1937 by Mahaim and Benatt and were initially described to extend from the His bundle into the ventricular myocardium.[22] Originally, these Mahaim fibers were classified into two main groups: nodoventricular fibers (between the AV node and ventricular myocardium) and fasicoventricular fibers (between the His bundle and the ventricular myocardium). In the 1980s, it became clear that the anatomic cause of Mahaim tachycardia is a slowly conducting AV AP with *decremental (or AV nodal) conduction properties*, in which case, conduction speed through the AP is negatively correlated to sinus node frequency.[23] Nowadays, two types of decremental right-sided APs, both arising from the tricuspid annulus but with different ventricular insertions, are recognized in Mahaim tachycardia: (1) *atriofascicular (AF) connections* and (2) *AV pathways (pseudo-Mahaim fibers)*. Atriofascicular (AF) pathways account for 80% of Mahaim fibers and have a long intracardiac course and insert into the distal right bundle branch (RBB) or right ventricle near its apex.[24] Atrioventricular (AV) pathways account for only 20% of Mahaim fibers and insert proximally into the right ventricle near the AV annulus close to the cardiac conduction system (CCS).[24]

Multiple APs or dual AV nodal pathways are found in approximately 40% of patients with Mahaim fibers.[25] Mahaim fibers are found in 10–40% of patients with Ebstein's anomaly of the tricuspid valve.[26]

16.2.3 Concealed Bypasses

Approximately 17–37% of diagnosed APs are concealed bypasses, which are capable of retrograde conduction only.[8] The true prevalence of concealed APs is unknown, since its presence cannot be established on resting ECGs. Concealed bypasses consist of myocardial strands that course through the epicardial fat. The vast majority of concealed bypasses are left-sided.[27] These pathways usually exhibit nondecremental retrograde conduction, which allows them to function as the retrograde arm in orthodromic tachycardia. Sometimes, concealed pathways can also exhibit decremental conduction, in which case, they are associated with Permanent Junctional Reciprocating Tachycardia (PJRT), an unusual form of SVT.

16.3 Pathogenesis in AVRT

16.3.1 Accessory Pathways (APs)

Accessory pathways are congenital in origin, resulting from failure of the normal developmental isolation process of the annulus fibrosus cordis, leaving persistent myocardial strands coursing through the tricuspid and mitral valve annulus connecting the atrial myocardium to the ventricular myocardium bypassing the AV node. Developmentally, these AV pathways derive from the embryonic AV junctional myocardium and histologically consist of myocardial and fibrous cells. Moreover, epicardium-derived-cells have recently been implicated in the normal developmental process of annulus fibrosis. [28–30]

The AP in WPW syndrome, known as the *bundle of Kent*, is usually located in the lateral rings of the annulus fibrosis and consists of a thin muscular segment that does not posses decremental properties.[31–33] Other less frequently encountered APs include substrates with decremental (nodal) properties mostly located in the septal or inferior half of the tricuspid valve and slow conducting APs giving rise to Permanent Junctional Reciprocating Tachycardia (PJRT).[3]

Mahaim fibers (mostly AF fibers) connect the atria to the right bundle branch or to the distal right ventricle and have decremental conduction properties.[3] Accessory pathways in Mahaim tachycardia conduct only in antegrade direction and most often arise in the anterior wall of the right atrium.[34] Sometimes, the Mahaim AP fibers arise in the His bundle or bundle branches and insert into the ventricular myocardium. These are generally not involved in tachycardias and are called fasciculoventricular tracts.[3] Histologically, APs in patients with Mahaim tachycardia demonstrate features similar to normal AV nodal tissue, not seen in normal APs as in WPW-syndrome. A possible morphologic and functional etiological explanation for the appearance of these pathways has been proposed, involving the embryonic development of the right ventricular inflow tract and the moderator band.[35]

16.3.2 Arrhythmia Mechanisms in AVRT

Accessory pathways in patients with the Wolff-Parkinson-White (WPW) syndrome give rise to ventricular preexcitation on ECG recordings during sinus rhythm.[3] In contrast, the ECG in a patient with a concealed AV bypass tract, which is capable of conducting solely in the retrograde direction (from the ventricles to the atria) during sinus rhythm is normal (no signs of ventricular preexcitation).

Mechanistically, besides an anatomical substrate (the AP), two additional conditions are required for functional preexcitation during sinus rhythm to occur: (1) electrical coupling between adjacent ventricular and atrial myocytes, and (2) a higher conduction velocity through the AP than in the normal ventricular conduction system. Subsequent reentry tachycardia can occur when (1) at least two functionally distinct conduction pathways are present, (2) unidirectional block is induced in one pathway, and (3) conduction time is slow enough over the nonblocked pathway to allow recovery of excitability in the blocked pathway, thereby permitting retrograde conduction over the blocked pathway and completing the reentry circuit.[36,37]

The macroreentry circuit in AVRT is produced in the atrial muscle, the AV node, the ventricular muscle, and the AP itself. This circuit facilitates continuous alternate depolarization of the atrial and ventricular myocardium. In typical *orthodromic AVRT*, the electrical impulse proceeds from the atria to the ventricles through the AV node and retrograde back up to the atria via the AP. In this type of tachycardia, the effective refractory period of the AP exceeds that of the normal AV nodal His-Purkinje conduction pathway.[38] In orthodromic AVRT, the QRS complexes are narrow and a retrograde P wave can be seen embedded in the early portion of the T wave (best seen in leads II and V2).[3]

Infrequently, the reentry circuit in AVRT can proceed in the opposite direction and produce *antidromic* (or preexcited) *AVRT*. In this type of tachycardia, the effective refractory period of the AP is shorter than that of the normal AV nodal His-Purkinje conduction pathway.[38] The ECG demonstrates very abnormally wide QRS complexes resembling those seen in ventricular tachycardia.[3] Reentrant tachycardia may also involve multiple APs, providing both antegrade and retrograde conduction.[36]

Because Mahaim fibers conduct only in antegrade direction, Mahaim tachycardia is always characterized by antidromic AVRT. In Mahaim tachycardia, the ECG findings during tachycardia are a normal or short PR interval and an abnormally wide QRS complex with a left bundle branch block appearance. Additional useful, but nondiagnostic, features suggesting Mahaim tachycardia instead of antidromic tachycardia in WPW syndrome include: (1) a QRS axis between 0° and minus 75°, (2) a QRS duration of ≤0.15 s, R-wave in lead I, (3) rS complex in lead V1, (4) precordial transition in lead V4 or later, and (5) cycle length between 220 and 450 ms (heart rates of 130–270 bpm). Less frequently, antidromic Mahaim tachycardia can use another AP as the retrograde limb of the reentry circuit instead of the AV node. Additionally, Mahaim fibers can also function as a bystander pathway, conducting anterogradely (instead of normal AV nodal antegrade conduction) during AVNRT, atrial flutter of atrial tachycardia.[39]

16.4 Clinical Aspects

16.4.1 AVRT in General

Episodes of AVRT are mostly characterized by paroxysmal regular palpitations with a sudden onset and termination, in contrast to sinus tachycardia, in which gradual acceleration and termination is observed. Arrhythmia-related symptoms in AVRT include palpitations, fatigue, lightheadedness, chest discomfort, dyspnea, presyncope, and syncope (in ~15% of patients). Polyuria is seen in sustained SVT and is caused by release of atrial natriuretic peptide in response to increased atrial pressures from contraction of the atria against a closed AV valve.[4]

Symptoms of AVRT in infancy differ from those in childhood or adolescence. Newborns may present with a history of fetal tachycardia, signs of left ventricular dysfunction, or even with *hydrops fetalis*, representing severe heart failure from persistent rapid fetal tachycardia.[40,41] Neonates can also present with new onset incessant and difficult to treat tachycardia at birth with no history of any fetal tachycardias.[3] Moreover, neonates with AVRTs are at high risk for sudden cardiac arrest as cardiac reserve in neonates is very small and typical AVRT with heart rates exceeding 200/min. can lead to life-threatening myocardial dysfunction within a few days.[42]

In infants, symptoms of AVRTs are inconspicuous and mimic those of many other common illnesses in infancy, including irritability, poor feeding, tachypnea, diaphoresis, and poor color. Most infants with AVRT have structurally normal hearts, while in 15% of patients, tachycardia is associated with heart disease, drug administration, or febrile illness.[3]

Older children and adolescents complain of palpitations (in excess of 150/min.), general malaise, indistinct pressure or discomfort in the throat, isolated headaches, fatigue, chest discomfort, shortness of breath, or lightheadedness. Syncope is unusual and may indicate life-threatening arrhythmia. In adolescence, typical SVTs are characterized by a sudden onset (often in rest) and sudden termination of tachycardia.

16.4.1.1 WPW Syndrome

Patients with WPW syndrome usually present with the same symptoms as of other forms of AVRT, but additionally, episodes of atrial fibrillation are reported in 20–30% of adult patients with WPW syndrome, while atrial fibrillation is uncommon in children.[43,44] These episodes of atrial fibrillation are clinically important as extremely rapid rates can occur over the bypass tract, leading to hemodynamic deterioration or ventricular fibrillation.

While sudden death (SD) in ventricular preexcitation syndrome is rare with an overall risk rate of 0.006 per patient-year, and mostly occurs in young and otherwise healthy individuals, it can be the first manifestation of the WPW syndrome in previously asymptomatic patients.[9] A clinical history of irregular and paroxysmal palpitations in a patient with baseline preexcitation suggests episodes of atrial fibrillation, which requires immediate electrophysiological evaluation as these patients are at risk for sudden death.[4]

Pathophysiologically, in most cases SD results from a rapid ventricular response to atrial fibrillation over an AP with a short refractory period.[45] In these cases, if atrial fibrillation (AF) develops, the normal rate-limiting effects of the AV node are bypassed as the impulse is conducted antegradely 1:1 through the AP, and the resultant excessive ventricular rates (sometimes 200–240 beats/min) may lead to ventricular fibrillation and, ultimately, sudden death. Atrial fibrillation might be secondary to AV reentrant tachycardia (AVRT)[45] or be triggered by a primary atrial pathology (such as atrial myocarditis), as was demonstrated in a clinicopathological series on 273 SDs in children and young adults (aged < 35 years), in which 50% of SD patients with ventricular preexcitation, additional to the AV bypass tract, showed isolated acute atrial myocarditis on histological examination.[46] Anamnestically, syncope or near-syncope in patients with WPW might indicate high risk of sudden death and would warrant risk stratification.

Risk stratification in patients with WPW includes exercise testing and programmed electrical stimulation

(PES) during invasive electrophysiologic testing. Individuals with WPW syndrome in whom the delta waves disappear with increases in the heart rate during exercise testing are considered at lower risk of SD. This is because the loss of the delta wave shows that the accessory pathway cannot conduct electrical impulses at a high rate (in the anterograde direction). These individuals will typically not have fast conduction down the accessory pathway during episodes of atrial fibrillation.[47] When delta waves do not abruptly disappear during exercise testing, an electrophysiological study to evaluate the properties of the AP and the propensity to develop AF or other tachycardia in WPW syndrome can be considered,[48] as recent studies have reported that asymptomatic young patients with induced tachycardia and short AP refractory period benefit from the preventive RF ablation of the AP.[49–51]

Patients are considered to be at high risk for ventricular fibrillation when the shortest interval between two subsequent preexcited ventricular beats during atrial fibrillation is less than 220 ms.[52,53] It is unclear whether invasive risk stratification (with programmed electrical stimulation) is necessary in the asymptomatic individual.[54] While some groups advocate PES for risk stratification in all individuals under 35 years of age, others offer it only to individuals who have history suggestive of a tachyarrhythmia as the incidence of SD is so low.[17,55]

16.5 Electrocardiographic Diagnosis

16.5.1 WPW Syndrome

While during sinus rhythm the ECG of patients with concealed bypasses is normal,[3] the typical ECG recorded during sinus rhythm in patients with WPW syndrome shows ventricular preexcitation manifested by: (1) a short PR interval, (2) a wide QRS complex, and (3) initial slurring (*delta wave*) of the QRS complex. In most patients, ventricular preexcitation is present at all times and all heart rates, while in some cases, preexcitation is intermittent. In the latter, preexcitation is usually present only at lower heart rates.[36] During tachycardia, in case of orthodromic AVRT, the ECG demonstrates narrow QRS complexes and a retrograde P wave embedded in the early portion of the T wave (best seen in leads II and V2), while the QRS complexes are very abnormally wide in antidromic AVRT, resembling those seen in ventricular tachycardia.[3] A fast, broad, irregular ("FBI") tachycardia on the ECG should trigger the awareness of atrial fibrillation with conduction over an accessory pathway (WPW).[56]

16.5.2 Mahaim Tachycardia

During sinus rhythm, the ECG in patients with Mahaim fibers is usually normal due to preferential ventricular activation via the AV node at normal rest heart rates. In contrast to the WPW syndrome, during sinus rhythm, no delta wave is present in patients with Mahaim fibers. Enhanced vagal tone or rapid pacing may however slow AV nodal conduction and favor conduction down the Mahaim fiber. Sometimes, a random discrete jump from AV nodal conduction to Mahaim conduction or a fusion complex of AV nodal and Mahaim conduction can be seen. In some patients, the relative refractoriness of the AV node and the Mahaim fiber may change with time, in which case, a variable degree of preexcitation may be seen on resting ECG. During tachycardia, the ECG demonstrates very abnormally wide QRS complexes (antidromic AVRT) with a left bundle branch block pattern, resembling those seen in ventricular tachycardia.[3,4]

16.6 Treatment

Regardless of SVT type, maneuvers that increase parasympathetic (vagal) tone, slow down conduction through the AV node, and break the reentry circuit responsible for SVT, apply to all patients. Different measures may be successful in different patients and include placing an ice-bag around the nose and mouth or abdomen, immersing in ice-cold water, applying the Valsalva maneuver (take a deep breath and bear down), pressuring on the abdomen, performing the carotid sinus massage, or blowing on a thumb.[3]

Acute treatment with intravenous administration of adenosine as a rapid bolus usually breaks the AVRT and is also safe in children of all ages. When symptoms resume, however, beta-blockers (e.g. propranolol), digoxin, procainamide, a calcium blocker (e.g.verapamil), or amiodarone should additionally be administered.[57] Patients with WPW syndrome should however not be treated with medications which enhance conduction along the bypass tract and hence increase the risk of atrial fibrillation degenerating in ventricular fibrillation, such as calcium channel blockers and digoxin.[10]

In chronic management of AVRTs, conservative treatment with class I anti-arrhythmic drugs, beta-blockers (class II), and/or amiodarone (class III) achieves a 70–80% efficacy in prevention of arrhythmia relapse.[58] General prophylactic treatment with antiarrhythmic medication is prescribed to infants younger than 1 year of age. At the age of 1 year, this prophylactic treatment is temporarily discontinued in order to see if tachycardia recurs.[59] After thorough evaluation of the risk/benefit ratio, radiofrequency catheter ablation (RFCA) has been shown to be a highly effective definitive treatment even in the pediatric age group.[4,13] Since the risk of fatal complications is estimated to be up to 0.3% in children less than 4–5 years of age, the Pediatric Radiofrequency Catheter Ablation Registry advises to restrict the indication for RF ablation to children older than 10–12 years.[13] Currently, main concerns in pediatric RF ablation focus on X-ray exposure and possible damage to the CCS,[13] in which respect, new systems of 3D mapping or the "non-contact" system and the introduction of cryoablation are very promising.[60–62]

During a RFCA procedure, the tissue is locally heated to 50–60°C through the small metal tip of a RF catheter by alternating current at 350 kHz to 1 MHz, producing a permanent small scar measuring approximately 4 mm in diameter and 4 mm in depth.[63] The initial success rates of RFCA exceed 95%.[4,11] Complications in RFCA are rare and result from radiation exposure, vascular access (hematomas, deep venous thrombosis, pneumothorax, and arterial perforation), catheter manipulation (valvular damage, micoremboli, perforation of the coronary sinus or myocardial wall, coronary artery dissection, thrombosis) or delivery of RF energy (AV-block, myocardial perforation, coronary artery spasm, transient ischemic attacks, cerebrovascular accidents).[4]

In acute management of patients with WPW and atrial fibrillation and rapid ventricular response, direct-current cardioversion is usually the treatment of choice. The usual rate-slowing drugs (e.g. beta blockers) used in AF are not effective in AF in WPW, and digoxin and the nondihydropyridine calcium channel blockers (e.g., verapamil or diltiazem) are contraindicated because they may increase the ventricular rate by enhancing conduction along the bypass tract and cause ventricular fibrillation. In emergencies, adenosine can be used to terminate AVRT in WPW syndrome. If cardioversion is impossible, drugs that prolong the refractory period of the accessory connection should be used. Intravenous procainamide or amiodarone is preferred, but any class Ia, class Ic, or class III antiarrhythmic can be used.[64] In chronic management of AF in WPW, medications that prolong AP refractory periods (flecainide, propafenone, and amiodarone), preventing rapid AP anterograde conduction (from atria to ventricles) in atrial tachycardias such as atrial fibrillation or flutter, are used.[65]

16.7 Genetics in AVRT

16.7.1 WPW Syndrome

16.7.1.1 Genes

In the majority of cases, WPW syndrome has no clear familial involvement (nonsyndromic or isolated preexcitation), while a minority of cases result from a clearly inherited disease or occur as part of a syndrome with a strong genetic basis (syndromic preexcitation). While in isolated WPW syndrome no disease genes have yet been identified, in syndromic preexcitation – WPW as a component of a more extensive cardiac disease or as part of a multisystem syndrome – multiple disease genes have been identified.[19]

The genetic basis of isolated WPW syndrome is still unkown. Vidaillet et al. however found that the prevalence of APs in first-degree relatives of 383 patients with proven ventricular preexcitation on the ECG was 3.4%, while the overall prevalence of WPW syndrome in the general population was estimated at 0.15%.[7] Patients with familial isolated ventricular preexcitation were found to have a higher incidence of multiple APs and an increased risk of SD compared to patients with ventricular preexcitation in the general population, and the pattern of inheritance appeared to be autosomal dominant.[7]

Syndromic preexcitation includes glycogen storage hypertrophic cardiomyopathy (HCM) associated with

WPW ventricular preexcitation syndrome and progressive CCS disease.[66] This has been identified as an autosomal dominant trait of which the gene has been localized on chromosome 7q34-q36.[67,68] A point mutation in the *PRKAG2 gene*, which encodes the regulatory γ-subunit of AMP-activated protein kinase (AMPK), results in the substitution of glutamine for arginine (R302Q).[67]

Another known syndrome associated with ventricular preexcitation includes Pompe disease, an autosomal recessive lysosomal glycogen storage disorder due to deficiency of acid alpha-1,4-glucosidase or acid maltase.[69,70] Danon disease is also a syndrome associated with ventricular preexcitation and is an X-linked lysosomal storage disorder with mutations in the Lysosome-associated protein 2 (LAMP2), which usually presents in young males with mental retardation, skeletal myopathy, and cardiac hypertrophy.[71] WPW is also frequently seen in patients with tuberous sclerosis associated with cardiac rhabdomyoma (autosomal recessive trait).[72] Mitochondrial syndromes, such as Leber's Hereditary Optic Neuropathy (LHON), are also associated with ventricular preexcitation.[73] Ebstein's anomaly, mapped to chromosome 11q in human, is yet another form of syndromic preexcitation inherited in autosomal dominant fashion, since WPW syndrome accompanies Ebstein's anomaly in at least 10% of patients.[16–18]

16.7.1.2 Animal Models for WPW Syndrome

Most electrophysiological characteristics of APs and their role in causing AVRT have been obtained from clinical studies in humans. Extensive electrophysiological experiments in dogs have provided additional detailed insights into ventricular preexcitation.[38] Moreover, since one of the major mutations for familial WPW syndrome was identified, transgenic technology has been used to generate transgenic models for WPW.[67,68]

Transgenic mice overexpressing the mutated PRKAG2 gene were recently found to nicely recapitulate the human phenotype of familial WPW syndrome and glycogen storage cardiac hypertrophy.[67,69] In this model, the mutated PRKAG2 gene, which encodes for the γ-2 subunit of AMP-activated-protein-kinase (AMPK), was cloned alongside a powerful alpha-myosin heavy chain promoter. Histopathology demonstrated glycogen-filled cardiomyocytes disrupting the annulus fibrosis and functionally giving rise to ventricular preexcitation. Additionally, in a similar transgenic mouse model overexpressing the PRKGA2 mutation, the preexcitation phenotype was reproduced and SVTs could be induced.[74]

Furthermore, deletion of the ALK3 gene (BMP receptor type IA) in mice, which encodes for the type 1a receptor for bone morphogenetic proteins (BMPs), in the AV canal cardiomyocytes during development causes ventricular preexcitation, possibly indicating an important role for the ALK3 gene in the etiology of the WPW syndrome.[75,76]

16.7.2 Genetics in Mahaim Tachycardia

While the role of Mahaim fibers in the genesis of arrhythmias has been extensively studied since their first description in the late 1930ss,[22] the cardiogenetics of Mahaim tachycardia remains unknown. Only one case description on the familial occurrence of Mahaim tachycardia in a mother and her son, suggesting genetic transmission, has been reported in the literature.[77] A possible morphologic and functional etiological explanation for the appearance of these Mahaim pathways involving the development of the right ventricular inflow tract and the *moderator band* in mouse cardiogenesis has been proposed for these right-sided APs.[35]

16.8 Summary

In this Chapter, the epidemiology, pathogenesis, clinical aspects, electrocardiographic diagnosis, therapy, and genetic aspects of atrioventricular reentrant tachycardia (AVRT), and in particular, the Wolff-Parkinson-White (WPW) syndrome, concealed bypasses, and Mahaim tachycardia are discussed. Atrioventricular reentrant tachycardia involves the presence of an accessory myocardial AV pathway.

- In Wolff-Parkinson-White (WPW) syndrome, the AP usually exhibits bidirectional conduction giving rise to ventricular preexcitation on ECG during sinus rhythm, involving a short PR interval, a wide QRS complex, and initial slurring of the QRS complex (delta wave).

- Mahaim fibers are capable of only antegrade conduction and usually do not give rise to preexcitation during sinus rhythm due to preferential conduction through the AV node.
- Concealed bypasses are usually capable of only retrograde nondecremental conduction. The ECG during sinus rhythm in a patient with a concealed accessory pathway is normal.
- Tachycardia in WPW syndrome can be both orthodromic and antidromic, while patients with Mahaim tachycardia demonstrate only antidromic AVRT and patients with concealed bypasses demonstrate only orthodromic AVRT.
- In general, treatment options in all types of AVRT include maneuvers that increase parasympathetic tone, pharmacological treatment, and radiofrequency catheter ablation (RFCA).
- While multiple genes for familial WPW syndrome (syndromic preexcitation) have been identified, the genetic basis of isolated WPW syndrome, Mahaim fibers, and other APs remains unknown.
- Risk stratification in patients with WPW including exercise testing and programmed electrical stimulation (PES) during invasive electrophysiologic testing should be considered in patients presenting with syncope and/or a family history of tachycardia or SD.
- Identification of WPW in more than one family member should prompt clinical evaluation of relatives for additional findings of ventricular hypertrophy or conduction abnormalities, since preexcitation and cardiomyopathy (e.g. PRKAG2, Danon disease, Pompe disese) and preexcitation and progressive AV conduction block (e.g. PRKAG2) is a rare combination and might imply an underlying mutation and autosomal dominant trait.

References

1. Ko JK, Deal BJ, Strasburger JF, Benson DW. Supraventricular tachycardia mechanisms and their age distribution in pediatric patients. *Am J Cardiol.* 1992;69(12):1028-1032.
2. Morady F. Catheter ablation of supraventricular arrhythmias: state of the art. *J Cardiovasc Electrophysiol.* 2004;15(1):124-139.
3. Kantoch MJ. Supraventricular tachycardia in children. *Indian J Pediatr.* 2005;72(7):609-619.
4. Blomström-Lundqvist C, Scheinman MM, Aliot EM, et al. European Society of Cardiology Committee N-HRS. ACC/AHA/ESC guidelines for the management of patients with supraventricular arrhythmias–executive summary. a report of the American college of cardiology/American heart association task force on practice guidelines and the European society of cardiology committee for practice guidelines (writing committee to develop guidelines for the management of patients with supraventricular arrhythmias) developed in collaboration with NASPE-Heart Rhythm Society. *J Am Coll Cardiol.* 2003;42(8):1493-1531.
5. Wellens HJ. Twenty-five years of insights into the mechanisms of supraventricular arrhythmias. *J Cardiovasc Electrophysiol.* 2003;14(9):1020-1025.
6. Blaufox AD, Rhodes JF, Fishberger SB. Age related changes in dual AV nodal physiology. *Pacing and clinical electrophysiology: PACE.* 2000;23(4 Pt 1):477-480.
7. Vidaillet HJ, Pressley JC, Henke E, Harrell FE, German LD. Familial occurrence of accessory atrioventricular pathways (preexcitation syndrome). *N Engl J Med.* 1987;317(2):65-69.
8. Miller JM. Therapy of Wolff-Parkinson-White syndrome and concealed bypass tracts: Part I. *J Cardiovasc Electrophysiol.* 1996;7(1):85-93.
9. Guize L, Soria R, Chaouat JC, Chrétien JM, Houe D, Le Heuzey JY. Prevalence and course of Wolf-Parkinson-White syndrome in a population of 138, 048 subjects. *Annales Méd Int.* 1985;136(6):474-478.
10. Goudevenos JA, Katsouras CS, Graekas G, Argiri O, Giogiakas V, Sideris DA. Ventricular pre-excitation in the general population: a study on the mode of presentation and clinical course. *Heart.* 2000;83(1):29-34.
11. Kolditz DP, Blom NA, Bökenkamp R, Bootsma M, Zeppenfeld K, Schalij MJ. Radiofrequency catheter ablation for treating children with cardiac arrhythmias: favourable results after a mean of 4 years. *Ned Tijdschr Geneeskd.* 2005;149(24):1339-1346.
12. Perry JC, Garson A. Supraventricular tachycardia due to Wolff-Parkinson-White syndrome in children: early disappearance and late recurrence. *J Am Coll Cardiol.* 1990;16(5):1215-1220.
13. Kugler JD, Danford DA, Houston KA, Felix G. Society PRARotPRARotPE. Pediatric radiofrequency catheter ablation registry success, fluoroscopy time, and complication rate for supraventricular tachycardia: comparison of early and recent eras. *J Cardiovasc Electrophysiol.* 2002;13(4):336-341.
14. Bauersfeld U, Pfammatter JP, Jaeggi E. Treatment of supraventricular tachycardias in the new millennium–drugs or radiofrequency catheter ablation? *Eur J Pediatr.* 2001;160(1):1-9.
15. Deal BJ, Keane JF, Gillette PC, Garson A. Wolff-Parkinson-White syndrome and supraventricular tachycardia during infancy: management and follow-up. *J Am Coll Cardiol.* 1985;5(1):130-135.
16. Mantakas ME, McCue CM, Miller WW. Natural history of Wolff-Parkinson-White syndrome discovered in infancy. *Am J Cardiol.* 1978;41(6):1097-1103.
17. Munger TM, Packer DL, Hammill SC, et al. A population study of the natural history of Wolff-Parkinson-White syndrome in Olmsted County, Minnesota, 1953–1989. *Circulation.* 1993;87(3):866-873.
18. Light PE. Familial Wolff-Parkinson-White Syndrome: a disease of glycogen storage or ion channel dysfunction? *J Cardiovasc Electrophysiol.* 2006;17(Suppl 1):S158-S161.

19. Ehtisham J, Watkins H. Is Wolff-Parkinson-White syndrome a genetic disease? *J Cardiovasc Electrophysiol.* 2005;16(11):1258-1262.
20. Braunwald E, Lambrew CT, Rockoff SD, Ross J, Morrow AG. Idiopathic hypertrophic subaortic stenosis. I. A description of the disease based upon an analysis of 64 patients. *Circulation.* 1964;30(Suppl 4):3-119.
21. Fananapazir L, Tracy CM, Leon MB, et al. Electrophysiologic abnormalities in patients with hypertrophic cardiomyopathy. A consecutive analysis in 155 patients. *Circulation.* 1989;80(5):1259-1268.
22. Mahaim I, Benatt A, JM M. Nouvelles recherches sur les connections superieures de la branche du fasceau de His-Tawara avec cloison intervntriculaire. *Cardiologica.* 1937;1:61–76.
23. Tchou P, Lehmann MH, Jazayeri M, Akhtar M. Atriofascicular connection or a nodoventricular Mahaim fiber? Electrophysiologic elucidation of the pathway and associated reentrant circuit. *Circulation.* 1988;77(4):837-848.
24. Haïssaguerre M, Cauchemez B, Marcus F, et al. Characteristics of the ventricular insertion sites of accessory pathways with anterograde decremental conduction properties. *Circulation.* 1995;91(4):1077-1085.
25. Bardy GH, Fedor JM, German LD, Packer DL, Gallagher JJ. Surface electrocardiographic clues suggesting presence of a nodofascicular Mahaim fiber. *J Am Coll Cardiol.* 1984;3(5):1161-1168.
26. Ellenborgen KARMPD. Accessory nodoventricular (Mahaim) fibers: a clinical review. *Pacing Clin Electrophysiol.* 1986;9:868.
27. Kuck KH, Friday KJ, Kunze KP. Basis for concealed accessory pathways. *Circulation.* 1990;82:407.
28. Kolditz DP, Wijffels MC, Blom NA, et al. Persistence of functional atrioventricular accessory pathways in postseptated embryonic avian hearts: implications for morphogenesis and functional maturation of the cardiac conduction system. *Circulation.* 2007;115(1):17-26.
29. Kolditz DP, Wijffels MC, Blom NA, van der Laarse A, Hahurij ND, Lie-Venema H, Markwald RR, Poelmann RE, Schalij MJ, Gittenberger-De Groot AC. Epicardium-derived cells in development of annulus fibrosis and persistence of accessory pathways. *Circulation.* 2008. 25;117(12):1508–1517
30. Gittenberger-de Groot AC, Vrancken Peeters MP, Mentink MM, Gourdie RG, Poelmann RE. Epicardium-derived cells contribute a novel population to the myocardial wall and the atrioventricular cushions. *Circ Res.* 1998;82(10):1043-1052.
31. Becker AE, Anderson RH, Durrer D, Wellens HJ. The anatomical substrates of wolff-parkinson-white syndrome A clinicopathologic correlation in seven patients. *Circulation.* 1978;57(5):870-879.
32. James TN, Puech P. De subitaneis mortibus. IX. Type A Wolff-Parkinson-White syndrome. *Circulation.* 1974;50(6):1264-1280.
33. Rossi L. Anatomic variants of AV junction and surgery of supraventricular arrhythmias. *Circulation.* 1980;62(4):916-917.
34. Gilette PC, GACD. Prolonged and decremental antegrade conduction properties in right anterior accessory connections: wide QRS antidromic tachycardia of left bundle branmch block pattern without Wolff=Parkinson-White configuration in sinus rhythm. *Am Heart J.* 1982;103:66.
35. Jongbloed MR, Wijffels MC, Schalij MJ, et al. Development of the right ventricular inflow tract and moderator band: a possible morphological and functional explanation for Mahaim tachycardia. *Circ Res.* 2005;96(7):776-783.
36. Obel OA, Camm AJ. Accessory pathway reciprocating tachycardia. *Eur Heart J.* 1998;19(Suppl E):E13–24, E50–11.
37. Prystowsky EN, Heger JJ, Zipes DP. The Wolff-Parkinson-White syndrome–diagnosis and treatment. *Heart Lung J Crit Care.* 1981;10(3):465-474.
38. Boineau JP, Moore EN. Evidence for propagation of activation across an accessory atrioventricular connection in types A and B pre-excitation. *Circulation.* 1970;41(3):375-397.
39. Gallagher JJ, Smith WM, Kasell JH, Benson DW, Sterba R, Grant AO. Role of Mahaim fibers in cardiac arrhythmias in man. *Circulation.* 1981;64(1):176-189.
40. Boldt T, Eronen M, Andersson S. Long-term outcome in fetuses with cardiac arrhythmias. *Obstet Gynecol.* 2003;102(6):1372-1379.
41. Kolditz DP, Blom NA, Bökenkamp R, Schalij MJ. Low-energy radiofrequency catheter ablation as therapy for supraventricular tachycardia in a premature neonate. *Eur J Pediatr.* 2005;164(9):559-562.
42. Juneja R, Shah S, Naik N, Kothari SS, Saxena A, Talwar KK. Management of cardiomyopathy resulting from incessant supraventricular tachycardia in infants and children. *Ind Heart J.* 2002;54(2):176-180.
43. Gallagher JJ, Svenson RH, Sealy WC, Wallace AG. The Wolff-Parkinson-White syndrome and the preexcitation dysrhythmias Medical and surgical management. *Med Clin North Am.* 1976;60(1):101-123.
44. Wellens HJ, Brugada P, Penn OC. The management of pre-excitation syndromes. *JAMA.* 1987;257(17):2325-2333.
45. Klein GJ, Bashore TM, Sellers TD, Pritchett EL, Smith WM, Gallagher JJ. Ventricular fibrillation in the Wolff-Parkinson-White syndrome. *N Engl J Med.* 1979;301(20):1080-1085.
46. Basso C, Corrado D, Rossi L, Thiene G. Ventricular preexcitation in children and young adults: atrial myocarditis as a possible trigger of sudden death. *Circulation.* 2001;103(2):269-275.
47. Bricker JT, Porter CJ, Garson A, et al. Exercise testing in children with Wolff-Parkinson-White syndrome. *Am J Cardiol.* 1985;55(8):1001-1004.
48. Garson A, Gillette PC. Electrophysiologic studies of supraventricular tachycardia in children. II. Prediction of specific mechanism by noninvasive features. *Am Heart J.* 1981;102(3 Pt 1):383-388.
49. Pappone C, Manguso F, Santinelli R, et al. Radiofrequency ablation in children with asymptomatic Wolff-Parkinson-White syndrome. *N Engl J Med.* 2004;351(12):1197-1205.
50. Pappone C, Santinelli V, Manguso F, et al. A randomized study of prophylactic catheter ablation in asymptomatic patients with the Wolff-Parkinson-White syndrome. *N Engl J Med.* 2003;349(19):1803-1811.
51. Pappone C, Santinelli V, Rosanio S, et al. Usefulness of invasive electrophysiologic testing to stratify the risk of arrhythmic events in asymptomatic patients with Wolff-Parkinson-White pattern: results from a large prospective long-term follow-up study. *J Am Coll Cardiol.* 2003;41(2):239-244.

52. Bromberg BI, Lindsay BD, Cain ME, Cox JL. Impact of clinical history and electrophysiologic characterization of accessory pathways on management strategies to reduce sudden death among children with Wolff-Parkinson-White syndrome. *J Am Coll Cardiol*. 1996;27(3):690-695.
53. Dubin AM, Collins KK, Chiesa N, Hanisch D, Van Hare GF. Use of electrophysiologic testing to assess risk in children with Wolff-Parkinson-White syndrome. *Cardiol Young*. 2002;12(3):248-252.
54. Campbell RM, Strieper MJ, Frias PA, Collins KK, Van Hare GF, Dubin AM. Survey of current practice of pediatric electrophysiologists for asymptomatic Wolff-Parkinson-White syndrome. *Pediatrics*. 2003;111(3):e245-247.
55. Fitzsimmons PJ, McWhirter PD, Peterson DW, Kruyer WB. The natural history of Wolff-Parkinson-White syndrome in 228 military aviators: a long-term follow-up of 22 years. *Am Heart J*. 2001;142(3):530-536.
56. Wellens HJJ. *ECG in Emergency Decision Making*. Philadelphia: W.B. Saunders; 2005.
57. Vaughan Williams EM. A classification of antiarrhythmic actions reassessed after a decade of new drugs. *J Clin Pharmacol*. 1984;24(4)):129-147.
58. Van Hare GF, Carmelli D, Smith WM, et al. Prospective assessment after pediatric cardiac ablation: design and implementation of the multicenter study. *PACE*. 2002; 25(3):332-341.
59. Weindling SN, Saul JP, Walsh EP. Efficacy and risks of medical therapy for supraventricular tachycardia in neonates and infants. *Am Heart J*. 1996;131(1):66-72.
60. Kirsh JA, Gross GJ, O'Connor S, Hamilton RM, Registry CIP. Transcatheter cryoablation of tachyarrhythmias in children: initial experience from an international registry. *J Am Coll Cardiol*. 2005;45(1):133-136.
61. Sporton SC, Earley MJ, Nathan AW, Schilling RJ. Electroanatomic versus fluoroscopic mapping for catheter ablation procedures: a prospective randomized study. *J Cardiovasc Electrophysiol*. 2004;15(3):310-315.
62. Van Hare GF, Dubin AM, Collins KK. Invasive electrophysiology in children: state of the art. *J Electrocardiol*. 2002;35(Suppl):165-174.
63. Haines DE. The biophysics of radiofrequency catheter ablation in the heart: the importance of temperature monitoring. *PACE*. 1993;16(3 Pt 2):586-591.
64. Fengler BT, Brady WJ, Plautz CU. Atrial fibrillation in the Wolff-Parkinson-White syndrome: ECG recognition and treatment in the ED. *Am J Emerg Med*. 2007;25(5):576-583.
65. Barlett HL, Loomis JL, Deno NS, Kollias J, Hodgson JL, Buskirk ER. A system for automatic end-tidal gas sampling at rest and during exercise. *J Appl Physiol*. 1973;35(2):301-303.
66. Wolf CMAM, Ahmad F, Sanbe A, et al. Reversibility of PRKAG2 glycogen-storage cardiomyopathy and electrophysiologic manifectations. *Circulation*. 2008;117(2): 144-154.
67. Gollob MH, Green MS, Tang AS, et al. Identification of a gene responsible for familial Wolff-Parkinson-White syndrome. *N Engl J Med*. 2001;344(24):1823-1831.
68. MacRae CA, Ghaisas N, Kass S, et al. Familial Hypertrophic cardiomyopathy with Wolff-Parkinson-White syndrome maps to a locus on chromosome 7q3. *J Clin Invest*. 1995;96(3):1216-1220.
69. Arad M, Moskowitz IP, Patel VV, et al. Transgenic mice overexpressing mutant PRKAG2 define the cause of Wolff-Parkinson-White syndrome in glycogen storage cardiomyopathy. *Circulation*. 2003;107(22):2850-2856.
70. Bulkley BH, Hutchins GM. Pompe's disease presenting as hypertrophic myocardiopathy with Wolff-Parkinson-White syndrome. *Am Heart J*. 1978;96(2):246-252.
71. Yang Z, McMahon CJ, Smith LR, et al. Danon disease as an underrecognized cause of hypertrophic cardiomyopathy in children. *Circulation*. 2005;112(11):1612-1617.
72. O'Callaghan FJ, Clarke AC, Joffe H, et al. Tuberous sclerosis complex and Wolff-Parkinson-White syndrome. *Arch Dis Child*. 1998;78(2):159-162.
73. Mashima Y, Kigasawa K, Hasegawa H, Tani M, Oguchi Y. High incidence of pre-excitation syndrome in Japanese families with Leber's hereditary optic neuropathy. *Clin Genet*. 1996;50(6):535-537.
74. Sidhu JS, Rajawat YS, Rami TG, et al. Transgenic mouse model of ventricular preexcitation and atrioventricular reentrant tachycardia induced by an AMP-activated protein kinase loss-of-function mutation responsible for Wolff-Parkinson-White syndrome. *Circulation*. 2005;111(1):21-29.
75. Gaussin V. Offbeat mice. *Anat Rec A Discov Mol Cell Evol Biol*. 2004;280(2):1022-1026.
76. Gaussin V, Morley GE, Cox L, et al. Alk3/Bmpr1a receptor is required for development of the atrioventricular canal into valves and annulus fibrosus. *Circ Res*. 2005;97(3):219-226.
77. Ott P, Marcus FI. Familial Mahaim syndrome. *Ann Noninvasive Electrocardiol*. 2001;6(3):272-275.

Hereditary Cardiac Conduction Diseases

17

Jean-Jacques Schott

17.1 Introduction

Cardiac conduction defects (CCD) are a group of serious and potentially life-threatening disorders.[1] CCD belongs to a group of pathologies with an alteration of cardiac conduction through the atrioventricular (AV) node, the His-Purkinje system with right or left bundle branch block, and widening of QRS complexes. CCD can lead to complete atrioventricular block (AV block) and cause syncope and sudden death.[1,2] Originally, CCD was considered as a structural disease of the heart, with anatomic changes in the conduction system underlying abnormal impulse propagation. In a substantial number of cases, however, conduction disturbances are found to occur in the absence of anatomical abnormalities. In these cases, functional, rather than structural alterations appear to underlie conduction disturbances. These functional defects are called 'primary electrical disease of the heart'.[3-6]

In the past decade, the field of molecular genetics has rapidly expanded and is becoming increasingly important in the area of cardiac electrophysiology and the genetic basis of cardiac conduction disturbances are now recognized as a critical factor in disease pathology.

The pathophysiological mechanisms underlying CCD are diverse, but the most frequent form of CCD is a degenerative form also called Lenègre-Lev disease[7,8] (idiopathic bilateral bundle-branch fibrosis), and two genes for this disorder have been identified. Besides isolated CCD, several other diseases, such as dilated or restrictive cardiomyopathies, congenital heart defects, Wolff-Parkinson-White syndrome, as well as Emery-Dreifus muscular dystrophy may be associated with CCD.[9,10]

The goal of this book is to provide the adult and pediatric cardiologist, geneticist, internist, pediatrician, and trainee, a comprehensive review and update of the field of cardiac conduction defects and to discuss the opportunity to propose genetic testing.

17.2 First Description of Cardiac Conduction Defect

Morgagni[11] was the first to report recurrent fainting episodes in a man occurring simultaneously with an observed slow pulse rate. In the nineteenth century, Adams[12] and Stokes[13] made similar observations. The first known report of an Adams-Stokes attack combined with ECG recordings came from van den Heuvel[14] who described a case of congenital heart block. Lenègre and Lev combined ECG findings, clinical observations, and detailed *postmortem* studies of the heart in the 1960s, thereby proving the direct relationship between complaints, CCD, and fibrosis of the conduction system. The names of Lenègre and Lev became synonymous with progressive cardiac conduction defect (PCCD). Lenègre-Lev disease is characterized by progressive alteration of cardiac conduction through the His-Purkinje system with right or left bundle branch block and widening of QRS complexes, eventually leading to complete atrioventricular block and causing syncope and sudden death.[15-17] In both diseases, a sclerodegenerative process causes fibrosis of the His Purkinje system. The severity and extent of the fibrotic lesions differ in Lev and Lenègre descriptions. For Lenègre, histological studies identified

J.-J. Schott
L'institut du thorax, Inserm U915, Faculte de Medecine,
Nantes, France
e-mail: jean-jacques.schott@univ-nantes.fr

In 2001, Tan et al.[37] provided a functional characterization of a *SCN5A* mutation that causes a sustained, isolated conduction defect with pathological slowing of the cardiac rhythm. ECG findings reported bradycardia, AV-nodal escape, broad P wave, long PR, and wide QRS. By analyzing the *SCN5A* coding region, they reported a *SCN5A*-G514C mutation in five affected family members. Biophysical characterization of the mutant channel showed abnormal voltage-dependent "gating" behavior.

In 2002, Wang et al.,[38] reported two new *SCN5A* mutations that result in AV conduction block presented during childhood. Molecular genetic studies revealed a first *SCN5A*-G298S mutation in a proband with progressive AV block (QRS = 135 ms at age 9 and QRS = 133 ms at age 20). A second *SCN5A*-D1595N mutation was identified in a proband with complete heart block at the age of 12 years. The functional consequences of the two mutations are impaired fast inactivation but not sustained noninactivating current. The mutations reduce the sodium current density and enhance slow inactivation components.

Since its first description in 1999, over a dozen reports have identified new *SCN5A* mutations causing PCCD or nonprogressive CCD sometimes associated with dilated cardiomyopathy or other arrhythmias[39–44].

17.3.1.3 SCN5A Overlap Syndrome

Cardiac sodium channel dysfunction caused by mutations in the *SCN5A* gene is associated with a number of relatively uncommon arrhythmia syndromes, including long-QT syndrome type 3 (LQT3), Brugada syndrome, conduction disease, sinus node dysfunction and atrial standstill which potentially leads to fatal arrhythmias in relatively young individuals. Although these various arrhythmia syndromes were originally considered separate entities, recent evidence indicates more overlap in clinical presentation and biophysical defects of associated mutant channels than previously appreciated. Various *SCN5A* mutations are now known to present with mixed phenotypes, a presentation that has become known as "overlap syndrome of cardiac sodium channelopathy." In many cases, multiple biophysical defects of single *SCN5A* mutations are suspected to underlie the overlapping clinical manifestations.[45]

17.3.1.4 Sodium Channel beta1 Subunit Mutations Associated with Cardiac Conduction Disease

Regulatory proteins of cardiac-specific sodium channels Nav1.5 are logical candidates for CCD. Sodium channels are multisubunit protein complexes composed not only of pore-forming α subunits but also of multiple other protein partners including auxiliary function-modifying β subunits. The β1 transcript arises from the splicing of exons 1–5 of *SCN1B* gene, and a second transcript has been described that arises from the splicing of exons 1–3, with retention of a segment of intron 3 (termed exon 3A) leading to an alternate 3′ sequence (β1B transcript).

Watanabe et al. (2008)[46] screened 44 probands with conduction disease for mutations in *SCN1B* and β1B transcript. They reported 1 *SCN1B* and 2 β1B mutations in three kindreds with conduction abnormalities.

A missense mutation, c.259G→C in exon 3, resulting in p.Glu87Gln within the extracellular immunoglobulin loop of the protein was identified in a Turkish kindred affected by conduction disease. None of the families had a history of syncope, sudden cardiac death, or epilepsy (also of relevance since SCN1B mutations may also cause generalized epilepsy). The proband was a 50-year-old woman who presented with palpitations and dizziness and had complete left bundle branch block. Electrophysiologic analysis revealed a prolonged His-ventricle interval of 80 ms and inducible atrioventricular nodal reentrant tachycardia, complete atrioventricular block occurred following atrial programmed stimulation and during induced tachycardia. Her brother was found to have bifascicular block (right bundle branch block and left anterior hemiblock), whereas their mother had a normal ECG.

Two nonsense mutations in the same codon were identified in two other families: A c.536G→A in exon 3A in a French kindred affected with Brugada syndrome and conduction disease. This mutation results in p.Trp179X and is predicted to generate a prematurely truncated protein that lacks the membrane-spanning segment and intracellular portion of the protein. The French proband was a 53-year-old man who presented with chest pain but had normal coronary angiography and echocardiography. ECG revealed ST segment elevation typical of Brugada syndrome as well as conduction abnormalities, including a prolonged PR interval of 220 ms and left anterior hemiblock. A baseline ECG in his brother, who had no history of palpitations or syncope, showed left anterior hemiblock and minor ST segment elevation suggestive of Brugada syndrome (type II saddleback) abnormalities. Their sister had a normal ECG, but her son was found to have right bundle branch block and type II Brugada syndrome after flecainide challenge.

A different nonsense mutation, c.537G → A in exon 3A resulting in p.Trp179X, affecting the same codon as in family 2, was identified in a Dutch kindred. The Dutch proband was a 17-year-old girl who had right bundle branch block and a prolonged PR interval of 196 ms on ECG; echocardiography was normal and a flecainide test for Brugada syndrome was negative. Her father had a normal ECG.

Both the canonical and alternately processed transcripts were expressed in the human heart and were expressed to a greater degree in Purkinje fibers than in heart muscle, consistent with the clinical presentation of conduction disease. Sodium current was lower when Nav1.5 was coexpressed with mutant β1 or β1B subunits than when it was coexpressed with WT β1 or β1B subunits. These findings implicate *SCN1B* as a disease gene for human arrhythmia susceptibility.

17.3.2 Clinical and Molecular Description of Lamin A/C Mutations in Dilated Cardiomyopathies Associated with Conduction Defects

Over the last several decades, it has become increasingly clear that the cause of many 'idiopathic' dilated cardiomyopathies is genetic. Familial Dilated Cardiomyopathy (FDC) is now thought to account for up to 50% of Idiopathic Dilated Cardiomyopathy (IDC) patients. Most of these cases (>90%) are thought to show autosomal dominant inheritance, although X-linked and autosomal recessive forms have been identified.[47]

More than 20 different genes have been identified in patients with FDC. These include genes that affect sarcomeric proteins, cytoskeletal proteins, nuclear proteins, ion channel proteins, and mitochondria[48, 49]. Although up to 50% of patients with IDC may have Familial Dilated cardiomyopathy (FDC) by history, a genetic test will identify the cause in only a small minority of patients.

The exception to this is lamin A/C mutation. Several mutations in *LMNA* have been identified in five families with DCM with conduction system disease by Fatkin et al. in 1999[50] and Lamin A/C mutations are thought to be the cause in up to 10% of FDC cases.[51]

Lamin A and C are type V intermediate filament proteins found in the nuclear membrane or lamina.[52, 53] They are expressed in terminally differentiated somatic cells and found in multiple different tissues, including skeletal and cardiac muscle. The protein is composed of a conserved rod domain with a globular head and tail region. The functions of lamin A and C are incompletely understood, but it is thought that they are important in maintaining nuclear architecture.

The earliest cardiac finding in patients with lamin A/C deficiency is usually conduction system disease. In a meta-analysis of 299 carriers of a lamin A/C mutation, 18% of patients less than 10 years of age had evidence of delayed intracardiac conduction. In patients over 30 years of age, 92% had conduction system disease, with 44% requiring pacemaker placement.[51]

In early stages, patients have a characteristic electrocardiogram with low amplitude P waves, prolonged PR interval, but a relatively normal QRS complex. By age 50, over 60% of patients have symptoms of cardiac heart failure. Meune et al.[54] found that, in a small cohort of patients with known lamin A/C defects, 42% of patients experienced SCD.

17.3.3 Clinical and Molecular Description of Secundum Atrial Septal Defects Associated with Atrioventricular Conduction Defects

Developmental aspects of the cardiac conduction system are complex and play a crucial role in the pathophysiology of CCD. Although several genes involved in mature conduction system function have been identified, their association with development of specific subcomponents of the cardiac conduction system remains challenging. Several transcription factors, including homeodomain proteins and T-box proteins, are essential for cardiac conduction system morphogenesis and activation or repression of key regulatory genes. In addition, several transcription factors modify expression of genes encoding the ion channel proteins that contribute to the electrophysiological properties of the conduction system and govern contraction of the surrounding myocardium.

Four families with secundum atrial septal defects and atrioventricular conduction defects[55] were found by Schott et al.[56] to show an autosomal dominant pedigree pattern and linkage to 5q35, and to have, in affected individuals, mutations in the *CSX* gene which encodes the cardiac-specific homeobox transcription factor NKX2–5. Since then, numerous mutations in *NKX2.5* have been reported in isolated or familial forms of congenital heart defects.[57-59]

17.3.4 Molecular Screening in Cardiac Conduction Defects

Cardiac conduction defect is a heritable heterogeneous condition of still mostly unknown origin. It is likely that the most common forms of CCD have a mutifactorial origin, combining environmental and genetic factors.

Screening for mutations in patients with cardiac conduction defects requires a high index of suspicion.

Today, systematic genetic testing is not recommended as the few genes identified do not allow risk stratification, except for lamin A/C mutations in idiopathic dilated cardiomyopathy (IDC), which predispose the patient to increased mortality. Patients with idiopathic dilated cardiomyopathy with at least one other first-degree relative with IDC should be considered for genetic screening in lamin A/C gene. Most patients with lamin A/C deficiency do have conduction system disease as one of the first features, and up to 30% of patients with DCM and conduction system disease are thought to have lamin A/C deficiency. Identifying a specific genetic mutation within a family can be important for family planning and directing follow-up for the family members. Family members of probands should undergo genetic counseling.

Genetic screening in *NKX2.5* gene in patients with congenital heart defects can also be envisioned. Genetic screening of *SCN5A* and *TRPM4* genes in isolated forms of conduction defects should not be undertaken unless a clear familial history of conduction disease is documented.

17.3.5 Conclusions

The most frequent form of progressive cardiac conduction disease is Lenegre-Lev disease, which is a frequently occurring degenerative disease with an expected increasing prevalence given the general aging of the population. It is a heritable heterogeneous condition of mostly unknown origin. It is likely that this most common form of PCCD has a multifactorial origin, combining environmental and genetic factors. However, in some cases, familial PCCD with a monogenetic origin and an autosomal dominant mode of inheritance has been described. In most cases, this is due to a pathogenic mutation in the alfa-subunit of the *SCN5A* gene. Progressive familial heart block is another form of hereditary cardiac conduction disease due to mutations in the *TRPM4* gene. In addition to isolated cases of cardiac conduction diseases, there are several types of conduction diseases that occur in conjunction with other pathologies (among others, dilated cardiomyopathy due to lamine A/C mutations and several congenital heart defects). In selected cases, molecular genetic testing is indicated.

Today, pacemaker implantation constitutes the only therapeutic treatment to prevent sudden death in CCD. It is conceivable that, in the near future, once the precise pathophysiological mechanism for PCCD is identified, less invasive approaches will become available, improving patient management.

References

1. Michaelsson M, Jonzon A, Riesenfeld T. Isolated congenital complete atrioventricular block in adult life A prospective study. *Circulation*. 1995;92(3):442-449.
2. Balmer C, Fasnacht M, Rahn M, et al. Long-term follow up of children with congenital complete atrioventricular block and the impact of pacemaker therapy. *Europace*. 2002;4(4): 345-349.
3. Tan HL, Bezzina CR, Smits JP, Verkerk AO, Wilde AA. Genetic control of sodium channel function. *Cardiovasc Res*. 2003;57(4):961-973.
4. Roden DM, Balser JR, George AL Jr, et al. Cardiac ion channels. *Annu Rev Physiol*. 2002;64:431-475.
5. Roden DM, George AL Jr. Structure and function of cardiac sodium and potassium channels. *Am J Physiol*. 1997;273 (2 Pt 2):H511-H525.

6. Benson DW. Genetics of atrioventricular conduction disease in humans. *Anat Rec A Discov Mol Cell Evol Biol.* 2004;280(2):934-939.
7. Lenegre J. The pathology of complete atrio-ventricular block. *Prog Cardiovasc Dis.* 1964;6:317-323.
8. Lev M, Kinare SG, Pick A. The pathogenesis of complete atrioventricular block. *Prog Cardiovasc Dis.* 1964;6:317-326.
9. Hanson EL, Jakobs PM, Keegan H, et al. Cardiac troponin T lysine 210 deletion in a family with dilated cardiomyopathy. *J Card Fail.* 2002;8:28-32.
10. Oropeza ES, Cadena CN. New phenotype of familial dilated cardiomyopathy and conduction disorders. *Am Heart J.* 2003;145:317-323.
11. Morgagni GB. De sedibus, et causis morborum per anatomen indagatis libri quinque. 2 volums. In 1.Venetis, typ. Remondiniana 1761.
12. Adams R. Cases of disease of the heart, accompanied with pathological observation. *Dublin Hospital reports.* 1827; 4:353-453.
13. Stokes W. Observations on some cases of permanently slow pulse. *Quat J Med Sci.* 1846;2:73-85.
14. van den Heuvel GCJ. Die ziekte van Stokes-Adams en een geval van aangeboren hart blok. Groningen 1908.
15. Lev M, Kinare SG, Pick A. The pathogenesis of atrioventricular block in coronary disease. *Circulation.* 197;42: 409-425.
16. Bharati S, Lev M, Dhingra RC, et al. Electrophysiologic and pathologic correlations in two cases of chronic second degree atrioventricular block with left bundle branch block. *Circulation.* 1975;52(2):221-229.
17. Lev M, Cuadros H, Paul MH. Interruption of the atrioventricular bundle with congenital atrioventricular block. *Circulation.* 1971;43(5):703-710.
18. Morquio L. Sur une maladie infantile et familiale caracterisee par des modifications permanentes du pouls, des attaques syncopales et epileptiformes et la mort subite. *Arch Med Enfants.* 1901;4:467-475.
19. Osler W. On the so-called Stokes-Adams disease. *Lancet.* 1903;II:516-524.
20. Fulton ZMK, Judson CF, Norris GW. Congenital heart block occurring in a father and two children, one an infant. *Am J Med Sci.* 1910;140:339-348.
21. Wallgren A, Winblad S. Congenital heart-block. *Acta Paediat.* 1937;20:175-204.
22. Wendkos MH. Familial congenital complete A-V heart blocks. *Am Heart J.* 1947;34:138-142.
23. Gazes PC, Culler RM, Taber E, et al. Congenital familial cardiac conduction defects. *Circulation.* 1965;32:32-34.
24. Combrink JMD, Snyman HW. Familial bundle branch block. *Am Heart J.* 1962;64:397-400.
25. Steenkamp WF. Familial trifascicular block. *Am Heart J.* 1972;84:758-760.
26. Brink AJ, Torrington M. Progressive familial heart block—two types. *S Afr Med J.* 1977;52:53-59.
27. Van der Merwe PL, Weymar HW, Torrington M, Brink AJ. Progressive familial heart block. Part II. Clinical and ECG confirmation of progression-report on 4 cases. *S Afr Med J.* 1986;70(6):356-357.
28. Van der Merwe PL, Weymar HW, Torrington M, Brink AJ. Progressive familial heart block (type I). A follow-up study after 10 years. *S Afr Med J.* 1988;73:275-276.
29. Stephan E. Hereditary bundle branch system defect: survey of a family with four affected generations. *Am Heart J.* 1978;95:89-95.
30. Stephan E, de Meeus A, Bouvagnet P. Hereditary bundle branch defect: right bundle branch blocks of different causes have different morphologic characteristics. *Am Heart J.* 1997;133:249-256.
31. Brink PA, Ferreira A, Moolman JC, et al. Gene for progressive familial heart block type I maps to chromosome 19q13. *Circulation.* 1995;91:1633-1640.
32. de Meeus A, Stephan E, Debrus S, et al. An isolated cardiac conduction disease maps to chromosome 19q. *Circ Res.* 1995;77(4):735-740.
33. Kruse M, Schulze-Bahr E, Corfield V, et al. Impaired endocytosis of the ion channel TRPM4 is associated with human progressive familial heart block type I. *J Clin Invest.* 2009; 119(9):2737-2744.
34. Hui Liu, Loubna EL ZEIN, Martin Kruse, et al. Gain-of-function mutations in TRPM4 cause autosomal dominant isolated cardiac conduction disease. *Circulation*: Cardiovascular Genetics 2010;3:374-385 Published online before print June 19, 2010, doi: 10.1161/CIRCGENETICS.109.930867
35. Schott JJ, Alshinawi C, Kyndt F, et al. Cardiac conduction defects associate with mutations in SCN5A. *Nat Genet.* 1999;23:20-21.
36. Probst V, Kyndt F, Potet F, et al. Haploinsufficiency in combination with aging causes SCN5A-linked hereditary Lenegre disease. *J Am Coll Cardiol.* 2003;41(4):643-652.
37. Tan HL, Bink-Boelkens MT, Bezzina CR, et al. A sodium-channel mutation causes isolated cardiac conduction disease. *Nature.* 2001;409:1043-1047.
38. Wang DW, Viswanathan PC, Balser JR, et al. Clinical, genetic, and biophysical characterization of SCN5A mutations associated with atrioventricular conduction block. *Circulation.* 2002;105:341-346.
39. Bezzina CR, Rook MB, Groenewegen WA, et al. Compound heterozygosity for mutations (W156X and R225W) in SCN5A associated with severe cardiac conduction disturbances and degenerative changes in the conduction system. *Circ Res.* 2003;92:159-168.
40. Niu DM, Hwang B, Hwang HW, et al. A common SCN5A polymorphism attenuates a severe cardiac phenotype caused by a non-sense SCN5A mutation in a Chinese family with an inherited cardiac conduction defect. *J Med Genet.* 2006;43(10):817-821.
41. Viswanathan PC, Benson DW, Balser JR. A common SCN5A polymorphism modulates the biophysical effects of an SCN5A mutation. *J Clin Invest.* 2003;111(3):341-346.
42. Makita N, Sasaki K, Groenewegen WA, et al. Congenital atrial standstill associated with coinheritance of a novel SCN5A mutation and connexin 40 polymorphisms. *Heart Rhythm.* 2005;2(10):1128-1134.
43. McNair WP, Ku L, Taylor MR, et al. SCN5A mutation associated with dilated cardiomyopathy, conduction disorder, and arrhythmia. *Circulation.* 2004;110(15):2163-2167.

44. Olson TM, Michels VV, Ballew JD, et al. Sodium channel mutations and susceptibility to heart failure and atrial fibrillation. *JAMA*. 2005;293(4):447-454.
45. Remme CA, Wilde AA, Bezzina CR. Cardiac sodium channel overlap syndromes: different faces of SCN5A mutations. *Trends Cardiovasc Med*. 2008 Apr;18(3):78-87.
46. Watanabe H et al. Sodium channel beta1 subunit mutations associated with Brugada syndrome and cardiac conduction disease in humans. *J Clin Invest*. 2008;118:2260-2268.
47. Mestroni L, Rocco C, Gregori D, et al. Familial dilated cardiomyopathy: evidence for genetic and phenotypic heterogeneity. *J Am Coll Cardiol*. 1999;34:181-190.
48. Karkkainen S, Peuhkurinen K. Genetics of dilated cardiomyopathy. *Ann Med*. 2007;39:91-107. A comprehensive review of the known genetic mutations that have been shown to cause FDC.
49. Burkett EL, Hershberger RE. Clinical and genetic issues in familial dilated cardiomyopathy. *J Am Coll Cardiol*. 2005;45:969-981.
50. Fatkin D, MacRae C, Sasaki T, et al. Missense mutations in the rod domain of the lamin A/C gene as causes of dilated cardiomyopathy and conduction system disease. *N Engl J Med*. 1999;341:1715-1724.
51. van Berlo JH, de Voogt WG, van der Kooi AJ, et al. Meta-analysis of clinical characteristics of 299 carriers of LMNA gene mutations: do lamin A/C mutations portend a high risk of sudden death? *J Mol Med*. 2005;83:79-83.
52. Gruenbaum Y, Goldman RD, Meyuhas R, et al. The nuclear lamina and its functions in the nucleus. *Int Rev Cytol*. 2006;226:1-62.
53. Shumaker DK, Kuczmarski ER, Goldman RD. The nucleoskeleton: lamins and actin are major players in essential nuclear functions. *Curr Opin Cell Biol*. 2003;15:358-366.
54. Meune C, van Berlo J, Anselme F, et al. Primary prevention of sudden death in patients with lamin A/C gene mutations. *N Engl J Med*. 2006;354:209-210.
55. Pease WE, Nordenberg A, Ladda RL. Genetic counselling in familial atrial septal defect with prolonged atrio-ventricular conduction. *Circulation*. 1976;53:759-762.
56. Schott J-J, Benson DW, Basson CT, et al. Congenital heart disease caused by mutations in the transcription factor NKX2–5. *Science*. 1998;281:108-111.
57. Watanabe Y, Benson DW, Yano S, Akagi T, Yoshino M, Murray JC. Two novel frameshift mutations in NKX2.5 result in novel features including visceral inversus and sinus venosus type ASD. *J Med Genet*. 2002;39:807-811.
58. Hirayama-Yamada K, Kamisago M, Akimoto K, et al. Phenotypes with GATA4 or NKX2.5 mutations in familial atrial septal defect. *Am J Med Genet*. 2005;135A:47-52.
59. Gutierrez-Roelens I, De Roy L, Ovaert C, Sluysmans T, Devriendt K, Brunner HG, Vikkula M. A novel CSX/NKX2–5 mutation causes autosomal-dominant AV block: are atrial fibrillation and syncopes part of the phenotype?

Part V

Other Hereditary Cardiac Diseases and Conditions

Connective Tissue Disorders and Smooth Muscle Disorders in Cardiology

K. van Engelen and B.J.M. Mulder

18.1 Introduction

The *hereditary disorders of connective tissue* encompass a spectrum of clinically and genetically heterogeneous conditions caused by genetic defects in structural connective tissue proteins, such as collagen, fibrillin, and elastin. Clinical manifestations of these disorders are variable, and include musculoskeletal, skin, ocular, cardiovascular, and other visceral pathologies. Substantial overlap in clinical features between different disorders of connective tissue is present. Many connective tissue disorders constitute a risk of aneurysm formation and dissection of the aorta and/or other arteries, due to a defect of connective tissue within the vessel wall. Marfan syndrome is the most common syndromic presentation of ascending aortic aneurysm, but other syndromes including vascular Ehlers–Danlos syndrome and Loeys–Dietz syndrome also show ascending aortic aneurysms and the associated risk of aortic dissection and rupture. Familial segregation of the risk for ascending aortic aneurysm can also occur in the absence of associated systemic findings of connective tissue abnormalities in patients with familial Thoracic Aortic Aneurysms and aortic Dissections (TAAD), with or without structural heart defects such as bicuspid aortic valve, aortic coarctation, or patent ductus arteriosus.

Over the past years, the molecular basis and pathogenesis of connective tissue disorders have been studied extensively, leading to an increased understanding of the underlying mechanisms of disease. Mutations in the fibrillin-1 (FBN1) gene are the major cause of Marfan syndrome, although mutations in this gene have also been found in the so-called Marfan-related disorders, such as TAAD, familial ectopia lentis, or MASS (Mitral valve prolapse, Myopia, borderline and nonprogressive Aortic root enlargement, and nonspecific Skin and Skeletal findings) phenotype. Mutations in the genes encoding the transforming growth factor-β (TGF-β) receptors 1 and 2 (TGFBR1 and TGFBR2, respectively) have also been found in patients with a spectrum of phenotypic manifestations, from relatively mild familial TAAD, on the one hand, to the severe Loeys–Dietz syndrome with aggressive dissections and ruptures in early childhood on the other hand, as well as in patients fulfilling the clinical criteria for Marfan syndrome.[1-3] The importance of TGF-β signaling in the pathogenesis of connective tissue disorders has now been established.

In this chapter, the focus will be on Marfan syndrome, the most common and most well-known hereditary disorder of connective tissue. In addition, other connective tissue disorders that may show cardiovascular involvement will be discussed shortly.

18.2 Marfan Syndrome

Marfan syndrome is a common yet under-recognized autosomal dominant systemic disorder of connective tissue, caused by heterozygous mutations in the FBN1 gene at 15q21.1, which encodes the extracellular matrix protein fibrillin-1. The disorder shows characteristic but highly variable manifestations in mainly the cardiovascular, ocular, and musculoskeletal systems. It was first reported in 1896, when Antoine-Bernard Marfan described a young girl with unusual musculoskeletal features.[4] The estimated

K. van Engelen (✉)
Department of Clinical Genetics and Department of Cardiology,
Academic Medical Center, Amsterdam, The Netherlands
e-mail: k.vanengelen@amc.uva.nl

prevalence of Marfan syndrome is one in 3,000–5,000 individuals, with no ethnic or gender predilection.

18.2.1 Etiology

18.2.1.1 Molecular Genetics

FBN1 is a large gene with 65 exons coding for *fibrillin-1*,[5,6] a 320- kD glycoprotein consisting of 2,871 amino acids. It is highly conserved among different species. The polypeptide comprises 47 repeated cysteine-rich motifs resembling epidermal growth factor (EGF-like) and an 8-cysteine motif (TB/8-cys). Forty-three of the 47 motifs contain a consensus sequence for calcium binding and are termed calcium-binding EGF-like motifs (cbEGF-like).[7]

Fibrillin molecules polymerize to form the *microfibrils*, a constituent of the *extracellular matrix*. Microfibrils can associate with elastin to form *elastic fibers*. The microfibrils and elastic fibers have a widespread distribution in connective tissue throughout the body, including the skin, vascular wall, tendons, fascia, alveolar wall, and ciliary zonules that suspend the ocular lens, where they provide force-bearing structural support needed by these individual organ systems. In addition, it has become increasingly clear that fibrillin-rich microfibrils have functions that are not directly related to structural integrity but rather have to do with homeostasis of the elastic matrix, matrix-cell attachments, and regulation of cytokines.[7]

To date, about 1,000 different mutations have been reported in FBN1. Most of the reported mutations are missense mutations affecting the conserved cysteine residues or residues of the consensus sequence of the cbEGF-like motifs. Nonsense mutations, mutations of splice sites associated with exon skipping, and more rarely, small deletions are also found.[8] The majority of mutations are private mutations that are unique to a family or an individual patient. Approximately 25% of cases of Marfan syndrome are caused by de novo mutations.

Many mutations in FBN1 are believed to adversely affect the normal, wild-type gene product, that is, they are thought to have a dominant-negative effect. However, Marfan syndrome and related disorders can also be caused by mutations that prevent or reduce the expression of the mutant allele. Haplo insufficiency may therefore also contribute to the pathogenesis.[9]

The mutation detection rate in patients fulfilling the diagnostic criteria for Marfan syndrome is currently about 90%.[10] In the remaining 10% of patients, no causal mutation or deletion in FBN1 can be identified. In those cases, the causative mutation in FBN1 might not be detectable with conventional techniques, or these patients may harbor a mutation in another (unknown) gene.

In a few FBN1-negative patients fulfilling the criteria of Marfan syndrome, as well as in patients with "incomplete" Marfan syndrome, mutations in the transforming growth factor-β receptor-2 (TGFBR2) and transforming growth factor-β receptor-1 (TGFBR1) genes have been reported.[3,11] Major cardiovascular involvement was almost always present in these patients, while ectopia lentis was rare.

18.2.1.2 Pathophysiology

Early theories of disease pathogenesis in Marfan syndrome assumed that the manifestations of the disease were caused by the loss of structural integrity of affected tissues, due to mutant fibrillin-1 in the microfibrils. This was thought to result in weak tissue that could not withstand enduring stress over time. Some features of Marfan syndrome could indeed be explained by these models, such as aortic aneurysms, ectopia lentis, and dural ectasia. However, other features, including bone overgrowth, craniofacial features, and myxomatous changes of the mitral valve did not seem compatible with these theories. It has become increasingly clear that other mechanisms contribute to the pathogenesis of Marfan syndrome.

Animal studies have shown that fibrillin-rich microfibrils have an essential role in homeostasis of the elastic matrix during postnatal life.[12] Elastic fibers have intimate connections with adjacent vascular endothelial cells and smooth muscle cells, mediated by fibrillin 1. As a result of defective fibrillin 1, these connections may be absent or inadequate. In mice, this resulted in abortive matrix remodeling, characterized by overproduction of multiple structural components and matrix-degrading enzymes, including *metalloproteinases* 2 and 9. Subsequent events are infiltration of inflammatory cells, intimal hyperplasia, elastic fiber calcification, and structural collapse of the vessel wall leading to aneurysm formation.[12] These manifestations have also been observed in pathologic specimens from patients with Marfan syndrome.[13]

Fibrillin-rich microfibrils also play a significant role in the regulation of *cytokines*. *Transforming growth factors-β (TGF-βs)* are multifunctional cytokines that can induce many cellular events including proliferation, differentiation, cell cycle arrest, programmed cell death, and matrix deposition.[14] The activation of TGF-β is limited by fibrillin 1. It was therefore hypothesized that abnormal fibrillin 1 or reduced levels of fibrillin 1 result in excessive amounts of active TGF-β.[15] Subsequently, an increased output of TGF-β-responsive genes, such as collagen and connective tissue growth factor, and altered cellular events lead to the phenotypic manifestations of Marfan syndrome. Support for this hypothesis has been found in several studies in mice and humans, with excess TGF-β signaling being observed in the developing lung, mitral valve, and ascending aorta.[15–17] However, the precise mechanism by which fibrillin-1 controls TGF-β activation remains unknown.

18.2.2 Clinical Aspects

Despite advances in molecular genetic screening, the diagnosis of Marfan syndrome is still primarily made on clinical grounds. The evolving understanding of the genetic and clinical features of Marfan syndrome has led to a revision of the Berlin diagnostic criteria of 1988, leading to the currently used *Ghent nosology* in 1996.[18] In the Ghent nosology, clinical features are assessed within six organ systems, and these are combined with familial or genetic findings (Table 18.1). The presence of a major criterion in two different systems and a minor criterion in a third system are required to make the diagnosis of Marfan syndrome.

Marfan syndrome shows a high penetrance, but marked inter- and intrafamilial variability. The disorder should be regarded as a spectrum of diverse and highly variable manifestations in the different organs, and not all patients display the classic habitus. Most of the manifestations have an age-dependent penetrance. Moreover, many of the physical findings are also encountered in the general population or in other connective tissue disorders. This variability in clinical expression, with manifestations that emerge from childhood onward, and the high rate of de novo mutations can pose difficulties in establishing the diagnosis in some patients, particularly in younger individuals with few symptoms.[19] The current Ghent criteria are now being revised worldwide to optimize the diagnostic process. A multidisciplinary approach, including clinical genetics, cardiology, ophthalmology, and radiology is essential in establishing the diagnosis, treatment, and follow-up. When the diagnosis has been established in an individual, first-degree relatives should be screened for the disorder as well.

18.2.2.1 Skeletal System

The most striking skeletal manifestation is often the overgrowth of the long bones, leading to the characteristic appearance of patients with Marfan syndrome. The extremities are disproportionately long for the size of the trunk (*dolichostenomelia*), which leads to an increased arm-span-to-height and upper-lower segment ratio (Fig. 18.1). The fingers and toes are long and thin (*arachnodactyly*) and in combination with hypermobility of the joints this leads to the characteristic wrist (Walker-Murdoch) sign and thumb (Steinberg) sign. The thumb sign is positive if the thumb, when completely opposed within the clenched hand, projects beyond the ulnar border of the hand. The wrist sign is positive if the distal phalanges of the first and fifth digits of one hand overlap when wrapped around the opposite wrist. Individuals with Marfan syndrome are taller than predicted based on their non-affected relatives; however, they are not necessarily tall compared to the general population (Fig. 18.2). Overgrowth of the rib cartilage can lead to pectus excavatum or pectus carinatum, by pushing the sternum posteriorly or anteriorly, respectively. Scoliosis, affecting around 60% of patients, may lead to deformity, pain, and even respiratory problems. Joint hypermobility is another common finding, and it may lead to pain, performance deficits, and premature osteoarthritis. Additional skeletal manifestations include pes planus and an abnormally deep acetabulum (protrusio acetabuli) with accelerated erosion, which can be confirmed on radiographs. Some patients surprisingly present with pes cavus or reduced extension of the elbow, however. Typical facial features of Marfan syndrome are a long and narrow face with underdeveloped cheekbones (malar hypoplasia), downward slanting palpebral fissures, enophtalmus, and retro- or micrognathia. A highly arched and narrow palate and tooth crowding are often present as well.

Skeletal abnormalities in Marfan syndrome emerge and may progress during childhood and adolescence, typically during periods of rapid growth. In the current Ghent nosology, Four of eight specific skeletal manifestations are required for a major criterion for the Marfan syndrome (Table 18.1).

Table 18.1 Marfan syndrome : diagnostic criteria. (Adapted from: www.genetests.org. Copyright University of Washington and Children's Health System, Seattle)

System	Criteria	
	Major	Minor
Cardiovascular	At least one of the following: • Dilatation of the ascending aorta involving the sinuses of Valsalva • Dissection of the ascending aorta	At least one of the following: • Mitral valve prolapse with or without mitral regurgitation • Dilatation of the main pulmonary artery, in the absence of obvious cause, before the age of 40 years • Calcification of the mitral annulus before the age of 40 years • Dilatation or dissection of the descending thoracic or abdominal aorta before the age of 50 years
Skeletal	Presence of at least four of the following components: • Pectus carinatum • Pectus excavatum requiring surgery • Scoliosis of >20° or spondylolisthesis • Reduced upper-to-lower segment ratio for age (<0.85 for older children or adults) or arm span-to-height ratio (>1.05)[a] • Wrist (Walker Murdoch) and thumb (Steinberg) signs[b] • Reduced extension at the elbow (<170°) • Medial rotation of the medial malleolus causing pes planus • Protrusio acetabulae (abnormally deep acetabulum with accelerated erosion) of any degree (ascertained on radiographs)	Two major components or one major component and at least two of the following: • Pectus excavatum of moderate severity • Joint hypermobility • Highly arched palate with tooth crowding • Facial appearance (dolichocephaly, malar hypoplasia, enophthalmos, retrognathia, down-slanting palpebral fissures)
Ocular	Ectopia lentis	At least two of the following: • Abnormally flat cornea (as measured by keratometry) • Increased axial length of the globe (as measured by ultrasound) • Hypoplastic iris or hypoplastic ciliary muscle causing decreased pupillary miosis
Pulmonary		At least one of the following: • Spontaneous pneumothorax • Apical blebs (ascertained by chest radiography)
Skin and Integument		At least one of the following: • Striae atrophicae without obvious cause • Recurrent or incisional herniae
Dura	Lumbosacral dural ectasia (ascertained by CT or MRI)	
Family/Genetic history	At least one of the following: • Having a parent, child, or sib who meets these diagnostic criteria independently • Presence of a mutation in *FBN1* known to cause Marfan syndrome • Presence of a haplotype around *FBN1*, inherited by descent, known to be associated with Marfan syndrome in the family (ascertained by linkage analysis)	

[a]The lower segment (LS) is measured from the top of the symphysis pubis to the floor; the LS is subtracted from the height to obtain the upper segment (US). The arm span is measured between the tips of the middle fingers with the arms outstretched
[b]Walker-Murdoch wrist sign is the overlapping of the complete distal phalanx of the thumb and fifth finger when wrapped around the opposite wrist. The "thumb sign" (Steinberg) is extension of the entire distal phalanx of the thumb beyond the ulnar border of the hand when apposed across the palm

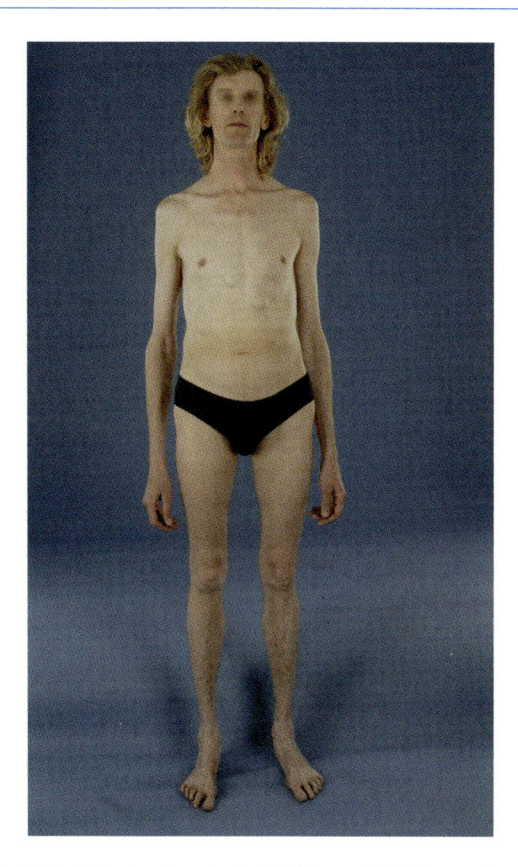

Fig. 18.1 Patient with a typical habitus of Marfan syndrome, with overgrowth of long bones and hypoplasia of skeletal muscle and adipose tissue

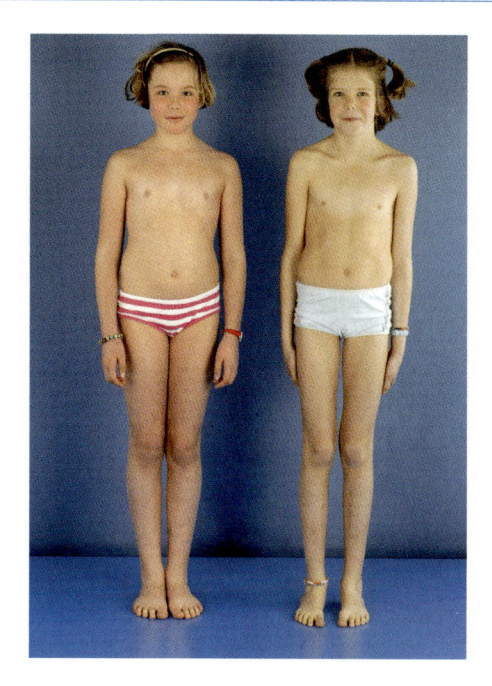

Fig. 18.2 The 8-year old girl on the right has Marfan syndrome. She is two years younger than her sister on the left, though the girls have the same height. (Courtesy of dr. J.M. Cobben)

18.2.2.2 Cardiovascular System

The major sources of morbidity and mortality in Marfan syndrome are due to manifestations in the cardiovascular system, of which *aortic aneurysm* and *dissection* are the most life threatening. *Dilatation of the aorta* occurs in the aortic root at the level of the *sinuses of Valsalva* (Fig. 18.3), resulting in a typical pear-shape of the aortic root. The onset and progression of aortic dilatation is highly variable, in rare cases beginning in utero, while other individuals never develop dilatation to dangerous diameters. Normal aortic dimensions are dependent on body surface area and age and therefore the dimensions measured in patients have to be compared with age-dependent normograms.[20,21] Reduced *aortic distensibility*, hypertension, and aortic regurgitation have been identified as predictors of rapid aortic growth.[22,23] Dissections in Marfan syndrome mostly constitute *type A aortic dissections* involving the aortic root, in some cases propagating along the descending aorta. The risk of aortic dissection in a patient is most importantly determined by the maximal aortic dimension, the rate of growth, and a family history of aortic dissection. However, dimensions within normal limits do not completely rule out the risk of dissection. Aortic dissection in children is rare. Dilatation of the ascending aorta involving the sinuses of Valsalva or dissection of the ascending aorta is a major criterion for the diagnosis of Marfan syndrome.

As Marfan patients currently survive longer due to surgical replacement of the aortic root, an increasing amount of patients develop aneurysms and/or dissections elsewhere in the arterial tree (Fig. 18.4).[24,25] In about 17% of patients, the aorta distal from the root is the first site of complications.[24,25] Predictors for aortic growth and adverse events in the distal aorta include a larger aortic diameter, lower aortic distensibility, and previous aortic root replacement.[22,25,26] The carotid or coronary arteries can also be involved, leading to cerebrovascular injury or myocardial infarction.

Within the heart, the atrioventricular valves are most often involved with thickening and prolapse of mitral and/or tricuspid valves and subsequent regurgitation. *Mitral valve prolapse* is a minor criterion for the clinical diagnosis. Aortic valve regurgitation usually arises in the context of stretching of the aortic annulus due to a dilated aortic root. Left ventricular

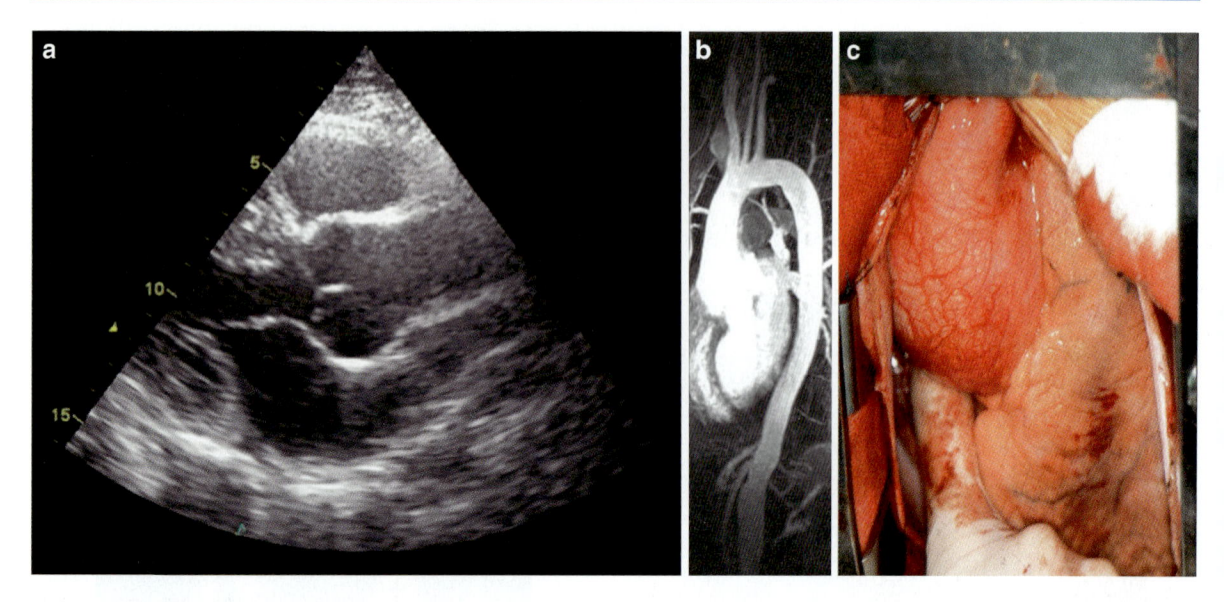

Fig. 18.3 (a) Echocardiogram showing a dilated aortic root at the level of the sinuses of Valsalva. Note the typical pear-shape of the aortic root. (b) Magnetic Resonance Angiography of aortic root aneurysm. (Courtesy of Groenink et al.[20]) (c) Aortic root aneurysm in a patient with Marfan syndrome at surgery (with permission by Bohn Stafleu van Loghum (Aangeboren hartafwijkingen bij volwassenen, 2nd edition, 2006, editors: prof dr. B.J.M. Mulder, dr. P.G. pieper, dr. F.J. Meijboom, dr. J.P.M. Hamer))

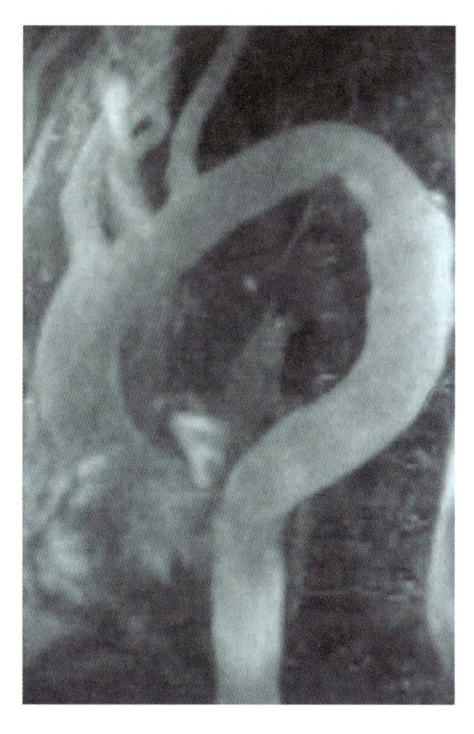

Fig. 18.4 Dilatation of the descending aorta after surgical repair for ascending aorta and aortic arch aneurysm (Magnetic Resonance Angiography). (Courtesy of Groenink et al.[20])

failure may be the consequence. *Calcification* of the aortic and atrioventricular valves can occur, and calcification of the mitral annulus under the age of 40 years is considered a minor criterion of Marfan syndrome. In the current Ghent nosology other minor criteria are dilatation of the main pulmonary artery at age younger than 40 years,[27] and dilatation or dissection of the descending aorta or more distal parts of the aorta.[28]

An increased frequency of venous varicosities, particularly of the superficial veins of the legs, is seen in Marfan syndrome, but it rarely causes significant problems.

18.2.2.3 Ocular System

Ocular lens dislocation (*ectopia lentis*), often bilateral and symmetric and mostly upward, is considered a major criterion and occurs in about 60% of patients with Marfan syndrome.[29] When dislocation of a lens is detected in the absence of a traumatic event (the most common cause), Marfan syndrome should always be considered. Subluxation usually develops in childhood, but may first appear later in life. A slit-lamp examination is an essential part of the diagnostic examination. *Myopia*, often rapidly progressing during childhood, is the most common ocular finding in patients with Marfan syndrome. It is associated with an increased length of the globe and an increased risk of *retinal detachment*.[30] The cornea can be flat and the iris or ciliary muscle may be hypoplastic.[29] A predisposition to *cataracts* and *glaucoma* exists. The lens dislocation, retinal detachment, cataract, and glaucoma may cause significant visual impairment.

18.2.2.4 Pulmonary System

In the lungs, widening of distal airspaces and *lung bullae* or blebs may be present, particularly in the upper lobes, which can predispose to spontaneous *pneumothorax*.[31] In addition, pectus deformities and scoliosis may lead to significantly reduced lung capacity.

18.2.2.5 Dural Sac

Stretching and ballooning of the dural sac (*dural ectasia*) in the lumbosacral region is seen in about two thirds of patients with Marfan syndrome. It can be assessed by lumbosacral imaging with MRI or CT (Fig. 18.5).[32] In the current Ghent nosology the presence of dural ectasia constitutes a major criterion in the Ghent nosology. However, dural ectasia can also be present in other connective tissue disorders,[33] and in healthy individuals. Possible symptoms include back pain and weakness, pain, and numbness in the proximal legs;[34] although it is often asymptomatic. Bone erosion and nerve entrapment may occur.

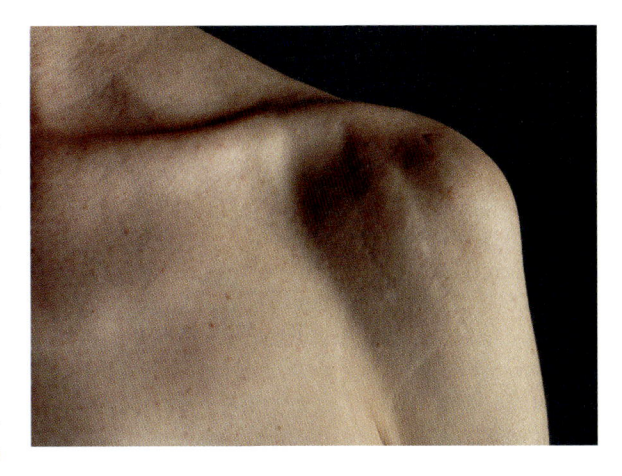

Fig. 18.6 Typical striae atrophicae on the anterior side of the shoulder of an adult patient with Marfan syndrome

18.2.2.6 Skin and Integuments

In contrast to many other connective tissue disorders, most patients with Marfan syndrome have a normal skin texture and elasticity, although in some people the skin is unusually thin or elastic. A common feature in Marfan syndrome is the presence of stretch marks (*striae distensae*) that are not associated with rapid weight gain, and at sites that are not typically stretched, such as the lumbar area and the anterior and posterior sides of the shoulders (Fig. 18.6). Inguinal and umbilical *hernias*, congenital or acquired, are also common.

18.2.2.7 Other

Hypoplasia of skeletal muscle and adipose tissue, contributes to the slender and asthenic appearance of some patients (Fig. 18.1).

18.2.3 Genetic Diagnosis

Genetic testing can aid in establishing the diagnosis of Marfan syndrome, as the presence of a mutation in FBN1 constitutes a major criterion in the Ghent nosology. However, its role remains limited, because sensitivity and specificity of genetic testing is not 100%.

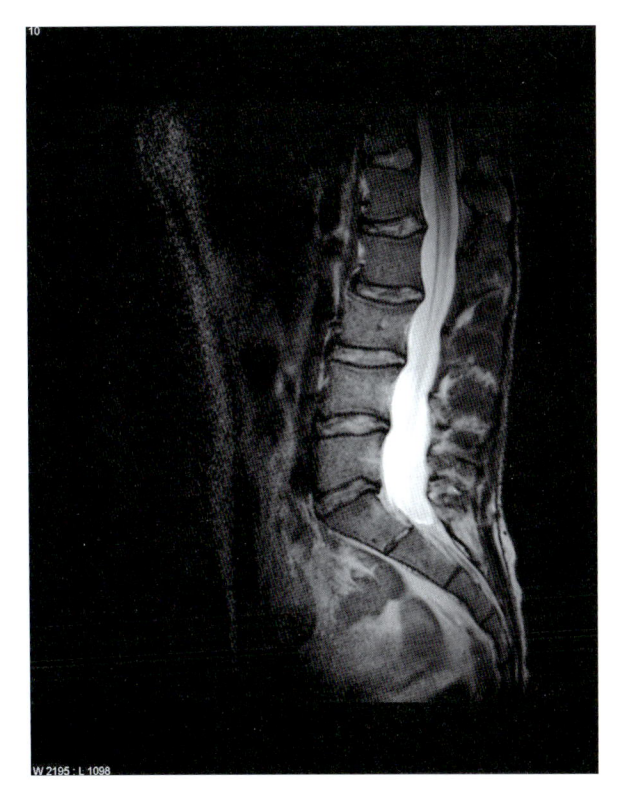

Fig. 18.5 Magnetic Resonance Imaging showing lumbar and sacral dural ectasia in a patient with Marfan syndrome. (Courtesy of Groenink et al.[20])

In about 10% of patients with classic Marfan syndrome, no mutation is detected using the current conventional screening methods.[10] In addition, other conditions than Marfan syndrome can be caused by mutations in FBN1, such as TAAD, familial ectopia lentis, or MASS phenotype. Genetic testing is useful for identification of relatives who have inherited the genetic predisposition within a family, if the causative mutation in the family is known. In addition, genetic testing can be used for prenatal diagnosis or *preimplantation genetic diagnosis*. Although requests for prenatal testing for Marfan syndrome are uncommon, the greater availability of mutation testing of the FBN1 gene might increase the number of requests for prenatal diagnosis.[35] The severity of disease in a child who inherits a mutant FBN1 is unpredictable, however.

18.2.4 Genotype–Phenotype Correlation

No definitive genotype–phenotype correlations seem to be present in Marfan syndrome.[36,37] Therefore, the identification of a particular mutation in a patient has little prognostic value and cannot determine individual management. However, some generalizations in genotype–phenotype correlations can be made. Mutations causing in-frame loss or gain of central coding sequence due to deletions, insertions, or splicing errors are associated with more severe disease. In contrast, mutations that create a premature termination codon leading to rapid degradation of the transcript can be associated with mild manifestations that may not fulfill the criteria for Marfan syndrome.[38–40] The mutations that have been found in patients with a neonatal presentation of severe and rapidly progressive Marfan syndrome, the so-called *neonatal Marfan syndrome* are located in a center portion of the FBN1 gene between exons 24 and 32, although many other patients with mutations in this region have a classic or mild phenotype.[36,41]

18.2.5 Differential Diagnosis

The differential diagnosis of Marfan syndrome is extensive because several disorders display similar skeletal, cardiovascular, or ocular manifestations. Diagnostic distinctions have prognostic value and may affect the clinical management and lifestyle of patients and are therefore of great importance. Table 18.2 is an overview of the clinical and genetic findings of the differential diagnoses of Marfan syndrome.

In an individual with Marfan-like skeletal manifestations, several disorders have to be considered: The *MASS phenotype*, an acronym of Mitral valve prolapse, Myopia, borderline and nonprogressive Aortic root enlargement, and nonspecific Skin and Skeletal findings, is a phenotypic continuum of Marfan syndrome.[42] It can occur sporadically or as a highly variable, autosomal dominant disorder. Important aortic and ocular problems rarely develop in MASS phenotype. Especially in children and adolescents it can be difficult to distinguish between Marfan syndrome and MASS phenotype, because of the age-dependant penetrance of many features in Marfan syndrome. In a few patients with MASS phenotype, mutations in FBN1 have been identified.[38,40]

Mitral valve prolapse has long been known to occur within families, and an autosomal dominant inheritance with reduced penetrance is assumed (see Chap. 15).[43] Long arms, joint hypermobility, pectus abnormalities, and spontaneous pneumothorax occur with higher frequency than in the general population.[44] This phenotype is called *mitral valve prolapse syndrome*. It is genetically heterogeneous, with several loci described in different families.

Familial ectopia lentis can also be inherited as an autosomal dominant trait, sometimes with mild, nonspecific signs of a systemic connective tissue disorder. Mutations in FBN1 have been described in several families.[45]

Congenital contractural arachnodactyly is a condition primarily affecting the skeleton with contractures of digits, elbows, and knees evident at birth, elongated long bones and kyphoscoliosis. In addition, the pinna of the ear is typically crumpled. Mitral valve prolapse and aortic root dilatation have been reported, with unknown frequency. Mutations in the fibrillin-2 gene account for about half of cases.[46,47] *Shprintzen–Goldberg syndrome* is a rare craniosynostosis syndrome characterized by marfanoid skeletal manifestations, exopthalmos, hypertelorism, downslanting palpebral fissures and other dysmorphic features, and developmental delay. The majority of patients do not show vascular involvement. In two patients, a mutation in FBN1 has been reported.[48] *Homocystinuria* is a disorder caused by a deficiency of

18 Connective Tissue Disorders and Smooth Muscle Disorders in Cardiology

Table 18.2 Differential diagnosis of Marfan syndrome

Syndrome	Inheritance	Genes involved	Cardiovascular	Ocular	Systemic
Marfan syndrome	AD	FBN1, TGFBR1[a], TGFBR2[a]	Sinus of Valsalva aneurysm, type A aortic dissection, mitral valve prolapse	Ectopia lentis, myopia	Typical skeletal manifestations: long bone overgrowth, scoliosis, pectus deformity, joint hypermobility and other. Dural ectasia, striae distensae
Loeys-Dietz syndrome	AD	TGFBR1, TGFBR2	Sinus of Valsalva aneurysm, other arterial aneurysm, dissections at younger age and at smaller aortic diameters, arterial tortuosity, patent ductus arteriosus, atrial septal defect, bicuspid aortic valve	Blue sclerae	Hypertelorism, craniosynostosis, bifid uvula/cleft palate, scoliosis, pectus deformity, joint hypermobility, developmental delay (rare)
Ehlers Danlos syndrome, vascular type	AD	COL3A1	Aneurysm and spontaneous rupture of mainly medium sized arteries, mitral valve prolapse		Gastrointestinal and uterine rupture, thin skin, characteristic face, easy bruising
Arterial tortuosity syndrome	AR	GLUT10	Arterial tortuosity, aneurysms of the large and medium sized arteries		Pectus deformity, joint hypermobility, hyperextensible/lax skin, elongated face, beaked nose, micrognatia
Familial thoracic aortic aneurysm and dissections	AD	ACTA2[a], MYH11[a], TGFBR1[a], TGFBR2[a], FBN1[a], NOTCH1[a]	Sinus of Valsalva aneurysm, ascending aorta aneurysm. In some families associated with patent ductus arteriosus (MYH11), bicuspid aortic valve (NOTCH1), aortic coarctation		ACTA2: livedo reticularis, iris flocculi
Familial ectopia lentis	AD	FBN1[a]		Ectopia lentis	Non specific skeletal features
MASS phenotype	AD	FBN1	Borderline and nonprogressive aortic root enlargement, mitral valve prolapse	Myopia	Non specific skeletal- and skin features
Shprintzen-Goldberg syndrome	Unknown	Unknown[b] (FBN1[a])	Sinus of Valsalva aneurysm (rare)		Craniosynostosis, facial dysmorphism, scoliosis, pectus deformity, long bone overgrowth, joint hypermobility, developmental delay
Homocystinuria	AR	CBS	Thromboembolism, coronary artery atherosclerosis, mitral valve prolapse	Ectopia lentis, myopia	Long bone overgrowth, scoliosis, pectus deformity, developmental delay
Congenital contractural arachnodactyly	AD	FBN2	Aortic root aneurysm[c]		Long bone overgrowth, joint contractures, crumpled pinnae of ears

[a] Mutation detected in a small subset of patients
[b] A mutation in FBN1 has been detected in a few atypical patients
[c] Has been reported, but frequency unknown

AD autosomal dominant, *AR* autosomal recessive, *CBS* cystathionine b-synthase, *MASS* Mitral valve prolapse, Myopia, borderline and non-progressive Aortic root enlargement, and non-specific Skin and Skeletal findings

the cystathionine β-synthase (CBS) enzyme. Clinical features are variable, and include developmental delay, ectopia lentis, severe myopia, skeletal abnormalities (excessive height and long bone overgrowth), and thromboembolism. Homocystinuria is inherited in an autosomal recessive manner, the causative gene being the CBS-gene, with mutations identified in over 95% of patients.[49,50]

In individuals or families with a history of aortic aneurysms and dissections, other differential diagnoses should be considered, such as *Loeys–Dietz syndrome*, the vascular type of *Ehlers–Danlos syndrome*, *Arterial Tortuosity syndrome*, as well as nonsyndromic forms of familial aortic aneurysms, including *familial Thoracic Aortic Aneurysms and aortic Dissections (TAAD)*, with or without the association with *patent ductus arteriosus* or *bicuspid aortic valve*. These disorders will be discussed in more detail in paragraphs 16.3–16.6.

18.2.6 Therapy/Follow-up/Prognosis

Type A aortic dissection is the most important cause of death in patients with Marfan syndrome (60%). Due to improved medical and surgical therapy of cardiovascular complications, the life expectancy of patients has increased importantly during the past decades.[51,52] A report from 1972 mentioned an average age of death in the fourth and fifth decades,[53] while in 1995 a median cumulative probability of survival of 72 years was reported.[52] Because of the multiple and variable manifestations of the disorder, an individualized and multidisciplinary approach is required for treatment and follow-up.

18.2.6.1 Cardiovascular Management

Because the risk of type A aortic dissection is related to the diameter of the aortic root as well as the rate of growth of the aorta, echocardiographic follow-up is of utmost importance to determine the optimal timing of surgery. If the echocardiographic measurements are unclear or imprecise, for example, due to thorax deformities, MRI can be of help.[54] In addition, MRI or CT scanning of the entire aorta is indicated in every Marfan patient because 20% of aortic dissections occur in the abdominal aorta. By means of MRI, decreasing aortic distensibility can be followed noninvasively. Yearly follow-up with echocardiography and less frequent MRI or CT scanning is sufficient for the majority of patients. If the aorta reaches a size that is close to the threshold for surgery, or when rapid aortic growth is present, a more frequent follow-up is indicated.

All Marfan patients should be treated with *β-adrenergic blockers*, since these have been shown to slow the dilatation of the aorta and to diminish the risk of dissection.[55–58] β blockers are probably beneficial through negative inotropic as well as negative chronotropic effects. Verapamil is commonly used when β blockers cannot be used, for example, in individuals with asthma. In addition to this medicamental therapy, *lifestyle modifications* are advised. To reduce hemodynamic stress, Marfan patients are advised to avoid situations that cause a substantial increase in blood pressure and heart rate. Because of the risk of acute aortic dissection, isometric exercise, contact sports, and competitive sports must be discouraged. However, light-to-moderate aerobic activities promote long-term cardiovascular and musculoskeletal as well as psychosocial health and, therefore, should be encouraged in patients with a lower risk of dissection.

18.2.6.2 Surgical Treatment

The two most important reasons for aortic replacement therapy are the prevention of aortic dissections and the prevention of left ventricular failure due to aortic regurgitation. Aortic root replacement therapy is currently recommended at a maximum diameter of 50 mm. In patients with a family history of dissection, a rapid increase in diameter (>2 mm/year) or severe aortic or mitral regurgitation requiring surgical intervention, aortic root replacement therapy is considered at smaller dimensions (45*50 mm). In addition, in women who wish to become pregnant, surgical intervention may be indicated at an aortic diameter below 45 mm.

The classical and most well-known procedure is the *Bentall procedure* (Fig. 18.7), which involves replacement of the aortic valve and ascending aorta with a graft and mechanical valve. The mortality of an elective Bentall procedure is less than 2%,[59] as opposed to an emergency procedure mortality rate of 15–25%. Currently, *valve sparing procedures* are increasingly performed, such as re-implantation of the native aortic valve in a Dacron tube or remodeling of the aortic root.[60–62] Excellent early outcome, favorable long-term results, and acceptable durability of

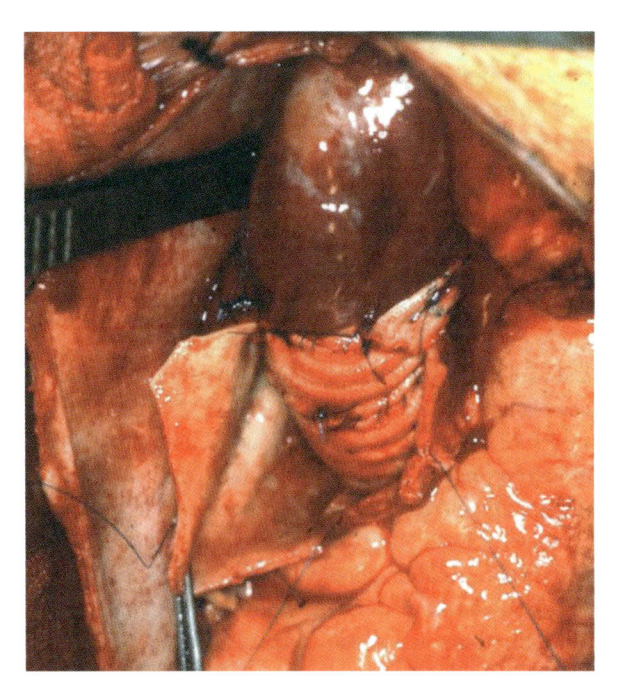

Fig. 18.7 Bentall procedure: the graft and mechanical valve have been incorporated and the coronary arteries have been reimplanted (with permission by Bohn Stafleu van Loghum (Aangeboren hartafwijkingen bij volwassenen, 2nd edition, 2006, editors: prof dr. B.J.M. Mulder, dr. P.G. pieper, dr. F.J. Meijboom, dr. J.P.M. Hamer))

the reimplanted valve have been reported for this technique in patients with Marfan syndrome.[63] The Bentall procedure requires lifelong anticoagulation, whereas valve-conserving techniques may avoid the need for anticoagulation.

After a Bentall procedure, 5- and 10-year survival rates are about 83% and 70%, respectively.[59,64] The limited prognosis is caused by late complications, mainly aneurysms and dissections as well as reoperations in the same or other localizations in the aorta or arterial tree. Therefore, also after surgical treatment a lifelong need for *endocarditis prophylaxis*, β blocking agents and regular screening of the entire aorta with echocardiography and MRI are warranted.

Effective repair or replacement procedures are also available for the treatment of mitral valve prolapse and regurgitation.[65]

18.2.6.3 Management of Other Manifestations

Periodic ophthalmic review is appropriate in childhood and early adolescence because ectopia lentis most often becomes evident in preschool years and may be slowly progressive. Adult patients should have ophthalmic screening at low frequency, because of the risk of glaucoma and cataracts. Most often, the myopia and lens dislocation can be managed with eyeglasses. Sometimes surgical intervention is necessary, including the implantation of artificial lenses. Growth should be monitored and the spine has to be evaluated for scoliosis. Surgical interventions are sometimes needed for stabilization of the spine or correction of severe pectus abnormalities, for medical as well as cosmetic reasons. The involvement of a skilled orthopedist is needed in these severe cases. Growth-reducing sex hormone therapy, starting before puberty, may be considered to limit adult height when an extreme height is anticipated.

18.2.6.4 Pregnancy

Pregnancy in women who have Marfan syndrome poses a problem, because a risk of aortic dissection during and shortly after pregnancy exists. The risk appears low in patients with an aortic root diameter of less than 40 mm,[66,67] but even in patients with normal aortic diameters there is a risk of aortic dissection. When the aortic diameter exceeds 45 mm, aortic surgery should be considered before pregnancy.[66,68] Another issue is the use of anticoagulation therapy in patients with an artificial valve. Coumarins may have teratogenic effects during the first trimester. Heparin, usually given in the form of low molecular weight heparin, is less effective against valve thrombosis and may lead to osteoporosis during longer-term use. The choice of antithrombotic therapy should be individualized. Pregnant women should be closely monitored by a cardiologist and obstetrician.

The risk of transmission of the disease to offspring is 50%, and genetic counseling should be offered to prospective parents to discuss this issue as well as the possibilities of prenatal diagnosis or preimplantation genetic diagnosis.

18.2.6.5 New Developments in Treatment of Marfan Syndrome

TGF-β antagonists seem to be a promising new treatment for aortic manifestations of Marfan syndrome. In fibrillin-1 deficient mice, TGF-β antagonists have been shown to rescue the pulmonary emphysema and myxomatous changes of the mitral valve. The administration of the angiotensin II type 1 (AT1) receptor blocker *losartan* (an antagonist of TGF-β signaling) in one of these animal models prevented aortic dilatation and improved disease manifestations in the aortic wall and the lungs.[69] Preliminary clinical data were provided by a cohort study of 18 pediatric patients in which losartan reduced the rate of aortic dilatation.[70]

Doxycycline, a matrix metalloproteinase inhibitor, was recently demonstrated to improve aortic wall architecture and delayed aortic dissection in mouse models.[71] Given that matrix metalloproteinases can contribute to the activation of TGF-β, it is possible that doxycycline and losartan will show synergistic effects. However, the effect of TGF-β antagonists in human patients still has to be confirmed.

18.3 Loeys–Dietz Syndrome

Loeys–Dietz syndrome is a recently described autosomal dominant condition characterized by *arterial tortuosity, hypertelorism,* and *bifid uvula*/cleft palate.[2,16] Patients are prone to aggressive *arterial aneurysms and dissections*, even in early childhood. The youngest patient described was 6 months of age. Moreover, dissections occur at aortic dimensions that are not considered hazardous in other connective tissue disorders such as Marfan syndrome. The mean age at death of patients who died was 26 years in one study.[2] Other features, such as *dural ectasia*, Arnold Chiari type I malformation, and *craniosynostosis* (premature fusion of the cranial sutures in children) are also part of the syndrome. Skeletal manifestations include club feet, pectus deformity, scoliosis, and arachnodactyly. Congenital heart defects, including bicuspid aortic valve, patent ductus arteriosus, and atrial septal defects do also occur. Developmental delay, hydrocephalus, spondylolistesis, congenital hip dislocation, cervical spine instability, submandibular branchial cysts, osteoporosis with fracture at young age, and defective tooth enamel have all been described.

A subset of patients with the Loeys–Dietz syndrome shows overlap with the vascular type of Ehlers–Danlos syndrome.[2] Physical findings in these patients include easy bruising, thin and wide scars, soft and translucent skin with easily visible veins, spontaneous rupture of bowel and spleen, uterine and arterial ruptures during pregnancy or in the postpartum period, and diffuse arterial aneurysms and dissections. These patients do not have cleft palate, craniosynostosis, or hypertelorism.

18.3.1 Molecular Background

Loeys–Dietz syndrome is caused by mutations in the genes encoding the receptors for *TGF-β (TGFBR1* and *TGFBR2)*. Most mutations are missense substitutions of evolutionary conserved residues that encode the intracellular kinase domain of the receptors.[3] No differences exist in clinical manifestation of patients with mutations in either gene. Mutations in TGFBR1 and TGFBR2 have also been reported in patients with familial thoracic aortic aneurysm and dissection (TAAD), without other features of Loeys–Dietz syndrome[1] as well as in patients fulfilling the current Ghent criteria for Marfan syndrome.[11] There are no apparent differences in the type of mutations in these disorders as opposed to those found in Loeys–Dietz syndrome.

There is considerable intrafamilial variability in phenotype in Loeys–Dietz syndrome, and multiple cases of apparent non-penetrance have been reported.[2,3] Most cases of severe Loeys-Dietz syndrome are due to de novo mutations.

Although the pathogenetic basis of Loeys–Dietz syndrome is not completely understood, increased TGF-β signaling has been demonstrated to be the underlying mechanism.[16] Altered gene activity downstream of TGF-β subsequently induces the phenotypic manifestations. The mutations in TGFBR1 and TGFBR2 that cause the vascular phenotypes are predicted to give rise to a mutant receptor protein that is able to traffic to the cell surface and bind extracellular ligand, but that lacks the capacity to propagate an intracellular signal. How the defect results in increased TGF-β signaling remains unclear.

18.3.2 Differential Diagnosis

There is extensive overlap in clinical features between Loeys–Dietz syndrome and Marfan syndrome, such as aortic root aneurysm and dissection, scoliosis, pectus deformity, and arachnodactyly. The main distinguishing features between Loeys–Dietz syndrome and Marfan syndrome are the presence of the typical triad of hypertelorism, cleft palate/bifid uvula, and arterial tortuosity. Moreover, patients with Loeys–Dietz syndrome do not have ectopia lentis and the majority does not have the typical overgrowth of long bones as seen in Marfan syndrome. It is therefore important to not only ask whether an individual meets the criteria for Marfan syndrome, but also to look at, often subtle, discriminating features to distinguish between the two disorders. Many other diseases mentioned in the differential diagnosis of Marfan syndrome have overlapping features with Loeys–Dietz syndrome, such as the *vascular Ehlers–Danlos syndrome, Shprintzen–Goldberg syndrome, congenital contractural arachnodactyly* and *nonsyndromic forms of aortic aneurysms*. In addition, *arterial tortuosity syndrome* and autosomal dominant and recessive forms of *cutis laxa* should be considered.

18.3.3 Treatment

Careful follow-up of patients with Loeys–Dietz syndrome is mandatory. Many of the measures recommended for patients with Marfan syndrome also apply to patients with Loeys–Dietz syndrome. Avoidance of isometric exercise, contact sports and competitive sports, as well as the use of β-adrenergic blockers to reduce hemodynamic stress are indicated. Because the vascular pathology can be seen throughout the entire arterial tree, routine surveillance includes imaging of the arterial tree from head through pelvis by magnetic resonance or computed tomography angiography. Ascending aortic surgery in adolescents and adults is recommended at lower diameters than in other aortic aneurysm syndromes, that is when maximal ascending aortic dimensions approach 4.0 cm, or when expanding more than 0.5 cm per year.[72] Surgical intervention at smaller dimensions may be indicated, based upon family history or personal risk assessments. Aneurysms distant from the aortic root are often amenable to surgical intervention as well.[2] In children, surgery should be considered when the maximal ascending aortic dimension exceeds a z-score of 3.0 (in patients with severe craniofacial features) or 4.0 (in the presence of mild craniofacial features), and should preferably be performed when the size of the aortic annulus allows the insertion of a sufficiently sized graft to accommodate growth.[72] The reported risk of aortic surgery is approximately 1.7% in Loeys–Dietz syndrome, and might be higher in the subset of patients with features overlapping with the Ehlers–Danlos syndrome.[2] Radiologic surveillance for cervical spine instability should be offered. The therapeutic use of TGF-β antagonists seems appealing considering their function in mouse models of Marfan syndrome; however, this has not been tested yet in Loeys–Dietz syndrome models.

18.4 Ehlers–Danlos Syndrome, Vascular Type

Ehlers–Danlos syndrome is an inherited heterogeneous group of connective tissue disorders, comprising several different clinical subtypes. The prevalence is estimated at one in 10,000 to one in 25,000 for all types. The most common types are the classical type and hypermobility type.

Ehlers–Danlos syndrome, vascular type (former EDS type IV), is estimated to constitute 5–10% of cases of Ehlers–Danlos syndrome. It is characterized by severe arterial and intestinal complications. Vascular Ehlers–Danlos syndrome is an autosomal dominant disorder, resulting from mutations in the gene for *type III procollagen (COL3A1)*.[73] There is a high rate of de novo mutations; sporadic cases account for approximately half of the cases.

The clinical diagnosis of the vascular type of Ehlers–Danlos syndrome is made on the basis of four clinical criteria: easy bruising; thin skin with visible veins; characteristic facial features; and *rupture of arteries, uterus, or intestines*.[74] The diagnosis is confirmed by the demonstration of cultured fibroblasts synthesizing abnormal type III procollagen molecules, or by the identification of a mutation in the COL3A1 gene. The facial features are often distinctive and consist of an "old-looking" face, with prominent cheekbones and sunken or bulging eyes. A thin and pinched nose, as well as thin lips, are present (Fig. 18.8). The skin on the extremities, especially the hands, appears aged (acrogeria) (Fig. 18.9). The skin appears translucent with visible veins, particularly on the chest and abdomen. *Hypermobility* of small joints can be present (Fig. 18.10), while hypermobility of large joints and

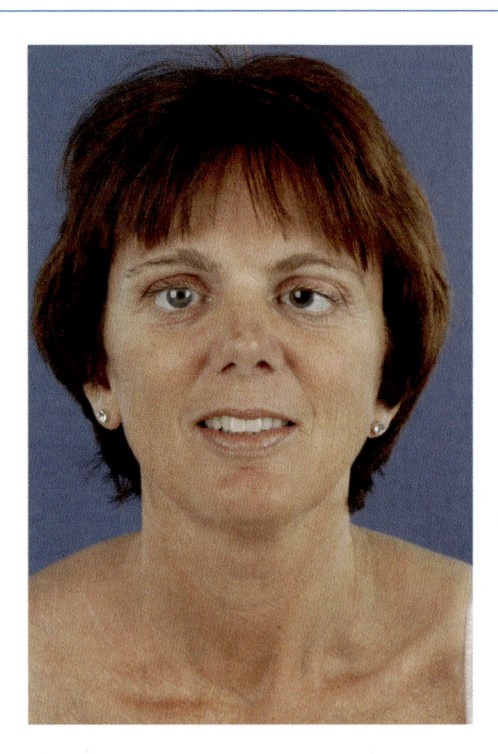

Fig. 18.8 43-year old woman with Ehlers Danlos syndrome, vascular type. Typical facial features including bulging eyes and a thin and pinched nose are present, although subtle in this patient. The left pupil is wide and unresponsive to light, resulting from a vascular complication and surgery. (Courtesy of dr. J.M. Cobben)

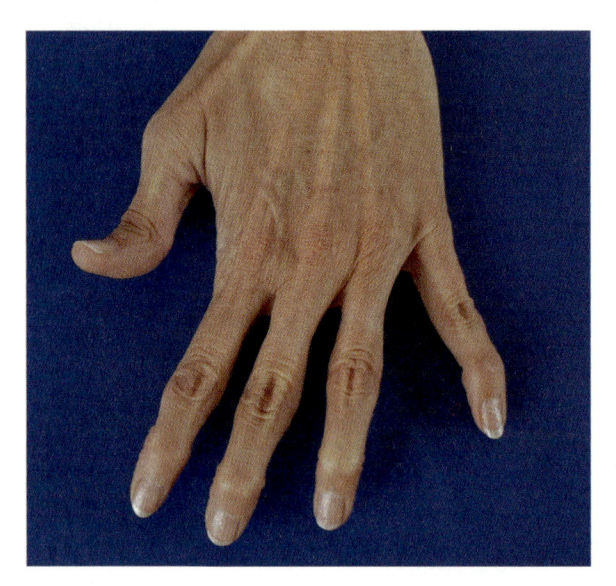

Fig. 18.9 Hand of the 43-year-old woman with vascular Ehlers Danlos syndrome shown in figure 18.8, showing acrogeria. (Courtesy of dr. J.M. Cobben)

hyperextensibility of the skin, characteristic of the more common forms of Ehlers–Danlos syndrome, are unusual in the vascular type. In many patients, the diagnosis is made only after a catastrophic complication or at postmortem examination. Affected patients are at risk for aneurysms and rupture or dissection, especially of medium-sized arteries.[75] The proximal and distal branches of the aortic arch, the descending thoracic aorta, and abdominal aorta are often affected, as well as vertebral and carotid arteries. Rupture of the gastrointestinal tract is another serious complication, occurring in about 25% of affected individuals and being lethal in 3% of cases.[75] It mostly occurs in the sigmoid colon, but the small intestine and stomach can also be affected. Pregnancy increases the risk of uterine and vascular rupture, particularly during the last 3 months, and the maternal mortality is reported to be as high as 12%.[75] Spontaneous ruptures of spleen, liver, and heart can also occur, although on much rarer occasions.[76] An increased prevalence of mitral valve prolapse has been reported in patients with vascular Ehlers–Danlos syndrome. Children may present with *pneumothorax*, inguinal *hernia*, and recurrent joint dislocation. Excessive bruising in children can be a presenting symptom to pediatricians.

Management of patients with vascular Ehlers–Danlos syndrome is focused on prophylactic measures and symptomatic treatment. A conservative approach is usually recommended for vascular complications. Surgical repair is frequently complicated by the presence of fragile tissue that does not heal well, although it might be needed to treat potentially fatal complications. Surgical mortality has been reported to be about 45%.[75,77] Premature death in vascular Ehlers–Danlos syndrome is due to complications of pregnancy, vascular disease, and difficulties during repair; the median life span is 48 years.[75]

18.5 Arterial Tortuosity Syndrome

Arterial tortuosity syndrome is a rare autosomal recessive connective tissue disorder, characterized by widespread arterial involvement with *tortuosity*, elongation, stenosis, and aneurysms of the large and middle-sized arteries.[78,79] Other clinical findings include pectus deformities, *joint hypermobility*, arachnodactyly, and hyperextensible and lax skin. Characteristic facial features

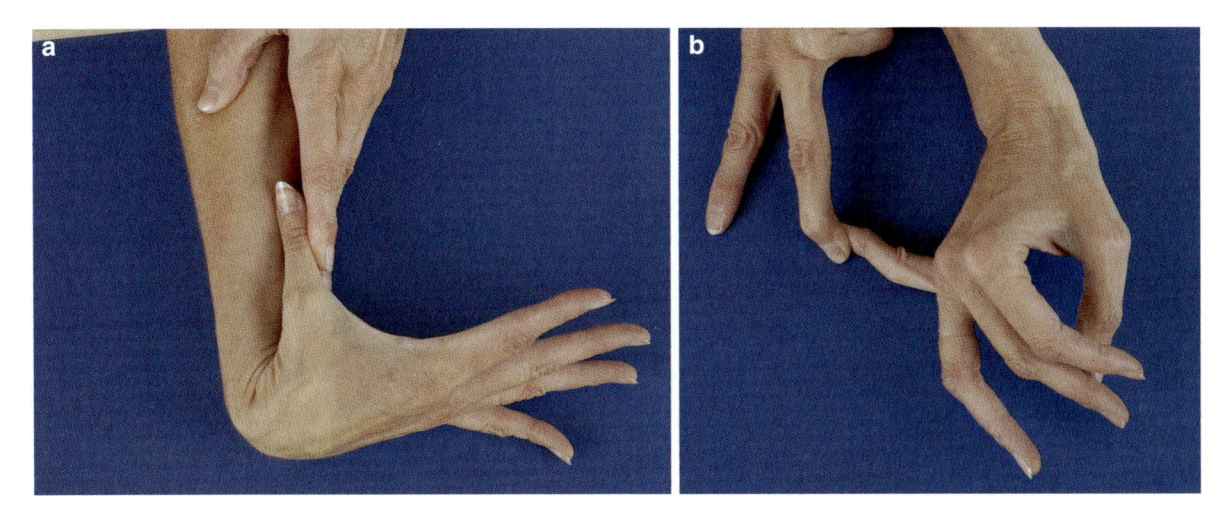

Fig. 18.10 Hypermobility of the wrist and fingers in Ehlers Danlos syndrome, vascular type. (Courtesy of dr. J.M. Cobben)

include elongated face, downslanting palpebral fissures, beaked nose, micrognatia, and high-arched palate. Mutations in the *SLC2A10 gene* on 20q13.1, coding for *GLUT10*, a member of the glucose transporter family, have been identified in several ATS families.[78–80] GLUT10 deficiency is associated with upregulation of the TGF-β pathway in the arterial wall.

18.6 Nonsyndromic Aortic Aneurysms and Dissections

Nonsyndromic thoracic aortic aneurysms can result from diverse etiologies, including infectious agents and hemodynamic forces. The vast majority of descending thoracic aneurysms are associated with atherosclerosis and the risk factors for aneurysm formation are the same as those for atherosclerosis (e.g., hypertension, hypercholesterolemia, smoking).[81] Atherosclerosis is an infrequent cause of ascending thoracic aortic aneurysms, however.[81]

About 20% of individuals with a thoracic aortic aneurysm or a type A aortic dissection without a history of connective tissue disorder have an affected first-degree relative.[82,83] *Familial Thoracic Aortic Aneurysms and Aortic Dissections (TAAD)* is the presence of familial dilatation and/or dissection of the thoracic aorta, in the absence of Marfan syndrome, Loeys–Dietz syndrome, and other connective tissue abnormalities, and in the absence of another cause of the aortic problems such as atherosclerosis. Familial TAAD is inherited in an autosomal dominant manner with decreased penetrance and variable expression. The thoracic aneurysms can be isolated, or may be associated with structural heart defects, such as bicuspid aortic valve, aortic coarctation, or patent ductus arteriosus within some families. Affected individuals have progressive *aortic dilatation* of the sinuses of Valsalva and/or ascending aorta and *aortic dissection*. In the majority of individuals with familial TAAD, enlargement of the aorta precedes dissection.[84] Aneurysms in more distal parts of the aorta, as well as cerebral aneurysms, may also be present. In TAAD, the onset and rate of progression of aortic dilatation is highly variable, with some individuals developing dilatation in childhood while others reach high age without aneurysms. Individuals with familial TAAD have a younger mean age at presentation than individuals with nonfamilial thoracic aortic aneurysms, but older than individuals with Marfan syndrome.[82] Aortic dissection in childhood is rare.

Familial TAAD exhibits significant genetic heterogeneity, although in most families the underlying genetic basis is not found. Mutations in *FBN1*,[85] *TGFBR1*, and *TGFBR2*[1,3] have been reported in a small proportion of patients with TAAD. The identification and characterization of these genes suggest that increased TGF-β signaling plays a role in pathogenesis. Only recently it was discovered that mutations in the vascular smooth muscle cell-specific *beta-myosin*

(MYH11)[86] and *alpha-actin (ACTA2)*[87] can also cause this disorder. These mutations were predicted to disrupt the contractile function of smooth muscle cells, leading to the hypothesis that decreased smooth muscle cell contractile function may be the underlying cause of the disease in these families. Mutations in ACTA2 have been found in 14 families and it is estimated that mutations in this gene account for 14% of familial TAAD.[87] Mutations in MYH11 have been found in two families with thoracic aneurysms and dissection associated with *patent ductus arteriosus* in some family members.[86]

Familial aortic aneurysms may also occur in association with bicuspid aortic valve. *Bicuspid aortic valve* is the most common congenital heart defect and occurs in approximately 1–2% of the general population. Bicuspid aortic valve is part of a spectrum of *left-sided obstructive lesions*, which also includes *coarctation of the aorta* and hypoplastic left heart syndrome, among others. Oligogenic or autosomal dominant inheritance with reduced penetrance has been suggested as the underlying genetic mechanism in some families.[88,89] Patients with bicuspid aortic valve are at risk of developing ascending aorta dilatation. Within families, some affected individuals only have the aortic dilatation, without associated structural cardiac abnormalities. Therefore, it is believed that in some families, bicuspid aortic valve, aortic coarctation, associated cardiac lesions, and *aortic aneurysms* are part of a phenotypic spectrum and variable manifestations of the same gene defect. Mutations in *NOTCH1* have been found in some patients who also had significant valve calcification and stenosis.[90,91]

Other loci associated with TAAD with or without bicuspid aortic valve have been identified, and further locus heterogeneity is assumed. The genetic basis for the disorder remains unknown in the majority of families.

18.6.1 Management

When a patient is diagnosed with a thoracic aortic aneurysm, clinical evaluation including family history, physical examination, and ocular assessment is recommended to exclude the diagnosis of Marfan syndrome or another systemic connective tissue disorder. Medications that reduce hemodynamic stress, such as *β-adrenergic blockers,* are recommended for individuals with TAAD. Careful follow-up is warranted. Aortic surgery is recommended at ascending aortic diameters similar to Marfan syndrome[92] and should be individualized for each patient, taking into account family history, rate of aortic growth, underlying gene defect, etc.

In families without an identified causal mutation, it is not possible to determine which individuals are at risk of developing aneurysms and dissection. Therefore, all first-degree relatives of affected patients are advised to undergo regular echocardiographic screening at low frequency.

18.7 Other Connective Tissue Disorders

Several other hereditary disorders displaying abnormalities of connective tissue exist, and in some of these there is an increased risk of cardiovascular manifestations. *Williams syndrome* is caused by a heterozygous *microdeletion on chromosome 7q11*.23. The commonly deleted interval encompasses 1.55 Mb and contains several genes, including the *elastin (ELN) gene*. Cardiovascular problems with narrowing of arteries are common: 75% of patients have a *supravalvular aortic stenosis*; stenoses of the pulmonary or other arteries may also be present. Abnormalities of connective tissue are present, including herniae, joint limitation or laxity, and soft, lax skin. The face is distinct and often described as "elfin-facies," with bitemporal narrowness, periorbital fullness, short nose, full nasal tipfull lips, wide mouth, and prominent earlobes. In most patients, mild mental retardation is present. Affected individuals have a unique personality, being overfriendly and empathic. Adult height is decreased and endocrine abnormalities (hypercalcemia, hypercalciuria, hypothyroidism) may be present.

Familial supravalvular aortic stenosis is an autosomal dominant disorder with the identical vascular phenotype of Williams syndrome, but without the associated abnormalities, although some individuals have a Williams-like mouth or connective tissue abnormalities such as herniae. Causative mutations or small gene deletions in ELN have been described in several families as well as sporadic cases of the disorder.

Mutations in ELN can also cause autosomal dominant *cutis laxa*, while autosomal recessive forms of this disorder are caused by mutation in *FBLN (Fibulin)-4* or *Fibulin-5*. This disorder is associated with loose skin, facial drooping, floppy airways, and inguinal and umbilical herniae. Cardiovascular manifestations include mild dilatation of the proximal aorta and great vessels, tortuous carotid arteries, peripheral pulmonary stenoses, and mitral or tricuspid valve abnormalities.

Osteogenesis imperfecta is a group of autosomal dominantly or recessively inherited disorders characterized by fractures with minimal or absent trauma, dentinogenesis imperfecta, and hearing loss. *Mitral valve prolapse, aortic dilatation*, and aortic and mitral regurgitation have been reported. *COL1A1* and *COL1A2* are the only two genes currently known to play a role in the disorder.

Stickler syndrome is a connective tissue disorder that can include ocular findings of myopia, cataract, and retinal detachment, hearing loss, midfacial underdevelopment, and cleft palate and mild spondyloepiphyseal dysplasia and/or precocious arthritis. *Mitral valve prolapse* has been reported to be more common than in the general population. Mutations in three collagen genes, *COL2A1, COL11A1,* and *COL11A2*, have been associated with Stickler syndrome.

18.8 Summary

The hereditary disorders of connective tissue constitute a clinically and genetically heterogeneous group of disorders, with substantial overlap in clinical features. Many connective tissue disorders constitute a risk of aneurysm and dissection of the aorta. The most common syndromic presentation of ascending aortic aneurysm is Marfan syndrome, which diagnosis is made using the Ghent criteria though these are currently being revised. These criteria are primarily based on clinical features, although the detection of a disease-causing FBN1 mutation contributes in establishment of the diagnosis. Other syndromes, including vascular Ehlers–Danlos syndrome and Loeys–Dietz syndrome also show ascending aortic aneurysms dissections. In addition, thoracic aortic aneurysm and dissection can occur in the absence of associated systemic findings of connective tissue abnormalities in TAAD, with or without structural heart defects such as bicuspid aortic valve, aortic coarctation, or patent ductus arteriosus.

Mutations in FBN1 are the major cause of Marfan syndrome, although mutations in this gene have also been found in several Marfan-related disorders, among which TAAD. Likewise, mutations in TGFBR1 and TGFBR2 can cause a spectrum of disorders, including Loeys–Dietz syndrome and TAAD, and mutations in these genes have also been described in patients fulfilling the criteria for Marfan syndrome. TAAD, however, can also be caused by mutations in MYH11 and ACTA2. Ehlers–Danlos syndrome, vascular type, is caused by mutations in COL3A1. Mutations in GLUT10 give rise to the autosomal recessive arterial tortuosity syndrome.

Because of the overlapping clinical features of the different disorders, a detailed clinical evaluation and family history is warranted to establish the right diagnosis in an individual. When a patient is diagnosed with a thoracic aortic aneurysm, ocular and physical examination is recommended to exclude the diagnosis of Marfan syndrome or another connective tissue disorder. In patients with ectopia lentis or skeletal manifestations, cardiovascular evaluation should be performed. In addition, molecular characterization aids in establishment of the correct diagnosis. Accurate diagnosis has important implications regarding treatment, prognosis, and recurrence risks for family members. A multidisciplinary team should therefore be involved in establishing the diagnosis, as well as in follow-up and treatment. In addition, family members of an affected individual should be screened for the disorder, clinically, or, if the disease-causing mutation in the family is known, by genetic screening. Due to the advances in medical and surgical therapy, life expectancy of patients with Marfan syndrome and other connective tissue disorders has improved substantially.

A major role for TGF-β signaling in the origin of Marfan syndrome and other disorders predisposing to aortic aneurysms and dissections has been demonstrated by the ongoing increase of insight in the molecular and pathogenetic basis of these disorders. This may provide a basis for future therapies to decrease the morbidity and mortality still associated with aortic aneurysms and dissections.

References

1. Pannu H, Fadulu VT, Chang J, et al. Mutations in transforming growth factor-beta receptor type II cause familial thoracic aortic aneurysms and dissections. *Circulation.* 2005 July 26;112(4):513-520.
2. Loeys BL, Schwarze U, Holm T, et al. Aneurysm syndromes caused by mutations in the TGF-beta receptor. *N Engl J Med.* 2006 August 24;355(8):788-798.
3. Stheneur C, Collod-Beroud G, Faivre L, et al. Identification of 23 TGFBR2 and 6 TGFBR1 gene mutations and genotype-phenotype investigations in 457 patients with Marfan syndrome type I and II, Loeys-Dietz syndrome and related disorders. *Hum Mutat.* 2008 November;29(11):E284-E295.
4. Marfan AB. Un cas de déformation congénitale des quatre membres, plus prononcée aux extrémités, caracterisée par l'allongement des os avec un certain degré d'aminicissement. *Bull mem Soc Med Hop Paris.* 1896;13:220-226.
5. Corson GM, Chalberg SC, Dietz HC, Charbonneau NL, Sakai LY. Fibrillin binds calcium and is coded by cDNAs that reveal a multidomain structure and alternatively spliced exons at the 5' end. *Genomics.* 1993 August;17(2):476-484.
6. Biery NJ, Eldadah ZA, Moore CS, Stetten G, Spencer F, Dietz HC. Revised genomic organization of FBN1 and significance for regulated gene expression. *Genomics.* 1999 February 15;56(1):70-77.
7. Hubmacher D, Tiedemann K, Reinhardt DP. Fibrillins: from biogenesis of microfibrils to signaling functions. *Curr Top Dev Biol.* 2006;75:93-123.
8. Collod-Beroud G, Le BS, Ades L, et al. Update of the UMD-FBN1 mutation database and creation of an FBN1 polymorphism database. *Hum Mutat.* 2003 September;22(3):199-208.
9. Judge DP, Biery NJ, Keene DR, et al. Evidence for a critical contribution of haploinsufficiency in the complex pathogenesis of Marfan syndrome. *J Clin Invest.* 2004 July;114(2):172-181.
10. Loeys B, De BJ, Van AP, et al. Comprehensive molecular screening of the FBN1 gene favors locus homogeneity of classical Marfan syndrome. *Hum Mutat.* 2004 August;24(2):140-146.
11. Mizuguchi T, Collod-Beroud G, Akiyama T, et al. Heterozygous TGFBR2 mutations in Marfan syndrome. *Nat Genet.* 2004 August;36(8):855-860.
12. Pereira L, Lee SY, Gayraud B, et al. Pathogenetic sequence for aneurysm revealed in mice underexpressing fibrillin-1. *Proc Natl Acad Sci U S A.* 1999 March 30;96(7):3819-3823.
13. Bunton TE, Biery NJ, Myers L, Gayraud B, Ramirez F, Dietz HC. Phenotypic alteration of vascular smooth muscle cells precedes elastolysis in a mouse model of Marfan syndrome. *Circ Res.* 2001 January 19;88(1):37-43.
14. Massague J. The TGF-beta family of growth and differentiation factors. *Cell.* 1987 May 22;49(4):437-438.
15. Neptune ER, Frischmeyer PA, Arking DE, et al. Dysregulation of TGF-beta activation contributes to pathogenesis in Marfan syndrome. *Nat Genet.* 2003 March;33(3):407-411.
16. Loeys BL, Chen J, Neptune ER, et al. A syndrome of altered cardiovascular, craniofacial, neurocognitive and skeletal development caused by mutations in TGFBR1 or TGFBR2. *Nat Genet.* 2005 March;37(3):275-281.
17. Ng CM, Cheng A, Myers LA, et al. TGF-beta-dependent pathogenesis of mitral valve prolapse in a mouse model of Marfan syndrome. *J Clin Invest.* 2004 December;114(11):1586-1592.
18. De Paepe A, Devereux RB, Dietz HC, Hennekam RC, Pyeritz RE. Revised diagnostic criteria for the Marfan syndrome. *Am J Med Genet.* 1996 April 24;62(4):417-426.
19. Faivre L, Masurel-Paulet A, Collod-Beroud G, et al. Clinical and molecular study of 320 children with Marfan syndrome and related type I fibrillinopathies in a series of 1009 probands with pathogenic FBN1 mutations. *Pediatrics.* 2009 January;123(1):391-398.
20. Groenink M, Rozendaal L, Naeff MS, et al. Marfan syndrome in children and adolescents: predictive and prognostic value of aortic root growth for screening for aortic complications. *Heart.* 1998 August;80(2):163-169.
21. Rozendaal L, Groenink M, Naeff MS, et al. Marfan syndrome in children and adolescents: an adjusted nomogram for screening aortic root dilatation. *Heart.* 1998 January;79(1):69-72.
22. Nollen GJ, Groenink M, Tijssen JG, Van Der Wall EE, Mulder BJ. Aortic stiffness and diameter predict progressive aortic dilatation in patients with Marfan syndrome. *Eur Heart J.* 2004 July;25(13):1146-1152.
23. Lazarevic AM, Nakatani S, Okita Y, et al. Determinants of rapid progression of aortic root dilatation and complications in Marfan syndrome. *Int J Cardiol.* 2006 January 13;106(2):177-182.
24. Finkbohner R, Johnston D, Crawford ES, Coselli J, Milewicz DM. Marfan syndrome. Long-term survival and complications after aortic aneurysm repair. *Circulation.* 1995 February 1;91(3):728-733.
25. Engelfriet PM, Boersma E, Tijssen JG, Bouma BJ, Mulder BJ. Beyond the root: dilatation of the distal aorta in Marfan's syndrome. *Heart.* 2006 September;92(9):1238-1243.
26. Mulder BJ. The distal aorta in the Marfan syndrome. *Neth Heart J.* 2008 November;16(11):382-386.
27. Nollen GJ, van Schijndel KE, Timmermans J, et al. Pulmonary artery root dilatation in Marfan syndrome: quantitative assessment of an unknown criterion. *Heart.* 2002 May;87(5):470-471.
28. van Karnebeek CD, Naeff MS, Mulder BJ, Hennekam RC, Offringa M. Natural history of cardiovascular manifestations in Marfan syndrome. *Arch Dis Child.* 2001 February;84(2):129-137.
29. Maumenee IH. The eye in the Marfan syndrome. *Birth Defects Orig Artic Ser.* 1982;18(6):515-524.
30. Dotrelova D, Karel I, Clupkova E. Retinal detachment in Marfan's syndrome Characteristics and surgical results. *Retina.* 1997;17(5):390-396.
31. Wood JR, Bellamy D, Child AH, Citron KM. Pulmonary disease in patients with Marfan syndrome. *Thorax.* 1984 October;39(10):780-784.
32. Oosterhof T, Groenink M, Hulsmans FJ, et al. Quantitative assessment of dural ectasia as a marker for Marfan syndrome. *Radiology.* 2001 August;220(2):514-518.
33. Villeirs GM, Van Tongerloo AJ, Verstraete KL, Kunnen MF, De Paepe AM. Widening of the spinal canal and dural ectasia in Marfan's syndrome: assessment by CT. *Neuroradiology.* 1999 November;41(11):850-854.
34. Foran JR, Pyeritz RE, Dietz HC, Sponseller PD. Characterization of the symptoms associated with dural ectasia in the Marfan patient. *Am J Med Genet A.* 2005 April 1;134A(1):58-65.

35. Loeys B, Nuytinck L, Van AP, et al. Strategies for prenatal and preimplantation genetic diagnosis in Marfan syndrome (MFS). *Prenat Diagn*. 2002 January;22(1):22-28.
36. Loeys B, Nuytinck L, Delvaux I, De BS, De PA. Genotype and phenotype analysis of 171 patients referred for molecular study of the fibrillin-1 gene FBN1 because of suspected Marfan syndrome. *Arch Intern Med*. 2001 November 12; 161(20):2447-2454.
37. De Backer J, Nollen GJ, Devos D, et al. Variability of aortic stiffness is not associated with the fibrillin 1 genotype in patients with Marfan's syndrome. *Heart*. 2006 July;92(7):977-978.
38. Nijbroek G, Sood S, McIntosh I, et al. Fifteen novel FBN1 mutations causing Marfan syndrome detected by heteroduplex analysis of genomic amplicons. *Am J Hum Genet*. 1995 July;57(1):8-21.
39. Dietz HC, McIntosh I, Sakai LY, et al. Four novel FBN1 mutations: significance for mutant transcript level and EGF-like domain calcium binding in the pathogenesis of Marfan syndrome. *Genomics*. 1993 August;17(2):468-475.
40. Tynan K, Comeau K, Pearson M, et al. Mutation screening of complete fibrillin-1 coding sequence: report of five new mutations, including two in 8-cysteine domains. *Hum Mol Genet*. 1993 November;2(11):1813-1821.
41. Faivre L, Collod-Beroud G, Callewaert B et al. Clinical and mutation-type analysis from an international series of 198 probands with a pathogenic FBN1 exons 24–32 mutation. Eur J Hum Genet 2008 November 12.
42. Glesby MJ, Pyeritz RE. Association of mitral valve prolapse and systemic abnormalities of connective tissue A phenotypic continuum. *JAMA*. 1989 July 28;262(4):523-528.
43. Grau JB, Pirelli L, Yu PJ, Galloway AC, Ostrer H. The genetics of mitral valve prolapse. *Clin Genet*. 2007 October;72(4): 288-295.
44. Roman MJ, Devereux RB, Kramer-Fox R, Spitzer MC. Comparison of cardiovascular and skeletal features of primary mitral valve prolapse and Marfan syndrome. *Am J Cardiol*. 1989 February 1;63(5):317-321.
45. Ades LC, Holman KJ, Brett MS, Edwards MJ, Bennetts B. Ectopia lentis phenotypes and the FBN1 gene. *Am J Med Genet A*. 2004 April 30;126A(3):284-289.
46. Gupta PA, Putnam EA, Carmical SG, et al. Ten novel FBN2 mutations in congenital contractural arachnodactyly: delineation of the molecular pathogenesis and clinical phenotype. *Hum Mutat*. 2002 January;19(1):39-48.
47. Nishimura A, Sakai H, Ikegawa S, et al. FBN2, FBN1, TGFBR1, and TGFBR2 analyses in congenital contractural arachnodactyly. *Am J Med Genet A*. 2007 April 1;143(7): 694-698.
48. Kosaki K, Takahashi D, Udaka T, et al. Molecular pathology of Shprintzen-Goldberg syndrome. *Am J Med Genet A*. 2006 January 1;140(1):104-108.
49. Kruger WD, Wang L, Jhee KH, Singh RH, Elsas LJ. Cystathionine beta-synthase deficiency in Georgia (USA): correlation of clinical and biochemical phenotype with genotype. *Hum Mutat*. 2003 December;22(6):434-441.
50. De Lucca M, Casique L. Characterization of cystathionine beta-synthase gene mutations in homocystinuric Venezuelan patients: identification of one novel mutation in exon 6. *Mol Genet Metab*. 2004 March;81(3):209-215.
51. Groenink M, Lohuis TA, Tijssen JG, et al. Survival and complication free survival in Marfan's syndrome: implications of current guidelines. *Heart*. 1999 October;82(4):499-504.
52. Silverman DI, Burton KJ, Gray J, et al. Life expectancy in the Marfan syndrome. *Am J Cardiol*. 1995 January 15;75(2): 157-160.
53. Murdoch JL, Walker BA, Halpern BL, Kuzma JW, McKusick VA. Life expectancy and causes of death in the Marfan syndrome. *N Engl J Med*. 1972 April 13;286(15):804-808.
54. Meijboom LJ, Groenink M, Van Der Wall EE, Romkes H, Stoker J, Mulder BJ. Aortic root asymmetry in marfan patients; evaluation by magnetic resonance imaging and comparison with standard echocardiography. *Int J Card Imaging*. 2000 June;16(3):161-168.
55. Salim MA, Alpert BS, Ward JC, Pyeritz RE. Effect of beta-adrenergic blockade on aortic root rate of dilation in the Marfan syndrome. *Am J Cardiol*. 1994 September 15;74(6): 629-633.
56. Rossi-Foulkes R, Roman MJ, Rosen SE, et al. Phenotypic features and impact of beta blocker or calcium antagonist therapy on aortic lumen size in the Marfan syndrome. *Am J Cardiol*. 1999 May 1;83(9):1364-1368.
57. Shores J, Berger KR, Murphy EA, Pyeritz RE. Progression of aortic dilatation and the benefit of long-term beta-adrenergic blockade in Marfan's syndrome. *N Engl J Med*. 1994 May 12;330(19):1335-1341.
58. Engelfriet P, Mulder B. Is there benefit of beta-blocking agents in the treatment of patients with the Marfan syndrome? *Int J Cardiol*. 2007 January 18;114(3):300-302.
59. Gott VL, Greene PS, Alejo DE, et al. Replacement of the aortic root in patients with Marfan's syndrome. *N Engl J Med*. 1999 April 29;340(17):1307-1313.
60. David TE, Feindel CM, Webb GD, Colman JM, Armstrong S, Maganti M. Aortic valve preservation in patients with aortic root aneurysm: results of the reimplantation technique. *Ann Thorac Surg*. 2007 February;83(2):S732-S735.
61. David TE, Feindel CM, Webb GD, Colman JM, Armstrong S, Maganti M. Long-term results of aortic valve-sparing operations for aortic root aneurysm. *J Thorac Cardiovasc Surg*. 2006 August;132(2):347-354.
62. Yacoub MH, Gehle P, Chandrasekaran V, Birks EJ, Child A, Radley-Smith R. Late results of a valve-preserving operation in patients with aneurysms of the ascending aorta and root. *J Thorac Cardiovasc Surg*. 1998 May;115(5):1080-1090.
63. Kallenbach K, Baraki H, Khaladj N, et al. Aortic valve-sparing operation in Marfan syndrome: what do we know after a decade? *Ann Thorac Surg*. 2007 February;83(2):S764-S768.
64. Favaloro RR, Casabe JH, Segura M, et al. Surgical treatment of ascending aortic complications in Marfan syndrome: early and long-term outcomes. *Rev Esp Cardiol*. 2008 August;61(8): 884-887.
65. Bhudia SK, Troughton R, Lam BK, et al. Mitral valve surgery in the adult Marfan syndrome patient. *Ann Thorac Surg*. 2006 March;81(3):843-848.
66. Meijboom LJ, Vos FE, Timmermans J, Boers GH, Zwinderman AH, Mulder BJ. Pregnancy and aortic root growth in the Marfan syndrome: a prospective study. *Eur Heart J*. 2005 May;26(9):914-920.
67. Rossiter JP, Repke JT, Morales AJ, Murphy EA, Pyeritz RE. A prospective longitudinal evaluation of pregnancy in the Marfan syndrome. *Am J Obstet Gynecol*. 1995 November; 173(5):1599-1606.
68. Expert consensus document on management of cardiovascular diseases during pregnancy. Eur Heart J 2003 April; 24(8):761–81.

69. Habashi JP, Judge DP, Holm TM, et al. Losartan, an AT1 antagonist, prevents aortic aneurysm in a mouse model of Marfan syndrome. *Science*. 2006 April 7;312(5770):117-121.
70. Brooke BS, Habashi JP, Judge DP, Patel N, Loeys B, Dietz HC III. Angiotensin II blockade and aortic-root dilation in Marfan's syndrome. *N Engl J Med*. 2008 June 26;358(26): 2787-2795.
71. Xiong W, Knispel RA, Dietz HC, Ramirez F, Baxter BT. Doxycycline delays aneurysm rupture in a mouse model of Marfan syndrome. *J Vasc Surg*. 2008 January;47(1): 166-172.
72. Williams JA, Loeys BL, Nwakanma LU, et al. Early surgical experience with Loeys-Dietz: a new syndrome of aggressive thoracic aortic aneurysm disease. *Ann Thorac Surg*. 2007 February;83(2):S757-S763.
73. Superti-Furga A, Gugler E, Gitzelmann R, Steinmann B. Ehlers-Danlos syndrome type IV: a multi-exon deletion in one of the two COL3A1 alleles affecting structure, stability, and processing of type III procollagen. *J Biol Chem*. 1988 May 5;263(13):6226-6232.
74. Beighton P, De PA, Steinmann B, Tsipouras P, Wenstrup RJ. Ehlers-Danlos syndromes: revised nosology, Villefranche, 1997. Ehlers-Danlos National Foundation (USA) and Ehlers-Danlos Support Group (UK). *Am J Med Genet*. 1998 April 28;77(1):31-37.
75. Pepin M, Schwarze U, Superti-Furga A, Byers PH. Clinical and genetic features of Ehlers-Danlos syndrome type IV, the vascular type. *N Engl J Med*. 2000 March 9;342(10): 673-680.
76. Ng SC, Muiesan P. Spontaneous liver rupture in Ehlers-Danlos syndrome type IV. *J R Soc Med*. 2005 July;98(7): 320-322.
77. Oderich GS, Panneton JM, Bower TC, et al. The spectrum, management and clinical outcome of Ehlers-Danlos syndrome type IV: a 30-year experience. *J Vasc Surg*. 2005 July;42(1):98-106.
78. Coucke PJ, Willaert A, Wessels MW, et al. Mutations in the facilitative glucose transporter GLUT10 alter angiogenesis and cause arterial tortuosity syndrome. *Nat Genet*. 2006 April;38(4):452-457.
79. Callewaert BL, Willaert A, Kerstjens-Frederikse WS, et al. Arterial tortuosity syndrome: clinical and molecular findings in 12 newly identified families. *Hum Mutat*. 2008 January; 29(1):150-158.
80. Faiyaz-Ul-Haque M, Zaidi SH, Wahab AA, et al. Identification of a p.Ser81Arg encoding mutation in SLC2A10 gene of arterial tortuosity syndrome patients from 10 Qatari families. *Clin Genet*. 2008 August;74(2):189-193.
81. Isselbacher EM. Thoracic and abdominal aortic aneurysms. *Circulation*. 2005 February 15;111(6):816-828.
82. Coady MA, Davies RR, Roberts M, et al. Familial patterns of thoracic aortic aneurysms. *Arch Surg*. 1999 April;134(4): 361-367.
83. Biddinger A, Rocklin M, Coselli J, Milewicz DM. Familial thoracic aortic dilatations and dissections: a case control study. *J Vasc Surg*. 1997 March;25(3):506-511.
84. Milewicz DM, Chen H, Park ES, et al. Reduced penetrance and variable expressivity of familial thoracic aortic aneurysms/dissections. *Am J Cardiol*. 1998 August 15;82(4): 474-479.
85. Milewicz DM, Michael K, Fisher N, Coselli JS, Markello T, Biddinger A. Fibrillin-1 (FBN1) mutations in patients with thoracic aortic aneurysms. *Circulation*. 1996 December 1;94(11):2708-2711.
86. Zhu L, Vranckx R, Van Khau KP, et al. Mutations in myosin heavy chain 11 cause a syndrome associating thoracic aortic aneurysm/aortic dissection and patent ductus arteriosus. *Nat Genet*. 2006 March;38(3):343-349.
87. Guo DC, Pannu H, Tran-Fadulu V, et al. Mutations in smooth muscle alpha-actin (ACTA2) lead to thoracic aortic aneurysms and dissections. *Nat Genet*. 2007 December;39(12): 1488-1493.
88. McBride KL, Pignatelli R, Lewin M, et al. Inheritance analysis of congenital left ventricular outflow tract obstruction malformations: Segregation, multiplex relative risk, and heritability. *Am J Med Genet A*. 2005 April 15;134A(2): 180-186.
89. Cripe L, Andelfinger G, Martin LJ, Shooner K, Benson DW. Bicuspid aortic valve is heritable. *J Am Coll Cardiol*. 2004 July 7;44(1):138-143.
90. Garg V, Muth AN, Ransom JF, et al. Mutations in NOTCH1 cause aortic valve disease. *Nature*. 2005 September 8;437(7056):270-274.
91. McKellar SH, Tester DJ, Yagubyan M, Majumdar R, Ackerman MJ, Sundt TM III. Novel NOTCH1 mutations in patients with bicuspid aortic valve disease and thoracic aortic aneurysms. *J Thorac Cardiovasc Surg*. 2007 August;134(2): 290-296.
92. Gleason TG. Heritable disorders predisposing to aortic dissection. *Semin Thorac Cardiovasc Surg*. 2005;17(3):274-281.

Genetics of Congenital Heart Defects

19

I.C. Joziasse and J.W. Roos-Hesselink

19.1 Introduction

The first reference in history to the presence of congenital heart defects comes from a Babylonian tablet which dates back to around 4,000 BC. The description mentions: "When a woman gives birth to an infant that has the heart open and has no skin, the country will suffer from calamities", which might refer to ectopia cordis.[1] Leonardo da Vinci then was the first to describe a congenital heart defect (atrial septal defect) in humans in his Quaderni de Anatomia[1] (Fig. 19.1).

As a group, congenital heart defects are the most common birth defect and occur in approximately 4–13 per 1000 live births (excluding bicuspid aortic valves and minor self-resolving ventricle septal defects).[2] This is certainly an underestimation of the total incidence of congenital heart disease, as the incidence of congenital heart disease is much higher in fetuses that die prenatally.[3] Since many liveborn children with congenital heart defects nowadays survive into adulthood, congenital heart defects are more frequently seen in adults which require specialized care (reproductive issues, heart failure, arrhythmias, etc.). Reproductive loss is common and a high proportion of early fetal loss is associated with chromosomal defects (20–70%), which often cause congenital heart defects.[4-6] The most common congenital heart defects at birth are ventricular septal defect (26–50.5%), atrial septal defect (3.4–14.3%), pulmonary stenosis (2.4–13.5%), tetralogy of Fallot (2–10.4%), and coarctation of aorta (2–9.8%).[2,7]

Since a long time, congenital heart defects have been considered to be multifactorial in origin. Special effort has been put into identifying modifiable risk factors for risk reduction. Identified modifiable risk factors are, for example, maternal diabetes (strict glucose control during pregnancy), maternal infections, and therapeutic drug use such as anticonvulsans. Several reports suggest that maternal multivitamin use, including folic acid, is associated with a reduction of congenital cardiovascular defects (for review, see Jenkins et al. Circulation 2007).[8-13] As a result, the American Academy of Pediatrics has made specific recommendations for prospective parents to reduce the risk of having a child with a congenital heart defect.[12] They recommend to take a multivitamin with folic acid daily, optimize preconception and prenatal care with specific attention to detection and management of maternal illness (diabetes, rubella vaccination etc.), discuss any medication use, avoid contact with people with febrile illnesses, and avoid exposure to organic solvents.[12] As maternal smoking and maternal alcohol use have been found to substantially increase the relative risk for an infant with a congenital heart defect, soon-to-be mothers should be advised to refrain from these adverse habits.[14-17]

Although *environmental factors* do play a role in the pathogenesis, the prevalence of heart defects among family members of patients with isolated congenital heart disease is higher than expected (2–16% vs 0.4–0.8%). Heritability estimates (percentage of phenotype explained by inherited factors) suggest that a genetic component is very likely to contribute.[18-22] Sanchez-Cascos et al. 1978, demonstrated that 50–95% of 1148 isolated cases of congenital heart disease could be explained by *heritable factors*.[22] The recurrence risk for a sibling or offspring of an affected patient to have a congenital heart defect varies between specific

I.C. Joziasse (✉)
Department of Cardiology, University Medical Center Utrecht, Utrecht, The Netherlands
e-mail: i.c.joziasse@umcutrecht.nl

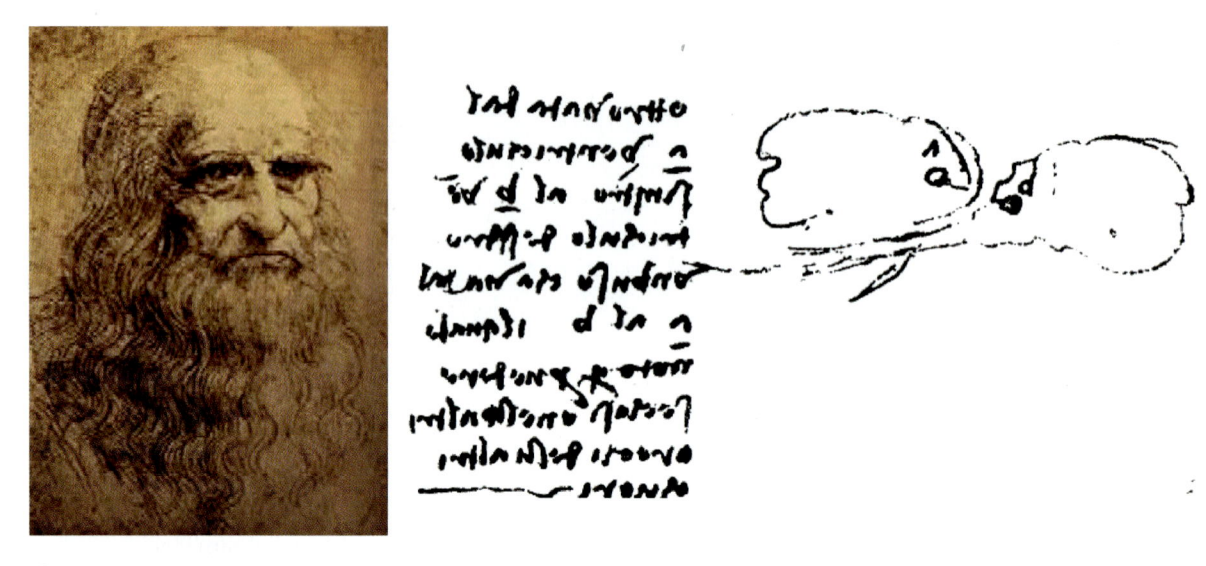

Fig. 19.1 Leonardo da Vinci, drawing of atrial septal defect in his Quaderni de Anatomia II published in 1513. The inscription read right to left: "I have found that a, left auricle, to b, right auricle, a perforating channel from a to b, which I not here to see whether this occurs in other auricles of other hearts.[1]

Table 19.1 Recurrence risks of congenital heart disease in offspring

Type of Heart disease	Total risk	Mother affected	Father affected
Acyanotic congenital heart disease			
• Atrial septal defect	3–5%	4.5–6%	1.5–3.5%
• Ventricular septal defect	4–8%	6–9.5%	2–3.6%
• Atrio-ventricular septal defect	10–15%	7.5–15%	1–7%
• Patent ductus arteriosus	3–4%	4%	2%
• Pulmonary Stenosis	4%	5.3–6.5%	2–3.5%
• Left ventricular obstruction	11–15%	10–11%	3%
• Coarctation of aorta	6%	4–6.3%	2.5–3%
Cyanotic congenital heart disease			
• Tetralogy of Fallot	2.2–3.1%	2–2.5%	1.5–2.2%
• Transposition of great vessel	0.5%		
Mendelian disorders			
• Holt-Oram syndrome	50%	50%	50%
• Noonan syndrome	50%	50%	50%
• Marfan syndrome	50%	50%	50%

cardiac defects, but the overall recurrence risk for non-Mendelian inherited congenital heart defects lies between 3–15% (Table 19.1).[23, 24] Surprisingly, for still unknown reasons, congenital heart defects occur more frequently in offspring when the mother is affected than when the father is affected.

For clinicians, it is important to identify whether a congenital heart defect has an underlying genetic component. Congenital heart disease is frequently associated with *syndromes*, which can be recognized by other dysmorphic features. In certain syndromes associated with congenital heart defects, other organ systems can be involved as well. An example is the 22q11.2 deletion syndrome (DiGeorge's syndrome/velo-cardio-facial syndrome). Children with 22q11.2 deletion syndrome most often have intellectual disabilities, and dysmorphic characteristics are usually minor (may be very difficult to detect especially in infancy). Frequently, outflow tract defects (conotruncal heart defects) such as tetralogy of Fallot are observed in these children.[25–28] Often, these children are immune compromised because of thymus hypoplasia,[25] therefore requiring irradiated erythrocytes after cardiac surgery in order to prevent graft versus host reaction. The presence of an underlying syndrome, such as a 22q11.2 deletion syndrome, may influence prognosis. Children with a microdeletion 22q11.2 and congenital heart defects are at relatively high risk for mortality and morbidity, as determined by both the severity of the cardiac lesions and the extracardiac anomalies associated with the microdeletion.[29] Moreover, establishing a syndrome diagnosis can change follow-up protocols. For example, special attention should be paid to the follow-up patients with Turner syndrome. These patients have a relatively high risk of aortic aneurysm, coarctation of the aorta, aortic valve disease, and coronary artery disease and should be monitored regularly. Additionally, identifying a syndrome or isolated genetic defect will have implications for reproductive risk (genetic counseling) and for family members (screening for genetic defects and/or cardiac anomalies). In summary, diagnosing a syndrome as the underlying cause of a heart defect can have implications for treatment and prognosis of the patient and the change on recurrence.

19.2 Congenital Heart Defects and Monogenic Disease

In isolated congenital heart disease a *monogenic* cause is rarely found. Only few families demonstrate clear Mendelian inheritance patterns. In these families, an autosomal dominant pattern or, rarely, autosomal recessive inheritance patterns are found. In families with an autosomal dominant inherited congenital heart defect, genetic research has been able to identify several mutations in genes important for the occurrence of congenital heart defects such as atrial septal defects and bicuspid aortic valves (further discussed below).[30–34] Penetrance in these families seems to be incomplete and these single gene mutations demonstrate a high phenotypic variability (*pleiotropy*) (Table 19.2). Conversely, mutations in different genes may also cause identical phenotypes, demonstrating the *genetic heterogeneity* of cardiac defects (Table 19.2). A possible explanation for this phenomenon comes from animal studies.[35] Cardiac development is controlled by multiple genetic pathways. Even in relatively small steps of development, for example, in endocardial cushion formation important to cardiac valve development, numerous genetic pathways are involved. Disturbances in each single element of a such a pathway, such as ligands, receptors, and transcription factors, could theoretically cause similar cardiac defects.

As the population frequency of recessive genetic mutations associated with congenital heart defects is small, recessive disease occurs rarely and is therefore difficult to identify. Only in regions where consanguinity is common, recessive disease has been found more frequently. Persistent Ductus Arteriosus (PDA) is such an example. PDA generally occurs sporadically, but in an Iranian population, PDA was mapped to a recessive locus at 12q24.[36]

19.3 Complex Role of Genes in Congenital Heart Defects

Most of the isolated congenital heart defects do not show a typical Mendelian inheritance pattern, but as mentioned earlier, a genetic component is very likely to contribute. This can be explained by applying the *polygenic threshold theory* for discontinuous traits. Consider a very large number of genes that are associated with heart development. Variation in all these genes may have relatively small deleterious or protective effects on the risk of congenital heart disease. All these effects added up create a certain individual disease susceptibility. This susceptibility follows a normal (Gaussian) distribution in the population. Embryos whose susceptibility exceeds a critical threshold will develop a congenital heart defect; those, whose susceptibility is below the threshold, will not. Affected people have inherited an unfortunate combination of risk-susceptibility genes. Their relatives who share many of these unfavorable genetic variants with them will therefore, as a group, also have an above-average susceptibility but will, usually, not exceed the threshold. In line with such a *multifactorial* model, in a recent study, Roesler et al. reported 35 different genetic variations or single nucleotide polymorphism (SNP's) in three genes from the Nodal pathway in 375 unrelated individuals with left–right asymmetry cardiac defects.[37] They detected only a few pathogenic mutations, but they did find functionally relevant common SNPs more frequently than expected. In addition, in several patients, they found more than one functionally relevant SNP. These common SNPs could

Table 19.2 Gene mutation and phenotype

Gene	Chromosomal location	Cardiac defect
ALK2	2q23-q24	Primum type ASD, MVP[63, 200]
BMPR2	2q33	AVSD, ASD, PDA, PAPVR+PAH[201]
CFC1	2q21.1	Heterotaxia, TGA, DORV, common AV canal, AA hypoplasia, pulmonary artresia, DIRV[202, 203]
CITED2	6q23.3	TOF, VSD, ASD, anomalous pulmonary venous return, RVOT obstruction[204]
CRELD1	22p13	AVSD, cleft mitral valve, ASD type I, heterotaxy[205, 206]
ELN	7q11.2	Supravalvular AoS[207]
FOG2	8q23	TOF[208]
GATA 4	8p23.1-p22	ASD, AVSD, pulmonary valve thickening, insufficiency of cardiac valves[30,33]
JAG1	20p12	TOF, VSD with aortic dextroposition, PPS[209]
KRAS	12p12.1	ASD, VSD, valvular PS, HCM, HOCM, MVP, TVP, LVH[210]
MYH6	14q12	Secundum ASD[211]
NKX2.5	5q34	ASD, VSD, TOF, AoS, VH Pulmonary atresia, Mitral valve anomalies, conduction disturbances[34, 212]
NKX2.6	8p21	TA[213]
NOTCH1	9q34.3	Bicuspid aortic valve, mitral valve stenosis, TOF, VSD[31, 32]
PROSIT240	12q24	TGA[214]
TBX1	22q11.2	Interrupted aortic arch, TA, other aortic arch anomalies[154]
TBX5	12q24.1	ASD, AVSD[67]
ZIC3	Xq26.2	Heterotaxy, TGA, DORV, ASD, AVSD[215]

AA aortic arch, *AoS* aortic stenosis, *ASD* atrial septal defect, *AV* atrioventricular, *AVSD* atrioventricular septal defect, *DIRV* double inlet right ventricle, *DORV* double outlet right ventricle, *HCM* hypertrophic cardiomyopathy, *HOCM* hypertrophic obstructive cardiomyopathy, *MVP* Mitral valve prolapse, *PAH* pulmonic artery hypertension, *PAPVR* partial anomalous venous return, *PDA* patent ductus arteriosus, *PPS* peripheral pulmonic stenosis, *PS* pulmonary stenosis, *RVOT* right ventricle outflow tract, *TA* truncus arteriosus, *TGA* transposition of great arteries, *TOF* tetralogy of fallot, *TVP* tricuspid valve prolapse, *VSD* ventricular septal defect, *VH* ventricular hypertrophy

theoretically act as unfavorable variants adding to the susceptibility for congenital cardiac defects Therefore, in most patients, cumulative impairment of a genetic pathway by several minor variants with a small effect seems more likely to explain the congenital heart defect than a single gene mutation.[37]

A multifactorial genesis congenital heart defects is also in line with the epidemiological principles of the *sufficient cause model* described by Dr. Kenneth Rothman.[38] A sufficient cause is a complete causal mechanism or a minimal set of conditions (components) and events that together are sufficient for the outcome, in this case, a congenital heart defect (Fig. 19.2). Each component (for example, a certain genetic variant) itself is unable to cause the defect, but when acting together with other component causes (e.g. other genetic variants or environmental factors) will result in the development of a congenital heart defect.[38] Several combinations of different components are possible to cause a congenital heart defect and thus several sufficient causes.

Genetic variation may, in addition, alter the response to our environment. If the genetic susceptibility is already high, only small additional *environmental factors* may induce disease, while in persons with low genetic susceptibility it may not. This is demonstrated in infants carrying the Reduced Folate Carrier (*RFC1* gene) polymorphism. A study of Shaw et al.

Fig. 19.2 Sufficient cause model. Constellation of components that combined has a sufficient cause to develop a congenital heart defect. *G* genetic variation, *E* environmental factor or exposure. Different combinations can lead to a sufficient cause. For example, in the first tart, a combination of two genetic variants and one environmental factor (e.g. maternal diabetes or no folic acid intake) are sufficient to cause the development of a heart defect, while in the second tart, different genetic variations altogether is sufficient to cause congenital heart disease. Also, a combination of only environmental factors could lead to congenital heart disease as depicted in tart III. Of course, for disease-causing mutations, only one component is sufficient to cause the defect as depicted in tart IV

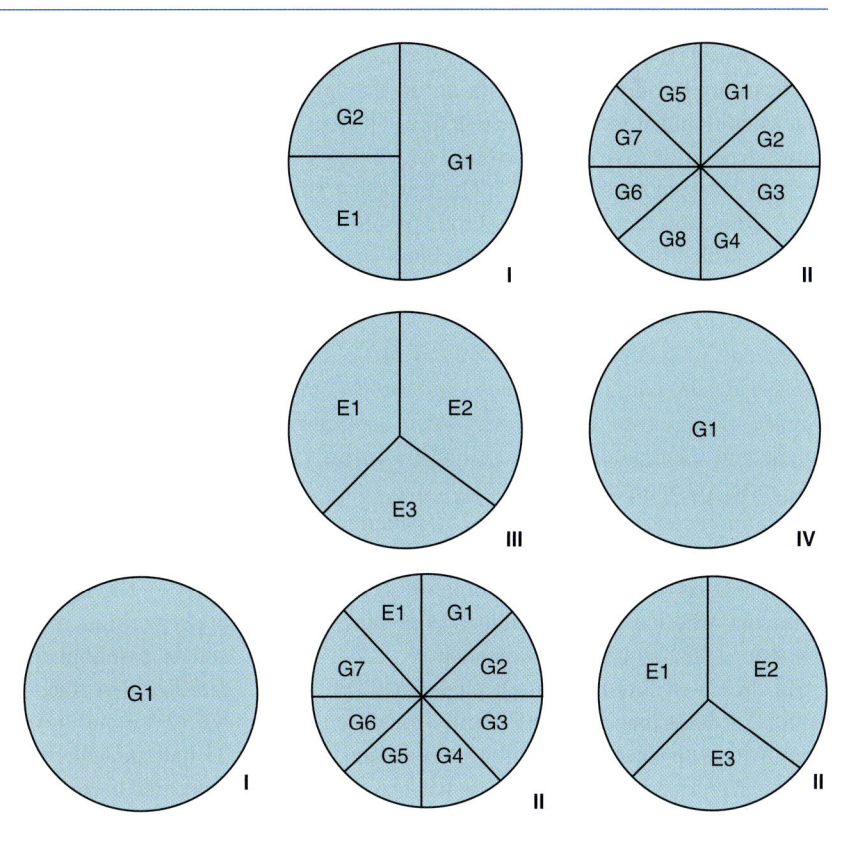

investigated whether an interaction existed between this SNP and maternal use of vitamin supplement containing folic acid on risk of conotruncal heart defects (and orofacial clefts). They found that infants who carried the *RFC1* G80 polymorphism were at increased risk (1.6–2.3 fold) of developing conotruncal heart defects. When taking into account maternal vitamin use (including folic acid), a substantially higher risk was observed for infants carrying the *RFC1* G80 polymorphism, whose mothers were nonusers of vitamin, versus infants both with and without the *RFC1* G80 polymorphism, whose mother did use vitamins. Moreover, infants with the *RFC1* A80 polymorphism seemed not at increased risk for conotruncal heart defects, even when their mother did not use vitamins (protective effect). This example nicely demonstrates how genetic variation may interact with environmental factors.[39]

19.4 Isolated Congenital Heart Defects

As described above, only in a minority of isolated congenital heart defects autosomal dominant inheritance patterns have been identified. Below, a description is given of some of the isolated congenital heart defects in which specific genes have been implicated in a Mendelian fashion.

19.4.1 Bicuspid Aortic Valves

The prevalence in the general population of a bicuspid aortic valve (BAV) is estimated to be 0.5–1.3%.[40–42] *Bicuspid aortic valves* progress more rapidly into regurgitation or stenosis of the valve.[43] This results in a higher occurrence of aortic valve replacement, especially at a younger age.[44] Patients with a bicuspid aortic valve are also more prone for endocarditis and ascending aorta dilatation which may ultimately result in aortic dissection or rupture (aortic dissection occurs 9 times more often with BAV than with tricuspid aortic valves).[45–47] Although infective endocarditis is observed more in patients with a bicuspid aortic valve, the last task force criteria from the American Heart Association (2007) have changed their recommendations for endocarditis prophylaxis.[48] They shifted their recommendation from all antibiotic prophylaxis for all patients with an increased lifetime risk for developing infective endocarditis to only patients with an underlying cardiac condition that increases the risk of an adverse outcome from infective endocarditis, which does not include bicuspid aortic valves.[48] The European guidelines (2004) still recommend antibiotic prophylaxis for patients with a bicuspid aortic valve, but it's suspected that the European guidelines will adopt the American guidelines, when they're revised.[49]

Patients with BAV are more likely to die because of cardiac death, aortic syndromes, and sudden unexplained death, even after aortic valve replacement (compared with patients with a tricuspid aortic valve and aortic valve replacement).[50]

Therefore, early identification of patients with a bicuspid aortic valve is warranted. Pedigree analysis suggests two types of inheritance patterns: autosomal dominant inheritance with reduced penetrance and a non-Mendelian pattern of inheritance.

In 36.7% of families with a bicuspid aortic valve patient, at least one additional family member can be identified with such a condition.[51] In a later report, *heritability* of a bicuspid aortic valve was estimated to be 0.89 (89% of the bicuspid aortic valve cases is explained by inherited factors).[52] Family-based genome-wide linkage analysis linked several loci with BAV such as a locus on chromosome 18 (LOD score 3.8: this means, the logarithm of the odds ratio in this case indicating the odds of ~6300:1, that the disease gene in this family is linked to this locus). Other interesting loci are on chromosome 9q34-35 and on chromosome 15q15-21 and 13q33-qter (suggestive linkage).[31,53] Thus, the origin of a bicuspid aortic valve displays genetic heterogeneity. Until now, two studies have found mutations in the *NOTCH1* gene (chromosome 9q34) in either sporadic patients or familial bicuspid aortic valve disease.[31,32] *NOTCH1* encodes a transmembrane receptor and is important for cellular differentiation, proliferation, and apoptotic programs during embryogenesis and could therefore influence organ formation and morphogenesis.[54] In approximately 4% of patients with a bicuspid aortic valve, mutations in *NOTCH1* are implicated.[32]

A bicuspid aortic valve can also occur in a family as part of a cluster of left ventricular outflow tract abnormalities, including hypoplastic left ventricle, aortic stenosis, coarctation of the aorta, mitral valve stenosis, supra valvular mitral valve stenosis, and a hypoplastic aortic arch (the *shone complex*).[55] As in most other familial isolated congenital heart defects, the phenotypic variability within families is considerably high.

19.4.2 Atrial Septal Defect (Type Secundum)

Secundum type *atrial septal defect* (ASD II) is a common congenital heart defect (3.4–14.3% of all CHD) and frequently occurs sporadically.[2,56] However, some families have been identified with an autosomal dominant inheritance pattern of ASD II both with and without conduction abnormalities.[57-60] Schott et al. was the first to link ASD II with atrioventricular (AV) conduction delay with mutations in *NKX2.5* (Chromosome 5q34).[34] NKX2.5 is a cardiac homeobox protein which plays a critical role in cardiac-specific gene expression and is essential for cardiac differentiation.[61] In four families described in the literature, all carriers of the *NKX2.5* mutations had AV conduction delay and 27 of 33 had an ASD. Other structural heart malformations identified in carriers included ventricular septal defects, tetralogy of Fallot, pulmonary atresia, and subvalvular aortic stenosis (Table 19.2).

Garg et al. and Okubo et al. identified loss of function mutations in *GATA4* (chromosome 8p23.1–p22) - a zinc-finger transcriptional factor that controls cardiac-specific gene expression and modulates cardiogenesis- segregating in 3 families with ASD II without conduction abnormalities.[30,33,62] Although, in these kindreds ASD II was the most common cardiac malformation, also atrioventricular septal defects, ventricular septal defects, pulmonic stenosis, and persistent ductus arteriosus were observed (Table 19.2). Though the mutations in *GATA4* explained the congenital heart defects observed in these families, only in a small percentage (0.5–3%) of ASD patients GATA4 mutations are detected.[63,64]

Additionally, mutations in *TBX5* have been found to be associated with ASD. *TBX5* is located on chromosome 12q21.3-q22 and has an essential role in cardiac specification and morphogenesis.[65] Mutations in *TBX5* were first found responsible for the Holt-Oram syndrome, a developmental disease affecting limbs and heart (described below).[66] Examination of heart tissue from 68 formalin fixed hearts of patients with complex heart disease (including ASD) identified 2 recurrent mutations in *TBX5* in the hearts of 16 unrelated patients with ASD.[67] It should be taken into account that these mutations were found in heart tissue and not in peripheral blood lymphocytes. This likely implies that these mutations are *somatic mutations* and not germline mutations. Somatic mutations are acquired in a somatic cell, for example, a cardiac progenitor cell, and passed on to the progeny of this mutated cell, resulting, for example, in mutations only in the heart and not in other organs/tissues. Somatic mutations are frequently caused by environmental factors such as radiation, UV light, and chemical agents, but can also occur spontaneously when aging. Germline mutations, on the other hand, are inherited genetic alterations that occur in the germ cells and can be passed on to offspring. Somatic mutations have also been described in *NKX2.5* in patients with ASD II.[68]

Conclusion: ASD II is a complex genetic heterogeneous disease (several genes are associated with one defect). Although most often the defect is isolated, a genetic component should be considered. If either a first-degree relative or two or more second-degree relatives are also known with a congenital heart defect, referral for further genetic analysis and counseling may be appropriate. Both a complex inheritance pattern and autosomal dominant inheritance pattern have been described. In addition, ASD II might be part of a syndrome. If suspicion of a syndromal or chromosomal disorder should arise, based on developmental or unexplained growth delay, or the presence of dysmorphic features, referral for syndrome diagnostics is indicated. In any case, a thorough family history for heart defects should be obtained.

19.4.3 Atrioventricular Septal Defect

Atrioventricular septal defects (AVSD) originate from an abnormal or inadequate fusion of the superior and inferior endocardial cushion with the atrial and muscular portion of the ventricular septum and can be divided in complete AVSD and partial AVSD. Of all liveborn children with a congenital heart defect, approximately 2.0–8.6% have an AVSD.[2, 69] Digilio et al. investigated the prevalence of congenital heart defects among relatives of patients with nonsyndromic AVSD and found in circa 12% of pedigrees one or more relatives with concordant or discordant congenital heart disease.[70] Congenital heart disease occurred in 1.9% of the 206 parents of an AVSD proband, in 3.6% of the 111 siblings, and in 0.8% of the 644 second-degree relatives.[70] Up to now, two genetic loci have been associated with isolated AVSD. First, analysis of a large pedigree consisting of 14 affected individuals with AVSD demonstrated linkage with a locus on chromosome 1 (1p31-21), now called AVSD1. The causative gene, however, has not yet been identified. A second locus was identified in a population of patients with **3p25 deletion syndrome**, a syndrome characterized by mental retardation, growth retardation, microcephaly, ptosis, and micrognathia. In a third of patients with 3p25 deletion syndrome congenital heart defects, mainly AVSD, are found.[71] Analysis of the critical region for congenital heart defects in this chromosomal syndrome led to the discovery of the second AVSD locus (AVSD2), located on chromosome 3p25.3, with ***CRELD1*** as the responsible gene (Cysteine-Rich protein with EGF-like domain 1).[71–74] Mutations in *CRELD1* were subsequently identified in patients with isolated AVSD as well (Table 19.2). Additionally, *GATA4* mutations have been described in patients with AVSD.[30, 33]

Most often AVSD is associated with chromosomal defects. The most frequently associated chromosomal defect (with both complete and partial AVSD) is trisomy 21 (Down's syndrome). In up to 76% of patients with AVSD, a chromosomal aberration of chromosome 21 is found (see below).[75] Other syndromes associated with AVSD are, for instance, Ivemark syndrome, Ellis van Creveld syndrome, and the CHARGE syndrome (Table 19.3).[76–79]

19.5 Congenital Heart Defects Associated with Syndromes

Frequently, congenital heart disease is associated with dysmorphic features, associated congenital malformations, mental retardation, or unexplained growth deficits, thus suggesting a syndrome diagnosis. Syndromes can be caused by both chromosomal aberrations as well as single gene mutations. Below, a description of some of the most well-known syndromes associated with congenital heart defects is given. Table 19.3 gives an overview of the most common syndromes associated with congenital heart disease.

Table 19.3 Syndromes associated with congenital heart defects (not exhaustive)

Syndrome	Chromosome	Gene	Congenital Heart defect	Other features
Adams-Oliver syndrome	unknown	unknown	Pulmonary vein stenosis, ToF, VSD, DORV, pulmonary HT [217]	Congenital aplasia cutis, limb defects, vascular defects
Alagille syndrome	20p12, 1p13-p11	JAG1, NOTCH2 [218, 219]	Pulmonic artery stenosis, peripheral arterial stenosis, ASD, VSD en CoA	Cholestasis, skeletal abnormalities, ocular abnormalities and characteristic facies
Bindewald syndrome	Unknown; AR	Unknown	ToF, DORV with subaortic VSD and PS [220]	minor facial anomalies, failure to thrive, mental retardation
Cardio-facio-cutaneous syndrome (CFC)	7q34, 12p12.1, 15q21, 7q32, 2p22-p21	BRAF, KRAS, MEK1, MEK2, SOS1 [131, 221, 222]	Valvular PS, ASD, hypertrophic cardiomyopathy	Similar features as NS, but more severe mental retardation and ectodermal abnormalities, friable hair, absent eyebrows
CHARGE syndrome	8q12, 22q11	CHD7	ASD, VSD, ToF, PDA, conotruncal and aortic arc anomalies	Coloboma, Atresia of Chonae, Retardation of growth, Genital hypoplasia and Ear abnormalities
Costello Syndrome	11p15.5, 7q34, 15q21	HRAS, BRAF, MEK1 [130, 223, 224]	Hypertrophic cardiomyopathy, VSD, PS, PDA, MVP, atrial arrhytimias [225]	Characteristic facies, short stature, distinctive hand posture and appearance, feeding difficulties, developmental delay
DiGeorge (VCF/ Takao/ 22q11 deletion/ CATCH22)	Microdeletion 22q11,18q21.33, 10p13 [145, 148]	TBX1 [152, 153]	VSD, ToF, pulmonary atresia with VSD, interrupted AA, TA, DORV, double AA, ASD	Neonatal hypocalcaemia (hypoplasia parathyroid gland) T-cell deficiency (hypoplasia thymus), low set ears, cleft palate, short stature, mental retardation, typical facies
Down syndrome/ Trisomy 21 syndrome	Trisomy 21		AVSD, VSD, ASD	Characteristic facies, mental retardation, short stature, duodenal atrasia, Hirschsprung disease, etc.
Ellis van Creveld syndrome	4p16	EVC1, EVC2 [79]	Common atrium	Dwarfism, polydactyly [226]
Holt-Oram	12q2:20q13.13-13.2	TBX5 [66, 163], SALL4	Single or multiple ASD, VSD, PDA sinus bradycardia and variable degrees of AV-block [159]	Upper limb anomalies (unilateral or bilateral), mostly on the radial side
Ivemark Syndrome	Unknown	Unknown	Dextrocardia, ASD, VSD, PS, endocardial cushion defects, conotruncal defects	Asplenia or polysplenia, malposition (abnormal lateralization) and maldevelopment of other abdominal organs, abnormal lobulation of lungs
Leopard syndrome	12q24.1, 3p25	PTPN11, RAF1 [128, 227]	Valvular PS, conduction abnormalities, left sided obstructive cardiomyopathy	Multiple lentigines, ocular hypertelorism, abn.genitalia, growth retardation, sensorineural deafness

Noonan Syndrome (NS)	12q24.1, 7q34, 12p12.1, 2p22-p21, 3p25, 15q21	*PTPN11*(~50%,), *BRAF, KRAS, SOS1, RAF1, MEK1* [123, 126, 127, 211, 227-229]	Valvular PS, ASD, hypertrophic cardiomyopathy	Short stature, characteristic facial features, high arched palate, webbed and/or short neck, pectus deformity, curly hair, coagulation defects, cryptorchidism
Opitz/GBBB Syndrome	Xp22	*MID1* [230]	ASD, VSD, ToF, HLV, PS	Characteristic face, hypertelorism, telecanthus, cleft lip/ palate, laryngomalacia, hypospadias, genitourinary anomalies
Thoraco-abdominal syndrome	Xq25-q26.1 [231]	Unknown	TGA, PDA [232]	Diaphragmatic and ventral hernia, hypoplasia of lungs, hydrocephalus, anencephaly, cleft lip, renal agenesis, hypospadias
Turner's syndrome	45,X		Coarctation of the aorta, BAV, HLHS	Growth retardation, skeletal defects, high arched palate, webbed neck, lymphoedema, ovarian failure, infertility, renal anomalies
Williams-Beuren syndrome	Deletion 7q11.23 (ELN locus) [183]	*ELN*	Supravalvular AoS, multiple pheripheral pulmonary artery stenosis, MVP, BAV coctail manner behaviour [191, 233]	Elfin face, mental and statural deficiency, dental malformation, infantile hypercalcemia

AA aortic arch, *AoS* aortic stenosis, *AR* autosomal recessive, *ASD* atrial septal defects, *AV* atrioventricular, *BAV* bicuspid aortic valve, *DORV* double outlet right ventricle, *HLV* hypoplastic left ventricle, *HT* hypertension, *NS* noonan syndrome, *MVP* mitral valve prolapse, *PDA* persistent ductus arteriosus, *PS* pulmonic stenosis, *TA* truncus arteriosus, *TGA* transposition of great arteries, *ToF* tetralogy of fallot

19.5.1 Down Syndrome/Trisomy 21

Down syndrome is the most frequent form of mental retardation. It is caused by a microscopically demonstrable chromosomal 21 aberration (**trisomy 21**) and occurs in one of 800–1000 live births.[80–82] This syndrome is characterized by well-defined and distinctive phenotypic features including characteristic facies, minor dysmorphisms of the limbs, hypotonia, and growth retardation.[83] Frequently, Down syndrome is associated with other specific congenital malformations such as Hirschprung's disease and duodenal abnormalities.[83] The incidence of congenital heart defects in Down syndrome is about 40–60%, most frequently **atrioventricular septal defects** (AVSD) and ventricular septal defects (VSD).[75, 84]

The risk for having a child with Down syndrome increases with maternal age.[80, 81, 85, 86] Besides, it is indicated that certain maternal genetic polymorphisms of genes involved in the **folate/homocysteine metabolism** or the combination of these polymorphisms can be associated with higher incidence of offspring with trisomy 21, although results from different reports have been conflicting and/or inconclusive.[87–92]

Additionally, it has been hypothesized and suggested in some reports that factors in the maternal grandmother such as age and altered folate/homocysteine metabolism can be implicated in chromosome 21 nondisjunction risk as well.[93–97]

In 95% of Down syndrome patients, three free copies of chromosome 21 are identified, in most instances caused by an error in maternal meiosis I. In most of the remaining patients, one copy of chromosome 21 is translocated to another acrocentric chromosome (**robertsonian translocation**).[81] Robertsonian translocations may be the cause of familial occurrence of Down syndrome. Although chromosome analysis is often not absolutely necessary to establish a diagnosis of Down syndrome, it should be performed to determine whether there may be an increased risk for future offspring.

Though it can be assumed that increased dosage of specific chromosome 21 genes plays a role, thus far, no chromosome 21 genes responsible for AVSD have been unambiguously identified. Importantly, only a percentage of Down syndrome patients exhibit a congenital heart defect. This suggests that factors other than overexpression of genes on chromosome 21 are needed for cardiac defects to occur. These may be environmental factors or genetic variation at other loci. Specific environmental risk factors have been implicated with the occurrence of Down syndrome with the cooccurrence of a congenital heart defect such as maternal diabetes (OR20.6) and smoking, but these findings have not been confirmed in other studies.[75, 98, 99] The presence of specific gene variants could, in addition to trisomy 21, further increase susceptibility for cardiac defects. Mutations in **CRELD1** have been found in Down syndrome patients with AVSD.[71, 100, 101]

19.5.2 Turner (Ullrich-Turner) Syndrome

Turner syndrome appears in 32–50 per 100.000 females and is cytogenetically characterized by the absence of one of both X-chromosomes (45,X) or a structurally abnormal X-chromosome, where a part has been deleted. Clinical manifestations include short stature, specific dysmorphic features (that are not always present), primary amenorrhea, absence of secondary sex characteristics, and cardiovascular malformations.[102–105] Additionally, there is an increased incidence of hypothyroidism, other congenital malformations (such as malformations of the urinary tract), deafness, and fractures (osteoporotic fractures in adulthood, nonosteoporotic fractures in childhood).[106, 107]

Congenital cardiac defects occur in about 22–55% of affected individuals. Most frequently, coarctation of the aorta and bicuspid aortic valves, and also anomalous pulmonary venous drainage and other types of aortic valve disease can be found.[108–110] Not completely unexpected, the presence of cardiac lesions seems to be dependent on the karyotype. In a report of Prandstraller et al. severe congenital heart defects and multiple lesions were found only in patients with a 45,X karyotype. Patients with an X-ring pattern had a higher prevalence of bicuspid aortic valves (BAV) while patients with deletions of the X-chromosome did not demonstrate any cardiac defects.[109] Specific phenotypic features may hint on the presence of congenital heart defects, such as neck webbing; 50% of patients with webbing of the neck were found to have congenital heart defects, versus 23% of the patients without neck webbing.[110]

Additionally, a specific ECG pattern can be found in Turner patients. More frequently, a left posterior fascicular Block and accelerated AV conduction is observed and the QTc is significantly longer (average 423 mm vs 397 ms).[111]

While Turner syndrome is a relatively common syndrome, there is a considerable delay in diagnosis of the syndrome. Although, many are ascertained at an earlier age because of short stature, it is not uncommon for these patients to come to medical attention because of primary amenorrhea (around 15 years of age).[104] Women with Turner syndrome have an increased cardiovascular risk as a result of an increased incidence of diabetes (both insulin dependent and noninsulin dependent), hypertension, and decreased estrogen levels.[106, 108] They need life-long estrogen replacement therapy starting from puberty onwards. Among Turner females, ischemic heart disease and stroke are more frequently observed than in the general population.[106]

All-cause mortality is increased considerably in patients with Turner syndrome (standardized mortality ratio of 3).[104, 112] Mortality risks are generally greatest in women with 45, X monosomy and in patients diagnosed young.[112] Mortality is specifically increased for circulatory diseases such as aortic aneurysm, aortic dilatation and dissection, aortic valve disease, hypertensive and ischemic heart disease, and cerebrovascular disease. Therefore, the aorta and aortic valve (aortic diameter, as well as the presence of aortic valve disease) should be monitored regularly. Furthermore, primary and secondary preventive strategies should be developed for further decreasing cardiovascular risk, such as myocardial infarction and strokes, in this patient population. Next to circulatory diseases, respiratory disease mortality (mainly pneumonia) and mortality from endocrine and associated disease (especially diabetes mellitus type 2) are raised.[104, 112]

19.5.3 Noonan Syndrome

In 1962, Dr. JA Noonan presented a new syndrome in 9 children with pulmonary stenosis. The syndrome was characterized by small stature, ptosis, hypertelorism, mild mental retardation, and in some instances, undescended testes and skeletal malformations.[113] Later, she described the typical facies, cardiac features, and clinical phenotype as resembling Turner syndrome (described above), but with normal chromosomes and occurring in both males and females.[114] Subsequently, more clinical details were identified in individuals with **Noonan syndrome**, including low set posteriorly rotated ears with a thick helix, deafness and pectus carinatum and excavatum, and blood clotting anomalies.[115] Phenotypic craniofacial appearance changes with age. Subtle features must be searched for in parents of affected children.[116] Congenital cardiac defects are found in two-third of patients. Common anomalies are pulmonary valvular abnormalities in 50–62% (both dysplastic and stenotic valves), hypertrophic cardiomyopathy in 10–25% (especially anterior septal hypertrophy), and atrial septal defect in 8–10%.[115, 117–119] This listing is not exhaustive and other congenital heart defects have been described in individuals with Noonan syndrome (Table 19.3).[119] Although exact birth prevalence of Noonan syndrome is unknown, it has been suggested to be between 1 in 1000 to 1 in 2500 livebirths.[115]

The syndrome is inherited in an autosomal dominant fashion and has been initially linked to the ***PTPN11*** gene (chromosome 12q24.1).[120–123] Approximately, 33–50% of Noonan patients have activating missense mutations in *PTPN11*, but also mutations in other genes of the **MAPK pathway** (e.g. *SOS1, KRAS, MEK1, MEK2, RAF1, HRAS, BRAF*) have been found in affected individuals (Table 19.3, Fig. 19.3).[124–127] Some of these genes have also been implicated in Cardiofaciocutaneous (CFC) syndrome, Costello's syndrome, and LEOPARD syndrome.[128–131] However, distinct genotype–phenotype correlations exist (RAF1 associates with hypertrophic cardiomyopathy, PTPN11/SOS1 dysplastic pulmonic valves, etc.). This, taken together with the fact that these syndromes phenotypically overlap and can occasionally be found within one family, indicates that the old syndromal nomenclature is not always adequate and stressed the need for molecular diagnosis.

19.5.4 Digeorge Syndrome/ Velocardiofacial Syndrome/22q11 Deletion

DiGeorge was the first who reported in 1965 congenital absence of the thymus and parathyroid glands in 4 children,[132] resulting in hypocalcaemia and defective cellular immunity. Subsequently, congenital heart

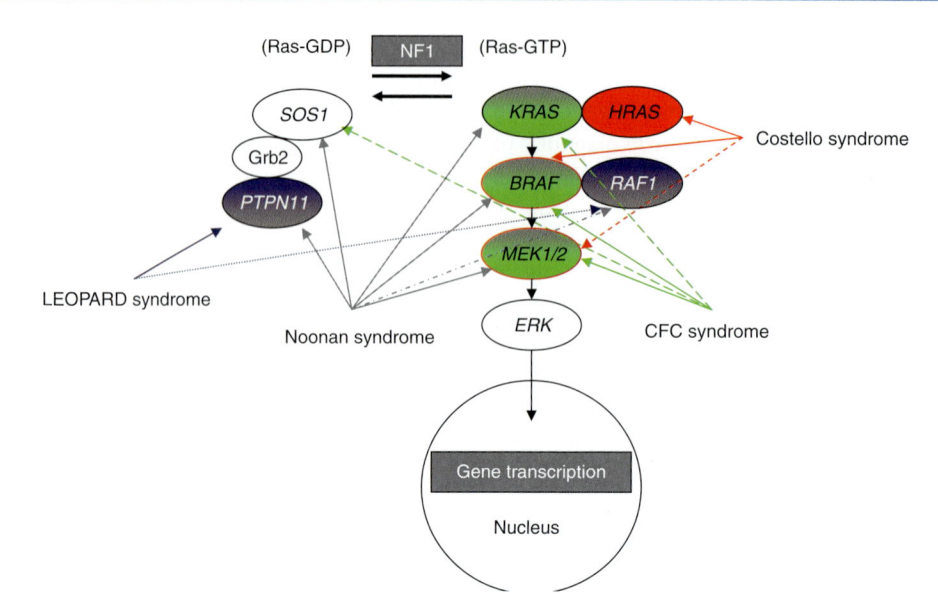

Fig. 19.3 MAPK-pathway and associated syndromes. Mutations in genes of the MAPK pathway have been found in individuals with Noonan syndrome, Cardiofaciocutaneous (*CFC*) syndrome, LEOPARD syndrome, and Costello's syndrome. These syndromes phenotypically overlap and as the same genes are implicated, these syndromes probably belong to the same disease entity. PTPN11 and RAF1 can be mutated in Noonan syndrome and in LEOPARD syndrome and are shown in both gray and blue. Mutations have been found in SOS1 and KRAS in patients with both Noonan syndrome and CFC syndrome and are shown in both gray and green, while BRAF and MEK mutations have been found in the latter two syndromes plus Costello syndrome (encircled in red). Mutations in HRAS have been found only in patients with Costello syndrome depicted in red here

defects and a characteristic facial appearance were found to be associated with this finding.[133, 134] Additional features found in **DiGeorge syndrome** are velopharyngeal dysfunction, or cleft palate, discrete dysmorphisms, developmental and behavioral problems, and psychiatric disorders in adulthood (mainly psychosis and schizophrenia).[28, 135–137] Congenital heart defects are detected in circa 80% of DiGeorge patients and consist primarily of **outflow tract anomalies** like interrupted aortic arch (IAA), truncus arteriosus (TA), and tetralogy of Fallot (TOF)(Table 19.3).[134, 137, 138]

The incidence of this syndrome is estimated to be at least 1 in 4000–6000 livebirths, but this might be an underestimation as many cases with mild features may remain undiagnosed.[138–140] Familial cases of DiGeorge have been described demonstrating an autosomal dominant inheritance pattern,[141–144] but usually DiGeorge syndrome occurs sporadically and results from a de novo 22q11.2 (micro)deletion.[142, 143, 145–148] Although the penetrance of a 22q11 deletion is nearly 100%, the severity of the disorders is variable.[28, 149] Importantly, 22q11.2 deletions supposedly are found in 6–11% of patients diagnosed with isolated **tetralogy of Fallot** (TOF) as well.[150, 151] Especially, when TOF is associated with pulmonary atresia, the chance of finding a 22q11.2 deletion is very high. Mutations in *TBX1*, located within the minimal deleted region, have been found responsible for the major features of this syndrome, but are a rare cause of this syndrome (Table 19.3).[152, 153] Other genes within the deletion definitely contribute to the phenotype, especially to the mental retardation and psychiatric symptoms. In rare cases of isolated aortic arch anomalies and truncus arteriosus mutations in TBX1 have also been found.[154]

In a minority of patients with a DiGeorge syndrome phenotype, other genetic chromosome defects have been associated with the defect such as deletions at 18q21.33 and 10p13.[148]

Various diagnostic terms have been assigned to the constellation of features of DiGeorge syndrome including **Velo-cardio-facial syndrome** (VCF), **22q11.2 deletion syndrome**, **Takao syndrome**, and **CATCH22**. All of these terms are now acknowledged to represent variant manifestations of the same entity, as all of these syndromes are caused by the same 22q11.2 microdeletion and demonstrate an extensive overlap of phenotypes.[138, 155]

For the clinician, it is important to be aware of the possibility of this syndrome in patients with **conotruncal cardiac malformations**. Mortality and morbidity after corrective surgery for congenital heart defects is higher in these patients than in those with isolated congenital heart defects. In addition, irradiated blood products and blood sero-negative for CMV should be given when blood transfusion is indicated (because of immuno incompetence due to thymus hypoplasia and T-cell dysfunction). Furthermore, because of parathyroid hypoplasia, there is always a chance of hypocalcemia, even in patients who have never experienced episodes of hypocalcemia before.[156] As dysmorphic features can be very subtle, especially in neonates newly identified with a heart defect, it is recommended to screen newborns with TOF, TA, and IAA for the presence of a 22q11.2 deletion.

19.5.5 Holt-Oram

The **Holt-Oram syndrome**, also called **heart-hand syndrome**, was first described in 1960.[157] It is an inherited disorder causing anomalies of the upper limb (mainly the thumbs) and heart with an incidence of approximately 1 per 100.000 births.[157, 158] Affected individuals exhibit skeletal abnormalities ranging from subclinical radiographic findings (abnormal carpal bones are almost always present) to more obvious radial defects (flattened thenar, fingerlike triphalangeal thumbs, or reduction defects).[157, 159–161] Heart defects occur in circa 76–95% of patients and are most often **atrial septal defects** (secundum type), ventricular septal defects, conduction abnormalities, and supraventricular arrhythmias.[159, 160, 162] But, as for most syndromes, other heart defects are sometimes observed as well (Table 19.3). The syndrome is transmitted as an autosomal dominant trait that is highly penetrant, although the clinical manifestations vary significantly, even within families.[157, 159, 162] Sporadic cases of Holt-Oram occur frequently as well, presumably due to *de novo* mutations.[160, 161, 163]

This syndrome maps to a gene named ***TBX5***, a transcription factor important for cardiac tissue specification and formation, located on chromosome 12q24.1.[66, 163–165] Mutations in *TBX5* (loss of function of the transcription factor *TBX5*) are found in circa 35–74% of affected individuals, depending on (strict) clinical criteria used for diagnosis.[166, 167] In a small minority of Holt-Oram cases, mutations in *SALL4* have been demonstrated and also deletions on chromosome 14 (14q23.3q31.1) and on chromosome 6 have been found.[168–170] Therefore, the Holt-Oram syndrome is genetically heterogeneous.

19.5.6 CHARGE Syndrome

The combination of choanal atresia and coloboma with other congenital, specific for the **CHARGE syndrome** was first described by Hall and Hittner in 1979 and has a birth prevalence of circa 1 in 8500 live births[25, 171–173]. The CHARGE acronym stands for Coloboma of the eye, Heart defects, Atresia of the choanae (congenital abnormality of the anterior base skull characterized by blockage of one or both of the posterior nasal cavities), Retardation of growth and development, Genital hypoplasia, and Ear abnormalities.[173, 174] Although this acronym helped identification of affected individuals, diagnostic uncertainties occurred in patients with only some CHARGE features warranting more specific diagnostic criteria which were developed by Blake and further specified by Verloes (Table 19.4).[175, 176] In approximately 83–85% of affected individuals, congenital heart defects can be observed (of any type), ranging from persistent ductus arteriosus, ventricle septal defects, atrial septal defects, and conotruncal abnormalities.[25, 78, 173] CHARGE usually appears sporadic but some familial cases have been described as well. In circa 65–73% of CHARGE patients, mutations or deletions of the chromodomain helicase DNA-binding 7 gene (**CHD7**) can be detected, mostly *de novo* mutations.[25, 177, 178] When the diagnostic criteria according to Blake and Verloes are strictly applied this percentage is much higher.[177] Therefore, CHD7 seems to be the major gene involved in the development of this syndrome. However, microdeletions in 22q11 have been detected in CHARGE patients as well.[179, 180]

19.6 Williams (Williams-Beuren) Syndrome

Williams syndrome is a developmental disorder affecting circa 1 in 7500 caused by a heterozygous deletion of circa 1.5–2 Mb of chromosome 7q11.23

Table 19.4 Recommendations for clinical care in patients with congenital heart defects

1. Obtain a thorough family history (pedigree analysis)
2. Consider physical examination with a special emphasis on cardiac murmurs in first-degree relatives, especially for defects known to cluster within families (bicuspid aortic valve and other left ventricular outflow tract obstructions)
3. Consider consultation of clinical geneticist if: – A congenital heart defect present in one first-degree or two second/third-degree relatives – Family history of other congenital defects or syndromes (e.g. laterality defects, mental retardation, choanal atresia, cleft lip/palate) – Suspicion of a syndrome (e.g. additional birth defects, learning disabilities/mental retardation, short stature, and dysmorphic features) – Certain cardiac defects (tetralogy of Fallot/truncus arteriosus/interrupted aortic arch type B (22q11del))
4. Offer prenatal evaluation during pregnancy if: – Congenital heart defect is present in parent – Congenital heart defect is present in one or more offspring(s) – Evidence for strong genetic predisposition for cardiac defects based on family history

including the **Elastin (ELN) locus** in 90–94%.[181–184] Usually, it occurs sporadically, but some families with an autosomal dominant inheritance pattern have been described as well.[182, 185] This syndrome is characterized by a characteristic facies (including an elfin face), developmental delay of variable severity, a friendly personality, infantile hypercalcemia, and congenital heart defects (in circa 80%).[185–188] **Supravalvular aortic stenosis** (SVAS) and peripheral pulmonary artery stenosis are most frequently associated with Williams syndrome. Especially SVAS has been related to Elastin hemizygosity.[182, 189] Also, other congenital heart defects have been described in patients with Williams syndrome such as atrial and ventricular septal defects, bicuspid aortic valve, pulmonary valve stenosis, coarctation of the aorta, and mitral valve prolapse.[188, 190–192]

Severity of SVAS can change over time. In one report, with a mean follow-up of 12.9 years in 59 patients with Williams syndrome and SVAS, it became evident that pressure gradients of less than 20 mmHg in infancy generally remained unchanged during the first 2 decades of life. Pressure gradients exceeding 20 mmHg, increased from an average of 36–53 mmHg in 13 patients. Additionally, after corrective surgery for SVAS, restenosis occurred frequently.[193] Conversely, in an Asian population, progression of SVAS was uncommon and the probability of regression was around 30%.[192] Interestingly, an increased prevalence and severity of SVAS and total congenital heart defects are observed in male patients with Williams syndrome. As age at diagnosis of congenital heart defects is strongly related with age at clinical diagnosis of Williams syndrome, this could explain why male patients are diagnosed, on average, at a younger age than female patients.[188]

Besides congenital heart defects, in circa 17–55% of both young and older patients hypertension can be diagnosed.[185, 187, 194, 195] Increased aortic stiffness and decreased arterial compliance have been described in William syndrome patients, which could explain most of the hypertension cases in these patients.[196] Also, renal artery stenosis and abdominal aortic narrowing, as underlying cause of hypertension, have been described in patients with Williams syndrome.[185, 197]

19.7 Summary and Future Research Projects

Isolated congenital heart defects are most frequently sporadic. Although sporadic, a genetic component is very likely to contribute to the occurrence of these defects.[22]

A Mendelian inheritance pattern is only rarely observed and therefore, up to now, **molecular genetic diagnostic tests**, apart from 22q11.2 deletion testing in newborns and small children, only occasionally play a role of importance. However, the clinician should be aware that congenital heart defects are frequently associated with **syndromes** and in these syndromic cases a genetic diagnosis is important. Syndromes associated with congenital heart defects may have very subtle dysmorphic features, and can easily be missed

(Table 19.4). Therefore, clinicians should be trained to be able to recognize at least the most common syndromes. Features that should raise suspicion of a syndrome are associated birth defects, learning disabilities or mental retardation, and short stature. Besides, the presence of family members with a syndrome diagnosis or congenital heart defect may indicate the possible occurrence of a syndrome in the patient. As phenotypical variability has been proven to be high, even the presence of another type of congenital heart defect (with or without dysmorphic features) in one first-degree relative or two second or third-degree relatives, or request of a familymember warrants consultation from a clinical geneticist. Therefore, special attention should be given to a thorough family history.

Finally, specific types of congenital heart defects may also indicate the possible presence of a genetic defect. In patients with Tetralogy of Fallot, for example, frequently a 22q11 deletion is found, and as explained above, this has consequences for treatment and reproductive risk, and might have implications for other family members as well (Table 19.4).[150,151] For family members of patients with a bicuspid aortic valve, the presence of such a bicuspid aortic valve or other obstructive left ventricular outflow tract anomaly should be suspected[52,55] (Table 19.4).

Additionally, **prenatal evaluation** should be offered to prospective parents when one parent has a congenital heart defect or when they previously had a child with a congenital heart defect. In specific cases, where there is strong evidence for a genetic contribution to cardiac defects within the families, prenatal evaluation can also be useful (Table 19.4). Prenatal evaluation is primarily needed to medically plan and prepare the delivery of a child with a congenital heart defect, such as delivery in a hospital with specialized neonatal care and prostaglandin administration to maintain an open ductus for cyanotic congenital heart defects depending on ductal flow.

Future research in isolated congenital heart defects will focus on unraveling the intricate role of genes in the development of congenital heart defects. We expect through both candidate gene sequencing screens and **genome-wide association studies**, (GWAs) that more and more genes will be discovered to be involved in human congenital heart defects. The advantage of GWAs is that it is hypothesis free. Using Single Nucleotide Polymorphism (SNP)-arrays, the whole genome is scanned for regions associated with the trait of interest. It's now used to detect genetic factors contributing to common complex diseases such as myocardial infarction. One important assumption is that the genetic factor one is looking for is a common variant, to be able to reliably detect an association if present.[198] In congenital heart disease the genetic variants associated are supposedly rare. Variants with a large effect on cardiac development will be selflimitting as a result of selection and therefore will not be consistently linked to a specific haplotype. To detect an association with a rare variant (with a small effect), very large numbers of patients and controls are needed. When interesting variants are picked up, a replication cohort should be sought for to confirm the findings.[198] As the traits of interest are also relatively rare, the number of affected persons is limited. Only through large worldwide collaborations, this type of research will be within reach in the near future.

Another hot topic at the moment is whole genome sequencing. A competition has been set up named "the 1000 dollar genome". Researchers all over the world are now trying to determine an individual's whole genome sequence for only 1000 dollars.[199] Therefore, in the near future, **whole genome sequencing** will not only be achievable but also come within reach for advanced clinical diagnostic test. Whole genome sequencing will generate a lot of new information and interpreting these results will be difficult. Tools for the interpretation of such huge amounts of data in the context of congenital heart defects are not readily available. The more complex and heterogeneous the disorder, the more difficult this will be. While determining which genetic variations are associated with the development of congenital heart defects, also interactions between genes (gene–gene interaction) and between genes and environmental factors (gene–environment interaction) should be further investigated. In summary, in the coming decade, a lot of new information will be generated and this will probably change the understanding of the pathogenesis of congenital heart defects tremendously.

As the genetic knowledge of the development of congenital heart defects progresses, collaboration between the clinician involved in the treatment of patients with congenital heart defects and the department of clinical genetics is warranted for both interpretation of new findings and genetic counseling.

References

1. Rashkind WJ. Pediatric Cardiology: A Brief Historical Perspective. *Pediatr Cardiol.* 1979;1:63-71.
2. Hoffman JI. Incidence of congenital heart disease: I. Postnatal incidence. *Pediatr Cardiol.* 1995;16:103-13.
3. Hoffman JI. Incidence of congenital heart disease: II. Prenatal incidence. *Pediatr Cardiol.* 1995;16:155-65.
4. Burgoyne PS, Holland K, Stephens R. Incidence of numerical chromosome anomalies in human pregnancy estimation from induced and spontaneous abortion data. *Hum Reprod.* 1991;6:555-65.
5. Hook EB. Prevalence of chromosome abnormalities during human gestation and implications for studies of environmental mutagens. *Lancet.* 1981;2:169-72.
6. Mikamo K. Anatomic and chromosomal anomalies in spontaneous abortion. *Possible correlation with overripeness of oocytes Am J Obstet Gynecol.* 1970;106:243-54.
7. Dadvand P, Rankin J, Shirley MD, Rushton S, Pless-Mulloli T. Descriptive epidemiology of congenital heart disease in Northern England. *Paediatr Perinat Epidemiol.* 2009;23:58-65.
8. Loffredo CA, Wilson PD, Ferencz C. Maternal diabetes: an independent risk factor for major cardiovascular malformations with increased mortality of affected infants. *Teratology.* 2001;64:98-106.
9. Ramos-Arroyo MA, Rodriguez-Pinilla E, Cordero JF. Maternal diabetes: the risk for specific birth defects. *Eur J Epidemiol.* 1992;8:503-8.
10. Botto LD, Lynberg MC, Erickson JD. Congenital heart defects, maternal febrile illness, and multivitamin use: a population-based study. *Epidemiology.* 2001;12:485-90.
11. Kelly TE, Edwards P, Rein M, Miller JQ, Dreifuss FE. Teratogenicity of anticonvulsant drugs. II: A prospective study. *Am J Med Genet.* 1984;19:435-443.
12. Jenkins KJ, Correa A, Feinstein JA, et al. Noninherited risk factors and congenital cardiovascular defects: current knowledge: a scientific statement from the American Heart Association Council on Cardiovascular Disease in the Young: endorsed by the American Academy of Pediatrics. *Circulation.* 2007;115:2995-3014.
13. Smedts HP, de Vries JH, Rakhshandehroo M, et al. High maternal vitamin E intake by diet or supplements is associated with congenital heart defects in the offspring. *BJOG.* 2009;116:416-23.
14. Källén K. Maternal smoking and congenital heart defects. *Eur J Epidemiol.* 1999;15:731-7.
15. Malik S, Cleves MA, Honein MA, et al. Maternal smoking and congenital heart defects. *Pediatrics.* 2008;121:e810-e816.
16. Webster WS, Germain MA, Lipson A, Walsh D. Alcohol and congenital heart defects: an experimental study in mice. *Cardiovasc Res.* 1984;18:335-8.
17. Autti-Ramo I, Fagerlund A, Ervalahti N, Loimu L, Korkman M, Hoyme HE. Fetal alcohol spectrum disorders in Finland: clinical delineation of 77 older children and adolescents. *Am J Med Genet Part A.* 2006;140:137-43.
18. Campbell M. Incidence of cardiac malformations at birth and later, and neonatal mortality. *Br Heart J.* 1973;35:189-200.
19. Samánek M. Congenital heart malformations: prevalence, severity, survival, and quality of life. *Cardiol Young.* 2000;10:179-85.
20. Loffredo CA, Chokkalingam A, Sill AM, et al. Prevalence of congenital cardiovascular malformations among relatives of infants with hypoplastic left heart, coarctation of the aorta, and d-transposition of the great arteries. *Am J Med Genet Part A.* 2004;124:225-30.
21. van der Velde ET, Vriend JW, Mannens MM, Uiterwaal CS, Brand R, Mulder BJ. CONCOR, an initiative towards a national registry and DNA-bank of patients with congenital heart disease in the Netherlands: rationale, design, and first results. *Eur J Epidemiol.* 2005;20:549-57.
22. Sanchez-Cascos A. The recurrence risk in congenital heart disease. *Eur J Cardiol.* 1978;7:197-210.
23. Burn J, Brennan P, Little J, et al. Recurrence risks in offspring of adults with major heart defects: results from first cohort of British collaborative study. *The Lancet.* 1998; 351:311-6.
24. Pradat P. Recurrence risk for major congenital heart defects in Sweden: a registry study. *Genet Epidemiol.* 1994;11:131-40.
25. Moerman P, Goddeeris P, Lauwerijns J, Van der Hauwaert LG. Cardiovascular malformations in DiGeorge syndrome (congenital absence of hypoplasia of the thymus). *Br Heart J.* 1980;44:452-9.
26. Webber SA, Hatchwell E, Barber JC, et al. Importance of microdeletions of chromosomal region 22q11 as a cause of selected malformations of the ventricular outflow tracts and aortic arch: a three-year prospective study. *J Pediatr.* 1996;129:26-32.
27. Khositseth A, Tocharoentanaphol C, Khowsathit P, Ruangdaraganon N. Chromosome 22q11 deletions in patients with conotruncal heart defects. *Pediatr Cardiol.* 2005;26:570-3.
28. Bassett AS, Chow EW, Husted J, et al. Clinical features of 78 adults with 22q11 Deletion Syndrome. *Am J Med Genet Part A.* 2005;138:307-13.
29. Kyburz A, Bauersfeld U, Schinzel A, et al. The fate of children with microdeletion 22q11.2 syndrome and congenital heart defect: clinical course and cardiac outcome. *Pediatr Cardio.* 2008;29:76-83.
30. Smith KA, Joziasse IC, Chocron S, et al. Identification of a dominant-negative ALK2 allele in a family with congenital heart defects. *Circulation.* 2009;119:3062-9.
31. Joziasse IC, Smith K, van der Smagt JJ, Mulder BJ, Bakkers J, Doevendans PA. Abstract 2133: Mutations In Alk2 Are Associated With Congenital Atrioventricular Valve-And Septal Defects. *Circulation.* 2007;116:II.
32. Roberts KE, McElroy JJ, Wong WPK, et al. BMPR2 mutations in pulmonary arterial hypertension with congenital heart disease. *Eur Respir J.* 2004;24:371-4.
33. Bamford RN, Roessler E, Burdine RD, et al. Loss-of-function mutations in the EGF-CFC gene CFC1 are associated with human left-right laterality defects. *Nat Genet.* 365;26:365-369.
34. Goldmuntz E, Bamford R, Karkera JD, Dela D, Roessler E, Muenke M. CFC1 mutations in patients with transposition of the great arteries and double-outlet right ventricle. *Am J Hum Genet.* 2002;70:776-780.

35. Sperling S, Grimm CH, Dunkel I, et al. Identification and functional analysis of CITED2 mutations in patients with congenital heart defects. *Hum Mutat*. 2005;26:575-82.
36. Sheffield VC, Pierpont ME, Nishimura D, et al. Identification of a complex congenital heart defect susceptibility locus by using DNA pooling and shared segment analysis. *Hum Mol Genet*. 1997;6:117-21.
37. Robinson SW, Morris CD, Goldmuntz E, et al. Missense mutations in CRELD1 are associated with cardiac atrioventricular septal defects. *Am J Hum Genet*. 2003;72:1047-52.
38. Metcalfe K, Rucka AK, Smoot L, et al. Elastin: mutational spectrum in supravalvular aortic stenosis. *Eur J Hum Genet*. 2000;8:955-63.
39. Pizzuti A, Sarkozy A, Newton AL, et al. Mutations of ZFPM2/FOG2 gene in sporadic cases of tetralogy of Fallot. *Hum Mutat*. 2003;22:372-377.
40. Garg V, Kathiriya IS, Barnes R, et al. GATA4 mutations cause human congenital heart defects and reveal an interaction with TBX5. *Nature*. 2003;424:443-7.
41. Okubo A, Miyoshi O, Baba K, et al. A novel GATA4 mutation completely segregated with atrial septal defect in a large Japanese family. *J Med Genet*. 2004;41:e97.
42. Eldadah ZA, Hamosh A, Biery NJ, et al. Familial Tetralogy of Fallot caused by mutation in the jagged1 gene. *Hum Mol Genet*. 2001;10:163-9.
43. Schubbert S, Zenker M, Rowe SL, et al. Germline KRAS mutations cause Noonan syndrome. *Nat Genet*. 2006;38:331-6.
44. Ching YH, Ghosh TK, Cross SJ, et al. Mutation in myosin heavy chain 6 causes atrial septal defect. *Nat Genet*. 2005;37:423-8.
45. Schott JJ, Benson DW, Basson CT, et al. Congenital heart disease caused by mutations in the transcription factor NKX2-5. *Science*. 1998;281:108-11.
46. König K, Will JC, Berger F, Müller D, Benson DW. Familial congenital heart disease, progressive atrioventricular block and the cardiac homeobox transcription factor gene NKX2.5:: identification of a novel mutation. *Clin Res Cardiol*. 2006;95:499-503.
47. Heathcote K, Braybrook C, Abushaban L, et al. Common arterial trunk associated with a homeodomain mutation of NKX2.6. *Hum Mol Genet*. 2005;14:585-93.
48. Garg V, Muth AN, Ransom JF, et al. Mutations in NOTCH1 cause aortic valve disease. *Nature*. 2005;437:270-4.
49. Mohamed SA, Aherrahrou Z, Liptau H, et al. Novel missense mutations (p.T596M and p.P1797H) in NOTCH1 in patients with bicuspid aortic valve. *Biochem Biophys Res Commun*. 2006;345:1460-5.
50. Muncke N, Jung C, Rüdiger H, et al. Missense Mutations and Gene Interruption in PROSIT240, a Novel TRAP240-Like Gene, in Patients With Congenital Heart Defect (Transposition of the Great Arteries). *Circulation*. 2003;108:2843-50.
51. Gong W, Gottlieb S, Collins J, et al. Mutation analysis of TBX1 in non-deleted patients with features of DGS/VCFS or isolated cardiovascular defects. *J Med Genet*. 2001;38:E45.
52. Reamon-Buettner SM, Borlak J. TBX5 mutations in non-Holt-Oram syndrome (HOS) malformed hearts. *Hum Mutat*. 2004;24:104.
53. Ware SM, Peng J, Zhu L, et al. Identification and functional analysis of ZIC3 mutations in heterotaxy and related congenital heart defects. *Am J Hum Genet*. 2004;74:93-105.
54. Joziasse IC, van de Smagt JJ, Smith K, et al. Genes in congenital heart disease: atrioventricular valve formation. *Basic Res Cardiol*. 2008;103:216-27.
55. Mani A, Meraji SM, Houshyar R, et al. Finding genetic contributions to sporadic disease: a recessive locus at 12q24 commonly contributes to patent ductus arteriosus. *Proc Natl Acad Sci USA*. 2002;99:15054-9.
56. Roessler E, Ouspenskaia MV, Karkera JD, et al. Reduced NODAL Signaling Strength via Mutation of Several Pathway Members Including FOXH1 Is Linked to Human Heart Defects and Holoprosencephaly. *Am J Hum Genet*. 2008;83:18-29.
57. Rothman KJGS, Poole C, Lash TL. *caustation and causal inference. Modern Epidemiology*. 3rd ed. Philadelphia: Lippincott Williams & Wilkins; 2008:5-9.
58. Shaw GM, Iovannisci DM, Yang W, et al. Endothelial nitric oxide synthase (NOS3) genetic variants, maternal smoking, vitamin use, and risk of human orofacial clefts. *Am J Epidemiol*. 2005;162:1207-14.
59. Movahed MR, Hepner AD, hmadi-Kashani M. Echocardiographic prevalence of bicuspid aortic valve in the population. *Heart Lung Circ*. 2006;15:297-9.
60. Tutar E, Ekici F, Atalay S, Nacar N. The prevalence of bicuspid aortic valve in newborns by echocardiographic screening. *Am Heart J*. 2005;150:513-5.
61. Hoffman JI, Kaplan S. The incidence of congenital heart disease. *J Am Coll Cardiol*. 2002;39:1890-900.
62. Beppu S, Suzuki S, Matsuda H, Ohmori F, Nagata S, Miyatake K. Rapidity of progression of aortic stenosis in patients with congenital bicuspid aortic valves. *Am J Cardiol*. 1993;7:322-7.
63. Roberts WC, Ko JM. Frequency by decades of unicuspid, bicuspid, and tricuspid aortic valves in adults having isolated aortic valve replacement for aortic stenosis, with or without associated aortic regurgitation. *Circulation*. 2005;111:920-5.
64. Lamas CC, Eykyn SJ. Bicuspid aortic valve–A silent danger: analysis of 50 cases of infective endocarditis. *Clin Infect Dis*. 2000;30:336-41.
65. Larson EW, Edwards WD. Risk factors for aortic dissection: a necropsy study of 161 cases. *Am J Cardiol*. 1984;53:849-55.
66. Puvimanasinghe JP, Takkenberg JJ, Edwards MB, et al. Comparison of outcomes after aortic valve replacement with a mechanical valve or a bioprosthesis using microsimulation. *Heart*. 2004;90:1172-8.
67. Wilson W, Taubert KA, Gewitz M, et al. Prevention of infective endocarditis: guidelines from the American Heart Association: a guideline from the American Heart Association Rheumatic Fever, Endocarditis, and Kawasaki Disease Committee, Council on Cardiovascular Disease in the Young, and the Council on Clinical Cardiology, Council on Cardiovascular Surgery and Anesthesia, and the Quality of Care and Outcomes Research Interdisciplinary Working Group. *Circulation*. 2007;116:1736-54.
68. Horstkotte D, Follath F, Gutschik E, et al. Guidelines on prevention, diagnosis and treatment of infective endocarditis executive summary; the task force on infective endocarditis

of the European society of cardiology. *Eur Heart J*. 2004; 25:267-76.
69. Russo CF, Mazzetti S, Garatti A, et al. Aortic complications after bicuspid aortic valve replacement: long-term results. *Ann Thorac Surg*. 2002;74:S1773-S1776.
70. Huntington K, Hunter AG, Chan KL. A prospective study to assess the frequency of familial clustering of congenital bicuspid aortic valve. *J Am Coll Cardiol*. 1997;30:1809-12.
71. Cripe L, Andelfinger G, Martin LJ, Shooner K, Benson DW. Bicuspid aortic valve is heritable. *J Am Coll Cardiol*. 2004;44:138-43.
72. Martin LJ, Ramachandran V, Cripe LH, et al. Evidence in favor of linkage to human chromosomal regions 18q, 5q and 13q for bicuspid aortic valve and associated cardiovascular malformations. *Hum Genet*. 2007;121:275-84.
73. Artavanis-Tsakonas S, Rand MD, Lake RJ. Notch signaling: cell fate control and signal integration in development. *Science*. 1999;284:770-6.
74. Wessels MW, Berger RM, Frohn-Mulder IM, et al. Autosomal dominant inheritance of left ventricular outflow tract obstruction. *Am J Med Genet Part A*. 2005;134:171-9.
75. Sanchez CA. Genetics of atrial septal defect. *Arch Dis Child*. 1972;47:581-8.
76. Benson DW, Sharkey A, Fatkin D, et al. Reduced penetrance, variable expressivity, and genetic heterogeneity of familial atrial septal defects. *Circulation*. 1998;97:2043-8.
77. Gelernter-Yaniv L, Lorber A. The familial form of atrial septal defect. *Acta Paediatr*. 2007;96:726-30.
78. Zuckerman HS, Zuckerman GH, Mammen RE, Wassermil M. Atrial septal defect: Familial occurrence in four generations of one family. *Am J Cardiol*. 1962;9:515-20.
79. Bosi G, Sensi A, Scorrano M, Croci G, Giusti S, Calzolari E. Atrial septal defect type Ostium secundum with and without prolonged atrioventricular conduction. *Eur Heart J*. 1995; 16:2014-5.
80. Shiojima I, Komuro I, Inazawa J, et al. Assignment of cardiac homeobox gene CSX to human chromosome 5q34. *Genomics*. 1995;27:204-6.
81. Arceci RJ, King AA, Simon MC, Orkin SH, Wilson DB. Mouse GATA-4: a retinoic acid-inducible GATA-binding transcription factor expressed in endodermally derived tissues and heart. *Mol Cell Biol*. 1993;13:2235-46.
82. Tomita-Mitchell A, Maslen CL, Morris CD, Garg V, Goldmuntz E. GATA4 sequence variants in patients with congenital heart disease. *J Med Genet*. 2007;44:779-83.
83. Horb ME, Thomsen GH. Tbx5 is essential for heart development. *Development*. 1999;126:1739-51.
84. Basson CT, Bachinsky DR, Lin RC, et al. Mutations in human TBX5 [corrected] cause limb and cardiac malformation in Holt-Oram syndrome. *Nat Genet*. 1997;15:30-5.
85. Reamon-Buettner SM, Borlak J. Somatic NKX2-5 mutations as a novel mechanism of disease in complex congenital heart disease. *J Med Genet*. 2004;41:684-90.
86. Carmi R, Boughman JA, Ferencz C. Endocardial cushion defect: further studies of "isolated" versus "syndromic" occurrence. *Am J Med Genet*. 1992;43:569-75.
87. Digilio MC, Marino B, Cicini MP, Giannotti A, Formigari R, Dallapiccola B. Risk of congenital heart defects in relatives of patients with atrioventricular canal. *Am J Dis Child*. 1993;147:1295-7.
88. Green EK, Priestley MD, Waters J, Maliszewska C, Latif F, Maher ER. Detailed mapping of a congenital heart disease gene in chromosome 3p25. *J Med Genet*. 2000;37:581-7.
89. Mowrey PN, Chorney MJ, Venditti CP, et al. Clinical and molecular analyses of deletion 3p25-pter syndrome. *Am J Med Genet*. 1993;46:623-9.
90. Phipps ME, Latif F, Prowse A, Payne SJ. etz-Band J, Leversha M, Affara NA, Moore AT, Tolmie J, Schinzel A, Molecular genetic analysis of the 3p- syndrome. *Hum Mol Genet*. 1994;3:903-8.
91. Rupp PA, Fouad GT, Egelston CA, et al. Identification, genomic organization and mRNA expression of CRELD1, the founding member of a unique family of matricellular proteins. *Gene*. 2002;293:47-57.
92. Loffredo CA, Hirata J, Wilson PD, Ferencz C, Lurie IW. Atrioventricular septal defects: possible etiologic differences between complete and partial defects. *Teratology*. 2001;63:87-93.
93. Hajdú J, Marton T, Papp C, Cesko I, Oroszne NJ, Papp Z. Prenatal diagnosis of atrioventricular septal defect and its prognostic significance. *Orv Hetil*. 1998;139:23-6.
94. Hutchins GM, Moore GW, Lipford EH, Haupt HM, Walker MC. Asplenia and polysplenia malformation complexes explained by abnormal embryonic body curvature. *Pathol Res Pract*. 1983;177:60-76.
95. Tellier AL, Cormier-Daire V, Abadie V, et al. CHARGE syndrome: report of 47 cases and review. *Am J Med Genet*. 1998;76:402-9.
96. Ruiz-Perez VL, Ide SE, Strom TM, et al. Mutations in a new gene in Ellis-van Creveld syndrome and Weyers acrodental dysostosis. *Nat Genet*. 2000;24:283-6.
97. Owens JR, Harris F, Walker S, McAllister E, West L. The incidence of Down's syndrome over a 19-year period with special reference to maternal age. *J Med Genet*. 1983;20:90-3.
98. Mikkelsen M, Fischer G, Stene J, Stene E, Petersen E. Incidence study of Down's syndrome in Copenhagen, 1960-1971; with chromosome investigation. *Ann Hum Genet*. 1976;40:177-82.
99. Bell R, Rankin J, Donaldson LJ. Down's syndrome: occurrence and outcome in the north of England, 1985-99. *Paediatr Perinat Epidemiol*. 2003;17:33-9.
100. Korenberg JR, Chen XN, Schipper R, et al. Down syndrome phenotypes: the consequences of chromosomal imbalance. *Proc Natl Acad Sci USA*. 1994;91:4997-5001.
101. Paladini D, Tartaglione A, Agangi A, et al. The association between congenital heart disease and Down syndrome in prenatal life. *Ultrasound Obstet Gynecol*. 2000;15:104-8.
102. Hook EB, Cross PK, Schreinemachers DM. Chromosomal abnormality rates at amniocentesis and in live-born infants. *JAMA*. 1983;249:2034-8.
103. Ingalls TH. Maternal health and mongolism. *Lancet*. 1972; 2:213-5.
104. Coppedè F, Marini G, Bargagna S, et al. Folate gene polymorphisms and the risk of Down syndrome pregnancies in young Italian women. *Am J Med Genet Part A*. 2006; 140:1083-91.
105. Scala I, Granese B, Lisi A, Mastroiacovo P, Andria G. Re: folate gene polymorphisms and the risk of Down syndrome pregnancies in young Italian women. *Am J Med Genet Part A*. 2007;143:1015-7.

106. Bosco P, Guéant-Rodriguez RM, Anello G, et al. Methionine synthase (MTR) 2756 (A ->G) polymorphism, double heterozygosity methionine synthase 2756 AG/methionine synthase reductase (MTRR) 66 AG, and elevated homocysteinemia are three risk factors for having a child with Down syndrome. *Am J Med Genet Part A.* 2003;121:219-24.
107. James SJ, Pogribna M, Pogribny IP, et al. Abnormal folate metabolism and mutation in the methylenetetrahydrofolate reductase gene may be maternal risk factors for Down syndrome. *Am J Clin Nutr.* 1999;70:495-501.
108. Zintzaras E. Maternal gene polymorphisms involved in folate metabolism and risk of Down syndrome offspring: a meta-analysis. *J Hum Genet.* 2007;52:943-53.
109. Pozzi E, Vergani P, Dalprà L, et al. Maternal polymorphisms for methyltetrahydrofolate reductase and methionine synthetase reductase and risk of children with Down syndrome. *Am J Obstet Gynecol.* 2009;200:636.
110. Aagesen L, Grinsted J, Mikkelsen M. Advanced grandmaternal age on the mother's side–a risk of giving rise to trisomy 21. *Ann Hum Genet.* 1984;48:297-301.
111. Malini SS, Ramachandra NB. Influence of advanced age of maternal grandmothers on Down syndrome. *BMC Med Genet.* 2006;7:4.
112. Papp Z, Váradi E, Szabó Z. Grandmaternal age at birth of parents of children with trisomy 21. *Hum Genet.* 1977;39:221-4.
113. Coppede F, Migheli F, Bargagna S, et al. Association of maternal polymorphisms in folate metabolizing genes with chromosome damage and risk of Down syndrome offspring. *Neuroscience Letters.* 2009;449:15-9.
114. Patterson D. Folate metabolism and the risk of Down syndrome. *Downs Syndr Res Pract.* 2008;12:93-7.
115. McKusick VA, Egeland JA, Eldridge R, Krusen DE. Dwarfism in the Amish I. The Elis-van Creveld Syndrome. *Bull Johns Hopkins Hosp.* 1964;115:306-36.
116. Verdyck P, Blaumeiser B, Holder-Espinasse M, Van HW, Wuyts W. Adams-Oliver syndrome: clinical description of a four-generation family and exclusion of five candidate genes. *Clin Genet.* 2006;69:86-92.
117. Bindewald B, Ulmer H, Müller U. Fallot complex, severe mental, and growth retardation: a new autosomal recessive syndrome? *Am J Med Genet.* 1994;50:173-6.
118. Siwik ES, Zahka KG, Wiesner GL, Limwongse C. Cardiac disease in Costello syndrome. *Pediatrics.* 1998;101:706-9.
119. Li L, Krantz ID, Deng Y, et al. Alagille syndrome is caused by mutations in human Jagged1, which encodes a ligand for Notch1. *Nat Genet.* 1997;16:243-51.
120. McDaniell R, Warthen DM, Sanchez-Lara PA, et al. NOTCH2 mutations cause Alagille syndrome, a heterogeneous disorder of the notch signaling pathway. *Am J Hum Genet.* 2006;79:169-73.
121. Niihori T, Aoki Y, Narumi Y, et al. Germline KRAS and BRAF mutations in cardio-facio-cutaneous syndrome. *Nat Genet.* 2006;38:294-6.
122. Rodriguez-Viciana P, Tetsu O, Tidyman WE, et al. Germline mutations in genes within the MAPK pathway cause cardio-facio-cutaneous syndrome. *Science.* 2006;311:1287-90.
123. Nyström AM, Ekvall S, Berglund E, et al. Noonan and cardio-facio-cutaneous syndromes: two clinically and genetically overlapping disorders. *J Med Genet.* 2008;45:500-6.
124. Aoki Y, Niihori T, Kawame H, et al. Germline mutations in HRAS proto-oncogene cause Costello syndrome. *Nat Genet.* 2005;37:1038-40.
125. Zenker M, Lehmann K, Schulz AL, et al. Expansion of the genotypic and phenotypic spectrum in patients with KRAS germline mutations. *J Med Genet.* 2007;44:131-5.
126. Gripp KW, Lin AE, Nicholson L, et al. Further delineation of the phenotype resulting from BRAF or MEK1 germline mutations helps differentiate cardio-facio-cutaneous syndrome from Costello syndrome. *Am J Med Genet Part A.* 2007;143:1472-80.
127. Lindsay EA, Vitelli F, Su H, et al. Tbx1 haploinsufficieny in the DiGeorge syndrome region causes aortic arch defects in mice. *Nature.* 2001;410:97-101.
128. Yagi H, Furutani Y, Hamada H, et al. Role of TBX1 in human del22q11.2 syndrome. *Lancet.* 2003;362:1366-73.
129. Maslen CL, Babcock D, Robinson SW, et al. CRELD1 mutations contribute to the occurrence of cardiac atrioventricular septal defects in Down syndrome. *Am J Med Genet Part A.* 2006;140:2501-5.
130. Driscoll DA, Budarf ML, Emanuel BS. A genetic etiology for DiGeorge syndrome: consistent deletions and microdeletions of 22q11. *Am J Hum Genet.* 1992;50:924-33.
131. Greenberg F, Elder FF, Haffner P, Northrup H, Ledbetter DH. Cytogenetic findings in a prospective series of patients with DiGeorge anomaly. *Am J Hum Genet.* 1988;43:605-11.
132. Carmi R, Barbash A, Mares AJ. The thoracoabdominal syndrome (TAS): a new X-linked dominant disorder. *Am J Med Genet.* 1990;36:109-14.
133. Hallidie-Smith KA, Karas S. Cardiac anomalies in Williams-Beuren syndrome. *Arch Dis Child.* 1988;63:809-13.
134. Jones KL, Smith DW. The Williams elfin facies syndrome. A new perspective. *J Pediatr.* 1975;86:718-23.
135. Digilio MC, Conti E, Sarkozy A, et al. Grouping of multiple-lentigines/LEOPARD and Noonan syndromes on the PTPN11 gene. *Am J Hum Genet.* 2002;71:389-94.
136. Pandit B, Sarkozy A, Pennacchio LA, et al. Gain-of-function RAF1 mutations cause Noonan and LEOPARD syndromes with hypertrophic cardiomyopathy. *Nat Genet.* 2007;39:1007-12.
137. Tartaglia M, Mehler EL, Goldberg R, et al. Mutations in PTPN11, encoding the protein tyrosine phosphatase SHP-2, cause Noonan syndrome. *Nat Genet.* 2001;29:465-8.
138. Roberts AE, Araki T, Swanson KD, et al. Germline gain-of-function mutations in SOS1 cause Noonan syndrome. *Nat Genet.* 2007;39:70-4.
139. Tartaglia M, Pennacchio LA, Zhao C, et al. Gain-of-function SOS1 mutations cause a distinctive form of Noonan syndrome. *Nat Genet.* 2007;39:75-9.
140. Sarkozy A, Carta C, Moretti S, et al. Germline BRAF mutations in Noonan, LEOPARD, and cardiofaciocutaneous syndromes: molecular diversity and associated phenotypic spectrum. *Hum Mutat.* 2009;30:695-702.
141. Razzaque MA, Nishizawa T, Komoike Y, et al. Germline gain-of-function mutations in RAF1 cause Noonan syndrome. *Nat Genet.* 2007;39:1013-7.
142. Quaderi NA, Schweiger S, Gaudenz K, et al. Opitz G/BBB syndrome, a defect of midline development, is due to

mutations in a new RING finger gene on Xp22. *Nat Genet.* 1997;17:285-91.
143. Parvari R, Carmi R, Weissenbach J, Pilia G, Mumm S, Weinstein Y. Refined genetic mapping of X-linked thoracoabdominal syndrome. *Am J Med Genet.* 1996;61:401-2.
144. Ewart AK, Morris CA, Atkinson D, et al. Hemizygosity at the elastin locus in a developmental disorder, Williams syndrome. *Nat Genet.* 1993;5:11-6.
145. Torfs CP, Christianson RE. Maternal risk factors and major associated defects in infants with Down syndrome. *Epidemiology.* 1999;10:264-70.
146. Fixler DE, Threlkeld N. Prenatal exposures and congenital heart defects in Down syndrome infants. *Teratology.* 1998; 58:6-12.
147. Zatyka M, Priestley M, Ladusans EJ, et al. Analysis of CRELD1 as a candidate 3p25 atrioventicular septal defect locus (AVSD2). *Clin Genet.* 2005;67:526-8.
148. Ford CE, Jones KW, Polani PE, De Almeida JC, Briggs JH. A sex-chromosome anomaly in a case of gonadal dysgenesis (Turner's syndrome). *Lancet.* 1959;1:711-3.
149. Gravholt CH, Juul S, Naeraa RW, Hansen J. Prenatal and postnatal prevalence of Turner's syndrome: a registry study. *BMJ.* 1996;312:16-21.
150. Stochholm K, Juul S, Juel K, Naeraa RW, Gravholt CH. Prevalence, incidence, diagnostic delay, and mortality in Turner syndrome. *J Clin Endocrinol Metab.* 2006; 91:3897-902.
151. Nielsen J, Wohlert M. Sex chromosome abnormalities found among 34, 910 newborn children: results from a 13-year incidence study in Arhus, Denmark. *Birth Defects Orig Artic Ser.* 1990;26:209-23.
152. Gravholt CH, Juul S, Naeraa RW, Hansen J. Morbidity in Turner syndrome. *J Clin Epidemiol.* 1998;51:147-58.
153. Chiovato L, Larizza D, Bendinelli G, et al. Autoimmune hypothyroidism and hyperthyroidism in patients with Turner's syndrome. *Eur J Endocrinol.* 1996;134:568-75.
154. Sybert VP. Cardiovascular malformations and complications in Turner syndrome. *Pediatrics.* 1998;101:E11.
155. Prandstraller D, Mazzanti L, Picchio FM, et al. Turner's syndrome: cardiologic profile according to the different chromosomal patterns and long-term clinical follow-Up of 136 nonpreselected patients. *Pediatr Cardiol.* 1999;20:108-12.
156. Sachdev V, Matura LA, Sidenko S, et al. Aortic valve disease in Turner syndrome. *J Am Coll Cardiol.* 2008;51:1904-9.
157. Bondy CA, Van PL, Bakalov VK, et al. Prolongation of the cardiac QTc interval in Turner syndrome. *Medicine.* 2006;85:75-81.
158. Schoemaker MJ, Swerdlow AJ, Higgins CD, Wright AF, Jacobs PA. Mortality in women with turner syndrome in Great Britain: a national cohort study. *J Clin Endocrinol Metab.* 2008;93:4735-42.
159. Midwest Society for Pediatric Research Cincinnati, Ohio, Oct. 25 and 26, 1962. *The Journal of Pediatrics.* 1963; 63:466-500.
160. Noonan JA. Hypertelorism with Turner phenotype. A new syndrome with associated congenital heart disease. *Am J Dis Child.* 1968;116:373-80.
161. Allanson JE. Noonan syndrome. *J Med Genet.* 1987; 24:9-13.
162. Allanson JE, Hall JG, Hughes HE, Preus M, Witt RD. Noonan syndrome: the changing phenotype. *Am J Med Genet.* 1985;21:507-14.
163. Sharland M, Burch M, McKenna WM, Paton MA. A clinical study of Noonan syndrome. *Arch Dis Child.* 1992;67:178-83.
164. Burch M, Sharland M, Shinebourne E, Smith G, Patton M, McKenna W. Cardiologic abnormalities in Noonan syndrome: phenotypic diagnosis and echocardiographic assessment of 118 patients. *J Am Coll Cardiol.* 1993;22:1189-92.
165. Marino B, Digilio MC, Toscano A, Giannotti A, Dallapiccola B. Congenital heart diseases in children with Noonan syndrome: An expanded cardiac spectrum with high prevalence of atrioventricular canal. *J Pediatr.* 1999;135:703-6.
166. Jamieson CR, van dB I, Brady AF, et al. Mapping a gene for Noonan syndrome to the long arm of chromosome 12. *Nat Genet.* 1994;8:357-60.
167. Brady AF, Jamieson CR, van der Burgt I, et al. Further delineation of the critical region for noonan syndrome on the long arm of chromosome 12. *Eur J Hum Genet.* 1997;5:336-7.
168. Legius E, Schollen E, Matthijs G, Fryns JP. Fine mapping of Noonan/cardio-facio cutaneous syndrome in a large family. *Eur J Hum Genet.* 1998;6:32-7.
169. Carta C, Pantaleoni F, Bocchinfuso G, et al. Germline missense mutations affecting KRAS Isoform B are associated with a severe Noonan syndrome phenotype. *Am J Hum Genet.* 2006;79:129-35.
170. Schubbert S, Zenker M, Rowe SL, et al. Germline KRAS mutations cause Noonan syndrome. *Nat Genet.* 2006; 38:331-6.
171. Araki T, Chan G, Newbigging S, Morikawa L, Bronson RT, Neel BG. Noonan syndrome cardiac defects are caused by PTPN11 acting in endocardium to enhance endocardial-mesenchymal transformation. *Proc Natl Acad Sci USA.* 2009;106:4736-41.
172. Cooper MD, Peterson RDA, Good RA. A new concept of the cellular basis of immunity. *The Journal of Pediatrics.* 1965;67:907-8.
173. Finley JP, Collins GF, de Chadarévian JP, Williams RL. DiGeorge syndrome presenting as severe congenital heart disease in the newborn. *Can Med Assoc J.* 1977;116: 635-40.
174. Karayiorgou M, Morris MA, Morrow B, et al. Schizophrenia susceptibility associated with interstitial deletions of chromosome 22q11. *Proc Natl Acad Sci USA.* 1995;92:7612-6.
175. Pulver AE, Nestadt G, Goldberg R, et al. Psychotic illness in patients diagnosed with velo-cardio-facial syndrome and their relatives. *J Nerv Ment Dis.* 1994;182:476-8.
176. Conley ME, Beckwith JB, Mancer JF, Tenckhoff L. The spectrum of the DiGeorge syndrome. *J Pediatr.* 1979; 94:883-90.
177. Botto LD, May K, Fernhoff PM, et al. A population-based study of the 22q11.2 deletion: phenotype, incidence, and contribution to major birth defects in the population. *Pediatrics.* 2003;112:101-7.
178. Devriendt K, Fryns JP, Mortier G, van Thienen MN, Keymolen K. The annual incidence of DiGeorge/velocardiofacial syndrome. *J Med Genet.* 1998;35:789-90.
179. Tézenas Du Montcel S, Mendizabai H, Aymé S, Lévy A, Philip N. Prevalence of 22q11 microdeletion. *J Med Genet.* 1996;33:719.
180. Strong WB. Familial syndrome of right-sided aortic arch, mental deficiency, and facial dysmorphism. *J Pediatr.* 1968;73:882-8.

181. Kelley RI, Zackai EH, Emanuel BS, Kistenmacher M, Greenberg F, Punnett HH. The association of the DiGeorge anomalad with partial monosomy of chromosome 22. *J Pediatr.* 1982;101:197-200.
182. de la Chapelle A, Herva R, Koivisto M, Aula P. A deletion in chromosome 22 can cause DiGeorge syndrome. *Hum Genet.* 1981;57:253-6.
183. Raatikka M, Rapola J, Tuuteri L, Louhimo I, Savilahti E. Familial third and fourth pharyngeal pouch syndrome with truncus arteriosus: DiGeorge syndrome. *Pediatrics.* 1981; 67:173-5.
184. Scambler PJ, Carey AH, Wyse RK, et al. Microdeletions within 22q11 associated with sporadic and familial DiGeorge syndrome. *Genomics.* 1991;10:201-6.
185. Wilson DI, Cross IE, Goodship JA, et al. A prospective cytogenetic study of 36 cases of DiGeorge syndrome. *Am J Hum Genet.* 1992;51:957-63.
186. Wilson DI, Goodship JA, Burn J, Cross IE, Scambler PJ. Deletions within chromosome 22q11 in familial congenital heart disease. *Lancet.* 1992;340:573-5.
187. Trainer AH, Morrison N, Dunlop A, Wilson N, Tolmie J. Chromosome 22q11 microdeletions in tetralogy of Fallot. *Arch Dis Child.* 1996;74:62-3.
188. Gioli-Pereira L, Pereira AC, Bergara D, Mesquita S, Lopes AA, Krieger JE. Frequency of 22q11.2 microdeletion in sporadic non-syndromic tetralogy of Fallot cases. *Int J Cardiol.* 2008;126:374-8.
189. Stevens CA, Carey JC, Shigeoka AO. Di George anomaly and velocardiofacial syndrome. *Pediatrics.* 1990;85: 526-30.
190. Cuneo BF, Driscoll DA, Gidding SS, Langman CB. Evolution of latent hypoparathyroidism in familial 22q11 deletion syndrome. *Am J Med Genet.* 1997;69:50-5.
191. Holt M, Oram S. Familial heart disease with skeletal malformations. *Br Heart J.* 1960;22:236-42.
192. Elek C, Vitéz M, Czeizel E. Holt-Oram syndrome. *Orv Hetil.* 1991;132:73-8.
193. Basson CT, Cowley GS, Solomon SD, et al. The clinical and genetic spectrum of the Holt-Oram syndrome (heart-hand syndrome). *N Engl J Med.* 1994;330:885-91.
194. Newbury-Ecob RA, Leanage R, Raeburn JA, Young ID. Holt-Oram syndrome: a clinical genetic study. *J Med Genet.* 1996;33:300-7.
195. Smith AT, Sack GH Jr, Taylor GJ. Holt-Oram syndrome. *J Pediatr.* 1979;95:538-43.
196. Basson CT, Huang T, Lin RC, et al. Different TBX5 interactions in heart and limb defined by Holt-Oram syndrome mutations. *Proc Natl Acad Sci USA.* 1999;96:2919-24.
197. Li QY, Newbury-Ecob RA, Terrett JA, et al. Holt-Oram syndrome is caused by mutations in TBX5, a member of the Brachyury (T) gene family. *Nat Genet.* 1997;15: 21-9.
198. Bonnet D, Pelet A, Legeai-Mallet L, et al. A gene for Holt-Oram syndrome maps to the distal long arm of chromosome 12. *Nat Genet.* 1994;6:405-8.
199. Terrett JA, Newbury-Ecob R, Cross GS, et al. Holt-Oram syndrome is a genetically heterogeneous disease with one locus mapping to human chromosome 12q. *Nat Genet.* 1994;6:401-4.
200. Brassington AM, Sung SS, Toydemir RM, et al. Expressivity of Holt-Oram syndrome is not predicted by TBX5 genotype. *Am J Hum Genet.* 2003;73:74-85.
201. McDermott DA, Bressan MC, He J, et al. TBX5 genetic testing validates strict clinical criteria for Holt-Oram syndrome. *Pediatr Res.* 2005;58:981-6.
202. Byth BC, Costa MT, Teshima IE, Wilson WG, Carter NP, Cox DW. Molecular analysis of three patients with interstitial deletions of chromosome band 14q31. *J Med Genet.* 1995;32:564-7.
203. Le Meur N, Goldenberg A, Michel-Adde C, et al. Molecular characterization of a 14q deletion in a boy with features of Holt-Oram syndrome. *Am J Med Genet Part A.* 2005; 134:439-42.
204. Adamopoulos S, Kokkinou S, Parissis JT, Kremastinos DT. New insight into "heart-hand" syndromes: a newly discovered chromosomal abnormality in a family with "heart-hand" syndrome. *Int J Cardiol.* 2004;97:129-32.
205. Hall BD. Choanal atresia and associated multiple anomalies. *J Pediatr.* 1979;95:395-8.
206. Hittner HM, Hirsch NJ, Kreh GM, Rudolph AJ. Colobomatous microphthalmia, heart disease, hearing loss, and mental retardation–a syndrome. *J Pediatr Ophthalmol Strabismus.* 1979;16:122-8.
207. Issekutz KA, Graham JM Jr, Prasad C, Smith IM, Blake KD. An epidemiological analysis of CHARGE syndrome: preliminary results from a Canadian study. *Am J Med Genet Part A.* 2005;133:309-17.
208. Lalani SR, Safiullah AM, Fernbach SD, et al. Spectrum of CHD7 mutations in 110 individuals with CHARGE syndrome and genotype-phenotype correlation. *Am J Hum Genet.* 2006;78:303-14.
209. Pagon RA, Graham JM Jr, Zonana J, Yong SL. Coloboma, congenital heart disease, and choanal atresia with multiple anomalies: CHARGE association. *J Pediatr.* 1981;99:223-7.
210. Blake KD, Davenport SL, Hall BD, et al. CHARGE association: an update and review for the primary pediatrician. *Clin Pediatr (Phila).* 1998;37:159-73.
211. Verloes A. Updated diagnostic criteria for CHARGE syndrome: a proposal. *Am J Med Genet Part A.* 2005;133:306-8.
212. Jongmans MC, Admiraal RJ, van der Donk KP, et al. CHARGE syndrome: the phenotypic spectrum of mutations in the CHD7 gene. *J Med Genet.* 2006;43:306-14.
213. Vissers LE, van Ravenswaaij CM, Admiraal R, et al. Mutations in a new member of the chromodomain gene family cause CHARGE syndrome. *Nat Genet.* 2004;36:955-7.
214. Devriendt K, Swillen A, Fryns JP. Deletion in chromosome region 22q11 in a child with CHARGE association. *Clin Genet.* 1998;53:408-10.
215. Emanuel BS, Budarf ML, Sellinger B, Goldmuntz E, Driscoll DA. Detection of microdeletions of 22q11 with fluorescence in situ hubridization (FISH): diagnosis fo DiGeorge syndrome (DGS), velo-cardio-facial (VCF) syndrome, CHARGE association and conotruncal cardiac malformations. *Am J Hum Genet.* 1992;51(Suppl):A3.
216. Strømme P, Bjornstad PG, Ramstad K. Prevalence estimation of Williams syndrome. *J Child Neurol.* 2002;17:269-71.
217. Nickerson E, Greenberg F, Keating MT, McCaskill C, Shaffer LG. Deletions of the elastin gene at 7q11.23 occur in approximately 90% of patients with Williams syndrome. *Am J Hum Genet.* 1995;56:1156-61.
218. Pérez Jurado LA, Peoples R, Kaplan P, Hamel BC, Francke U. Molecular definition of the chromosome 7 deletion in Williams syndrome and parent-of-origin effects on growth. *Am J Hum Genet.* 1996;59:781-92.

219. Morris CA, Demsey SA, Leonard CO, Dilts C, Blackburn BL. Natural history of Williams syndrome: physical characteristics. *J Pediatr*. 1988;113:318-26.
220. Preus M. The Williams syndrome: objective definition and diagnosis. *Clin Genet*. 1984;25:422-8.
221. Eronen M, Peippo M, Hiippala A, et al. Cardiovascular manifestations in 75 patients with Williams syndrome. *J Med Genet*. 2002;39:554-8.
222. Sadler LS, Pober BR, Grandinetti A, et al. Differences by sex in cardiovascular disease in Williams syndrome. *J Pediatr*. 2001;139:849-53.
223. Ewart AK, Jin W, Atkinson D, Morris CA, Keating MT. Supravalvular aortic stenosis associated with a deletion disrupting the elastin gene. *J Clin Invest*. 1994;93:1071-7.
224. Sugayama SM, Moisés RL, Wagënfur J, et al. Williams-Beuren syndrome: cardiovascular abnormalities in 20 patients diagnosed with fluorescence in situ hybridization. *Arq Bras Cardiol*. 2003;81:462-73.
225. Wang CC, Hwu WL, Wu ET, Lu F, Wang JK, Wu MH. Outcome of pulmonary and aortic stenosis in Williams-Beuren syndrome in an Asian cohort. *Acta Paediatr*. 2007;96:906-9.
226. Wessel A, Pankau R, Kececioglu D, Ruschewski W, Bürsch JH. Three decades of follow-up of aortic and pulmonary vascular lesions in the Williams-Beuren syndrome. *Am J Med Genet*. 1994;52:297-301.
227. Amenta S, Sofocleous C, Kolialexi A, et al. Clinical manifestations and molecular investigation of 50 patients with Williams syndrome in the Greek population. *Pediatr Res*. 2005;57:789-95.
228. Cherniske EM, Carpenter TO, Klaiman C, et al. Multisystem study of 20 older adults with Williams syndrome. *Am J Med Genet Part A*. 2004;131:255-64.
229. Salaymeh KJ, Banerjee A. Evaluation of arterial stiffness in children with Williams syndrome: Does it play a role in evolving hypertension? *Am Heart J*. 2001;142:549-55.
230. Scheiber D, Fekete G, Urban Z, et al. Echocardiographic findings in patients with Williams-Beuren syndrome. *Wien Klin Wochenschr*. 2006;118:538-42.
231. Iles MM. What Can Genome-Wide Association Studies Tell Us about the Genetics of Common Disease? *PLoS Genet*. 2008;4:e33.
232. Mardis ER. Anticipating the 1,000 dollar genome. *Genome Biol*. 2006;7:112.

Genetic Disorders of the Lipoprotein Metabolism; Diagnosis and Management

20

A. Bakker, L. Jakulj, and J.J.P. Kastelein

Abbreviations

ABC	Adenosine tri-phosphate (ATP) binding cassette
ACAT	Acyl-coenzyme A: cholesterol O-acyltransferase
ApoA1	Apolipoprotein A1
BAS	Bile acid sequestrants
CAD	Coronary artery disease
CE	Cholesterylester
CETP	Cholesterylester transfer protein
cIMT	Carotid intima media thickness
CHD	Coronary heart disease
CVD	Cardio vascular disease
FCH	Familial combined hyperlipidemia
FD	Familial dysbetalipoproteinemia
FDB	Familial defective apolipoprotein B
FH	Familial hypercholesterolemia
FHTG	Familial hypertriglyceridemia
HDL-C	High density lipoprotein cholesterol
HL	Hepatic lipase
HMG-CoA	3-Hydroxyl-3-methylglutaryl coenzyme A
IDL	Intermediate density lipoprotein
LCAT	Lecithin:cholesteryl acyltransferase
LDL-C	Low density lipoprotein cholesterol
LDL-R	Low density lipoprotein receptor
LDLRAP	LDL-receptor adapting protein
LIPC	Gene encoding hepatic lipase
LIPG	Gene encoding endothelial lipase
LPL	Lipoprotein lipase
NPC1L1	Niemann-Pick C1 like 1
PCSK9	Proprotein convertase subtilisin/kexin type 9
PLTP	Phospholipid transfer protein
RCT	Reverse cholesterol transport
SNP	Single nucleotide polymorphism
SR-B1	Scavenger receptor B1
TC	Total cholesterol
VLDL	Very low density lipoprotein

20.1 Introduction

Atherosclerosis, leading to ischemic manifestations in different vascular beds, is the leading cause of morbidity and mortality worldwide. It is a multifactorial disease, driven by a combination of genetic, environmental, and behavioral factors. The process of atherosclerosis accelerates in the presence of classical risk factors such as *dyslipidemia*, hypertension, diabetes mellitus, obesity, and smoking. Dyslipidemia is one of the major contributors to atherosclerosis and includes both elevated low-density-lipoprotein cholesterol (*LDL-C*) levels, as well as decreased high-density-lipoprotein cholesterol (*HDL-C*) levels.[1] The crucial role of increased plasma LDL-C levels in the pathogenesis of atherosclerosis has been well established. This also applies to the pharmacological reduction of plasma LDL-C levels accomplished by hydroxyl–methyl–glutaryl coenzyme A (HMG-CoA) reductase inhibitors or *statins*. A large prospective meta-analysis including over 90,000 individuals demonstrated that a LDL-C reduction of 1 mmol/L is associated with a 21% reduction in major cardiovascular events.[2] In addition, decreased plasma HDL-C levels are an independent

J.J.P. Kastelein (✉)
Department of Vascular Medicine, Academic Medical Center, University of Amsterdam, Meibergdreef 9, room F4-159.2, 1105 AZ Amsterdam, The Netherlands
e-mail: j.j.kastelein@amc.nl/j.s.jansen@amc.nl

predictor of cardiovascular disease (CVD), as has been unequivocally established by numerous epidemiological studies. Almost 40% of patients with premature CAD have low HDL-C levels, either alone or in conjunction with hypertriglyceridemia or combined hyperlipidemia.[3] Furthermore, it has been estimated that each 0.03 mmol/L (1 mg/dL) increase in HDL-C is associated with a 2% reduction CAD risk in men and a 3% reduction in women.[4] However, whether raising HDL-C by pharmacological means will result in cardiovascular benefit is disputable. A recent meta-regression analysis of 108 randomized controlled trials, including more than 300,000 patients using several lipid-modifying interventions, did not show a relationship between treatment-induced increases in HDL-C and a decrease in coronary heart disease events or deaths when corrected for concurrent LDL-C reductions.[5] Nevertheless, this study does not prove that increasing HDL-C in selected patients with low HDL-C levels has no value.[6] In addition, these studies evaluated only HDL-C concentrations and did not address HDL functionality.

Finally, the relationship between *hypertriglyceridemia* and CVD risk remains to be elucidated.[7] Although several studies have indicated elevated plasma triglyceride levels as an independent risk factor for CVD,[8–10] it has also been suggested that plasma triglycerides merely represent a biomarker of CVD risk rather than an independent risk factor, as this trait is frequently accompanied by additional risk factors for atherosclerosis, such as insulin resistance or low HDL-C levels.[11] Ongoing studies will have to determine whether triglyceride lowering confers additional CVD benefit beyond that achieved by LDL-C lowering.[12]

Although dyslipidemia has a largely polygenic background, a number of *monogenetic disorders* have been identified. This chapter provides an overview of genetic causes underlying disturbances in lipid and lipoprotein metabolism, in which the focus will be primarily on these monogenetic disorders. The chapter starts with a global overview of lipid and lipoprotein metabolism, followed by the genetic background of disturbances in LDL-C and HDL-C levels, respectively. Finally, genetic causes of disorders in triglyceride metabolism are also discussed. For each of these categories, genetics, clinical phenotype, diagnosis, and management will be addressed.

20.2 Structure of Lipids and Lipoproteins

Cholesterol and *triglycerides* exert essential functions in body cell membranes and in hormone and energy homeostasis. Due to their hydrophobic properties, cholesterol and triglycerides are transported in large macromolecular complexes, so-called lipoproteins. *Lipoproteins* contain a core of hydrophobic lipids surrounded by hydrophilic molecules such as phospholipids, unesterified cholesterol, and *apolipoproteins*. The latter are proteins that provide structural integrity to the lipoprotein and serve as ligands for binding to specific receptors. Based on their relative density, lipoproteins can be categorized into five major classes: chylomicrons, very low-density lipoproteins (*VLDL*), intermediate-density lipoproteins (*IDL*), low-density lipoproteins (*LDL*), and high-density lipoproteins (*HDL*). The first two categories are large, buoyant triglyceride-rich particles, whereas the latter three are dense, cholesterol-rich particles. When fasting, plasma cholesterol levels are usually a reflection of the amount of LDL in the plasma, whereas plasma triglyceride levels reflect the amount of VLDL.

20.3 Lipid and Lipoprotein Metabolism

The liver and the intestine are the most important sources of lipoproteins. Their transport and metabolism is generally divided into three systems: absorption of exogenous and endogenous lipids and lipoproteins, endogenous synthesis of lipids and lipoproteins, and *reverse cholesterol transport (RCT)*. These processes are depicted in Fig. 20.1.

20.3.1 Absorption of Exogenous and Endogenous Lipids

The average Western diet consists of a daily intake of approximately 100 g of fat and 500 mg of cholesterol. Phospholipids and bile acids, present in hepatic bile emulsify lipids from food to form micelles within the intestinal lumen. Hepatic bile also delivers significant amounts of unesterified cholesterol to these micelles.

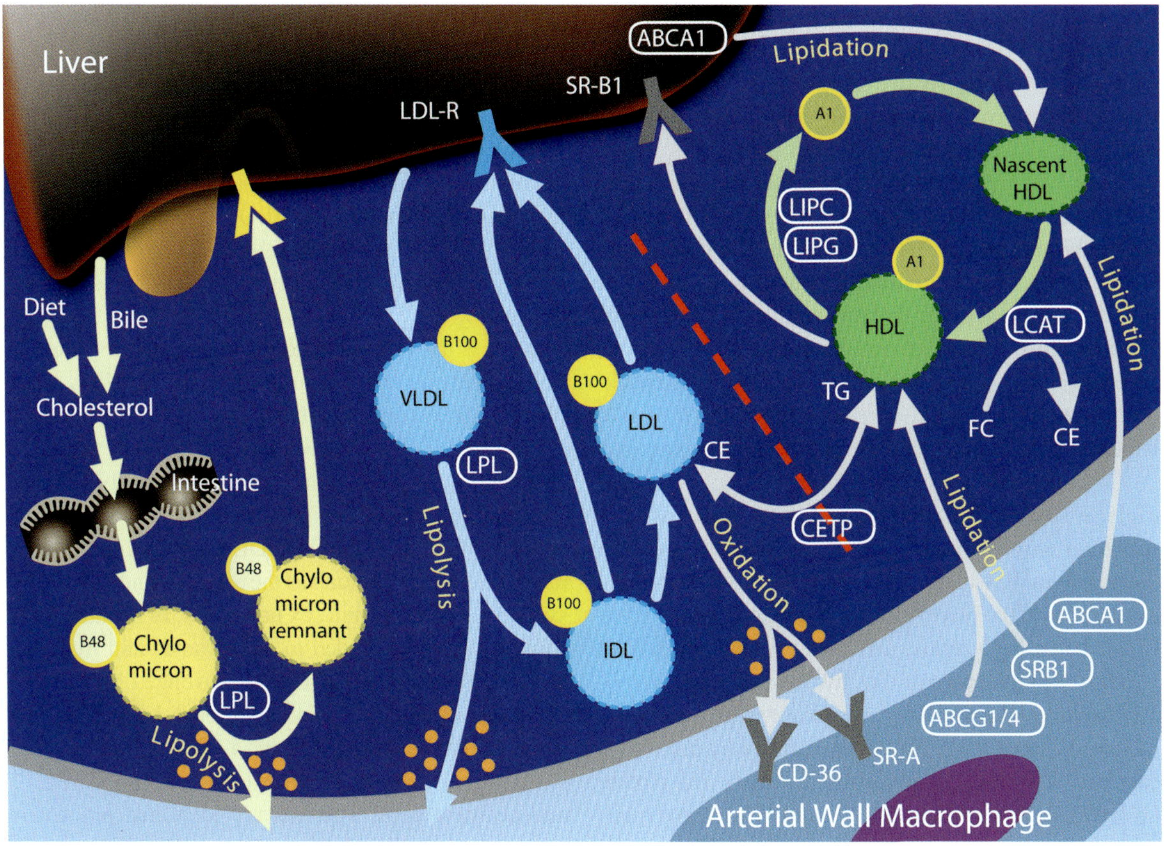

20 Genetic Disorders of the Lipoprotein Metabolism; Diagnosis and Management

Fig 20.1 Overview of lipoprotein metabolism Dietary lipids and cholesterol from the hepatic bile are absorbed in the intestine, packaged into chylomicrons, and secreted into the lymph, which drains into the systemic circulation. In the bloodstream, the triglyceride-rich (*TG*) chylomicrons are hydrolyzed through action of lipoprotein lipase (*LPL*) and the removed TGs and free fatty acids are taken up by extrahepatic tissues like the arterial endothelial wall. The chylomicrons remnants are taken up by the liver for further processing. In the fasting state, the liver assembles TG-rich very low density lipoprotein (*VLDL*). Also VLDL are hydrolyzed by LPL and thereby transformed to smaller VLDL-remnants, IDL. Half of the IDL are directly taken up by the liver through binding of the LDL-R, whereas the other half is converted to cholesterol-rich LDL. Most of the plasma LDL-C is cleared from the circulation by binding to the LDL-R of the liver. Of the remaining LDL, some subfractions are especially prone to oxidative modification and then taken up by scavenger receptors (SR-A and CD-36) of arterial wall macrophages resulting into foam cells and atherosclerotic plaques. High-density-lipoprotein (*HDL*) is responsible for the reverse cholesterol transport from extrahepatic tissues to the liver. Nascent HDL is formed from lipid-poor apo-A1, which is secreted by the liver and intestine and which is lipidated through interaction with ABCA1. Nascent HDL is also generated from surface components shed during lipolysis of TG-rich lipoproteins by LPL (not depicted). After lipidation, LCAT esterifies free cholesterol (*FC*) to cholesterylesters (*CE*) which migrates into the core of the HDL making them larger spherical particles. These larger HDL particles acquire additional lipids from extrahepatic tissues, including arterial wall macrophages, by receptor mediated pathways like ABCG1, ABCG4, and SR-B1, as well as from lipolysis of TG-rich lipoproteins and passive diffusion (not depicted). The HDL particles can be metabolized in several ways. First, they can deliver CE to the liver by binding to SR-B1 on the hepatocyte surface. In the liver the cholesterol can be processed and eliminated. Alternatively, CE in HDL can be exchanged for TG in apoB-containing lipoproteins, by the action of CETP. The TG-enriched HDL is hydrolyzed by LIPC and LIPG to smaller HDL and lipid-poor ApoA1 particles. These can either be recycled to acquire cholesterol or apoA1 is excreted from the body through the kidneys

Pancreatic lipases secreted into the intestinal lumen digest dietary lipids to chemical entities that can be absorbed by enterocytes. Fatty acids and monoacylglycerides are almost entirely absorbed through both passive diffusion and carrier-mediated processes.[13] In contrast, *cholesterol absorption* is an active process, mediated by several transporter proteins which are located at the intestinal brush border membrane. Cholesterol and sterols derived from plants are taken up by the enterocyte through the recently identified

Niemann-Pick C Like 1 (NPC1L1) transporter,[14] whereas the ATP-binding cassette transporters (ABC) G5 and G8 actively secrete plant sterols, and to lesser extent cholesterol, back into the intestinal lumen.[15] Of note, NPC1L1 and ABCG5 and G8 are also located in the liver, where they are involved in hepatic cholesterol trafficking to the bile.[15,16] Intestinal cholesterol absorption exhibits on average about 50% efficiency, with large interindividual variation, ranging from 20 to 80%.[17] Free cholesterol that has entered the enterocyte is either re-esterified intracellularly by Acyl-coenzyme A: cholesterol acyltransferase (ACAT) 2 and then packaged into chylomicrons, or trafficked toward the basolaterally located ABCA1 protein for HDL formation.

Chylomicrons consist for approximately 80–95% of triglycerides and apolipoprotein B48 (apoB48) as their structural surface protein. They are secreted into the thoracic lymph, which drains directly into the systemic circulation. In the bloodstream, chylomicrons are hydrolyzed, that is, triglycerides and free fatty acids (FFAs) are removed from the core of the chylomicrons, by *lipoprotein lipase (LPL)*, thereby generating remnant particles. LPL is anchored to the endothelial surface by proteoglycans and/or by the recently identified anchoring protein GPIHBP1 (glycosylphosphatidylinositol anchored high-density lipoprotein-binding protein 1).[18] LPL requires apolipoprotein CII as a cofactor for adequate hydrolysis. The removed triglycerides and FFAs are taken up by peripheral cells, whereas the chylomicrons remnants, are taken up by the liver for further processing, as described below. Chylomicrons have a short half-life in the circulation, averaging approximately 10–20 min, provided that clearance is undisturbed. Hence, chylomicrons are not present in the bloodstream in the fasting state. However, when postprandial levels of chylomicrons and their remnants remain high, due to intestinal overproduction or delayed clearance, this can promote delivery of these particles to the arterial endothelium, with subsequent generation of *foam cells* and fatty streaks and eventually *atherosclerotic plaque formation*.

subsequent esterification by ACAT2 or from remnant particles that have been taken up from the circulation. Like chylomicrons, VLDL are triglyceride-rich particles secreted into the bloodstream, where they are hydrolyzed by LPL and thereby transformed to smaller and denser VLDL-remnants, IDL, and finally, LDL particles. In general, half of the VLDL-remnants are directly taken up by the liver through binding to the *LDL receptor* (LDL-R), whereas the other half is converted to LDL.

LDL is the most abundant cholesterol-carrying particle in humans and accounts for more than 75% of the plasma cholesterol. Mediated by *apoB100*, most of plasma LDL is cleared from the circulation by the LDL-R, which is located at the surface of hepatocytes and internalized entirely (lipoprotein + receptor) upon binding of LDL. The remaining LDL particles are delivered to peripheral tissues such as the adrenals and gonads for synthesis of steroid and sex hormones. In hepatic endosomes, LDL is degraded to amino acids and free cholesterol, whereas the LDL-R is scavenged back to the cell surface for uptake of additional LDL particles. Approximately 70–80% of the LDL catabolism takes place via the LDL-R. The remaining part is cleared via nonspecific routes. The recently identified proprotein convertase subtilisin/kexin type 9 (*PCSK9*) protein is thought to promote degradation of the LDL-R, although the exact mechanism of action is not known.[19]

Finally, LDL is not a homogeneous lipoprotein fraction, as it consists of several subfractions with varying mass and density. *Small-dense LDL* is particularly associated with atherosclerotic disease. This subfraction is mostly prevalent in subjects with elevated triglyceride levels. Small dense LDL particles are prone to oxidative modification, resulting in uptake by scavenger receptors of arterial macrophages, which express a strong affinity for these so-called ox-LDL particles. Since a negative feedback system for these scavenger receptors is lacking, unlimited amounts of ox-LDL can be taken up by these macrophages, which transform into foam cells and atherosclerotic plaques.

20.3.2 Endogenous Synthesis of Lipids and Lipoproteins

In the fasting state, the liver assembles VLDL-C by combining triglycerides, phospholipids, apolipoprotein B100 (apoB100), and cholesteryl esters (CE). The latter originate either from de novo synthesis and

20.3.3 HDL Metabolism and Reverse Cholesterol Transport

HDL is a highly heterogeneous class of lipoprotein particles that differ in protein component and lipid composition, size, shape, density, and charge. In addition to the observational support for the atheroprotective role of HDL, numerous in-vitro and in-vivo animal studies have demonstrated various mechanisms through which HDL exerts its beneficial effects on the arterial wall. The most widely acknowledged mechanism is its role in *reverse cholesterol transport (RCT)*. This involves the ability of HDL to stimulate efflux of cholesterol from peripheral tissues, transport in the plasma, and uptake by the liver, followed by biliary excretion and elimination via the feces. Specifically, the efflux of cholesterol from macrophage foam cells in the artery wall is thought to be central to the anti-atherogenic properties of HDL. In addition, putative atheroprotective properties of HDL include its ability to improve endothelial function, inhibit LDL oxidation and induce several anti-apoptotic, anti-inflammatory, and anti-thrombotic effects. However, whether stimulation of these processes results in clinical benefit in humans is not clear.[20]

The process of RCT starts by lipidation of *apolipoprotein A1* (apo-AI) through interaction with the ATP-binding cassette transport protein A1 (ABCA1). Apolipoprotein-AI is the most important structural protein of HDL and comprises approximately 70% of the proteins in HDL-C. It is synthesized by the liver and intestine and released into the circulation either in a free non-lipidated form or incorporated in small discoid particles, rich in phospholipids and poor in cholesterol, so-called nascent or pre-β-HDL. Nascent HDL is also generated from redundant surface components shed during lipolysis of triglyceride-rich lipoproteins like chylomicrons and VLDL by LPL. The *ABCA1 transporter* resides at the cellular membrane and facilitates the transfer of free cholesterol and phospholipids from intracellular lipid pools to apo-AI. New insights suggest that, in contrast to previous opinion, hepatic ABCA1 appears to be critical for the initial lipidation of lipid-poor apo-AI, protecting it from rapid degradation and allowing it to go on to form mature HDL. Conversely, macrophage ABCA1 appears to contribute little to bulk lipidation of HDL and therefore to plasma HDL-C levels, but does seem to be important for protection against atherosclerosis.[21] After lipidation, lecithin:cholesteryl acyltransferase (*LCAT*) subsequently esterifies the externalized free cholesterol to cholesterylesters on the surface of HDL on activation by its cofactor apo-AI. The esterified cholesterol then migrates into the core of the HDL and as larger amounts of cholesterylesters become incorporated, HDL becomes a larger spherical particle. These larger so-called HDL-3 and HDL-2 particles acquire additional free cholesterol and phospholipids from extrahepatic tissues, including macrophage foam cells, by means of passive diffusion or receptor-mediated pathways, like ABCG1, ABCG4, and scavenger receptor B1 (SR-B1), as well as from lipolysis of triglyceride-rich lipoproteins.[20] The HDL-3 and HDL-2 particles can be metabolized in several ways. First, they can directly deliver cholesterylesters to the liver by binding to *SR-B1* on the hepatocyte surface. In the liver the cholesterol can be processed and eliminated as bile or converted to cholesterol-containing steroids. Once the HDL particle is delipidated, it dissociates from SR-B1 and can then reinitiate another cycle of RCT. Alternatively, cholesterylesters in HDL can be exchanged for triglycerides in apoB–containing lipoproteins such as LDL, by the action of *cholesterylester transfer protein (CETP)*, after which these cholesterylesters are available for hepatic clearance via the LDL receptor. However, if a population of apoB-containing lipoproteins enriched with cholesterylesters by CETP interacts with macrophages in arterial walls and promotes net cholesterol uptake, this process is potentially atherogenic. Whether the sum effect of CETP activity in humans is pro- or antiatherogenic is not clear. Most experimental evidence in animals favors a proatherogenic role for CETP. The triglyceride-enriched HDL is a substrate for hydrolysis by *hepatic lipase* (LIPC) while phospholipids are mainly hydrolyzed by *endothelial lipase* (LIPG). In this way, HDL is remodeled to lipid poor ApoA1 and smaller HDL particles which can either be recycled to acquire cholesterol from extrahepatic tissues or dissociated apoA1 is excreted from the body through the kidneys. *Phospholipid transfer protein (PLTP)* also plays a major role in HDL metabolism in various ways. PLTP facilitates the transfer of phospholipids from triglyceride-rich lipoproteins during lipolysis and evidence has accumulated over the years that PLTP can also remodel HDL particles.[22]

20.4 Genetic Causes of Elevated LDL-C Levels

Mutations in genes involved in LDL metabolism can result in increased plasma LDL-C concentrations. Three genes have been characterized: the *LDL-R* gene, the *ApoB* gene, and most recently, the *PCSK9* gene.[19]

These genes are involved in *autosomal dominant hypercholesterolemia*. The single known *autosomal recessive* form of *hypercholesterolemia* (ARH) is caused by failing internalization of the LDL-R/LDL particle complex in the hepatocytes and is caused by a mutation in the ARH gene.[23,24]

20.4.1 Familial Hypercholesterolemia

Familial hypercholesterolemia (FH) is the most common autosomal dominant inherited disorder of metabolism. Approximately 1:500 people are affected, resulting in almost ten million patients worldwide. In some populations this prevalence is higher due to a founder effect.[25] Homozygosity is rare, with an average frequency of one per million. FH subjects are characterized plasma LDL-C levels above the 95th percentile for age and gender, due to impaired internalization of LDL particles caused by structural alterations in the LDL receptor.[26] Moreover, the decrease in the hepatic cholesterol pool stimulates cholesterol synthesis, resulting in increased production of VLDL, which further increases LDL-C levels.

20.4.1.1 Genetics

The molecular defect underlying FH consists of a mutation in the *LDL-R gene*, located on chromosome 19p13.[26] At present, over 1,000 different mutations in the LDL-R or promoter region leading to an FH phenotype have been described,[27] 91% of which are point mutations.[28]

In addition, variations in *PCSK9 gene* on chromosome 1 are a rare cause of the FH phenotype. To date, eight hypercholesterolemic missense mutations in PCSK9 have been reported.[29] These "gain-of-function" mutations are thought to cause hypercholesterolemia due to PCSK9-induced enhanced degradation of LDL receptors. Yet, as stated previously, the precise function of PCSK9 in cholesterol metabolism remains to be elucidated. Finally, mutations in the LDL-R-binding domain of apoB100 can result in a phenotype which resembles FH. This is outlined in the next paragraph on familial defective apolipoprotein B (FDB).

20.4.1.2 Clinical Characteristics

A hallmark of FH are plasma LDL-C levels above the 95th percentile for age and gender. This induces accelerated deposition of cholesterol in arterial walls and other tissues, resulting in the clinical hallmarks of FH: premature atherosclerosis, *tendon xanthomas, xanthelasmata*, and corneal arcus.[30] However, these clinical characteristics (Figs. 20.2 and 20.3) may not be present in every patient with FH. If untreated, approximately 50% of male and 30% of female heterozygous FH patients will develop symptomatic CVD before the age of 50 years.[31] However, onset and progression of atherosclerotic disease varies considerably between FH individuals. It was shown that event-free survival depends more on actual LDL-C levels caused by the mutation, rather than the type of mutation itself.[32]

Patients with homozygous FH have plasma cholesterol levels > 13 mmol/L. If untreated, patients suffer from CVD before 20 years of age and generally do not survive past 30 years of age.

Although cardiovascular events are rare in children heterozygous for FH, affected children were shown to have an impaired endothelial function,[33] as well as an increased *carotid intima-media thickness (cIMT)*,[34] when compared to their unaffected siblings, which is indicative for an early onset of subclinical atherosclerosis. Based on these findings, current guidelines advise early pharmacological cholesterol-lowering in children with FH, as discussed below.

20.4.1.3 Diagnosis

Genetic analysis provides the only unequivocal diagnosis. Nevertheless, FH is usually diagnosed on the basis of clinical features. Several clinical tools have been developed, with different diagnostic criteria, some of which combined with DNA analysis (reviewed in Reference[35]). In the Netherlands, we use an algorithm of the Dutch Lipid Network as shown in Fig. 20.4. The primary clinical diagnostic criteria are elevated LDL-C levels above the 95th percentile for age and gender, presence of tendon xanthomata in the patient or a first degree relative, and a pattern of autosomal dominant inheritance of premature coronary heart disease or hypercholesterolemia.

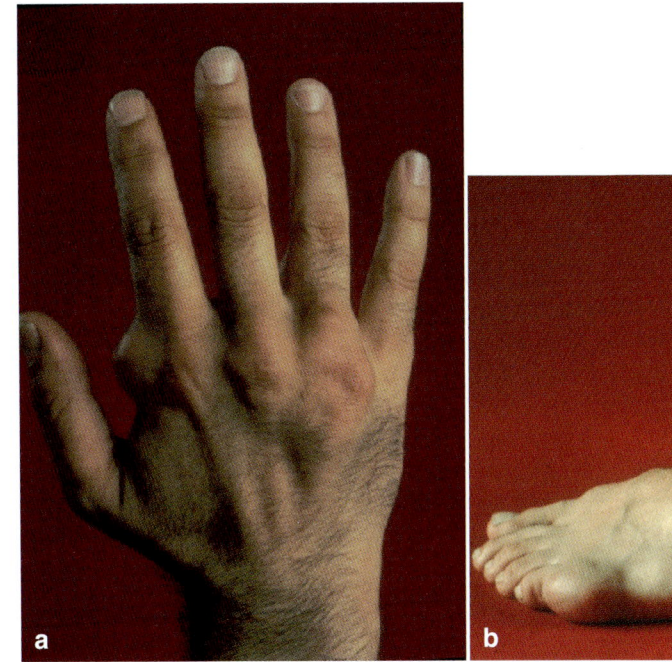

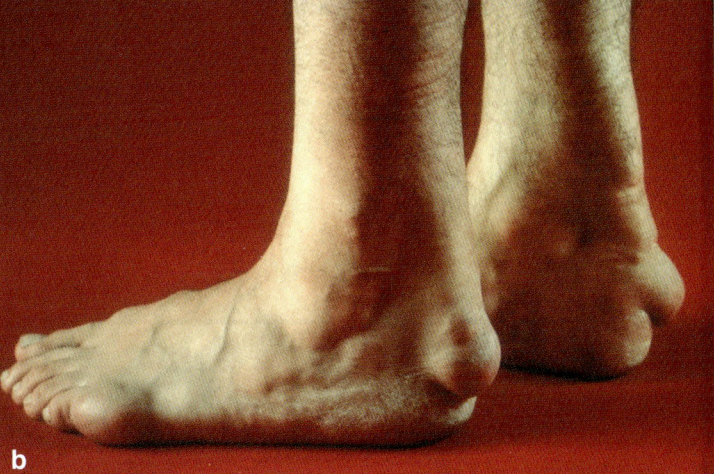

Fig. 20.2 Skin xanthomas (**a**) Xanthomas of the hand (**b**) Achilles tendon xanthomas

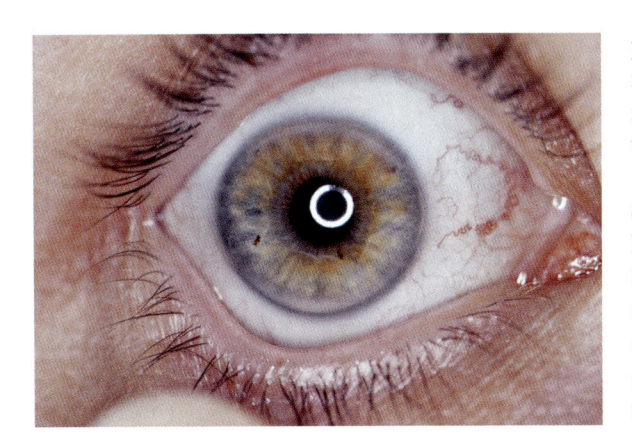

Fig. 20.3 Arcus lipoides

20.4.1.4 Management

Treatment with *high-dose statins* is currently the most effective strategy to reduce CVD risk in FH patients.[36] The recommended LDL-C targets in FH were recently reviewed.[35] In addition, patients are treated in combination with lifestyle modifications aimed to reduce the risk of other atherogenic factors. Finally, several new pharmacological agents have been developed to optimize cholesterol-lowering treatment in those who do not reach their LDL-C targets or who are unable to tolerate high doses of statins.

Ezetimibe is a cholesterol absorption inhibiting compound, which acts by blocking the intestinal NPC1L1 protein. Although the beneficial cholesterol-lowering effects of ezetimibe combined with statin therapy have been established, it is still unclear whether ezetimibe provides the same cardiovascular benefits as statins.[37] Hence, the results of an ongoing end-point trial have to be awaited.[38]

Bile acid sequestrants (BAS) bind bile acids in the intestine and subsequently increase hepatic conversion of cholesterol into bile acids. The concomitant decrease in hepatic cholesterol content results in decreased LDL-C levels, mediated by increased hepatic LDL-R expression. *Colesevelam* is a novel BAS with a more favorable side-effect profile, as it is thought to bind with higher affinity compared to other BAS.[39] It is currently being evaluated in patients with FH not at their target LDL-C level, despite a maximally tolerated and stable dose of statin and ezetimibe.

Family history*

I	First-degree relative with CVD < 60 years of age	1
II	First-degree relative with plasma LDL-cholesterol levels > 5 mmol/l	
III	First-degree relative with an corneal arcus < 45 years and/or tendon xanthomas	2
IV	Children < 18 years of age with plasma LDL-cholesterol levels > 3.5 mmol/l	

Personal Medical History

I	First-degree relative with CHD < 60 years of age	2
II	Cerebro-vascular event or peripheral arterial disease < 60 years of age	1

Physical Examination

I	Presence of tendon xanthomas	6
II	Presence of corneal arcus < 45 years of age	4

Laboratory parameters

I	LDL-cholesterol > 8.5 mmol/l	8
II	LDL-cholesterol 6.5 - 8.4 mmol/l	5
III	LDL-cholesterol 5.0 - 6.4 mmol/l	3
IV	LDL-cholesterol 4.0 - 4.9 mmol/l	1

FH DIAGNOSIS

Almost certain   score = 8

Likely   score= 6-7

Possible   score= 3-5

Additional DNA testing is advised if the score > 6

* In this category, only the highest applicable number should be scored; the highest score for family history is 2.

Fig. 20.4 Diagnostic algorithm for familial hypercholesterolemia

Selective inhibition of apo-B100 mRNA synthesis by *antisense oligonucleotides (ASOs)* is an entirely new approach to lower cholesterol levels. ASOs bind to a complementary mRNA sequence by Watson-Crick hybridization, resulting in selective degradation of the targeted mRNA sequence and thereby in a reduction in apoB100 synthesis. The drug is administered subcutaneously once a week or less and induces an approximately 50% LDL-C reduction in FH.[40] The most common adverse events are mild injection site reactions, as well as modest increases in liver enzymes, as seen with all other lipid-lowering drugs. Other studies are evaluating *antisense PCSK9* as a LDL-C lowering strategy, with promising results.[41]

Finally, with respect to treatment of *children with FH*, several statin trials have been performed over the past decade.[42] On the basis of these studies, showing that statin treatment lowers LDL-C safely and effectively in children with FH,[42,43] and a study demonstrating reduced cIMT progression in FH adolescents on statin therapy,[44] current guidelines of the American Heart Association recommend initial statin treatment in children with FH from the age of 10 years in males and after onset of the menses in females.[45]

In addition to pharmacological treatment of hypercholesterolemia, management should also comprise screening of first-degree relatives.

20.4.2 *Familial Defective Apolipoprotein B*

Familial Defective Apolipoprotein B (FDB) is an autosomal dominant disorder which resembles the clinical phenotype of FH. The mechanism underlying the hypercholesterolemia is defective binding of apo-B100 of the LDL particle to the LDL receptor. The estimated prevalence is 1:500 in Central Europe and 1:700 in Northern America. Due to founder effects, prevalence up to 1:200 was observed in certain regions of Europe.[46] However, the exact prevalence remains unknown, since it frequently overlaps with that of FH.

20.4.2.1 Genetics

FDB is caused by mutations in the apo-B100 gene located on chromosome 2p23–24. So far, 11 true mutations at the apoB locus have been identified. The R3500Q mutation is the most frequent one, with a prevalence of 1:600–700 in Caucasians.[28]

20.4.2.2 Clinical Characteristics, Diagnosis, and Management

FDB is clinically indistinguishable from FH,[47] although with slightly lower LDL-C levels.[48] FDB is diagnosed by genotyping or according to clinical diagnostic criteria for FH. Equal to FH, patients with FDB are treated with lipid-lowering medication combined with lifestyle modifications.

20.4.3 *Autosomal Recessive Hypercholesterolemia*

ARH is the single known recessive disorder causing hypercholesterolemia. Only about 50 individuals with ARH have been identified worldwide. In ARH, hepatic endocytosis of the LDL-R/LDL particle complex, mediated by the *LDL-R adapting protein (LDLRAP)* is disrupted.[24,49]

20.4.3.1 Genetics

To date, 17 mutations in the ARH gene, located on chromosome 1p35–36.1 have been identified, the great majority being truncating mutations.[23]

20.4.3.2 Clinical Characteristics

ARH is characterized by a phenotype which resembles homozygous FH, consisting of severe hypercholesterolemia, large xanthomas, and premature CVD, although the phenotype in ARH is slightly milder, since patients tend to have higher HDL-C levels and are more responsive to lipid-lowering therapy and express a longer event-free survival when compared to homozygous FH patients.[50] The presence of residual LDL-R activity, as demonstrated in skin fibroblasts might explain the favorable plasma cholesterol concentrations and the response to cholesterol-lowering medication in patients with ARH.[51] In general, no clinical symptoms before the age of 20 years are present in ARH. Heterozygous carriers of the ARH-mutation have slightly elevated lipid levels within the normal range.

20.4.3.3 Diagnosis

ARH is diagnosed by genetic testing. Affected individuals meet clinical criteria for homozygous FH, as described in paragraph 20.4.1.2; however, based on clinical evaluation of first-degree relatives, a lipid disorder of recessive origin, rather than homozygous FH, should be considered.

20.4.3.4 Management

Patients with ARH are sensitive to treatment with statins and a cholesterol-lowering diet.[52]

20.4.4 Familial Combined Hyperlipidemia

Familial combined hyperlipidemia (FCH) is a relatively common lipoprotein disorder, with a prevalence of 1:200. The disease is based on increased *VLDL synthesis*, due to an overproduction of apo-B100, sometimes combined with delayed hepatic clearance of VLDL.[53]

20.4.4.1 Genetics

Initially, FCH was considered an autosomal dominant monogenetic disorder; however, the hereditary background might be *polygenetic*, as some families display a convincingly autosomal dominant mode of inheritance, whereas in others, a multifactorial basis is considered to be more likely. Although the causative gene has not been indentified to date, several candidate genes and loci have been associated with FCH.[54]

20.4.4.2 Clinical Characteristics

FCH is phenotypically heterogeneous and, in most individuals, not manifest until adulthood. It is characterized by elevated LDL-C and/or triglyceride levels, a tendency to decreased HDL-C levels, and is often accompanied by central obesity, insulin resistance, and hypertension. In addition, there are no clinical stigmata such as in FH. Different phenotypes can be expressed within members of one affected family. In most cases, apoB100 levels are elevated above 1.2 g/L and plasma triglycerides are mildly to moderately increased; however, cholesterol and triglyceride levels can vary over time within affected individuals. FCH patients have an increased risk of premature CVD.[55]

20.4.4.3 Diagnosis

FCH is diagnosed based on a combination of plasma lipid abnormalities and a positive family history of dyslipidemia: either solitary elevated LDL-C, triglycerides or both, with or without premature CVD in the index patient or family members. The diagnosis of FCH can accurately be calculated by a nomogram, which has shown to be a good predictor of cardiovascular risk in FCH.[56]

20.4.4.4 Management

Untreated FCH patients are prone to premature CVD; therefore, aggressive lipid-lowering treatment equal to that for FH is required. Most FCH patients are treated with high-dose statins, with a target LDL-C of 2.5 mmol/L or 1.8 mmol/L in subjects with CVD. In case triglyceride levels are elevated as well, patients can be treated with a *fibrate*. In addition, FCH patients should be treated with lifestyle modification, also in order to target the accompanying symptoms of obesity, insulin resistance, and hypertension. Finally, since FCH patients have functioning LDL receptors, the response to dietary interventions and pharmacological cholesterol-lowering is generally better than observed in FH.

20.4.5 Sitosterolemia

Sitosterolemia is a rare autosomal recessive disorder characterized by premature atherosclerosis. Although hypercholesterolemia is not obligatory, elevations in LDL-C levels may be observed. The underlying defect in sitosterolemia is hyperabsorption of *plant sterols* and decreased biliary secretion of both cholesterol and plant sterols.[57] Plant sterols are structurally similar to cholesterol and are derived solely from the diet. Normally, plasma plant sterol levels in humans are extremely low due to active efflux, as achieved by the *ABCG5/G8 transporters*. In sitosterolemia, this mechanism is disrupted. The exact prevalence is unknown; approximately 50 cases have been described worldwide.

20.4.5.1 Genetics

The ABCG5 and ABCG8 transporter genes are arranged in a head-to-head configuration on chromosome 2p21.[58] Mutations in either the ABCG5 or G8 gene can cause sitosterolemia.[57,59] All of the missense mutations in either ABCG5 or ABCG8 studied to date, either prevent formation of the obligate heterodimer or block the efficient trafficking of the heterodimer to the plasma membrane.[15]

20.4.5.2 Clinical Characteristics

Sitosterolemia is characterized by xanthomas, arthralgias, anemia, and premature atherosclerosis.[60] Plasma cholesterol levels are not necessarily elevated; however, affected individuals are highly sensitive to dietary cholesterol and become markedly hypercholesterolemic when fed a high-cholesterol diet.[61]

20.4.5.3 Diagnosis

The disease should be suspected in patients who develop xanthomas in early childhood, despite normal or only moderately elevated plasma cholesterol concentrations.

Sitosterolemia can be diagnosed by genetic analysis or by plasma plant sterol levels exceeding 0.024 mM (1 mg/dL).

20.4.5.4 Management

Affected individuals should be restricted from a cholesterol- and plant sterol-rich diet, as well as from plant sterol-enriched food products. In addition, subjects benefit from treatment with ezetimibe, a cholesterol absorption inhibitor which also inhibits intestinal absorption of plant sterols,[62] alone or combined with a BAS.[63] Statins are not effective in sitosterolemia.

20.5 Genetic Causes of HDL-C Disorders

Disorders of HDL-C identified in humans may result from interaction between genetic and environmental factors. Plasma HDL-C levels are under pronounced genetic influence, with heritability estimates ranging between 40 and 60%.[64] Several monogenetic defects in various proteins involved in HDL metabolism have been identified in humans to date. The genes encoding apolipoprotein-AI, ABCA1, and LCAT are essential for the de novo synthesis of HDL. A complete lack of any of these factors confers severe HDL deficiency, which is referred to as familial hypoalphalipoproteinemia. In contrast, CETP deficiency mostly induces accumulation of HDL in the circulation, so-called hyperalphalipoproteinemia. However, the vast majority of cases with HDL-C deficiency, defined as an age- and sex-adjusted plasma HDL-C concentration below the 10th percentile, are polygenic and/or multifactorial in origin. Decreased HDL levels are often found in patients with genetically disturbed metabolic pathways like hypertriglyceridemia, diabetes mellitus type 2, and obesity and metabolic syndrome.[65] In addition, multiple other factors have been identified to negatively influence HDL-C levels, such as smoking, physical inactivity, and certain medication or diseases like rheumatoid arthritis and systemic lupus erythematosus.[66]

The first step in the diagnostic workup of HDL deficiency consists of exclusion of these underlying conditions. Patients with a virtual absence of HDL must undergo careful physical examination to unravel the clinical characteristics of certain HDL deficiency syndromes as described below. In addition, family studies should be initiated to show segregation of low HDL in the family. Definitive diagnosis requires specialized biochemical tests and the demonstration of a functionally relevant mutation in an HDL gene.[65]

To date, no routinely used drug is able to increase HDL-C in patients with specific familial HDL deficiency syndromes so that prevention of cardiovascular disease in these patients must be focused on the avoidance and treatment of additional risk factors. In general, several lifestyle and pharmacologic interventions have shown to modestly increase HDL, although the impact of these interventions on the functional quality of HDL is unclear. Lifestyle modifications such as weight reduction, exercise, and smoking cessation can increase HDL levels by approximately 10–15%. In addition, pharmacologic therapies with niacin, fibrates, or statins, alone or in combination, raises HDL. Niacin therapy is the most effective pharmacological agent currently available and results in significant HDL increases of 15–35%. Several mechanisms have been suggested, although it is not exactly clear how niacin raises HDL. The most common reason for treatment failure is inability to tolerate cutaneous flushing. This can be reduced by prescribing the long-acting form, or by administering premedication with aspirin, or may diminish spontaneously after several days of therapy, as patients develop tolerance. Also, administration of niacin in combination with laropiprant (MK-0524A) significantly reduces flushing and therefore complicated introduction periods can be avoided.[67] Niacin used as monotherapy has shown benefit with regard to CHD risk reduction.[68] Fibric acid derivatives increase synthesis of apo-AI, enhancing the formation of new HDL particles and raising HDL by 5–20%, with the largest increases seen in patients with hypertriglyceridemia. Triglycerides are reduced by 20–50%, but LDL is changed minimally, if at all. A recently published meta-analysis of randomized placebo-controlled clinical trials using fibrates as monotherapy for the primary and secondary prevention of cardiovascular events, showed a significant reduction in nonfatal myocardial infarction. However, there was no significant impact on stroke, fatal myocardial infarction, or cardiovascular mortality. Most worrying was the trend toward an increase in all-cause mortality.[69] Statins induce an increase in hepatic apo-AI production and HDL levels of approximately 10–15%. What is not clear from available data is the role for combination therapy with statins and either fibrates or niacin. Given that statin therapy represents a well-tolerated, evidence-based approach for the primary and secondary prevention of cardiovascular events in patients with low HDL, the question arises as to whether there is incremental value to adding either a fibrate or niacin if HDL remains low. Ongoing studies such as ACCORD, AIM-HIGH, and HPS2-THRIVE should assist in answering these questions, hopefully within the next 3 years. To date, no formal HDL target goals or treatment guidelines have been implemented yet, as there is a lack of strong clinical evidence to support effective pharmacologic therapy for CVD risk reduction.[70] However, promising new agents which target both quantity and quality of HDL particles are currently under development including CETP inhibitors, apo-AI and HDL mimetics, intravenous apo-AI (Milano) infusion, and agonists of PPAR-alpha, LRH-1 and LXR.[71] CETP inhibitors, like torcetrapib, dalcetrapib (JTT-705), and anacetrapib, are powerful HDL raisers and decrease LDL. However, all torcetrapib clinical trials have been discontinued because there were significantly more major cardiovascular events in the groups receiving torcetrapib in combination with atorvastatin than in the group receiving atorvastatin monotherapy. The increased mortality and morbidity may in part be explained by the significant increase in systolic blood pressure induced by torcetrapib.[72] Animal models demonstrated that torcetrapib induces an increase in blood pressure in all evaluated species, accompanied by increased circulating levels of aldosterone, independent of CETP inhibition.[73] In addition, in rats torcetrapib increased mean arterial pressure and expression of renin–angiotensin–aldosterone–system (RAAS)-related as well as endothelin mRNAs were found in adrenals and aorta. Conversely, no effects on blood pressure and RAAS were observed with anacetrapib and dalcetrapib in these studies. Furthermore, early clinical data with anacetrapib[74] and dalcetrapib,[75,76] demonstrating a respectively, 129 and 28–34% increase in HDL-C, reported so far no effect on systemic blood pressure.

In this section we will focus on the established monogenetic disorders of HDL metabolism including ApoA1, ABCA1, LCAT. Also genetic disorders of CETP will be discussed.

20.5.1 Apolipoprotein AI Deficiency

Apo-AI is the major protein component of HDL-C in plasma and plays a central role in cholesterol efflux from tissues to the liver for excretion. Apo-AI deficiency is a rare autosomal recessive inherited disorder characterized by decreased HDL-C levels.

20.5.1.1 Genetics

The *apo-AI gene* is located on the long arm of chromosome 11, adjacent to the genes encoding the apolipoproteins C-III and IV. Of the approximately 70 reported distinctive mutations of this gene, mostly found in heterozygous state, some are functionally relevant, that is, are associated with reduced levels of apo-AI and HDL-C.[77]

20.5.1.2 Clinical Characteristics

Heterozygous carriers of a functionally relevant mutation usually present with half normal apo-AI and HDL-C levels. Some mutations can even lead to more pronounced decreases. In most cases, heterozygous carriers of apo-AI variants do not present with specific clinical symptoms. An important exception are some structural apo-AI variants with amino acid substitutions in the N-terminus, which have been detected in patients with familial amyloidosis.[78] Surprisingly, susceptibility for premature coronary heart disease has been shown to differ markedly between apo-AI variants. Low HDL-C levels due to heterozygosity for a specific apo-AI mutation (L178P) were associated with vascular dysfunction, accelerated carotid arterial wall thickness, and an increased incidence of premature vascular events compared with their family controls.[79] In contrast, despite very low HDL levels, carriers of the apo-AI (R173C) Milano mutant did not differ from controls in terms of vascular function[80] and arterial wall thickness.[81] These differences are likely due to the profoundly different effects of the mutations at the protein level.

Patients with complete apo-AI deficiency, due to homozygosity or compound heterozygosity, present with a virtually absent HDL-C. In adult patients variable clinical manifestations have been described, such as abnormalities of the skin (xanthelasmata and xanthomas) and/or eyes (corneal opacities)[65] (see Table 20.1). Remarkably, only 11 of the 25 reported cases with complete apoA-1 deficiency, suffered from premature cardiovascular events. However, the remaining 14 cases were almost all below the age of 50 and may have been too young for clinical manifestations of atherosclerosis to occur. In addition, this small number of cases and differences in the type of apoA-1 gene defect make conclusions on the susceptibility to premature coronary heart disease in these specific patients difficult.[82]

20.5.1.3 Diagnosis

The diagnosis of apo-AI deficiency requires sequencing of the apo-AI gene and the demonstration of a functionally relevant mutation.

20.5.1.4 Management

Since no routinely used drug is able to increase HDL-C levels in patients with familial low HDL cholesterol, prevention of cardiovascular disease in these patients must be focused on the avoidance and treatment of

Table 20.1 Clinical hallmarks of familial HDL deficiency syndromes[65]

	Apo A1 deficiency	Tangier disease	Fish eye disease	Familial LCAT deficiency
Affected gene	APOA1	ABCA1	LCAT	LCAT
Enlarged tonsils	No	Occasionally	No	No
Hepato/splenomegaly	No	Occasionally	No	No
Neuropathy	No	Occasionally	No	No
Corneal opacities	+++	+	+++	+++
Xanthomas	Occasionally	No	No	Occasionally
Xantelasmata	Occasionally	No	No	No
Nephropathy	No	No	No	Yes
Hemolytic anemia	No	No	No	Yes

additional risk factors and the use of statins to obtain very low LDL-C levels.[65]

20.5.2 ABCA1 Deficiency and Tangier Disease

ATP-binding cassette transporter A1 (ABCA1) mediates the efflux of cholesterol and phospholipids from peripheral tissues to lipid-poor apo-AI in plasma and thereby plays a central role in forming HDL. Functionally relevant mutations in the ABCA1 gene lead to cholesterol efflux defects, which subsequently cause low HDL-C and apo-AI levels. Complete ABCA1 deficiency is the underlying cause of *Tangier disease*. This rare autosomal recessive disorder has been diagnosed in about 70 patients worldwide.

20.5.2.1 Genetics

The ABCA1 gene resides on chromosome 9q31. To date, more than 70 mutations and several common and rare variants have been described in the ABCA1 gene, with a wide range of biochemical and clinical phenotypes.[83] Several recent genome-wide association studies have identified common variants in ABCA1 as a significant source of variation in plasma HDL cholesterol levels across multiple ethnic groups [84,85] establishing ABCA1 as a major gene influencing HDL levels in humans.[86]

20.5.2.2 Clinical Characteristics

Heterozygote carriers of functionally relevant ABCA1 mutations can present with a broad range of plasma HDL-C levels ranging from 30% to 83% of age- and sex-matched controls.[86] However, the majority of these mutations are associated with an approximately 50% reduction in serum HDL-C and apo-AI levels and increased triglycerides. LDL levels are typically within the normal range.

Tangier disease, which is caused by complete ABCA1 deficiency due to homozygosity or compound heterozygosity, is characterized by profoundly decreased HDL-C plasma and apo-AI levels. Frequently, serum levels of total and LDL cholesterol are also low, whereas serum levels of triglycerides are mildly elevated. The clinical presentation of Tangier disease varies considerably and if present, clinical symptoms can be isolated or combined (see Table 20.1). It is likely that this phenotypic heterogeneity might at least in part be accounted for by the nature of the mutation and its effect on the protein.[87] Presenting features of Tangier disease include enlarged orange tonsils, hepatomegaly, and splenomegaly. Also, lymph nodes can have bright yellow streaks and morphologic characteristics as those present in the tonsils. A symptom with significant implications for quality of life is a peripheral neuropathy, which, however, has a highly variable expression. These clinical symptoms result from accumulation of cholesterylesters in reticuloendothelial cells, that is, macrophages, Kupffer cells or histiocytes, leading to accumulation of these cells in various organs.[65] Despite the known role of *ABCA1* in determining plasma HDL levels, the impact of *ABCA1* on atherosclerosis remains controversial and incompletely understood. Prior to the identification of ABCA1 mutations as the genetic basis of Tangier Disease in 1999, patients were identified based on their clinical phenotype, that is, extremely low HDL cholesterol in homozygotes, with the offspring and parents of homozygotes being obligate heterozygotes. Considering the wide variation in phenotype, misclassification of patients was likely and this complicated accurate CAD risk estimation. Since the assignment of disease has been based on genotype, allowing a more unambiguous diagnosis, several studies have addressed the risk for CAD in these patients. Large family studies, studying several mutations, showed a more than threefold excess of coronary artery disease (CAD) and increased carotid arterial wall thickness in affected family members when compared to unaffected members.[88,89] In both studies, levels of cholesterol efflux correlated well with HDL-C levels and there was a strong correlation between levels of cholesterol efflux and CAD and/or carotid arterial wall thickness. However, these family studies potentially suffer from selection bias, as only families with the most severe phenotypes may have presented at clinics. Also CAD risk estimates were based on few individuals and were not adjusted for age and other cardiovascular risk factors. Bypassing these problems, seven different ABCA1 mutations were studied in two different population cohorts and a large case-control study, including a total of 109 heterozygotes, 6,666 ischemic heart disease cases, and a total of 41,961 participants.[90] Four mutations were found to be associated with an on average 30% reduction in HDL-C and decreased cholesterol efflux. Carriers of these 4 mutations, however, did not display an increased risk of CVD. However, this conclusion should also be interpreted with caution as the variants studied were mild mutations, giving relatively small reductions in HDL cholesterol levels and cholesterol efflux.[91] The findings are also conflicting with several reports showing that common genetic variations of the ABCA1 gene influences risk of CAD in the general population.[90,92,93] Interestingly, these associations with atherosclerosis are independent of effects on HDL levels. These findings, of an altered risk for CAD but without corresponding differences in lipid levels, suggest that although ABCA1 may be an important atherosclerosis susceptibility locus, the mechanism by which it exerts this effect is not necessarily by altering steady-state HDL-C levels. In conclusion, any specific ABCA1 variant must be considered in relation to its impact on protein function, as different variants will have different effects on HDL and susceptibility to atherosclerosis.[94]

20.5.2.3 Diagnosis

The findings of virtually absent HDL-C and low levels of apo-AI are not sufficient to diagnose Tangier disease which ultimately requires ABCA1 gene sequence analysis. Cholesterol efflux defects can be demonstrated with the cholesterol efflux assay on cultivated skin fibroblasts. However, even in the absence of coding sequence mutations in ABCA1, cellular cholesterol efflux defects are a common feature in subjects with low HDL.[95] Foam cell formation, responsible for the clinical symptoms in Tangier disease, can be detected in the rectal mucosa by endoscopic examination as pale mucosa with studded 1–2 mm discrete orange brown spots.[65]

20.5.2.4 Management

To date, no specific treatment for Tangier disease exists. It is advised to identify and tightly regulate modifiable cardiovascular risk factors and possibly institute statin therapy as a means to drive LDL-C levels down even further.

20.5.3 Familial LCAT Deficiency and Fish Eye Disease

Lecithin:cholesterol acyltransferase (LCAT) plays a key role in the maturation of small HDL by means of esterification of free cholesterol, primarily at the surface of the HDL particle (so-called alpha-LCAT activity) but also on lipids transported by apoB containing lipoproteins (so-called beta-LCAT activity). After esterification, the cholesteryl ester molecules migrate to the inner core of the lipoprotein, promoting further cholesterol efflux from peripheral tissues and leading to larger, cholesteryl-ester-enriched HDL particles. Mutations in the LCAT gene causing LCAT deficiency represent another rare autosomal recessive disorder that underlies HDL deficiency. Low HDL-C values result from defective HDL maturation followed by rapid clearance of nascent HDL particles from the circulation. Depending on the mutation, patients with complete LCAT deficiency present with one of the two clinical phenotypes, *familial LCAT deficiency* (FLD) or *fish-eye disease* (FED).

20.5.3.1 Genetics

The gene encoding LCAT is located on chromosome 16, locus 16q22.1. Mutations in LCAT account for approximately 4% of low HDL.[95] Thus far, over 40 mutations in the LCAT gene have been described in reports that predominantly investigated single cases or small nuclear families.[82]

20.5.3.2 Clinical Characteristics

Heterozygous carriers of LCAT mutations lack clinical symptoms, although frequently they present with half normal HDL-C levels and mild hypertriglyceridemia.[82] Homozygous or compound heterozygous patients with mutations in the LCAT gene present with one of two clinical phenotypes, familial LCAT deficiency (FLD) or fish-eye disease (FED). In FLD both alpha-LCAT, which is specific for HDL, and beta-LCAT, which is specific for VLDL and LDL, is deficient, that is, the deficient esterification is generalized. In contrast, patients with FED have a selective alpha-LCAT deficiency. Because LCAT is still partly active, these patients have, in general, a less severe phenotype. Both FLD as FED are characterized by corneal opacifications, which become apparent after the third decade (see Table 20.1). In addition, FLD is characterized by hemolytic anemia, and deposition of foam cells in bone marrow, spleen, and particularly in kidneys. Progressive renal disease, with proteinuria and hematuria, which progresses to terminal renal insufficiency, has been described in a high percentage of these patients[65] (see Table 20.1). Biochemically, FLD and FED are both characterized by variable loss of LCAT activity and *HDL deficiency* (5–10% of normal HDL-C levels). Serum levels of apoA-I are usually decreased but not as low as in patients with apoA-I deficiency or Tangier disease. Additionally, hypertriglyceridemia is observed.[82]

Because of the limited number of carriers of LCAT gene mutations it is difficult to draw conclusions regarding the risk for atherosclerosis. A recent 25-year follow-up of nine heterozygote family members,[96] as well as a large family study, including 68 carriers of LCAT defects of which 59 heterozygotes and 74 family controls[82] which measured carotid arterial wall thickness indicated that heterozygous carriers of LCAT defects may have an increased risk of atherosclerotic vascular disease.

20.5.3.3 Diagnosis

The identification of LCAT deficiency needs either genetic testing or measurement of LCAT activity. Depending on the kind of mutation, immunoassays of LCAT detect either no, or slightly reduced concentrations of LCAT protein in plasma. Routine lipid and lipoprotein analyses do not help to distinguish patients with FLD and FED. However, patients with FLD show an increased proportion of unesterified cholesterol in plasma (80–100% instead of normal <30%). By contrast, the plasma of patients with FED have a normal or slightly elevated (up to 70%) unesterified cholesterol/cholesterol ester ratio.[65]

20.5.3.4 Management

Only symptomatic treatments exists for LCAT deficiency. Because deposition of highly abnormal apoB-containing lipoproteins in the kidneys of FLD patients has been implicated as the pathogenetic factor in the formation of renal disease, therapies which reduce the concentration of apoB-containing lipoproteins (such as a fat-restricted diet and statins) are at least theoretically useful.[65]

20.5.4 Genetic Disorders of CETP

As a regulator of cholesterol flux through the RCT system, cholesteryl ester transfer protein (CETP), may be viewed as potentially having both pro-atherogenic and anti-atherogenic properties (see Fig. 20.1). By facilitating the exchange of cholesteryl esters for triglycerides between HDL and Apo-B containing lipoproteins (LDL and VLDL), CETP may decrease direct RCT via the HDL/hepatic SR-B1 route. In addition, pro-atherogenic effects of CETP activity may result from a reduction in overall HDL levels, potentially reducing cellular cholesterol efflux from the arterial wall, and from an increase in atherogenic LDL levels. However, the potentially proatherogenic activities of CETP may, to a large extent, be neutralized by an increase in indirect RCT via the LDL/hepatic LDL receptor route.[97]

20.5.4.1 Genetics

CETP is encoded by a gene located on the long arm of chromosome 16. Several mutations of the CETP gene have been associated with altered CETP activity and HDL-c levels. Recent genome-wide association studies have reported that CETP genotypes are associated with HDL-C levels more strongly than any other locus across the genome.[84,85]

20.5.4.2 Clinical Characteristics

Significant differences between ethnic groups with regard to allele frequencies of CETP polymorphisms exist.[98] Particularly in Japan, CETP gene defects are common and there are appreciable numbers of individuals who are homozygous for mutations in the CETP gene. Not surprisingly, functional mutations of the CETP gene can produce significant changes in lipid and lipoprotein metabolism. For example, homozygous individuals for a G to A mutation at the +1 position of intron 14 have no measurable CETP activity and exhibit up to five times normal HDL-C levels.[99] Another mutation, D442G, retains partial CETP activity and, consequently, displays less markedly raised HDL-C levels in homozygous individuals.[100] Not all CETP gene mutations have an as dramatic effect on CETP protein levels as the ones described above. Various single nucleotide polymorphisms of the CETP gene are associated with only small changes in plasma CETP levels and subsequently HDL-C levels, in either direction.[98] This demonstrates clearly that the precise nature of changes in lipid and lipoprotein metabolism can vary between different mutations.[97] Consequently, the role of CETP mutations on cardiovascular risk profiles is complex. Although sparse, there is evidence emerging from clinical trials that elevated CETP levels, regardless of the cause, are associated with increased risk of CVD.[101–103] However, studies on individuals with CETP protein deficiency, arising from different genetic mutations, have reported ambiguous findings on the relationship between CETP protein deficiency and CAD risk. In some studies, CETP-deficient patients were thought to have a increased CAD risk[104] but, conversely, this concept was not supported by others.[105,106] In addition, a recent meta-analysis was published that involved a total of more than 113,000 individuals and six CETP polymorphisms.[107] Three common CETP gene variants (TaqIB, I405V, and −629C>A) were consistently associated with a decreased CETP concentrations, modestly increased HDL-C and apoA-I levels, and weakly decreased triglycerides and coronary risk. Data were insufficient for informative per-allele estimates in relation to three uncommon CETP variants (D442G, −631C>A, and R451Q). However, they were associated with mean differences in HDL-C of 13.4%, −0.7%, and −8.8%, respectively, compared with controls. Furthermore, a recent study showed that the 373P and 451Q polymorphisms, which are associated with higher CETP activity and concentration and lower HDL-C concentration, are associated with atherogenic effects as manifested by a greater presence of coronary artery calcification as measured by computed tomography.[98] Thus, from the results of studies on individuals with CETP protein deficiency arising from genetic mutations, the relationship between CETP and the risk of CVD is not entirely conclusive. Overall cardiovascular risk is presumably not only dependent on the effect of CETP deficiency on overall levels of HDL-C but also on the effect on functionality of the HDL particles. Moreover, additional factors affecting the metabolic setting of the CETP gene mutation probably also play an important role. It was shown that high HDL-C resulting from simultaneous presence of CETP- and LIPC gene variants did not protect against coronary artery disease. In contrast, an increased risk for coronary artery disease was found in these patients.[108] In addition high triglyceride levels have been suggested to enhance the effect of CETP concentration on CHD risk.[103] Also potential joint effects of CETP genotypes with environmental determinants of HDL-C levels (e.g., exercise and alcohol) on risk of coronary disease have been reported.[109]

20.6 Genetic Causes of Elevated Triglycerides

Severely elevated *triglyceride* concentrations are a risk factor for developing *pancreatitis* and in the absence of other causes such as diabetes mellitus, alcohol abuse, chronic renal failure, or hypothyroidism, and generally point to genetic disorders of triglyceride-rich lipoprotein-modulating enzymes or apolipoproteins. Mutations in several genes have been described, of which the *LPL*, *apo-CII*, and *Apo-E* genes are the most important ones. Very recently, the *GPIHBP1 gene* has been introduced as a contributor to *primary hypertriglyceridemia*.[18,110] These genetic defects will be briefly discussed.

Regardless of its origin, management of hypertriglyceridemia consists of therapeutic lifestyle changes aiming at dietary and weight control, as well pharmacological treatment with fibrates, niacin, or high doses of fish oil, alone or in various combinations. In case triglyceride levels exceed 10 mmol/L (800 mg/dL), combinations of different drugs are usually required, in order to reduce the risk of pancreatitis.[111] The benefit of treating mild-to-moderate triglyceride elevations is less clear.[112] If hypertriglyceridemia is a comorbidity, statins can lower triglyceride levels by 20–40%.[113] Fibrates lower triglyceride levels by approximately 40–60% and modestly raise HDL-C levels by approximately 15–25%.[113] Patients who do not respond to *fibrates* can be treated with *niacin*, which lowers triglyceride levels by 30–50%, raises HDL-C levels by

20–30%, and lowers LDL-C levels by 5–25%.[113,114] However, the use of niacin is limited, due to its relatively frequent vasomotor side effects and liver enzyme elevations. Finally, fish oil with 2–4 g of *omega-3 fatty acids* daily can reduce triglyceride levels by 30–50%.[115] Over-the-counter preparations usually contain far less than these required amounts.[116] Of note, in patients with *diabetes mellitus*, optimizing glycemic control might help to lower triglyceride levels without additional medication for hypertriglyceridemia.

20.6.1 LPL-Deficiency and Apo-CII Deficiency

Plasma lipoprotein lipase (LPL) and its cofactor apo-CII are involved in the hydrolysis of triglyceride-rich particles such as chylomicrons and VLDL. Genetic *LPL deficiency* is a rare autosomal recessive disorder causing severe hypertriglyceridemia. Estimations of prevalence vary between 1:1,000,000 in the general population and 1:5,000 in French Quebec. The incidence of apo-CII deficiency is even lower than that of LPL deficiency.

20.6.1.1 Genetics

The LPL gene is located on chromosome 8p22.[117] Nearly 100 mutations have been described.[118] The apo-CII gene is located on chromosome 19, in which at least 13 mutations have been described.[119]

20.6.1.2 Clinical Characteristics

Affected individuals have insufficient capacity to hydrolyze triglycerides, resulting in extremely high plasma triglyceride concentrations, often accompanied by recurrent episodes of *pancreatitis*. LPL deficiency typically manifests itself in early childhood with severe and repetitive colicky pain in the abdomen, acute pancreatitis, and failure to thrive. Eruptive xanthomas (Fig. 20.4), lipemia retinalis, and hepatosplenomegaly can also be present. Plasma is lipemic, reflecting increased plasma levels of both chylomicrons and VLDL.

Whether LPL deficiency is associated with an increased incidence of CVD remains unclear. Only two studies have demonstrated premature atherosclerosis in LPL deficient patients.[120,121] To date, the only apo-CII mutation described to cause early atherosclerosis is the apo-CII St Michel mutation.[122]

20.6.1.3 Diagnosis

Genetic LPL- and *apo-CII deficiency* are diagnosed by genotyping, combined with the phenotype as described above. Apo-CII deficiency can also be diagnosed by a post-heparin LPL activity assay, in which the patient's post-heparin plasma is mixed with that of a non-affected individual. In this experiment, triglyceride levels will decrease rapidly in an apo-CII deficient patient, in contrast to subjects with LPL deficiency.

20.6.1.4 Treatment

Treatment consists of *dietary fat restriction*. Hypertriglyceridemia is treated as described above, however, in genetic LPL and apo-CII deficiency, most of these strategies do not result in a substantial reduction in triglyceride levels. Nevertheless, promising new compounds for treatment of this patient group are under investigation, such as *ibrolipim*, a pharmacological stimulator of tissue LPL formation, *LPL gene therapy*,[123] and *antisense apo-CIII therapy*.[124]

20.6.2 Familial Dysbetalipoproteinemia (Apo E2/E2 Deficiency)

Familial dysbetalipoproteinemia (FD) is characterized by the defective clearance of VLDL- and *chylomicron-remnant particles* caused by homozygosity for apoE2, the type of apoE unable to bind to its receptor. There are three common apoE isoforms: apoE3, apoE2, and apoE4.[125] Although approximately 0.5% of the population worldwide is homozygous for apoE2, only a small minority develops FD with an estimated prevalence of 1–2:10,000. This is due to the necessity of concomitant environmental, hormonal, and possibly genetic factors, inducing VLDL or chylomicron overproduction, such as a high caloric diet or alcohol abuse, diabetes mellitus, obesity, hypothyroidism, renal disease, or estrogen deficiency.

20.6.2.1 Genetics

Most people have an apoE3/E3 genotype, with a ~55% prevalence ; however, apoE4 and apoE2 also exist, with an estimated frequency of 0.5% for *apoE2/E2*, 15% for apoE2/E3, 25% for apoE3/E4, 1–2% for apoE4/E4, and 3–4% for apoE2/E4.

ApoE2 differs from ApoE3 by a single substitution of cysteine for arginine at residue 158.

Less common, dominant negative mutations may also cause the disorder (ApoE3-Leiden or ApoE2-Lys146 > Gln).[126]

20.6.2.2 Clinical Characteristics

Clinically, apoE2/E2 patients present with *tubero-eruptive xanthomas* (see Fig. 20.5), palmar streaks, elevated TC and triglyceride concentrations, and are at high risk for premature CVD and peripheral vascular disease.[127] Tubero-eruptive xanthomas begin as clusters of small papules on elbows, knees, or buttocks and can grow to the size of small grapes. Palmar xanthomas are orange yellow discolorations of palm and wrist creases. Both are pathognomonic for FD, but their absence does not exclude the disorder. Plasma TC concentrations usually exceed 8.0 mmol/L (300 mg/dL) and may approach 26.0 mmol/L (1,000 mg/dL).

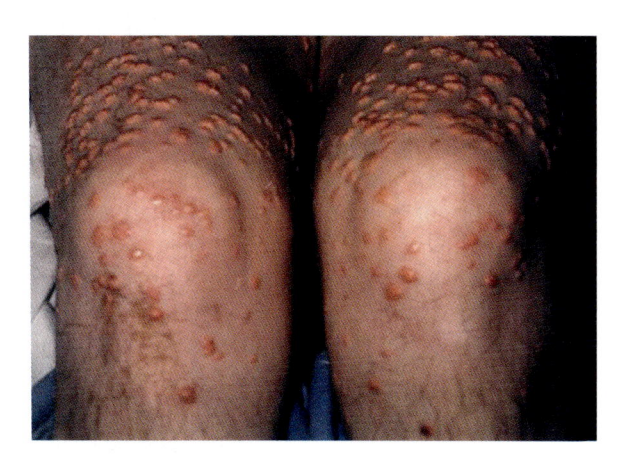

Fig. 20.5 Eruptive xanthomas

Triglyceride concentrations are within the same range. Dyslipidemia in FD rarely manifests before adulthood. The average age of clinically overt vascular disease is approximately 40 years in men and 59 in women.

20.6.2.3 Diagnosis

FD can be diagnosed either by lipoprotein ultracentrifugation and electrophoresis with a *VLDL/triglyceride ratio* >0.3 or by apoE genotyping. However, the absence of apoE2/E2 does not rule out the disease, as other genetic causes might also give rise to this trait.

20.6.2.4 Management

Treatment of FD is aimed at reducing overproduction of VLDL and/or chylomicrons, by means of dietary restrictions, including alcohol intake and weight reduction, combined with pharmacological treatment with statins, alone or combined with other compounds, as described above.

20.6.3 Familial Combined Hyperlipidemia

FCH is discussed in paragraph 20.4.4

20.6.4 Familial Hypertriglyceridemia

Familial hypertriglyceridemia (FHTG) is a common disorder causing hypertriglyceridemia with prevalence of 1:500. Although the genetic basis is unknown and the onset of disease depends on the presence of certain lifestyle factors, FHTG is discussed due to its high prevalence. The metabolic defect is a combination of hepatic VLDL overproduction and decreased catabolism of both VLDL and chylomicrons.

20.6.4.1 Clinical Characteristics

Typically, patients have moderately elevated plasma triglycerides, 3–10 mmol/L, often accompanied by low HDL-C levels. FHTG is associated with obesity, insulin resistance, hypertension, and hyperuricemia. The onset of hypertriglyceridemia is usually in adult age, when lifestyle factors which increase triglyceride levels, such as obesity, become more prominent. When the hypertriglyceridemia becomes more severe, the clinical picture can resemble that of LPL-deficiency. The association with CVD is weak, at most.

20.6.4.2 Diagnosis

FHTG is diagnosed by exclusion of other causes of hypertriglyceridemia. A first-degree family member with the same disorder is useful for the diagnosis. FCH and FD should definitely be excluded, since these disorders are associated with a more pronounced CVD risk and therefore require a more stringent therapy.

20.6.4.3 Management

The first line of treatment is lifestyle modification, possibly combined with pharmacological treatment in case of more severe hypertriglyceridemia.

Finally, mutations in the recently identified *GPIHBP1 protein* might be a cause of severe hypertriglyceridemia, resembling LPL or apo-CII deficiency.[18] This protein is thought to anchor LPL on the luminal surface of capillaries, where lipolysis of triglyceride-rich particles takes place. At present, two mutations have been described in humans: the G56R [128,129] and Q115P mutation,[110] of which only the latter was proven to be causal. In addition, the LMF1 gene has recently been identified as an interesting candidate gene for severe *hypertriglyceridemia*.[130]

20.7 Future Candidates

Recently, three *genome-wide association studies* reported several genes or loci that contribute significantly to variations in LDL-C, HDL-C, and triglycerides.[84,85,131] A total of seven new associated genes or loci and more than a dozen of previously suggested genes were found. For the newly identified loci, the causal variants and genes are not yet clear and warrant further investigation to establish causality in lipoprotein disorders.

20.8 Summary

Disorders of lipoprotein metabolism are major contributors to cardiovascular disease (CVD), a leading cause of mortality and morbidity worldwide. Dyslipidemia includes both elevated low-density-lipoprotein cholesterol (LDL-C) levels, as well as decreased high-density-lipoprotein cholesterol (HDL-C) levels.

LDL mediates cholesterol transport from the liver to peripheral tissues, including macrophages in the arterial wall, which, after uptake and accumulation of cholesterol, can transform into foam cells and atherosclerotic plaques. Conversely, HDL is thought to exert beneficial effects on the arterial wall through its role in the reverse cholesterol transport (RCT), which involves the transport of cholesterol from peripheral tissues to the liver followed by biliary excretion and elimination via the feces.

The crucial role of increased plasma LDL-C levels in the pathogenesis of atherosclerosis has been firmly established, as well as the beneficial effects of LDL-C reduction accomplished by hydroxy-methyl-glutaryl coenzyme A (HMG-CoA) reductase inhibitors or statins. In addition, decreased plasma HDL-C levels are an established independent predictor of CVD. However, pharmacological raising of plasma HDL levels has failed to reduce cardiovascular events thus far. It is therefore uncertain whether HDL plays a causative role in CVD protection or if it is merely an epiphenomenon or nonfunctional biomarker. Finally, the relationship between hypertriglyceridemia and CVD risk also remains to be elucidated.

Most cases of CVD are multifactorial and/or polygenic in origin. However, when cardiovascular disease occurs at young age, a number of monogenetic disorders of lipoprotein are frequently seen. These monogenetic disorders of lipoproteins are the primary focus of this chapter.

Regarding LDL metabolism, mutations in four genes are currently identified to result in increased plasma LDL-C concentrations, namely the LDL-R gene, ApoB gene, ARH gene, and most recently, the PCSK9 gene. Clinical hallmarks of these disorders, of which familial hypercholesterolemia is the most frequent and well-known, are elevated plasma

LDL-C levels and, consequently, premature atherosclerosis.

To date, several rare monogenetic defects in various proteins involved in HDL metabolism have been identified in humans. The genes encoding apolipoprotein-AI, ABCA1 and LCAT respectively, are essential for the de novo synthesis of HDL. A complete lack of any of these factors confers severe HDL deficiency, which is referred to as familial hypoalphalipoproteinemia. In contrast, CETP deficiency mostly induces accumulation of HDL in the circulation. Although FHA patients display extremely low plasma HDL-C levels, the association of these genetic disorders with atherosclerosis is disputed. Since HDL is a heterogeneous class of lipoprotein particles, these different classes may have different associations with disease. Furthermore, the functionality of the HDL particles, rather than their abundance may be an important determinant of their anti-atherogenic effects.

Finally, severely elevated triglyceride concentrations can be induced by mutations in several genes, of which the LPL, apo-CII, and Apo-E genes are the most important ones. Despite the unclear role of hypertriglyceridemia in atherogenesis, severely elevated triglyceride levels confer a health risk due to the increased risk of pancreatitis.

In general, the vast majority of dislipidemias are polygenic and/or multifactorial in origin.

The first step in the diagnostic workup of dyslipidemias consists of the exclusion of underlying conditions through careful medical history taking, physical examination, and biochemical testing (see Table 20.2). Suggestive for a genetic cause are the presence of specific clinical hallmarks (see text and Table 20.1) and/or the presence of familial dyslipidemias/premature atherosclerosis. In these cases, specialized biochemical tests and/or the demonstration of a functionally relevant mutation in the involved genes should be performed to obtain a definitive diagnosis. In addition, family studies should be initiated to evaluate the inheritance pattern of the phenotype (see Table 20.3).

Treatment consists of lifestyle modifications such as weight reduction, exercise, and smoking cessation to reduce the risk of other atherogenic risk factors, possibly in combination with pharmacological agents. High-dose statins are currently the most effective pharmacological strategy to reduce CVD risk. Also in case of low HDL and hypertriglyceridemia, statin monotherapy, possibly in combination with other agents, reduces the risk of CVD.

Table 20.2 Underlying causes of dyslipidemias

Elevated LDL-C levels	Decreased HDL-C levels	Elevated TG levels
Hypothyreoidism	Obesity	Obesity
Kidney diseases	Diabetes mellitus	Diabetes mellitus
Certain medications, e.g.,	Metabolic syndrome	Metabolic syndrome
• Corticosteroids	Certain medication	Alcohol abuse
• Thiazide diuretics	Underlying diseases, e.g.,	Chronic renal failure
	• Rheumatoid arthritis	Hypothyroidism
	• Systemic lupus erythematosus	

Table 20.3 Steps in the diagnostic workup of dyslipidemias[65]

1. Exclude underlying conditions
2. Suspected genetic cause?
– Profoundly decreased HDL-C levels? (<5th percentile adjusted for age and sex)
– Presence of specific clinical hallmarks? (see text and Table 20.1)
– Presence of familiar dyslipidemias/premature atherosclerosis?
3. Perform specialized biochemical tests and/or specific HDL gene sequencing
4. Initiate family studies

References

1. Yusuf S, Hawken S, Ounpuu S, et al. Effect of potentially modifiable risk factors associated with myocardial infarction in 52 countries (the INTERHEART study): case-control study. *Lancet*. 2004 September 11;364(9438):937-952.
2. Baigent C, Keech A, Kearney PM, et al. Efficacy and safety of cholesterol-lowering treatment: prospective meta-analysis of data from 90, 056 participants in 14 randomised trials of statins. *Lancet*. 2005 October 8;366(9493):1267-1278.
3. Genest JJ Jr, Martin-Munley SS, McNamara JR, et al. Familial lipoprotein disorders in patients with premature coronary artery disease. *Circulation*. 1992 June;85(6):2025-2033.
4. Gordon DJ, Probstfield JL, Garrison RJ, et al. High-density lipoprotein cholesterol and cardiovascular disease. Four prospective American studies. *Circulation*. 1989 January; 79(1):8-15.

5. Briel M, Ferreira-Gonzalez I, You JJ, et al. Association between change in high density lipoprotein cholesterol and cardiovascular disease morbidity and mortality: systematic review and meta-regression analysis. *BMJ.* 2009; 338:b92.
6. Ghali WA, Rodondi N. HDL cholesterol and cardiovascular risk. *BMJ.* 2009;338:a3065.
7. Sarwar N, Danesh J, Eiriksdottir G, et al. Triglycerides and the risk of coronary heart disease: 10, 158 incident cases among 262, 525 participants in 29 Western prospective studies. *Circulation.* 2007 January 30;115(4):450-458.
8. Hokanson JE, Austin MA. Plasma triglyceride level is a risk factor for cardiovascular disease independent of high-density lipoprotein cholesterol level: a meta-analysis of population-based prospective studies. *J Cardiovasc Risk.* 1996 April;3(2):213-219.
9. Bansal S, Buring JE, Rifai N, Mora S, Sacks FM, Ridker PM. Fasting compared with nonfasting triglycerides and risk of cardiovascular events in women. *JAMA.* 2007 July 18;298(3):309-316.
10. Nordestgaard BG, Benn M, Schnohr P, Tybjaerg-Hansen A. Nonfasting triglycerides and risk of myocardial infarction, ischemic heart disease, and death in men and women. *JAMA.* 2007 July 18;298(3):299-308.
11. Taskinen MR. Diabetic dyslipidaemia: from basic research to clinical practice. *Diabetologia.* 2003 June;46(6):733-749.
12. Gandotra P, Miller M. The role of triglycerides in cardiovascular risk. *Curr Cardiol Rep.* 2008 November;10(6): 505-511.
13. Neeli I, Siddiqi SA, Siddiqi S, et al. Liver fatty acid-binding protein initiates budding of pre-chylomicron transport vesicles from intestinal endoplasmic reticulum. *J Biol Chem.* 2007 June 22;282(25):17974-17984.
14. Davies JP, Levy B, Ioannou YA. Evidence for a Niemann-Pick C (NPC) gene family: identification and characterization of NPC1L1. *Genomics.* 2000 April 15;65(2):137-145.
15. Graf GA, Cohen JC, Hobbs HH. Missense mutations in ABCG5 and ABCG8 disrupt heterodimerization and trafficking. *J Biol Chem.* 2004 June 4;279(23):24881-24888.
16. Temel RE, Tang W, Ma Y, et al. Hepatic Niemann-Pick C1-like 1 regulates biliary cholesterol concentration and is a target of ezetimibe. *J Clin Invest.* 2007 July;117(7): 1968-1978.
17. Grundy SM. Absorption and metabolism of dietary cholesterol. *Annu Rev Nutr.* 1983;3:71-96.
18. Beigneux AP, Davies BS, Gin P, et al. Glycosylphosphatidylinositol-anchored high-density lipoprotein-binding protein 1 plays a critical role in the lipolytic processing of chylomicrons. *Cell Metab.* 2007 April;5(4):279-291.
19. Abifadel M, Varret M, Rabes JP, et al. Mutations in PCSK9 cause autosomal dominant hypercholesterolemia. *Nat Genet.* 2003 June;34(2):154-156.
20. Tall AR. Cholesterol efflux pathways and other potential mechanisms involved in the athero-protective effect of high density lipoproteins. *J Intern Med.* 2008 March;263(3): 256-273.
21. Lewis GF, Rader DJ. New insights into the regulation of HDL metabolism and reverse cholesterol transport. *Circ Res.* 2005 June 24;96(12):1221-1232.
22. Van TA. Phospholipid transfer protein. *Curr Opin Lipidol.* 2002 April;13(2):135-139.
23. Quagliarini F, Vallve JC, Campagna F, et al. Autosomal recessive hypercholesterolemia in Spanish kindred due to a large deletion in the ARH gene. *Mol Genet Metab.* 2007 November;92(3):243-248.
24. Garcia CK, Wilund K, Arca M, et al. Autosomal recessive hypercholesterolemia caused by mutations in a putative LDL receptor adaptor protein. *Science.* 2001 May 18;292(5520): 1394-1398.
25. Leitersdorf E, Tobin EJ, Davignon J, Hobbs HH. Common low-density lipoprotein receptor mutations in the French Canadian population. *J Clin Invest.* 1990 April;85(4):1014-1023.
26. Goldstein JL, Hobbs HH, Brown MS. *The metabolic and molecular bases of inherited disease.* New York: McGraw-Hill; 2001:2863-2913.
27. Leigh SE, Foster AH, Whittall RA, Hubbart CS, Humphries SE. Update and analysis of the University College London low density lipoprotein receptor familial hypercholesterolemia database. *Ann Hum Genet.* 2008 July;72(Pt 4):485-498.
28. Varret M, Abifadel M, Rabes JP, Boileau C. Genetic heterogeneity of autosomal dominant hypercholesterolemia. *Clin Genet.* 2008 January;73(1):1-13.
29. Allard D, Amsellem S, Abifadel M, et al. Novel mutations of the PCSK9 gene cause variable phenotype of autosomal dominant hypercholesterolemia. *Hum Mutat.* 2005 November; 26(5):497.
30. Goldstein JL, Hobbs HH, Brown MS. *The metabolic and molecular bases of inherited disease.* New York: McGraw-Hill; 2001:2863-2913.
31. Slack J. Risks of ischaemic heart-disease in familial hyperlipoproteinaemic states. *Lancet.* 1969 December 27;2(7635): 1380-1382.
32. Souverein OW, Defesche JC, Zwinderman AH, Kastelein JJ, Tanck MW. Influence of LDL-receptor mutation type on age at first cardiovascular event in patients with familial hypercholesterolaemia. *Eur Heart J.* 2007 February;28(3): 299-304.
33. De JS, Lilien MR, Op't RJ, Stroes ES, Bakker HD, Kastelein JJ. Early statin therapy restores endothelial function in children with familial hypercholesterolemia. *J Am Coll Cardiol.* 2002 December 18;40(12):2117-2121.
34. Wiegman A, De GE, Hutten BA, et al. Arterial intima-media thickness in children heterozygous for familial hypercholesterolaemia. *Lancet.* 2004 January 31;363(9406):369-370.
35. Huijgen R, Vissers MN, Defesche JC, Lansberg PJ, Kastelein JJ, Hutten BA. Familial hypercholesterolemia: current treatment and advances in management. *Expert Rev Cardiovasc Ther.* 2008 April;6(4):567-581.
36. Lewington S, Whitlock G, Clarke R, et al. Blood cholesterol and vascular mortality by age, sex, and blood pressure: a meta-analysis of individual data from 61 prospective studies with 55, 000 vascular deaths. *Lancet.* 2007 December 1;370(9602):1829-1839.
37. Kastelein JJ, Akdim F, Stroes ES, et al. Simvastatin with or without ezetimibe in familial hypercholesterolemia. *N Engl J Med.* 2008 April 3;358(14):1431-1443.
38. Cannon CP, Giugliano RP, Blazing MA, et al. Rationale and design of IMPROVE-IT (IMProved Reduction of Outcomes: Vytorin Efficacy International Trial): comparison of ezetimbe/simvastatin versus simvastatin monotherapy on cardiovascular outcomes in patients with acute coronary syndromes. *Am Heart J.* 2008 November;156(5):826-832.

39. Florentin M, Liberopoulos EN, Mikhailidis DP, Elisaf MS. Colesevelam hydrochloride in clinical practice: a new approach in the treatment of hypercholesterolaemia. *Curr Med Res Opin*. 2008 April;24(4):995-1009.
40. Kastelein JJ, Wedel MK, Baker BF, et al. Potent reduction of apolipoprotein B and low-density lipoprotein cholesterol by short-term administration of an antisense inhibitor of apolipoprotein B. *Circulation*. 2006 October 17;114(16):1729-1735.
41. Visser ME, Jakulj L, Kastelein JJ, Stroes ES. LDL-C-lowering therapy: current and future therapeutic targets. *Curr Cardiol Rep*. 2008 November;10(6):512-520.
42. Avis HJ, Vissers MN, Stein EA, et al. A systematic review and meta-analysis of statin therapy in children with familial hypercholesterolemia. *Arterioscler Thromb Vasc Biol*. 2007 August;27(8):1803-1810.
43. Arambepola C, Farmer AJ, Perera R, Neil HA. Statin treatment for children and adolescents with heterozygous familial hypercholesterolaemia: a systematic review and meta-analysis. *Atherosclerosis*. 2007 December;195(2): 339-347.
44. Rodenburg J, Vissers MN, Wiegman A, et al. Statin treatment in children with familial hypercholesterolemia: the younger, the better. *Circulation*. 2007 August 7;116(6): 664-668.
45. Kavey RE, Allada V, Daniels SR, et al. Cardiovascular risk reduction in high-risk pediatric patients: a scientific statement from the American Heart Association Expert Panel on Population and Prevention Science; the Councils on Cardiovascular Disease in the Young, Epidemiology and Prevention, Nutrition, Physical Activity and Metabolism, High Blood Pressure Research, Cardiovascular Nursing, and the Kidney in Heart Disease; and the Interdisciplinary Working Group on Quality of Care and Outcomes Research: endorsed by the American Academy of Pediatrics. *Circulation*. 2006 December 12;114(24):2710-2738.
46. Miserez AR, Muller PY. Familial defective apolipoprotein B-100: a mutation emerged in the mesolithic ancestors of Celtic peoples? *Atherosclerosis*. 2000 February; 148(2):433-436.
47. Defesche JC, Pricker KL, Hayden MR, van der Ende BE, Kastelein JJ. Familial defective apolipoprotein B-100 is clinically indistinguishable from familial hypercholesterolemia. *Arch Intern Med*. 1993 October 25;153(20):2349-2356.
48. Miserez AR, Keller U. Differences in the phenotypic characteristics of subjects with familial defective apolipoprotein B-100 and familial hypercholesterolemia. *Arterioscler Thromb Vasc Biol*. 1995 October;15(10):1719-1729.
49. Eden ER, Sun XM, Patel DD, Soutar AK. Adaptor protein disabled-2 modulates low density lipoprotein receptor synthesis in fibroblasts from patients with autosomal recessive hypercholesterolaemia. *Hum Mol Genet*. 2007 November 15;16(22):2751-2759.
50. Cohen JC, Kimmel M, Polanski A, Hobbs HH. Molecular mechanisms of autosomal recessive hypercholesterolemia. *Curr Opin Lipidol*. 2003 April;14(2):121-127.
51. Wilund KR, Yi M, Campagna F, et al. Molecular mechanisms of autosomal recessive hypercholesterolemia. *Hum Mol Genet*. 2002 November 15;11(24):3019-3030.
52. Rodenburg J, Wiegman A, Vissers MN, Kastelein JJ, Stalenhoef AF. A boy with autosomal recessive hypercholesterolaemia. *Neth J Med*. 2004 March;62(3):89-93.
53. Wierzbicki AS, Graham CA, Young IS, Nicholls DP. Familial combined hyperlipidaemia: under - defined and under - diagnosed? *Curr Vasc Pharmacol*. 2008 January;6(1):13-22.
54. Gaddi A, Cicero AF, Odoo FO, Poli AA, Paoletti R. Practical guidelines for familial combined hyperlipidemia diagnosis: an up-date. *Vasc Health Risk Manag*. 2007;3(6):877-886.
55. Schaefer EJ, Genest JJ Jr, Ordovas JM, Salem DN, Wilson PW. Familial lipoprotein disorders and premature coronary artery disease. *Atherosclerosis*. 1994 August;108(Suppl): S41-S54.
56. Veerkamp MJ, De GJ, Hendriks JC, Demacker PN, Stalenhoef AF. Nomogram to diagnose familial combined hyperlipidemia on the basis of results of a 5-year follow-up study. *Circulation*. 2004 June 22;109(24):2980-2985.
57. Berge KE, Tian H, Graf GA, et al. Accumulation of dietary cholesterol in sitosterolemia caused by mutations in adjacent ABC transporters. *Science*. 2000 December 1;290(5497): 1771-1775.
58. Lu K, Lee MH, Yu H, et al. Molecular cloning, genomic organization, genetic variations, and characterization of murine sterolin genes Abcg5 and Abcg8. *J Lipid Res*. 2002 April;43(4):565-578.
59. Hubacek JA, Berge KE, Cohen JC, Hobbs HH. Mutations in ATP-cassette binding proteins G5 (ABCG5) and G8 (ABCG8) causing sitosterolemia. *Hum Mutat*. 2001 October; 18(4):359-360.
60. Bhattacharyya AK, Connor WE. Beta-sitosterolemia and xanthomatosis. A newly described lipid storage disease in two sisters. *J Clin Invest*. 1974 April;53(4):1033-1043.
61. Salen G, Shefer S, Nguyen L, Ness GC, Tint GS, Shore V. Sitosterolemia. *J Lipid Res*. 1992 July;33(7):945-955.
62. Salen G, Von BK, Lutjohann D, et al. Ezetimibe effectively reduces plasma plant sterols in patients with sitosterolemia. *Circulation*. 2004 March 2;109(8):966-971.
63. Salen G, Starc T, Sisk CM, Patel SB. Intestinal cholesterol absorption inhibitor ezetimibe added to cholestyramine for sitosterolemia and xanthomatosis. *Gastroenterology*. 2006 May;130(6):1853-1857.
64. Wang X, Paigen B. Genetics of variation in HDL cholesterol in humans and mice. *Circ Res*. 2005 January 7;96(1):27-42.
65. Von EA. Differential diagnosis of familial high density lipoprotein deficiency syndromes. *Atherosclerosis*. 2006 June;186(2):231-239.
66. Khovidhunkit W, Memon RA, Feingold KR, Grunfeld C. Infection and inflammation-induced proatherogenic changes of lipoproteins. *J Infect Dis*. 2000 June;181(Suppl 3): S462-S472.
67. Paolini JF, Mitchel YB, Reyes R, et al. Effects of laropiprant on nicotinic acid-induced flushing in patients with dyslipidemia. *Am J Cardiol*. 2008 March 1;101(5):625-630.
68. Canner PL, Berge KG, Wenger NK, et al. Fifteen year mortality in Coronary Drug Project patients: long-term benefit with niacin. *J Am Coll Cardiol*. 1986 December;8(6):1245-1255.
69. Saha SA, Kizhakepunnur LG, Bahekar A, Arora RR. The role of fibrates in the prevention of cardiovascular disease – a pooled meta-analysis of long-term randomized placebo-controlled clinical trials. *Am Heart J*. 2007 November; 154(5):943-953.
70. Link JJ, Rohatgi A, de Lemos JA. HDL cholesterol: physiology, pathophysiology, and management. *Curr Probl Cardiol*. 2007 May;32(5):268-314.

71. Hausenloy DJ, Yellon DM. Targeting residual cardiovascular risk: raising high-density lipoprotein cholesterol levels. *Heart.* 2008 June;94(6):706-714.
72. Barter PJ, Caulfield M, Eriksson M, et al. Effects of torcetrapib in patients at high risk for coronary events. *N Engl J Med.* 2007 November 22;357(21):2109-2122.
73. Forrest MJ, Bloomfield D, Briscoe RJ, et al. Torcetrapib-induced blood pressure elevation is independent of CETP inhibition and is accompanied by increased circulating levels of aldosterone. *Br J Pharmacol.* 2008 August;154(7):1465-1473.
74. Krishna R, Anderson MS, Bergman AJ, et al. Effect of the cholesteryl ester transfer protein inhibitor, anacetrapib, on lipoproteins in patients with dyslipidaemia and on 24-h ambulatory blood pressure in healthy individuals: two double-blind, randomised placebo-controlled phase I studies. *Lancet.* 2007 December 8;370(9603):1907-1914.
75. De Grooth GJ, Kuivenhoven JA, Stalenhoef AF, et al. Efficacy and safety of a novel cholesteryl ester transfer protein inhibitor, JTT-705, in humans: a randomized phase II dose-response study. *Circulation.* 2002 May 7;105(18):2159-2165.
76. Kuivenhoven JA, de Grooth GJ, Kawamura H, et al. Effectiveness of inhibition of cholesteryl ester transfer protein by JTT-705 in combination with pravastatin in type II dyslipidemia. *Am J Cardiol.* 2005 May 1;95(9):1085-1088.
77. Pisciotta L, Fasano T, Calabresi L, et al. A novel mutation of the apolipoprotein A-I gene in a family with familial combined hyperlipidemia. *Atherosclerosis.* 2008 May;198(1):145-151.
78. Joy T, Wang J, Hahn A, Hegele RA. APOA1 related amyloidosis: a case report and literature review. *Clin Biochem.* 2003 November;36(8):641-645.
79. Hovingh GK, Brownlie A, Bisoendial RJ, et al. A novel apoA-I mutation (L178P) leads to endothelial dysfunction, increased arterial wall thickness, and premature coronary artery disease. *J Am Coll Cardiol.* 2004 October 6;44(7):1429-1435.
80. Gomaraschi M, Baldassarre D, Amato M, et al. Normal vascular function despite low levels of high-density lipoprotein cholesterol in carriers of the apolipoprotein A-I(Milano) mutant. *Circulation.* 2007 November 6;116(19):2165-2172.
81. Sirtori CR, Calabresi L, Franceschini G, et al. Cardiovascular status of carriers of the apolipoprotein A-I(Milano) mutant: the Limone sul Garda study. *Circulation.* 2001 April 17;103(15):1949-1954.
82. Hovingh GK, De GE, der SW Van, et al. Inherited disorders of HDL metabolism and atherosclerosis. *Curr Opin Lipidol.* 2005 April;16(2):139-145.
83. Brunham LR, Singaraja RR, Hayden MR. Variations on a gene: rare and common variants in ABCA1 and their impact on HDL cholesterol levels and atherosclerosis. *Annu Rev Nutr.* 2006;26:105-129.
84. Kathiresan S, Melander O, Guiducci C, et al. Six new loci associated with blood low-density lipoprotein cholesterol, high-density lipoprotein cholesterol or triglycerides in humans. *Nat Genet.* 2008 February;40(2):189-197.
85. Willer CJ, Sanna S, Jackson AU, et al. Newly identified loci that influence lipid concentrations and risk of coronary artery disease. *Nat Genet.* 2008 February;40(2):161-169.
86. Singaraja RR, Visscher H, James ER, et al. Specific mutations in ABCA1 have discrete effects on ABCA1 function and lipid phenotypes both in vivo and in vitro. *Circ Res.* 2006 August 18;99(4):389-397.
87. Singaraja RR, Brunham LR, Visscher H, Kastelein JJ, Hayden MR. Efflux and atherosclerosis: the clinical and biochemical impact of variations in the ABCA1 gene. *Arterioscler Thromb Vasc Biol.* 2003 August 1;23(8):1322-1332.
88. Clee SM, Kastelein JJ, Van DM, et al. Age and residual cholesterol efflux affect HDL cholesterol levels and coronary artery disease in ABCA1 heterozygotes. *J Clin Invest.* 2000 November;106(10):1263-1270.
89. Van Dam MJ, De GE, Clee SM, et al. Association between increased arterial-wall thickness and impairment in ABCA1-driven cholesterol efflux: an observational study. *Lancet.* 2002 January 5;359(9300):37-42.
90. Frikke-Schmidt R, Nordestgaard BG, Stene MC, et al. Association of loss-of-function mutations in the ABCA1 gene with high-density lipoprotein cholesterol levels and risk of ischemic heart disease. *JAMA.* 2008 June 4;299(21):2524-2532.
91. Brunham LR, Singaraja RR, Duong M, et al. Tissue-specific roles of ABCA1 influence susceptibility to atherosclerosis. *Arterioscler Thromb Vasc Biol.* 2009 April;29(4):548-554.
92. Clee SM, Zwinderman AH, Engert JC, et al. Common genetic variation in ABCA1 is associated with altered lipoprotein levels and a modified risk for coronary artery disease. *Circulation.* 2001 March 6;103(9):1198-1205.
93. Zwarts KY, Clee SM, Zwinderman AH, et al. ABCA1 regulatory variants influence coronary artery disease independent of effects on plasma lipid levels. *Clin Genet.* 2002 February;61(2):115-125.
94. Brunham LR, Kastelein JJ, Hayden MR. ABCA1 gene mutations, HDL cholesterol levels, and risk of ischemic heart disease. *JAMA.* 2008 November 5;300(17):1997-1998.
95. Kiss RS, Kavaslar N, Okuhira K, et al. Genetic etiology of isolated low HDL syndrome: incidence and heterogeneity of efflux defects. *Arterioscler Thromb Vasc Biol.* 2007 May;27(5):1139-1145.
96. Ayyobi AF, McGladdery SH, Chan S, John Mancini GB, Hill JS, Frohlich JJ. Lecithin: cholesterol acyltransferase (LCAT) deficiency and risk of vascular disease: 25 year follow-up. *Atherosclerosis.* 2004 December;177(2): 361-366.
97. Shah PK. Inhibition of CETP as a novel therapeutic strategy for reducing the risk of atherosclerotic disease. *Eur Heart J.* 2007 January;28(1):5-12.
98. Tsai MY, Johnson C, Kao WH, et al. Cholesteryl ester transfer protein genetic polymorphisms, HDL cholesterol, and subclinical cardiovascular disease in the Multi-Ethnic Study of Atherosclerosis. *Atherosclerosis.* 2008 October;200(2):359-367.
99. Inazu A, Brown ML, Hesler CB, et al. Increased high-density lipoprotein levels caused by a common cholesteryl-ester transfer protein gene mutation. *N Engl J Med.* 1990 November 1;323(18):1234-1238.
100. Inazu A, Jiang XC, Haraki T, et al. Genetic cholesteryl ester transfer protein deficiency caused by two prevalent mutations as a major determinant of increased levels of high density lipoprotein cholesterol. *J Clin Invest.* 1994 November;94(5):1872-1882.

101. Klerkx AH, de Grooth GJ, Zwinderman AH, Jukema JW, Kuivenhoven JA, Kastelein JJ. Cholesteryl ester transfer protein concentration is associated with progression of atherosclerosis and response to pravastatin in men with coronary artery disease (REGRESS). *Eur J Clin Invest*. 2004 January;34(1):21-28.
102. Smilde TJ, Van WS, Wollersheim H, Trip MD, Kastelein JJ, Stalenhoef AF. Effect of aggressive versus conventional lipid lowering on atherosclerosis progression in familial hypercholesterolaemia (ASAP): a prospective, randomised, double-blind trial. *Lancet*. 2001 February 24;357(9256):577-581.
103. Boekholdt SM, Kuivenhoven JA, Wareham NJ, et al. Plasma levels of cholesteryl ester transfer protein and the risk of future coronary artery disease in apparently healthy men and women: the prospective EPIC (European Prospective Investigation into Cancer and nutrition)-Norfolk population study. *Circulation*. 2004 September 14;110(11):1418-1423.
104. Gerholm-Larsen B, Tybjaerg-Hansen A, Schnohr P, Steffensen R, Nordestgaard BG. Common cholesteryl ester transfer protein mutations, decreased HDL cholesterol, and possible decreased risk of ischemic heart disease: the Copenhagen City Heart Study. *Circulation*. 2000 October 31;102(18):2197-2203.
105. Curb JD, Abbott RD, Rodriguez BL, et al. A prospective study of HDL-C and cholesteryl ester transfer protein gene mutations and the risk of coronary heart disease in the elderly. *J Lipid Res*. 2004 May;45(5):948-953.
106. Moriyama Y, Okamura T, Inazu A, et al. A low prevalence of coronary heart disease among subjects with increased high-density lipoprotein cholesterol levels, including those with plasma cholesteryl ester transfer protein deficiency. *Prev Med*. 1998 September;27(5 Pt 1):659-667.
107. Thompson A, Di AE, Sarwar N, et al. Association of cholesteryl ester transfer protein genotypes with CETP mass and activity, lipid levels, and coronary risk. *JAMA*. 2008 June 18;299(23):2777-2788.
108. van Acker BA, Botma GJ, Zwinderman AH, et al. High HDL cholesterol does not protect against coronary artery disease when associated with combined cholesteryl ester transfer protein and hepatic lipase gene variants. *Atherosclerosis*. 2008 September;200(1):161-167.
109. Mukherjee M, Shetty KR. Variations in high-density lipoprotein cholesterol in relation to physical activity and Taq 1B polymorphism of the cholesteryl ester transfer protein gene. *Clin Genet*. 2004 May;65(5):412-418.
110. Beigneux AP, Franssen R, Bensadoun A, Gin P, Melford K, Peter J, Walzem RL, Weinstein MM, Davies BS, Kuivenhoven JA, Kastelein JJ, Fong LG, linga-Thie GM, Young SG. Chylomicronemia with a mutant GPIHBP1 (Q115P) that cannot bind lipoprotein lipase. *Arterioscler Thromb Vasc Biol*. 2009 March 19;29:956–962.
111. Brunzell JD. Clinical practice. Hypertriglyceridemia. *N Engl J Med*. 2007 September 6;357(10):1009-1017.
112. Birjmohun RS, Hutten BA, Kastelein JJ, Stroes ES. Efficacy and safety of high-density lipoprotein cholesterol-increasing compounds: a meta-analysis of randomized controlled trials. *J Am Coll Cardiol*. 2005 January 18;45(2):185-197.
113. Grundy SM, Cleeman JI, Merz CN, et al. Implications of recent clinical trials for the National Cholesterol Education Program Adult Treatment Panel III guidelines. *Circulation*. 2004 July 13;110(2):227-239.
114. McKenney J. New perspectives on the use of niacin in the treatment of lipid disorders. *Arch Intern Med*. 2004 April 12;164(7):697-705.
115. Harris WS, Ginsberg HN, Arunakul N, et al. Safety and efficacy of Omacor in severe hypertriglyceridemia. *J Cardiovasc Risk*. 1997 October;4(5–6):385-391.
116. Hopper L, Ness A, Higgins JP, Moore T, Ebrahim S. GISSI-prevenzione trial. *Lancet*. 1999 October 30;354(9189):1557.
117. Merkel M, Eckel RH, Goldberg IJ. Lipoprotein lipase: genetics, lipid uptake, and regulation. *J Lipid Res*. 2002 December;43(12):1997-2006.
118. Wang J, Cao H, Ban MR, et al. Resequencing genomic DNA of patients with severe hypertriglyceridemia (MIM 144650). *Arterioscler Thromb Vasc Biol*. 2007 November;27(11):2450-2455.
119. Lam CW, Yuen YP, Cheng WF, Chan YW, Tong SF. Missense mutation Leu72Pro located on the carboxyl terminal amphipathic helix of apolipoprotein C-II causes familial chylomicronemia syndrome. *Clin Chim Acta*. 2006 February;364(1–2):256-259.
120. Benlian P, De Gennes JL, Foubert L, Zhang H, Gagne SE, Hayden M. Premature atherosclerosis in patients with familial chylomicronemia caused by mutations in the lipoprotein lipase gene. *N Engl J Med*. 1996 September 19;335(12):848-854.
121. Saika Y, Sakai N, Takahashi M, et al. Novel LPL mutation (L303F) found in a patient associated with coronary artery disease and severe systemic atherosclerosis. *Eur J Clin Invest*. 2003 March;33(3):216-222.
122. Connelly PW, Maguire GF, Little JA. Apolipoprotein CIISt. Michael. Familial apolipoprotein CII deficiency associated with premature vascular disease. *J Clin Invest*. 1987 December;80(6):1597-1606.
123. Stroes ES, Nierman MC, Meulenberg JJ, et al. Intramuscular administration of AAV1-lipoprotein lipase S447X lowers triglycerides in lipoprotein lipase-deficient patients. *Arterioscler Thromb Vasc Biol*. 2008 December;28(12):2303-2304.
124. Franssen R, Visser ME, Kuivenhoven JA, Kastelein JJP, Dallinga-Thie GM, Stroes ESG. Role of lipoprotein lipase in triglyceride metabolism: potential therapeutic target. *Future Lipidol*. 2008;3(4):385-397.
125. Yuan G, Al-Shali KZ, Hegele RA. Hypertriglyceridemia: its etiology, effects and treatment. *CMAJ*. 2007 April 10;176(8):1113-1120.
126. Smelt A, de Beer F. Apolipoprotein E and familial dysbetalipoproteinemia: clinical, biochemical, and genetic aspects. *Semin Vasc Med*. 2009;4(3):249-257.
127. Walden CC, Hegele RA. Apolipoprotein E in hyperlipidemia. *Ann Intern Med*. 1994 June 15;120(12):1026-1036.
128. Wang J, Hegele RA. Homozygous missense mutation (G56R) in glycosylphosphatidylinositol-anchored high-density lipoprotein-binding protein 1 (GPI-HBP1) in two siblings with fasting chylomicronemia (MIM 144650). *Lipids Health Dis*. 2007;6:23.
129. Gin P, Beigneux AP, Davies B, et al. Normal binding of lipoprotein lipase, chylomicrons, and apo-AV to GPIHBP1

containing a G56R amino acid substitution. *Biochim Biophys Acta*. 2007 December;1771(12):1464-1468.
130. Peterfy M, Ben-Zeev O, Mao HZ, et al. Mutations in LMF1 cause combined lipase deficiency and severe hypertriglyceridemia. *Nat Genet*. 2007 December;39(12):1483-1487.
131. Kooner JS, Chambers JC, Guilar-Salinas CA, et al. Genome-wide scan identifies variation in MLXIPL associated with plasma triglycerides. *Nat Genet*. 2008 February;40(2):149-151.

Novel Insights into Genetics of Arterial Thrombosis

Joke Konings, José W.P. Govers-Riemslag, and Hugo ten Cate

Abbreviations

ACS	Acute coronary syndrome
ADP	Adenosine diphosphate
(A)MI	(Acute) myocardial infarction
Bp	Base pair
CAD	Coronary artery disease
CHD	Coronary heart disease
EPCR	Endothelial protein C receptor
F	Coagulation factor
GP	Glycoprotein
HMWK	High molecular weight kininogen
HVR	Hypervariable region
PAI-1	Plasminogen activator inhibitor-1
PAR	Protease-activated receptor
PCI	Percutaneous coronary intervention
PlA	Platelet antigen
(s)TM	(Soluble) Thrombomodulin
STE-ACS	ST-segment elevation ACS
TAFI	Thrombin activatable fibrinolysis inhibitor
TF	Tissue factor
tPA	Tissue type plasminogen activator
uPA	Urokinase plasminogen activator
VNTR	Variable number of tandem repeats
vWF	von Willebrand factor
vWD	von Willebrand disease

H. ten Cate (✉)
Laboratory for Clinical Thrombosis and Haemostasis,
Department of Internal Medicine, Cardiovascular Research
Institute Maastricht, Maastricht University Medical Center
Maastricht, The Netherlands
e-mail: h.ten.cate@mumc.nl

21.1 Introduction

The vast majority of arterial thrombotic complications develop either on the top of an atherosclerotic lesion, or as thromboemboli originating in the heart, mostly related to atrial fibrillation. The origin of these two entities is quite distinct. While atherothrombosis is largely dependent on vessel wall characteristics (plaque lesion size and composition), emboli from the heart are more dependent on flow and blood composition characteristics. In all the cases, the three components that were indicated by Rudolph Virchow, abnormalities in the blood flow, hypercoagulability of the blood, and injury to the vessel wall, are operational and there is no black-and-white distinction between the different mechanisms of thromboembolic disease.

The involvement of the blood coagulation system in atherothrombosis as well as atrial fibrillation is evident from the efficacy of antithrombotic medication. While platelet inhibitors are particularly effective in preventing atherothrombotic complications, such as myocardial infarction, plasmatic coagulation inhibition is an effective pharmacological intervention in patients with atrial fibrillation. These differences in efficacy of medication also point to the different weight of blood coagulation components in the pathophysiology of arterial thrombosis. In this chapter we discuss the contributing role of the blood coagulation mechanism to arterial thromboembolism from a genetic perspective. While the relationship between genetics and *venous* thrombosis has been extensively studied, this is less so in arterial thrombosis. Still, the coagulation genotype may contribute to thrombosis risk as recently established in a large meta-analysis showing that factors that are considered specific for venous thrombosis, such as factor V Leiden, also contribute to the risk of myocardial infarction.[1] In atrial fibrillation there have been several studies clearly demonstrating an independent association between coagulation activity (e.g., D-dimer levels) and thromboembolic risk.[2] Hence, it is likely that the genetic influence of determinants of coagulation may also be of importance in embolic stroke.

There is another reason for looking at the interaction between (genetic) coagulation activity and arterial thrombosis: several lines of evidence indicate that hypercoagulability and in particular thrombin generation are linked to atherosclerosis.[3, 4] It is also likely that genetic determinants of coagulation contribute to atherogenesis and may even influence plaque phenotype, such that the risk of a thrombotic event is changed.

The genotype may affect concentration and/or activity of a coagulation protein, or may have no apparent effect at all. Some genetic variations, such as those underlying hemophilia A, have a major inhibitory impact on blood coagulation; others, like factor V Leiden have very limited coagulation-enhancing effects. Thus, the direction and severity of the biological effects on coagulation, and hence on thrombosis, may vary considerably, depending on the specific genetic variation.

21.2 Hemostatic System

The primary role of the hemostatic system is to confine bleeding without causing thrombotic complications. This system consists of platelets, coagulation, and fibrinolysis. Platelet activation and coagulation are complementary processes. Platelets provide a procoagulant surface and additional coagulation proteins, whereas thrombin, a product of coagulation, is one of the most potent platelet activators. Furthermore, fibrin fibers stabilize the platelet aggregate to form a hemostatic plug and prevent blood loss.

Upon injury to the vessel wall, platelets adhere to the site of trauma and form an aggregate, which serves as a primary plug that stops bleeding.[5] Adhesion of platelets is characterized by several stages: tethering, rolling, activation, and stable adhesion. Receptor–ligand interactions mediate the adhesion and activation of platelets. Platelets express a high density of receptors on their membrane surface. The platelet receptors interact with ligands expressed on the surface of endothelial cells, within the subendothelial matrix, or as soluble proteins in the circulation[6] (see Fig. 21.1).

During tethering, the interaction of platelets with the subendothelium is dependent on flow conditions. Blood flowing in a vessel generates stress forces, called shear stress. Low shear stress is present in veins and larger arteries, whereas high shear stress is especially found in the microvasculature and at places of stenosis.[7] At high shear rates, von Willebrand factor (vWF) is essential to bridge the receptors on the platelet membrane and components in the vessel wall such as collagen: the glycoprotein (GP) Ib-IX-V receptor on platelets adheres to immobilized vWF bound to collagen in the vessel wall. This interaction is quite unstable, but allows GP VI and GPIa/IIa (*integrin $\alpha_2\beta_1$*) to interact with collagen in the subendothelium. Furthermore, these interactions activate the GPIIb/IIIa *(integrin αIIb/β3)* receptor, which binds to vWF and fibrinogen. These interactions allow stable platelet adhesion.[8] At low shear rates, vWF is not essential for platelet adhesion, and other receptors such as GP Ia/IIa and GP VI mediate the adhesion of platelets to the vessel wall.[9]

After adhesion, platelets are activated. Several activation pathways exist, all requiring platelet receptors. Agonist-activating platelets include collagen, adenosine diphosphate (ADP), serotonin, thromboxane A2, and thrombin. Upon activation, ADP, serotonin, Ca^{2+} ions, and several proteins such as fibrinogen, vWF, and coagulation factors are secreted by the platelet. Furthermore, thromboxane A2 is formed by the platelet. ADP is able to activate platelets via the $P2Y_1$ and the $P2Y_{12}$ receptors, and thromboxane A2 via the thromboxane receptor. Thrombin, a key enzyme in blood coagulation, is an important platelet activator. Platelet activation via thrombin involves two protease-activated receptors (PAR), PAR-1 and PAR-4, and triggers platelet secretion and aggregation. The main receptor involved in aggregation is GPIIb/IIIa, and this receptor assists in the mutual interaction of platelets and between platelets and other substrates. These substrates include fibrinogen and vWF, and they help to form bridges between the platelets. Platelet aggregates are an essential constituent of the arterial thrombus. The second critical role of platelets in hemostasis is to expose a platform for activation of coagulation proteins. Phosphatidylserine provides a procoagulant surface for the assembly of enzymatic complexes of coagulation factors and promotes thrombin generation.

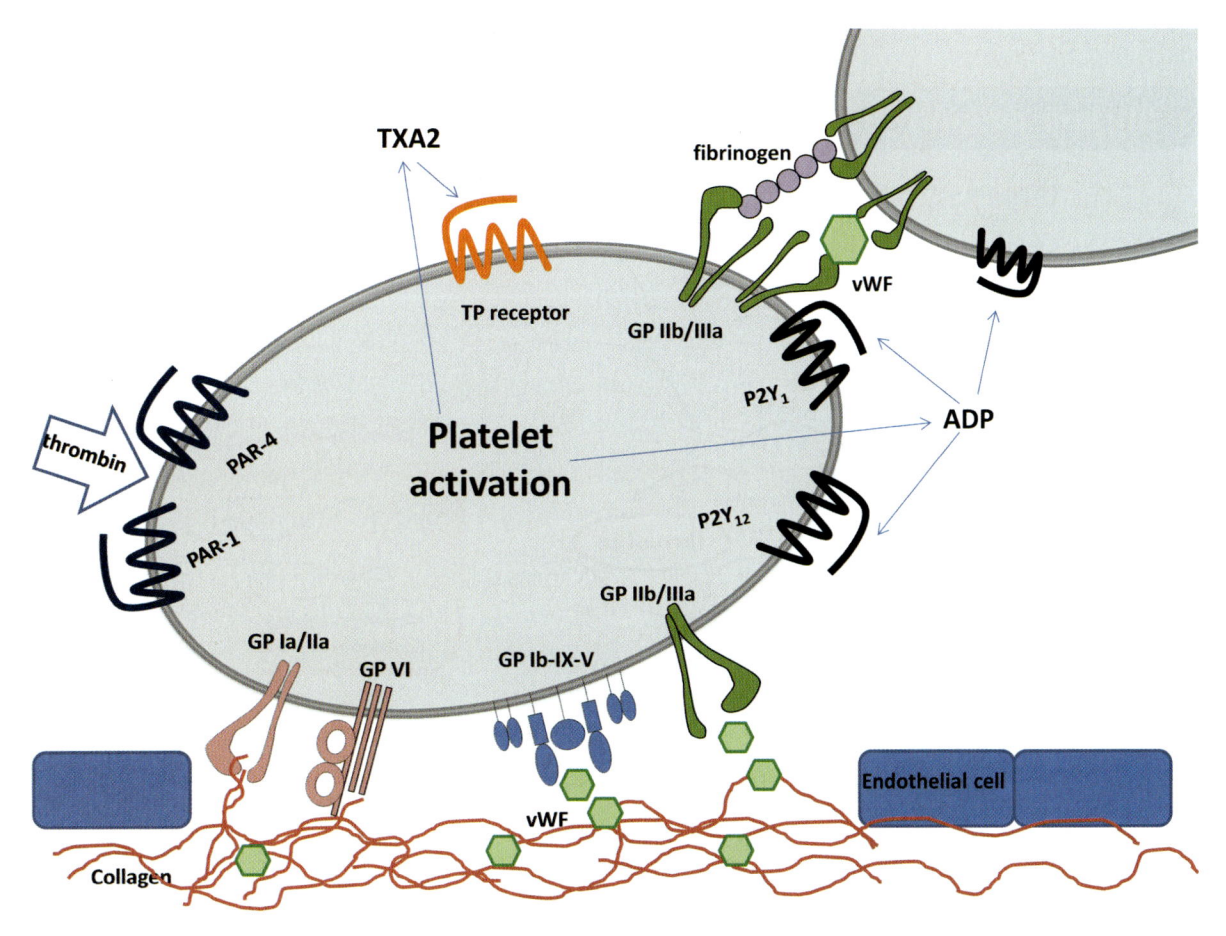

Fig. 21.1 Platelet and its receptors. Platelets express several receptors, which are important for platelet adhesion, activation, and aggregation. The GP Ib-IX-V receptor is essential for platelet adhesion under high shear conditions. The GP Ibα subunit interacts with von Willebrand factor (vWF) in the exposed subendothelium. This is an unstable interaction, but allows the more stable adhesion of the GP Ia/IIa receptor and the GP VI receptor to collagen in the exposed subendothelium. Platelets are activated by the interaction of these receptors with their ligand and the binding of several agonists to platelet receptors. Thrombin is one of the most powerful platelet-activating agonists. Activation of platelets via thrombin is mediated by proteinase-activated receptor (PAR)-1 and PAR-4. Activated platelets release adenosine diphosphate (ADP), which binds to the $P2Y_1$ and $P2Y_{12}$ receptors and produce thromboxane A2 (TXA2), which binds to thromboxane receptors (TP). The main receptor involved in platelet aggregation is GP IIb/IIIa, binding of this receptor to vWF or fibrinogen bridges between platelets. vWF, von Willebrand factor, GP glycoprotein, TXA2 thromboxane A2, TP thromboxane receptor, PAR proteinase-activated receptor

The combination of collagen and thrombin is an effective trigger for a procoagulant response.[10]

The proteins of the coagulation system are activated in a cascade of sequentially activated plasma serine proteases (see Fig. 21.2). Limited proteolytic activation of the coagulation proteins yields active enzymes and cofactors that catalyze the next reaction in the cascade to form fibrin, the protein that forms the meshwork of the clot. Major coagulation reactions occur at phospholipid surfaces on platelets, particularly after "flip-flop" exposure of phosphatidylserine. The formation of fibrin can be initiated through the initiation of two cascades of the coagulation system: the extrinsic pathway and the intrinsic pathway.[11,12]

The extrinsic pathway is triggered by the exposure of tissue factor (TF), constitutively present in the subendothelium, to blood upon vessel injury. TF binds to factor (F) VII(a) and the TF/FVIIa complex then cleaves factor X (FX) into its active form FXa. FXa is a key enzyme common to both the extrinsic and the intrinsic coagulation pathways. FXa associates with factor Va (FVa) onto the exposed procoagulant surface containing phosphatidylserine of the platelets via Ca^{2+}-ions to form the enzymatic prothrombinase complex.

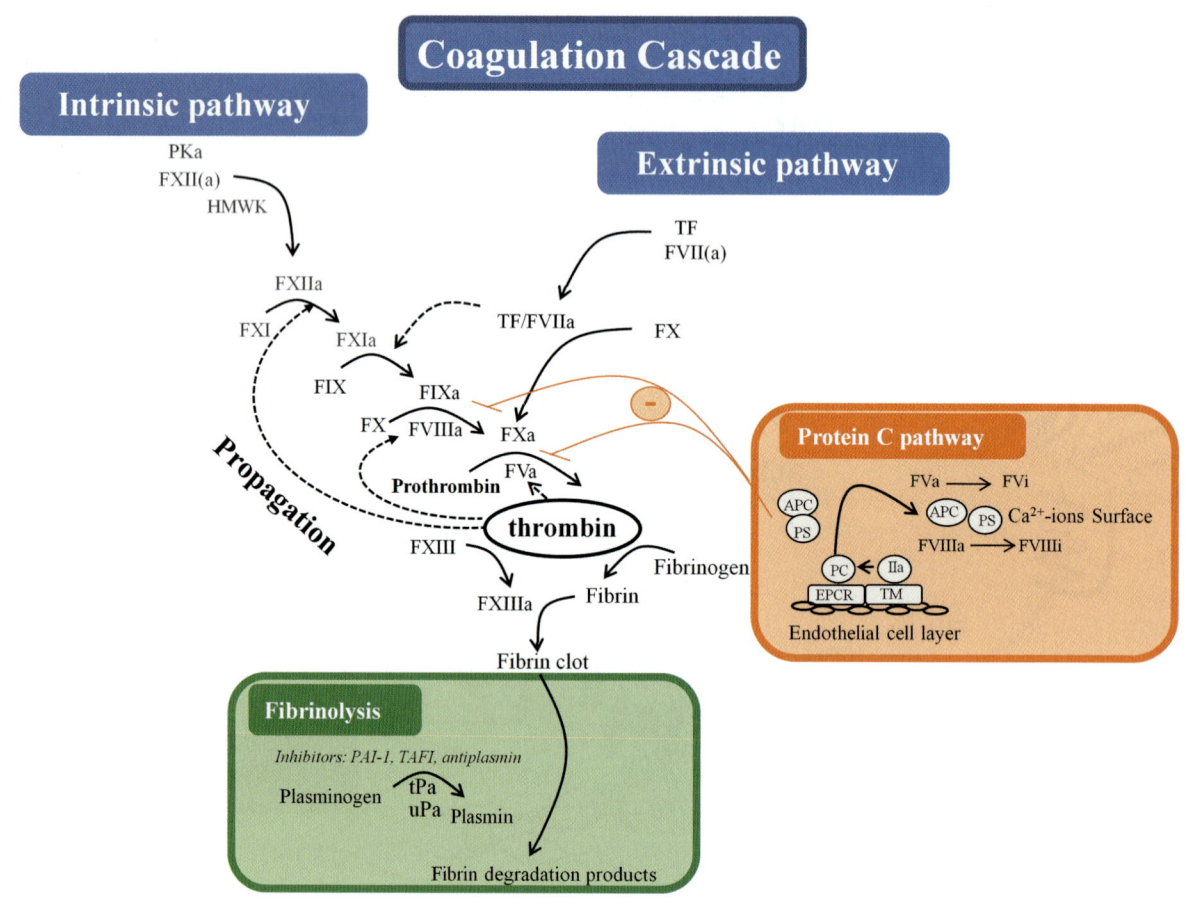

Fig. 21.2 Current view of coagulation and related systems. The coagulation cascade is a sequential activation of coagulation proteins and consists of two pathways initiated by different triggers: the intrinsic pathway and the extrinsic pathway. FXa is the central enzyme common to both the extrinsic and the intrinsic coagulation pathway. After injury, the extrinsic pathway is initiated by the formation of the TF/FVIIa complex and cleaves FX to form FXa. The intrinsic pathway is triggered by activation of FXII in a process called contact activation; FXIIa will also produce FXa via FXI, FIX, and FX activation. FXa associates with FV(a) on a phospholipid membrane and converts prothrombin (FII) into thrombin. Thrombin will form fibrin out of fibrinogen and will also activate FXIII. FXIIIa polymerizes the fibrin fibers. Clot formation is accelerated by thrombin activation of the cofactors FV and FVIII and by feedback activation of FXI. The TF/FVIIa complex can also activate FIX next to FX. Thrombin slows down its own formation by the protein C pathway (see inset). The formed activated protein C (APC) will inactivate the cofactors FVa and FVIIIa. After 1 week the process of fibrinolysis (see inset) will dissolve the fibrin clot. Roman numerals indicate unactivated coagulation factors, and activated factors are indicated by a lower case "a." TF tissue factor, PKa kallikrein, HMWK high molecular weight kininogen, APC-activated protein C, PS protein S, EPCR endothelial protein C receptor, TM thrombomodulin, tPa tissue-type plasminogen activator, uPa urokinase plasminogen activator, PAI-1 plasminogen activator-1, TAFI thrombin-activatable fibrinolysis inhibitor

This prothrombinase complex (FXa, FVa, Ca^{2+}, and phospholipid membrane) converts prothrombin (FII) to thrombin (FIIa). Thrombin, in turn, catalyses the conversion of fibrinogen to fibrin. Fibrin molecules polymerize to fibers that are covalently linked by FXIIIa.[13] The intrinsic pathway is triggered by activation of factor XII (FXII) on a negative surface in a process called "contact activation." The contact activation system consists of four proteins: FXII, FXI, prekallikrein, and high molecular weight kininogen (HMWK). FXIIa can either activate factor XI (FXI) into its active form FXIa, leading to activation of the intrinsic coagulation pathway or can stimulate the kallikrein–kinin system (leading to the formation of bradykinin). In the coagulation system, FXIa then converts factor IX (FIX) into FIXa. FIXa associates with cofactor VIIIa (FVIIIa) in the tenase complex (FIXa, FVIIIa, Ca^{2+}, and phospholipid membrane), which in turn activates FX. FXa is the starting point of the common pathway already described above, which will form thrombin after activation of prothrombin (prothrombinase complex). In the last 2 decades there were two main adaptations of the original cascade/waterfall model. It is now generally accepted that (1) the TF/FVIIa complex not only activates FX, but is also able to activate FIX. (2) FXIa is not only formed by activation through FXIIa, but thrombin can also activate FXI (a positive feedback loop of thrombin). Thrombin plays a key role in coagulation and mediates more procoagulant feedback reactions, including the activation of the cofactors V and VIII, as well as FXIII. Thrombin also initiates an important inhibitory mechanism involving the thrombomodulin-mediated activation of protein C by thrombin. Activated protein C (APC) attenuates in concert with the cofactor protein S, the intrinsic route of coagulation by the inactivation of the cofactors FVIIIa and FVa. Other important natural anticoagulants include antithrombin (which directly inhibits serine proteases such as thrombin) and tissue factor pathway inhibitor (TFPI). The detailed discussion of these systems is beyond the scope of this chapter.[14, 15]

The fibrinolytic system plays an important role in breaking down cross-linked fibrin clots. The main enzyme responsible for the dissolution of fibrin is plasmin. The inactive zymogen plasminogen is converted to plasmin, mainly by the enzymes tissue-type plasminogen activator (tPA) and urokinase plasminogen activator (uPA). This process is enhanced in the presence of fibrin. The fibrinolytic system is inhibited by plasminogen activator inhibitor-1 (PAI-1), thrombin-activatable fibrinolysis inhibitor (TAFI), and α2-antiplasmin. While PAI-1 is able to inhibit tPA and uPA, TAFI prevents the binding of tPA and plasminogen to fibrin, which slows down the conversion of plasminogen into plasmin and plasmin is inhibited by the action of α2-antiplamin.

In concert, these complex mechanisms maintain a delicate balance: defects in any of these systems may contribute to a bleeding or clotting (thrombosis) phenotype, depending on the nature and severity of the defect. In the following we specifically address the role of genotypical variation in hemostatic proteins in relation to arterial thromboembolic disease. Most clinical studies that we refer to include coronary artery disease (CAD) or myocardial infarction as an endpoint; fewer studies addressed stroke and atherosclerosis as major clinical endpoints.

21.3 Platelet-Membrane Glycoproteins and Von Willebrand Factor Gene Polymorphisms

Current treatment of atherothrombotic complications focuses on the inhibition of platelets. Therefore, one would expect polymorphisms that influence platelet count, the function of platelets, or vWF to affect the risk and/or the progression of arterial thrombosis. Many polymorphisms have been identified in platelet receptors; however, their contribution to arterial thrombosis seems minor. Polymorphisms in the receptors GPIb-IX-V, GPIa/IIa, GPIIb/IIIa, GP VI, and vWF are discussed.

21.3.1 Platelet Receptors

The GPIb-IX-V receptor is crucial for platelet adhesion at places of high shear. It is a complex of four proteins coded by four different genes. GPIb consists of two proteins, GPIbα and Ibβ, linked to each other via a disulfide bridge and noncovalently associated with GPIX and GPV. Especially GPIbα is capable of binding to vWF and thrombin. Three polymorphisms have been identified in the gene coding for GPIbα. Two of these occur in the coding sequence of the gene and affect the structure of the protein: (1) a 434 C/T (Thr145Met) substitution[16, 17] and (2) a variable number of tandem repeats (VNTR) of 39 base pairs (bp) resulting in four variants with different length termed D, C, B, A (1, 2, 3, or 4 repeats).[18, 19] A third polymorphism, a −5C/T substitution, occurs in the Kozak sequence of GPIbα gene.[20] This sequence occurs in the promoter region and is essential for efficient protein translation. The C-variant is associated with increased surface level of GPIb-IX-V receptors, and the highest expression is found in individuals with the CC genotype.[20, 21]

Most research has focused on the −5C/T substitution in the promoter region; however, the data are inconsistent. The −5C allele of CC genotype was found to be associated with acute coronary syndrome (ACS), thrombotic complications after percutaneous coronary intervention (PCI),[21] and myocardial infarction.[22] In contrast, others did not find any association between this polymorphism and (premature) MI, stroke, or CAD.[23–26] A large meta-analysis, including more than 5,000 cases and 5,000 controls from 14 different studies showed only a marginal, not statistically significant effect on coronary heart disease (CHD) of the −5C/T substitution, with a per-allele relative risk for the −5C allele of 1.05 (95% confidence interval (CI) 0.96–1.13).[1] Pooled analysis indicated that the Kozak sequence of −5C/T polymorphism is strongly associated with the risk of ischemic stroke; however, the direction of this association is highly variable.[27] The contribution of the VNTR and Thr145Met polymorphisms to arterial thrombosis is not evident, and the data from several clinical studies are inconsistent.[16, 28–31] However, meta-analysis did show an increased risk of ischemic stroke for carriers of at least one 145Met-allele.[27]

The GPIIb/IIIa (integrin $\alpha 2\beta 3$) receptor is constituted of two glycoproteins, GPIIIa and GPIIb, and is the main receptor involved in aggregation. A common variation is the 1565C/T (Leu33Pro; Leu33:PlA1), Pro33:PlA2) substitution in exon 2 of the GPIIIa gene. In fact, 25% of individuals of northern European ancestry have the PlA1 variant, with 2% being homozygous.[32] Biologically, PlA2-positive platelets display a lower threshold for activation, theoretically reducing the thrombosis threshold.[33]

Many studies investigated the relation between the Leu33Pro polymorphism and arterial thrombosis, with conflicting results. The PlA2 genotype has been associated with an increased risk of arterial thrombosis and unstable angina or MI, especially in patients younger than 60,[34] sudden cardiac death due to coronary thrombosis,[35] and stent thrombosis following catheter-based procedures.[36] The prevalence of the PlA2 allele was increased in siblings of patients with a history of premature ischemic heart disease,[37] as well as in subjects with atherothrombotic stroke.[31] However, a substantial number of reports do not confirm the association between arterial disease and the PlA2 genotype.[38–44] Furthermore, the Framingham investigators conclude on the basis of functional testing in combination with genotyping that although heritable factors play a role, the GPIIIa genotype only makes a small contribution to platelet aggregation.[45] Several meta-analyses addressed the influence of the PlA2 allele on CAD, with different results. Burr et al.[46] did find an association between the PlA2 allele and CHD, the effect of the polymorphism however differs substantially between study populations. Other meta-analyses did not observe a statistically significant contribution of the PlA2 polymorphism to the risk of MI,[43] cerebrovascular disease,[44] or CHD.[1]

GPIa/IIa is a complex of two proteins and assists in platelet adhesion by binding to collagen. The expression of this receptor is subject to high variation, which results in a variable response to collagen.[28] Two silent polymorphisms in the GPIa gene, 807C/T (Phe224) and 873G/A (Thr246), are linked and related to the density of this receptor on the platelet surface. The 807T/873A variant is associated with high receptor density, and consequently with a faster rate of platelet adhesion to type I collagen.[47] Contrasting data have been published,[48–50] however, meta-analyses did not reveal any significant associations between the polymorphisms and CAD[1, 51] or ischemic stroke.[52]

The structure and function of the collagen receptor GPVI is affected by the 13254T/C (Ser219Pro) substitution.[53] The 13254C-allele was associated with an increased risk for (premature) MI and coronary thrombosis,[38, 53, 54] especially in association with the −148 T allele of the β-fibrinogen gene (this polymorphism will be discussed in the section about fibrinogen) indicating an interaction between these genes.[53]

21.3.2 Von Willebrand Factor

The presence of vWF plays an important role in platelet aggregation at sites of high shear, such as coronary plaque lesions. vWF supports the adhesion of platelets to the subendothelium at the site of injury, enhancing platelet aggregation. Furthermore, it acts as a carrier for FVIII, supports local accumulation of FVIII, and behaves as an acute phase protein. vWF protein concentration is a marker of cardiovascular risk and vWF levels are elevated in patients with acute coronary syndrome.[55] vWF levels predict CAD in initially healthy individuals, but this association disappears after adjustment for conventional risk factors.[55]

Many mutations have been identified in the vWF gene that cause von Willebrand disease (vWD), an inherited bleeding disorder. In animal studies, pigs and mice with complete vWF deficiency were protected against atherosclerosis.[56–59] In humans, evidence for such a protective effect is less clear. Patients with vWD, or hemophilia A or B, had fewer carotid plaques, a smaller degree of carotid stenosis,[60] and a lower number and grade of atherosclerotic plaques of the legs and abdominal aorta.[61] However, the effects appeared to be vascular bed-dependent: whereas the intima-media thickness in the femoral artery was minimally reduced, that in the carotid artery was not.[62] In these investigations, however, both patients with hemophilia and vWD were included and the type of hemophilia was not specified. Patients with type-3 vWD (absence of detectable vWF and reduced levels of FVIII) were not protected from the development of early and advanced atherosclerotic lesions,[63] although the patient population may have been too young to assess the effect.[64] Type 2B vWD (characterized by a qualitative abnormality in vWF levels) did not protect from atherosclerosis.[65]

Several polymorphisms have been identified in the promoter region of the vWF gene: −1793G/C, −1234C/T, −1185A/G, and −1051G/A. These polymorphisms are in strong linkage disequilibrium, resulting in two haplotypes: haplotype 1 (GCAG) and haplotype 2 (CTGA). Homozygotes for haplotype 1 have the highest levels of vWF to antigen (vWF:Ag), homozygotes for haplotype 2 the lowest and heterozygotes intermediate vWF:Ag levels.[66, 67] The Sma I polymorphism, a T to C substitution, is situated in intron 2 of the vWF gene and is not associated with vWF/Ag levels.[68] The 2365A/G (Thr789Ala) polymorphism is associated with increased levels of vWF.[69, 70] The effect of these polymorphisms is not entirely clear. Haplotype 1 was found to be associated with an increased risk of CAD in subjects with advanced atherosclerosis,[71] but not with MI[72] or CAD in subjects undergoing coronary angiography.[73] The CC genotype of the Sma I polymorphism was associated with higher risk for ischemic stroke, but not with MI.[68] Thr789Ala polymorphism might affect the risk for CHD in type-1 diabetic patients,[69], but not in type-2 diabetes,[70] through modulation of the plasma vWF level.[69]

Levels of vWF are strongly influenced by blood group: individuals who are carriers of blood groups A, B, or AB have mean vWF levels that are 25–30% higher than carriers of blood group O. Carriers of non-O blood group demonstrate an increased risk of arterial thrombosis, most likely due to elevated vWF levels.[74]

21.4 Coagulation Gene Polymorphisms

21.4.1 Initiation of the Extrinsic Pathway of Coagulation

21.4.1.1 Factor VII

The relation between FVII concentration and arterial thrombosis is controversial. Increased levels of FVII zymogen and of FVII-clotting activity (FVII:C) have been associated with an increased risk of CAD in the Northwick Park Heart Study-II (NPHS-II) and the Prospective Cardiovascular Münster (PROCAM) study,[75, 76] but not consistently.[77, 78] In these studies, different methods have been used to determine the level of FVII (activated FVII, FVII:C, FVII antigen), possibly explaining some of the discrepancies between these studies.[79–81]

Five polymorphic sites of the FVII gene have been described that account for up to 30% of the variance in FVII levels in plasma.[82–90] These polymorphic sites include a 10976A/G (Arg353Gln) substitution in the catalytic region, a decanucleotide insertion/deletion (−323ins10) in the promoter region at nucleotide −323 (A1: deletion, A2: insertion); a polymorphism in the hypervariable region 4 (HVR4) of intron 7 with three alleles of different length: H5, H6, and H7; and two promoter polymorphisms −401G/T and −402G/A.[90] The most studied polymorphisms are the Arg353Gln and −323ins10. The Gln353 and A2 variants are associated with a reduction in FVII plasma levels of 20–25%.[83, 91, 92] Carriers of the FVII Gln353 have lower FVII levels due to lower secretion efficacy of this variant.[84, 93] The polymorphisms in the promoter region alter the transcriptional activity: the −401T is associated with reduced transcription and reduced levels of FVII, the −402A allele is associated with increased transcriptional activity and higher FVII levels.[90] The H7H7 genotype is associated with lower FVII:C and FVII antigen levels.[94]

Even though there is a strong correlation between the FVII polymorphisms and FVII levels, the relationship with arterial thrombosis is not so clear. Conflicting data are published. The 353Gln and the −323ins10 A2 variant and H7H7 genotype were reported to protect from MI.[94–96] The −402A allele was associated with a higher risk of MI[96]; homozygotes for the Arg353 allele had more complications after PCI.[97] In contrast, several other studies did not find a significant effect of polymorphisms in the FVII gene on the risk of arterial thrombosis. The West of Scotland Coronary Prevention (WOSCOP) study failed to confirm an association between the Arg353Gln polymorphism and MI.[81] In a Dutch case-control study, patients with the Arg353 variant had a nonsignificant lower risk of MI.[98] In a large meta-analysis there was no association between CAD and the Arg353Gln polymorphism in the FVII gene.[1]

21.4.1.2 Tissue Factor

Atherosclerotic plaques contain high levels of TF, making them trombogenic. Furthermore, levels of circulating TF are increased in patients with acute coronary syndrome and the expression of TF on monocytes is increased in patients with unstable angina.[99]

In the promoter region of the gene coding for TF four polymorphisms (−1812 C/T, −1322 C/T, −1208 deletion/insertion, −603 A/G) have been identified that are in complete linkage disequilibrium and code for two haplotypes: −1208D and −1208I.[99] The −1208D haplotype is associated with lower levels of circulation TF. These haplotypes are in linkage disequilibrium with a 5466A/G polymorphism situated in intron 2 of the TF gene: −1208D is linked to 5466G.[100] The data regarding the association of the polymorphism and disease are contradictory. The −1208I haplotype has been associated with an increased risk of MI,[101] −1208D with an increased risk of cardiac death in patients with acute coronary syndrome.[100] The polymorphism did not influence the risk of CAD or recurrent cardiovascular events in patients with a MI.[99, 102]

21.4.2 Contact Activation and the Intrinsic Pathway of Coagulation

21.4.2.1 Factor XII

The initiation of the intrinsic pathway via FXII is thought to occur only under pathologic conditions,[103] which may involve activators including activated platelets, collagen,[104] and polyphosphates.[105] Furthermore, FXII is involved in other biological processes including complement activation and inflammation.

Elevated levels of prekallikrein and HMWK have been associated with MI,[106, 107] however little is known about the function of prekallikrein and HMWK in arterial thrombosis. The relation between FXII and arterial thrombosis is not straightforward. In the NPHS-II high levels of FXIIa and low levels of FXIIa in complex with C1-esterase inhibitor were associated with increased risk of CHD.[108, 109] High FXIIa levels were also associated with a higher and earlier recurrence rate of ACS in patients who survived a first MI[110] and were positively associated with coronary calcifications.[111] In the Study of Myocardial Infarction Leiden (SMILE), levels of FXII:C were decreased in men that had developed an MI,[112] whereas Merlo et al.[106] did not find an effect on the levels of FXII:C in CAD. Furthermore, there are no convincing arguments for an effect in arterial thrombosis of FXII deficiency.[113]

The most investigated polymorphism in the FXII gene is a −4C/T (previous notation: 46C/T) substitution in the 5′ untranslated region. This polymorphism is associated with lower plasma levels of FXII.[114]

The association between this polymorphism and arterial thrombosis is not clear. The TT-genotype was associated with an increased risk of CAD in men with high cholesterol[115] and with both CAD and ischemic stroke in the Spanish population[116, 117]; however, in patients with preexisting CAD the TT-genotype had a protective effect on the development of ACS.[118] Furthermore, several studies did not observe an effect of the polymorphism on arterial thrombosis.[119–123]

The association of the −4C/T polymorphism and arterial thrombosis is influenced by cholesterol levels. Pravastatin treatment reduced the risk of CAD in men with a CC or CT genotype, but not in men with a TT genotype.[115] Furthermore, in patients with CAD before the age of 45, the presence of the −4T allele increased the risk of MI, especially in the presence of hypercholesterolemia. If both risk factors were present, the risk was increased 2.26-fold.[124]

21.4.2.2 Factor XI, Factor IX, and Factor VIII

Both environmental and genetic factors contribute to the levels of circulating FXI, FIX, and FVIII. The impact of genetic factors on plasma levels is unclear. Even though elevated levels of FXI as well as FIX and FVIII have consistently been associated with MI and ACS, no polymorphisms have been described that correlate with plasma levels of these factors.[89]

While deficiency in FXI is associated with a rather mild bleeding diathesis, deficiency in FIX (hemophilia B) or FVIII (hemophilia A) is associated with a severe bleeding phenotype. Deficiency in FXI may protect against ischemic stroke; the incidence of ischemic stroke was reduced in a cohort of FXI-deficient patients when compared to the general population.[125] In contrast, it did not protect from MI.[126] Deficiency in FIX or FVIII appears to protect against the occurrence of MI: in a long follow-up study, patients with hemophilia A and B had an 80% reduction in the risk of fatal ischemic heart disease.[127] Comparison of causes of death in patients with hemophilia in the United States yielded similar results.[128] Female carriers of hemophilia A or B also showed a reduced mortality from ischemic heart disease.[129] It is uncertain whether hemophilia A and B have protective properties against ischemic stroke.[130]

21.4.3 Common Coagulation Pathway

21.4.3.1 Factor V and Prothrombin

The prothrombin 20210 G/A variant and FV 1691G/A (Arg506Gln; FV Leiden) polymorphisms are relatively common in Caucasian persons with a frequency of ~2% and 5% respectively. The biological effect of the FII 20210 G/A variant results from a higher transcription rate of prothrombin resulting in somewhat higher prothrombin protein concentrations and a higher rate of thrombin generation. The effect of FV Leiden resides in the impaired inactivation of FVa by activated protein C. These effects have both been characterized as risk factors for venous thrombosis in many studies. In arterial thrombosis, however, the influence has been more difficult to establish. In the meta-analysis by Ye et al., both variants appear to contribute to the risk of a MI.[1] In contrast, an effect on stroke risk, large or small vessel based has not been detected.[131] However, the effects are small and, as single-risk factors, probably negligible in persons with premature vascular disease without other apparent risk factors.[132]

Interestingly, in animal experiments the FV Leiden mutation is associated with an increased burden of atherosclerosis.[133, 134] A number of studies indeed suggests that atherosclerosis in the carotid artery is influenced by FV Leiden carriership,[135, 136] whereas also other prothrombotic gene mutations including the prothrombin 20210 G/A variant are associated with severity of atherosclerotic burden.[137] One of the mechanisms involved may be an interaction between hypercoagulability and LDL cholesterol transfer to the vessel wall. Such interactions may contribute to a greater risk of MI.[138, 139] Thus, although the effects of the polymorphisms as single factors may be very small, the interaction with commonly recognized risk mechanisms for atherosclerosis may give unexpected strong effects on the risks of atherosclerosis and thrombosis, at least in animal models.[135]

Finally, the concentration of FV may also play a role in the risk of thrombosis. While carriers of the FV Leiden mutation do not have an altered protein concentration, an increased FV:C level has also been identified as a risk factor for AMI.[140] So far, no genetic basis for elevated FV levels has been published. In contrast, a common R2 haplotype in FV has been shown to lower the FV concentration, but any (protective) effect on arterial thrombosis risk has not been described.[141]

21.4.4 Protein C Pathway

The protein C anticoagulant pathway has antithrombotic activities and limits inflammatory responses. The essential components of this pathway include thrombin, thrombomodulin (TM), the endothelial cell protein C receptor (EPCR), protein C, and protein S. A retrospective cohort study showed that deficiency in protein S or protein C is associated with arterial thrombosis.[142]

Thrombomodulin is a transmembrane protein, expressed in endothelial cells. In healthy people, high levels of soluble thrombomodulin (sTM) were associated with a decreased risk of coronary heart disease.[143] However, a cross-sectional analysis found a positive association between levels of sTM and carotid atherosclerosis. A possible explanation for this difference is that in healthy individuals elevated levels of sTM correlated with the expression level of membrane bound TM, whereas in patients with atherosclerotic disease elevated levels indicate damage to the endothelium.[144] Different polymorphisms have been identified in the thrombomodulin gene. Two polymorphisms in the 5′ untranslated region, −133C/A and −33G/A, are associated with reduced expression of TM.[145] Furthermore, a 127G/A (Ala25Thr) and 1418C/T substitution have been identified in the TM gene. Polymorphisms in the promoter region of the TM gene were more common in patients diagnosed with MI than in matched controls.[146] The −33A allele has been associated with an increased risk of AMI and carotid atherosclerosis,[147–149] however, not consistently.[150] In the SMILE, the 25Thr allele was associated with MI, especially in young men and in the presence of additional risk factors (such as smoking and the presence of a metabolic risk factor: obesity, hypertension, or hypercholesterolemia).[151] The 455Val allele was more common in patients with a MI and the 455Val genotype in patients with ischemic stroke compared to healthy controls.[152, 153] The Ala455Val predicted the risk of CHD in the black population, but not in the white population.[154] Other studies failed to find an association.[155, 156] TM accelerates the inactivation of the complement factors C3b and C5a, through the activation of TAFI by the thrombin-TM complex. Delvaeye et al.[157] identified mutations in the TM gene that impaired the activation of TAFI and therefore were less protected from activated complement.

A polymorphism in the gene coding for EPCR, 6936G/A (Ser219Gly), contributes to a higher basal release of soluble EPCR, causing increased levels of soluble EPCR in plasma. In the NPHS-II, homozygotes for the 219Gly allele had an increased risk for coronary heart disease.[158]

21.4.5 Fibrinogen and Factor XIII

21.4.5.1 Fibrinogen

The level of fibrinogen has consistently been associated with the risk of arterial thrombotic disorders. High levels of fibrinogen have been associated with MI, ischemic stroke, and peripheral arterial disease.[75, 159–162] Fibrinogen levels cluster with other risk markers for arterial thrombosis, such as hypertension, diabetes, and smoking.[90] Both genetic and environmental factors influence the levels of circulating fibrinogen. It is estimated that genetic factors account for approximately 20–50% of variation in fibrinogen levels.[89, 163] Even though the level of fibrinogen is a risk marker for an arterial thrombotic event, it is not clear if there is a causal relationship.[164] Some authors argue that elevated fibrinogen levels merely reflect the presence of inflammation, atherosclerosis, or other risk factors.[164]

Several polymorphisms have been identified in the genes that code for the polypeptide chains of fibrinogen. Most of the interest has been on polymorphisms on the Bβ chain because in vitro studies suggest that the production of the β chain is the rate-limiting step in the formation of a fibrinogen molecule.[90] The main polymorphisms include the −148C/T, −455G/A, −854G/A polymorphisms in the 5′ promoter region, the BclI polymorphism in the 3′ region and the Arg448Lys substitution in the C-terminal region of the Bβ chain, and the Thr312Ala substitution in the Aα chain. The rare allele of the −148C/T, −455G/A, −854G/A, and BclI polymorphisms are associated with higher levels of fibrinogen.[90] The 448Lys allele is associated with a clot with lower permeability and a tighter structure with thinner fibers than the 448Arg variant,[165] although this has not been confirmed.[166] The Thr312Ala polymorphism is situated near FXIII cross-linking sites, causing increased FXIII cross-linking, and formation of thicker fibrin fibers.[167] This polymorphism is associated with decreased plasma fibrinogen levels.[168]

The association between arterial thrombosis and polymorphisms in the fibrinogen genes is not so clear. Even though some studies suggest that there is an association between these polymorphisms and atherosclerosis,[169, 170] the susceptibility to CAD,[171] MI,[169, 172] and ischemic stroke,[168] others have not foud an association.[168, 173, 174]

Fibrinogen γ′, an alternative splicing variant of the γ-chain of fibrinogen, has been associated with susceptibility to thrombotic disease. A minor fraction of circulating fibrinogen contains the γ′ chain: 7–15% circulates as a heterodimeric fibrinogen molecule (AαBβγA) (AαBβγ′), referred to as γA/γ′, and 1% circulates as a homodimeric molecule (AαBβγ′) (AαBβγ′), referred to as γ′/γ′. Coronary artery disease has been associated with elevated levels of γA/γ′, independent from total fibrinogen levels.[166] Fibrinogen γ′ contains binding sites for thrombin and FXIII. Clots formed from γA/γ′ fibrinogen are highly cross-linked, have thinner fibers with more branch points, and are more resistant to lysis.[175] Variations in the fibrinogen genes are associated with the level of γ′ and the γ′/total fibrinogen ratio[176]: the FFG-haplotype 2 (H2) is associated with lower levels of γA/γ′ in plasma; the C-allele of the 9340T/C polymorphism and the G-allele of the 2224G/A polymorphism are associated with higher γ′ levels.[177]

Mutations in the fibrinogen gene may lead to quantitative and/or qualitative disorders. The quantitative disorders include afibrinogenemia and hypofibrinogenemia and cause reduced levels of circulating fibrinogen. Quantitative disorders are mostly associated with a bleeding diathesis, however arterial thrombosis does occur in these patients.[178, 179] Besides its procoagulant activity, fibrin also has antithrombotic properties: it contains non-substrate-binding places for thrombin. Hereby, fibrin sequesters and downregulates thrombin activity: thrombin is trapped in the clot and is not available for platelet activation in the arterial wall.[180] Qualitative disorders include dysfibrinogenemia and hypodysfibrinogenemia and cause normal or reduced levels of fibrinogen with an abnormal functional activity.[180] Most qualitative disorders are clinically silent, however they can also cause a bleeding diathesis, lead to a hypercoagulable state or a combination of bleeding and thromboembolic symptoms. It is believed that there are two mechanisms responsible for thrombosis associated with dysfibrinogenemia: (1) abnormal fibrinogen is defective in binding thrombin resulting in elevated thrombin levels; (2) formed fibrin clots are more resistant to plasmin degradation.[181] Maybe some of these effects contribute to incident cases of arterial thrombosis in individuals with dysfibrinogenemia.[182, 183]

21.4.5.2 Factor XIII

Only a few studies have addressed the relation between plasma levels of FXIII and CAD, showing inconsistent results.[184] FXIII deficiency is extremely rare (1 in 1–3 million), leads to a bleeding diathesis and may theoretically protect from (arterial) thrombosis.[178] Several polymorphisms have been identified. The most extensively studied polymorphism is the Val34Leu (G163T) substitution in the A subunit. The 34Leu allele enhances the activation of FXIII by thrombin two- to threefold. This affects the structure of the formed fibrin clot.[166] The Tyr204Phe polymorphism in the A subunits is associated with lower levels and lower activity of FXIII. A polymorphism in the B subunit, His95Arg (A8259G), is associated with an increased dissociation rate of the factor XIII subunits following activation by thrombin.[185]

Kohler et al.[186] were the first to find a protective effect of the Leu34 variant for AMI. Several authors confirmed these findings,[187–189] however not all.[190, 191] Two meta-analyses both suggest that the FXIII-A Val34Leu polymorphism exerts a moderate, but statistically significant protective effect against CAD.[192, 193] The effect of this polymorphism on the risk of ischemic stroke is not clear, it did not influence the risk of peripheral artery disease.[184] In young women, the Tyr204Phe polymorphism was associated with a ninefold increased risk for the occurrence of ischemic stroke. The risk was higher in homozygotes than in heterozygotes, indicating a dose–response relationship.[194] The Arg95 variant alone did not have an effect on the risk for MI, however in the presence of Leu34 variant, the risk for MI was reduced in postmenopausal women.[195] The risk for ischemic stroke was not associated with this polymorphism in young women.[194] In patients with non-disabling cerebral ischemia of arterial origin the Arg95 variant was more common in patients with large artery disease compared to patients with small artery disease.[131]

21.5 Fibrinolytic System

21.5.1 Fibrinolysis

Based on the functions of the fibrinolytic system, one may postulate important regulatory roles for each of the players in the fibrinolytic cascade. Overall, reduced fibrinolytic activity is associated with increased risk for arterial thrombosis. However, only for some proteins a clear significance for atherogenesis and arterial thrombosis has been established, although with remaining uncertainties regarding the magnitude of the effects. Plasminogen deficiency does not seem to be a thrombotic risk factor.[196] Elevated levels of circulating tPA have been shown to be associated with arterial thrombosis.[197] This might seem contradictory; however, most circulating tPA is bound to PAI-1, and high levels of tPA in plasma could reflect disturbance of the endothelium.[197] No clear associations have been found between tPA-polymorphisms and arterial thrombosis.[196] The −7531 C/T substitution and the Alu repeat insertion/deletion (I/D) in the eight intron are associated with changed release rates[198] The −7531 C/T polymorphism was associated with MI,[199] but not with ischemic stroke.[200] In the Rotterdam study, the I allele had significantly more presence than the D allele in patients with nonfatal MI.[201] However, other studies did not find an association between I/D polymorphism and MI.[202, 203]

21.5.2 Fibrinolytic Inhibitors

High levels of TAFI antigen and activity have been associated with increased risk of arterial thrombosis in multiple studies.[196] The TAFI plasma levels are determined for approximately 25% by TAFI gene polymorphisms.[204] Several polymorphisms have been identified among which 1040C/T (Thr325Ile), 505A/G (Ala147Thr) in the coding region and 1542C/G in the 3′ untranslated region. The 1040TT genotype and the 1542C allele are associated with lower levels of circulating TAFI.[205–207] Furthermore, the Thr325Ile substitution changes the stability of activated TAFI, causing an altered fibrinolytic activity, while the fibrinolytic activity is enhanced for the Ile325 allele.[196] The 505A/G substitution is associated with increased TAFI levels.[208] These polymorphisms were investigated in several studies, however with inconsistent results.[206, 209–212]

The clinically most relevant factor in the interplay of fibrinolysis and thrombosis is PAI-1. Concentrations of PAI-1 are an independent risk factor for CAD.[213] There is a common polymorphism known as 4G/5G in the −675 promoter region. The 5G allele is slightly less transcriptionally active than the 4G. The PAI-1 4G allele frequency is 0.58 in Caucasians and 0.13 in Africans.[214] PAI-1 levels were strongly dependent on body mass index (BMI), showing highest levels in the highest BMI quartile,[215] with BMI being a much stronger determinant of PAI-1 levels than genotype. This metabolic effect is more obvious in African than Caucasian persons, possibly also related to morphometric determinants.[214] The result is that plasma levels are highest in lean persons with a 4G/4G genotype as compared to the 5G/5G genotype (about twofold difference in two population based cohorts).[215]

Regulation of PAI-1 production is complex. PAI-1 is synthesized in different cell types, including hepatocytes, adipocytes, and endothelial cells, and PAI-1 transcription is regulated by several factors, including inflammatory mediators such as Il-1 and TNFα, as well as hormones like insulin and glucocorticoids; it also acts as an acute phase protein.[216] The contributing effect of the 4G/5G genotype appears at the cellular level, e.g., in the amount of PAI-1 produced under influence of Il-1[217] in vivo. An example is the effect of a diet rich in unsaturated fatty acids, which lowers PAI-1 in 4G carriers but has no detectable effect in 5G carriers that have a lower PAI-1 level to begin with.[218] The 4G genotype is also associated with high cholesterol levels.[219] It is likely that the interplay between inflammatory mediators, metabolic factors, and PAI-1 genotype, depending on ethnic background, affects fibrinolysis, and hence, thrombosis risk. In addition to thrombosis, PAI-1 has different modulating effects on the vessel wall, including remodeling and atherosclerosis, but many of these effects have been quite inconsistent, such that the biological significance remains unclear. Recent data are however challenging, showing that a recombinant PAI-1 based peptide inhibits vasa vasorum and atherogenesis[220] and that overexpressing PAI-1 attenuates aortic aneurysm formation in mice.[221]

The clinical effect of the PAI-1 4G/5G genotype has been mainly studied with regard to atherothrombosis and stroke. As a single factor the PAI-1 polymorphism emerged as an independent risk factor for recurrent ischemic events in nonhyperlipidemic postinfarction patients, suggesting that in the absence of major risk factors like hypercholesterolemia, the effect of a single genotype can be clinically relevant.[222] This observation deviates from the more commonly held belief that the contribution of single SNPs in the hemostatic mechanism does not contribute to the risk of arterial thrombosis.[223] The absence of evident risk influence may indeed be much dependent on the population studied. However, also in the pooled analysis of seven hemostatic gene polymorphisms the PAI-1 4 G variant yielded a per allele relative risk of coronary disease of 1.06 (1.02–1.10), albeit that a concern remains regarding heterogeneity of the studied populations.[1] As part of a series of common SNPs, PAI-1 4G contributed to risk of AMI in patients with advanced coronary atherosclerosis, suggesting the additive effect of this mutation.[224]

With regard to ischemic stroke, the evidence for a risk association is highly significant in pooled analysis, but there is the concern of extreme heterogeneity.[225] The most recent meta-analysis does not show any differential effects of PAI-1 genotype with regard to type of stroke (small or large vessel).[226]

21.6 Conclusions

There is no clear-cut association between polymorphisms in genes coding for hemostatic proteins and arterial thrombosis (see Table 21.1). Reported results are often contradictory due to heterogeneity of patient populations, deficiencies in study design, and so on. Meta-analyses show only small or even no association of these polymorphisms and arterial thrombosis. This may seem remarkable, especially for those factors that are established risk markers of arterial thrombosis. Elevated levels of fibrinogen are consistently associated with arterial thrombosis, polymorphisms in these genes are at best contributing, by modifying fibrinogen structure or concentration. Platelet inhibitors are extensively used in the treatment of arterial thrombosis; however, the studied polymorphisms were not related to CHD.[1]

Table 21.1 Polymorphism in hemostatic factors and association with arterial thrombosis

Protein	Polymorphism/mutation	Phenotype	Association with arterial thrombosis
Platelet receptors			
GPIb-IX-V	GPIbα 434C/T (Thr145Met)	Structural change of receptor	Inconsistent results for CAD, risk factor for ischemic stroke
	GPIbα VNTR	Structural change of receptor	Inconsistent results
	GPIbα −5C/T in the Kozak sequence	Increased surface level of GP Ib-IX-V receptor	Marginal, nonsignificant effect on MI, risk factor for ischemic stroke
GPIIb/IIIa	1565C/T (Leu33Pro; Leu33: PlA1, Pro33:PlA2)	Lower activation threshold	No risk factor for MI, cerebrovascular disease or CAD
GPIa/IIa	807C/T (Phe224); 873G/A (Thr246)	Increased receptor density and faster rate of platelet adhesion to type-1 collagen	No significant association with CAD or ischemic stroke
GPVI	13254T/C (Ser219Pro)	Altered structure-function	Probable risk factor for MI
Von Willebrand Factor (vWF)	−1793G/C, −1234C/T, −1185A/G, and −1051G/A	Decreased vWF:Ag levels	Probably no effect on risk for CAD
	Sma I polymorphism	No effect on vWF levels	Suggestive risk factor for ischemic stroke
	2365A/G (Thr789Ala) substitution	Increased vWF levels	Suggestive risk factor for CHD in type-1 diabetes
Factor VII	10976A/G (Arg353Gln)	Lower secretion efficiency	Inconsistent results
	−323ins10 (A1: deletion, A2: insertion)	Reduced transcription rates	Inconsistent results
	HVR4 (H5, H6 and H7)	Lower levels of FVII:C and FVII antigen	Inconsistent results
	−401G/T	Reduced transcription	Inconsistent results
	−402G/A	Increased transcriptional activity	Inconsistent results
Tissue factor	−1812 C/T, −1322 C/T, −1208 deletion/insertion, −603 A/G	Higher levels of circulating TF	Inconsistent results
	5466A/G	Lower TF mRNA and basal TF activity in monocytes, an increased relative increase of TF activation upon stimulation of monocytes with lipopolysaccharide	Suggestive risk factor for cardiac death in patients with acute coronary syndrome
Factor XII	−4C/T (previous notation:46C/T)	Lower levels of FXII	Inconsistent results

(continued)

Table 21.1 (continued)

Protein	Polymorphism/mutation	Phenotype	Association with arterial thrombosis
Factor V	1691G/A (Arg506Gln) (FV Leiden)	APC resistance	Risk factor for CHD
Prothrombin	20210G/A	Higher prothrombin levels	Risk factor for CHD
Thrombomodulin	−33G/A	Reduced transcription and expression of TM	Inconsistent results
	127G/A (Ala25Thr)	Does not alter the surface expression of TM	Suggestive risk factor, especially in the presence of additional risk factors
	1418C/T (Ala455Val)	Unknown	Inconsistent results
Endothelial cell protein C receptor (EPCR)	6936G/A (Ser219Gly)	Increased levels of soluble EPCR	Suggestive risk factor CHD
Fibrinogen	−148C/T	Higher levels of fibrinogen	Inconsistent results
	−455G/A (HaeIII)	Higher levels of fibrinogen	Inconsistent results
	BclI	Higher levels of fibrinogen	Inconsistent results
	Arg448Lys	Altered clot structure	Inconsistent results
	−854G/A	Higher levels of fibrinogen	Inconsistent results
	Thr312Ala	Altered clot structure, lower fibrinogen levels	Inconsistent results
Factor XIII	Val34Leu	Altered clot structure	Protective against AMI
	Tyr204Phe	Lower levels and activity of FXIII	Suggestive risk factor for ischemic stroke
	His95Arg	Increased dissociation rate of FXIII subunits following activation by thrombin	Inconsistent results
tPA	−7351 C/T	Reduced release rate T allele, no effect on plasma levels	Suggestive risk factor for MI, no risk factor for ischemic stroke
	Alu repeat I/D	Release rates higher in subjects homozygous for the I allele	Inconsistent results
Thrombin activatable fibrinolysis inhibitor (TAFI)	1040C/T (Thr325Ile)	Lower TAFI levels, enhanced fibrinolytic activity	Inconsistent results
	505A/G (Ala147Thr)	Increased TAFI levels	Inconsistent results
	1542C/G	Lower TAFI levels	Inconsistent results
Plasminogen activator inhibitor-1 (PAI-1)	4G/5G	Reduced PAI-1 levels	4G variant risk factor for coronary disease

Clearly, arterial thrombosis is a multifactorial disease and single polymorphisms likely contribute in a limited manner. From that perspective it is surprising that a common polymorphism such as factor V Leiden is a contributing factor for CAD.[1] Such observations are important from a mechanistic perspective, showing that plasma proteins such as factor V may play a role in (athero)thrombosis. The recent experimental data that link hypercoagulability to atherosclerosis could explain such a gene effect on the atherosclerosis phenotype. Clearly, more research is needed to further explore this exciting field.

21.7 Clinical Relevance

From a clinical perspective, even more so than in venous thromboembolism, there is no indication for genotyping or thrombophilia screening in arterial thrombosis.

Even though there is evidence from clinical studies that some SNPs do contribute to the development of arterial thrombosis, the contribution is only minor. Therefore, screening for these SNPs is not useful in most cases, particularly because the identification of risk alleles does not alter current management of the patient. Even in venous thromboembolic disease, thrombophilia is not recommended in current CBO guidelines. This may seem remarkable but is also based on the lack of management consequences. Nevertheless, from the perspective of the patients, their partners, and family, it may be unsatisfactory to refrain from any kind of additional thrombophilia testing, when techniques are available. The authors would indeed consider thrombophilia screening in individual cases, such as in young patients without any other obvious cardiovascular risk factors. One should, however, keep in mind that such analyses are primarily aimed at satisfying the doctor's or the patient's curiosity and one should also consider the potential negative consequences of genotyping, including effects on (life) insurance premiums.

Acknowledgments The authors thank the Netherlands Heart Foundation for their financial support (NHF grant number 2008B120).

References

1. Ye Z et al. Seven haemostatic gene polymorphisms in coronary disease: meta-analysis of 66, 155 cases and 91, 307 controls. *Lancet*. 2006;367(9511):651-658.
2. Watson T, Shantsila E, Lip GY. Mechanisms of thrombogenesis in atrial fibrillation: Virchow's triad revisited. *Lancet*. 2009;373(9658):155-166.
3. Seehaus S et al. Hypercoagulability inhibits monocyte transendothelial migration through protease-activated receptor-1-, phospholipase-Cbeta-, phosphoinositide 3-kinase-, and nitric oxide-dependent signaling in monocytes and promotes plaque stability. *Circulation*. 2009;120(9):774-784.
4. Borissoff JI et al. Is thrombin a key player in the "coagulation-atherogenesis" maze? *Cardiovasc Res*. 2009;82(3):392-403.
5. Davi G, Patrono C. Platelet activation and atherothrombosis. *N Engl J Med*. 2007;357(24):2482-2494.
6. Kiefer TL, Becker RC. Inhibitors of platelet adhesion. *Circulation*. 2009;120(24):2488-2495.
7. Rivera J et al. Platelet receptors and signaling in the dynamics of thrombus formation. *Haematologica*. 2009;94(5):700-711.
8. Varga-Szabo D, Pleines I, Nieswandt B. Cell adhesion mechanisms in platelets. *Arterioscler Thromb Vasc Biol*. 2008;28(3):403-412.
9. Surin WR, Barthwal MK, Dikshit M. Platelet collagen receptors, signaling and antagonism: emerging approaches for the prevention of intravascular thrombosis. *Thromb Res*. 2008;122(6):786-803.
10. Bevers EM et al. Generation of prothrombin-converting activity and the exposure of phosphatidylserine at the outer surface of platelets. *Eur J Biochem*. 1982;122(2):429-436.
11. Macfarlane RG. An enzyme cascade in the blood clotting mechanism, and its function as a biochemical amplifier. *Nature*. 1964;202:498-499.
12. Davie EW, Ratnoff OD. Waterfall sequence for intrinsic blood clotting. *Science*. 1964;145:1310-1312.
13. Standeven KF, Ariens RA, Grant PJ. The molecular physiology and pathology of fibrin structure/function. *Blood Rev*. 2005;19(5):275-288.
14. Esmon CT. The protein C pathway. *Chest*. 2003;124(3 Suppl):26S-32S.
15. Hackeng TM, Rosing J. Protein S as cofactor for TFPI. *Arterioscler Thromb Vasc Biol*. 2009;29(12):2015-2020.
16. Murata M et al. Coronary artery disease and polymorphisms in a receptor mediating shear stress-dependent platelet activation. *Circulation*. 1997;96(10):3281-3286.
17. Kuijpers RW et al. NH2-terminal globular domain of human platelet glycoprotein Ib alpha has a methionine 145/threonine145 amino acid polymorphism, which is associated with the HPA-2 (Ko) alloantigens. *J Clin Invest*. 1992;89(2):381-384.
18. Meyer M, Schellenberg I. Platelet membrane glycoprotein Ib: genetic polymorphism detected in the intact molecule and in proteolytic fragments. *Thromb Res*. 1990;58(3):233-242.
19. Lopez JA, Ludwig EH, McCarthy BJ. Polymorphism of human glycoprotein Ib alpha results from a variable number of tandem repeats of a 13-amino acid sequence in the mucin-like macroglycopeptide region. Structure/function implications. *J Biol Chem*. 1992;267(14):10055-10061.
20. Afshar-Kharghan V et al. Kozak sequence polymorphism of the glycoprotein (GP) Ibalpha gene is a major determinant of the plasma membrane levels of the platelet GP Ib-IX-V complex. *Blood*. 1999;94(1):186-191.
21. Meisel C et al. Role of Kozak sequence polymorphism of platelet glycoprotein Ibalpha as a risk factor for coronary artery disease and catheter interventions. *J Am Coll Cardiol*. 2001;38(4):1023-1027.
22. Kenny D et al. Platelet glycoprotein Ib alpha receptor polymorphisms and recurrent ischaemic events in acute coronary syndrome patients. *J Thromb Thrombolysis*. 2002;13(1):13-19.
23. Frank MB et al. The Kozak sequence polymorphism of platelet glycoprotein Ibalpha and risk of nonfatal myocardial infarction and nonfatal stroke in young women. *Blood*. 2001;97(4):875-879.
24. Ishida F et al. Genetic linkage of Kozak sequence polymorphism of the platelet glycoprotein Ib alpha with human platelet antigen-2 and variable number of tandem repeats polymorphism, and its relationship with coronary artery disease. *Br J Haematol*. 2000;111(4):1247-1249.
25. Sucker C et al. Are prothrombotic variants of platelet glycoprotein receptor polymorphisms involved in the pathogenesis of thrombotic microangiopathies? *Clin Appl Thromb Hemost*. 2009;15(4):402-407.
26. Croft SA et al. Kozak sequence polymorphism in the platelet GPIbalpha gene is not associated with risk of myocardial infarction. *Blood*. 2000;95(6):2183-2184.

27. Maguire JM et al. Polymorphisms in platelet glycoprotein 1balpha and factor VII and risk of ischemic stroke: a meta-analysis. *Stroke*. 2008;39(6):1710-1716.
28. Gonzalez-Conejero R et al. Polymorphisms of platelet membrane glycoprotein Ib associated with arterial thrombotic disease. *Blood*. 1998;92(8):2771-2776.
29. Sonoda A et al. Association between platelet glycoprotein Ibalpha genotype and ischemic cerebrovascular disease. *Stroke*. 2000;31(2):493-497.
30. Ardissino D et al. Prothrombotic genetic risk factors in young survivors of myocardial infarction. *Blood*. 1999;94(1):46-51.
31. Carter AM et al. Platelet GP IIIa PlA and GP Ib variable number tandem repeat polymorphisms and markers of platelet activation in acute stroke. *Arterioscler Thromb Vasc Biol*. 1998;18(7):1124-1131.
32. von dem Borne AE, Decary F. Nomenclature of platelet-specific antigens. *Hum Immunol*. 1990;29(1):1-2.
33. Michelson AD et al. Platelet GP IIIa Pl(A) polymorphisms display different sensitivities to agonists. *Circulation*. 2000;101(9):1013-1018.
34. Weiss EJ et al. A polymorphism of a platelet glycoprotein receptor as an inherited risk factor for coronary thrombosis. *N Engl J Med*. 1996;334(17):1090-1094.
35. Mikkelsson J et al. Glycoprotein IIIa Pl(A) polymorphism associates with progression of coronary artery disease and with myocardial infarction in an autopsy series of middle-aged men who died suddenly. *Arterioscler Thromb Vasc Biol*. 1999;19(10):2573-2578.
36. Walter DH et al. Platelet glycoprotein IIIa polymorphisms and risk of coronary stent thrombosis. *Lancet*. 1997;350(9086):1217-1219.
37. Goldschmidt-Clermont PJ et al. Higher prevalence of GPIIIa PlA2 polymorphism in siblings of patients with premature coronary heart disease. *Arch Pathol Lab Med*. 1999;123(12):1223-1229.
38. Motovska Z et al. Platelet glycoprotein GP VI 13254C allele is an independent risk factor of premature myocardial infarction. *Thromb Res*. 2010;125(2):e61-4.
39. Ridker PM et al. PIA1/A2 polymorphism of platelet glycoprotein IIIa and risks of myocardial infarction, stroke, and venous thrombosis. *Lancet*. 1997;349(9049):385-388.
40. Marian AJ, Brugada R, Kleiman NS. Platelet glycoprotein IIIa PlA polymorphism and myocardial infarction. *N Engl J Med*. 1996;335(14):p. 1071-p. 1072. author reply 1073-4.
41. Herrmann SM et al. The Leu33/Pro polymorphism (PlA1/PlA2) of the glycoprotein IIIa (GPIIIa) receptor is not related to myocardial infarction in the ECTIM Study. Etude Cas-Temoins de l'Infarctus du Myocarde. *Thromb Haemost*. 1997;77(6):1179-1181.
42. Carter AM et al. Association of the platelet Pl(A) polymorphism of glycoprotein IIb/IIIa and the fibrinogen Bbeta 448 polymorphism with myocardial infarction and extent of coronary artery disease. *Circulation*. 1997;96(5):1424-1431.
43. Samani NJ, Lodwick D. Glycoprotein IIIa polymorphism and risk of myocardial infarction. *Cardiovasc Res*. 1997;33(3):693-697.
44. Brenner B. Are platelet membrane glycoprotein polymorphisms predictive of arterial thrombosis? *Isr Med Assoc J*. 2002;4(6):458-459.
45. O'Donnell CJ et al. Genetic and environmental contributions to platelet aggregation: the Framingham heart study. *Circulation*. 2001;103(25):3051-3056.
46. Burr D et al. A meta-analysis of studies on the association of the platelet PlA polymorphism of glycoprotein IIIa and risk of coronary heart disease. *Stat Med*. 2003;22(10):1741-1760.
47. Kunicki TJ et al. Hereditary variation in platelet integrin alpha 2 beta 1 density is associated with two silent polymorphisms in the alpha 2 gene coding sequence. *Blood*. 1997;89(6):1939-1943.
48. Moshfegh K et al. Association of two silent polymorphisms of platelet glycoprotein Ia/IIa receptor with risk of myocardial infarction: a case-control study. *Lancet*. 1999;353(9150):351-354.
49. Santoso S et al. Association of the platelet glycoprotein Ia C807T gene polymorphism with nonfatal myocardial infarction in younger patients. *Blood*. 1999;93(8):2449-2453.
50. Croft SA et al. The GPIa C807T dimorphism associated with platelet collagen receptor density is not a risk factor for myocardial infarction. *Br J Haematol*. 1999;106(3):771-776.
51. Tsantes AE et al. Lack of association between the platelet glycoprotein Ia C807T gene polymorphism and coronary artery disease: a meta-analysis. *Int J Cardiol*. 2007;118(2):189-196.
52. Nikolopoulos GK et al. Integrin, alpha 2 gene C807T polymorphism and risk of ischemic stroke: a meta-analysis. *Thromb Res*. 2007;119(4):501-510.
53. Croft SA et al. Novel platelet membrane glycoprotein VI dimorphism is a risk factor for myocardial infarction. *Circulation*. 2001;104(13):1459-1463.
54. Ollikainen E et al. Platelet membrane collagen receptor glycoprotein VI polymorphism is associated with coronary thrombosis and fatal myocardial infarction in middle-aged men. *Atherosclerosis*. 2004;176(1):95-99.
55. Spiel AO, Gilbert JC, Jilma B. von Willebrand factor in cardiovascular disease: focus on acute coronary syndromes. *Circulation*. 2008;117(11):1449-1459.
56. Fuster W et al. Resistance to arteriosclerosis in pigs with von Willebrand's disease. Spontaneous and high cholesterol diet-induced arteriosclerosis. *J Clin Invest*. 1978;61(3):722-730.
57. Nichols TC et al. von Willebrand factor and occlusive arterial thrombosis. A study in normal and von Willebrand's disease pigs with diet-induced hypercholesterolemia and atherosclerosis. *Arteriosclerosis*. 1990;10(3):449-461.
58. Methia N et al. Localized reduction of atherosclerosis in von Willebrand factor-deficient mice. *Blood*. 2001;98(5):1424-1428.
59. Griggs TR et al. Susceptibility to atherosclerosis in aortas and coronary arteries of swine with von Willebrand's disease. *Am J Pathol*. 1981;102(2):137-145.
60. Bilora F et al. Do hemophilia A and von Willebrand disease protect against carotid atherosclerosis? A comparative study between coagulopathics and normal subjects by means of carotid echo-color Doppler scan. *Clin Appl Thromb Hemost*. 1999;5(4):232-235.
61. Bilora F et al. Hemophilia A, von Willebrand disease, and atherosclerosis of abdominal aorta and leg arteries: factor VIII and von Willebrand factor defects appear to protect abdominal aorta and leg arteries from atherosclerosis. *Clin Appl Thromb Hemost*. 2001;7(4):311-313.

62. Sramek A et al. Decreased coagulability has no clinically relevant effect on atherogenesis: observations in individuals with a hereditary bleeding tendency. *Circulation*. 2001; 104(7):762-767.
63. Sramek A et al. Patients with type 3 severe von Willebrand disease are not protected against atherosclerosis: results from a multicenter study in 47 patients. *Circulation*. 2004; 109(6):740-744.
64. Leebeek FW, van der Meer IM, Witteman JC. Genetic variability of von Willebrand factor and atherosclerosis. *Circulation*. 2004;110(5):e57.
65. Bilora F et al. Type IIb von Willebrand disease: role of qualitative defects in atherosclerosis and endothelial dysfunction. *Clin Appl Thromb Hemost*. 2007;13(4):384-390.
66. Keightley AM et al. Variation at the von Willebrand factor (vWF) gene locus is associated with plasma vWF:Ag levels: identification of three novel single nucleotide polymorphisms in the vWF gene promoter. *Blood*. 1999;93(12): 4277-4283.
67. Harvey PJ et al. A single nucleotide polymorphism at nucleotide -1793 in the von Willebrand factor (VWF) regulatory region is associated with plasma VWF:Ag levels. *Br J Haematol*. 2000;109(2):349-353.
68. Dai K, Gao W, Ruan C. The Sma I polymorphism in the von Willebrand factor gene associated with acute ischemic stroke. *Thromb Res*. 2001;104(6):389-395.
69. Lacquemant C et al. Association between high von willebrand factor levels and the Thr789Ala vWF gene polymorphism but not with nephropathy in type I diabetes. The GENEDIAB Study Group and the DESIR Study Group. *Kidney Int*. 2000;57(4):1437-1443.
70. Klemm T et al. Impact of the Thr789Ala variant of the von Willebrand factor levels, on ristocetin co-factor and collagen binding capacity and its association with coronary heart disease in patients with diabetes mellitus type 2. *Exp Clin Endocrinol Diabetes*. 2005;113(10):568-572.
71. van der Meer IM et al. Genetic variability of von Willebrand factor and risk of coronary heart disease: the Rotterdam Study. *Br J Haematol*. 2004;124(3):343-347.
72. Di Bitondo R et al. The -1185 A/G and -1051 G/A dimorphisms in the von Willebrand factor gene promoter and risk of myocardial infarction. *Br J Haematol*. 2001;115(3):701-706.
73. Simon D et al. Association studies between -1185A/G von Willebrand factor gene polymorphism and coronary artery disease. *Braz J Med Biol Res*. 2003;36(6):709-714.
74. Wu O et al. ABO(H) blood groups and vascular disease: a systematic review and meta-analysis. *J Thromb Haemost*. 2008;6(1):62-69.
75. Meade TW et al. Haemostatic function and ischaemic heart disease: principal results of the Northwick Park Heart Study. *Lancet*. 1986;2(8506):533-537.
76. Hoffman CJ et al. Elevation of factor VII activity and mass in young adults at risk of ischemic heart disease. *J Am Coll Cardiol*. 1989;14(4):941-946.
77. Smith FB et al. Hemostatic factors as predictors of ischemic heart disease and stroke in the Edinburgh Artery Study. *Arterioscler Thromb Vasc Biol*. 1997;17(11):3321-3325.
78. Lane A et al. Factor VII Arg/Gln353 polymorphism determines factor VII coagulant activity in patients with myocardial infarction (MI) and control subjects in Belfast and in France but is not a strong indicator of MI risk in the ECTIM study. *Atherosclerosis*. 1996;119(1):119-127.
79. Humphries SE et al. Factor VII coagulant activity and antigen levels in healthy men are determined by interaction between factor VII genotype and plasma triglyceride concentration. *Arterioscler Thromb*. 1994;14(2):193-198.
80. Miller GJ et al. Factor VII-deficient substrate plasmas depleted of protein C raise the sensitivity of the factor VII bio-assay to activated factor VII: an international study. *Thromb Haemost*. 1994;71(1):38-48.
81. Lowe GD et al. Interleukin-6, fibrin D-dimer, and coagulation factors VII and XIIa in prediction of coronary heart disease. *Arterioscler Thromb Vasc Biol*. 2004;24(8):1529-1534.
82. Green F et al. A common genetic polymorphism associated with lower coagulation factor VII levels in healthy individuals. *Arterioscler Thromb*. 1991;11(3):540-546.
83. Marchetti G et al. A polymorphism in the 5' region of coagulation factor VII gene (F7) caused by an inserted decanucleotide. *Hum Genet*. 1993;90(5):575-576.
84. Bernardi F et al. Factor VII gene polymorphisms contribute about one third of the factor VII level variation in plasma. *Arterioscler Thromb Vasc Biol*. 1996;16(1):72-76.
85. Sacchi E et al. Plasma factor VII levels are influenced by a polymorphism in the promoter region of the FVII gene. *Blood Coagul Fibrinolysis*. 1996;7(2):114-117.
86. Pollak ES et al. Functional characterization of the human factor VII 5'-flanking region. *J Biol Chem*. 1996;271(3): 1738-1747.
87. Van't Hooft FM et al. wo common functional polymorphisms in the promoter region of the coagulation factor VII gene determining plasma factor VII activity and mass concentration. *Blood*. 1999;93(10):3432-3441.
88. Di Castelnuovo A et al. The decanucleotide insertion/deletion polymorphism in the promoter region of the coagulation factor VII gene and the risk of familial myocardial infarction. *Thromb Res*. 2000;98(1):9-17.
89. Endler G, Mannhalter C. Polymorphisms in coagulation factor genes and their impact on arterial and venous thrombosis. *Clin Chim Acta*. 2003;330(1–2):31-55.
90. Lane DA, Grant PJ. Role of hemostatic gene polymorphisms in venous and arterial thrombotic disease. *Blood*. 2000;95(5): 1517-1532.
91. Lane A et al. Genetic and environmental determinants of factor VII coagulant activity in ethnic groups at differing risk of coronary heart disease. *Atherosclerosis*. 1992;94(1):43-50.
92. Quek SC et al. The effects of three factor VII polymorphisms on factor VII coagulant levels in healthy Singaporean Chinese, Malay and Indian newborns. *Ann Hum Genet*. 2006;70(Pt 6):951-957.
93. Hunault M et al. The Arg353Gln polymorphism reduces the level of coagulation factor VII. In vivo and in vitro studies. *Arterioscler Thromb Vasc Biol*. 1997;17(11):2825-2829.
94. Iacoviello L et al. Polymorphisms in the coagulation factor VII gene and the risk of myocardial infarction. *N Engl J Med*. 1998;338(2):79-85.
95. Girelli D et al. Polymorphisms in the factor VII gene and the risk of myocardial infarction in patients with coronary artery disease. *N Engl J Med*. 2000;343(11):774-780.
96. Bozzini C et al. Influence of polymorphisms in the factor VII gene promoter on activated factor VII levels and on the risk

of myocardial infarction in advanced coronary atherosclerosis. *Thromb Haemost.* 2004;92(3):541-549.
97. Mrozikiewicz PM et al. Reduced procedural risk for coronary catheter interventions in carriers of the coagulation factor VII-Gln353 gene. *J Am Coll Cardiol.* 2000;36(5):1520-1525.
98. Doggen CJ et al. A genetic propensity to high factor VII is not associated with the risk of myocardial infarction in men. *Thromb Haemost.* 1998;80(2):281-285.
99. Arnaud E et al. Polymorphisms in the 5' regulatory region of the tissue factor gene and the risk of myocardial infarction and venous thromboembolism: the ECTIM and PATHROS studies. Etude Cas-Temoins de l'Infarctus du Myocarde. Paris Thrombosis case-control Study. *Arterioscler Thromb Vasc Biol.* 2000;20(3):892-898.
100. Malarstig A et al. Genetic variations in the tissue factor gene are associated with clinical outcome in acute coronary syndrome and expression levels in human monocytes. *Arterioscler Thromb Vasc Biol.* 2005;25(12):2667-2672.
101. Ott I et al. Tissue factor promotor polymorphism -603 A/G is associated with myocardial infarction. *Atherosclerosis.* 2004;177(1):189-191.
102. Campo G et al. Tissue factor and coagulation factor VII levels during acute myocardial infarction: association with genotype and adverse events. *Arterioscler Thromb Vasc Biol.* 2006;26(12):2800-2806.
103. Renne T et al. Defective thrombus formation in mice lacking coagulation factor XII. *J Exp Med.* 2005;202(2):271-281.
104. van der Meijden PE et al. Dual role of collagen in factor XII-dependent thrombus formation. *Blood.* 2009;114(4):881-890.
105. Smith SA et al. Polyphosphate modulates blood coagulation and fibrinolysis. *Proc Natl Acad Sci USA.* 2006;103(4):903-908.
106. Merlo C et al. Elevated levels of plasma prekallikrein, high molecular weight kininogen and factor XI in coronary heart disease. *Atherosclerosis.* 2002;161(2):261-267.
107. Vaziri ND et al. Coagulation, fibrinolytic, and inhibitory proteins in acute myocardial infarction and angina pectoris. *Am J Med.* 1992;93(6):651-657.
108. Cooper JA et al. Comparison of novel hemostatic factors and conventional risk factors for prediction of coronary heart disease. *Circulation.* 2000;102(23):2816-2822.
109. Govers-Riemslag JW et al. The plasma kallikrein-kinin system and risk of cardiovascular disease in men. *J Thromb Haemost.* 2007;5(9):1896-1903.
110. Grundt H et al. Activated factor 12 (FXIIa) predicts recurrent coronary events after an acute myocardial infarction. *Am Heart J.* 2004;147(2):260-266.
111. Colhoun HM et al. Activated factor XII levels and factor XII 46C>T genotype in relation to coronary artery calcification in patients with type 1 diabetes and healthy subjects. *Atherosclerosis.* 2002;163(2):363-369.
112. Doggen CJ, Rosendaal FR, Meijers JC. Levels of intrinsic coagulation factors and the risk of myocardial infarction among men: opposite and synergistic effects of factors XI and XII. *Blood.* 2006;108(13):4045-4051.
113. Girolami A et al. Myocardial infarction and arterial thrombosis in severe (homozygous) FXII deficiency: no apparent causative relation. *Clin Appl Thromb Hemost.* 2005;11(1):49-53.
114. Kanaji T et al. A common genetic polymorphism (46 C to T substitution) in the 5'-untranslated region of the coagulation factor XII gene is associated with low translation efficiency and decrease in plasma factor XII level. *Blood.* 1998;91(6):2010-2014.
115. Zito F et al. Association of the factor XII 46C>T polymorphism with risk of coronary heart disease (CHD) in the WOSCOPS study. *Atherosclerosis.* 2002;165(1):153-158.
116. Santamaria A et al. Homozygosity of the T allele of the 46 C–>T polymorphism in the F12 gene is a risk factor for acute coronary artery disease in the Spanish population. *Haematologica.* 2004;89(7):878-879.
117. Santamaria A et al. Homozygosity of the T allele of the 46 C->T polymorphism in the F12 gene is a risk factor for ischemic stroke in the Spanish population. *Stroke.* 2004;35(8):1795-1799.
118. Endler G et al. Homozygosity for the C–>T polymorphism at nucleotide 46 in the 5' untranslated region of the factor XII gene protects from development of acute coronary syndrome. *Br J Haematol.* 2001;115(4):1007-1009.
119. Kohler HP, Futers TS, Grant PJ. FXII (46C–>T) polymorphism and in vivo generation of FXII activity–gene frequencies and relationship in patients with coronary artery disease. *Thromb Haemost.* 1999;81(5):745-747.
120. Bach J et al. Coagulation factor XII (FXII) activity, activated FXII, distribution of FXII C46T gene polymorphism and coronary risk. *J Thromb Haemost.* 2008;6(2):291-296.
121. Athanasiadis G et al. Polymorphism FXII 46C>T and cardiovascular risk: additional data from Spanish and Tunisian patients. *BMC Res Notes.* 2009;2:154.
122. Oguchi S et al. Genotype distribution of the 46C/T polymorphism of coagulation factor XII in the Japanese population: absence of its association with ischemic cerebrovascular disease. *Thromb Haemost.* 2000;83(1):178-179.
123. Yazdani-Biuki B et al. The functional promoter polymorphism of the coagulation factor XII gene is not associated with peripheral arterial disease. *Angiology.* 2010;61(2):211-215.
124. Roldan V et al. Synergistic association between hypercholesterolemia and the C46T factor XII polymorphism for developing premature myocardial infarction. *Thromb Haemost.* 2005;94(6):1294-1299.
125. Salomon O et al. Reduced incidence of ischemic stroke in patients with severe factor XI deficiency. *Blood.* 2008;111(8):4113-4117.
126. Salomon O et al. Inherited factor XI deficiency confers no protection against acute myocardial infarction. *J Thromb Haemost.* 2003;1(4):658-661.
127. Rosendaal FR et al. Mortality and causes of death in Dutch haemophiliacs, 1973-86. *Br J Haematol.* 1989;71(1):71-76.
128. Aronson DL. Cause of death in hemophilia A patients in the United States from 1968 to 1979. *Am J Hematol.* 1988;27(1):7-12.
129. Sramek A, Kriek M, Rosendaal FR. Decreased mortality of ischaemic heart disease among carriers of haemophilia. *Lancet.* 2003;362(9381):351-354.
130. Darby SC et al. Mortality rates, life expectancy, and causes of death in people with hemophilia A or B in the United Kingdom who were not infected with HIV. *Blood.* 2007;110(3):815-825.

131. Pruissen DM et al. Prothrombotic gene variation in patients with large and small vessel disease. *Neuroepidemiology*. 2008;31(2):89-92.
132. Martini CH et al. No effect of polymorphisms in prothrombotic genes on the risk of myocardial infarction in young adults without cardiovascular risk factors. *J Thromb Haemost*. 2005;3(1):177-179.
133. Eitzman DT et al. Homozygosity for factor V Leiden leads to enhanced thrombosis and atherosclerosis in mice. *Circulation*. 2005;111(14):1822-1825.
134. Kerlin BA et al. Survival advantage associated with heterozygous factor V Leiden mutation in patients with severe sepsis and in mouse endotoxemia. *Blood*. 2003;102(9):3085-3092.
135. Volzke H et al. Interaction between factor V Leiden and serum LDL cholesterol increases the risk of atherosclerosis. *Atherosclerosis*. 2005;180(2):341-347.
136. Pruissen DM et al. Prothrombotic genetic variants and atherosclerosis in patients with cerebral ischemia of arterial origin. *Atherosclerosis*. 2009;204(1):191-195.
137. Marcucci R et al. Thrombophilic risk factors in patients with severe carotid atherosclerosis. *J Thromb Haemost*. 2005;3(3):502-507.
138. Inbal A et al. Synergistic effects of prothrombotic polymorphisms and atherogenic factors on the risk of myocardial infarction in young males. *Blood*. 1999;93(7):2186-2190.
139. Doggen CJ et al. Interaction of coagulation defects and cardiovascular risk factors: increased risk of myocardial infarction associated with factor V Leiden or prothrombin 20210A. *Circulation*. 1998;97(11):1037-1041.
140. Redondo M et al. Coagulation factors II, V, VII, and X, prothrombin gene 20210G–>A transition, and factor V Leiden in coronary artery disease: high factor V clotting activity is an independent risk factor for myocardial infarction. *Arterioscler Thromb Vasc Biol*. 1999;19(4):1020-1025.
141. Ameziane N et al. No association between the R2 factor V gene and acute coronary events. *Thromb Haemost*. 2001;85(3):566-567.
142. Mahmoodi BK et al. Hereditary deficiency of protein C or protein S confers increased risk of arterial thromboembolic events at a young age: results from a large family cohort study. *Circulation*. 2008;118(16):1659-1667.
143. Salomaa V et al. Soluble thrombomodulin as a predictor of incident coronary heart disease and symptomless carotid artery atherosclerosis in the Atherosclerosis Risk in Communities (ARIC) Study: a case-cohort study. *Lancet*. 1999;353(9166):1729-1734.
144. Wu KK. Soluble thrombomodulin and coronary heart disease. *Curr Opin Lipidol*. 2003;14(4):373-375.
145. Weiler H, Isermann BH. Thrombomodulin. *J Thromb Haemost*. 2003;1(7):1515-1524.
146. Ireland H et al. Thrombomodulin gene mutations associated with myocardial infarction. *Circulation*. 1997;96(1):15-18.
147. Li YH et al. G-33A mutation in the promoter region of thrombomodulin gene and its association with coronary artery disease and plasma soluble thrombomodulin levels. *Am J Cardiol*. 2000;85(1):8-12.
148. Li YH et al. Functional mutation in the promoter region of thrombomodulin gene in relation to carotid atherosclerosis. *Atherosclerosis*. 2001;154(3):713-719.
149. Park HY et al. Association of G-33A polymorphism in the thrombomodulin gene with myocardial infarction in Koreans. *Hypertens Res*. 2002;25(3):389-394.
150. Zhao J et al. Association study of the thrombomodulin -33G>A polymorphism with coronary artery disease and myocardial infarction in Chinese Han population. *Int J Cardiol*. 2005;100(3):383-388.
151. Doggen CJ et al. A mutation in the thrombomodulin gene, 127G to A coding for Ala25Thr, and the risk of myocardial infarction in men. *Thromb Haemost*. 1998;80(5):743-748.
152. Norlund L et al. A common thrombomodulin amino acid dimorphism is associated with myocardial infarction. *Thromb Haemost*. 1997;77(2):248-251.
153. Cole JW et al. Thrombomodulin Ala455Val Polymorphism and the risk of cerebral infarction in a biracial population: the Stroke Prevention in Young Women Study. *BMC Neurol*. 2004;4(1):21.
154. Wu KK et al. Thrombomodulin Ala455Val polymorphism and risk of coronary heart disease. *Circulation*. 2001;103(10):1386-1389.
155. Faioni EM et al. Mutations in the thrombomodulin gene are rare in patients with severe thrombophilia. *Br J Haematol*. 2002;118(2):595-599.
156. van der Velden PA et al. A frequent thrombomodulin amino acid dimorphism is not associated with thrombophilia. *Thromb Haemost*. 1991;65(5):511-513.
157. Delvaeye M et al. Thrombomodulin mutations in atypical hemolytic-uremic syndrome. *N Engl J Med*. 2009;361(4):345-357.
158. Ireland H et al. EPCR Ser219Gly: elevated sEPCR, prothrombin F1+2, risk for coronary heart disease, and increased sEPCR shedding in vitro. *Atherosclerosis*. 2005;183(2):283-292.
159. Kelleher CC. Plasma fibrinogen and factor VII as risk factors for cardiovascular disease. *Eur J Epidemiol*. 1992;8(Suppl 1):p. 79-p. 82.
160. Scarabin PY et al. Associations of fibrinogen, factor VII and PAI-1 with baseline findings among 10, 500 male participants in a prospective study of myocardial infarction – the PRIME Study. Prospective Epidemiological Study of Myocardial Infarction. *Thromb Haemost*. 1998;80(5):749-756.
161. Danesh J et al. Plasma fibrinogen level and the risk of major cardiovascular diseases and nonvascular mortality: an individual participant meta-analysis. *Jama*. 2005;294(14):1799-1809.
162. Meijer WT et al. Determinants of peripheral arterial disease in the elderly: the Rotterdam study. *Arch Intern Med*. 2000;160(19):2934-2938.
163. Green FR. Fibrinogen polymorphisms and atherothrombotic disease. *Ann N Y Acad Sci*. 2001;936:549-559.
164. van der Bom JG et al. Elevated plasma fibrinogen: cause or consequence of cardiovascular disease? *Arterioscler Thromb Vasc Biol*. 1998;18(4):621-625.
165. Lim BC et al. Genetic regulation of fibrin structure and function: complex gene-environment interactions may modulate vascular risk. *Lancet*. 2003;361(9367):1424-1431.
166. Scott EM, Ariens RA, Grant PJ. Genetic and environmental determinants of fibrin structure and function: relevance to clinical disease. *Arterioscler Thromb Vasc Biol*. 2004;24(9):1558-1566.

The Pharmacogenetics of Atherosclerosis

22

P.S. Monraats and J.W. Jukema

22.1 Introduction

Atherosclerosis is a type of arteriosclerosis. It comes from the Greek words *athero* (meaning gruel or paste) and *sclerosis* (hardness). Atherosclerosis is the pathophysiological basis of the majority of morbidity and mortality in Western societies. In recent years, progression has been made in unraveling the exact cause for atherosclerosis. It is considered to have a complex pathophysiology, and is a disease of the arterial intima leading to the formation of fibrous (atheromatous) plaques and to stenosis/occlusion of the lumen (Fig. 22.1).[1] Atherosclerosis affects large and medium-sized arteries. The artery and position of the plaque varies with each person. In general, atherosclerosis is a slowly progressive disease that may start in childhood. In some people this disease progresses rapidly in their third decade. In others, it does not become threatening until they are in their 50s or 60s.[1,2] As mentioned above, the development of atherosclerosis is a complex process, and several major risk factors[3] and a growing number of novel risk markers[4] have been reported to play a role in its development. Evidence accumulated over decades convincingly demonstrates that family history of atherosclerosis in a parent or a sibling is an important risk factor associated with atherosclerotic coronary artery disease (CAD).[5]

The contribution of genetic factors to CAD is well illustrated by studies conducted on twins. In a study of monozygotic and dizygotic twins, death from CAD at an early age of one twin was a strong predictor of the risk of death of the other twin (Fig. 22.2). The fact that this risk was greatest in monozygotic compared to dizygotic twins indicates a strong genetic contribution to CAD.[6] The genes that contribute to this common disease are without doubt many, and they code for proteins involved in multiple metabolic pathways. Such genes include those controlling lipoprotein metabolism, vascular tone and reactivity, macrophage structure and function, and haemostatic as well as fibrinolytic pathways.

Advances have been made in both primary and secondary prevention of complications of atherosclerosis. Treatment of individuals with cholesterol-lowering therapy, like 3-hydroxy-3-methyl-glutaryl coenzyme A reductase inhibitors, has been shown to effectively reduce coronary events and even reduce the progression of atherosclerosis.[7] However, it remains questionable if all individuals have the same benefit of this treatment, since it is well recognized that considerable interindividual variation exists in response to statin therapy, in terms of both low-density lipoprotein (LDL)-lowering and clinical outcomes. The individual response to specific drugs is to a large extent inherited: this genetically determined variance in drug response defines the research area known as "pharmacogenetics." In this chapter, further clarification of the field of pharmacogenetics, focused on atherosclerosis, is provided. Furthermore, genetic markers and their relevance to clinical practice are discussed.

22.2 Pharmacogenetics: Tailored Therapy to Fit Individual Profiles

With the Human Genome Project nearing completion, genetic association studies are stimulated enormously. The purpose of genetic association studies is to investigate the relation of genetic variations or polymorphisms

P.S. Monraats (✉)
Department of Cardiology, Leiden University Medical Center, Leiden, The Netherlands
e-mail: p.s.monraats@lumc.nl

Fig. 22.1 The development of atherosclerosis

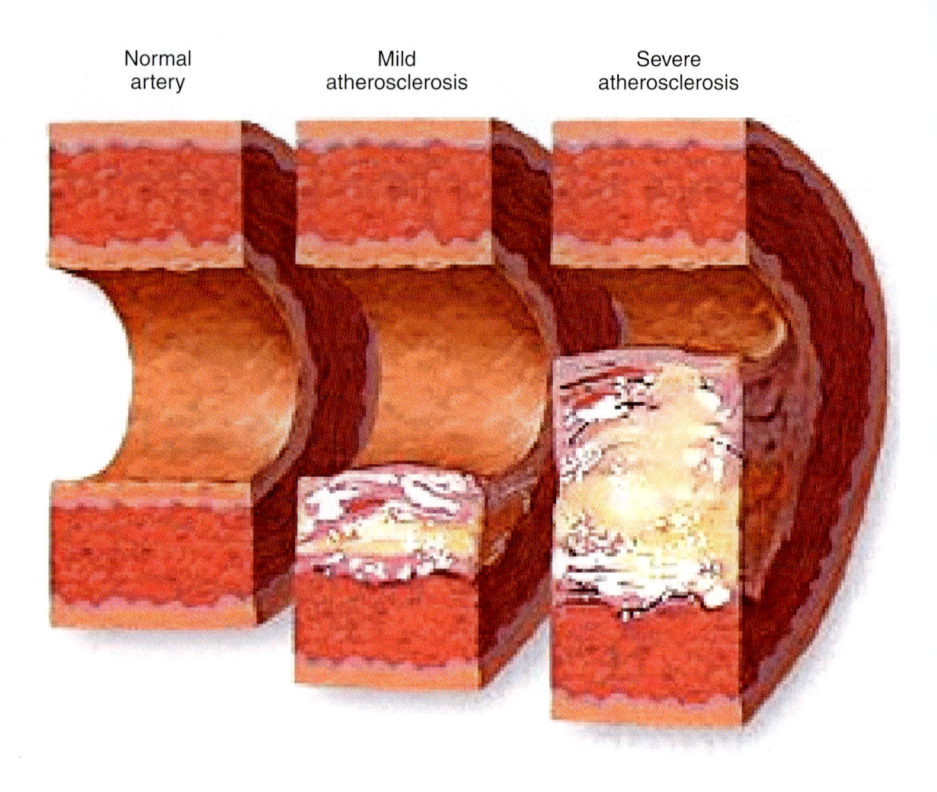

Age at Death	Relative Hazard (95% CI)*	
	MEN	WOMEN†
Monozygotic twins		
36–55 yr	13.4 (5.1–35.1)	14.9 (7.5–29.6)‡
56–65 yr	8.1 (4.9–13.2)‡	
66–75 yr	4.3 (3.0–6.1)‡	3.9 (2.3–6.6)‡
76–85 yr	1.9 (1.4–2.7)	2.2 (1.5–3.3)
≥86 yr	0.9 (0.4–1.8)	1.1 (0.6–2.1)
Dizygotic twins		
36–55 yr	4.3 (1.8–10.6)	2.2 (0.8–5.8)
56–65 yr	2.6 (1.7–3.9)	
66–75 yr	1.7 (1.3–2.2)	1.9 (1.3–2.8)
76–85 yr	1.4 (1.0–1.8)	1.4 (1.1–1.9)
≥86 yr	0.7 (0.4–1.4)	1.0 (0.6–1.7)

*CI denotes confidence interval.

†The age categories of 36 to 55 years and 56 to 65 years were merged into the age category of 36 to 65 years for women because of the small number of deaths.

‡$P<0.05$ for the comparison of the monozygotic twins with the dizygotic twins, by the Wald test.

Fig. 22.2 Relative hazard of death from coronary heart disease in subjects according to the age of their twins at death from coronary heart disease[6]

with disease and/or environment. Ecogenetics embodies the study of gene–environment interactions in the broadest sense. These environmental factors include among others food, chemicals, radiation, smoking, and drugs. For example, studies in humans demonstrate dietary hypo- and hyper-responders [8] and the presence of individuals at exaggerated risk of CAD from tobacco smoke due to underlying mutations in the nitric oxide synthase gene.[9] More specifically, the study of inheritable responses to pharmaceutical agents is known as pharmacogenetics.[10] The position of pharmacogenetics in the field of genetic testing in general is illustrated in Fig. 22.3.

Pharmaceutical treatment in the individual might be either beneficial, have no effect at all, or result in toxicity or adverse drug reactions. The first would be the desired effect. Identification of individuals in the second group, the nonresponders, would facilitate both a reduction of the costs and of the number of individuals at risk of toxicity. Furthermore, identification of persons at increased risk of adverse reactions to a specific drug or treatment would enable the physician to reduce costs, morbidity, and perhaps even mortality due to toxicity. Patterns of genetic markers or polymorphisms are highly individual and thus provide the theoretical possibility to make this distinction.

The genetic markers used in association studies are most often single nucleotide polymorphisms (SNPs), variants of a single nucleotide in DNA. SNPs are frequent throughout the genome. It is estimated that there is about one SNP in every 1,000 base pairs. These SNPs occur in the coding region (coding region SNP or cSNP), in noncoding regions in or in the immediate vicinity of the gene (perigenic SNP or pSNP) and between the genes (intergenic SNP or iSNP).[10] Individuals that carry a particular allele of a gene are likely to carry specific variants of several SNP markers, which are not necessarily located in the coding region of the gene. This is due to linkage disequilibrium. Linkage disequilibrium is present if certain polymorphisms occur together more frequently than might be expected by chance alone. This suggests that the combination of these polymorphisms is inherited together. Therefore, the association of a certain SNP with disease does not necessarily implicate that the SNP itself is associated with the pathophysiological mechanism.[11] Still this marker for disease may be useful and further study of the region of linked alleles might indicate the key functional mutation. However, due to these biological complexities, one of the problems is the poor reproducibility, which is often experienced in the field of genetic association studies, which are the main form of genetic studies. In addition to true biological effects that influence the relationship of interest, a number of methodological and epidemiological factors may distort the results of genetic association studies as well, such as differences in patient and control definition, population heterogeneity, and limited statistical power. These limitations have cast doubt on this type of study and some biomedical journals have even adopted a policy of not publishing the results of association studies related to complex diseases, or only publish positive associations which leads to publication bias.[12]

Although insight in disease mechanisms requires the detection of actual mutations responsible for the difference in protein levels, this is not absolutely necessary in pharmacogenetics. As long as it is possible to identify groups of individuals with a specific genetic marker profile that responds positively or negatively to medication, the goal of pharmacogenetics can be achieved. However, a plausible pathophysiologic link will have to provide the hypothesis for every project studying gene–environment interactions.[11,13]

Candidate genes for pharmacogenetic studies can be classified into the following three categories:

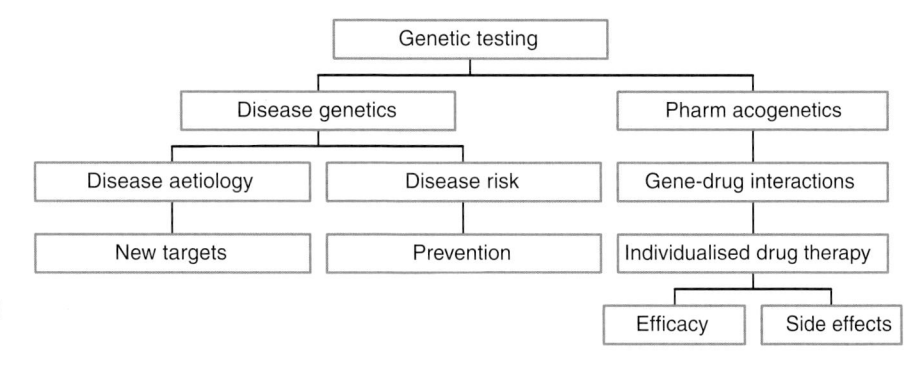

Fig. 22.3 The position of pharmacogenetics in the field of genetic testing

(1) genes encoding for drug-metabolizing enzymes and/or transporters that influence pharmacokinetics (e.g., liver P450 enzymes), (2) genes that influence pharmacodynamics by encoding for targets and other elements of a pathway in which a drug acts (e.g., molecules involved in cholesterol metabolism); and (3) genes that are involved in the underlying disease condition or intermediate phenotype (factors influencing atherosclerotic vascular disease development).[14]

22.3 Thiopurine Methyltransferase

An example of the use of pharmacogenetics comes from the study of the enzyme thiopurine methyltransferase (TPMT), which is involved in the breakdown of certain antitumor agents used in childhood leukemia. An inherited deficiency of this enzyme results in higher serum levels of the chemotherapeutic agent 6-mercaptopurine and thus increases toxicity, especially on the hematopoeietic system.[15–17] On the other hand, a certain gene polymorphism at this gene locus is associated with difficulty in achieving an effective dose of these drugs. Although the TPMT polymorphism is exceedingly rare (1% of the white population is homozygous for the polymorphism), the importance of the phenotype of TPMT deficiency has persuaded some centers to provide diagnostic services.

22.4 The Cytochrome P450

An example, which is important for cardiovascular medicine, is provided by variation in the gene encoding cytochrome P450 (CYP2D6). The cytochrome P450 system includes a number of enzymes involved in the metabolic elimination of drugs used in medicine today, such as those used for treating psychiatric, neurological, and cardiovascular diseases. Amiodarone, propafenone, flecainide, carvedilol, timolol, and mexiletine are examples of drugs used in cardiovascular medicines that are substrates for the cytochrome P450 CYP2D6 system.[18,19] CYP2D6 is highly polymorphic, and is inactive in about 6% of Caucasian people. In contrast, some individuals inherit up to 13 copies of the gene. These people have a more active drug metabolism so that therapeutic effects cannot be obtained with conventional doses of the prescribed drugs. On the other hand, in some individuals, simultaneously administered drugs inhibiting or competing for P450 may render their metabolism nonactive. These individuals would turn out to be poor metabolizers and thus are at an increased risk for drug toxicity. For instance, one study found that the clearance of the R(+) enantiomer of carvedilol was 66% lower and the area under the concentration-versus-time curve 156% higher among poor metabolizers than extensive metabolizers.[20]

In a recent study among persons treated with clopidogrel, carriers of a reduced-function CYP2C19 allele had significantly lower levels of the active metabolite of clopidogrel, diminished platelet inhibition, and a higher rate of major adverse cardiovascular events, including stent thrombosis, than did noncarriers.[21] When these results are confirmed in other studies, screening patients for this genotype could lead to a better risk stratification of patients at increased risk of developing a cardiovascular event due to a reduced effect of clopidogrel and thereby adjust treatment and by doing so individualize treatment.

22.5 Prediction of Efficacy in the Treatment of Atherosclerosis

Identifying genetic variants associated with drug efficacy can provide simple improvements in patient care. Polymorphisms of genes-encoding enzymes involved in the metabolism of lipids can influence the clinical response to cholesterol-lowering therapy. For example, certain polymorphisms in the lipoprotein lipase gene, the HMG-CoA reductase gene, the PCSK9 gene, the cholesterylester transfer protein gene, and the apolipoprotein E gene have been implicated to effect efficacy of statin therapy, as clarified below. However, another example, that of the angiotensin-converting enzyme insertion/deletion polymorphism, demonstrates the potential fallacy of methodological shortcomings.

22.6 Lipoprotein Lipase

In most people, plasma triglycerides are in dynamic equilibrium, mediated by a balance between very low-density lipoprotein (VLDL) and chylomicron (CM) synthesis, lipolysis of triglyceride (TG)-rich lipoproteins, and by uptake of remnant particles through appropriate receptors. However, with abnormalities involving synthesis, lipolysis, or remnant uptake, hypertriglyceridemia may ensue. Lipoprotein lipase (LPL) is a multifunctional protein. After synthesis in parenchymal cells, primarily in adipose tissue and skeletal muscle, LPL is transported to the intimal surface of the vascular endothelium where it is anchored by the heparin sulfate side chains of membrane proteoglycans. LPL plays a pivotal role in the hydrolysis of lipoprotein triacylglycerols to monoacylglycerols and fatty acids. The enzyme has also been shown to have other important functions where it acts as a ligand for the LDL receptor-related protein (LRP) and influences the secretion and uptake of LDL-cholesterol.

It is possible that mutations in the gene encoding LPL may underlie mild-to-moderate hypertriglyceridemia, but that their phenotype will not be recognized until other environmental or genetic factors are present. We have shown that a mutation in the human LPL gene, Asn291Ser, which results in a partial deficiency of lipolytic activity, is present with increased frequency in individuals with angiographically proven premature CAD, where it is associated with significantly decreased HDL cholesterol and increased TG levels.[22] Another common mutation in the LPL gene, the Ser447stop mutation, which is due to a C to G transversion at nucleotide 1595 in exon 9, which converts the serine 447 codon (*TCA*) to a premature termination codon (*TGA*) was significantly increased in those individuals in the highest quartile of HDL cholesterol in the Regression Growth Evaluation Statin Study (REGRESS – a lipid-lowering regression trial), pointing toward a positive association of this premature stop codon with elevated levels of HDL cholesterol and thus subsequently with protection against the development of CAD.[23] These results are confirmed in a recent meta-analysis of Sagoo et al. and in an earlier review article.[24,25]

In individuals with combined hyperlipidemia and CAD, another mutation, that is, an aspartic acid to an asparagine residue at position 9 (Asp9Asn) in the mature LPL protein was also identified. This mutation, too, was accompanied by high-TG/low-HDL cholesterol phenotype.[26] In addition, we investigated whether the presence of this Asp9Asn mutation could confer increased susceptibility to atherosclerosis and therefore be associated with more rapid progression of CAD. The mutation was identified in 4.8% of the patients who participated in the lipid-lowering angiographic clinical study REGRESS.[7,27] Carriers of this Asp9Asn mutation more frequently had a positive family history of CAD and exhibited lower HDL cholesterol levels than noncarriers. Indeed, it could be shown that these patients – with only subtle disturbances of lipoprotein metabolism due to the presence of this mutation – exhibited accelerated progression of coronary atherosclerosis and a diminished clinical event-free survival.[27] Of specific interest was the observation that the carriers of this mutation seemed particularly sensitive to pravastatin therapy. On average, progression was abolished in the treated group.

22.7 HMG-CoA-Reductase Gene

One thousand five hundred and thirty-six participants of the Pravastatin Inflammation/CRP Evaluation (PRINCE) study were free of statin use for 6 months prior to enrollment of this study and received pravastatin 40 mg daily. Changes in lipid levels after a 24-week therapy were related to all analyzed SNPs. The authors identified two common and tightly linked intronic SNPs (SNP12 and SNP29) in HMGCR, which were significantly associated with reduced efficacy of pravastatin therapy. Compared to homozygotes for the major (wild-type) allele in SNP12, which had a mean change of −42.0 mg/dL total cholesterol with pravastatin, patients with a single copy of the minor allele only reached a mean change of −32.8 mg/dL. Thus, carriers of the variant allele experienced a 22% smaller relative decrease in total cholesterol levels. This effect was largely due to a 19% smaller relative reduction in LDL-cholesterol in heterozygote individuals. Interestingly, no significant difference in basal lipid levels or HDL-cholesterol levels with pravastatin was seen between the genotypes. For SNP29 in the HMG-CoA reductase gene, nearly identical results were presented. In the remaining nine genes (i.e., *ABCG5*, *ABCG8*, *APOB*, *APOE*, *CETP*, *CYP3A4*, *CYP3A5*, *FDFT1*, and *LDLR*), three other SNPs, including SNP4 in the squalene synthase gene (*FDFT1*), were associated with differential effects of pravastatin on lipid reduction. However, most of these variations were also significantly associated with baseline lipid levels. Although the absolute difference in total cholesterol reduction of 9 mg/dL linked with the HMG-CoA reductase SNPs is relatively low compared to environmental factors such as diet, these genetic variations combined with other pharmacogenomic markers may affect health care on a population basis.[28]

22.8 PCSK9

Proprotein convertase subtilisin kexin-like 9 (*PCSK9*) is a secreted glycoprotein that is transcriptionally regulated by cholesterol status. It modulates levels of circulating low-density lipoprotein cholesterol (LDLC) by negatively regulating low-density lipoprotein receptor (LDLR) levels. *PCSK9* is a highly polymorphic gene with over 40 non-synonymous, exonic SNPs reported in humans. Variants resulting in "a gain-of-function" predispose carriers to hypercholesterolemia, whereas *PCSK9* variants resulting in a "loss-of-function" associate with hypocholesterolemia. Genetic variation at *PCSK9* has been reported to significantly affect LDL cholesterol levels in plasma, LDLC lowering response to statins, and risk for premature coronary heart disease.[29–33] The Prospective Study of Pravastatin in the Elderly at Risk (PROSPER) examined in 5,783 elderly, of whom 43% had a history of vascular disease at baseline, and who were randomized to pravastatin or placebo with follow up, two SNPs in *PCSK9*, namely R46L and E640G. In this population 3.5% were carriers of the T allele at R46L, and these subjects had significantly ($p<0.001$) lower levels of LDLC, no difference in LDLC-lowering response to pravastatin, and a nonsignificant 19% unadjusted and 9% adjusted decreased risk of vascular disease at baseline.[34] Berge et al. sequenced *PCSK9* exons in 38 unrelated hypocholesterolemic subjects, 25 unrelated patients with heterozygous familial hypercholesterolemia (FH) with LDL receptor mutations, who were "hyper-responders" to statin therapy, and 441 hypercholesterolemic patients who did not have an LDL receptor mutation. They report that 15.8% (6 of 38) of the hypocholesterolemic patients were heterozygous for one of three *PCSK9* variants, R46L, G106R, or R237W. They also noted that 8.8% (2 of 25) of the heterozygous hyper-responders were carriers of either the R46L or the N157K alleles. They found none of the rare alleles in the 441 hypercholesterolemic patients without LDL receptor mutations. They concluded that these *PCSK9* variants are linked to hypocholesterolemia and possibly to hyper-response to statins.[29] Rashid et al. reported hypersensitivity to statins in mice lacking PCSK9.[35]

In the PROSPER trial the other SNP, the E670G, showed no significant relationship with the endpoints.[34] However other studies did find that the E670G allele at PCSK9 was associated with increased LDLC.[30,32]

To further investigate the effect of statin therapy in relation to PCSK9, one study examined the effect of statins and fibrates on *PCSK9* plasma levels in human subjects. Changes in plasma *PCSK9* following fenofibrate or gemfibrozil treatments (fibric acid derivatives) were inversely correlated with changes in LDLC levels ($p=0.012$). Atorvastatin administration significantly increased plasma *PCSK9* ($p=0.033$) and these changes were inversely correlated with changes in LDLC levels ($p=0.012$). They conclude that plasma *PCSK9* can be used as an accessible and meaningful measurement of its intracellular regulation. In this way, individuals who would benefit most from combination therapy with *PCSK9* inhibitors could be identified.[36]

22.9 Cholesterylesther Transfer Protein

In another exemplary study from the REGRESS trial, we could demonstrate that variation at the cholesterylesther transfer protein (CETP) gene locus is associated with changes in CETP activity and changes in lipid and lipoprotein activity.[37] In the placebo arm of this lipid-lowering regression trial it became obvious that homozygosity for a certain genotype (B1B1) was associated with the highest CETP protein mass and activity and consequently with the lowest HDL cholesterol. In addition, this high-activity CETP genotype led to a faster rate of progression of coronary atherosclerosis in this particular patient group. The notion that high CETP activity, particularly when genetically determined, confers a high risk for the development of atherosclerosis was recently confirmed by the Framingham Offspring Study (FOS) research group.[38] In the FOS cohort, almost a complete Caucasian population, B1B1 CETP genotype was again associated with low HDL cholesterol and a higher incidence of cardiovascular disease. The associations, both in a CAD and in the general population, suggest that variation at the CETP gene influences the risk for premature CAD, at least in Caucasians.

Far more important, in the pravastatin arm of the REGRESS trial a different picture emerged. Patients carrying the B1B1 genotype reacted more favorably to statin therapy when compared to their counterparts with the B2B2 genotype. This pharmacogenomic observation led to the concept that lowering CETP activity that is already low, that is, in the case of a B2B2 genotype, is possibly unwanted and does not lead to regression of coronary atherosclerosis. This notion was strengthened by the observation that pravastatin influenced lipids and lipoproteins to a similar extent in all CETP genotypes and decreased CETP mass activity by approximately 20% in all REGRESS patients.[37] Furthermore, recent results of the 10-year follow-up study in the REGRESS cohort showed that the CETP B2B2 genotype adversely affects clinical outcome when a statin was used, when compared to B1B1 carriers. The hazard ratio per B2 copy was 1.59 for atherosclerotic disease death.[39] If these data are confirmed in other trials with clinical endpoints, CETP genotyping could become an important decision point in the prescription of statins.

Variations in genes encoding microsomal triglyceride transfer protein (MTP), lecithin–cholesterol acyltransferase (LCAT), hepatic lipase (HL), and LPL are currently studied to extend on these pharmacogenomic observations in those patients eligible for statin therapy.

22.10 Apolipoprotein E

Another intensively investigated gene in lipid metabolism is the apolipoprotein E gene (*ApoE*). *ApoE* is 299 amino acids long, and it has multiple roles in lipid and lipoprotein metabolism: for example, it binds to LDLR and mediates the liver uptake of chylomicron remnants, VLDL, and intermediate-density lipoproteins (IDL). It is synthesized principally in the liver, but has also been found in other tissues such as the brain, kidneys, and spleen. *ApoE* is a polymorphic protein with three major isoforms, ε2, ε3, ε4, resulting from cysteine–arginine interchanges at residues 112 and 158. The most common isoform, ε3, contains cysteine at position 112 and arginine at 158, while ε2 contains cystein at both positions and ε4 contains arginine at both positions. The three *ApoE* isoforms have different affinities for the *ApoE*-binding receptors of cell surfaces. The binding of ε 2 to LDLR is ineffective, owing to cysteine at amino acid position 158. This leads to a delayed clearance of *ApoE* ε2-bearing particles from the circulation, which in turn results in upregulation of the synthesis of 3-hydroxy-3-methylglutaryl coenzyme A (HMG-CoA) and LDLR, with a net effect of typically lower-plasma LDL cholesterol (LDL-C). Conversely, ε4 has a faster clearance from plasma than ε3, leading to downregulation of both HMG-CoA and LDLR, causing in aggregate an increase in the plasma LDL-C concentration, which is associated with a higher risk of atherosclerosis and cardiovascular mortality. Therefore, statins may be less effective in reducing cholesterol levels in *ApoE* ε4 carriers, as they already have low HMG-CoA reductase levels. Conversely, patients with the *ApoE* ε2 genotype may especially profit from statin therapy. A meta-analysis reviewing 48 clinical studies stated that *ApoE* ε4 carriers had a 42% higher risk of CAD in comparison with ε3/ε3 individuals, whereas the ε2 allele did not significantly associate with CAD risk. The detrimental role of the *ApoE* ε4 allele is mainly thought to be due to its association with a poor lipid profile (high LDL-C and low HDL-C). In general, the literature supports the notion that ApoE ε3 homozygotes get a larger percentage of benefit from statin treatment than *ApoE* ε4 homozygotes in terms of LDL-C decrease. In addition, those with the *ApoE* ε2 alleles have an even greater reduction in LDL-C during statin medication than ε3 homozygotes. These differences remain significant even after adjustment for age and baseline LDL concentration, both of which are proven important factors in the response to statins. ε2 carriers more frequently achieve treatment goals set than ε4 carries. As mentioned earlier, ε2 carriers possess enhanced activity of HMG-CoA activity and blockage of this activity by statins logically leads to a greater treatment response than for ε3 homozygotes. Conversely, the lower basal HMG-CoA activity of ε4 carriers diminishes the potential of statin therapy. Interestingly, less compliance and a 2.3-fold higher risk of discontinuation has been associated with ε4 homozygote patients, which could be explained by less therapeutic efficacy in these individuals.[40] In conclusion, it can be assumed that ApoE genotyping may help to identify patients with good treatment response and filter subgroups prone to discontinue statin therapy.[41]

In contrast to these results however, a substudy of the Scandinavian Simvastatin Survival Study (4 S) demonstrated that myocardial infarction survivors with the ε4 allele have a nearly twofold increased risk of dying compared with other patients, but that this excess mortality could be abolished by treatment with simvastatin. This effect was probably not due to differences in lipid levels, but could be due to the inhibition of the progression of lesions and/or the stabilization of vulnerable plaques.[42]

22.11 Adverse Drug Reactions

As mentioned above, identification of persons at increased risk of adverse reactions to a specific drug or treatment would enable the physician to reduce costs, morbidity, and perhaps even mortality due to toxicity. In rare cases, statins can cause muscle pain or weakness in association with elevated creatinine kinase levels (myopathy), and occasionally, this leads to muscle breakdown and myoglobin release (rhabdomyolysis) with the risk of renal failure and death. Recently, the SEARCH Collaborative Group has identified common variants in SLCO1B1 that are strongly associated with an increased risk of statin-induced myopathy.[43] SCLO1B1 encodes the organic anion-transporting polypeptide OATP1B1, which has been shown to regulate the hepatic uptake of statins. This genome-wide association study was performed using approximately 300,000 markers in 85 subjects with definite or incipient myopathy and 90 controls, all of whom were taking 80 mg of simvastatin daily as part of a trial involving 12,000 participants. This study showed a single strong association of myopathy with the rs4363657 SNP located within SLCO1B1 on chromosome 12. The noncoding rs4363657 SNP was in nearly complete linkage disequilibrium with the nonsynonymous rs4149056 SNP, which has been linked to statin metabolism. The prevalence of the rs4149056 C-allele in the population was 15%. The odds ratio for myopathy was 4.5 (95% C.I. 2.6–7.7) per copy of the C-allele, and 16.9 (95%C. I. 4.7–61.1) in CC as compared with TT homozygotes. More than 60% of the myopathy cases could be attributed to the C variant. The association of rs4149056 with myopathy was replicated in another trial where patients received 40 mg of simvastatin daily, which also showed an association between the rs4149056 and the cholesterol-lowering effects of simvastatin. None of the SNPs in any other region was clearly associated with myopathy. So genotyping these variants may help to achieve the benefits of statin therapy more safely and effectively.[43]

22.12 ACE Insertion/Deletion Polymorphism

The renin–angiotensin–aldosterone system (RAAS) plays a major role in the development and progression of cardiovascular diseases by promoting vasoconstriction, sodium reabsorption, cardiac remodeling, norepinephrine release, and other potential detrimental effect. Angiotensin-converting enzyme (ACE) is the regulating enzyme in the formation of angiotensin II (Fig. 22.4).

ACE-inhibitors and angiotensin II type 1-receptor (AR1R) blockers are recommended for managing cardiovascular diseases, such as hypertension and heart failure, since inhibitors of ACE have been shown to effectively lower blood pressure and improve prognosis in heart failure patients. Furthermore, certain ACE inhibitors seem to have an anti-atherosclerotic effect. However, there is substantial variability in individual responses to these agents. For example, fewer than 50% of hypertensive patients achieve adequate blood pressure control with ACE inhibitor monotherapy.

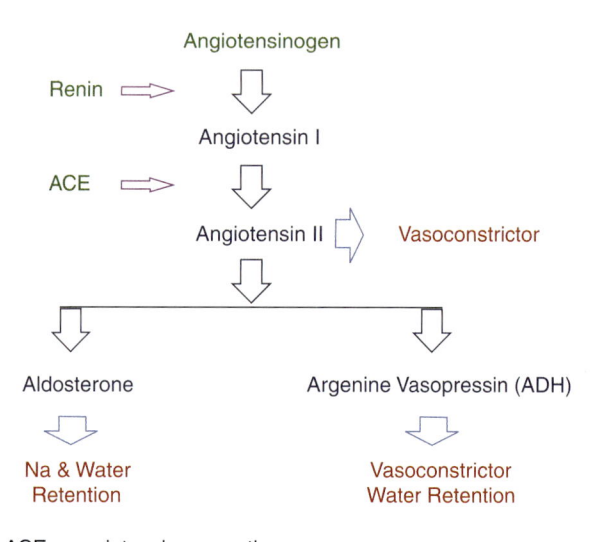

ACE = angiotensin converting enzyme
ADH = antidiuretic hormone

Fig. 22.4 Schematic overview of the renin angiotensin aldosterone system

Variations in the genes encoding, ACE, angiotensinogen, and AT1R have been associated with RAAS activity, suggesting that they may influence responses to RAAS antagonists.[37]

In 1992, Cambien et al. reported on an insertion/deletion (I/D) in intron 16 of the DCP1 or ACE gene.[44] In this polymorphism, a fragment of 287 base pairs is either present or absent. The DD genotype emerged as a potent risk factor for myocardial infarction, since the ACE D allele has been associated with higher plasma and tissue angiotensin II levels and greater AT1R expression than the I-allele. Since then, the DD genotype has been implicated as a risk factor for hypertension, ischemic cerebrovascular disease, multivessel coronary artery disease, left ventricular hypertrophy, elite athletic performance, and response to physical training and restenosis after percutaneous coronary interventions, particularly after coronary stenting.

However, a meta-analysis by Agema et al. concluded that a clinically significant association of the ACE polymorphism with restenosis after percutaneous transluminal coronary revascularization in patients is unlikely.[45] Most likely the results are distorted by publication bias.[45] Furthermore, recently, another meta-analysis of the ACE gene polymorphism in cardiovascular disease did not show any association of the DD genotype with myocardial infarction, ischemic heart disease, or ischemic cerebrovascular disease.[46] However, the ACE gene did affect plasma ACE activity. Although it seems likely that an effect would be more evident in highly selected subgroups (e.g., young people), much doubt has been generated whether or not this polymorphism is the causative mutation or just a mere genetic marker in linkage disequilibrium with another mutation. In fact, recently Rieder et al. sequenced 24kb of the ACE gene in 11 individuals, identifying 78 varying sites. It became evident that 17 variant sites were in absolute linkage disequilibrium with the insertion/deletion polymorphism, implicating that each of these sites might be the actual pathophysiologically important mutation.[47] Thus, cautious interpretation of a phenotypic association with a single SNP is warranted.

Despite the wobbly theoretical basis, pharmacogenetic studies have been undertaken to evaluate the efficacy of ACE inhibitors in prevention of in-stent restenosis after percutaneous coronary interventions. A Japanese, a Danish, and an Italian study have been presented, in which there was diversity of the genotype exhibiting the highest risk of in-stent restenosis (II genotypes bearing the highest risk in Japanese, DD genotypes in Caucasians) and the efficacy of ACE inhibitors (not effective in Danish, effective in Italians).[48] The role of this gene in hypertension also showed conflicting results. In Greek hypertensive patients, reductions in systolic and diastolic blood pressures were significantly greater for patients with the DD genotype than for patients with the ID and II genotypes. In contrast, among Japanese hypertensive patients treated with ACE inhibition treatment, diastolic blood pressure tended to decrease more for the ACE II genotype than for other ACE genotypes. Differences in agents used and in the duration of drug exposure and differences in study population may all have influenced the results.

Another small study showed, in a cohort of patients with systolic dysfunction, that the ACE D allele was associated with a significantly poorer transplant-free survival. The effect was primarily evident in patients not treated with B-blockers and was not seen in patients receiving therapy. Therapy with ACE inhibitors has been shown to increase myocardial B-receptor density, demonstrating the interdependence of sympathetic and renin–angiotensin activation. This finding suggests a potential pharmacogenetic interaction between the ACE ID polymorphism and therapy with B-blockers in the determination of heart failure survival.[49]

Other studies investigated different genes in the RAAS system. They showed that the aniotensinogen 235Met/Thr polymorphism also appears to affect RAAS activity and drug responses. Different studies found the best response to ACE inhibitor monotherapy for hypertension in carriers of the 235Thr allele, which had the highest angiotensinogen concentrations.

In addition to an association with drug efficacy, also for the RAAS polymorphisms ACE inhibitor related adverse effects have been studied. For example, decline in renal function during ACE inhibitor treatment tended to be greater in heart failure patients with the ACE II genotype. The II genotype has also been associated with increased susceptibility to development of cough during ACE inhibitor therapy.[20,37]

Much of the data on RAAS gene polymorphisms and drug effectiveness, particularly with respect to the ACE ID polymorphism, are inconsistent and even conflicting. However, these discordant results are not surprising in view of the complex signaling pathway of the RAAS. Numerous proteins are involved in the RAAS pathway and many of these proteins exhibit functional polymorphisms. While studies of individual RAAS gene polymorphisms are simple to perform it is unlikely that they will yield predictive accuracy sufficient to influence therapeutic regimens.[20] However, since it has been the subject of such a vast amount of research important lessons can be deducted from its results. These lessons result in the guidelines provided below.

22.13 Future Expectations

The above-mentioned examples are only a few in the rapidly increasing number of candidate genes. Genes involved in different processes like vascular wall homeostasis, inflammation, hemostasis, smooth muscle cell proliferation, extracellular matrix homeostasis, and many others will be evaluated in the future. Other interesting aspects of the lipid metabolism, like paraoxonase polymorphisms, will be part of future pharmacogenetic investigations.

If polymorphisms that clearly influence the response to medications become known, it will also be important to evaluate the cost-effectiveness of screening patients for these genotypes. Genotyping should be considered in the following scenarios: (1) when it is difficult to monitor drug response, (2) when there is a strong association between the gene variant and the relevant clinical outcome, (3) when the assay is inexpensive, (4) when the variant allele is common. Future studies with large-scale population will be required before genotyping information can be used to customize patient treatment. These future studies are important since pharmacogenetics has the potential to eliminate the trial-and-error approach to prescribing medications. Ideally, the most effective drug with the best safety profile for a given patient will be chosen on the basis of the patient's genotype. Before pharmacogenetics can routinely be applied to drug therapy decision making in cardiovascular disease management, the genetic polymorphisms or polymorphism combinations that best determine drug therapy response must be identified. Once these associations are discovered and confirmed, the next step will be to design rapid assays for the detection of polymorphisms that are predictive of drug responses and the use of these assays for tailoring drug therapy during clinic visits or at the bedside.

Pharmacogenetics is likely to increase the complexity of drug prescription, since genotype will be considered along with other factors in determining the most appropriate agents for a patient. The potential interactions among multiple polymorphisms, influencing the overall response to each agent, will further complicate pharmacotherapy. To optimize the outcomes of drug therapy, clinicians must know which genetic polymorphisms to test for and how to interpret the results before instituting therapy for a particular patient.[20]

Given the large number of drugs currently available for the treatment and prevention of CVD and the large number of patients eligible to receive these drugs, even small sources of variation in drug efficacy and safety have important implications for public health. For example, withholding a drug from individuals predisposed to a diminished response and/or adverse effect could reduce the financial and personal costs of ineffective or dangerous therapy while ensuring that the remaining drug-eligible subjects receive greater benefit and safety. Pharmacogenetics also has the potential to influence drug development and clinical trial design to achieve greater efficiency, thereby allowing a larger number of more effective and cheaper drugs to be available for treatment and prevention of CVD.

22.14 Pharmacogenetics and Current Practice

What is the current position of pharmacogenetics in clinical practice? This question can be addressed from different viewpoints. Pharmacogenetics involves clinical medicine and genetics but also ethical issues and industrial interest will play a role. The way to handle current knowledge therefore depends on one's position.

We have summarized several genetic polymorphisms that may play a role in variable response to specific classes of cardiovascular agents. Polymorphisms in genes encoding CETP and ApoE are associated with patients' response to statins. In studies of other polymorphisms, such as those in the genes encoding ACE, findings have been less consistent regarding the relationship between these genetic variations and CAD. Although more data are needed before genetic factors influence treatment decisions, these examples are important. Targeted individualized treatment based on genetic composition may be used to maximize the efficacy and minimize the potential toxicity of medications in the future. Once polymorphisms that best determine responses to particular drugs are known, patients may

22.15 The Pharmaceutical Industry's View

The current knowledge can be used to make clinical trials more effective. Knowledge of responders would make trials more effective thus increasing cost-effectiveness. Identification of drug-sensitive genotypes in phase II studies would enable the design of a phase-III study with responders only, thus allowing smaller trials and preventing nonresponders to be exposed unnecessarily to a risk of adverse drug reactions. On the other hand, the industry often was not interested in pharmacogenetic research, since the results might limit sales if a particular drug turns out to have a strong effect in a certain subpopulation only, with no or little effect in the remaining population (Table 22.1).

One step further, considering that rare adverse drug reactions frequently emerge only after introduction of a drug to the market, enhanced surveillance systems after market introduction might incorporate DNA storage of all individuals, who receive this newly introduced drug. This would quickly and reliably enable the identification of SNP marker profiles once a rare adverse drug reaction has been encountered.[11,13,50]

At present, these stimulating ideas are only theory. Industry, regulatory authorities, and governments should explore the details. As long as the fear for genetic research in medical ethic committees and among politicians prevails above the insight that this approach both enhances safety and cost-effectiveness, and thus the cost of health care, pharmacogenomic clinical trials are elusive. Furthermore, there is a need for incentives to promote the development and implementation of a pharmacogenetic approach to clinical therapeutics because agencies may be uninterested in strategies designed to restrict the use of products to a select subgroup of patients.

22.16 The Geneticist's View

Polymorphisms are frequent throughout the genome and occur in both coding and regulatory segments of a gene as in noncoding areas of which the significance is unknown.[10] Multiple single gene disorders have been identified in which single mutations by themselves may be sufficient to cause disease. In multifactorial or multigenetic disorders like atherosclerosis the impact

Table 22.1 Common pharmacogenetics in human drug metabolizing enzymes (Adapted from Weber ref.[56])

Gene	Phenotype	Frequency in different ethnic groups	Total no of drugs	Examples
Thiopurine methyltransferase (S-methylation)	Poor metabolizer	Low in all populations	<10	6-Mercaptopurine, 6-thioguanine, azathioprine
Cytochrome P450 (drug oxidation):				
CYP2D6	Poor metabolizer	Caucasian 6%, African-American 2%, Asian 1%	>100	Codeine, nortryptiline, dextromethorphan
	Ultra-rapid metabolizer	Ethiopian 20%, Spanish 7%, Scandinavian 1.5%		
CYP2C9	Reduced activity		>60	Tolbutamide, diazepam, ibuprofen, warfarin
CYP2C19	Poor metabolizer	Asian 23%, Caucasian 4%	>50	Mephenytoin, omeprazole, proguanil, citalopram, clopidogrel
N-acetyltransferase	Poor metabolizer	Caucasian 60%, African-American 60%, Asian 20%, Inuit 5%	>15	Isoniazid, procainamide, sulphonamides, hydralazines

of a single polymorphism might be small. However, the same rules of genetics should apply as outlined below. In some genetic association studies confusing results have been found, not in concurrence with accepted genetic theories. To overcome these drawbacks in genetic epidemiological research criteria have been suggested to be met in establishing useful links between genetic variations and diseases.[51,52] First, the change in the gene must cause a relevant alteration in the function or level of a gene product (which is not always a protein). However, most of the time this cannot readily be deduced from sequence alone. It is not always possible to perform functional studies, since the relevant proteins may only expressed in tissues with limited access, and not in blood. Furthermore, effects of genetic variants in inhibitors of certain proteins are not necessarily reflected by serum protein levels. Second, the beneficial and harmful phenotypes must have apparent clinical differences. Third, the hypothesis linking the genotype to disease must be physiologically plausible. Fourth, the number of cases investigated in an individual study linking a genotype to disease must be sufficient to detect even small relative risks. These small differences may still be important since they exert their influence over a lifetime.

In practice it is not always possible to meet with all these requirements. For example, following these criteria the study of the impact of the nitric oxide (NO) synthase gene in the development of atherosclerosis would require the measurement of NO.[53] However, NO is immediately degraded and cannot be measured.

22.17 The Clinician's View

Until now pharmacogenetics has been merely a fascinating field of research. The examples presented above are exemplary studies, which need to be extended into clinical trials and to clear phenotypic endpoints like death, myocardial infarction, or need for revascularization.[54] A nonexhaustive summary of the practical status of these gene polymorphisms is provided in Table 22.2.

Despite the promise of pharmacogenetics, several barriers to translation of pharmacogenetic findings into routine clinical CVD care remain. One of these is that in clinical practice, to date, detailed knowledge of genetics is often insufficient; many practitioners are not familiar with the principles of genetics and the mechanisms by which gene variants might influence clinical decision-making. The clinician should be aware of the promises and above all the drawbacks of pharmacogenetic and genetic association studies in

Table 22.2 Practical use of pharmacogenetics and atherosclerosis treatment

Enzyme	Polymorphism	Allele frequency in healthy Caucasians	Classification[a] Atherosclerosis	Other diseases
Cytochrome P450	CYP2D6	6%	D	A
	CYP2C9	6%	D	
	CYP2C19	4%	D	
N-acetyltransferase		60%	D	
Lipoprotein lipase	Asn291Ser	1–7%	D	
	Asp9Asn	2–4%	D	
	Ser447stop	17–22%	D	
PCSK9	E670G	2–6%	D	
	R46L	3.5%	D	
Cholesterylesther transfer protein	TaqIB (B1 allele)	59.4%	D	
SLCO1B1	rs4149056	15%	D	
Angiotensin-converting enzyme	Deletion allele	52–61%	D	D

[a]"A" denotes molecular diagnostics available and diagnosis relevant for treatment. "D" denotes polymorphism with implications for predisposition, but minor phenotypic consequences

general. Furthermore, at present, technology does not provide the capacity or the cost-effectiveness to test complete patient populations. However, technological advances like high-throughput sequencing studies and the development of DNA-microarray analysis with computerized detection methods are already facilitating the implementation of genome-wide SNP marker screening in clinical trials.[55] Moreover, development of bioinformatics programs will allow combined marker analysis in patients who differ in their response to therapy, both in a positive and negative way.

To definitively show that pharmacogenetic testing has value in clinical practice, it is not enough to simply demonstrate that drug response varies by genotype. There must also be an alternative treatment strategy that could be triggered by knowledge of genotype and proof that testing for the genotype and subsequently tailoring the treatment strategy based on genetic information is clinically more effective or more cost-effective (or both) than merely treating everyone in the usual manner. These will be difficult and costly studies to perform. On the other hand, the total costs of CVD and its treatment in our society are of such magnitude that even modest gains achieved through application of pharmacogenetics to CVD care could have substantial impact on attributable risk and cost.[5]

Finally, the power of genetic techniques and possible implications for insurances demand strict guidelines for the use of genetic specimen and profound ethical consideration. Coding of databases and other safety features to warrant privacy to the individual are of paramount importance. Therefore, genetic studies are frequently hampered by difficult negotiations with ethical committees. The lack of an effective treatment is a frequent argument for limiting diagnostic and prognostic tests. However, disease susceptibility gene polymorphisms are much less discussed and certainly do not have the same impact. Moreover, restrictive regulations for genetic studies might inadvertently impede our ability to prescribe safe and effective medication. Therefore, clinicians should understand the importance to stress the necessity of pharmacogenetic research.

22.18 Summary

Atherosclerosis is a multifactorial disease with a multigenetic background. Pharmacogenetics is the search for genetic polymorphisms that affect responses to drug therapy. So far many associations have been found between genetic polymorphisms and responses to cardiovascular drugs. In this chapter, we have given examples of polymorphisms that have been shown to play a role in the response to drug therapy. Once the polymorphisms that best determine the response to a particular drug are known, and tests to rapidly identify these variations are available, individual patients may be screened for genetic polymorphism before drug therapy is begun and the information can be used to choose agents with the greatest potential for efficacy and least potential for toxicity. Much more must be learned, however, before phamacogenetic factors can be routinely incorporated in the therapeutic decisions.

References

1. Murray CJ, Lopez AD. Alternative projections of mortality and disability by cause 1990–2020: Global Burden of Disease Study. *Lancet.* 1997;349:1498-1504.
2. Ross R. Atherosclerosis–an inflammatory disease. *N Engl J Med.* 1999;340:115-126.
3. Wilson PW, D'Agostino RB, Levy D, Belanger AM, Silbershatz H, Kannel WB. Prediction of coronary heart disease using risk factor categories. *Circulation.* 1998; 97:1837-1847.
4. Ridker PM. Evaluating novel cardiovascular risk factors: can we better predict heart attacks? *Ann Intern Med.* 1999; 130:933-937.
5. Arnett DK, Baird AE, Barkley RA, et al. Relevance of genetics and genomics for prevention and treatment of cardiovascular disease: a scientific statement from the American Heart Association Council on Epidemiology and Prevention, the Stroke Council, and the Functional Genomics and Translational Biology Interdisciplinary Working Group. *Circulation.* 2007;115:2878-2901.
6. Marenberg ME, Risch N, Berkman LF, Floderus B, De Faire U. Genetic susceptibility to death from coronary heart disease in a study of twins. *N Engl J Med.* 1994;330: 1041-1046.
7. Jukema JW, Bruschke AV, van Boven AJ, et al. Effects of lipid lowering by pravastatin on progression and regression of coronary artery disease in symptomatic men with normal to moderately elevated serum cholesterol levels. The Regression Growth Evaluation Statin Study (REGRESS). *Circulation.* 1995;91:2528-2540.
8. Katan MB, Beynen AC, de Vries JH, Nobels A. Existence of consistent hypo- and hyperresponders to dietary cholesterol in man. *Am J Epidemiol.* 1986;123:221-234.
9. Wang XL, Sim AS, Badenhop RF, McCredie RM, Wilcken DE. A smoking-dependent risk of coronary artery disease associated with a polymorphism of the endothelial nitric oxide synthase gene. *Nat Med.* 1996;2:41-45.

Genetics of (Premature) Coronary Artery Disease

Heribert Schunkert and Jeanette Erdmann

23.1 Introduction

Coronary artery disease (CAD) and its major complication myocardial infarction (MI) mark the clinical manifestation of a chronic pathomorphological process in the vascular wall that originates in the interaction of multiple genetic and environmental factors. Likewise, a multifactorial etiology applies to many of the underlying cardiovascular risk factors including hypercholesterolemia, hypertension, diabetes mellitus, and smoking addiction. Thus, endogenous (genetic) and exogenous (nutrition, physical activity, therapy, etc.) mechanisms all affect the development of atherosclerotic lesions either directly in the arterial wall, or indirectly via traditional risk factors, or interactively by augmentation or amelioration of other players in the game. On a cellular level, atherosclerosis is also a complex process characterized by endothelial dysfunction, lipid and matrix accumulation, migration and local transformation of circulating cells, smooth muscle cell (SMC) proliferation, calcification, inflammation, and, finally, thrombus formation. In this scenario, the potential involvement of genetically modulated mechanisms may occur at multiple facets of the disease (Fig. 23.1).

For long, evaluation of the family history served as a guide to approach a patient's genetic risk for coronary events. While a positive family history is found in 20–30% of cases, modern molecular genetics reveal that genetic variants affecting the risk of coronary artery disease may be fairly common in our population. Indeed, the increasing number of risk alleles identified since the year 2007 implies that basically all individuals share a variable number of genetic risk factors. For example, 75% of Western Europeans carry at least one variant of the chromosome 9p21.3 risk allele, which increases the probability of coronary artery disease by 25% irrespective of whether the family history is positive.[1] Thus, genetic factors may play a variable role in basically all cases of coronary artery disease even when the family history is negative. Nevertheless, rare families with multiple affected members may allow identification of specific molecular gene defects with profound implications and thereby novel targets for risk prediction and a better understanding of pathophysiology.

23.2 Importance of Family History

For the time being, the assessment of family history is fundamental for approaching the genetic components in the complex disease processes leading to coronary artery disease. Particularly, a familial predisposition is assumed, when myocardial infarction is diagnosed before 55 years in a male first-degree relative or before 65 years in a female first-degree relative. The Framingham Heart Study revealed that such positive family history for premature myocardial infarction increases the risk by a slightly different extent depending on parental premature CAD (1.45-fold) or sibling CAD (1.99-fold). Moreover, the familial risk was found to increase with a decrease in the ages of onset of disease in the affected relatives.[2–4] To a lesser degree, genetic effects regarding myocardial infarction risk can be traced in affected second-degree relatives.[5]

H. Schunkert (✉)
Universität zu Lübeck, Medizinische Klinik II, Ratzeburger Allee 160, D-23538 Lübeck, Germany
e-mail: heribert.schunkert@uk-sh.de

Fig. 23.1 Key players in coronary artery disease and potential interference of genetic variants. Chromosomal loci listed have been demonstrated to be highly significantly associated with coronary artery disease ($P < 5 \times 10^{-8}$, i.e. genome-wide significance). *symbolizes modulation by multiple genetic factors

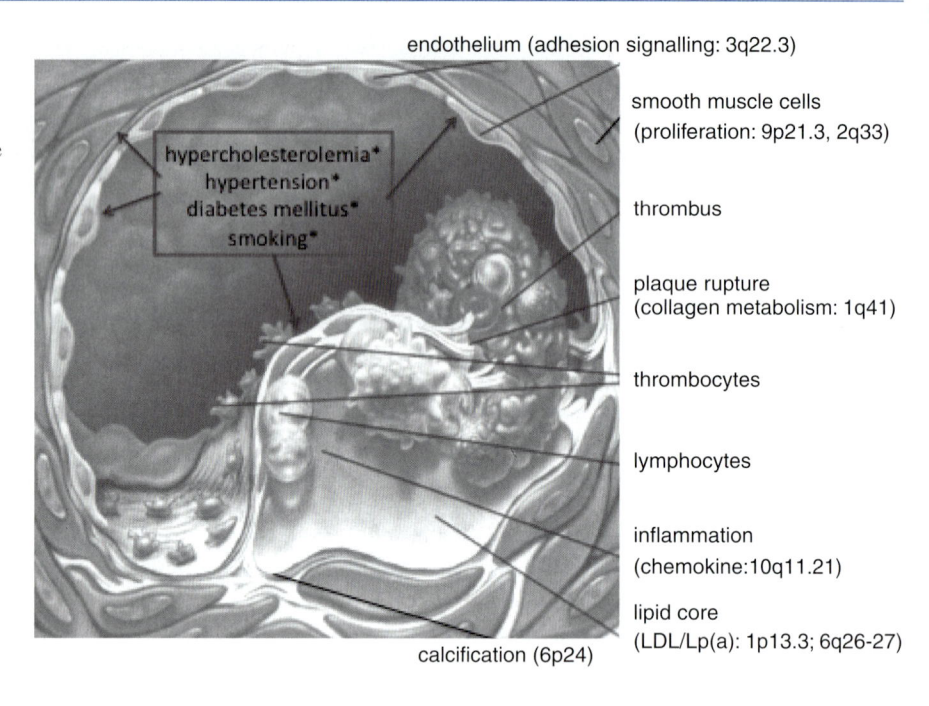

endothelium (adhesion signalling: 3q22.3)

hypercholesterolemia*
hypertension*
diabetes mellitus*
smoking*

smooth muscle cells (proliferation: 9p21.3, 2q33)

thrombus

plaque rupture (collagen metabolism: 1q41)

thrombocytes

lymphocytes

inflammation (chemokine: 10q11.21)

lipid core (LDL/Lp(a): 1p13.3; 6q26-27)

calcification (6p24)

Fig. 23.2 The relative increase in risk of myocardial infarction/CAD is shown in relation to different familial backgrounds. The risk for identical and nonidentical twins is based on the hypothesis that the partner twin had died of myocardial infarction at an age of 55 years

In families with several affected family members, traditional cardiovascular risk factors are often found with increased frequency.[6] Furthermore, lifestyle habits associated with a raised incidence of myocardial infarction (e.g. smoking) are more frequently shared in affected family members. Interestingly, the Northwick Park Heart Study as well as the Reykjavik Cohort Study both revealed that the increase in risk in terms of a positive family history remains to be highly significant (odds ratio 1.5–1.8) even after adjustment for traditional risk factors.[7,8] Thus, the excess risk related to a positive family history is partially independent of traditional risk factors suggesting that unrelated mechanisms may be causative in this respect.[9]

Furthermore, a high rate for reoccurrence of myocardial infarction was found in identical twins of myocardial infarction patients. Such form of a positive family history was related to an eightfold increased probability to die of myocardial infarction before the age of 55 years, when the twin was affected at an early age as well[10] (Fig. 23.2). The highest risk related to family history, however, is found in rare families with an autosomal-dominant pattern of inheritance for myocardial infarction.[11, 12]

23.2.1 Familial Forms of Coronary Artery Disease

Some families present with an extremely high prevalence of CAD/MI. With the exception of two large families studied by Wang et al.[11] and Mani et al.[12] (see next section), many of such families could not be systematically analyzed genetically due to the high lethality of the disease. In the German Myocardial Infarction Family Study, we specifically looked for myocardial infarction in large families with at least four surviving affected individuals. Overall, these families represent less than 0.1% of cases with myocardial infarction.[13] On the basis of analyses on 19 family pedigrees and statistical simulations, the presence of an autosomal-dominant inheritance pattern was probable in all cases.[13, 14] The family pedigrees will hopefully extend the knowledge of genes involved in myocardial infarction in the near future (Fig. 23.3).

23.2.2 MEF2A

Wang and co-workers recently succeeded in identifying a mutation in the gene of the transcription factor MEF2A in a family with an autosomal-dominant form of myocardial infarction. For the first time, a familial genetic defect was shown to give rise to myocardial infarction in humans.[11] A 21-bp deletion in the gene appeared to result in alterations of the coronary walls, thus favoring plaque deposition, which ultimately may lead to myocardial infarction. Interestingly, the same pathway is crucial in preventing apoptosis in endothelial cells and death due to vascular obstruction in mice. However, at present the significance of this gene with respect to the CAD/MI morbidity in humans is still unclear, given the fact that several studies showed no association between single nucleotide polymorphisms (SNPs) in the *MEF2A* gene and coronary artery disease/myocardial infarction in other families or large case-control studies.[14]

23.2.3 LRP6

In 2007, Mani et al. described a large Iranian family segregating autosomal dominant coronary artery disease with hyperlipidemia, hypertension, type 2 diabetes, and osteoporosis. They identified a C-to-T transition in exon 9 of the LRP6 gene, resulting in an arginine-to-cysteine substitution at codon 611 (R611C). The index case in this family was found to be homozygous for this mutation, while all other affected family members were heterozygous. However, even heterozygous subjects manifested early coronary artery disease and metabolic syndrome. Expression of LRP6 containing this mutation in NIH3T3 cells showed a 49% reduction of Wnt signaling compared with that of wild-type

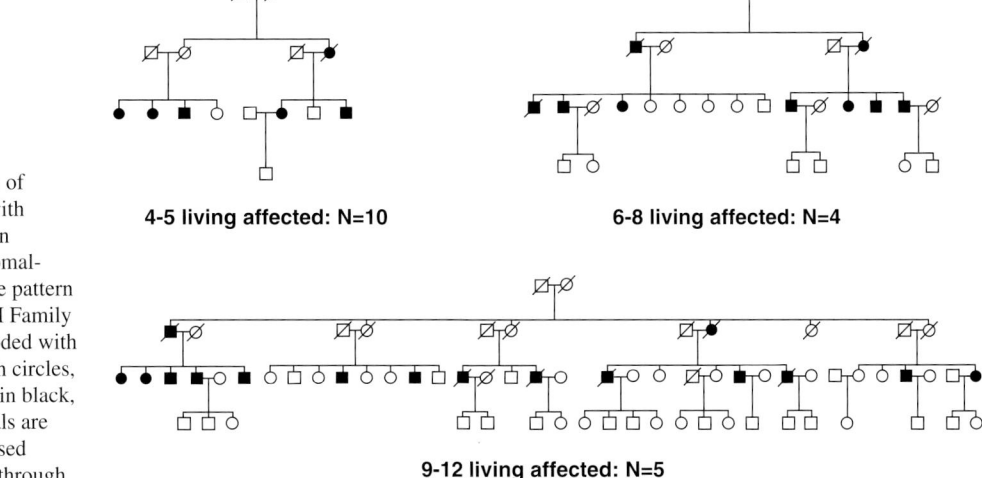

Fig. 23.3 Examples of multiplex families with myocardial infarction presenting an autosomal-dominant inheritance pattern from the German MI Family Study; men are encoded with squares, women with circles, affected individuals in black, unaffected individuals are bordered, and deceased persons are crossed through

4-5 living affected: N=10

6-8 living affected: N=4

9-12 living affected: N=5

LRP6. The addition of low doses of Wnt3a also demonstrated markedly reduced signaling of LRP6 carrying the R611C mutation.[12]

23.3 Heritability Estimates of Coronary Artery Disease

The classic measure of the genetic component for a phenotype (trait), termed "heritability" is defined as the percentage of the total variance of the trait that is explained by inheritance. By examining the increased similarity of trait values in related individuals as compared with unrelated or less-related individuals, one can estimate the heritability. The simplest conceptual study design in man is the comparison of monozygotic (MZ) and dizygotic (DZ) twins. Thus, MZ twins share 100% of their genes, whereas DZ twins share, on average, 50% of their genes. If a trait has a genetic component, MZ twins are more likely to resemble each other than DZ twins. Table 23.1 lists heritability estimates for myocardial infarction and various risk factors. Because of assumptions that are required for estimates of heritability, the calculated numbers must be considered to be rough proxies. Particularly, the high prevalence of risk alleles in apparently healthy subjects within a population may result in an underestimation of the true role of genetic factors involved.

23.4 Heritability of Coronary Anatomy and Pathology

We recently demonstrated that the heritability estimates of coronary artery disease depend in part on the pattern of coronary morphology. Particularly, left main disease and proximal coronary artery stenoses displayed high reoccurrence rates in affected siblings. The heritability estimate for ostial and proximal coronary stenoses was found as $h^2=0.32$, indicating that about one third of the variability of this phenotype is explained by genetic factors ($p=0.008$). Likewise, a highly significant heritability was found for the ecstatic form of coronary atherosclerosis and extraluminal calcification of the coronary arteries as well as the abdominal aorta.[15] Thus, in addition to family history, knowledge of the coronary pathology in an affected family member may enhance risk prediction in first-degree relatives from this patient.[16]

23.5 Genes Affecting Coronary Artery Disease

Over the past three decades, a great deal of research has focused on defining the genetic components of myocardial infarction, coronary artery disease, and their risk factors. Initially, this research focused on

Table 23.1 Heritability estimates for myocardial infarction and various risk factors. Heritability estimates, in most cases based on multiple studies, are taken from Jee et al. 2002,[66] King et al. 2002,[67] and Lusis et al. 1998[68]

Band	SNP	Risk-allele frequency	OR (95% CI)	p-value	Gene(s)	References
9p21.3	rs1333049	52%	1.20 (1.16–1.25)	$2.8 \cdot 10^{-21}$	CDKN2A, CDKN2B	20, 22, 23, 41, 24
1p13.3	rs599839	77%	1.13 (1.08–1.19)	$1.1 \cdot 10^{-14}$	PSRC1	20, 41
1q41	rs3008621	72%	1.10 (1.04–1.17)	$1.4 \cdot 10^{-9}$	MIA3	20, 19, 41
10q11	rs501120	84%	1.11 (1.05–1.18)	$9.5 \cdot 10^{-8}$	CXCL12	20, 41
2q33	rs6725887	14%	1.17 (1.11–1.23)	$1 \cdot 10^{-8}$	ALS2CR3	19
3q22.3	rs9818870	15%	1.15 (1.11–1.19)	$7.4 \cdot 10^{-13}$	MRAS	18
6p24	rs12526453	65%	1.13 (1.08–1.17)	$1 \cdot 10^{-9}$	PHACTR1	19
6q26–27	rs2048327 rs3127599 rs7767084 rs10755578	18%	1.20 (1.13–1.28)	$1.2 \cdot 10^{-9}$	LPA	25
12q24.3	rs2259816	36%	1.08 (1.05–1.11)	$4.8 \cdot 10^{-7}$	TCF1	18
21q22	rs9982601	13%	1.19 (1.14–1.27)	$6 \cdot 10^{-11}$	MRPS6, KCNE2	19

23 Genetics of (Premature) Coronary Artery Disease

candidate genes that hypothetically might affect known traits involved in the atherosclerotic process including the renin-angiotensin system, lipoprotein metabolism, inflammation, or coagulation. Many of these candidate gene studies failed replication in consecutive studies. Consequently, novel strategies for gene identification were explored with the beginning of the twenty-first century that allowed the exploration of the entire genome. Without a priori hypothesis, genome-wide linkage analysis searches the entire genome for chromosomal regions shared in affected family members. While these studies allowed identification of several chromosomal loci harboring myocardial infarction genes, these regions were too large for the elucidation of causative genes or molecular variants.[17]

The success came with technological and methodological advances that initiated the advent of genome-wide association (GWA) studies and thus a new era of the exploration of coronary artery disease and myocardial infarction. Most recently, these investigations identified multiple gene variants reproducibly to be associated with coronary heart disease, hypercholesterolemia, or diabetes mellitus. Surprisingly, most of the genes identified thus far were not expected to play a role in the development of atherosclerosis.[18-25] Thus, an important task for the immediate future is to understand the fundamental pathophysiological mechanisms affected by these genes. Another difficulty in this research is that, unlike Mendelian traits, genetic studies on complex cardiovascular disorders are complicated by variable cosegregation between the risk allele and the disease. In fact, many genetic variants associated with the disorders were found to be relatively common in the overall population and therefore – albeit to a variable degree – prevalent in both healthy and affected individuals.

Accordingly, functional information on these genetic factors and related gene expression as well as protein expression patterns is very much in need. Subsequently, genetic research may enhance diagnostic testing and development of new treatment targets.

Table 23.2 Compilation of CAD/MI risk variants identified by genome-wide association studies

Trait	Heritability estimate
Myocardial infarction	25–60%
Total cholesterol	40–60%
Low-density lipoprotein cholesterol	55–80%
High-density lipoprotein cholesterol	45–75%
Total triglycerides	40–80%
Body mass index	25–60%
Systolic blood pressure	50–70%
Diastolic blood pressure	50–65%
Lipoprotein [a] levels	90%
Homocysteine levels	45%
Type 2 diabetes	40–80%
Fibrinogen	20–50%

completed.[18-20, 22, 23, 25] These studies identified a total of ten chromosomal loci unequivocally associated with this phenotype (Table 23.2, Fig. 23.4).[18-20, 25] Most excitingly, all studies revealed uniformly that a chromosomal locus on 9p21.3 confers the strongest association with CAD and MI (Table 23.2).

23.7 Chromosome 9p21.3

Each C-allele of SNP rs1333049 (G/C, MAF 0.46 in HapMap CEU), representing the CAD/MI locus on 9p21.3, is associated with a 25% higher risk for CAD/MI (95% CI [1.16 to 1.35]). The high frequency of the risk allele (about 75% of all individuals in a Caucasian population carry at least one risk allele) explains why the proportion of CAD/MI risk that can be attributed to carrying the rs1333049 C-allele is fairly high (22% for rs1333049), even after adjustment for cardiovascular risk scores.[20]

23.6 Genome-Wide Association Studies for CAD and MI – Novel Insights

Currently, commercially used DNA-arrays allow evaluation of up to 2,400,000 SNPs for statistical analysis. Since 2007, seven independent GWA studies on coronary artery disease and myocardial infarction were

23.7.1 Pleiotropic Effects of Chromosome 9p21.3

Further data show that this locus not only affects CAD/MI risk but also affects the risk of abdominal aortic aneurysm (AAA), intracranial aneurysm, peripheral

Fig. 23.4 Schematic drawing of the human chromosomes. Genetic loci identified by GWAS for CAD and MI are marked

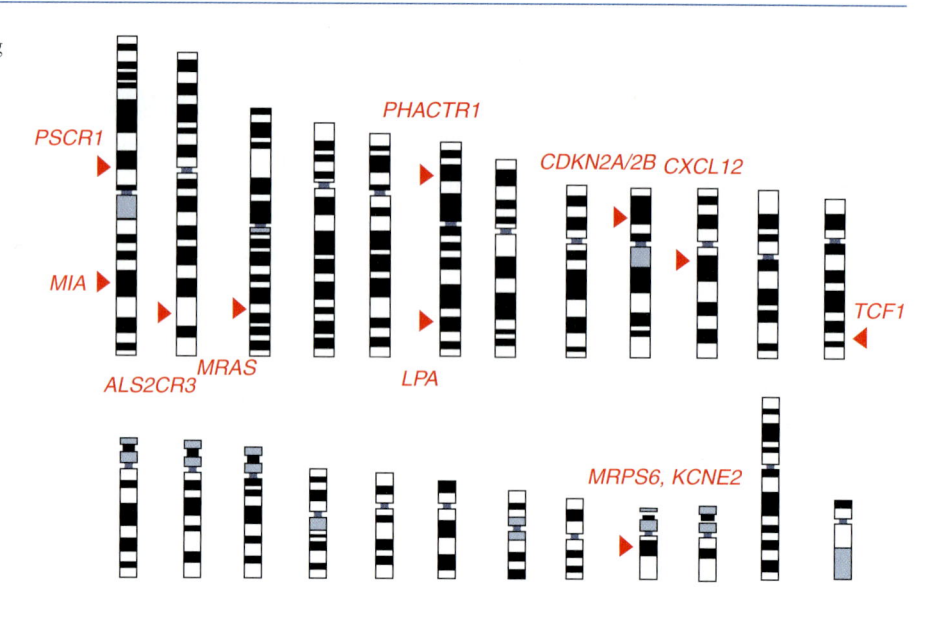

arterial disease, and cardioembolic stroke in many populations.[26] In addition, Matarin et al. (2008) and Geschwendtner et al. (2009) reported both that 9p21.3 region represents a major risk locus for atherosclerotic stroke.[27, 28] The effect of this locus on stroke appears to be independent of its relation to coronary artery disease and other stroke risk factors and further support a broad role of the 9p21 region in arterial disease.

Recently, additional associations have been reported between common variants located in 9p21.3 and a broad range of phenotypes not directly connected to atherosclerosis. The spectrum of these diseases reaches from periodontitis to different human cancers including glioma, basal cell carcinoma, and familial melanoma.[29, 30] Indeed, the CDKN2A/2B tumor suppressor genes encode critical regulators of cell cycle and/or apoptosis.[31] Of note, the CAD/MI risk haplotype seems not to overlap completely with the cancer locus. While the associations of multiple phenotypes with 9p21.3 are only descriptive by nature, functional studies will help to unravel the secrets behind these findings.

The same locus on chromosome 9p21.3 was also reported to be associated with type 2 diabetes mellitus (T2DM) in 3 out of 5 GWA studies on this phenotype.[32–34] However, more detailed studies revealed that indeed neighboring linkage disequilibrium (LD) blocks but not the same SNPs are responsible for T2DM (a risk factor for coronary artery disease) and CAD/MI.[35]

23.7.2 Pathophysiology behind Chromosome 9p21.3

First insights into the pathophysiological mechanisms of 9p21.3 in CAD/MI came from Broadbent and coworkers (2007). They reported about a large antisense noncoding RNA gene (ANRIL), which collocates with the CAD high-risk haplotype at chromosome 9p21.3. This gene is expressed in tissues and cell types that are affected by atherosclerosis and is therefore a prime candidate gene for the chromosome 9p21.3 CAD/MI locus.[35] Liu et al. (2009) analyzed the expression of 9p21 transcripts in purified peripheral blood T-cells (PBTL) from healthy probands.[36] They found a significantly reduced expression of all INK4/ARF transcripts (p15(INK4b), p16(INK4a), ARF, and ANRIL) in subjects with CAD, stroke, and aortic aneurysm. Expression of MTAP was not influenced by the genotype. A more detailed analysis by Jarinova et al. (2009) revealed that a conserved sequence within the 9p21.3 locus has enhancer activity shown by reporter gene expression analysis in primary aortic smooth muscle.[37] Furthermore, whole blood RNA expression of short ANRIL variants was increased by 2.2-fold, whereas expression of the long ANRIL variant was decreased by 1.2-fold in healthy subjects homozygous for the risk allele, respectively. Moreover, relevant to atherosclerosis, genome-wide expression profiling demonstrated upregulation of gene sets modulating cellular proliferation in carriers of the risk allele. These

Fig. 23.5 Current understanding of the pathogenetic mechanism underlying the chromosome 9p21 CAD-risk locus. A 58 kb intergenic region is strongly associated with CAD in humans. Visel et al. (2010) deleted a 70 kb orthologous interval in mice using a conventional knock-out strategy (chr4$^{\Delta 70k}$). Subsequently, the expression of two neighboring genes, Cdkn2a and Cdkn2b, was determined and it has been shown that the expression of both genes was significantly decreased due to the deletion of a regulatory element located within the 70 kb interval. The decrease of Cdkn2a and Cdkn2b gene expression leads to an increase of cell proliferation in smooth muscle cells [38]

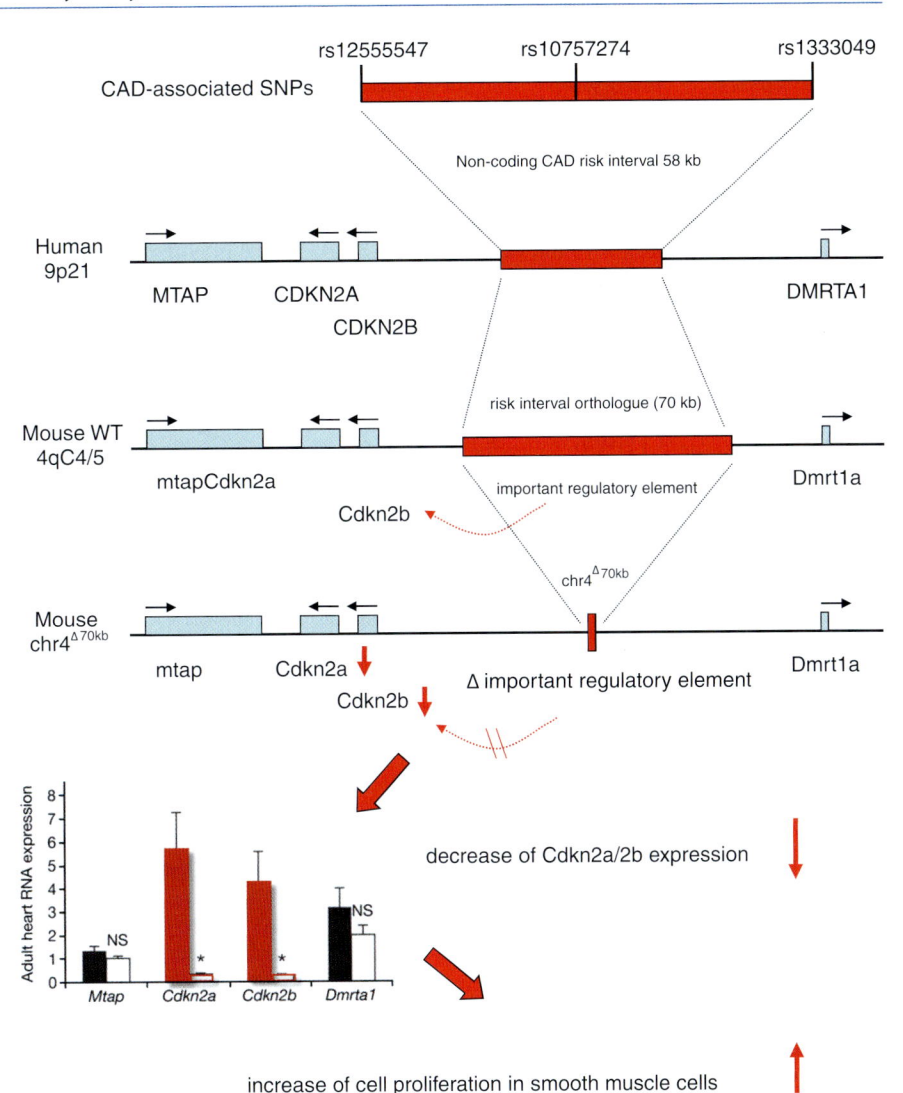

results allow hypothesizing that in risk-allele carriers, the activity of an enhancer element is altered promoting atherosclerosis by regulating expression of ANRIL, which in turn leads to altered expression of genes controlling cellular proliferation pathways.

More definite insights into the pathogenetic mechanism behind the chromosome 9p21 locus were reported very recently by Visel et al. (2010).[38] They showed that deletion of the orthologous 70-kb noncoding CAD risk interval on mouse chromosome 4 affects cardiac expression of neighboring genes, as well as proliferation properties of vascular cells. Particularly, cardiac expression of two genes near the noncoding interval, Cdkn2a and Cdkn2b, is severely reduced in chr4Δ70kb/Δ70kb mice, indicating that distant-acting gene regulatory functions are located in the noncoding CAD risk interval. Primary cultures of chr4Δ70kb/Δ70kb aortic smooth muscle cells exhibited excessive proliferation and diminished senescence, a cellular phenotype consistent with accelerated CAD pathogenesis,[38] (Fig. 23.5).

23.8 Chromosome 1p13.3 – Genetic Elevation of LDL Levels Translates to Risk of CAD/MI

The second most often replicated CAD/MI locus (represented by SNP rs599839) was initially identified through a GWA study for CAD[20] and is located on chromosome 1p13.3.[39–41] Interestingly, this locus has also consistently been associated to LDL-C in several studies.[19, 39, 42–45] The minor allele in European populations (A/G, MAF 0.28 in HapMap CEU) is associated with a lower risk of CAD and lower levels of LDL-C. SNP rs599839 explains about 1% of the variation in circulating LDL-cholesterol levels, equivalent to more established genes for LDL regulation, particularly APOE.

SNP rs599839 lies intergenic within an approximately 97 kb large haplotype block on 1p13.3. This chromosomal region harbors four genes: proline/serine-rich coiled coil protein 1 (PSRC1); cadherin, EGF LAG seven-pass G-type receptor 2 (CELSR2), myosin binding protein H-like (MYBPHL), and sortilin 1 (SORT1).

The hepatic mRNA expression levels of PSRC1, CELSR2, and SORT1 have been shown to correlate with LDL-C plasma levels in a mouse model of cardiovascular disease as well as in a human cohort.[45, 46] The CAD risk allele (A) was associated with lower levels of CELSR2 and SORT1 expression and with higher LDL-C levels. Both genes fall into the category of cell surface receptor linked signal transduction.[46] SORT1 is a transmembrane receptor protein that binds to a variety of different ligands and has been shown to be involved in the endocytosis and intracellular degradation of lipoprotein lipase (LPL)[47] which is a rate-limiting enzyme of triglyceride hydrolysis in lipoproteins. Recently, SORT1 has also been linked to the endocytosis of APO-A-V containing chylomicrons.[48] Recent studies from our group confirmed the association of the G allele of SNP rs588839 with higher sortilin mRNA levels in whole blood RNA.[49] Furthermore, we showed that overexpression of sortilin in transfected cells leads to increased uptake of LDL-particles into these cells. One possible explanation for the association of the chromosome 1p13 variant with LDL-C and CAD might therefore be increased sortilin expression leading to enhanced LDL-uptake into tissues, which in turn results in lower LDL-C levels and subsequently lower risk of CAD.[49]

23.9 Chromosome 1q41 – Link to Collagen Processing

The SNP rs17465637 (C/A, MAF 0.27 in HapMap CEU) located on chromosome 1q41 was identified through a GWA study to be associated with CAD/MI.[20] The common C-allele is associated with a higher risk (OR = 1.14 (95% CI (1.10–1.19)). Three studies were able to replicate this CAD/MI locus in independent samples.[18, 19, 41]

The signal lies within the melanoma inhibitory activity family member 3 (MIA3) gene, alias TANGO. SNP rs17465637, which shows the strongest association at this locus, is located in intron 4 of the gene.[42, 43, 44] MIA3 is required for collagen VII (COL7A1) secretion by loading COL7A1 into transport carriers. MIA3 may participate in cargo loading of COL7A1 at endoplasmic reticulum exit sites by binding to COPII coat subunits Sec23/24 and guiding SH3-bound COL7A1 into a growing carrier. It seems that MIA3 does not play a role in global protein secretion and is apparently specific to COL7A1 cargo loading. However, it may participate in secretion of other proteins in cells that do not secrete COL7A1.[50] The exact role in the pathophysiology of CAD/MI needs further elaboration.

23.10 Chromosome 10q11.21 – Link to EPC Recruitment and Inflammation

GWA studies have also identified genetic variation at the SDF1 gene locus as a risk factor for MI.[20] The lead SNP rs501120 (T/C, MAF 0.17 in HapMap CEU) is located 5′ of the SDF-1-gene with the common T-allele showing a higher risk on MI (OR = 1.33 (95% CI 1.20–1.48)).

SDF-1 (Stromal derived factor 1) is a member of the chemokine family and its presence has been confirmed in atherosclerotic plaques.[51] Overexpression of SDF-1 and its receptor CXCR-4 lead to a stabilization of atherosclerotic plaques potentially through an anti-inflammatory effect and by inhibiting the recruitment of leukocytes into lesion sites.[52] SDF1 is the only known ligand of the CXCR4-receptor, which is predominantly expressed in monocytes and endothelial cells. Biological effects of the SDF1/CXCR4 interaction are the recruitment of endothelial progenitor cells to regions of tissue hypoxemia and the regulation of monocyte inflammatory response. However, so far it is still not definite if the association of the 10q11 CAD variant is causally related to the SDF1/C×CR4 system. Hypothetical explanations for such a functional relationship could be an association of the genetic variant with SDF1-expression levels or receptor affinity, which in turn might have an impact on EPC recruitment under hypoxic conditions or cytokine release in the setting of acute MI.

If one applies the stringent threshold of genome-wide significance of $p < 5 \times 10^{-8}$[53] as the main criterion for the definition of a CAD/MI risk locus, the recent GWA studies identified five additional loci.[18, 19, 25] These loci are located on chromosomes 2q33, 3q22.3, 6p24, 6q26–27, and 12q22; however, one must keep in mind that the independent replication for these loci has to be awaited.

23.11 Chromosome 2q33 – Link to Cell Cycle Progression, Apoptosis

The rare C-allele of SNP rs6725887 (T/C, MAF 0.16 in HapMap CEU), located on chromosome 2q33 within intron 12 of the WDR12 gene, is associated with higher risk for CAD/MI (OR = 1.17 (95%CI (1.11–1.23)).[19] The WDR12 gene encodes a member of the WD repeat protein family (WD40), which are minimally conserved regions of approximately 40 amino acids typically bracketed by gly-his and trp-asp (GH-WD). Members of this gene-family are involved in a variety of cellular processes including cell cycle progression, signal transduction, apoptosis, and gene regulation. Especially, WDR12 is crucial for processing of the 32S precursor ribosomal RNA (rRNA) and cell proliferation.[54] The mechanisms by which WDR12 confers the increased risk of CAD/MI remains to be defined.

23.12 Chromosome 3q22.3 – Link to Adhesion Signaling

Recently, we identified the MRAS gene as a new susceptibility gene for CAD and MI.[18] SNP rs9818870 (C/T, MAF 0.17 in HapMap CEU) is located in the 3´ UTR of MRAS in proximity to a cluster of miRNA binding sites. The rare T-allele is associated with an increased risk of CAD/MI (OR = 1.15, 95% CI (1.11–1.19)).

The M-ras protein belongs to the ras superfamily of GTP-binding proteins and is widely expressed in all tissues, with a very high expression in the cardiovascular system, especially in the heart (SymAtlas). Previous work has shown that M-ras is involved in TNF-α–stimulated LFA-1 activation in splenocytes by using mice deficient in this process.[55] These findings suggest a possible role for M-ras in adhesion signalling, which is important in the atherosclerotic process.

23.13 Chromosome 6p24 – Link to Coronary Calcification

The common C-allele of SNP rs12526453 is associated with higher risk of MI (OR = 1.12 (95% CI 1.08–1.17)).[19] This SNP is located on chromosome 6p24 in intron 5 of PHACTR1 (phophastase and actin regulator 1) gene (C/G, MAF 0.37 in HapMap CEU). PHACTR1 is an inhibitor of protein phosphatase 1, an enzyme that dephosphorylates serine and threonine residues on a range of proteins.[56] Interestingly, the PHACTR1 locus may lead to MI by directly promoting the development of atherosclerosis in the coronary arteries. In an independent GWAS for coronary artery calcification in >10,000 participants from six prospective cohort studies, PHACTR1 SNPs were also found to be associated with coronary artery calcification at genome-wide significance (C.J. O'Donnell, National Heart, Lung and Blood Institute, personal communication 2009). These results could be replicated in our own data from the GerMIFS I (data not shown).

23.14 Chromosome 6q26–27 – Link to Lp(a)

Using a genome-wide haplotype approach, we were able to identify the SLC22A3-LPAL2-LPA gene cluster as a strong susceptibility locus for CAD.[25] Two haplotypes consisting of four SNPs (rs2048327 in the SLC22A3 gene, rs3127599 in the LPAL2 gene, and rs7767084 and rs10755578 in the LPA gene) are consistently associated with CAD/MI risk (CTTG haplotype, OR = 1.2 (95% CI 1.13–1.28); CCTC haplotype, OR = 1.82 (95% CI 1.57–2.12)).

Interestingly, this locus has not been identified from previous genome-wide association (GWA) studies that focused on univariate analyses of SNPs. The proposed approach in the paper by Tregouet et al. may have wide utility for analyzing GWA data for other complex traits. The haplotype association analysis was performed using a sliding-windows approach. While these analyses need enormous computer capacity, the analysis was carried out using the European EGEE grid of computers (http://www.eu-egee.org).

The locus partly overlaps the LPA gene, which encodes apolipoprotein(a), the main protein of lipoprotein(a) (Lp(a)), a well-known risk factor for CAD.[57, 58] Indeed, Tregouet et al. showed that the haplotypes associated with CAD also associated with the highest Lp(a) levels, and after adjustment for Lp(a) levels, these were no longer associated with CAD, suggesting that their relation to risk is mediated by an effect on Lp(a) levels.

Genetic variants, particularly a kringle repeat polymorphism, also affect the size of lp(a) particles, and recent studies suggested that small Lp(a) particle size may be an independent risk marker.[57] In the light of these findings, Clarke and coworkers identified new risk variants (rs10455872 and rs3798220) with low allele frequency but strong effects on CAD, i.e. a 2.5-fold risk increase in individuals who carried at least two of these risk alleles.[59]

23.15 Chromosome 21q22 – No Functional Link Yet Established

At chromosome 21q22, the rare T-allele of SNP rs9982601 (C/T, MAF 0.20 in HapMap CEU) increases the CAD/MI risk (OR = 1.20, 95%C (1.14–1.27)).[19] This SNP is located in an intergenic region between MRPS6 (mitochondrial ribosomal protein S6), SLC5A3 (solute carrier family 5 (inositol transporters) member 3), and KCNE2 (potassium voltage-gated channel, Isk-related family, member 2). MRPS6 encodes a subunit of the mitochondrial ribosomal protein 28S.[60] SLC5A3 is a gene embedded within MRPS6 and encodes a protein that transports sodium and myo-inositol in response to hypertonic stress.[61] KCNE2 encodes a subunit of a potassium channel, and mutations in this gene cause inherited arrhythmias.[62] Currently, it is not yet established which gene in this region is responsible for the increased risk for CAD/MI. However, in the the ENCODE freeze from April 2009 (http://www.genome.gov/10005107), a noncoding transcript has been annotated in this region spanning SNP rs9982601 (AP000318.2). This opens the possibility that the pathophysiological link between the association of rs9982601 and CAD/MI risk is due to changes in this noncoding transcript and not due to changes in any of the three protein-coding genes mentioned earlier.

23.16 Chromosome 12q24.3 – Link to LDL Metabolism

Recently, we identified a genetic variant at chromosome 12q24.3 (rs2259816, G/T, MAF 0.38 in HapMap CEU) to be associated with CAD/MI risk.[18] The rare T-allele increases the risk for CAD/MI (OR = 1.08, 95% CI (1.05,1.11)). SNP rs2259816 is located in intron 7 of HNF1A gene. This and two further associated SNPs (rs1169313 and rs2258287) are in an LD block that covers the coding region of HNF1A and C12orf43. HNF1A (also known as TCF1) encodes a transcription factor that binds to promoters of a variety of genes that are expressed exclusively in the liver.[63] Variants in HNF1A may cause maturity-onset diabetes of the young (MODY) (MIM600496) and affect plasma concentrations of C-reactive protein (CRP), a powerful risk marker for cardiovascular disease.[64] Moreover, a risk allele at the HNF1A locus (rs2258287) has been mapped to higher plasma levels of low-density lipoprotein cholesterol.[40]

23.17 General Lessons from Modern CAD Genetics

A summary of the wealth of new data on genes affecting the risk of MI/CAD loci could read as follows:

1. The precise mechanisms linking chromosomal loci and disease manifestation are often unclear, e.g. the chromosome 9p21.3 region carries no known protein-coding gene.[1]
2. Traditional risk factors do not mediate the risk since the majority of loci display no association with such intermediate phenotypes.[18–23]
3. The genetic risk conferred by the newly discovered loci is independent of the risk conferred by a positive family history. Thus, the molecular-genetic information for risk prediction goes beyond that of all traditional risk factors.
4. All currently identified risk alleles are relatively frequent.[1] It needs to be awaited whether these

high-frequency alleles are causative or tag rare variants with strong effects.
5. Each of the currently known risk alleles increases the probability of CAD by a relatively small margin, i.e. 10–30% per allele. In other words, individuals who are homozygous for the risk allele on chromosome 9p21.3 carry a 50% risk increase when compared with the 25% of the European population, who do not carry this allele.[1]
6. The high frequency of risk alleles, on the other hand, explains why the implications of the recently identified genetic factors at the population level are substantial, even though an affected individual carries only a relative moderate risk increase.[1] This wealth of new information on heritable aspects of coronary artery disease obviously opens multiple windows for scientific exploration. From a clinical point of view, the immediate needs concern risk prediction and (preventive) therapy for atherosclerosis.

23.18 Is Genetic Risk Prediction feasible?

While coronary artery disease is a chronic process, its clinical manifestation may occur suddenly and fatally in the form of a myocardial infarction. Thus, there is a strong clinical demand for predicting the disease onset. A simplified view on genetic risk calculation could be a count of risk alleles similar to the quantitative assessment of cholesterol levels in a population. The underlying assumption, i.e. that risk conferred by some alleles can be balanced by "protective" alleles at other loci, does not take into account that biological mechanisms as well as effect sizes at various loci are likely to be different. Thus, the development of genetic risk scores needs careful prospective testing.

An open question for estimating genetic risk is the definition of a "control sample" free of genetic predisposition. The number of known chromosomal loci for CAD is growing constantly (at the time of writing >10) and the frequency of most such alleles is high. Even in a "healthy" Western European population sample, the average number of currently known risk alleles is >10. Thus, if a group of individuals does not carry a specific risk allele, it cannot be expected that their genetic risk is "zero" but rather at the "population average." This population average is, however, heavily inflated by a multitude of perhaps untested genetically predisposing factors, minus the effect of the tested allele. Vice versa, what is the effect of a traditional risk factor, e.g. smoking, in a person who luckily does not carry any genetic predisposition, i.e. none of the CAD risk alleles? Is it outrageous to hypothesize that coronary artery disease could be eliminated altogether if the effects of susceptibility genes could be entirely neutralized? In this respect, it may be noteworthy that some mammalian species (mice) or vascular beds (internal mammary artery) do not develop atherosclerosis; in other words, even the presence of multiple established risk factors does not automatically result in the manifestation of CAD.

Currently, only a relative limited fraction (<10%) of the overall genetic risk (heritability) of the disease is explained by the identified loci. A part explanation for this relates to the limited power of individual GWA studies to detect such loci. A global consortium (CARDIoGRAM) is analyzing genome-wide information from more than 20,000 cases of CAD and over 60,000 controls and this will undoubtedly identify additional loci harboring even more common variants. Other studies are investigating the role of other forms of genomic variation such as copy number variation (CNV) or with a specific focus on candidate gene (IBC Consortium). Moreover, an increasing effort is being made on elucidating the role of rare variants, which will be aided by novel information on such variants coming out of the 1,000 genome project (http://www.1000genomes.org). In parallel, statistical methods have been developed that make use of SNPs for risk prediction even when their statistical level of association with disease does not reach the conservative genome-wide significance threshold of $p<5\times10^{-8}$. These algorithms take into account that analysis of all SNPs with association at significance levels of $p<5\times10^{-6}$ or $p<5\times10^{-5}$ will include multiple falsely associated SNPs. Yet, the predictive information derived from the large number of remaining truly associated SNP may go far beyond the information derived from the relatively few "established" SNPs.[65] Together, genetic susceptibility for myocardial infarction as well as for related risk factors will soon become more transparent.

In practical terms, the challenge is to utilize the genomic information for the refinement of clinically utilized risk scores. These scores are largely dominated by the predictive information of age and gender and based on prediction of short-term risks. It is obvious that a man in his 70s has a higher risk than a young woman, over the next 10 years, whatever genetic risk burden in these two subjects may carry. The clinically relevant question is, what difference genetic factors will make in refining risk prediction in for example two middle-aged men in order to better target future preventative measures. Epidemiological studies with prospective DNA and data collection are ongoing to address these clinically important issues. These prospective studies need to be awaited to understand by which degree molecular-genetic prediction of coronary artery disease can improve personalized risk assessment beyond that approximated from family history or risk scores such as the Framingham or Euro Score.

23.19 Cardiovascular System Biology

Systems biology offers the potential to provide new insights into our understanding of the pathogenesis of CAD. The rationale for this approach is based on the hypothesis that an interacting network precipitates complex diseases such as atherosclerosis. Players in this network modulate each other at multiple levels including the genome, transcriptome (mRNA and miRNAs), methylome/epigenome, proteome, and metabolome. The challenge is to comprehend the connections and interactions between individual constituents of this network. Specifically, it is of great interest to understand the communication between genetic (SNPs, CNVs) on this place, traditional, and environmental risk factors (SNPs, CNVs) at the level of the cell, tissue, and organ to ultimately describe the entire organism as a system prone to develop disease (Fig. 23.6). The intention is to identify the biological networks that connect the differing system elements, thereby defining the characteristics that describe the overall system. This information can then be used to derive mechanistic information on biological processes as well as identify potential target sites for therapeutic intervention.

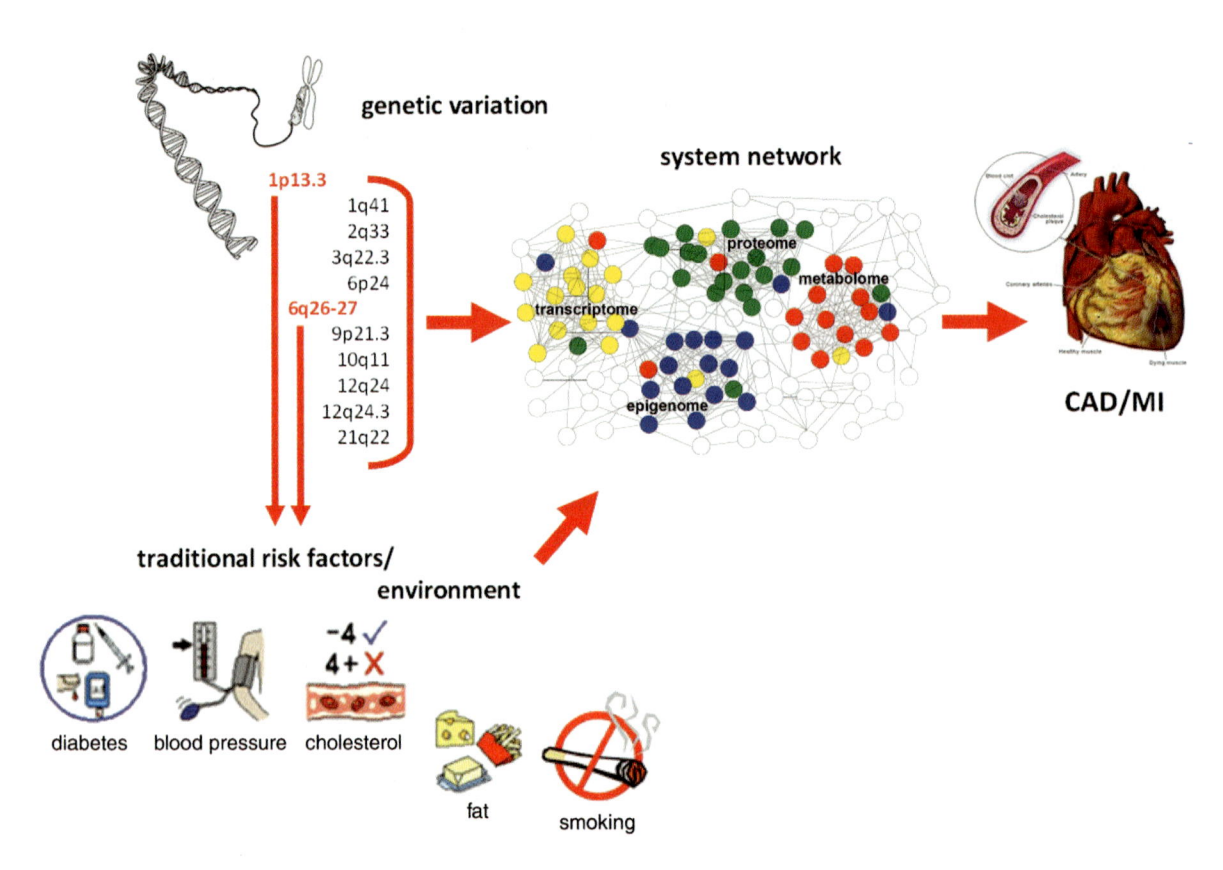

Fig. 23.6 Integrative view of genetic risk variants affecting gene expression or function in the context of traditional risk factors and hitherto unspecified environmental cofactors.[46] Ultimately, biological networks may malfunction resulting in the precipitation of coronary artery disease.

23.20 Conclusion

Molecular genetic approaches applied to coronary artery disease will continue to identify genes and pathways involved in predisposition and pathophysiology of this often life-threatening condition. Moreover, gene expression profiling studies will refine the understanding of the nature of atherosclerotic lesions within vascular wall and promises discovery and validation of targets for therapeutic intervention. Opportunities to transform genetic, genomic, proteomic, and metabolomic information into cardiovascular clinical practice have never been greater, but their implementation into clinical practice requires validation in large independent cohorts, achieved only through collaborative effort. Their continued success will depend on ongoing cooperation within the cardiovascular research community.

References

1. Schunkert H, Götz A, Braund P, et al. Repeated replication and a prospective meta-analysis of the association between chromosome 9p21.3 and coronary artery disease. *Circulation*. 2008;117(13):1675-1684.
2. Murabito JM, Pencina MJ, Nam BH, et al. Sibling cardiovascular disease as a risk factor for cardiovascular disease in middle-aged adults. *JAMA*. 2005;294:3117-3123.
3. Myers RH, Kiely DK, Cupples LA, Kannel WB. Parental history is an independent risk factor for coronary artery disease: the Framingham Study. *Am Heart J*. 1990;120:963-969.
4. Lloyd-Jones DM, Nam BH, D'Agostino RB Sr, et al. Parental cardiovascular disease as a risk factor for cardiovascular disease in middle-aged adults: a prospective study of parents and offspring. *Jama*. 2004;291:2204-2211.
5. Horne BD, Camp NJ, Muhlestein JB, Cannon-Albright LA. Identification of excess clustering of coronary heart diseases among extended pedigrees in a genealogical population database. *Am Heart J*. 2006;152:305-311.
6. Hengstenberg C, Holmer SR, Mayer B, et al. Siblings of myocardial infarction patients are overlooked in primary prevention of cardiovascular disease. *Eur Heart J*. 2001;22:926-933.
7. Andresdottir MB, Sigurdsson G, Sigvaldason H, Gudnason V. Fifteen percent of myocardial infarctions and coronary revascularizations explained by family history unrelated to conventional risk factors. The Reykjavik Cohort Study. *Eur Heart J*. 2002;23:1655-1663.
8. Hawe E, Talmud PJ, Miller GJ, Humphries SE. Family history is a coronary heart disease risk factor in the Second Northwick Park Heart Study. *Ann Hum Genet*. 2003;67:97-106.
9. Boer JM, Feskens EJ, Verschuren WM, Seidell JC, Kromhout D. The joint impact of family history of myocardial infarction and other risk factors on 12-year coronary heart disease mortality. *Epidemiology*. 1999;10:767-770.
10. Marenberg ME, Risch N, Berkman LF, Floderus B, de Faire U. Genetic susceptibility to death from coronary heart disease in a study of twins. *N Engl J Med*. 1994;330:1041-1046.
11. Wang L, Fan C, Topol SE, Topol EJ, Wang Q. Mutation of MEF2A in an inherited disorder with features of coronary artery disease. *Science*. 2003;302:1578-1581.
12. Mani A, Radhakrishnan J, Wang H, et al. LRP6 mutation in a family with early coronary disease and metabolic risk factors. *Science*. 2007;315:1278-1282.
13. Mayer B, Fischer M, Erdmann J, et al. dentification of rare forms of autosomal dominant heritability of myocardial infarction. *Circulation*. 2002;106:II-290.
14. Lieb W, Mayer B, Konig IR, et al. Lack of association between the MEF2A gene and myocardial infarction. *Circulation*. 2008;117:185-191.
15. Fischer M, Broeckel U, Holmer S, et al. Distinct heritable patterns of angiographic coronary artery disease in families with myocardial infarction. *Circulation*. 2005;111:855-862.
16. Fischer M, Mayer B, Baessler A, et al. Familial aggregation of left main coronary artery disease and future risk of coronary events in asymptomatic siblings of affected patients. *Eur Heart J*. 2007;28:2432-2437.
17. Broeckel U, Hengstenberg C, Mayer B, et al. A comprehensive linkage analysis for myocardial infarction and its related risk factors. *Nat Genet*. 2002;30:210-214.
18. Erdmann J, Grosshennig A, Braund PS, et al. New susceptibility locus for coronary artery disease on chromosome 3q22.3. *Nat Genet*. 2009;41:280-282.
19. Kathiresan S, Voight BF, Purcell S, Musunuru K, Ardissino D, et al. Genome-wide association of early-onset myocardial infarction with single nucleotide polymorphisms and copy number variants. *Nat Genet*. 2009;41:334-341.
20. Samani NJ, Erdmann J, Hall AS, et al. Genomewide association analysis of coronary artery disease. *N Engl J Med*. 2007;357:443-453.
21. Soranzo N, Rendon A, Gieger C, et al. A novel variant on chromosome 7q22.3 associated with mean platelet volume, counts, and function. *Blood*. 2009;113:3831-3837.
22. McPherson R, Pertsemlidis A, Kavaslar N, et al. A common allele on chromosome 9 associated with coronary heart disease. *Science*. 2007;316:1488-1491.
23. Helgadottir A, Thorleifsson G, Manolescu A, et al. A common variant on chromosome 9p21 affects the risk of myocardial infarction. *Science*. 2007;316:1491-1493.
24. Wellcome Trust Case Control Consortium. Genome-wide association study of 14, 000 cases of seven common diseases and 3, 000 shared controls. *Nature*. 2007;447:661-678.
25. Tregouet DA, Konig IR, Erdmann J, et al. Genome-wide haplotype association study identifies the SLC22A3-LPAL2-LPA gene cluster as a risk locus for coronary artery disease. *Nat Genet*. 2009;41:283-285.
26. Helgadottir A, Thorleifsson G, Magnusson KP, et al. The same sequence variant on 9p21 associates with myocardial infarction, abdominal aortic aneurysm and intracranial aneurysm. *Nat Genet*. 2008;40:217-224.

27. Matarin M, Brown WM, Hardy JA, et al. Association of integrin alpha2 gene variants with ischemic stroke. *J Cereb Blood Flow Metab.* 2008;28:81-89.
28. Gschwendtner A, Bevan S, Cole JW, et al. Sequence variants on chromosome 9p21.3 confer risk for atherosclerotic stroke. *Ann Neurol.* 2009;65:531-539.
29. Shete S, Hosking FJ, Robertson LB, et al. Genome-wide association study identifies five susceptibility loci for glioma. *Nat Genet.* 2009;41:899-904.
30. Wrensch M, Jenkins RB, Chang JS, et al. Variants in the CDKN2B and RTEL1 regions are associated with high-grade glioma susceptibility. *Nat Genet.* 2009;41:905-908.
31. Sherr CJ. Cell cycle control and cancer. *Harvey Lect.* 2000; 96:73-92.
32. Saxena R, Voight BF, Lyssenko V, et al. Genome-wide association analysis identifies loci for type 2 diabetes and triglyceride levels. *Science.* 2007;316:1331-1336.
33. Zeggini E, Scott LJ, Saxena R, et al. Meta-analysis of genome-wide association data and large-scale replication identifies additional susceptibility loci for type 2 diabetes. *Nat Genet.* 2008;40:638-645.
34. Scott LJ, Mohlke KL, Bonnycastle LL, et al. A genome-wide association study of type 2 diabetes in Finns detects multiple susceptibility variants. *Science.* 2007;316:1341-1345.
35. Broadbent HM, Peden JF, Lorkowski S, et al. Susceptibility to coronary artery disease and diabetes is encoded by distinct, tightly linked SNPs in the ANRIL locus on chromosome 9p. *Hum Mol Genet.* 2008;17:806-814.
36. Liu Y, Sanoff HK, Cho H, et al. INK4/ARF transcript expression is associated with chromosome 9p21 variants linked to atherosclerosis. *PLoS One.* 2009;4:e5027.
37. Jarinova O, Stewart AF, Roberts R, et al. Functional analysis of the chromosome 9p21.3 coronary artery disease risk locus. *Arterioscler Thromb Vasc Biol.* 2009;29:1671-1677.
38. Visel A, Zhu Y, May D, et al. Targeted deletion of the 9p21 non-coding coronary artery disease risk interval in mice. *Nature.* 2010;464:409-412.
39. Samani NJ, Braund PS, Erdmann J, et al. The novel genetic variant predisposing to coronary artery disease in the region of the PSRC1 and CELSR2 genes on chromosome 1 associates with serum cholesterol. *J Mol Med.* 2008;86:1233-1241.
40. Kathiresan S, Willer CJ, Peloso GM, et al. Common variants at 30 loci contribute to polygenic dyslipidemia. *Nat Genet.* 2009;41:56-65.
41. Samani NJ, Deloukas P, Erdmann J, et al. Large scale association analysis of novel genetic loci for coronary artery disease. *Arterioscler Thromb Vasc Biol.* 2009;29:774-780.
42. Muendlein A, Geller-Rhomberg S, Saely CH, et al. Significant impact of chromosomal locus 1p13.3 on serum LDL cholesterol and on angiographically characterized coronary atherosclerosis. *Atherosclerosis.* 2009;206(2):494-499.
43. Huang H, Pan L, Zhang L, Chen Y, Zeng Z. Association of single nucleotide polymorphism rs599839 on chromosome 1p13.3 with premature coronary heart disease in a Chinese Han population. *Zhonghua Yi Xue Yi Chuan Xue Za Zhi.* 2008;25:686-689.
44. Sandhu MS, Waterworth DM, Debenham SL, et al. LDL-cholesterol concentrations: a genome-wide association study. *Lancet.* 2008;371:483-491.
45. Kathiresan S, Melander O, Guiducci C, et al. Six new loci associated with blood low-density lipoprotein cholesterol, high-density lipoprotein cholesterol or triglycerides in humans. *Nat Genet.* 2008;40:189-197.
46. Schadt EE, Molony C, Chudin E, et al. Mapping the genetic architecture of gene expression in human liver. *PLoS Biol.* 2008;6:e107.
47. Nielsen MS, Jacobsen C, Olivecrona G, Gliemann J, Petersen CM. Sortilin/neurotensin receptor-3 binds and mediates degradation of lipoprotein lipase. *J Biol Chem.* 1999;274:8832-8836.
48. Nilsson SK, Christensen S, Raarup MK, Ryan RO, Nielsen MS, Olivecrona G. Endocytosis of apolipoprotein A-V by members of the low density lipoprotein receptor and the VPS10p domain receptor families. *J Biol Chem.* 2008;283:25920-25927.
49. Linsel-Nitschke P, Heeren J, Aherrahrou Z, et al. Genetic variation at chromosome 1p13.3 affects sortilin mRNA expression, cellular LDL-uptake and serum LDL levels which translates to the risk of coronary artery disease. *Atherosclerosis.* 2009;208(1):183-189.
50. Saito K, Chen M, Bard F, et al. TANGO1 facilitates cargo loading at endoplasmic reticulum exit sites. *Cell.* 2009; 136:891-902.
51. Abi-Younes S, Sauty A, Mach F, Sukhova GK, Libby P, Luster AD. The stromal cell-derived factor-1 chemokine is a potent platelet agonist highly expressed in atherosclerotic plaques. *Circ Res.* 2000;86:131-138.
52. Poznansky MC, Olszak IT, Foxall R, Evans RH, Luster AD, Scadden DT. Active movement of T cells away from a chemokine. *Nat Med.* 2000;6:543-548.
53. Dudbridge F, Gusnanto A. Estimation of significance thresholds for genomewide association scans. *Genet Epidemiol.* 2008;32:227-234.
54. Holzel M, Rohrmoser M, Schlee M, et al. Mammalian WDR12 is a novel member of the Pes1-Bop1 complex and is required for ribosome biogenesis and cell proliferation. *J Cell Biol.* 2005;170:367-378.
55. Yoshikawa Y, Satoh T, Tamura T, et al. The M-Ras-RA-GEF-2-Rap1 pathway mediates tumor necrosis factor-alpha dependent regulation of integrin activation in splenocytes. *Mol Biol Cell.* 2007;18:2949-2959.
56. Allen PB, Greenfield AT, Svenningsson P, Haspeslagh DC, Greengard P. Phactrs 1–4: A family of protein phosphatase 1 and actin regulatory proteins. *Proc Natl Acad Sci U S A.* 2004;101:7187-7192.
57. Holmer SR, Hengstenberg C, Kraft HG, et al. Association of polymorphisms of the apolipoprotein(a) gene with lipoprotein(a) levels and myocardial infarction. *Circulation.* 2003;107:696-701.
58. Boerwinkle E, Leffert CC, Lin J, Lackner C, Chiesa G, Hobbs HH. Apolipoprotein(a) gene accounts for greater than 90% of the variation in plasma lipoprotein(a) concentrations. *J Clin Invest.* 1992;90:52-60.
59. Clarke R, Peden JF, Hopewell JC, et al. Genetic variants associated with Lp(a) lipoprotein level and coronary disease. *N Engl J Med.* 2009;361:2518-2528.
60. Cavdar Koc E, Burkhart W, Blackburn K, Moseley A, Spremulli LL. The small subunit of the mammalian mitochondrial ribosome. Identification of the full complement of ribosomal proteins present. *J Biol Chem.* 2001;276:19363-19374.
61. Kwon HM, Yamauchi A, Uchida S, et al. Cloning of the cDNa for a Na+/myo-inositol cotransporter, a hypertonicity stress protein. *J Biol Chem.* 1992;267:6297-6301.

62. Abbott GW, Sesti F, Splawski I, et al. MiRP1 forms IKr potassium channels with HERG and is associated with cardiac arrhythmia. *Cell*. 1999;97:175-187.
63. Courtois G, Morgan JG, Campbell LA, Fourel G, Crabtree GR. Interaction of a liver-specific nuclear factor with the fibrinogen and alpha 1-antitrypsin promoters. *Science*. 1987;238:688-692.
64. Reiner AP, Barber MJ, Guan Y, et al. Polymorphisms of the HNF1A gene encoding hepatocyte nuclear factor-1 alpha are associated with C-reactive protein. *Am J Hum Genet*. 2008;82:1193-1201.
65. Purcell SM, Wray NR, Stone JL, et al. Common polygenic variation contributes to risk of schizophrenia and bipolar disorder. *Nature*. 2009;460:748-752.
66. Jee SH, Song KS, Shim WH, et al. Major gene evidence after MTHFR-segregation analysis of serum homocysteine in families of patients undergoing coronary arteriography. *Hum Genet*. 2002;111:128-135.
67. King RA, Rotter JI, Motulsky AG. *The Genetic Basis of Common Diseases*. Oxford: Oxford University Press; 2002.
68. Lusis AJ, Weinreb A, Drake TA. *Genetics of Atherosclerosis*. Philadelphia: Lippincott-Raven; 1998.

24
Hereditary Neuromuscular Diseases and Cardiac Involvement

A.J. van der Kooi and M. de Visser

24.1 Introduction

Neuromuscular disorders comprise a large group of diseases caused by dysfunction of motor neurons, peripheral nerves and skeletal muscles. A fair proportion of neuromuscular disorders have a genetic cause. The incidence and prevalence of cardiomyopathies associated with inherited neuromuscular diseases, particularly with muscular dystrophies, have until recently been underestimated, even though heart involvement is either the direct or indirect cause of death in many of these diseases.

This chapter focuses on the primary cardiac involvement in neuromuscular diseases, i.e. the primary cardiac changes which are caused by the same genetic anomalies that damage skeletal muscle or nerves. Cardiac involvement can manifest itself as impulse generation or conduction defects, focal or diffuse myocardial thickening, dilation of the cardiac cavities, relaxation abnormality, hypertrophic, dilated, non-compaction or restrictive cardiomyopathy, Takotsubo phenomenon, secondary valve insufficiency, intra-cardiac thrombus formation, or heart failure with systolic or diastolic dysfunction.[1] Secondary cardiac involvement in neuromuscular disorders ultimately manifesting with cor pulmonale is not addressed in detail. Chest wall disorders (e.g., in polio survivors, spinal muscular atrophy type 2 or congenital myopathies/dystrophies) or respiratory muscle weakness (e.g., in motor neuron disease or *Pompe*'s disease) reduce the pulmonary vascular bed and cause pulmonary hypertension, chronic hypoxia and hypercapnia. Respiratory muscle weakness is confirmed by pulmonary function tests that show a significant reduction of maximal respiratory pressures and vital capacity (VC) consistent with a restrictive ventilatory defect. In such cases, nocturnal ventilatory support is indicated.

The chapter reviews the probability and severity of cardiac disease in each type of neuromuscular disease, it offers guidelines for the clinical evaluation of cardiac involvement in the different disorders, and proposes useful forms of therapy.

24.2 Muscle Disorders

24.2.1 Muscular Dystrophies

Muscular dystrophies represent a clinically and genetically heterogeneous group of disorders, characterised by dystrophic changes in skeletal muscle and muscle wasting and weakness of variable distribution and severity. They can be caused by mutations in genes encoding sarcolemma-associated proteins, such as dystrophin and the dystrophin-associated glycoprotein complex, and genes encoding proteins of the nuclear envelope.

24.2.1.1 Sarcolemma-Associated Proteins

Dystrophinopathic cardiomyopathy

Dystrophinopathic cardiomyopathy is caused by a defect in the *dystrophin* gene on the X-chromosome encoding for the protein dystrophin. Dystrophin is

A.J. van der Kooi (✉)
Neurologists, Academic Medical center (AMC), Amsterdam, The Netherlands
e-mail: a.j.kooi@amc.uva.nl

expressed in heart, skeletal muscle, neural tissues, and smooth muscle, but progressive tissue damage is confined to heart and skeletal muscle. Dystrophin, together with other cytoskeletal proteins, provides mechanical support for the sarcolemma. A change in the amount, size or function of dystrophin causes a structurally weak sarcolemma, which ruptures under mechanical stress, allowing uncontrolled focal ingress of extracellular fluid components, especially calcium, into the muscle fibre interior.[2,3] Mutations of the dystrophin gene can result in different disorders manifesting with skeletal muscle involvement and/or cardiomyopathy: *Duchenne* muscular dystrophy (complete absence of dystrophin), *Becker* muscular dystrophy (qualitative and/or quantitative abnormalities of dystrophin), *X-linked dilated cardiomyopathy* (dystrophin abnormalities confined to the myocardium), and carrier (DMD/BMD) manifesting with cardiomyopathy. The lack of dystrophin in cardiac muscle leads to progressive cardiomyocyte degeneration and fibrosis. The posterior basal segment of the left ventricle is consistently the first site in which hypokinesis is detected on echocardiography. The characteristic electrocardiographic (ECG) alterations consist of tall right precordial R waves (R/S ratio greater than 1 in V1) and deep Q waves (greater than 3 mm) in the left precordial and limb leads. A progressive, global hypokinesia with ventricular dilatation then evolves. The right side of the heart is seldom affected, and in particular, the right atrium rarely demonstrates any echocardiographic abnormality. The most frequent cardiac abnormality in DMD is sinus tachycardia, occurring in childhood and persisting through life (Gilroy circulation 1963). Another rhythm abnormality frequently found is atrial premature beats. The incidence of cardiac conduction defects remained 6–13% throughout life.[4] AV blocks are only rarely found. Ventricular arrhythmias are infrequent at the early stages, but their incidence increases with the progression of the disease.[5,6]

Duchenne Muscular Dystrophy

Duchenne muscular dystrophy is caused by frame-shift mutations in the dystrophin gene, causing complete absence of the dystrophin protein. DMD has an incidence of one in 2,500 to 3,500 live male births.[7,8] Predominantly boys are affected in this X-linked disorder. The onset of disease is usually between 2 to 5 years of age with progressive symmetrical proximal weakness, legs more than arms, a characteristic hypertrophic appearance of the calves, and marked elevation of serum creatine kinase activity. Wheelchair dependency occurs around the age of 10 years. The introduction of ventilatory support led to a considerable extension of the life expectancy, from a mean age of death of 14.4 years in the 1960s to 25.3 years for those who were ventilated since 1990.[9] From a clinical perspective, a progressive dilated cardiomyopathy eventually occurs in all boys with DMD. Symptoms of overt cardiac failure are rare, probably because boys typically have a severely restricted physical ability. Left untreated, cardiomyopathy makes a significant contribution to early mortality.[9]

Management. Patients should have a cardiac investigation (ECG and echocardiography) at diagnosis. Cardiac investigations should be performed every 2 years up to the age of 10 years, before any surgery, and annually after age 10 years, or more frequently, if an abnormal echocardiogram is identified.[10] Assessment and treatment of respiratory function should be performed in parallel with the cardiological investigations.[9] There is a case for continued evaluation of more sophisticated tools (echo tissue Doppler imaging, cardiac magnetic resonance imaging, etc.) for earlier detection of abnormalities, but these are not required for routine management. Initial treatment with Angiotensin Converting Enzyme (ACE) inhibitors and/or beta blockers is indicated in case of impaired left ventricular function. Data from retrospective studies on the effect of corticosteroids showed that 5% of treated patients versus 58% of untreated patients developed cardiomyopathy, indicating a beneficial effect of corticosteroid treatment on the development and progression of cardiomyopathy.[11] There is also evidence indicating that treatment even before any impairment of ventricular function is detectable on echocardiogram may delay the onset and progression of cardiomyopathy (92.9% of the treated group were alive at 10 years versus 65.5% of the untreated group).[12,13] There are concerns about the possible impact of ACE inhibition on left ventricular development in very young children[10] and therefore, another trial was initiated. Anticoagulant therapy should be considered in patients with severe cardiac dysfunction to prevent systemic thromboembolic events.[14] DMD patients are rarely eligible for cardiac transplantation due to other complications including scoliosis and respiratory insufficiency.[10]

Becker Muscular Dystrophy

Becker muscular dystrophy is caused by in-frame mutations in the dystrophin gene which lead to reduced or otherwise altered dystrophin protein expression. The incidence of BMD is one third of that of DMD, a much higher figure than was previously thought, implying that BMD has been under-diagnosed in the past.[15] The clinical picture is characterised by later age of onset and slower rate of progression as compared to DMD. However, the spectrum of BMD encompasses a variety of phenotypes, including an intermediate form between BMD and DMD ('outliers'), a 'quadriceps only' form, and a very mild form in which BMD may manifest itself with myalgias and muscle cramps, exercise intolerance and myoglobinuria, or asymptomatic elevation of the serum CK activity. In most cases, the first symptoms were noticed between the 6th and 18th year of life with a mean age of onset of 11.1 years.[16] The age of loss of ambulation varies from 10 to 78 years (mean age is in the fourth decade). Becker cardiomyopathy evolves in the same manner as DMD cardiomyopathy. In BMD, the severity of cardiac disease does not correlate with that of skeletal muscle weakness.[17] A severe dilated cardiomyopathy can occur in patients with BMD with relatively preserved muscle function.

Management. Patients with Becker muscular dystrophy should have cardiac evaluation (ECG and echocardiography) at diagnosis. After that, they should be screened for the development of cardiomyopathy at least every 5 years, but preferably every 2 years.[10] When progressive abnormality is found, they should be seen more regularly and treated with ACE inhibitors and, if indicated, beta blockers. Cardiac transplantation may be a viable treatment option in this group of patients.[18,19]

X-Linked Dilated Cardiomyopathy

There are descriptions of male patients who present with early onset dilated cardiomyopathy and do not develop or have only mild skeletal muscle weakness.[20–23] Many but not all affected patients have an increased serum CK activity.[23,24] The disease is being referred to as X-linked dilated cardiomyopathy (X-LDC). X-LDC may be caused by the presence of a single point mutation at the first exon–intron boundary or a nonsense mutation in exon 29, by a rearrangement downstream from the 5' end of intron 11 or by a deletion in the mid-rod domain of the dystrophin gene. What all these mutations have in common is that they show a different pattern of expression in cardiac as compared to skeletal muscle.[78]

Management. Patients should have cardiac evaluation (ECG and echocardiography) at diagnosis. After that, they should be screened for the development of cardiomyopathy every 5 years, but preferably every 2 years. When progressive abnormality is found, they should be seen more regularly and treated with ACE inhibitors and, if indicated, beta blockers. Cardiac transplantation may be a viable treatment in this group of patients.

Female Carriers of Duchenne and Becker Muscular Dystrophy

Carriers of DMD and BMD are at risk of developing dilated cardiomyopathy. The cumulative risk of cardiomyopathy is estimated to be between 7% and 10%, irrespective of any manifest skeletal muscle weakness.[25–27]

Management. All carriers of DMD and BMD should have echocardiography and ECG at diagnosis, (there is no indication to test them presymptomatically before the age of 16 years) and at least every 5 years thereafter, or more frequently, in patients with abnormalities on investigation.[10] Clinical geneticists should refer women for cardiac evaluation when carriership is diagnosed. Carriers manifesting severe skeletal muscle symptoms or cardiac symptoms require more frequent investigation. Once significant abnormalities are detected, patients may benefit from treatment with ACE inhibitors and beta blockers, if indicated. Ultimately cardiac transplantation may be appropriate.[10,28]

Dystrophin-Associated Glycoprotein Complex Cardiomyopathies

Sarcoglycanopathies (limb girdle muscular dystrophy (LGMD) types 2C, D, E and F) constitute about 25% of the LGMD group and are inherited in an autosomal recessive way. Limb girdle muscular dystrophies constitute a heterogeneous group of disorders characterised by progressive weakness of the limb girdle muscles, i.e. the muscles of hip region and upper leg, and shoulder region and proximal arm. LGMD 2C-F are caused by defects in alpha, beta, gamma and delta sarcoglycan, which are part of the sarcoglycan transmembrane component of the dystrophin-associated glycoprotein complex. All types of sarcoglycanopathies can be associated with cardiomyopathy. The clinical course is comparable to that of patients with Duchenne muscular dystrophy or severe Becker muscular dystrophy.

Management. It is recommended to investigate sarcoglycanopathy patients with the same frequency as patients with DMD/BMD (see above).[10] Present perception is that the incidence of tachy- or brady-arrhythmias in sarcoglycanopathies is low but the issue has not been fully resolved. Arrhythmia surveillance with Holter ECG or other ambulatory ECG registrations is justified. Standard therapy should be effective in these patients with evidence of cardiomyopathy, but trial-based evidence of efficacy is lacking.

Other Plasma Membrane Proteins

Caveolinopathies. Caveolins are the structural proteins that are necessary for the formation of caveolae membrane domains. Caveolae are vesicular organelles (50–100 nm in diameter) that are particularly abundant in cells of the cardiovascular system. In these cell types, caveolae function both in protein trafficking and signal transduction.[29] The gene encoding for caveolin-3, the muscle-specific form of the caveolin family, is located on chromosome 3. Cardiac myocytes and skeletal muscle fibres express caveolin-3. In skeletal muscle, caveolin-3 is partly associated with the complex of dystrophin-associated proteins. Caveolin-3 mutations, predominantly autosomal dominant but occasionally autosomal recessive, may cause a variety of phenotypes, including LGMD1C, distal myopathy, rippling muscle disease, myoglobinuria and asymptomatic hyperCKemia. The incidence is unknown. There seems to be no evidence to suggest that cardiac surveillance is indicated routinely in LGMD1C.[10, 30] However, several recent observations of familial hypertrophic cardiomyopathy,[30] sudden death, possibly due to arrhythmia,[31] and long QT syndrome[32] associated with caveolin-3 mutations suggest that cardiac involvement is a feature of caveolin-3 deficiency, and careful cardiac assessment of these patients seems reasonable.[33]

24.2.1.2 Proteins with Enzymatic Activity

LGMD2I/MDC1C

LGMD2I, caused by mutations in the *fukutin-related protein* (FKRP) gene, is an autosomal recessive disorder. The *FKRP*-gene is a homologue of the fukutin gene encoding for the fukutin-related protein. FKRP is a putative glycosyltransferase whose precise function is uncertain. It has been localised in the Golgi apparatus and is involved in the glycosylation processing of α-dystroglycan, an indispensable molecule for binding laminin alpha2. FKRP is ubiquitously expressed. Mutations in the fukutin-related protein gene (FKRP) located on chromosome 19q13 give rise to a spectrum of phenotypes including a form of *congenital muscular dystrophy (MDC1C), Walker-Warburg* phenotype and a relatively mild form of limb-girdle muscular dystrophy (LGMD2I). The most common mutation is the C826A mutation. Patients with a homozygous C826A mutation generally exhibit milder and late-onset muscular dystrophy, whereas the compound heterozygous mutations are associated with more severe and early-onset type of muscular dystrophy phenotypically related to Duchenne muscular dystrophy.[34, 35] In the Netherlands, LGMD2I was diagnosed in only 8% of all LGMD families, whereas in the United Kingdom and Denmark, LGMD2I is considered the most frequent cause of LGMD.[35–37]

Left ventricular hypokinesis, dilated cardiomyopathy, and heart failure have been reported in about one third of LGMD2I patients, regardless of the gene mutation and the severity of the muscular disease, suggesting that all patients should be referred for cardiac evaluation.[38]

Management. All patients with LGDM2I should be evaluated for cardiac involvement (ECG and echocardiography) at diagnosis. After that, 2-year screening seems reasonable.[10]

Fukuyama Congenital Muscular Dystrophy

Fukuyama congenital muscular dystrophy (FCMD) is an autosomal recessive disorder, caused by mutations in the fukutin gene on chromosome 9q31.[39] Its protein product, fukutin, has sequence homologies with bacterial glycosyltransferase, but its precise function is unknown. FCMD also belongs to the group of disorders associated with glycosylation defects of α-dystroglycan. The disorder is particularly frequent in Japan where its incidence is 40% of that of Duchenne muscular dystrophy and is rare in Western countries.[40] FCMD is clinically characterised by a triad of mental retardation, brain deformities, and congenital muscular dystrophy.

In contrast with the severe dystrophic involvement of skeletal muscle, cardiac insufficiency is quite rare.

Typically, patients are able to sit but never attain independent ambulation. In contrast, the mildest fukutin-related phenotype, *LGMD2M*, presents with minimal muscle weakness, dilated cardiomyopathy and normal intelligence.[41] This suggests that late-onset LGMD patients with mutations in the fukutin protein should be evaluated with ECG and echocardiography at diagnosis. After that, 2-year screening seems reasonable.

24.2.1.3 Inner Nuclear Membrane Proteins

Emery-Dreifuss muscular dystrophy (EDMD) can present as an X-linked or autosomal dominant disorder. The disease is characterised by early contractures, and a humeroperoneal distribution of muscle weakness. Both emerin and lamin A/C, the causative genes in X-linked and autosomal dominant Emery-Dreifuss muscular dystrophy, respectively, are nuclear lamina genes. Defects in these genes cause conduction disorders and cardiomyopathy. In case of more prominent limb-girdle muscle weakness in the presence of a lamin A/C mutation the disorder is called LGMD1B.

X-Linked Emery Dreifuss Muscular Dystrophy

The gene locus for this entity is located at Xq28, and the *gene (STA)* which is 2,100 bp in length and consists of six exons encoding a 254 amino acid serine-rich protein, is called emerin.[42] Emerin mutations identified to date include a few missense mutations, and the majority are nonsense, splice site or small deletions/insertions that ultimately result in premature translation termination and complete absence of emerin expression on both Western blotting and immunohistochemistry. The function of the *emerin* protein which is expressed in different tissues[43] and in all vertebrates remains to be elucidated.

Clinically, the disorder is characterised by early contractures of the Achilles tendons, elbows and post cervical muscles, often before there is any significant weakness. Subsequently, limitation of neck flexion develops, but later forward flexion of the entire spine becomes limited.[44, 45] Muscle wasting and weakness with a distinctive humeroperoneal distribution early in the course of the disease is slowly progressive. Weakness later extends to the proximal limb girdle musculature, but is rarely profound. Onset in the first few years of life is not exceptional.[46, 47] The variability of the clinical severity in individual members of the same family appears to be much greater as compared to other forms of muscular dystrophy (even compared to Becker muscular dystrophy). Only very rarely, ambulation is lost as a result of muscle weakness or contractures.[46] Very rare cases seem to be completely asymptomatic still in the 4th decade of life.[46]

Cardiac features usually occur in patients' (early) teens or twenties, but a boy as young as age 5 years, in whom the heart was involved, has been reported.[46] Cardiac involvement is characterised by cardiac conduction defects, ranging from sinus bradycardia, prolongation of the PR interval on electrocardiography to complete heart block. Atrial paralysis is almost pathognomonic of EDMD. The finding of a dilated right atrium on echocardiography and isolated atrial paralysis with absent "p" waves on electrocardiography should always prompt the exclusion of EDMD.[48] The severity of heart disease does not correlate with the degree of skeletal muscle involvement. Sakata et al. (2005) described a Japanese family with X-linked recessive Emery-Dreifuss muscular dystrophy in which patients died suddenly without prior cardiac or neuromuscular complaints.[49] The living patients with cardiac involvement showed either no or mild muscle weakness and contractures were not present. EDMD affects the atria, and right heart involvement predominates. There is progressive replacement of the normal myocardium by fibrous and adipose tissue, which results in the loss of atrial contractility (atrial paralysis) and atrial dilatation. Evidence of left ventricular dysfunction (in addition to the invariable involvement of the conduction system) was reported by some groups but not by others.[46]

As with DMD, there may be some female carriers of this X-linked disease who manifest cardiac disease, in particular, atrial paralysis, albeit usually at a later age than male subjects.[49, 50] No association with any sign of muscle weakness, wasting or contractures appears to be present.[46] Published cases of manifesting carriers may have been diluted by cases of dominant disease.

Management. Cardiological evaluation at diagnosis and annually thereafter using 12 lead ECG (preferably at 50 mm/s) requires expert assessment as ECG changes may be subtle and difficult to interpret.[10] Holter monitoring should be recommended annually for tachy- or brady-arrhythmias. Echocardiography can be done on a less regular basis. Permanent pacemaker implantation is justified, even in asymptomatic patients[10] when ECG begins to show abnormalities of sinus node or AV node disease. However, nocturnal AV-Wenkebach may be a normal finding in young people. In the presence of sino-atrial or AV-nodal conduction abnormalities on surface ECG, invasive electrophysiology testing probably adds little to the decision to or timing of pacemaker implantation. However, such testing may have a role in determining the optimum mode of and sites for pacing.[10] Whether implantable defibrillators may be a more appropriate form of management than pacemakers when anti-bradycardia pacing is indicated for these patients is unclear.

It is recommended to establish the carrier status in females at risk and to offer them periodic ECG surveillance including 24 h ambulatory Holter monitoring to detect atrial or AV-nodal conduction disease.[51] There is a need for more systematic study of the natural history of cardiac involvement in X-linked EDMD carriers.

Autosomal Dominant EDMD/LGMD1B

Mutations in the *LMNA* gene on chromosome 1q11-q23[52] encoding *lamins A and C* by alternative splicing cause primary laminopathies including various types of lipodystrophies, muscular dystrophies (EDMD2 and LGMD1B) and *progeroid syndromes*, *mandibuloacral dysplasia*, dilated cardiomyopathies, *neuropathy*, restrictive *dermopathy*, and *arthropathy* with tendinous calcifications. When looking at LMNA/C-related muscular dystrophies, most cases will have an EDMD phenotype, but in some instances, a limb girdle phenotype, referred to as LGMD1B, is found.

Lamins are the main components of the intermediate filament lamina, which lines the inner nuclear membrane. Lamin proteins have been shown to bind to chromatin and to several inner nuclear membrane proteins.

The pattern and severity of cardiac disease is thought to be more severe in the autosomal dominant form as compared to the X-linked EDMD. Among patients with AD-EDMD, 35% will develop a progressive and potentially life-limiting dilated cardiomyopathy by middle age. Ventricular dysrhythmias are also significant in laminopathy patients and are an important cause of sudden death, despite pacing.[5,53]

Management. ECG at diagnosis and yearly thereafter. Holter monitoring for tachy or brady-arrhythmias annually. Echocardiography on a less regular basis. Management of patients with ventricular dysrhythmia is complex and the complications of implantable defibrillators may be greater than with pacemakers.[54] These patients should be managed in specialised centres and their data collated to contribute to further evidence in the future. In the meantime, there is a strong indication for defibrillator implantation to be considered when anti-bradycardia pacing is indicated. This recommendation needs to be validated over time through the collection of high-quality prospective data.[10]

24.2.2 Nucleotide Repeat Disorders with Myotonia

24.2.2.1 Myotonic Dystrophy

Myotonic dystrophy type 1 (*DM1*, also known as *dystrophia myotonica* or *Steinert's* disease) is an autosomal dominant multi-system disorder, and the most common myopathy presenting in adults (incidence one in 8,000 live births, prevalence is approximately five per 100 000 in most American and European populations).

DM1 is caused by an aberrantly expanded CTG repeat in the 3'-untranslated region of the DM protein kinase (DMPK) gene on chromosome 19q13.3. The mutated DMPK gene produces an altered version of messenger RNA, which interacts with certain proteins to form clumps within the cell. The abnormal clumps interfere with the production of many other proteins. The severity of the disease is related to the repeat length, which can expand from generation to generation (anticipation), and varies from very severe, often lethal congenital DM to late-onset mild muscle weakness, myotonia and cataract. Patients with adolescent-onset DM1 characteristically manifest with myotonia (delayed muscle relaxation after contraction), progressive weakness and atrophy of the skeletal muscles, with predominant distal weakness, and facial involvement with involvement of systems other than skeletal muscle, such as the heart, endocrine glands, central nervous system and smooth muscle.

Myocardial fibrosis and degeneration of the cardiac-conduction system occur in the majority of patients. Approximately, 90% show electrocardiographic (ECG) abnormalities, commonly, prolongation of the PR interval, and QRS duration. ECG changes do not correlate either with disease severity or with CTG repeat length. Arrhythmias can occur, including sinus-node dysfunction, progressive heart block, atrial tachycardia, flutter or fibrillation, and ventricular tachycardia or fibrillation.[55] Patients with adult DM1 are at high risk for arrhythmias and sudden death.[56,57] A rhythm other than sinus, PR interval of 240 msec or more, QRS duration of 120 msec or more, or second-degree or third-degree atrioventricular block and a diagnosis of atrial tachyarrhythmia (sustained atrial tachycardia, flutter or fibrillation) predict sudden death.[56] Cardiomyopathy and congestive heart failure occur far less frequently than conduction disturbances. The most prevalent echocardiographic changes are mitral valve prolapse and septal and myocardial fibrosis.

Management. Cardiac evaluation includes annual ECG, and Holter monitoring if annual ECG shows increasing PR interval or other evidence of increased risk of bradycardia. Echocardiogram should be performed at diagnosis in congenital myotonic dystrophy. Invasive measurement of the HV interval may help decide the need for pacing in borderline cases. If atrial tachyarrhythmias (atrial flutter, fibrillation) become symptomatic, anti-arrhythmic treatment may be justified.[10] However, anti-arrhythmic drugs may aggravate any pre-existing tendency to bradycardia or ventricular tachyarrhythmias. As a result of accumulating evidence of sudden death in DM1 patients who have been paced, the consensus recommendation at present is that implantable defibrillators may be a more appropriate form of management than pacemakers when anti-bradycardia pacing is indicated for these patients.[56,57]

24.2.2.2 Myotonic Dystrophy Type 2

Myotonic dystrophy type 2 (DM2), also called *Proximal myotonic myopathy (PROMM)*, is present in a large number of families of northern European ancestry. In Germany it has the same prevalence as DM1. DM2 is caused by an expanded CCTG tetra-nucleotide repeat in the first intron of the zinc finger protein 9 (ZNF9) gene on chromosome 3q21.[58]

DM2 shares many features with DM1, but the patients have less symptomatic distal, facial, and bulbar weakness, and less pronounced clinical myotonia. Important other differences include the absence of a congenital form of DM2, an apparent lack of mental retardation in juvenile cases, and less evident excessive daytime sleepiness.

The heart involvement is comparable to that in DM1.

24.2.3 Ion Channel Disorder Associated with Periodic Paralysis and Heart Involvement

24.2.3.1 Anderson Syndrome

Anderson syndrome is a very rare disease, and characterised by the clinical triad of dyskalemic paralytic attack, ventricular ectopy and potential dysmorphic features. It is inherited as an autosomal dominant trait. Mutations in the potassium channel gene *KCNJ2*, which encodes for the Kir2.1 potassium channel generating the I_{K1} current, have been found. Cardiac disturbances may comprise the long QT syndrome (type 7), ventricular extra-systoles, or tachycardia.[59] Tachydysrhythmia may cause syncopal attacks and sudden death. The cardiac symptoms are provoked or worsened by hypokalemia and digitalis. The *paralytic attacks* may be hyperkalemic or hypokalemic, and therefore the response to oral potassium is unpredictable.

24.2.4 Myofibrillar Myopathies

The term *myofibrillar myopathies (MFM)* was proposed as a noncommittal designation for a group of chronic neuromuscular diseases associated with common morphologic features, consisting of a distinct pathologic pattern of myofibrillar disorganisation that begins at the Z-disk and is followed by accumulation of myofibrillar degradation products and ectopic expression of diverse proteins that include desmin, αB-crystallin, dystrophin, myotilin, sarcoglycans, neural cell adhesion molecule, plectin, gelsolin, ubiquitin, filamin C, Xin and congophilic amyloid material. These disorders are transmitted by autosomal dominant inheritance, and typically manifest as distal myopathies, but may also affect proximal muscles.[60] Median age of onset is 55 years (range 7–77). Serum CK activity is normal or slightly elevated. Cardiomyopathy, often of the arrhythmogenic type, is a frequent associated feature.[61] Mutations in the *desmin, αB-crystallin, myotilin, ZASP, filamin C, and BAG3* genes have been identified in about half of the patients.[60] Mutations in myotilin cause *LGMD1A*. Because cardiac involvement in myofibrillar myopathies resembles cardiac involvement in laminopathies, the same guidelines for follow-up and management seem to be appropriate.

24.2.5 Congenital Myopathies

24.2.5.1 Central Core Disease

Central core disease (CCD) is caused by mutations in the *ryanodin receptor (RYR1), selenoprotein (SEPN1), and titin (TTN)* genes.[62] CCD is characterised by hypotonia and weakness in the neonatal period and a non-progressive course.[63] It is inherited in an autosomal dominant fashion in the vast majority of cases. Dysmorphic features may develop secondary to muscle weakness. CCD is a rare condition, but its true incidence is not known. An association with the potentially fatal malignant hyperthermia syndrome is well known. Serum CK activity is usually normal. Muscle biopsies reveal well-demarcated cores within most muscle fibres.

Cardiac involvement in CCD is rare. Recessive mutations in titin have been associated with congenital myopathy and fatal cardiomyopathy.[64]

24.2.5.2 Nemaline Rod Myopathy

Defects in six thin filament protein genes, including skeletal *α-actin (ACTA1), nebulin (NEB), α-tropomyosin (TPM3), β-tropomyosin (TPM2), troponin T (TNNT1), and cofilin-2 (CFL2)*, have so far been shown to result in nemaline myopathy,[62] 2008). Nemaline myopathy is characterised by the presence of rod-shaped structures in the muscle fibres.

The clinical spectrum of nemaline myopathies is wide, ranging from severe, often fatal conditions with prenatal onset to early childhood-onset conditions of varying severity. Disproportionately severe axial and respiratory muscle involvement is common in all variants, and is often the long-term prognostic determinant. The condition is otherwise essentially stable, though in a few patients with mutations in ACTA1, severe progression of weakness in late childhood has been noticed.

Primary cardiac involvement is rare. However, several cases with nemaline myopathy and predominantly, hypertrophic cardiomyopathy have been described.[65, 66]

24.2.5.3 Myosin Storage Myopathy

Mutations in the myosin heavy chain gene MYH7 cause myosin storage myopathy. MYH7 is the most frequent cause of hypertrophic cardiomyopathy. Cardiomyopathy has been described in combination with this disorder.[67, 68]

24.2.6 Metabolic Disorders Affecting Muscle

24.2.6.1 Lysosomal Glycogenosis

Pompe's Disease or Glycogen Storage Disease Type II

Pompe's disease is a rare autosomal recessive disorder caused by mutations in the gene that encodes for *α-glucosidase*. Alpha-glucosidase deficiency causes glycogen to accumulate in various tissues and disrupt function of skeletal and cardiac muscle in particular. Presentation in infancy is associated with respiratory failure, cardiomyopathy, and severe muscle weakness. Juvenile or adult-onset cases typically present with proximal muscle weakness, and often develop respiratory insufficiency or exertional dyspnoea due to diaphragmatic involvement.[69] Cardiac involvement in glycogenosis type II comprises cardiomyopathy, arrhythmias, and cardiac decompensation. The cardiac involvement depends on the residual acid alpha-glucosidase activity and the age at symptom onset. In the late-onset forms, cardiac involvement is rare.

Management. Until recently, treatment was focused on supportive measures, and infants diagnosed with classical Pompe's disease usually died within the first year of life. The recent introduction of enzyme replacement therapy (ERT) with recombinant α-glucosidase has dramatically improved the life expectancy of infantile-onset disease with anecdotal improvements in respiratory and motor function observed in juvenile or adult-onset cases.[70] Cardiac assessment in infants with glycogenosis type II should involve an echocardiogram at diagnosis, followed by check-ups at quarterly intervals during the first 2 years of treatment with ERT, and then at 6 monthly intervals. For adult patients, it is advocated to perform an electrocardiogram at least once in routine clinical follow-up. Additional echocardiography seems indicated only in those patients with abnormal electrocardiographic findings, a history of cardiac disease or evident cardiac symptoms.[71]

Danon Disease

Danon disease is caused by a primary deficiency of a major lysosomal membrane glycoprotein, *LAMP2 (lysosome-associated membrane protein 2)*. This is an extremely rare X-linked dominant disorder, characterised by hypertrophic cardiomyopathy, skeletal myopathy, and variable degree of mental retardation, with autophagic vacuoles in skeletal and cardiac muscle. Males are more affected than females. In probands, cardiac symptoms, such as exertional dyspnoea, start in teenage years. The association of hypertrophic cardiomyopathy and cardiac arrhythmia is common, and patients typically die of cardiac failure or cardiac arrest in their fourth decade. Problems with the electrical conduction in the heart can occur, presenting as 'Wolff-Parkinson-White' syndrome. The myopathy is usually mild. Serum CK activities are 5–10 times elevated.[72]

Management. Patients should have a cardiac investigation, including ECG and echocardiography at diagnosis. Cardiac investigations should be performed every 1–2 years or more frequently, if an abnormal echocardiogram is identified.

24.2.6.2 Mitochondrial Disorders

Primary Disorders of Mitochondrial Function

These are caused by mutations in both mitochondrial and nuclear genes encoding mitochondrial proteins. They are an increasingly recognised cause of multisystem diseases that have disorders of the central nervous system and skeletal muscle as their predominant manifestations. Because of its dependence on oxidative metabolism, the heart is also frequently involved in *mitochondrial* disease (see chapter 24). Several mitochondrial syndromes that involve the heart include *Kearns-Sayre syndrome, MELAS* (mitochondrial myopathy, encephalopathy, lactic acidosis, and stroke-like episodes, and *MERFF* (myoclonic epilepsy with ragged red fibres). Several types of cardiac abnormality including hypertrophic cardiomyopathy, dilated cardiomyopathy, *Wolff-Parkinson-White syndrome*, and cardiac arrhythmia have been described.[79] Conduction disturbances may be an important cause of mortality in patients with Kearns-Sayre syndrome.

Management. Patients should have cardiac evaluation (ECG and echocardiography) at diagnosis. After that, they should be screened for the development of cardiomyopathy every 2–5 years. Timely placement of a pacemaker can be life-saving in the presence of conduction block. In patients with isolated cardiomyopathy cardiac transplantation may be required.[73]

Carnitine Deficiency

Carnitine plays an essential role in the transfer of long-chain fatty acids across the inner mitochondrial membrane. This transfer requires enzymes and transporters that accumulate carnitine within the cell (*OCTN2 carnitine transporter*), conjugate it with long chain fatty acids (*carnitine palmitoyl transferase 1, CPT1*), transfer the acylcarnitine across the inner plasma membrane (*carnitine-acylcarnitine translocase, CACT*), and conjugate the fatty acid back to Coenzyme A for subsequent beta oxidation (*carnitine palmitoyl transferase 2, CPT2*). Deficiency of the OCTN2 carnitine transporter causes primary carnitine deficiency, characterised by increased loss of carnitine in the urine and decreased carnitine accumulation in tissues. Patients can present with hypoketotic hypoglycemia and hepatic encephalopathy, or with skeletal and cardiomyopathy. This disease responds to carnitine supplementation.

CACT deficiency presents in most cases in the neonatal period with hypoglycemia, hyperammonemia, and cardiomyopathy with arrhythmia leading to cardiac arrest. Plasma carnitine levels are extremely low.

In CPT1 deficiency the skeletal muscle and heart are usually unaffected. In adults with deficiency of CPT 2 rhabdomyolysis triggered by prolonged exercise may occur. More severe variants of CPT2 deficiency present in the neonatal period similar to CACT deficiency. Treatment for deficiency of CPT2 and CACT consists of a low-fat diet supplemented with medium chain triglycerides that can be metabolised by mitochondria.

Friedreich's ataxia

Friedreich's ataxia is an autosomal recessive disorder, in most cases caused by a homozygous expanded GAA-repeat (55–1700, normal 7–33) localised in the intron of the frataxin gene on chromosome 9q13. There is an inverse correlation between the length of the GAA-repeat and onset of the disease, progression and the occurrence of cardiomyopathy.[74] The *frataxin gene* encodes for the frataxin protein located at the inner mitochondrial membrane the function of which remains to be elucidated.

The estimated prevalence is 2–3 per 100.000 inhabitants. Onset of the disease is usually between 5 and 25 years. Progressive gait ataxia and ataxia of the legs are the first manifestations of the disease. Subsequently, cerebellar dysarthria, ataxia of the arms, oculomotor disturbances, pyramidal features and sensory abnormalities due to involvement of the posterior columns and the peripheral nerves occur. Hypertrophic cardiomyopathy is observed in 60–70% and can even precede cerebellar ataxia. In later stages dilated cardiomyopathy may develop. Most patients are wheelchair-bound after a disease duration of 8–15 years. There is a great range in age of death (30–70 years), dependent on the occurrence of cardiac involvement.

Management. Cardiac evaluation should take place at diagnosis, with re-screening every 3–5 years when no abnormalities are found. There is accumulating evidence that idebenone can prevent progression of cardiomyopathy, albeit long-term randomized controlled trials are still needed.

Barth Syndrome

Barth syndrome is an extremely rare X-linked cardioskeletal myopathy caused by a deficiency in tafazzin. *Tafazzin*, a phospholipid acyltransferase, is involved in acyl-specific remodelling of cardiolipin, which promotes structural uniformity and molecular symmetry among the cardiolipin molecular species. Inhibition of this pathway leads to changes in mitochondrial architecture and function.[75] Patients have variable clinical findings, often including heart failure, myopathy, cyclic neutropenia, growth retardation and organic aciduria. Affected boys usually die of heart failure in infancy or early childhood, but there may be relative improvement in those who survive to later childhood.[76]

Management. Patients should have a cardiac investigation, including ECG and echocardiography at diagnosis and thereafter every 1–2 years, or more frequently, if an abnormal echocardiogram is identified.

24.3 Neuropathies

24.3.1 Familial Amyloid Polyneuropathy

Familial amyloid polyneuropathy designates a group of dominantly inherited neuropathies, with extracellular deposition of amyloid substance in various tissues. The three main precursor proteins encountered in these disorders are *transthyretin (TTR), apolipoprotein A1 or not gelsolin*. TTR-associated neuropathies are by far the most frequent type with a severe sensorimotor and autonomic neuropathy as the hallmark of the disease, most often associated with cardiac manifestations. First described in Portugal, the disorder was subsequently reported across the world, although Portugal, Japan and Sweden are the three main areas of prevalence. In the past years, an increasing number of mutations have been identified in the TTR gene, along with a larger clinical spectrum than initially thought. Variable age of onset and penetrance are also largely reported with unclear phenotypic–genotypic correlations. Over the last 15 years, liver transplantation has enabled improved prognosis of this devastating condition. However, in some patients with substantial cardiac involvement prior to liver transplantation, the cardiac condition continues to worsen, as measured by left ventricular wall thickness and ejection fraction.[77] These findings have led to a very small number of combined liver and heart transplantations in cases of hereditary amyloidosis with cardiac involvement.

24.3.2 Charcot-Marie-Tooth Disease Type 2 Caused by Lamin A/C Mutations

Charcot-Marie-Tooth (CMT) disease comprises a group of clinically and genetically heterogeneous hereditary motor and sensory neuropathies, which are clinically characterised by distal muscle weakness and wasting, sensory disturbances, and foot and hand deformities. An axonal subtype, CMT2, is defined by (near-) normal nerve conduction velocities in combination with the loss of large myelinated fibres and axonal degeneration on nerve biopsy. CMT2 phenotypes are characterised by a large genetic heterogeneity. In one autosomal recessive subtype of CMT2 mutations in the

Table 24.1 Frequency, type and implications of cardiac involvement in different neuromuscular disorders (Adapted from Bushby et al.[10])

Disease (gene)	Cardiac involvement	% of patients with cardiac involvement	Age at onset	Morbidity/ Mortality	Evaluation	Management
Duchenne muscular dystrophy (dystrophin)	ECG abnormalities; DCM	Abnormal ECG > 90%; abnormal ECHO > 90%	Detectable from the age of 6 years onwards	Cardiac death 10–20%, usually in teens	ECG and ECHO at diagnosis, 2-yearly before 10 years, annually thereafter	ACE inhibitors, beta blockers
Becker muscular dystrophy (dystrophin)	ECG abnormalities; HCM and DCM	ECG abnormal – 90%, ECHO abnormal – 65%	Variable, may be disproportionate to skeletal involvement	Cardiac death in up to 50%	ECG and ECHO every 2–5 years	ACE inhibitors, beta blockers, heart transplantation in end-stage DCM in patients with relatively preserved skeletal muscle function
DMD/BMD carriers (dystrophin)	ECG abnormalities; DCM	7–10% dilated cardiomyopathy, ECG abnormalities 21–90	Variable, may be disproportionate to skeletal muscle involvement		ECHO and ECG at diagnosis, or after the age of 16 years and at least every 5 years thereafter	ACE inhibitors, beta blockers, cardiac transplantation in end-stage DCM
X-DCM (dystrophin)	ECG abnormalities; DCM	100% by definition	No evident muscle weakness	Cardiac transplantation sometimes necessary	ECG and ECHO every 2–5 years	ACE inhibitors, beta blockers. Cardiac transplantation in end-stage DCM
Sarcoglycanopathies (LGMD2C-F) (sarcoglycans)	ECG abnormalities; DCM	20–25%	Variable	Cardiac transplantation sometimes necessary	ECG and ECHO every 2–5 years	ACE inhibitors, beta blockers. Cardiac transplantation in end-stage DCM
LGMD1C (Caveolin-3)	HCM, long QT	Case-reports				
LGMD2I/MDC1C (FKRP)	ECG abnormalities, DCM	1/3 of adult-onset Cases	Possibly related to severity of overall disease		ECG and ECHO at diagnosis, every 2-year thereafter	ACE inhibitors, beta blockers
Fukutin congenital muscular dystrophy/ LGMD2M (fukutin)	DCM	Case-reports			Late-onset cases: ECG and ECHO at diagnosis, every 2-year thereafter	ACE inhibitors, beta blockers
X-EDMD (emerin)	AV block, atrial paralysis, atrial flutter and fibrillation	>95% by the age of 30 years	10–39	SCD common in non-paced individuals	ECG and Holter at diagnosis and annually thereafter	Pacemaker

(continued)

Table 24.1 (continued)

Disease (gene)	Cardiac involvement	% of patients with cardiac involvement	Age at onset	Morbidity/Mortality	Evaluation	Management
AD-EDMD/LGMD1B/myofibrillar myopathies (lamin A/C desmin, alpha-B-crystallin, myotilin)	AV block, atrial flutter and fibrillation; DCM	Rhythm and conduction disturbances >95% by the age of 30 years, DCM 35%		SCD despite pacing, heart failure	ECG and Holter at diagnosis and annually thereafter	ICD, ACE inhibitors, beta blockers, cardiac transplantation in end-stage DCM
DM1/DM2 (DMPK)	AV conduction disturbances, atrial flutter and fibrillation, ventricular tachy-arrhythmias	90% ECG abnormalities		SCD despite pacing.	ECG and Holter at diagnosis and annually thereafter, ECHO in congenital myotonic dystrophy	Pacemaker/ICD
Anderson syndrome (KCNJ2)	Long QT syndrome, ventricular extra-systoles, or tachycardia			Syncopal attacks, sudden death; provoked by hypokalemia and digitalis		
Congenital myopathy RYR1, SEPN1, TTN	HCM, fatal Cardiomyopathy	Rare, only case reports				
Pompe's disease (alpha-glucosidase)	Cardiomyopathy in juvenile cases	Rare in adult onset Cases			In infants ECHO with 3–6 months interval; ECG at least once in adult-onset cases	Enzyme replacement therapy
Danon disease (LAMP2)	HCM			Cardiac failure or cardiac arrest in 4th decade	ECG and ECHO at diagnosis. every 1–2 years thereafter	Cardiac transplantation in end-stage HCM
Mitochondrial	HCM, DCM Wolff-Parkinson-White syndrome, and cardiac arrhythmia, conduction disturbances				ECG and ECHO at diagnosis and every 2 to 5 years thereafter	Pacemaker, cardiac transplantation in end-stage DCM

Carnitine deficiency	Cardiomyopathy, arrhythmia		Dietary treatment
Barth syndrome (tafazzin)	DCM, HCM	Heart failure in infancy or early childhood	ECG and ECHO at diagnosis and thereafter every 1–2 years
Familial amyloid polyneuropathy (TTR, apolipoprotein A1, gelsolin)	HCM, DCM		Liver (+heart) transplantation
Refsum disease	HCM, DCM	Later stages of the disease	Dietary treatment
Friedreich's ataxia (frataxin)	HCM, DCM	60–70%	Cardiac involvement determines age of death	ECG and ECHO at diagnosis, re-screening every 3–5 years	Idebenone

DCM denotes dilated cardiomyopathy; HCM denotes hypertrophic cardiomyopathy; ACE denotes angiotensin converting enzyme; SCD denotes sudden cardiac death; AV denotes atrio ventricular; X denotes X-linked; AD denotes autosomal dominant; DM1 denotes myotonic dystrophy type 1; DM2 denotes myotonic dystrophy type 2.

lamin A/C gene have been found. Mutations in this gene also cause AD EDMD, LGMD1B, and dilated cardiomyopathy with conduction defects, and therefore, similar cardiac involvement may be anticipated, although, as yet, it has not been described.

24.3.3 Refsum's Disease

This is a rare autosomal recessive peroxisomal disorder. The classic triad encompasses ataxia, retinitis pigmentosa, and polyneuropathy. Refsum's disease is caused by an inborn error in the metabolism of a fatty acid, called phytanic acid. All patients have markedly increased serum concentrations of phytanic acid. Cardiomyopathy can occur in the course of the disease, mostly at an advanced stage of the disease. Chronic dietary treatment by restricting the exogenous sources of phytanic acid and its precursor phytol results in clinical improvement.

24.4 Summary

A fair proportion of the neuromuscular disorders have a genetic cause. Molecular genetics evaluation can reveal a pathogenic mutation in many cases. Heart involvement is either the direct or indirect cause of death in many of these diseases. It is also of importance to consider the presence of cardiac abnormalities in patients with inherited neuromuscular disease who are to be given a general anaesthetic, because arrhythmias and conduction abnormalities may be precipitated peri-operatively.

Cardiac involvement related to the primary skeletal muscle disorder can manifest itself as impulse generation or conduction defects, or cardiomyopathy.

Patients with neuromuscular disorders, known to be associated with cardiac pathology, should be referred to a cardiologist for extensive evaluation of ventricular function, impulse formation and conduction diseases. Vice verse patients presenting with dilated cardiomyopathies should be investigated for neuromuscular disease.

References

1. Finsterer J, Stollberger C. Cardiac involvement in primary myopathies. *Cardiology*. 2000;94(1):1-11.
2. Mokri B, Engel AG. Duchenne dystrophy: electron microscopic findings pointing to a basic or early abnormality in the plasma membrane of the muscle fiber. *Neurology*. 1975 Dec;25(12):1111-1120.
3. Head SI, Williams DA, Stephenson DG. Abnormalities in structure and function of limb skeletal muscle fibres of dystrophic mdx mice. *Proc Biol Sci*. 1992 May 22;248(1322): 163-169.
4. Nigro G, Comi LI, Politano L, Bain RJI. The incidence and evolution of cardiomyopathy in Duchene muscular dystrophy. *Int J Cardiol*. 1990;26:271-277.
5. Goodwin FC, Muntoni F. Cardiac involvement in muscular dystrophies: molecular mechanisms. *Muscle Nerve*. 2005 Nov;32(5):577-588.
6. Finsterer J, Stollberger C. The heart in human dystrophinopathies. *Cardiology*. 2003;99(1):1-19.
7. Emery AEH. Differential diagnosis. In: Emery AEH, ed. *Duchenne muscular dystrophy*. 2nd ed. Oxford: Oxford medical publications; 1993:80-107.
8. van Essen AJ, Busch HF, te Meerman GJ, ten Kate LP. Birth and population prevalence of Duchenne muscular dystrophy in The Netherlands. *Hum Genet*. 1992;88(3):258-266.
9. Eagle M, Baudouin SV, Chandler C, Giddings DR, Bullock R, Bushby K. Survival in Duchenne muscular dystrophy: improvements in life expectancy since 1967 and the impact of home nocturnal ventilation. *Neuromuscul Disord*. 2002 Dec;12(10):926-929.
10. Bushby K, Muntoni F, Bourke JP. 107th ENMC international workshop: the management of cardiac involvement in muscular dystrophy and myotonic dystrophy. 7th-9th June 2002, Naarden, the Netherlands. *Neuromuscul Disord*. 2003 Feb;13(2):166-172.
11. Silversides CK, Webb GD, Harris VA, Biggar DW. Effects of deflazacort on left ventricular function in patients with Duchenne muscular dystrophy. *Am J Cardiol*. 2003 Mar 15;91(6):769-772.
12. Duboc D, Meune C, Lerebours G, Devaux JY, Vaksmann G, Becane HM. Effect of perindopril on the onset and progression of left ventricular dysfunction in Duchenne muscular dystrophy. *J Am Coll Cardiol*. 2005 Mar 15; 45(6):855-857.
13. Duboc D, Meune C, Pierre B, et al. Perindopril preventive treatment on mortality in Duchenne muscular dystrophy: 10 years' follow-up. *Am Heart J*. 2007 Sep;154(3):596-602.
14. Cardiovascular health supervision for individuals affected by Duchenne or Becker muscular dystrophy. *Pediatrics*. 2005 Dec;116(6):1569–1573.
15. Bushby KM, Thambyayah M, Gardner-Medwin D. Prevalence and incidence of Becker muscular dystrophy. *Lancet*. 1991;337(8748):1022-1024.
16. Emery AE, Skinner R. Clinical studies in benign (Becker type) X-linked muscular dystrophy. *Clin Genet*. 1976;10(4): 189-201.
17. Hoogerwaard EM, De Voogt WG, Wilde AAM, et al. Evolution of cardiac abnormalities in Becker muscular dystrophy over a 13-year period. *J Neurol*. 1997;244:657-663.
18. Casazza F, Brambilla G, Salvato A, Morandi L, Gronda E, Bonacina E. Dilated cardiomyopathy and successful cardiac transplantation in Becker's muscular distrophy. Follow-up after two years. *G Ital Cardiol*. 1988;18(9):753-757.

19. Connuck DM, Sleeper LA, Colan SD, et al. Characteristics and outcomes of cardiomyopathy in children with Duchenne or Becker muscular dystrophy: a comparative study from the Pediatric Cardiomyopathy Registry. *Am Heart J.* 2008 Jun;155(6):998-1005.
20. Muntoni F, Cau M, Ganau A, et al. Deletion of the dystrophin muscle-promotor region associated with X-linked dilated cardiomyopathy. *N Engl J Med.* 1993;329:921-925.
21. Palmucci L, Doriguzzi C, Mongini T, et al. Dilating cardiomyopathy as the expression of Xp21 Becker type muscular dystrophy. *J Neurol Sci.* 1992;111(2):218-221.
22. Towbin JA. Fielding Hejtmancik J, Brink P, et al. *X-linked dilated cardiomyopathy Circulation.* 1993;87:1854-1865.
23. Milasin J, Muntoni F, Severini GM, et al. A point mutation in the 5' splice site of the dystrophin gene first intron responsible for X-linked dilated cardiomyopathy. *Hum Mol Genet.* 1996 Jan;5(1):73-79.
24. Muntoni F, Di LA, Porcu M, et al. Dystrophin gene abnormalities in two patients with idiopathic dilated cardiomyopathy. *Heart.* 1997 Dec;78(6):608-612.
25. Hoogerwaard EM, van der Wouw PA, Wilde AAM, et al. Cardiac involvement in carriers of Duchenne and Becker muscular dystrophy. *Neuromusc Disord.* 1999;9:347-351.
26. Politano L, Nigro V, Nigro G, et al. Development of cardiomyopathy in female carriers of Duchenne and Becker muscular dystrophy. *JAMA.* 1996;275:1335-1338.
27. Grain L, Cortina-Borja M, Forfar C, Hilton-Jones D, Hopkin J, Burch M. Cardiac abnormalities and skeletal muscle weakness in carriers of Duchenne and Becker muscular dystrophies and controls. *Neuromuscul Disord.* 2001 Mar;11(2):186-191.
28. Rees W, Schuler S, Hummel M, Hetzer R. Heart transplantation in patients with muscular dystrophy associated with end-stage cardiomyopathy. *J Heart Lung Transplant.* 1993 Sep;12(5):804-807.
29. Williams TM, Lisanti MP. The Caveolin genes: from cell biology to medicine. *Ann Med.* 2004;36(8):584-595.
30. Hayashi T, Arimura T, Ueda K, et al. Identification and functional analysis of a caveolin-3 mutation associated with familial hypertrophic cardiomyopathy. *Biochem Biophys Res Commun.* 2004 Jan 2;313(1):178-184.
31. Cronk LB, Ye B, Kaku T, et al. Novel mechanism for sudden infant death syndrome: persistent late sodium current secondary to mutations in caveolin-3. *Heart Rhythm.* 2007 Feb;4(2):161-166.
32. Vatta M, Ackerman MJ, Ye B, et al. Mutant caveolin-3 induces persistent late sodium current and is associated with long-QT syndrome. *circ.* 2006 Nov 14;114(20):2104-2112.
33. Goodwin FC, Muntoni F. Cardiac involvement in muscular dystrophies: Molecular mechanisms. *Muscle Nerve.* 2005 Jun 2;32(5):577-588.
34. Brockington M, Yuva Y, Prandini P, et al. Mutations in the fukutin-related protein gene (FKRP) identify limb girdle muscular dystrophy 2I as a milder allelic variant of congenital muscular dystrophy MDC1C. *Hum Mol Genet.* 2001 Dec 1;10(25):2851-2859.
35. Poppe M, Cree L, Bourke J, et al. The phenotype of limb-girdle muscular dystrophy type 2I. *Neurology.* 2003 Apr 22;60(8):1246-1251.
36. van der Kooi AJ, Frankhuizen WS, Barth PG, et al. Limb-girdle muscular dystrophy in the Netherlands: gene defect identified in half the families. *Neurology.* 2007 Jun 12;68(24):2125-2128.
37. Sveen ML, Schwartz M, Vissing J. High prevalence and phenotype-genotype correlations of limb girdle muscular dystrophy type 2I in Denmark. *Ann Neurol.* 2006 May;59(5):808-815.
38. Wahbi K, Meune C, Hamouda eH, et al. Cardiac assessment of limb-girdle muscular dystrophy 2I patients: an echography, Holter ECG and magnetic resonance imaging study. *Neuromuscul Disord.* 2008 Aug;18(8):650-655. doi: Hamouda eH.
39. Kobayashi K, Nakahori Y, Miyake M, et al. An ancient retrotransposal insertion causes Fukuyama-type congenital muscular dystrophy. *Nature.* 1998 Jul 23;394(6691):388-392.
40. Toda T, Kobayashi K, Kondo-Iida E, Sasaki J, Nakamura Y. The Fukuyama congenital muscular dystrophy story. *Neuromuscul Disord.* 2000 Mar;10(3):153-159.
41. Murakami T, Hayashi YK, Noguchi S, et al. Fukutin gene mutations cause dilated cardiomyopathy with minimal muscle weakness. *Ann Neurol.* 2006 Nov;60(5):597-602.
42. Bione S, Maestrini E, Rivella S, et al. Identification of a novel X-linked gene responsible for Emery-Dreifuss muscular dystrophy. *Nature Genetics.* 1994;8:323-327.
43. Manilal S, Nguyen TM, Sewry CA, Morris GE. The Emery-Dreifuss muscular dystrophy protein, emerin, is a nuclear membrane protein. *Hum Mol Genet.* 1996 Jun;5(6):801-808.
44. Yates JRW. Workshop report: European workshop on Emery-Dreifuss muscular dystrophy. *Neuromuscul Disord.* 1991;1:393-396.
45. Emery AE. Emery-Dreifuss muscular dystrophy - a 40 year retrospective. *Neuromuscul Disord.* 2000 Jun;10(4–5):228-232.
46. Wehnert M, Muntoni F. 60th ENMC International Workshop: non X-linked Emery-Dreifuss Muscular Dystrophy 5–7 June 1998, Naarden, The Netherlands. *Neuromuscul Disord.* 1999 Mar;9(2):115-121.
47. Talkop UA, Talvik I, Sonajalg M, et al. Early onset of cardiomyopathy in two brothers with X-linked Emery-Dreifuss muscular dystrophy. *Neuromuscul Disord.* 2002 Nov;12(9):878-881.
48. Buckley AE, Dean J, Mahy IR. Cardiac involvement in Emery Dreifuss muscular dystrophy: a case series. *Heart.* 1999 Jul;82(1):105-108.
49. Sakata K, Shimizu M, Ino H, et al. High incidence of sudden cardiac death with conduction disturbances and atrial cardiomyopathy caused by a nonsense mutation in the STA gene. *circ.* 2005 Jun 28;111(25):3352-3358.
50. Fishbein MC, Siegel RJ, Thompson CE, Hopkins LC. Sudden death of a carrier of X-linked Emery-Dreifuss muscular dystrophy. *Ann Intern Med.* 1993 Nov 1;119(9):900-905.
51. Anderson LVB. Multiplex Western Blot Analysis of the Muscular Dystrophy Proteins. In: Bushby KMD, Anderson LVB, eds. *Muscular Dystrophy: Methods and Protocols.* Totowa: Humana Press; 2001:369-386.
52. Bonne G, Di Barletta MR, Varnous S, et al. Mutations in the gene encoding lamin A/C cause autosomal dominant Emery-Dreifuss muscular dystrophy. *Nat Genet.* 1999 Mar;21(3):285-288.

53. van Berlo JH, de Voogt WG, van der Kooi AJ, et al. Meta-analysis of clinical characteristics of 299 carriers of LMNA gene mutations: do lamin A/C mutations portend a high risk of sudden death? *J Mol Med.* 2005 Jan;83(1):79-83.
54. Curtis JP, Luebbert JJ, Wang Y, et al. Association of physician certification and outcomes among patients receiving an implantable cardioverter-defibrillator. *JAMA.* 2009 Apr 22;301(16):1661-1670.
55. Sovari AA, Bodine CK, Farokhi F. Cardiovascular manifestations of myotonic dystrophy-1. *Cardiol Rev.* 2007 Jul;15(4):191-194.
56. Groh WJ, Groh MR, Saha C, et al. Electrocardiographic abnormalities and sudden death in myotonic dystrophy type 1. *N Engl J Med.* 2008 Jun 19;358(25):2688-2697.
57. Hermans MC, Faber CG, Pinto YM. Sudden death in myotonic dystrophy. *N Engl J Med.* 2008 Oct 9;359(15):1626-1628.
58. Ref Type: Generic
59. Sansone V, Griggs RC, Meola G, et al. Andersen's syndrome: a distinct periodic paralysis. *Ann Neurol.* 1997 Sep;42(3):305-312.
60. Selcen D. Myofibrillar myopathies. *Curr Opin Neurol.* 2008 Oct;21(5):585-589.
61. Selcen D, Ohno K, Engel AG. Myofibrillar myopathy: clinical, morphological and genetic studies in 63 patients. *Brain.* 2004 Feb;127(Pt 2):439-451.
62. Sewry CA, Jimenez-Mallebrera C, Muntoni F. Congenital myopathies. *Curr Opin Neurol.* 2008 Oct;21(5):569-575.
63. Jungbluth H. Central core disease. *Orphanet J Rare Dis.* 2007;2:25.
64. Carmignac V, Salih MA, Quijano-Roy S, et al. C-terminal titin deletions cause a novel early-onset myopathy with fatal cardiomyopathy. *Ann Neurol.* 2007 Apr;61(4):340-351.
65. Feng JJ, Marston S. Genotype-phenotype correlations in ACTA1 mutations that cause congenital myopathies. *Neuromuscul Disord.* 2009 Jan;19(1):6-16.
66. D'Amico A, Graziano C, Pacileo G, et al. Fatal hypertrophic cardiomyopathy and nemaline myopathy associated with ACTA1 K336E mutation. *Neuromuscul Disord.* 2006 Oct;16(9–10):548-552.
67. Tajsharghi H, Oldfors A, Macleod DP, Swash M. Homozygous mutation in MYH7 in myosin storage myopathy and cardiomyopathy. *Neurology.* 2007 Mar 20;68(12):962.
68. Uro-Coste E, Arne-Bes MC, Pellissier JF, et al. Striking phenotypic variability in two familial cases of myosin storage myopathy with a MYH7 Leu1793pro mutation. *Neuromuscul Disord.* 2009 Feb;19(2):163-166.
69. Hagemans ML, Hop WJ, Van Doorn PA, Reuser AJ, Van der Ploeg AT. Course of disability and respiratory function in untreated late-onset Pompe disease. *Neurology.* 2006 Feb 28;66(4):581-583.
70. Van der Beek NA, Hagemans ML, Van der Ploeg AT, Reuser AJ, Van Doorn PA. Pompe disease (glycogen storage disease type II): clinical features and enzyme replacement therapy. *Acta Neurol Belg.* 2006 Jun;106(2):82-86.
71. Van der Beek NA, Soliman OI, van Capelle CI, et al. Cardiac evaluation in children and adults with Pompe disease sharing the common c.-32–13T>G genotype rarely reveals abnormalities. *J Neurol Sci.* 2008 Dec 15;275(1–2):46-50.
72. Danon MJ, Oh SJ, DiMauro S, et al. Lysosomal glycogen storage disease with normal acid maltase. *Neurology.* 1981 Jan;31(1):51-57.
73. Tranchant C, Mousson B, Mohr M, et al. Cardiac transplantation in an incomplete Kearns-Sayre syndrome with mitochondrial DNA deletion. *Neuromuscul Disord.* 1993 Sep;3(5–6):561-566.
74. Bit-Avragim N, Perrot A, Schols L, et al. The GAA repeat expansion in intron 1 of the frataxin gene is related to the severity of cardiac manifestation in patients with Friedreich's ataxia. *J Mol Med.* 2001;78(11):626-632.
75. Barth PG, Valianpour F, Bowen VM, et al. X-linked cardioskeletal myopathy and neutropenia (Barth syndrome): an update. *Am J Med Genet A.* 2004 May 1;126A(4):349-354.
76. Barth PG, Wanders RJ, Vreken P. X-linked cardioskeletal myopathy and neutropenia (Barth syndrome)-MIM 302060. *J Pediatr.* 1999 Sep;135(3):273-276.
77. Stangou AJ, Hawkins PN, Heaton ND, et al. Progressive cardiac amyloidosis following liver transplantation for familial amyloid polyneuropathy: implications for amyloid fibrillogenesis. *Transplantation.* 1998 Jul 27;66(2):229-233.
78. Ferlini A, Galié N, Merlini L, Sewry C, Branzi A, Muntoni F. A novel Alu-like element rearranged in the dystrophin gene causes a splicing mutation in a family with X-linked dilated cardiomyopathy. *Am J Hum Genet.* 1998 Aug;63(2):436-46.
79. Ozawa T, Tanaka M, Sugiyama S, Hattori K, Ito T, Ohno K, Takahashi A, Sato W, Takada G, Mayumi B, et al. Multiple mitochondrial DNA deletions exist in cardiomyocytes of patients with hypertrophic or dilated cardiomyopathy. *Biochem Biophys Res Commun.* 1990 Jul 31;170(2):830-6.

Sudden Cardiac Death in the Young; Epidemiology and Cardiogenetic Evaluation of Victims and Their Relatives

Anneke Hendrix, Michiel L. Bots, and Arend Mosterd

25.1 Introduction

Sudden death (SD) in young persons can be divided into natural (death due to a disease) and nonnatural (e.g. car accident or suicide) causes of death. The natural deaths can be subdivided into noncardiac deaths, cardiac deaths, and unexplained deaths. The latter two categories comprise the cardiac deaths due to inherited diseases and are discussed further in this chapter (see Fig. 25.1).

The *sudden cardiac death (SCD)* or *sudden unexplained death (SUD)* of a young person has an enormous impact on those who are left behind. It evokes severe feelings of anxiety and incomprehension because such a dramatic event was not anticipated. During the last 10–15 years, it has become clear that in 50–70% of SCD and SUD victims aged 40 years or younger, potential inherited cardiac disease can be identified as the cause of sudden death.[1,2] *Cardiomyopathies* (e.g. hypertrophic cardiomyopathy [HCM]) or *primary arrhythmia syndromes* (e.g. congenital long-QT syndrome [LQTS]) can cause fatal arrhythmias that may lead to sudden death. Premature *coronary artery disease*, as observed in *familial hypercholesterolemia* (FH), is another cause of sudden death in the young. Relatives of young SCD/SUD victims have an increased risk of carrying the inherited predisposition to develop cardiac disease.[3–8] Furthermore, a family history of sudden death is associated with an increased risk of sudden death among adult family members.[9–11] Increasingly, genetic testing is available for inherited cardiac diseases and new mutations, accounting for specific phenotypes, are being discovered.[12,13]

Diagnostic evaluation of first-degree relatives followed by early treatment may reduce the risk of SCD in patients with inherited cardiac diseases.[14–17] However, as *sudden cardiac arrest* in the young is often the first "symptom" of inherited cardiac disease, early identification is difficult in apparently healthy individuals.[18–20] To prevent SCD in the young, presymptomatic *cardiogenetic evaluation* of high risk groups has been proposed, aiming primarily at first-degree *relatives* of SCD/SUD victims with possible inherited diseases.

In this chapter, we shall present an overview of the epidemiology of SCD/SUD in the young and discuss the potential benefits of presymptomatic cardiogenetic evaluation of first-degree relatives. In addition, we shall discuss *preparticipation screening* of young *athletes*.

25.2 Definitions

The terminology that is used to describe SCD is often confusing due to the variety of definitions that are used in the literature. A distinction can be made between sudden death in the young and the sudden infant death syndrome. The term "*sudden death* in the young" covers a broad spectrum of different definitions, but in general, comprises victims between 1 and 40 years of age. The general *definition* of sudden infant death syndrome (SIDS) is 'sudden unexpected death of an infant <1 year of age, with onset of the fatal episode apparently occurring during sleep, that remains unexplained after a thorough investigation, including performance of a complete autopsy and review of the circumstances of death and the clinical history'.[21] The distinction

---

A. Hendrix (✉)
Interuniversity Cardiology Institute of the Netherlands, Utrecht, The Netherlands and
Julius Center for Health Sciences and Primary Care, University Medical Center Utrecht, Utrecht, The Netherlands
e-mail: a.hendrix@umcutrecht.nl

Fig. 25.1 Flow chart, sudden death and inherited cardiac diseases (cardiogenetic evaluation of relatives is explained in Fig. 4)

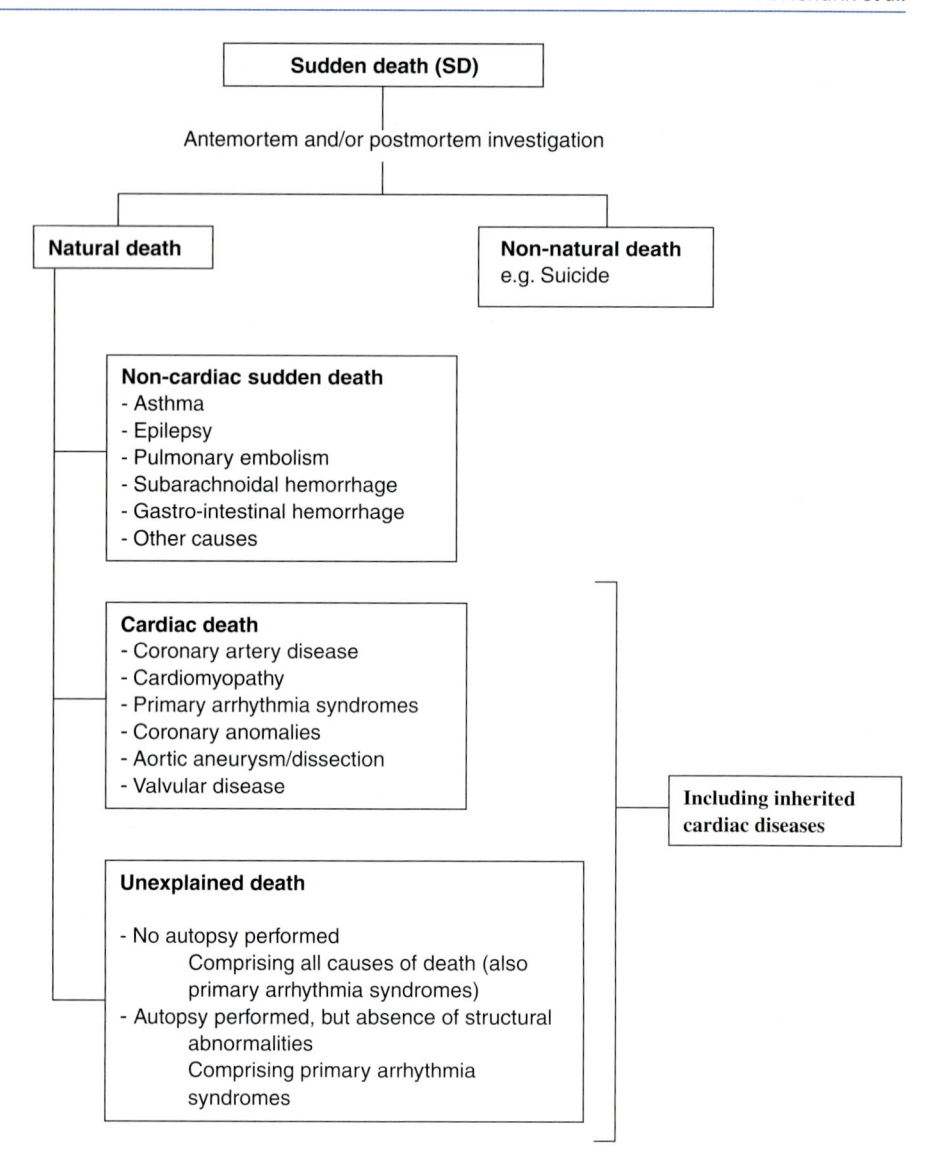

```
                        ┌─────────────────────┐
                        │  Sudden death (SD)  │
                        └──────────┬──────────┘
                                   │
                   Antemortem and/or postmortem investigation
                   ┌───────────────┴───────────────┐
          ┌────────┴────────┐              ┌───────┴─────────┐
          │  Natural death  │              │ Non-natural death│
          └────────┬────────┘              │   e.g. Suicide  │
                   │                       └─────────────────┘
                   │
   ┌───────────────┴─────────────────────────┐
   │  Non-cardiac sudden death               │
   │  - Asthma                               │
   │  - Epilepsy                             │
   │  - Pulmonary embolism                   │
   │  - Subarachnoidal hemorrhage            │
   │  - Gastro-intestinal hemorrhage         │
   │  - Other causes                         │
   └─────────────────────────────────────────┘

   ┌─────────────────────────────────────────┐        ┌──────────────────────┐
   │  Cardiac death                          │        │ Including inherited  │
   │  - Coronary artery disease              │        │  cardiac diseases    │
   │  - Cardiomyopathy                       │        └──────────────────────┘
   │  - Primary arrhythmia syndromes         │
   │  - Coronary anomalies                   │
   │  - Aortic aneurysm/dissection           │
   │  - Valvular disease                     │
   └─────────────────────────────────────────┘

   ┌─────────────────────────────────────────┐
   │  Unexplained death                      │
   │                                         │
   │  - No autopsy performed                 │
   │      Comprising all causes of death     │
   │      (also primary arrhythmia syndromes)│
   │  - Autopsy performed, but absence of    │
   │    structural abnormalities             │
   │      Comprising primary arrhythmia      │
   │      syndromes                          │
   └─────────────────────────────────────────┘
```

between these two age categories (<1 year and >1 year) is based on the differences in incidence and causes. For example, the occurrence of SIDS is strongly associated with sleeping position (after an international public health campaign to place infants on their back while sleeping, the incidence of SIDS decreased with 50–90%).[22] Although there is an overlap between the causes of death in SIDS (e.g. the congenital LQTS) and sudden death in the young, we shall focus on SCD victims older than one year in this chapter.

SCD was recently defined as 'death due to any cardiac disease that occurs out of hospital, in an emergency department, or in an individual reported death on arrival at a hospital'.[23] The term *'sudden unexplained death'* (SUD) is used when no diagnosis can be established based on antemortem and postmortem investigation (*autopsy-negative sudden deaths*).[24] However, SUD might also refer to cases, in which the cause of death remains uncertain because no autopsy is performed. In the absence of structural abnormalities on autopsy, the term *'sudden arrhythmic death syndrome'* has been proposed.[3,6] Although the assumption is that many of the autopsy-negative sudden deaths are due to an electrical heart disease or a cardiomyopathy with minor histopathologic changes, other causes cannot be excluded automatically (e.g. epileptic seizure, electrolyte abnormalities, intoxication).

In the literature, the 1, 6, and 24 h definitions are used in the SCD definition.[25–27] This distinction in duration of symptoms before the victim's death directly influences the established proportion of (cardiac) causes of death. In studies that use the '1 h definition', the reported proportion of cardiac deaths due to an arrhythmia will be higher in comparison to studies that use the '24 h definition'. The use of definitions that are based on the duration of preceding symptoms has limitations. In the first place, nonspecific symptoms (e.g. fatigue or atypical chest pain) are often present for a period longer than 24 h. Second, in daily practice, it can be difficult to determine the exact duration of symptoms before dying, especially because this information needs to be collected retrospectively and no witness is present in 40% of the sudden cardiac arrests.[27] To distinguish the 'sudden deaths' from deaths due to other terminal diseases, it may therefore be better to use 'unexpected' death instead of 'sudden' death. In the third place, victims may die a few days after a successful resuscitation because of irreversible damage (e.g. brain death).[28] These victims are often not included in sudden death studies while the causes of death and the clinical implications for relatives are the same. Surviving a sudden cardiac arrest depends mainly on external factors like early and adequate cardiopulmonary resuscitation and early defibrillation.

25.3 Incidence

In the general population (over all ages), SCD accounts for approximately one death per 1,000 person-years.[29] In the young (<40 years), the *incidence* of *SCD* is estimated to be 100-fold lower and lies between 0.8 and 1.6 per 100.000 person-years (Fig. 25.2).[30, 31] The population-based incidence of *autopsy-negative sudden death* is estimated to be 0.16–0.43/100.000 person-years.[26, 32] The incidence of SCD in *athletes* ranges from 0.6 to 3.6/100.000 person-years.[31,33]

Incidence estimates vary considerably between studies. The collection of data on this topic is complicated, because most cases of sudden death occur out of hospital and often information needs to be collected retrospectively.[27] In addition, *traumatic deaths* like car accidents or drownings can initially be caused by a cardiac arrhythmia, but are often not taken into account in incidence estimates. As no nationwide registrations of victims of SCD or SUD of 1–40 years exist, present studies are often restricted to regional observations, where socio-economical status, racial differences, and the presence of founder mutations predisposing to a specific cardiogenetic disorder might influence the occurrence of SCD.[25,34] From studies based on *death certificate diagnosis,* absolute numbers of sudden deaths in the young population may be adequately derived, but the proportion of cardiovascular deaths may be unreliable due to misclassification.[35, 36]

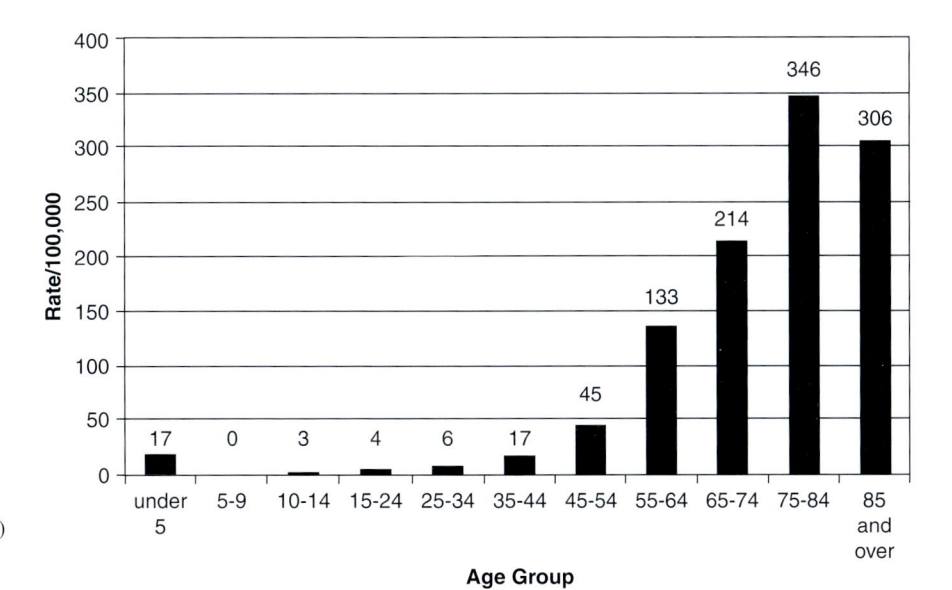

Fig. 2 Annual incidence of sudden cardiac death among residents of Multnomah County by age-groups, Oregon (population 660,486) (Adapted from J Am Coll Cardiol, 2004[37])

25.4 Causes

SCD in persons older than 40 years is mainly due to coronary artery disease that can result in myocardial ischemia and fatal arrhythmias. It has been estimated that 80% of the cardiac deaths in victims over 40 years of age are caused by coronary artery disease, 10–15% by cardiomyopathies, and 5% by other (less common) causes.[38] However, in the young (1–40 years), inherited cardiac causes are more frequently observed. Recently, a review was published that included articles from 1980 to 2007 on *causes* of death in the young. All studies were included in which autopsy was performed in >70% of the sudden death victims.[30] 17 publications were identified, including 3,528 cases of SCD in the age-group of 1–40 years that were collected between 1967 and 2004. The most common causes of SCD in persons aged 1–40 years in the general population were atherosclerotic coronary artery disease (accounting for 23% of the cases), followed by cardiomyopathies (13%) and myocarditis (6%). In athletes, cardiomyopathies were the most common causes of SCD (accounting for 48% of the cases), followed by (non-atherosclerotic) coronary pathology (e.g. coronary artery aneurysm and vasculitis) (16%) and atherosclerotic coronary artery disease (7%). A considerable proportion of the sudden deaths remained unexplained (autopsy-negative sudden deaths), accounting for 16% of the SCD in the general population and 4% in the athlete population. This latter group comprises primary arrhythmia syndromes (e.g. LQTS, catecholaminergic polymorphic ventricular tachycardia (CPVT) and the Brugada syndrome [BS]).[39, 40] The proportion of deaths due to coronary artery disease increases with age. In addition, *myocarditis* and *primary arrhythmia syndromes* were relatively more common in the younger population (1–25 years).[18, 30, 41]

25.5 Demographics

Information on regional, racial, and gender differences in SCD in the young is scarce. However, it seems that the incidence and causes of SCD (over all ages) differ among regions and populations.[42,43] This might be due

Causes of death	General population		Athletes	
	N	% (95%CI)	N	% (95%CI)
Sudden cardiac death				
Atherosclerotic disease	726	23 (22–25)	27	7 (5–10)
Conduction disorders	44	1 (1–2)	5	1 (0–3)
Myocarditis	195	6 (5–7)	16	4 (2–6)
Cardiomyopathy[a]	397	13 (11–14)	181	48 (43–53)
Coronary pathology (non-ischemic)[b]	73	2 (2–3)	61	16 (12–20)
Congenital cardiac diseases[c]	37	1 (1–2)	2	1 (0–1)
Valve abnormalities[d]	120	4 (3–5)	22	6 (4–8)
Other cardiovascular diseases	230	7 (6–8)	8	2 (1–4)
Sudden death with unknown cause	519	16 (15–18)	16	4 (2–6)
Sudden non-cardiac death				
Respiratory[e]	244	8 (7–9)	8	2 (1–4)
Neurologic[f]	289	9 (8–10)	2	1 (0–1)
Other non cardiac deaths	249	8 (7–9)	22	6 (4–8)
Abdominal aneurysma[g]	27	1 (1–1)	8	2 (1–4)
Total sudden deaths	3,150	100	378	100

[a] Arrhythmogenic right ventricular cardiomyopathy, dilated cardiomyopathy, left ventricular hypertrophy, diffuse fibrosis, endocardial fibroelastosis, myocardial fibrosis, idiopathic myocardial scarring, right ventricular dysplasia, fibroelastosis cordis
[b] Coronary abnormalities, coronary bridging, vasculitis, coronary artery aneurysm
[c] Marfan syndrome, tetralogy of Fallot
[d] Mitral valve prolapse, mitral valve insufficiency, aortic valve insufficiency
[e] Asthma, pulmonary embolism
[f] Epilepsy, subarachnoidal haemorrhage, intracranial haemorrhage, meningitis

to several factors, including the regional distribution of age and gender and the prevalence of inherited cardiac diseases and coronary artery disease. Several studies reported clustering of inherited diseases in populations (and regions); in southeast Asia for example, especially in Cambodia, Philippines, Thailand, and Japan, the incidence of nocturnal sudden death among young men was estimated to be as high as 26–38 per 100.000 person-years. Cardiogenetic evaluation suggested that a *primary arrhythmia syndrome* similar to the BS is underlying these sudden deaths.[44–47] Furthermore, SCD occurs more frequently in African Americans than in white Americans.[41] HCM is the most common cause of SCD in the athletes in the United States, while in the Veneto region in Italy, *arrhythmogenic right ventricular dysplasia/cardiomyopathy (ARVD/C)* is accounting for the majority of deaths among athletes.[33,48,49]

Overall, *SCD* in the young (1–40 years) is more common in men (2.27 per 100,000 person-years) than in women (0.95 per 100,000 person-years).[30] This difference can partly be explained by the high proportion of deaths due to coronary artery disease which is increasing over age (especially >30 years)[30,41] and perhaps because women are relatively protected for the development of atherosclerosis in the premenopausal period.[50–52]

25.6 Postmortem Diagnosis

A dedicated and focused *postmortem investigation* as compared to a 'routine' cardiac postmortem investigation is essential in detecting potential inherited cardiac diseases in sudden death victims. Postmortem investigation includes investigation of the circumstances of death, verification of the victim's medical history and family history, autopsy, and DNA storage.[28,53]

25.6.1 Circumstances of Death, Verification of the victim's Medical History and Family History

An effort should be made to obtain relevant information from health care professionals (e.g. resuscitation team) and other witnesses of the event regarding the circumstances of death (e.g. occurring during sleep, emotional stress, or exercise), the type and duration of *preceding symptoms* (e.g. chest pain, dizziness, nausea, fever, or headache), and the location of the fatal event. Relatives and general practitioners can be a useful source of information on medical and family history. Circumstances of death can provide important clues to the underlying causes of death, since triggers of sudden cardiac arrest might be specific for the underlying disease. Arrhythmias may be triggered by exercise in patients with *HCM, ARVD/C, LQTS type 1,* or *CPVT,* while in patients with *BS* or *LQTS* type 3, fatal arrhythmias more often occur during sleep. Furthermore, information on the victim's medical history should be collected (e.g. recent infections, surgical operations, or comorbidities (e.g. hypertension, neuromuscular diseases, asthma, or epilepsy). Medication use as well as cigarette exposure, alcohol and substance abuse should be reported.

Besides, the victim's *family history* with respect to sudden death and inherited cardiac or neuromuscular diseases can reveal important additional information. When available, *antemortem investigations* (e.g. electrocardiogram, echocardiography, results of exercise testing, or CT-scan) should be done.[28,53]

A systematic review and analysis to establish which specific clinical information is useful to determine the cause of death is yet to be performed. These types of studies are however desperately needed to enhance our knowledge and improve the clinical care of sudden death victims and their relatives.

25.6.2 Autopsy

It is difficult to compare *autopsy* studies of sudden death in the young as autopsies have not been performed according to a standardized protocol.[28] When autopsy is limited to macroscopic examination of the heart without histological sampling of cardiac tissue or *toxicological examination*, focal cardiac abnormalities or extracardiac causes such as an intoxication may remain undetected. In 79% of the cases of SCD without macroscopic abnormalities, *histopathological examination* of the heart revealed local pathology like focal *myocarditis* or the presence of conduction system abnormalities.[25] Revision of the heart by a cardiac expert pathologist is recommended when no cause of death can be established. A toxicological examination for drugs (e.g. opiates, amphetamine), alcohol, and medication may be considered necessary in cases with autopsy-negative findings.[54,55] Recently, international *guidelines* for autopsy in SCD/SUD in the young have been published.[28] Standardization of autopsy is an important tool in research on causes of SCD and SUD and facilitates comparison and pooling of data from various sources.

25.6.2.1 DNA Storage

Storage of the victim's *DNA* enables genetic testing when relatives consult a cardiologist or clinical-geneticist for cardiogenetic evaluation. Autopsy allows tissue to be collected and stored in a tissue bank. Usually, only *paraffin-embedded tissue* is stored when a SCD/SUD victim is autopsied, which is not an optimal source for extensive genetic testing. Guidelines have supported the notion that it is desirable to store EDTA-blood and/or frozen muscle, liver or spleen tissue that can be obtained during autopsy.[24,56] When no autopsy is performed, a skin *biopsy* might be taken (with permission from the victim's relatives). Prior to the biopsy, the skin needs to be disinfected with alcohol. The obtained tissue can be temporarily stored in a sterile vial with physiological isotonic saline before sending it to a DNA laboratory.[57]

25.7 Cardiogenetic Evaluation of First-Degree Relatives of Young Sudden (Cardiac) Death Victims

25.7.1 Cardiogenetic Clinic

When an inherited cardiac disease is suspected or when the cause of death remains unknown, relatives should be referred for *cardiogenetic evaluation* which comprises cardiological assessment and/or genetic testing. Cardiogenetic evaluation entails several aspects that need careful consideration upfront. These include difficulties in establishing a final diagnosis, the interpretation of genetic test results, and the ethical considerations concerning genetic testing.[58] Relatives should be informed about the advantages and disadvantages of cardiogenetic evaluation (see chapter clinical genetics). Therefore, dedicated *cardiogenetic outpatient clinics* have been established that provide integrated cardiogenetic care by combining expertise from the fields of ethics, genetics, and cardiology.

25.7.2 Genetic Testing

Recent cardiogenetic developments have resulted in the identification of many different mutated genes related to specific cardiac pathology, and consequently lead to a better understanding of the pathophysiology of clinical syndromes such as HCM and LQTS.[59] However, *genetic testing* does not always reveal a genetic mutation when an inherited cardiac disease is suspected. Not all causative mutations have been discovered yet and many clinical syndromes show genetic heterogeneity. In individuals with *LQTS*, the current yield of genetic testing is estimated to be 50–70%.[60] For many other diseases (e.g. *dilated cardiomyopathy* [DCM]) the yield of genetic testing is less.[61] In *autopsy-negative sudden death victims*, genetic testing is complicated, expensive, and time-consuming, especially when no specific clinical diagnosis is considered based on careful evaluation of circumstances of death and past medical history of the victim. Genetic testing in 49 autopsy-negative sudden death victims revealed a mutation in LQTS-associated genes in 20% of the victims and a mutation in *CPVT*-associated genes in 14% of the victims.[39,40] Yet, since such extensive genetic testing is not feasible in daily practice, it seems logical that the detection rate of heritable diseases is lower in the clinical setting. In the future, a molecular autopsy might become possible for victims of autopsy-negative sudden death. Molecular autopsy comprises testing of a large amount of genes that underlie primary arrhythmia syndromes.[62]

Because most inherited heart diseases show an *autosomal dominant* pattern of inheritance, first-degree relatives of *SCD* victims with genetic disease have a 50% risk of being a carrier of the same disease.[59] As mentioned earlier, recent studies showed that, with thorough clinical assessment of first-degree *relatives* of SUD victims, a cause of death can be established in 22–53% of the families.[3–6] A Dutch investigation of 43 families of *SUD* victims, of whom 22 were autopsied, found an inherited cardiac disease in 17 of the 43

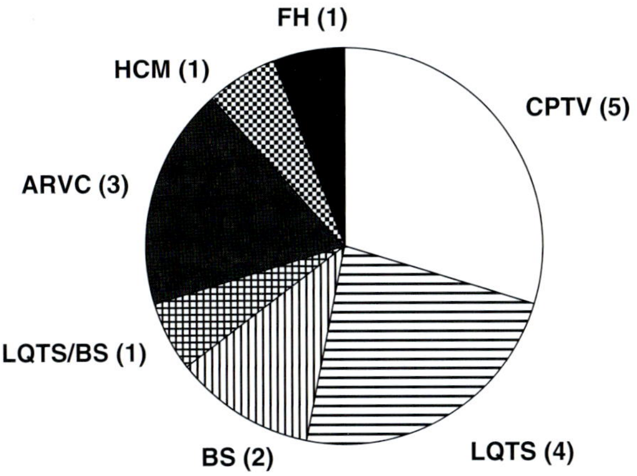

Fig. 25.3 Diagnosis after evaluation of families with ≥1 SUD victim *FH* familiar hypercholesterolemia, *HCM* hypertrophic cardiomyopathy, *ARVC* arrythmogenic right ventricular cardiomyopathy, *LQTS* long-QT syndrome, *BS* Brugada syndrome, *CPVT* catecholaminergic polymorphic ventricular tachycardia (Adapted from Tan HL et al., Circulation 2005[4])

families that explained the sudden death of the victims (Fig. 25.3).[4] Furthermore, a study executed in the United Kingdom revealed an inherited disease in 53% of the 57 families of autopsy-negative sudden death victims aged 4–64 years.[6]

The yield of genetic testing in relatives is high when the causative mutation is known.[63] However, it should be realized that a relative carrying the mutation of an inherited cardiac disease will not necessarily develop signs or symptoms of the clinical syndrome that is associated with the mutation. *Penetrance* of genetic mutations may vary among individuals of the same family and sometimes secondary factors (or genes) can influence the *phenotype expression*.[59,61]

25.7.3 Evaluation of First-Degree Relatives

Three scenarios for the cardiogenetic evaluation of first-degree *relatives* of young *SCD/SUD* victims can be distinguished (Fig. 25.4):

(a) The causative mutation in the SCD victim is known.
(b) An inherited cardiac disease is suspected in the victim, but not established by genetic testing.
(c) The cause of the victim's death is unknown (with or without extensive autopsy)

Scenario (a) allows for a targeted approach by genetic testing of the victim's relatives. Cascade screening, starting with genetic testing of the (genetically) first-degree relatives (which include the parents, children, brothers, and sisters) of an affected individual will genetically identify the causative mutation in one or more of the relatives. Subsequently, the screening can be extended to the connecting branch of the pedigree.[61] Consequently, absence of the mutation rules out the presence of the disease, and no further investigation of the pedigree is needed. In case a causative mutation is present in a relative, cardiologic evaluation and/or diagnostic follow-up is usually indicated.

In scenario (b), the *cardiogenetic evaluation* of relatives is more complicated. The postmortem findings in the victim may raise the suspicion of an inherited cardiac disease. Based on this, targeted cardiologic evaluation of the relatives can be performed. Based on the results of the cardiologic evaluation of relatives, targeted genetic testing can be performed in the victim (if the victim's DNA is available) or in the clinically affected relative (with cardiac abnormalities).

If a mutation is found, cascade screening in the pedigree is recommended (see scenario a). In case no mutation can be detected, cardiologic evaluation (guided by the findings in the affected relative or postmortem findings in the sudden death victim) of all first-degree relatives may be considered. As the penetrance of a causative mutation may differ among individuals, the absence of abnormalities on cardiologic evaluation does not automatically rule out the presence of an inherited disease. In some diseases, symptoms develop only at older age (e.g. HCM, DCM, and ARVD/C), which may mandate follow-up in these individuals.

In scenario (c), the cause of sudden death is unknown and no clues are available for a specific diagnosis, which makes *cardiogenetic evaluation* less feasible. A cardiac examination of the relatives may reveal a relevant diagnosis. Examination should include the following aspects; (1) medical history, (2) physical examination, (3) standard resting 12 lead electrocardiogram and 12 lead electrocardiogram with specific right precordial positioning of the leads (leads –V1, -V2, -1V1, and -1V2), (4) echocardiography, (5) Holter recording, (6) exercise testing, and (7) measurement of serum lipid levels.[64] If the initial examination raises the possibility of a specific genetic disorder, additional investigations may be indicated, which may include provocation

Fig. 25.4 Flow chart, cardiogenetic evaluation of first-degree relatives

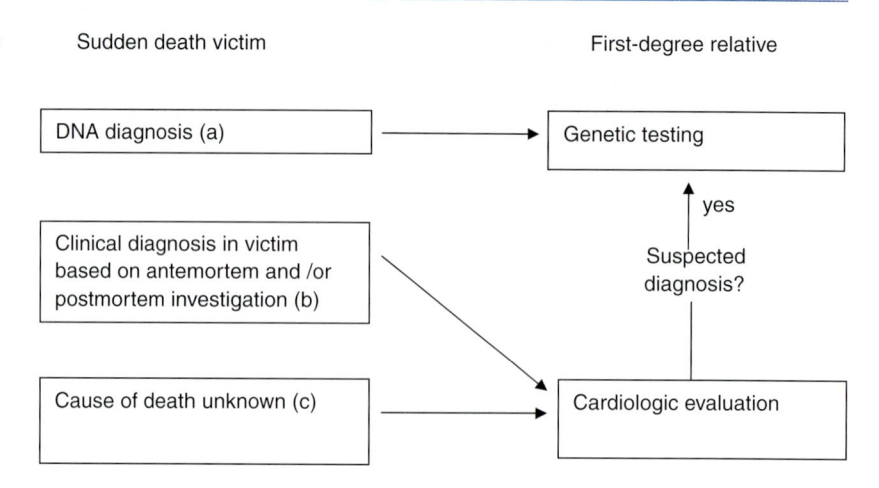

testing (e.g. ajmaline challenging), cardiac MRI, and genetic testing.[3–6]

25.7.3.1 Cost-Effectiveness of Cardiogenetic Evaluation of First-Degree Relatives

A formal evaluation of the above-mentioned scenarios to identify inherited cardiac diseases in relatives of young sudden death victims has not been performed. The *cost-effectiveness* of cardiogenetic evaluation of relatives of SCD/SUD victims depends on the balance between the probability of identifying the causative mutation and its associated therapeutic (risk reduction through treatment) and prognostic consequences (risk when no treatment is given). This can differ from situation to situation. Since the costs of *genetic testing* are associated with the number and size of the analyzed genes, a targeted genetic evaluation of relatives (e.g. scenarios a and b) is likely to result in a higher yield of genetic testing and lower costs.[62] When the cause of death is unknown (scenario c), genetic evaluation is less (cost)-effective.[65] Limited analyses of only those genes that are responsible for the major part of the clinical syndromes seem to increase efficiency. To date, no studies on the yield and cost-effectiveness of *cardiogenetic evaluation* of relatives of SCD/SUD victims have been published. A cost-effectiveness analysis through modeling is necessary to provide additional information on the value of cardiogenetic evaluation of relatives for different diseases and scenarios.[30] Based on these future studies, recommendations can be drafted regarding the cardiogenetic evaluation of relatives of SCD/SUD victims.

25.8 Preparticipation Screening of Athletes

Being physically active is generally regarded as the best way to prevent (cardiovascular) disease, but vigorous activity can also acutely and transiently increase the risk of acute cardiac events in susceptible persons.[66] *Physical activity* can trigger fatal arrhythmias in persons with cardiomyopathies (e.g. *HCM* or *ARVD/C*), CPVT, and LQTS or provoke coronary plaque rupture in those with coronary artery disease.[66,67] The sudden death of an apparently healthy *athlete* inevitably leads to discussions if death could have been prevented by preparticipation screening.

Approximately, 5–14% of all SCDs in the young occur during physical activity.[68–70] However, it is still largely unclear whether young athletes have an increased risk of SCD compared to nonathletes. An Italian study reported that the risk of SD was 2.5 (CI 1.8–3.4) times higher in young athletes than in nonathletes.[71] In this study, events not directly associated with physical activity were also taken into account. The physician's Health Study reported a 16.9 times higher risk of sudden death (CI 10.5 to 27.0) during physical activity and the 30 min after physical activity than during episodes of low activity, but an association between the frequency of physical activity and the long-term risk of sudden death could not be established.[70] A moderate to intensive level of physical activity was associated with a significant decrease in SCD in a study of 7,735 middle-aged men.[72]

In 2005, the Study Group of Sport Cardiology of the European Society of Cardiology issued recommendations for routine preparticipation cardiovascular screening of young competitive athletes. The evaluation consists of a questionnaire, physical examination, and an electrocardiogram.[73] An electrocardiogram is not part of the *preparticipation screening* recommended by the American Heart Association that essentially relies on medical history and physical examination.[74] The European recommendations are largely based on the Italian experience, suggesting that the introduction of a screening program in young athletes led to a decline in the incidence of SCD. In Italy, 55 cases of SCD aged 12–35 years among screened athletes were registered between 1997 and 2004. The incidence of SCD dropped from 4.19 (CI 1.78–7.59) to 0.87 (CI 0.46–1.28) per 100,000 athletes per year.[31]

The European preparticipation screening recommendations have led to a stream of pro-contra discussions.[75,76] The main criticisms of the recommendations are the lack of randomized studies to support the recommendations, the absence of validated questionnaires and electrocardiographic criteria, and the potential of false positive findings among screened athletes, especially given the rare occurrence of SCD in young athletes.[75] The interpretation of the electrocardiogram of athletes is hampered by physiological adaptation to systematic training, known as the athlete's heart.[77] Up to 40% of the electrocardiograms taken in athletes demonstrate variations that can be deemed abnormal, such as sinus bradycardia, atrial fibrillation, and ST-segment changes in the right precordial leads.[78] Furthermore, it has been estimated that only a small proportion of the athletes who suddenly died would have been previously identified as being at increased risk for SCD by the preparticipation program.[75]

The Netherlands Institute for Public Health and Environment calculated that a randomized study of the effects of *preparticipation screening* (assuming a 50% reduction in the rate of SCD from four to two per 100,000 athletes per year, with 80% power) would mandate two groups with 1,200,000 person-years of follow-up. Taking this into account, it is unlikely that definitive evidence supporting the use of preparticipation screening will ever be presented.

Raising awareness of the potential consequences of symptoms during exercise (e.g. collapse, chest discomfort), improving the availability of automated external defibrillators, and careful cardiogenetic evaluation of young SCD victims and their relatives may constitute sound alternatives to mandatory preparticipation screening. Given the higher absolute number of acute cardiac arrests in older athletes, in whom coronary artery disease is far more common, this group should not be neglected.[79]

25.9 Key Messages

- The *incidence* of *sudden cardiac death* in persons 40 years or younger is approximately 0.8–1.6 per 100,000 person-years.
- Sudden cardiac death or *sudden unexplained death* in the young is frequently caused by inherited cardiac diseases.
- A thorough *postmortem investigation* of young SCD/SUD victims is desirable to establish the cause of death.
- Standardization of *autopsy* is important to compare studies on the causes of sudden cardiac death and sudden unexplained death and to enable pooling of information from various sources.
- *Relatives* of sudden cardiac death victims with diagnosed inherited diseases have a high risk of being a carrier of an inherited cardiac disease, because most inherited cardiac diseases show an autosomal dominant pattern of inheritance.
- If postmortem investigation does not reveal any structural abnormalities in young victims (so called autopsy-negative sudden death) or when no postmortem examination is performed in SUD victims, *cardiogenetic evaluation* of relatives should be considered.
- The *cost-effectiveness* of screening of relatives of sudden cardiac and sudden unexplained death victims needs to be analyzed in future studies.
- Given the rare occurrence of SCD in young athletes, the preventive effects of routine *preparticipation screening* is likely to be limited.
- Raising awareness of the potential consequences of symptoms during exercise (e.g. collaps, chest discomfort), improving the availability of automated external defibrillators, and careful cardiogenetic evaluation of young SCD/SUD victims and their relatives may constitute sound alternatives to mandatory preparticipation screening.

References

1. Wilde AA, Bezzina CR. Genetics of cardiac arrhythmias. *Heart*. 2005 October;91(10):1352-1358.
2. Wilde AA, van Langen IM, Mannens MM, Waalewijn RA, Maes A. Sudden death at young age and the importance of molecular-pathologic investigation. *Ned Tijdschr Geneeskd*. 2005 July 16;149(29):1601-1604.
3. Behr E, Wood DA, Wright M, et al. Cardiological assessment of first-degree relatives in sudden arrhythmic death syndrome. *Lancet*. 2003 November 1;362(9394):1457-1459.
4. Tan HL, Hofman N, van Langen IM, van der Wal AC, Wilde AA. Sudden unexplained death: heritability and diagnostic yield of cardiological and genetic examination in surviving relatives. *Circulation*. 2005 July 12;112(2):207-213.
5. Hofman N, Tan HL, Clur SA, Alders M, van Langen IM, Wilde AA. Contribution of inherited heart disease to sudden cardiac death in childhood. *Pediatrics*. 2007 October;120(4): e967-e973.
6. Behr ER, Dalageorgou C, Christiansen M, et al. Sudden arrhythmic death syndrome: familial evaluation identifies inheritable heart disease in the majority of families. *Eur Heart J*. 2008 July;29(13):1670-1680.
7. Schwartz K, Carrier L, Guicheney P, Komajda M. Molecular basis of familial cardiomyopathies. *Circulation*. 1995 January 15;91(2):532-540.
8. Garson A Jr, Dick M, Fournier A, et al. The long QT syndrome in children. An international study of 287 patients. *Circulation*. 1993 June;87(6):1866-1872.
9. Friedlander Y, Siscovick DS, Weinmann S, et al. Family history as a risk factor for primary cardiac arrest. *Circulation*. 1998 January 20;97(2):155-160.
10. Jouven X, Desnos M, Guerot C, Ducimetiere P. Predicting sudden death in the population: the Paris Prospective Study I. *Circulation*. 1999 April 20;99(15):1978-1983.
11. Elliott PM, Poloniecki J, Dickie S, et al. Sudden death in hypertrophic cardiomyopathy: identification of high risk patients. *J Am Coll Cardiol*. 2000 December;36(7):2212-2218.
12. Alders M, Koopmann TT, Christiaans I, et al. Haplotype-sharing analysis implicates chromosome 7q36 harboring DPP6 in familial idiopathic ventricular fibrillation. *Am J Hum Genet*. 2009 April;84(4):468-476.
13. Lehnart SE, Ackerman MJ, Benson DW, et al. Inherited arrhythmias: a National Heart, Lung, and Blood Institute and Office of Rare Diseases workshop consensus report about the diagnosis, phenotyping, molecular mechanisms, and therapeutic approaches for primary cardiomyopathies of gene mutations affecting ion channel function. *Circulation*. 2007 November 13;116(20):2325-2345.
14. Maron BJ, Spirito P, Shen WK, et al. Implantable cardioverter-defibrillators and prevention of sudden cardiac death in hypertrophic cardiomyopathy. *JAMA*. 2007 July 25; 298(4):405-412.
15. Watanabe H, Chopra N, Laver D, et al. Flecainide prevents catecholaminergic polymorphic ventricular tachycardia in mice and humans. *Nat Med*. 2009 April;15(4):380-383.
16. Hobbs JB, Peterson DR, Moss AJ, et al. Risk of aborted cardiac arrest or sudden cardiac death during adolescence in the long-QT syndrome. *JAMA*. 2006 September 13;296(10): 1249-1254.
17. Moss AJ, Zareba W, Hall WJ, et al. Effectiveness and limitations of beta-blocker therapy in congenital long-QT syndrome. *Circulation*. 2000 February 15;101(6):616-623.
18. Drory Y, Turetz Y, Hiss Y, et al. Sudden unexpected death in persons less than 40 years of age. *Am J Cardiol*. 1991 November 15;68(13):1388-1392.
19. Amital H, Glikson M, Burstein M, et al. Clinical characteristics of unexpected death among young enlisted military personnel: results of a three-decade retrospective surveillance. *Chest*. 2004 August;126(2):528-533.
20. Wisten A, Forsberg H, Krantz P, Messner T. Sudden cardiac death in 15-35-year olds in Sweden during 1992-99. *J Intern Med*. 2002 December;252(6):529-536.
21. Krous HF, Beckwith JB, Byard RW, et al. Sudden infant death syndrome and unclassified sudden infant deaths: a definitional and diagnostic approach. *Pediatrics*. 2004 July;114(1): 234-238.
22. Moon RY, Horne RS, Hauck FR. Sudden infant death syndrome. *Lancet*. 2007 November 3;370(9598):1578-1587.
23. Goldberger JJ, Cain ME, Hohnloser SH, et al. American Heart Association/American College of Cardiology Foundation/Heart Rhythm Society scientific statement on noninvasive risk stratification techniques for identifying patients at risk for sudden cardiac death: a scientific statement from the American Heart Association Council on Clinical Cardiology Committee on Electrocardiography and Arrhythmias and Council on Epidemiology and Prevention. *Circulation*. 2008 September 30;118(14):1497-1518.
24. Carturan E, Tester DJ, Brost BC, Basso C, Thiene G, Ackerman MJ. Postmortem genetic testing for conventional autopsy-negative sudden unexplained death: an evaluation of different DNA extraction protocols and the feasibility of mutational analysis from archival paraffin-embedded heart tissue. *Am J Clin Pathol*. 2008 March;129(3): 391-397.
25. Corrado D, Basso C, Thiene G. Sudden cardiac death in young people with apparently normal heart. *Cardiovasc Res*. 2001 May;50(2):399-408.
26. Morentin B, Suarez-Mier MP, Aguilera B. Sudden unexplained death among persons 1-35 years old. *Forensic Sci Int*. 2003 August 27;135(3):213-217.
27. de Vreede-Swagemakers JJ, Gorgels AP, Dubois-Arbouw WI, et al. Out-of-hospital cardiac arrest in the 1990's: a population-based study in the Maastricht area on incidence, characteristics and survival. *J Am Coll Cardiol*. 1997 November 15;30(6):1500-1505.
28. Basso C, Burke M, Fornes P, et al. Guidelines for autopsy investigation of sudden cardiac death. *Virchows Arch*. 2008 January;452(1):11-18.
29. Straus SM, Bleumink GS, Dieleman JP, van der Lei J, Stricker BH, Sturkenboom MC. The incidence of sudden cardiac death in the general population. *J Clin Epidemiol*. 2004 January;57(1):98-102.
30. Vaartjes I, Hendrix A, Hertogh EM, et al. Sudden death in persons younger than 40 years: causes and incidence. *Eur J Cardiovasc Prev and Prehab*. 2009 October;16(5):592-596.
31. Corrado D, Basso C, Pavei A, Michieli P, Schiavon M, Thiene G. Trends in sudden cardiovascular death in young competitive athletes after implementation of a preparticipation screening program. *JAMA*. 2006 October 4;296(13): 1593-1601.

32. Behr ER, Casey A, Sheppard M, et al. Sudden arrhythmic death syndrome: a national survey of sudden unexplained cardiac death. *Heart*. 2007 May;93(5):601-605.
33. Maron BJ, Doerer JJ, Haas TS, Tierney DM, Mueller FO. Sudden deaths in young competitive athletes: analysis of 1866 deaths in the United States, 1980-2006. *Circulation*. 2009 March 3;119(8):1085-1092.
34. Maron BJ. Sudden death in young athletes. *N Engl J Med*. 2003 September 11;349(11):1064-1075.
35. Iribarren C, Crow RS, Hannan PJ, Jacobs DR Jr, Luepker RV. Validation of death certificate diagnosis of out-of-hospital sudden cardiac death. *Am J Cardiol*. 1998 July 1;82(1):50-53.
36. Fox CS, Evans JC, Larson MG, et al. A comparison of death certificate out-of-hospital coronary heart disease death with physician-adjudicated sudden cardiac death. *Am J Cardiol*. 2005 April 1;95(7):856-859.
37. Chugh SS, Jui J, Gunson K, et al. Current burden of sudden cardiac death: multiple source surveillance versus retrospective death certificate-based review in a large U.S. community. *J Am Coll Cardiol*. 2004 September 15;44(6):1268-1275.
38. Huikuri HV, Castellanos A, Myerburg RJ. Sudden death due to cardiac arrhythmias. *N Engl J Med*. 2001 November 15;345(20):1473-1482.
39. Tester DJ, Ackerman MJ. Postmortem long QT syndrome genetic testing for sudden unexplained death in the young. *J Am Coll Cardiol*. 2007 January 16;49(2):240-246.
40. Tester DJ, Spoon DB, Valdivia HH, Makielski JC, Ackerman MJ. Targeted mutational analysis of the RyR2-encoded cardiac ryanodine receptor in sudden unexplained death: a molecular autopsy of 49 medical examiner/coroner's cases. *Mayo Clin Proc*. 2004 November;79(11):1380-1384.
41. Zheng ZJ, Croft JB, Giles WH, Mensah GA. Out-of-hospital cardiac deaths in adolescents and young adults in the United States, 1989 to 1998. *Am J Prev Med*. 2005 December;29(5 Suppl 1):36-41.
42. Soo L, Huff N, Gray D, Hampton JR. Geographical distribution of cardiac arrest in Nottinghamshire. *Resuscitation*. 2001 February;48(2):137-147.
43. Nichol G, Thomas E, Callaway CW, et al. Regional variation in out-of-hospital cardiac arrest incidence and outcome. *JAMA*. 2008 September 24;300(12):1423-1431.
44. Roberts R, Brugada R. Genetics and arrhythmias. *Annu Rev Med*. 2003;54:257-267.
45. Vatta M, Dumaine R, Varghese G, et al. Genetic and biophysical basis of sudden unexplained nocturnal death syndrome (SUNDS), a disease allelic to Brugada syndrome. *Hum Mol Genet*. 2002 February 1;11(3):337-345.
46. Tungsanga K, Sriboonlue P. Sudden unexplained death syndrome in north-east Thailand. *Int J Epidemiol*. 1993 February;22(1):81-87.
47. Tatsanavivat P, Chiravatkul A, Klungboonkrong V, et al. Sudden and unexplained deaths in sleep (Laitai) of young men in rural northeastern Thailand. *Int J Epidemiol*. 1992 October;21(5):904-910.
48. Corrado D, Basso C, Schiavon M, Thiene G. Screening for hypertrophic cardiomyopathy in young athletes. *N Engl J Med*. 1998 August 6;339(6):364-369.
49. Maron BJ, Shirani J, Poliac LC, Mathenge R, Roberts WC, Mueller FO. Sudden death in young competitive athletes. Clinical, demographic, and pathological profiles. *JAMA*. 1996 July 17;27(3):199-204.
50. Gordon T, Kannel WB, Hjortland MC, McNamara PM. Menopause and coronary heart disease. The Framingham Study. *Ann Intern Med*. 1978 August;89(2):157-161.
51. Atsma F, Bartelink ML, Grobbee DE, van der Schouw YT. Postmenopausal status and early menopause as independent risk factors for cardiovascular disease: a meta-analysis. *Menopause*. 2006 March;13(2):265-279.
52. Colditz GA, Willett WC, Stampfer MJ, Rosner B, Speizer FE, Hennekens CH. Menopause and the risk of coronary heart disease in women. *N Engl J Med*. 1987 April 30;316(18):1105-1110.
53. de la Grandmaison GL. Is there progress in the autopsy diagnosis of sudden unexpected death in adults? *Forensic Sci Int*. 2006 January 27;156(2–3):138-144.
54. Safranek DJ, Eisenberg MS, Larsen MP. The epidemiology of cardiac arrest in young adults. *Ann Emerg Med*. 1992 September;21(9):1102-1106.
55. van Ingen G, van Loenen AC, Voortman M, Zweipfenning PG, Meijer CJ. Recommendations for toxicologic studies in sudden, unexpected death. *Ned Tijdschr Geneeskd*. 1996 January 27;140(4):179-181.
56. Ackerman MJ, Tester DJ, Driscoll DJ. Molecular autopsy of sudden unexplained death in the young. *Am J Forensic Med Pathol*. 2001 June;22(2):105-111.
57. Christiaans I, Langen IM, Wilde AA. Plotselinge dood op jonge leeftijd, de noodzaak van obductie en erfelijkheidsonderzoek. *Medisch Contact*. 2009;61:1253-1256.
58. Hendriks KS, Hendriks MM, Birnie E, et al. Familial disease with a risk of sudden death: a longitudinal study of the psychological consequences of predictive testing for long QT syndrome. *Heart Rhythm*. 2008 May;5(5):719-724.
59. Priori SG, Barhanin J, Hauer RN, et al. Genetic and molecular basis of cardiac arrhythmias: impact on clinical management parts I and II. *Circulation*. 1999 February 2;99(4):518-528.
60. Priori SG, Napolitano C. Role of genetic analyses in cardiology: part I: mendelian diseases: cardiac channelopathies. *Circulation*. 2006 February 28;113(8):1130-1135.
61. Clinical indications for genetic testing in familial sudden cardiac death syndromes: an HRUK position statement. Heart 2008 April;94(4):502–507.
62. Wilde AA, Pinto YM. Cost-effectiveness of genotyping in inherited arrhythmia syndromes, are we getting value for the money? *Circ Arrhythmia electrophysiol*. 2009;2:1-3.
63. Basso C, Calabrese F, Corrado D, Thiene G. Postmortem diagnosis in sudden cardiac death victims: macroscopic, microscopic and molecular findings. *Cardiovasc Res*. 2001 May;50(2):290-300.
64. Sangwatanaroj S, Prechawat S, Sunsaneewitayakul B, Sitthisook S, Tosukhowong P, Tungsanga K. New electrocardiographic leads and the procainamide test for the detection of the Brugada sign in sudden unexplained death syndrome survivors and their relatives. *Eur Heart J*. 2001 December;22(24):2290-2296.
65. Bai R, Napolitano C, Bloise R, Monteforte N, Priori SG. Yield of genetic screening in inherited cardiac channelopathies. *Circ Arrhythmia Electrophysiol*. 2009;2:6-15.
66. Thompson PD, Franklin BA, Balady GJ, et al. Exercise and acute cardiovascular events placing the risks into perspective:

a scientific statement from the American Heart Association Council on Nutrition, Physical Activity, and Metabolism and the Council on Clinical Cardiology. *Circulation.* 2007 May 1;115(17):2358-2368.
67. Mittleman MA, Maclure M, Tofler GH, Sherwood JB, Goldberg RJ, Muller JE. Triggering of acute myocardial infarction by heavy physical exertion. Protection against triggering by regular exertion. Determinants of Myocardial Infarction Onset Study Investigators. *N Engl J Med.* 1993 December 2;329(23):1677-1683.
68. Burke AP, Farb A, Virmani R, Goodin J, Smialek JE. Sports-related and non-sports-related sudden cardiac death in young adults. *Am Heart J.* 1991 February;121(2 Pt 1):568-575.
69. Quigley F, Greene M, O'Connor D, Kelly F. A survey of the causes of sudden cardiac death in the under 35-year-age group. *Ir Med J.* 2005 September;98(8):232-235.
70. Albert CM, Mittleman MA, Chae CU, Lee IM, Hennekens CH, Manson JE. Triggering of sudden death from cardiac causes by vigorous exertion. *N Engl J Med.* 2000 November 9;343(19):1355-1361.
71. Corrado D, Basso C, Rizzoli G, Schiavon M, Thiene G. Does sports activity enhance the risk of sudden death in adolescents and young adults? *J Am Coll Cardiol.* 2003 December 3;42(11):1959-1963.
72. Wannamethee G, Whincup PH, Shaper AG, Walker M, MacFarlane PW. Factors determining case fatality in myocardial infarction "who dies in a heart attack"? *Br Heart J.* 1995 September;74(3):324-331.
73. Corrado D, Pelliccia A, Bjornstad HH, et al. Cardiovascular pre-participation screening of young competitive athletes for prevention of sudden death: proposal for a common European protocol. Consensus Statement of the Study Group of Sport Cardiology of the Working Group of Cardiac Rehabilitation and Exercise Physiology and the Working Group of Myocardial and Pericardial Diseases of the European Society of Cardiology. *Eur Heart J.* 2005 March;26(5):516-524.
74. Maron BJ, Araujo CG, Thompson PD, et al. Recommendations for preparticipation screening and the assessment of cardiovascular disease in masters athletes: an advisory for healthcare professionals from the working groups of the World Heart Federation, the International Federation of Sports Medicine, and the American Heart Association Committee on Exercise, Cardiac Rehabilitation, and Prevention. *Circulation.* 2001 January 16;103(2):327-334.
75. Viskin S. Antagonist: routine screening of all athletes prior to participation in competitive sports should be mandatory to prevent sudden cardiac death. *Heart Rhythm.* 2007 April;4(4):525-528.
76. Corrado D, Thiene G. Protagonist: routine screening of all athletes prior to participation in competitive sports should be mandatory to prevent sudden cardiac death. *Heart Rhythm.* 2007 April;4(4):520-524.
77. Maron BJ, Pelliccia A. The heart of trained athletes: cardiac remodeling and the risks of sports, including sudden death. *Circulation.* 2006 October 10;114(15):1633-1644.
78. Pelliccia A, Maron BJ, Culasso F, et al. Clinical significance of abnormal electrocardiographic patterns in trained athletes. *Circulation.* 2000 July 18;102(3):278-284.
79. Mohlenkamp S, Lehmann N, Breuckmann F, et al. Running: the risk of coronary events: Prevalence and prognostic relevance of coronary atherosclerosis in marathon runners. *Eur Heart J.* 2008 August;29(15):1903-1910.

The Outpatient Clinic for Cardiogenetics

Irene M. van Langen and Arthur A.M. Wilde

26.1 Introduction

Until the nineties of the last century, genetic counseling and testing concentrated on reproductive issues, dysmorphology syndromes, neurogenetic diseases, and hereditary aspects of mental retardation. Due to the discovery of a rapidly increasing number of genes predisposing to heritable cancer syndromes and heart diseases a lot has changed since then. Currently, more than half of genetic counseling sessions in the Netherlands are dedicated to either genetic forms of cancer or genetic heart disease. Specialized outpatient clinics were founded, in which clinical geneticists, genetic nurses, and psychosocial workers from the departments of Clinical Genetics collaborate with oncologists and cardiologists, respectively. Molecular geneticists belonging to the labs performing diagnostic and predictive testing are involved as well. The principle goals of these clinics are, in the first place, early detection of individuals at risk for either inherited cancer or inherited potentially life-threatening cardiac disease. The disorders concerned are predominantly inherited in an autosomal dominant fashion. Consequently, as many high risk carrying relatives as possible are traced and surveillance in a risk stratification program is offered, whenever appropriate. Of course, such an undertaking has to be embedded into state-of-the-art *genetic counseling* guaranteeing informed free choice for every individual. The ultimate goal is to detect and inform everyone with an increased morbidity and mortality risk as a result of these monogenetic diseases and to prevent arrhythmias and premature death in those who appreciate this. Of course, this should be performed in the most cost-effective way, which, at this moment, appears to be *cascade screening*.[1] Besides, as a result of centralized data collection, these cardiogenetic clinics can also serve an important research purpose.

Most *outpatient cardiogenetic clinics* in the Netherlands focus on cardiomyopathies and inherited primary electrical diseases, although premature atherosclerosis, connective tissue diseases, and sometimes (isolated or syndromic) congenital heart disease may be covered as well. In addition, families in which *(unexplained) sudden death* at young age (below the age of 45, arbitrarily) has occurred are eligible for referral, because in the majority of such families, a heritable cardiovascular disease would have been the cause. Even if this cause is still unknown at time of referral, cardiac screening of close relatives is useful to identify those with an elevated risk.[2]

While the genetic predisposition cannot yet be cured, prevention of dangerous arrhythmias and especially of the most feared complication of sudden cardiac death is possible to a large extent by life style changes, and when appropriate, the prophylactic use of medication, pacemakers, or implantable cardioverter defibrillators (ICD's). Therefore, screening of yet unidentified people at risk seems useful.[3]

I.M. van Langen (✉)
Department of Genetics, University Medical Centre Groningen,
Groningen, The Netherlands
e-mail: i.m.van.langen@medgen.umcg.nl

A.A.M. Wilde
Department of Cardiology, University Medical Centre,
University of Amsterdam, Amsterdam,
The Netherlands
e-mail: a.a.wilde@amc.uva.nl

mutation (e.g. cascade screening) and/or have similar cardiac findings
(e) Counselees with multiple cases of heart disease in their families (congenital and/or structural or with onset later in life) who want to know if this condition increases the risk of such a disease in themselves and their children
(f) Counselees with one or more cases of unexplained sudden cardiac death in their family who want to know if they (and their children) are at risk

26.2.2 Genetic Testing

Genetic testing can be performed in the context of genotype-specific treatment and management or implementation of preventive measures in disease to modify the severity of the disorder. It can also be used to supply information on the risks for offspring or to enable prenatal diagnosis. In primary arrhythmias and cardiomyopathies, prenatal testing is currently seldom requested, most probably because preventive treatment can be offered and for most heritable cardiac disorders penetrance is far from complete. Predictive testing should always be in the context of *pretest and posttest counseling*, as described in the guidelines on testing for Huntington's disease and cancer susceptibility, independent by whom or in which way the genetic information is obtained.

26.2.2.1 Diagnostic Testing

The term diagnostic molecular testing is being used when a person or foetus has or is suspected of having a particular disorder on clinical grounds, and molecular techniques (DNA-tests) are being used for making, confirming, or refining of the diagnosis. An example is the use of genetic investigations in a person with recurrent syncopes and a prolonged QT-interval. By screening the eligible genes and finding a mutation, the diagnosis can be confirmed, the type of LQTS can be established with absolute certainty, and thereby, the most effective treatment and more detailed information on the phenotype becomes clear and family studies are much facilitated.

26.2.2.2 Predictive Testing

When healthy individuals (including the foetus in prenatal testing) undergo genetic (molecular) testing for a disease running in their family, this is a predictive or presymptomatic genetic test. The term presymptomatic is generally reserved for situations where an abnormal test result almost invariably implies the development of the disorder later in life. An example is Huntington's disease; when the characteristic (familiar) lengthened repeat in the Huntington gene is detected on one allele in a young person, it is certain that he or she will get all neurological symptoms of this devastating disease in the distant future. With the results of predictive testing, the risk of developing (the symptoms of) a disorder can generally be specified. DNA testing in a healthy person closely related to a LQTS patient with a proven mutation can be considered a predictive test; the result of the test can reduce the risk of dying suddenly at young age to, nearly, zero (not having the familiar mutation) or considerably enlarge the risk in carriers. However, it is not easy to predict if this carrier will ever get symptoms, because life-long penetrance is far from 100%. Besides, it is not possible to predict the severity or age of onset of the disease, based on the outcome of molecular testing although ECG- and other characteristics may give some clue. The same accounts for most other primary cardiac arrhythmias and cardiomyopathies.

In general, predictive molecular testing is possible only in families where the causal mutation in the index patient has already been identified. In other cases, predictive testing in cardiogenetics is possible by cardiac screening. Not finding clinical signs of the disease does not imply that the mutation is absent, however, due to the reduced (and age-dependent) penetrance in these disorders.

The initiative for predictive testing most often lies with the individual counselee or his or her parents, but may be initiated by a relative or a medical worker.

The justification of *predictive testing in children* has always been a point of debate. As with any other medical intervention, when children do not have the capacity to provide voluntary, informed consent, the decisive consideration in genetic testing in children should be based on the best interest of the child and not on anxiety in the parents.[12] In cardiogenetics, symptoms can start at young age (in LQTS and CPVT in particular) and effective prophylaxis is possible, so testing in children for these diseases seems clinically and ethically justified. Psychological research showed that the quality of life of children who proved to be carriers for LQTS, HCM, or familial hypercholesterolemia (FH) is not threatened in the majority of them, although LQTS-carriership, in particular, may be surrounded by anxiety in some.[13]

26.2.2.3 Genetic (Population) Screening

Genetic population screening differs from genetic testing in individuals or families. Here, a certain population is actively invited by medical or governmental organizations for testing to identify persons at risk, independent of family history or manifestation of symptoms. Population screening is never organized by cardiogenetics outpatient clinics. Primary goals of genetic screening are the timely prevention of deleterious effects of the disorder by prophylactic treatment or lifestyle rules (e.g. medication and dietary measures in familial hypercholesterolemia). *Prevention of sudden death* at young age could be a goal of genetic population screening as well. Examples are *preparticipation testing* in competition or top sports and ECG-testing in neonates to prevent sudden infant death due to Long QT Syndrome or congenital structural defects. DNA testing, and also cardiac testing could be used to achieve this goal.

26.2.2.4 Cascade Screening

Cascade screening is an intermediate form between individual genetic testing and population screening: In families with genetic disorders in which some kind of prevention is possible and in which a causal mutation is identified, all first-degree relatives (and second degree, in case of deceased first-degree relatives) are actively invited, by relatives or by medical workers, to undergo predictive testing. When a mutation carrier is identified, his or her first-degree relatives are invited for counseling and testing, etc. Cascade screening has proven to be feasible in families with familial hypercholesterolemia[14–16] with very high *uptake* of DNA testing. Cascade screening in cardiomyopathies and primary electrical diseases is also actively performed at the cardiogenetic outpatient clinics. Uptake in hypertrophic cardiomyopathy is lower than in familial hypercholesterolemia, which may be caused by the differences in counseling as well as by disease characteristics and more limited preventive treatment options.[17]

26.3 State of the Art in Cardiogenetic Counseling and Testing in the Netherlands

26.3.1 Background of Genetic Counseling and Testing in Inherited Cardiovascular Disease

26.3.1.1 From a Clinical Genetic Point of View

The guidelines for presymptomatic counseling and testing in Huntington's disease, as described above, are aimed at supporting the patient's autonomy and at self-selection of those with sufficient ego-strength to handle possible unfavorable results. In total, about 25% of those at high risk ask for genetic counseling and of this group, only about 50% eventually opt for DNA testing.

Guidelines for counseling and testing in oncogenetics are heavily based on *the "Huntington paradigms,"* which is remarkable, because in contrast to Huntington's chorea in these disorders preventive strategies, following cascade screening and identification of high risk individuals, are of great health benefit. Still, reinforcement of autonomy and (self) selection of those capable of handling bad outcomes remain important goals of counseling. In cardiogenetics, the Huntington's guidelines are followed as well, but with more emphasis on prevention, and less than in oncogenetics on self-selection. Reinforcement of autonomy remains important in all situations. The psychosocial negative effects of cascade screening in oncogenetics are reported to be small, probably because, as in Huntington's disease, only motivated and psychologically strong applicants start and proceed with testing. Pretest emotional state was predictive of subsequent distress in 14 of 27 analyses. Test results (being a carrier or not) were rarely predictive of distress. Therefore testing protocols, in neurogenetics and also in onco- and cardiogenetics should include a pretest assessment of emotional state, in order to single out the counselees who will need more specialized psychosocial care.[18]

26.3.1.2 From a Public Health Point of View

Testing for the presence of risk factors for cardiovascular morbidity and mortality like high blood pressure or high cholesterol has been common practice for years. The context varies from screening in an apparently healthy population (primary prevention) to screening patients who suffered from an acute myocardial infarction or cerebrovascular accident (secondary prevention). The prerequisites for the success of a screening program using genetic testing are (a) the possibility to diagnose the disease early, (b) the possibility to effectively treat the disease, and (c) a high prevalence of the disease within a certain population[19] In order to establish that screening programs are truly beneficial for the health of a certain (high risk) population, the psychological and social consequences of being tested and of the follow-up should also be evaluated. Furthermore, the implementation of a screening program depends on the prevailing organizational setting for genetic test services. Evaluation of genetic screening programs is possible by using well-known *criteria of Wilson and Jungner*, published by the WHO in 1968. Adaptation was made for the use in genetic screening for heritable cancer syndromes by the Crossroads 99 Group in 2001[20] (see Table 26.1). The use of these criteria proved to be valuable in the evaluation of the current screening for FH in the Netherlands.[15,16,21,22]

The British National Screening Committee recommended a paradigm of individual informed choice for participants in screening programs. This may lead to a more "clinical genetic counseling" way of informing those eligible, in which uncertainties and possible harms are emphasized just like the benefits. Only those who are really motivated for screening are asked to participate. This probably leads to a decreased uptake, but also to an increased compliance to changes in life style and/or taking medications of those proven to be at risk.[23]

Table 26.1 Criteria for assessment of screening adapted from the Crossroads 99 Group, based on the Wilson and Jüngner criteria[24]

1. Knowledge of population and disease
   (a) Important burden of the disease
   (b) Target population identifiable
   (c) Considerable level of risk
   (d) Preclinical phase of the disease existent
   (e) Natural course (from susceptibility to precursor, early disease and advanced disease) understood

2. Feasibility of screening procedure
   (a) Suitable test or examination available
   (b) Entire screening procedure acceptable to screened population
   (c) Screening process is a continuing process and encompasses all elements of screening procedures

3. Interventions and follow-up
   (a) Physical net benefit of the intervention likely
   (b) Psychological net benefit likely
   (c) Social net benefit likely
   (d) Facilities for adequate follow-up available
   (e) Consensus on accepted management for those with a positive test result

4. Societal and health system issues
   (a) Economic and medical costs balanced
   (b) Psychological costs balanced
   (c) Societal costs balanced
   (d) Appropriate screening services accessible to entire population without adverse consequences for nonparticipants
   (e) Appropriate confidentiality procedures and antidiscrimination provisions available for participants and nonparticipants

26.4 Current Genetic Counseling and Testing for Genetic Cardiovascular Diseases in the Netherlands

26.4.1 Multidisciplinary Outpatient Clinics for Cardiogenetics

Multidisciplinary outpatient clinics for cardiogenetics currently exist in all academic hospitals in the Netherlands and in some nonacademic clinics. Here cardiologists, clinical geneticists, genetic nurses, and psychosocial workers cooperate in the "genetic workup" and diagnostic and predictive DNA testing of families with (possible) genetic cardiac arrhythmias and cardiomyopathies and with sudden death at young age in relatives. In a mean of 50–90%, depending on the disease and the certainty of clinical diagnosis, the causative mutation in the index patient can be found, facilitating cascade screening in relatives.

Although referrals are increasing rapidly, a survey among Dutch cardiologists revealed that only 59% of cardiologists inform their HCM patients about the genetic nature of their disease and only 54% discuss the consequences for offspring.[24] Recently, a Dutch *guideline for genetic counseling and testing in HCM* has been approved by the scientific societies involved (clinical genetics, cardiology, pediatric cardiology).[25] Our group proposed a *guideline for the use in cardiogenetic counseling and testing in hereditary arrhythmias* (see Table 26.2).

In general, compared to the Huntington and oncogenetic guidelines, the built in time for reflection before actually embarking upon predictive testing in cardiogenetics is reduced or omitted (especially in LQTS), because the risk of sudden death may require expeditious identification and institution of prophylactic treatment to mutation carriers. Moreover, in our clinical experience, anxiety is raised, particularly in LQTS-testing, when applicants are forced to postpone testing for the sake of reflection. Also, while psychosocial support is actively offered to each applicant, it is not a prerequisite before testing, except in the predictive testing of children.

Counseling sessions combine the consultation of a cardiologist and a clinical geneticist (or genetic nurse), different from most predictive testing sessions in Huntington's disease and in cancer syndromes. Besides, in contrast to Huntington's disease, where in absence of preventive measures active cascade screening is deemed to be inappropriate, in cancer genetics and cardiogenetics cascade screening is usually facilitated by the distribution of so-called "*family- information-letters*" (supplied by a clinical geneticist). Usually, these family-information-letters are distributed in the family by the index patient (thus in a way circumventing the issue of patient confidentiality), but always in close collaboration with the clinical geneticist or genetic nurse, thus ensuring that the proper high-risk bearing relatives are addressed. Whether or not these letters could also be distributed directly by the medical team (if the index patient consents to this) is still a matter of debate (while on one hand, it seems strange that medical professionals send letters containing medical information to individuals who are not their patients and whom they do not know, on the other hand, it may be judged equally strange to transfer the burden of this responsibility to a diseased relative). These family-information-letters contain basic clinical information on the disease running in the family, its consequences and health risks, and what adequate preventive treatment is available, all in easy to understand plain language; these

Table 26.2 Proposed guidelines for genetic cascade screening in heritable arrhythmias (adapted from the Huntington guidelines and used by the authors since 1996)

1. Genetic counseling (including the drawing of a extended pedigree), extension of cardiologic evaluation (if necessary), and DNA testing of the index patient in a multidisciplinary outpatient cardiogenetics clinic
2. If a mutation is detected, education of the index patient and initiation of cascade screening
3. Information of first and second-degree relatives by the index patient (if necessary by the medical specialists), using an information letter written by the medical team
4. Genetic counseling of relatives prior to testing, during a family meeting and/or individually
5. Clinical testing of relatives at first consultation (mainly in LQTS and in relatives complaining of symptoms, irrespective of the familial disease)
6. DNA testing of relatives at first consultation (or at a second appointment, if desired by an individual needing more time to consider testing)
7. Psychosocial care (psychologist, social worker) mandatory for all families in whom minors are tested. If this is not the case, actively offered psychosocial care, but not mandatory
8. Results given personally, by telephone, or by letter, dependent on the preference of the individual
9. Actively offered follow-up appointments (including psychosocial care) for mutation carriers, especially those having children
10. Cardiologic testing and follow-up in mutation carriers, or referral for testing and follow-up to a cardiologist familiar with the disease in the neighborhood of the residence of the mutation carrier

informative letters necessarily reveal, as little as possible, specific medical information on the index patient, and always contain a name address and telephone number of a genetics service where more adequate information can be obtained or where they can be referred to. The distribution of family-information-letters is encouraged by the medical team. In some (large) families, the predictive counseling process can be initiated with a *family session*, in which up to 30 people can be informed at the same time, followed by (short) individual session. Test results are not necessarily revealed in a personal consultation, but may be given (based on the preferences of applicants) by e-mail or by telephone.[26]

Mutation carriers are always offered the opportunity of a *follow-up consultation* (including *psychosocial care*) immediately after receiving the results and are actively referred for follow-up cardiac care. Satisfaction and quality of life of HCM-carriers counseled in this way are very acceptable.[27]

In the population with primary inherited arrhythmia syndromes (LQTS, CPVT, Brugada syndrome, see chapters 9, 10, and 12 respectively) from our cardiogenetics database, we collected data on cascade screening and prophylactic treatment in carriers. From 1996 to 2007 in 130 probands, a disease-causing mutation in one of the involved genes was identified. Five hundred and nine relatives of these probands tested positive for the familial mutation. These subjects underwent cardiologic investigation. After a mean follow-up of 69±31 (LQTS), 60±19 (CPVT), and 56±21 (Brugada syndrome) months, treatment was initiated and ongoing in 65%, 71%, and 6% respectively. Treatment included drug treatment in 249, implantation of pacemakers in 26, and of cardioverter defibrillators (ICD's) in 14. Therefore, cascade screening resulted in immediate prophylactic treatment in a substantial proportion of carriers, in LQTS and CPVT in particular.[3] We are not aware of the exact number of first and second-degree relatives who did not participate in predictive testing. In LQTS, the amount of nonparticipants seems to be lower than in HCM, which is undoubtedly partly related to the perceived risk of sudden death in those eligible for testing, being higher in LQTS than in most HCM families. Extension of cascade screening to more distant relatives sometimes is problematic, due to the lack of personal contacts. In LQTS families with a known mutation a mean of 4.6 mutation-carrying relatives have been identified, of whom the majority is taking prophylactic medication. In HCM, an uptake of predictive testing of 39% in the first year after the detection of the mutation in the index patient has been reported from our center.[17] In the north of Holland (Groningen), the uptake of testing for HCM was 45, 1% (mean 1–5 year) and for primary arrhythmias 80%.[28]

26.4.2 Current Cascade Screening in Familial Hypercholesterolemia

Familial hypercholesterolemia (FH) is another type of cardiovascular disease predisposing tot sudden death. Similar to the inherited arrhythmias, FH is an autosomal dominantly inherited disorder with *reduced penetrance* and *variable expressivity* in which prophylactic measures reduce morbidity and mortality. The estimated frequency of FH is 1 in 400. Predictive DNA testing has long been possible, but achieved clinical relevance only after effective lipid lowering therapies became available in the last decade. In the Netherlands, cascade screening in FH is not yet part of the activities of outpatient clinics for cardiogenetics but is performed by a separate organisation, the "Foundation for tracing hereditary hypercholesterolemia" (Dutch acronym StOEH), in the setting of population screening (current funding until the end of 2013), which actively approaches and visits relatives of index patients at home, where information is given and blood samples may be taken. Mutation carriers receive the results by mail and are advised to visit their general practitioner with a letter stressing the need for treatment and follow-up through lipid clinics. Systematic evaluation studies revealed that psychological consequences are relatively small, but societal adverse effects do exist. Follow-up, prophylactic treatment, and adherence to life style advice of those who are eligible appear disappointing.[15,21,22,29] When funding ends, cascade screening for FH will have to be taken over by regular medical care, like in other inherited diseases with preventive treatment options.

26.4.3 Diagnostic Testing and Counseling in Diseases Associated with Aortic Dissections

In four of eight clinical genetic centers in the Netherlands, so-called "Marfan outpatient diagnostic clinics" are being held. In a multidisciplinary setting (clinical geneticist, cardiologist, ophthalmologist, in children also endocrinologist and orthopedist) people with signs of Marfan's syndrome and/or coming from Marfan families are being investigated for signs of this disease through physical examination and additional MRI and DNA testing if indicated. A minority is found to have Marfan's syndrome though, but many are diagnosed with, autosomal dominantly, inherited forms of (ascending) aortic dissections (FAAD, TAAD). Regular aortic screening of relatives at risk and timely surgical therapy prevents sudden death from aortic dissection in these families. DNA testing and therefore genetic cascade screening is currently feasible only in Marfan's syndrome, because only in this "aortic disease" a mutation in the fibrillin 1 gene can be found in the great majority of index patients. Of course, in those rare cases in which a causative mutation can be established in one of the other genes (i.e. TGFBR2, ACTA2, MYH11), genetic cascade screening is possible too.

26.4.4 Diagnostic Testing and Counseling in Congenital Heart Defects

In most cases, congenital heart defects are not monogenetically inherited; their background is *multifactorial* and DNA testing is not yet possible or does not contribute to the counseling process or policy making. A few exceptions exist (see Chap 17) and also syndromic forms should be distinguished from isolated heart defects. Referral is often asked in a reproductive genetic counseling setting, with a future parent having a heart defect or with parents having already had a former child with such a defect. These referrals are mostly handled by clinical dysmorphologists, not necessarily in the setting of an outpatient clinic for cardiogenetics.

26.4.5 Expected Future Developments

Until now, in the Netherlands, cardiogenetic test services (counseling, testing, follow-up) have to a large extent been supplied by departments of clinical genetics and cardiology of teaching hospitals. It has been recognized internationally that master degree *genetic counselors* are increasingly taking over the work of clinical geneticists, in particular in cascade screening.

This is due to the explosion of demands for predictive testing, the relatively low number of medical specialists in this field, and the good experiences with this new profession. Currently in the Netherlands, a professional masters educational program is being developed for genetic nurses in training. In the future, though, it is expected that even this will not be sufficient to meet the demands of the continuously growing number of people who opt for (predictive) testing in cardiovascular genetics. Therefore, regular medical specialists like cardiologists, who treat index patients with inherited arrhythmias, will be asked to perform part of this testing in the future, most preferably backed up by the professionals currently in charge of cascade screening for these and other genetic disorders. Cardiologists as well as clinical geneticist support this future model.[30] Education of cardiologists regarding this part of their (future) work is essential, especially in view of the results of a survey, indicating a substantial deficit in knowledge and skills in this specific field.[24] Which model for cascade screening should be chosen ultimately depends on outcomes of current research on the magnitude of the beneficial effect of early identification of mutation carriers in the different cardiogenetic diseases and on the way and to which extent active cascade screening will be reimbursed in the future.

26.4.6 Summary

In the last 15 years, the cardiogenetic outpatient clinics (in the Netherlands) have grown to maturity. Primary electrical diseases, cardiomyopathies, familial sudden death at young age and sometimes premature atherosclerosis, connective tissue diseases, and structural congenital defects are covered. These clinics are currently concentrated in the teaching hospitals and clinical geneticists, genetic counselors, cardiologists, social workers, and molecular geneticists are involved. Goals of these clinics are diagnosing, genetic counseling and testing of index patients, and active cascade screening in relatives, if appropriate. Research from bench to bedside is performed to fill the gaps in knowledge. Uptake of predictive testing is not yet optimal, in the cardiomyopathies in particular. If the *utility of cascade screening* in the different diseases can definitely be proven, more efficient ways of reaching all people at risk for sudden death at young age could be investigated and implemented.

References

1. Hadfield SG, Horara S, Starr BJ, et al. Steering Group for the Department of Health Familial Hypercholesterolaemia Cascade Testing Audit Project. Family tracing to identify patients with familial hypercholesterolaemia: the second audit of the Department of Health Familial Hypercholesterolaemia Cascade Testing Project. *Ann Clin Biochem.* 2009 Jan;46 (Pt 1):24-32. Epub 2008 Nov 21.
2. Tan HL, Hofman N, van Langen IM, van der Wal AC, Wilde AA. Sudden unexplained death: heritability and diagnostic yield of cardiological and genetic examination in surviving relatives. *Circulation.* 2005 Jul 12;112(2):207-13. Epub 2005 Jul 5.
3. Hofman N, Tan HL, Alders M, van Langen IM, Wilde A. Active cascade screening in inherited arrhythmia syndromes; does it lead to prophylactic treatment? JACC, revised version submitted.
4. Committee on Genetic Counseling. Genetic counseling. Am J Hum Genet 1975;27:240–63.
5. Proposed international guidelines on ethical issues in medical genetics and genetic services. World Health Organization. Human Genetics Programme,1998.
6. http://www4.od.nih.gov/oba/SACGHS/reports/CR_report_public_comment_draft.pdf, accessed May 12, 2005.
7. Proposed international guidelines on ethical issues in medical genetics and genetic services. World Health Organization. Human Genetics Programme.
8. Almqvist EW, Bloch M, Brinkman R, Craufurd D, Hayden MR. Worldwide assessment of the frequency of suicide, suicide attempts, or psychiatric hospitalization after predictive testing for Huntington disease. *Am J Hum Genet.* 1999;64: 1293-304.
9. Wright C, Kerzin-Storrar L, Williamson PR, et al. Comparison of genetic services with and without genetic registers: knowledge, adjustment, and attitudes about genetic counseling among probands referred to three genetic clinics. *J Med Genet.* 2002;39:e84-e84.
10. Skirton H, Barnes C, Curtis G, Walford-Moore J. The role and practice of the genetic nurse: report of theAGNC Working Party. *J Med Genet.* 1997;34:141-7.
11. Braithwaite D, Emery J, Walter F, Prevost AT, Sutton S. Psychological impact of genetic counseling for familial cancer: a systematic review and meta-analysis. *Fam Cancer.* 2006;5(1):61-75.
12. Borry P, Goffin T, Nys H, Dierickx K. Attitudes regarding predictive genetic testing in minors: A survey of European clinical geneticists. *Am J Med Genet Part C: Semin Med Genet.* 2010;148C(1):78-83.
13. Meulenkamp TM, Tibben A, Mollema ED, et al. Predictive genetic testing for cardiovascular diseases: impact on carrier children. *Am J Med Genet A.* 2008 Dec 15;146A(24): 3136-46.
14. Leren TP. Cascade genetic screening for familial hypercholesterolemia. *Clin Genet.* 2004;66:483-7.
15. van Maarle MC, Stouthard ME, Marang-vd Mheen PJ, Klazinga NS, Bonsel GJ. How disturbing is it to be approached for a genetic cascade screening programme for familial hypercholesterolaemia? Psychological impact and screenees' views. *Community Genet.* 2001;4:244-52.
16. van Maarle MC, Stouthard ME, Bonsel GJ. Quality of life in a family based genetic cascade screening programme for familial hypercholesterolaemia: a longitudinal study among participants. *J Med Genet.* 2003;40:e3.
17. Christiaans I, Birnie E, Bonsel GJ, Wilde AA, van Langen IM. Uptake of genetic counseling and predictive DNA testing in hypertrophic cardiomyopathy. *Eur J Hum Genet.* 2008 Oct;16(10):1201-7. Epub 2008 May 14.
18. Broadstock M, Michie S, Marteau T. Psychological consequences of predictive genetic testing: a systematic review. *Eur J Hum Genet.* 2000;8:731-8.
19. Wilson JMG, Jüngner G. Principles and practice of screening for disease. 34. 1968. Geneva. World Health Organization. Public Health Papers.
20. Goel V for Crossroads 99 Group. Appraising organised screening programmes for testing for genetic susceptibility to cancer. *BMJ.* 2001;322:1174-78.
21. van Maarle MC, Stouthard MEA, Marang-vd Mheen PJ, Klazinga NS, Bonsel GJ. Follow-up after a family-based genetic screening programme on Familial Hypercholesterolemia: Screening alone is not enough. *BMJ.* 2002; 324:1367-68.
22. Mharang-vd Mheen PJ, van Maarle MC, Stouthard MEA. Getting insurance after genetic screening on familial hypercholesterolemia: the need to educate both insurers and the public to increase adherence to national guidelines in The Netherlands. *J Epidemiol Community Health.* 2002;56: 145-47.
23. Marteau TM, Kinmonth AL. Screening for cardiovascular risk: public health imperative or matter for individual informed choice? *BMJ.* 2002;325:78-80.
24. van Langen M, Birnie E, Leschot NJ, Bonsel GJ, Wilde AAM. Genetic knowledge and counseling skills of Dutch cardiologists: sufficient for the genomics era? *I Eur Heart J.* 2003;24:560-566.
25. http://www.nvvc.nl/UserFiles/Richtlijnen/Richtlijnen.htm
26. Christiaans I, van Langen IM, Birnie E, Bonsel GJ, Wilde AA, Smets EM. Genetic counseling and cardiac care in predictively tested hypertrophic cardiomyopathy mutation carriers: the patients' perspective. *Am J Med Genet A.* 2009 Jul;149A(7): 1444-51.
27. Christiaans I, van Langen IM, Birnie E, Bonsel GJ, Wilde AA, Smets EM. Quality of life and psychological distress in hypertrophic cardiomyopathy mutation carriers: a cross-sectional cohort study. *Am J Med Genet A.* 2009 Feb 15;149A(4):602-12.
28. van der Roest WP, Pennings JM, Bakker M, van den Berg MP, van Tintelen JP. Family letters are an effective way to inform relatives about inherited cardiac disease. *Am J Med Genet Part A.* 2009;149A:357-363.
29. van Maarle MC, Stouthard MEA, Bonsel GJ. Quality of Life in a family-based genetic cascade screening programme on Familial Hypercholesterolemia: A longitudinal study among participants. *J Med Genet.* 2002;40:e3.
30. van Langen IM, Birnie E, Schuurman E, et al. Preferences of cardiologists and clinical geneticists for the future organization of genetic care in hypertrophic cardiomyopathy: a survey. *Clin Genet.* 2005;68:360-368.

Abdominal Aortic Aneurysm

A.F. Baas and S.E. Kranendonk

27.1 Introduction

Aneurysms and dissections are the major diseases affecting the aorta. Aneurysms are defined as arterial dilatations of approximately 1.5× the normal diameter and can be localized in the thoracic and abdominal aorta. The frequency of abdominal aortic aneurysms (AAA) is 3–7 times higher than the frequency of thoracic aortic aneurysms (TAA).[1]

AAA generally develops in older, smoking men, and is considered to be a silent killer. The condition is usually asymptomatic, but in case of AAA rupture the mortality risk is extremely high (up to 80%).[2] Rupture rates are exponentially related to the size of the AAA,[3] and they can be prevented by surgical intervention. Therefore, detection of an AAA at an early stage can be life-saving.

The pathogenesis of AAA is associated with atherosclerosis, extracellular matrix changes, and a chronic inflammation of the vessel wall, but molecular details are poorly understood.[4] Mendelian inherited patterns have been described in families with several affected family members.[5] But for the major part, AAA is considered as a multifactorial disorder in which both environmental and genetic factors contribute to its formation. The discovery of genes that are involved in AAA development will greatly enhance the understanding of its pathology. Linkage analysis and candidate gene association studies have attempted to identify genetic key players in AAA development, and genome wide association studies are currently being conducted.[6]

In this chapter, we summarize the most recent research on the genetic aspects of AAA, including familial studies and the identification of predisposing single nucleotide polymorphisms (SNPs). Their future implications are discussed as well. Furthermore, we highlight the clinical diagnosis and therapies, especially in the light of familial AAA. We conclude with an algorithm for familial forms of AAA. Based on this, we strongly support an AAA screening program for first degree family members of AAA patients.

27.2 Epidemiology

An AAA is generally defined as an infrarenal aorta with a diameter of at least 3 cm.[7] Rupture risks are related to the size of the AAA, and cause roughly 15,000 deaths per year in the USA, making it the 14th leading cause of death.[8] Especially, men above their sixties are at high risk of developing an AAA, with prevalences ranging from 5% to 8%.[7,9] The prevalence in women of the same age group is 1–2%.[7,9] In addition to male gender and age, smoking is a major risk factor,[10] which not only increases the risk of developing an AAA, but also causes more rapid enlargement, and thereby increases the chance of rupture.[11,12] Another important, but non-modifiable AAA risk factor, is a family history of AAA, increasing the risk up to tenfold.[13] Familial clustering of AAA was first reported in 1977.[14] Environmental factors can explain the familial occurrence, since smoking and unhealthy diet habits can run in a family. However, after multifactorial adjustments, familial AAA remained an independent AAA risk factor, suggesting a genetic basis.[15] Ultrasound screening studies among first-degree relatives of AAA patients have demonstrated an increased

A.F. Baas (✉)
Division of Biomedical Genetics,
Department of Medical Genetics, and
Julius Center for Health Science and Primary Care, University Medical Center Utrecht, Utrecht, The Netherlands
e-mail: a.f.baas@umcutrecht.nl

AAA prevalence in different populations, varying from 8.9% to 28.6% in brothers of AAA patients.[16] A systematic review of case-control studies showed that odds ratios for a family history of AAA vary between 2.6 (95% CI 1.0–7.1) and 9.7 (95% CI 4.1–23.0) based on the study design.[17] The mode of inheritance of genetically determined AAA has been controversial, and both autosomal recessive[18,19] and autosomal dominant[19,20] patterns have been described. However, the lack of a consistent inheritance pattern in most cases is considered to be consistent with a multigenetic origin.

Three other AAA risk factors have been described: white race;[21] chronic obstructive pulmonary disease;[22] and the presence of other atherosclerotic diseases, like cerebrovascular disease, coronary artery disease, and peripheral artery disease.[10] In addition, hypertension increases the risk of AAA rupture.[23]

with less than 25% expanding more than 0.5 cm per year.[24] Initially, the enlargement is slower; the speed of expansion is related to the size of the AAA. Rupture rates are exponentially related to AAA diameter. A population-based screening program reported cumulative rupture rates in small AAAs (3–4.4 cm) of 2.1% per year, while for large AAAs (4.5–5.9) this was 10.2% per year.[25]

27.3 Pathogenesis

27.3.1 Natural History

The natural history of an AAA is enlargement over several years, ultimately leading to rupture. In general, the AAA expands between 0.25 and 0.75 cm per year,

27.3.2 Vascular Pathology

As all blood vessels, the aortic wall consists of three layers: (1) the intima, a monolayer of endothelial cells; (2) the media, consisting of vascular smooth muscle cells (VSMCs) embedded in extracellular matrix made of elastic fibers formed by elastin and associated proteins; and (3) the adventitia, consisting of fibroblasts and collagen (Fig. 27.1). Elastic fibers are responsible for the viscoelastic properties, while collagens provide tensile strength to the vascular wall.[7] In the course of AAA formation, degenerative changes in each of these layers are observed; intimal atherosclerosis, medial extracellular matrix changes, and a chronic inflammatory process within the media and adventitia.

Inflammatory cells that are responsible for a chronic inflammatory state are thought to be recruited to the

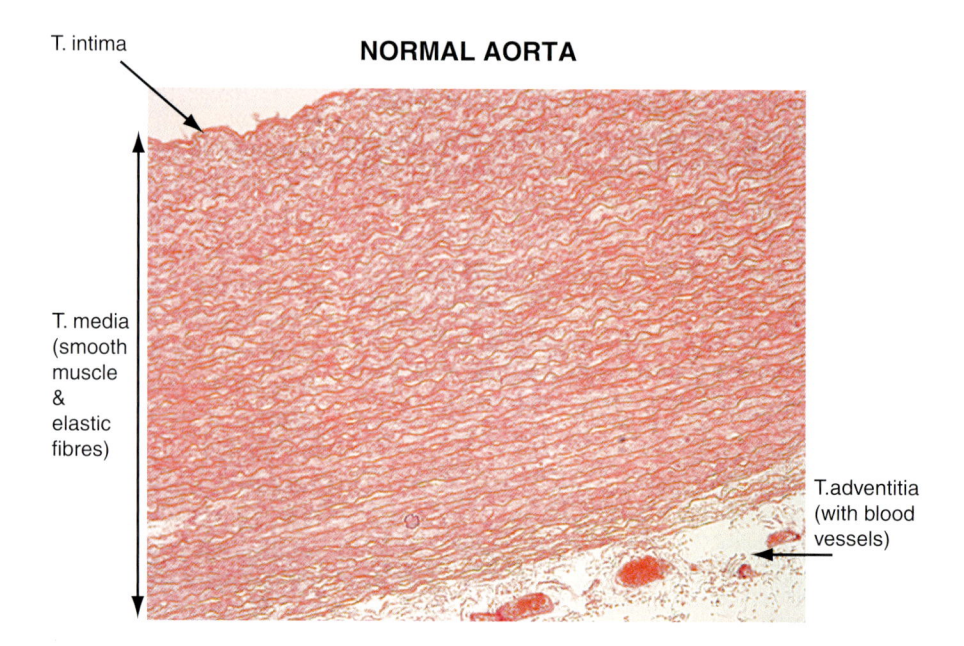

Fig. 27.1 Histology of the normal aorta

Fig. 27.2 Schematic representation of the processes involved in abdominal aortic aneurysm (AAA) formation with their associated risk factors

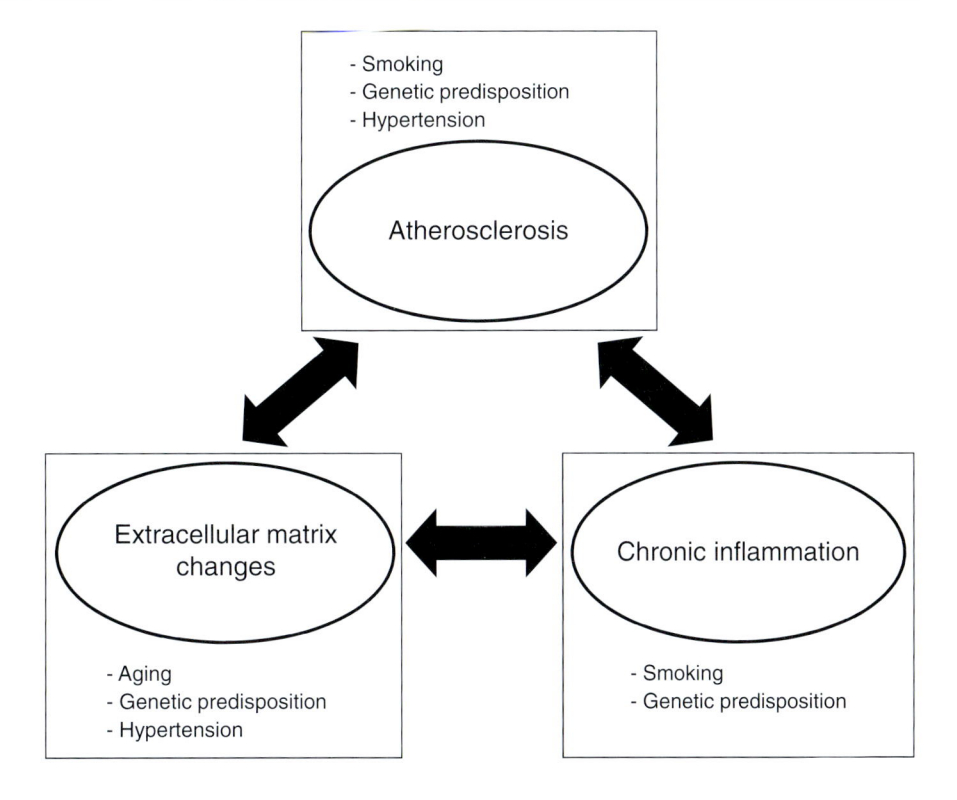

media and adventitia by different stimuli; oxidized lipoproteins, hypoxia, and localized cytokine production.[26] The most prominent medial change is fragmentation of the elastic fibers.[27,28] Proteolytic proteins that are highly expressed within the AAA, specifically metallo-matrix proteinases (MMPs), are responsible for elastin and collagen degradation.[29] Furthermore, within the media of AAAs, a lower density of VSMCs is observed.[30] These VSMCs are thought to play a major role in vascular remodeling by inducing localized expression of various extracellular matrix proteins and by protecting against inflammation and proteolysis.[31] The infrarenal aorta is more vulnerable to atherosclerosis than other parts of the aorta. Differences in hemodynamic conditions and structure are thought to be responsible for this.[32] Atheroma formation is induced by damaged endothelial cells, which can be caused by chemical (i.e., smoking) or mechanical (i.e., hypertension) triggers.[33] Aortic wall weakening caused by an expanded, penetrating, atherosclerotic lesion has previously been assumed to initiate AAA formation. However, it has also been proposed that atherosclerosis is a phenomenon occurring secondary to arterial dilation.[32]

Figure 27.2 shows the interplay of the different pathological processes that characterize the AAA, with their associated risk factors. The successive order in which these events play a role in AAA development is currently unknown. The identification of causal genetic factors will hopefully shed more light on this issue.

27.4 Molecular Genetics

27.4.1 Segregation and Linkage Analysis

Although the multifactorial nature of AAA is most obvious, segregation studies have described families in which the condition follows a Mendelian inherited pattern. Both autosomal dominant and autosomal recessive modes have been proposed.[5] Linkage studies are complicated by the late onset of the AAA. When a patient and maybe subsequently some of the siblings are diagnosed with an AAA, the parents are often deceased and their children are too young to have developed an AAA. Therefore, the study group usually comprises one generation. In a combined effort, a whole genome linkage scan was carried out using 119 families from different countries with at least two affected siblings. Strong evidence of linkage to chromosome 19q13 was found when the covariates sex and number of affected individuals were included (logarithm of odds [LOD] score 4.64).[34] Linkage to this region was replicated by a genome-wide linkage scan in three Dutch families with four or five affected siblings (LOD score 3.95),[35] although the peaks of the two studies did not overlap. The 95% confidence intervals (determined by taking the maximum LOD score minus one unit of each site of the peak) span approximately from 38.1 to 39.9 Mb[34] and from 54.0 to 58.0 Mb[35] and contain around 20–100 genes, respectively. To this date, no causal gene has been identified from the 19q13 loci. Gene mapping for AAA has been less successful compared to T AA, in which several causal genes, including the transforming beta receptor genes (*TGFBRI*, and *TGFBRII, ACTA2 and MYH11*), were identified from large families.[36-38] The main reason for this is that no large AAA families could be pursued, and therefore using a single family to map a disease locus has been impossible. The families in which AAA inheritance follows a clear Mendelian pattern are considered to be rare, and the main focus in genetic AAA research lies in finding SNPs that predispose to the disease in association studies.

27.4.2 Association Studies

There are roughly two different approaches in association studies: (1) the hypothesis-driven approach, in which SNPs in candidate genes or biological pathways are tested; and (2) the hypothesis-free approach, in which SNPs throughout the entire genome, or in parts of the genome, are analyzed for association with the disease of interest. Obviously, the latter approach has a major advantage. No prior knowledge of pathogenesis is required, and the main reason for performing this kind of genetic research is to unravel largely unknown pathological pathways. However, both logistically and financially, the so-called Genome Wide Association Studies (GWAS) are very challenging. More than 500,000 SNPs are tested simultaneously, and therefore extreme low p-values are required to remain significantly associated after correction for multiple testing ($P < 1 \times 10-7$)[6]. To obtain this kind of significance, substantial sample sizes are required, consisting of at least 2,000 cases and 2,000 controls. Furthermore, to verify the GWAS, replication of the results in additional large independent cohorts is recommended. The costs of analyzing the SNPs in the genome of one sample are currently approximately 500 euros, but this will likely be less costly in the future. For now, however, this means that for analyzing sufficient sample sizes considerable funding is required, as well as establishing collaborations between international groups.[6] A drawback of GWAS is that a large part of the inherited component of complex traits is apparently not accounted for in the analysis.[39] This is partly due to limited power to detect small genetic contributions, but most importantly due to the inability of GWAS to capture rare genetic variants. These variants have a minor allele frequency of less than 1%, compared to more than 5% for the common variants that are detected by GWAS.[39] Identification of these rare variants requires deep sequencing of multiple patient samples, and this will be a next challenge for unraveling the genetic contribution of complex diseases. Novel sequencing technologies will likely enable the future detection of rare variants.[40]

A wide set-up collaboration has recently led to the identification of an AAA-associated genetic variant on chromosome 9p21,[41,] which was previously found to predispose to coronary artery disease in GWAS.[42-44] An association study with AAA cohorts from seven different countries, consisting of 2,836 cases and 16,732 controls, yielded an odds ratio of 1.31 (95% 1.22–1.42) with a combined p-value of 1.2×10^{-12} [42]. Interestingly, the 9p21 variant is also associated with different intracranial aneurysm cohorts.[41] Since the latter is not specifically associated with atherosclerosis, the causal variant within this locus may influence another process leading to vascular pathology, like vascular remodeling. The associated SNP is not localized within a known gene, but lies within close proximity to a cluster of genes, *CDKN2A-ARF-CDKN2B*. The encoded proteins play a general role in cell proliferation and apoptosis. The mechanism by which the, not yet identified, causal variant predisposes to vascular pathology remains to be elucidated. Recent insights suggest that reduced expression levels of genes nearby the locus could be responsible for the phenotype.[45,46]

A way to circumvent the challenging requirements for GWAS is to pursue the investigation of candidate genes. Multiple associations in different genes have been described.[5,47] As expected, the tested SNPs are often localized within genes functioning in processes that are known to play a role in AAA pathogenesis;

extracellular matrix degradation, inflammatory and immune responses, and hypertension. A problem with these candidate gene studies is that replication of the results in different cohorts has proven to be often inconsistent. This may be caused by genetic heterogeneity, but may also indicate that the results are regularly false-positive, since for genetic analysis consistent replication addressing the same variant and phenotype is considered as the best test to correct for false-positive results.[48] Table 27.1 represents an overview of the genes investigated for association with AAA in the last decade.

Taken together, different approaches have been attempted to identify genetic AAA key-players, with the ultimate goal to (1) enhance the understanding of AAA pathology; (2) expose targets for future pharmaceutical therapies; and (3) allow the generation of AAA risk predicting systems, consisting of an array of AAA-associated genetic variants. So far, only one consistent replicable associated variant on chromosome 9p21 has been identified. The causal variant, and the mechanism by which the increased predisposition to AAA is generated, remains to be elucidated. Novel sequencing technologies will hopefully enable us to detect the causal variants.

Table 27.1 Overview of the candidate genes investigated for association with abdominal aortic aneurysms (AAA) in the last decade. No consistent replicable variants have been identified through this approach

Category	Gene	References
Proteolytic enzymes and inhibitors	MMP-1	73
	MMP-2	73–75
	MMP-3	73, 76, 77
	MMP-9	73, 77–80
	MMP-10	73
	MMP-12	73
	MMP-13	73
	TIMP-1	73, 81
	TIMP-2	73
	TIMP-3	73
	PAI-1	61, 77, 82
	PAF-AH	83
Extracellular matrix components	ELN	73, 84
	COL3A1	73
	CSPG2	85
	HSPG2	85
Cytokines/Chemokines and receptors	IL-1A	86
	IL-1B	86, 87
	IL-1RN	86
	IL-6	87, 88
	IL-10	87, 89
	CCR5	90
	TNFalpha	87
	PPARG	91
Renin-angiotensin system	ACE	92–97
	AT1R	93–95
	AGT	94
	BDKRB2	94

Table 27.1 (continued)

Category	Gene	References
Homocysteine metabolism	MTHFR	98–100
	MTR	98
	MTRR	98
	CBS	98
	MTHFD1	98
	SLC19A1	98
	NNMT	98
	TCN2	98
	AHCY	98
	BHMT	98
	BHMT2	98
	FOLH1	98
	TYMS	98
	ENOSF1	98
	SHMT1	98
	PON1	98, 101
	PON2	98
TGF-beta pathway	TGFBRI	93, 102, 103
	TGFBRII	102, 103
	TGFB1	73, 84
Inflammatory/immune responses	HO-1	104
	CRP	105
	COX-2	106
	HLA-DQA1	107
	HLA-DQB1	107
	HLA-DRB1	107–109
	HLA-DRB3	107
Others	ABCC6	110
	eNOS	111–113
	ERalpha	84
	ERbeta	84
	PR	84
	XYLT1	114

27.5 Clinical Diagnosis

In order to describe the diagnostic tools for AAA detection, we have to separate patients with unruptured AAAs from patients with ruptured AAAs.

27.5.1 Diagnosis of Unruptured AAA

Usually, unruptured AAAs are asymptomatic. Therefore, they are most commonly diagnosed by coincidence during physical examination, or while obtaining radiologic images for other purposes. A physician may consider the presence of an AAA based on age, gender, smoking habits, the presence of other atherosclerotic lesions or manifest cardiovascular diseases, and family history of the patient. It may be worthwhile to perform targeted physical examination on older, smoking, male patients, especially when AAA is known to run in the family. An AAA may be detected by abdominal palpation; a "pulsatile mass" is suggestive for its presence. Abdominal palpation is fast and safe, but both sensitivity and specificity for AAA detection are low, and vary with AAA size.[49] Furthermore, in the obese patient it is often impossible to detect an AAA by abdominal palpation. To confirm or exclude the presence of an AAA, abdominal ultrasonography is the golden standard (Fig. 27.3). This is a completely safe, accurate, and relatively cheap diagnostic tool, with a sensitivity and a specificity approaching 100%.[50]

Because of its efficiency, ultrasonographic screening strategies have been considered for years. The United States Preventive Services Task Force (USPSTF) recommends one-time screening for AAA using ultrasonograpy, in men aged 65–75 years who have never smoked,[51] and did not advice to screen women. These recommendations are based on the results of a meta-analysis of four randomized controlled trials, including the Multicenter Aneurysm Screening Study (MASS).[52] It appears that there is a significant reduction in AAA-related mortality in men who were invited for AAA screening (odds ratio 0.57, 95% confidence interval 0.45–0.74).[53] Additionally, a modestly significant difference in all-cause mortality has been observed in favor of men invited for AAA screening (odds ratio 0.97, 95% confidence interval 0.94–0.99).[54] The American College of Cardiology and the American Heart Association's peripheral arterial disease guideline recommend screening in men aged 65–75 years with a first-degree relative with AAA as well.[55] Several studies have addressed the efficiency of screening siblings of AAA patients. One of the most recent reports on this issue combined the results from all previously published studies, and found that 17.2%

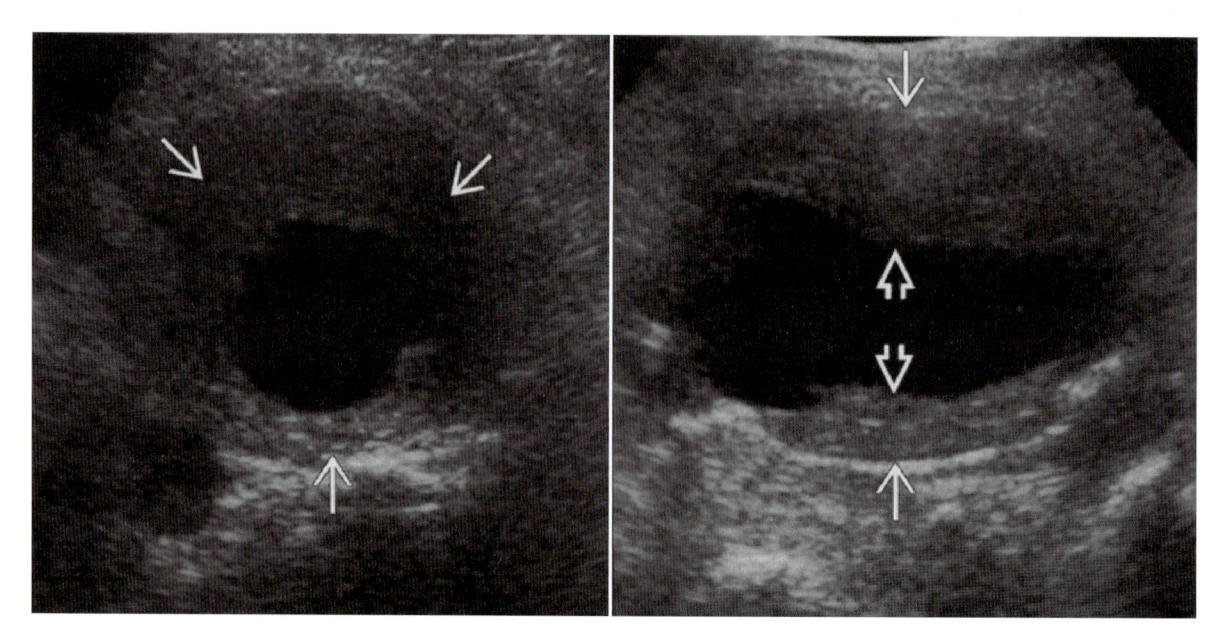

Fig. 27.3 (*Left*): transverse transabdominal ultrasound shows an abdominal aortic aneurysms (AAA) with a moderate amount of circumferential mural thrombus (→). (*Right*): Corresponding longitudinal transabdominal ultrasound shows that the outer wall (→) is substantially larger in diameter when compared with the lumen (⇒)

of the brothers, and 4.2% of the sisters of AAA patients harbored an AAA.[16] An investigation of 542 consecutive AAA patients revealed that familiar AAAs tends to develop at an earlier age.[56] Therefore, we recommend screening first-degree relatives of AAA patients by ultrasonography at the age of 45. Additionally, if no AAA is initially discovered, screening should be repeated every 5 years, until the age of 75.

Based on the results of two different reports,[57,58] England and Wales have recently implemented a national AAA screening program as well.

27.5.1.1 Future Genetic Diagnostic Tools

As discussed, GWAS, in combination with rare variant identification, may lead to the development of genetic risk predicting systems by testing an array of AAA-associated genetic variants in a single individual. Based on these systems, high-risk patients could enter specialized surveillance programs. Several reports have addressed the effect of different SNPs on AAA growth.[59–61] These genotype–phenotype studies may contribute to the decision making for follow-up and surgical intervention.

Since no single gene defect responsible for familiar AAA development has been identified so far, a genetic test in clear cases of familial AAA is not yet available. The importance of pursuing this research, including the personal importance for the family in question and the general importance of learning more about AAA pathogenesis, justifies referral of patients from families with at least three known cases of AAA to a clinical genetics department. Here, detailed pedigree construction may lead to the possibility of performing linkage analysis, in which a single gene defect may be uncovered.

27.5.2 Diagnosis of Ruptured AAA

As mentioned, a ruptured AAA is often fatal. Accurate, quick diagnosis is essential for increasing the chance of survival. A ruptured AAA can present itself with different symptoms, including pain and tenderness in the abdomen, flank, groin, or back.[50] Also urinary and gastrointestinal symptoms, like retentions and bleeding, can be present. In hemodynamically unstable patients, hypotension, syncope, and shock may be prominent. For diagnosing a ruptured AAA, with huge amounts of blood in the abdomen or retroperitoneum, computed tomography (CT) scanning or CT-angio is required. If the patient is hemodynamically unstable, radiological imaging may be impossible. In this case, emergency surgery (open abdominal repair or stent grafting) is the only change of survival.

Considering the life-threatening consequences, it is extremely important to suspect ruptured AAA in high-risk patients who present themselves in the emergency room with the symptoms mentioned above, so that appropriate actions can be undertaken as soon as possible. Therefore, it is mandatory to inquire after the risk factors, specifically tobacco use and family history of AAA.

27.6 Follow-up/Therapy

In this paragraph, we will focus on unruptured AAAs and thereby on elective therapy.

27.6.1 Follow-up

Once an AAA is diagnosed, it is important to survey growth, because the decision of a surgical intervention depends on the aneurysm size and on the velocity of expansion. Generally spoken, surgical correction of the AAA is considered when the risk of rupture outweighs the risk of surgery. The United Kingdom Small Aneurysm Trial (UKSAT) has been essential for current guidelines and recommendations.[24] In this trial, 1,049 patients with AAAs varying from 4.0 to 5.4 cm in diameter were randomly assigned to immediate elective AAA repair or surveillance, until the diameter of 5.5 cm was reached or until the AAA became symptomatic. Beyond the 5.5 cm threshold, elective repair was offered. The primary trial outcome, all-cause mortality, did not differ between the two groups. Based on these findings, the current threshold for surgical intervention is 5.5 cm. AAAs smaller than 4 cm can be monitored every 1–2 years.[50] For AAAs sized between 4.0 and 5.4 cm, surveillance imaging intervals between 6 and 12 months should be applied.[50] There appears to be no difference in AAA morphology between sporadic and familiar forms.[53] Therefore, these surveillance and repair guidelines are also applicable for familiar AAA cases. However, when family history reveals multiple AAA ruptures, specifically at younger ages, the surgeon should consider imaging at shorter intervals, and to intervene at a lower diameter threshold.

27.6.2 Therapy

Prevention of AAA growth and/or rupture can be achieved at different levels; by lifestyle adjustments, pharmacological treatment, or surgical intervention. The first two are specifically important for small AAAs, of which the frequency will increase by early detection in the screening programs. All three treatment options will be discussed successively.

27.6.2.1 Lifestyle

When an AAA is diagnosed, the physician should strongly recommend smoking cessation, because of its implications on growth and rupture rates. It is conceivable that gene–environment interactions exist. Therefore it is specifically important to advise family members of AAA patients to quit smoking as well.

27.6.2.2 Pharmacological Treatment

A couple of pharmaceutical agents have shown to be able to inhibit AAA progression in mouse models and in small human trials. These include statins, ACE-inhibitors, and anti-inflammatory agents.[62] Because of the small sample sizes and short follow-up periods of these studies, no firm conclusion can be drawn. Large randomized controlled trials are required to establish the effects of these treatments.

The former discussed medications are generally used for cardiovascular diseases, because of their lipid-lowering, blood pressure-controlling, or anti-inflammatory effects, leading to an improved condition of the vasculature at different levels. However, to treat small AAAs, the challenge lies within specifically targeting processes that cause aneurysm formation and/or expansion. Genetic research may expose the molecular key-players that are suitable for this purpose.

27.6.2.3 Surgical Intervention

AAAs are considered for elective surgery when they exceed the 5.5 cm threshold, above which the risk of rupture is considered higher than the mortality risk of surgery.[24] At present, there are two distinct surgical options to repair an AAA: conventional open repair (OR) and endovascular repair (EVAR). In the conventional method, after full midline laparotomy, a prosthesis is placed to replace the aneurysmal part of the aorta. During the anastomosis period, the abdominal aorta and the iliac arteries are clamped. This procedure has a relatively high mortality risk, especially in older and cardiological compromised patients.[63] In 1991, the first endovascular treatment was described.[64] During this procedure, an endograft is positioned within the AAA through a transfemoral or transiliac route. This device then replaces the aneurysm and excludes it from the circulation. This technique avoids laparotomy and clamping of the arteries. Additionally, milder and shorter anesthetics are required. EVAR was originally designed for patients that are unfit to undergo invasive OR. However, because of the initial positive results, EVAR was quickly implemented in the clinical routine and was also frequently used for low-risk patients. A drawback of the EVAR treatment is the relatively frequent occurrence of complications that arise due to graft problems (endoleaks, migration, kinking)[65] and require reinterventions (up to 20% within 4 years post-operatively).[66] Because of this, frequent follow-up using radiological imaging is required.[50] This leads to a substantial radiation exposure and is expensive. A second disadvantage of EVAR is that more than 30% of AAA patients are anatomically unsuitable to receive an endograft.[50] Furthermore, the devices are costly and therefore cost-effectiveness is not obvious.[67]

The Dutch Randomized Endovascular Aneurysm Management (DREAM) trial and the British EVAR-1 trial were initiated to determine which procedure is superior.[68,69] The DREAM trial included 351 low- and medium-risk patients and the EVAR-1 trial included 1,082 comparable patients. The trials had very similar perioperative and mid-term results. EVAR has an initial survival advantage, with a perioperative mortality rate of 1.2–1.7% against 4.6–4.7% for the OR treatment.[70,71] However, this benefit disappears after 1 year, and the 2-year survival rate is approximately 90% for both procedures.[66,72] Currently, it is therefore not possible to point out which procedure is best. Long-term results of

Fig. 27.4 Follow-up and therapy algorithm in familiar AAA patients

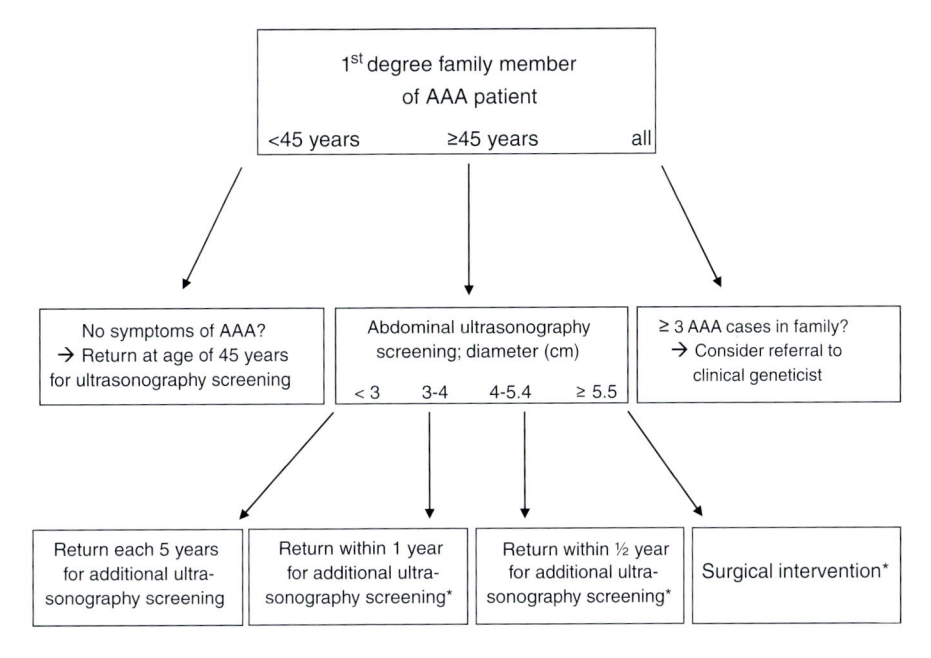

*Consider imaging at shorter intervals, and intervention at a lower diameter threshold, when family history reveals multiple AAA ruptures, specifically at younger ages.

the trials are awaited for and required to support the decision making.

27.7 Summary

In this chapter, we have outlined the most recent insights on genetic and clinical aspects of the AAA. It is obvious that genetic research may greatly contribute to the understanding of AAA pathology, which in the future may lead to the generation of risk-predicting systems based on genetic information and may aid to the generation of specific pharmaceutical therapies. To accomplish this, GWAS are required, and these are currently conducted by two different collaborative programs. The results are eagerly awaited for. Additionally, deep-sequencing multiple AAA patients, in order to identify rare genetic variants, will be essential to capture the major part of the heritable component. Meanwhile, it is equally important to pursue pedigree construction of large AAA families. This will enable the identification of AAA causal genes in single families by linkage and subsequent sequence analysis.

For the clinician it is important to realize that a family history of AAA is a major risk factor, and that familiar forms of AAA develop at a younger age. Otherwise, the condition is thought to progress similarly in sporadic and familial cases. Based on the information given in this chapter, we have constructed an algorithm to support decision making for follow-up and therapy in familiar AAA (Fig. 27.4). We strongly recommend ultrasound screening in first degree family members of AAA patients.

References

1. Estrera AL, Miller CC, Azizzadeh A, Safi HJ. Thoracic aortic aneurysms. *Acta Chir Belg*. 2006 May;106(3):307-316.
2. Budd JS, Finch DR, Carter PG. A study of the mortality from ruptured abdominal aortic aneurysms in a district community. *Eur J Vasc Surg*. 1989 August;3(4):351-354.
3. Lederle FA, Johnson GR, Wilson SE, et al. Rupture rate of large abdominal aortic aneurysms in patients refusing or unfit for elective repair. *JAMA*. 2002 June 12;287(22):2968-2972.
4. Annambhotla S, Bourgeois S, Wang X, Lin PH, Yao Q, Chen C. Recent advances in molecular mechanisms of abdominal aortic aneurysm formation. *World J Surg*. 2008 June;32(6):976-986.
5. Sandford RM, Bown MJ, London NJ, Sayers RD. The genetic basis of abdominal aortic aneurysms: a review. *Eur J Vasc Endovasc Surg*. 2007 April;33(4):381-390.

6. Genome Wide Association Studies: identifying the genes that determine the risk of abdominal aortic aneurysm. Eur J Vasc Endovasc Surg 2008 October;36(4):395–396.
7. Sakalihasan N, Limet R, Defawe OD. Abdominal aortic aneurysm. *Lancet*. 2005 April 30;365(9470):1577-1589.
8. Gillum RF. Epidemiology of aortic aneurysm in the United States. *J Clin Epidemiol*. 1995 November;48(11):1289-1298.
9. Pleumeekers HJ, Hoes AW, van der DE. Aneurysms of the abdominal aorta in older adults. The Rotterdam Study. *Am J Epidemiol*. 1995 December 15;142(12):1291-1299.
10. Cornuz J, Sidoti PC, Tevaearai H, Egger M. Risk factors for asymptomatic abdominal aortic aneurysm: systematic review and meta-analysis of population-based screening studies. *Eur J Public Health*. 2004 December;14(4): 343-349.
11. Smoking, lung function and the prognosis of abdominal aortic aneurysm. The UK Small Aneurysm Trial Participants. *Eur J Vasc Endovasc Surg* 2000 June;19(6):636–642.
12. Brady AR, Thompson SG, Fowkes FG, Greenhalgh RM, Powell JT. Abdominal aortic aneurysm expansion: risk factors and time intervals for surveillance. *Circulation*. 2004 July 6;110(1):16-21.
13. Johansen K, Koepsell T. Familial tendency for abdominal aortic aneurysms. *JAMA*. 1986 October 10;256(14): 1934-1936.
14. Clifton MA. Familial abdominal aortic aneurysms. *Br J Surg*. 1977 November;64(11):765-766.
15. Blanchard JF, Armenian HK, Friesen PP. Risk factors for abdominal aortic aneurysm: results of a case-control study. *Am J Epidemiol*. 2000 March 15;151(6):575-583.
16. Ogata T, MacKean GL, Cole CW, et al. The lifetime prevalence of abdominal aortic aneurysms among siblings of aneurysm patients is eightfold higher than among siblings of spouses: an analysis of 187 aneurysm families in Nova Scotia, Canada. *J Vasc Surg*. 2005 November;42(5): 891-897.
17. Van Vlijmen-Van Keulen CJ, Pals G, Rauwerda JA. Familial abdominal aortic aneurysm: a systematic review of a genetic background. *Eur J Vasc Endovasc Surg*. 2002 August; 24(2):105-116.
18. Majumder PP, St Jean PL, Ferrell RE, Webster MW, Steed DL. On the inheritance of abdominal aortic aneurysm. *Am J Hum Genet*. 1991 January;48(1):164-170.
19. Kuivaniemi H, Shibamura H, Arthur C, et al. Familial abdominal aortic aneurysms: collection of 233 multiplex families. *J Vasc Surg*. 2003 February;37(2):340-345.
20. Verloes A, Sakalihasan N, Koulischer L, Limet R. Aneurysms of the abdominal aorta: familial and genetic aspects in three hundred thirteen pedigrees. *J Vasc Surg*. 1995 April;21(4): 646-655.
21. Johnson G Jr, Avery A, McDougal EG, Burnham SJ, Keagy BA. Aneurysms of the abdominal aorta. Incidence in blacks and whites in North Carolina. *Arch Surg*. 1985 October; 120(10):1138-1140.
22. van Laarhoven CJ, Borstlap AC, van Berge Henegouwen DP, Palmen FM, Verpalen MC, Schoemaker MC. Chronic obstructive pulmonary disease and abdominal aortic aneurysms. *Eur J Vasc Surg*. 1993 July;7(4):386-390.
23. Brown LC, Powell JT. Risk factors for aneurysm rupture in patients kept under ultrasound surveillance. UK Small Aneurysm Trial Participants. *Ann Surg*. 1999 September; 230(3):289-296.
24. Mortality results for randomised controlled trial of early elective surgery or ultrasonographic surveillance for small abdominal aortic aneurysms. The UK Small Aneurysm Trial Participants. *Lancet* 1998 November 21;352(9141): 1649–1655.
25. Scott RA, Tisi PV, Ashton HA, Allen DR. Abdominal aortic aneurysm rupture rates: a 7-year follow-up of the entire abdominal aortic aneurysm population detected by screening. *J Vasc Surg*. 1998 July;28(1):124-128.
26. Thompson RW, Curci JA, Ennis TL, Mao D, Pagano MB, Pham CT. Pathophysiology of abdominal aortic aneurysms: insights from the elastase-induced model in mice with different genetic backgrounds. *Ann N Y Acad Sci*. 2006 November;1085:59-73.
27. Sakalihasan N, Heyeres A, Nusgens BV, Limet R, Lapiere CM. Modifications of the extracellular matrix of aneurysmal abdominal aortas as a function of their size. *Eur J Vasc Surg*. 1993 November;7(6):633-637.
28. Baxter BT, McGee GS, Shively VP, et al. Elastin content, cross-links, and mRNA in normal and aneurysmal human aorta. *J Vasc Surg*. 1992 August;16(2):192-200.
29. Thompson RW, Parks WC. Role of matrix metalloproteinases in abdominal aortic aneurysms. *Ann N Y Acad Sci*. 1996 November 18;800:157-174.
30. Lopez-Candales A, Holmes DR, Liao S, Scott MJ, Wickline SA, Thompson RW. Decreased vascular smooth muscle cell density in medial degeneration of human abdominal aortic aneurysms. *Am J Pathol*. 1997 March;150(3):993-1007.
31. Allaire E, Muscatelli-Groux B, Mandet C, et al. Paracrine effect of vascular smooth muscle cells in the prevention of aortic aneurysm formation. *J Vasc Surg*. 2002 November; 36(5):1018-1026.
32. Lindsay J Jr. Diagnosis and treatment of diseases of the aorta. *Curr Probl Cardiol*. 1997 October;22(10):485-542.
33. Guo DC, Papke CL, He R, Milewicz DM. Pathogenesis of thoracic and abdominal aortic aneurysms. *Ann N Y Acad Sci*. 2006 November;1085:339-352.
34. Shibamura H, Olson JM, van Vlijmen-Van KC, et al. Genome scan for familial abdominal aortic aneurysm using sex and family history as covariates suggests genetic heterogeneity and identifies linkage to chromosome 19q13. *Circulation*. 2004 May 4;109(17):2103-2108.
35. Van Vlijmen-Van Keulen CJ, Rauwerda JA, Pals G. Genome-wide linkage in three Dutch families maps a locus for abdominal aortic aneurysms to chromosome 19q13.3. *Eur J Vasc Endovasc Surg*. 2005 July;30(1):29-35.
36. Loeys BL, Schwarze U, Holm T, et al. Aneurysm syndromes caused by mutations in the TGF-beta receptor. *N Engl J Med*. 2006 August 24;355(8):788-798.
37. Guo DC, Pannu H, Tran-Fadulu V, et al. Mutations in smooth muscle alpha-actin (ACTA2) lead to thoracic aortic aneurysms and dissections. *Nat Genet*. 2007 December; 39(12):1488-1493.
38. Zhu L, Vranckx R, Van Khau KP, et al. Mutations in myosin heavy chain 11 cause a syndrome associating thoracic aortic aneurysm/aortic dissection and patent ductus arteriosus. *Nat Genet*. 2006 March;38(3):343-349.
39. Frazer KA, Murray SS, Schork NJ, Topol EJ. Human genetic variation and its contribution to complex traits. *Nat Rev Genet*. 2009 April;10(4):241-251.
40. Shendure J, Ji H. Next-generation DNA sequencing. *Nat Biotechnol*. 2008 October;26(10):1135-1145.

41. Helgadottir A, Thorleifsson G, Magnusson KP, et al. The same sequence variant on 9p21 associates with myocardial infarction, abdominal aortic aneurysm and intracranial aneurysm. *Nat Genet.* 2008 February;40(2):217-224.
42. Helgadottir A, Thorleifsson G, Manolescu A, et al. A common variant on chromosome 9p21 affects the risk of myocardial infarction. *Science.* 2007 June 8;316(5830):1491-1493.
43. Samani NJ, Raitakari OT, Sipila K, et al. Coronary artery disease-associated locus on chromosome 9p21 and early markers of atherosclerosis. *Arterioscler Thromb Vasc Biol.* 2008 September;28(9):1679-1683.
44. McPherson R, Pertsemlidis A, Kavaslar N, et al. A common allele on chromosome 9 associated with coronary heart disease. *Science.* 2007 June 8;316(5830):1488-1491.
45. Liu Y, Sanoff HK, Cho H, et al. INK4/ARF transcript expression is associated with chromosome 9p21 variants linked to atherosclerosis. *PLoS One.* 2009;4(4):e5027.
46. Jarinova O, Stewart AF, Roberts R et al. Functional analysis of the chromosome 9p21.3 coronary artery disease risk locus. *Arterioscler Thromb Vasc Biol* 2009 July 10.
47. Thompson AR, Drenos F, Hafez H, Humphries SE. Candidate gene association studies in abdominal aortic aneurysm disease: a review and meta-analysis. *Eur J Vasc Endovasc Surg.* 2008 January;35(1):19-30.
48. Vieland VJ. The replication requirement. *Nat Genet.* 2001 November;29(3):244-245.
49. Lederle FA, Simel DL. The rational clinical examination. Does this patient have abdominal aortic aneurysm? *JAMA.* 1999 January 6;281(1):77-82.
50. Lederle FA. In the clinic. Abdominal aortic aneurysm. *Ann Intern Med.* 2009 May 5;150(9):ITC5-ITC15.
51. Screening for abdominal aortic aneurysm: recommendation statement. *Ann Intern Med* 2005 February 1;142(3):198–202.
52. Fleming C, Whitlock EP, Beil TL, Lederle FA. Screening for abdominal aortic aneurysm: a best-evidence systematic review for the U.S. Preventive Services Task Force. *Ann Intern Med.* 2005 February 1;142(3):203-211.
53. Multicentre aneurysm screening study (MASS): cost effectiveness analysis of screening for abdominal aortic aneurysms based on four year results from randomised controlled trial. *BMJ* 2002 November 16;325(7373):1135.
54. Lindholt JS, Norman P. Screening for abdominal aortic aneurysm reduces overall mortality in men. A meta-analysis of the mid- and long-term effects of screening for abdominal aortic aneurysms. *Eur J Vasc Endovasc Surg.* 2008 August;36(2):167-171.
55. Hirsch AT, Haskal ZJ, Hertzer NR, et al. ACC/AHA Guidelines for the Management of Patients with Peripheral Arterial Disease (lower extremity, renal, mesenteric, and abdominal aortic): a collaborative report from the American Associations for Vascular Surgery/Society for Vascular Surgery, Society for Cardiovascular Angiography and Interventions, Society for Vascular Medicine and Biology, Society of Interventional Radiology, and the ACC/AHA Task Force on Practice Guidelines (writing committee to develop guidelines for the management of patients with peripheral arterial disease)--summary of recommendation. *J Vasc Interv Radiol.* 2006 September;17(9):1383-1397.
56. Darling RC III, Brewster DC, Darling RC, et al. Are familial abdominal aortic aneurysms different? *J Vasc Surg.* 1989 July;10(1):39-43.
57. Thompson SG, Ashton HA, Gao L, Scott RA. Screening men for abdominal aortic aneurysm: 10 year mortality and cost effectiveness results from the randomised Multicentre Aneurysm Screening Study. *BMJ.* 2009;338:b2307.
58. Ehlers L, Overvad K, Sorensen J, Christensen S, Bech M, Kjolby M. Analysis of cost effectiveness of screening Danish men aged 65 for abdominal aortic aneurysm. *BMJ.* 2009;338:b2243.
59. Thompson AR, Golledge J, Cooper JA, Hafez H, Norman PE, Humphries SE. Sequence variant on 9p21 is associated with the presence of abdominal aortic aneurysm disease but does not have an impact on aneurysmal expansion. *Eur J Hum Genet.* 2009 March;17(3):391-394.
60. Eriksson P, Jormsjo-Pettersson S, Brady AR, Deguchi H, Hamsten A, Powell JT. Genotype-phenotype relationships in an investigation of the role of proteases in abdominal aortic aneurysm expansion. *Br J Surg.* 2005 November;92(11):1372-1376.
61. Jones K, Powell J, Brown L, Greenhalgh R, Jormsjo S, Eriksson P. The influence of 4G/5G polymorphism in the plasminogen activator inhibitor-1 gene promoter on the incidence, growth and operative risk of abdominal aortic aneurysm. *Eur J Vasc Endovasc Surg.* 2002 May;23(5):421-425.
62. Miyake T, Morishita R. Pharmacological treatment of abdominal aortic aneurysm. *Cardiovasc Res.* 2009 August 1;83(3):436-443.
63. Henebiens M, Vahl A, Koelemay MJ. Elective surgery of abdominal aortic aneurysms in octogenarians: a systematic review. *J Vasc Surg.* 2008 March;47(3):676-681.
64. Parodi JC, Palmaz JC, Barone HD. Transfemoral intraluminal graft implantation for abdominal aortic aneurysms. *Ann Vasc Surg.* 1991 November;5(6):491-499.
65. Leurs LJ, Buth J, Laheij RJ. Long-term results of endovascular abdominal aortic aneurysm treatment with the first generation of commercially available stent grafts. *Arch Surg.* 2007 January;142(1):33-41.
66. Endovascular aneurysm repair versus open repair in patients with abdominal aortic aneurysm (EVAR trial 1): randomised controlled trial. *Lancet* 2005 June 25;365(9478):2179–2186.
67. Prinssen M, Buskens E, de Jong SE, et al. Cost-effectiveness of conventional and endovascular repair of abdominal aortic aneurysms: results of a randomized trial. *J Vasc Surg.* 2007 November;46(5):883-890.
68. Prinssen M, Buskens E, Blankensteijn JD. The Dutch Randomised Endovascular Aneurysm Management (DREAM) trial. Background, design and methods. *J Cardiovasc Surg (Torino).* 2002 June;43(3):379-384.
69. Brown LC, Epstein D, Manca A, Beard JD, Powell JT, Greenhalgh RM. The UK Endovascular Aneurysm Repair (EVAR) trials: design, methodology and progress. *Eur J Vasc Endovasc Surg.* 2004 April;27(4):372-381.
70. Greenhalgh RM, Brown LC, Kwong GP, Powell JT, Thompson SG. Comparison of endovascular aneurysm repair with open repair in patients with abdominal aortic aneurysm (EVAR trial 1), 30-day operative mortality results: randomised controlled trial. *Lancet.* 2004 September 4;364(9437):843-848.
71. Prinssen M, Verhoeven EL, Buth J, et al. A randomized trial comparing conventional and endovascular repair of abdominal aortic aneurysms. *N Engl J Med.* 2004 October 14;351(16):1607-1618.

72. Blankensteijn JD, de Jong SE, Prinssen M, et al. Two-year outcomes after conventional or endovascular repair of abdominal aortic aneurysms. *N Engl J Med.* 2005 June 9;352(23):2398-2405.
73. Ogata T, Shibamura H, Tromp G, et al. Genetic analysis of polymorphisms in biologically relevant candidate genes in patients with abdominal aortic aneurysms. *J Vasc Surg.* 2005 June;41(6):1036-1042.
74. Smallwood L, Warrington N, Allcock R, et al. Matrix metalloproteinase-2 gene variants and abdominal aortic aneurysm. *Eur J Vasc Endovasc Surg.* 2009 August;38(2):169-171.
75. Hinterseher I, Bergert H, Kuhlisch E, et al. Matrix metalloproteinase 2 polymorphisms in a caucasian population with abdominal aortic aneurysm. *J Surg Res.* 2006 June 15;133(2): 121-128.
76. Deguara J, Burnand KG, Berg J, et al. An increased frequency of the 5A allele in the promoter region of the MMP3 gene is associated with abdominal aortic aneurysms. *Hum Mol Genet.* 2007 December 15;16(24):3002-3007.
77. Yoon S, Tromp G, Vongpunsawad S, Ronkainen A, Juvonen T, Kuivaniemi H. Genetic analysis of MMP3, MMP9, and PAI-1 in Finnish patients with abdominal aortic or intracranial aneurysms. *Biochem Biophys Res Commun.* 1999 November 19;265(2):563-568.
78. Smallwood L, Allcock R, van BF, et al. Polymorphisms of the matrix metalloproteinase 9 gene and abdominal aortic aneurysm. *Br J Surg.* 2008 October;95(10):1239-1244.
79. Jones GT, Phillips VL, Harris EL, Rossaak JI, van Rij AM. Functional matrix metalloproteinase-9 polymorphism (C-1562T) associated with abdominal aortic aneurysm. *J Vasc Surg.* 2003 December;38(6):1363-1367.
80. Armani C, Curcio M, Barsotti MC, et al. Polymorphic analysis of the matrix metalloproteinase-9 gene and susceptibility to sporadic abdominal aortic aneurysm. *Biomed Pharmacother.* 2007 June;61(5):268-271.
81. Hinterseher I, Krex D, Kuhlisch E, et al. Tissue inhibitor of metalloproteinase-1 (TIMP-1) polymorphisms in a Caucasian population with abdominal aortic aneurysm. *World J Surg.* 2007 November;31(11):2248-2254.
82. Rossaak JI, van Rij AM, Jones GT, Harris EL. Association of the 4G/5G polymorphism in the promoter region of plasminogen activator inhibitor-1 with abdominal aortic aneurysms. *J Vasc Surg.* 2000 May;31(5):1026-1032.
83. Unno N, Nakamura T, Mitsuoka H, et al. Association of a G994 –>T missense mutation in the plasma platelet-activating factor acetylhydrolase gene with risk of abdominal aortic aneurysm in Japanese. *Ann Surg.* 2002 February;235(2): 297-302.
84. Massart F, Marini F, Menegato A, et al. Allelic genes involved in artery compliance and susceptibility to sporadic abdominal aortic aneurysm. *J Steroid Biochem Mol Biol.* 2004 December;92(5):413-418.
85. Baas AF, Medic J, van 't SR et al. The Intracranial Aneurysm Susceptibility Genes HSPG2 and CSPG2 Are Not Associated With Abdominal Aortic Aneurysm. *Angiology* 2010 January Online Available.
86. Marculescu R, Sodeck G, Domanovits H, et al. Interleukin-1 gene cluster variants and abdominal aortic aneurysms. *Thromb Haemost.* 2005 September;94(3):646-650.
87. Bown MJ, Burton PR, Horsburgh T, Nicholson ML, Bell PR, Sayers RD. The role of cytokine gene polymorphisms in the pathogenesis of abdominal aortic aneurysms: a case-control study. *J Vasc Surg.* 2003 May;37(5):999-1005.
88. Smallwood L, Allcock R, van BF, et al. Polymorphisms of the interleukin-6 gene promoter and abdominal aortic aneurysm. *Eur J Vasc Endovasc Surg.* 2008 January;35(1): 31-36.
89. Bown MJ, Lloyd GM, Sandford RM, et al. The interleukin-10-1082 'A' allele and abdominal aortic aneurysms. *J Vasc Surg.* 2007 October;46(4):687-693.
90. Ghilardi G, Biondi ML, Battaglioli L, Zambon A, Guagnellini E, Scorza R. Genetic risk factor characterizes abdominal aortic aneurysm from arterial occlusive disease in human beings: CCR5 Delta 32 deletion. *J Vasc Surg.* 2004 November;40(5):995-1000.
91. Moran CS, Clancy P, Biros E et al. Association of PPARgamma Allelic Variation, osteoprotegerin and abdominal aortic aneurysm. *Clin Endocrinol (Oxf)* 2009 April 25.
92. Korcz A, Mikolajczyk-Stecyna J, Gabriel M, et al. Angiotensin-converting enzyme (ACE, I/D) gene polymorphism and susceptibility to abdominal aortic aneurysm or aortoiliac occlusive disease. *J Surg Res.* 2009 May 1;153(1):76-82.
93. Lucarini L, Sticchi E, Sofi F, et al. ACE and TGFBR1 genes interact in influencing the susceptibility to abdominal aortic aneurysm. *Atherosclerosis.* 2009 January;202(1):205-210.
94. Jones GT, Thompson AR, van Bockxmeer FM, et al. Angiotensin II type 1 receptor 1166C polymorphism is associated with abdominal aortic aneurysm in three independent cohorts. *Arterioscler Thromb Vasc Biol.* 2008 April;28(4):764-770.
95. Fatini C, Pratesi G, Sofi F, et al. ACE DD genotype: a predisposing factor for abdominal aortic aneurysm. *Eur J Vasc Endovasc Surg.* 2005 March;29(3):227-232.
96. Pola R, Gaetani E, Santoliquido A, et al. Abdominal aortic aneurysm in normotensive patients: association with angiotensin-converting enzyme gene polymorphism. *Eur J Vasc Endovasc Surg.* 2001 May;21(5):445-449.
97. Hamano K, Ohishi M, Ueda M, et al. Deletion polymorphism in the gene for angiotensin-converting enzyme is not a risk factor predisposing to abdominal aortic aneurysm. *Eur J Vasc Endovasc Surg.* 1999 August;18(2):158-161.
98. Giusti B, Saracini C, Bolli P, et al. Genetic analysis of 56 polymorphisms in 17 genes involved in methionine metabolism in patients with abdominal aortic aneurysm. *J Med Genet.* 2008 November;45(11):721-730.
99. Strauss E, Waliszewski K, Gabriel M, Zapalski S, Pawlak AL. Increased risk of the abdominal aortic aneurysm in carriers of the MTHFR 677T allele. *J Appl Genet.* 2003;44(1):85-93.
100. Jones GT, Harris EL, Phillips LV, van Rij AM. The methylenetetrahydrofolate reductase C677T polymorphism does not associate with susceptibility to abdominal aortic aneurysm. *Eur J Vasc Endovasc Surg.* 2005 August;30(2):137-142.
101. Strauss E, Waliszewski K, Pawlak AL. The different genotypes of MTHFR 1298A>C and PON1–108C>T polymorphisms confer the increased risk of the abdominal aortic aneurysm in the smoking and nonsmoking persons. *Przegl Lek.* 2005;62(10):1023-1030.
102. Golledge J, Clancy P, Jones GT, et al. Possible association between genetic polymorphisms in transforming growth factor beta receptors, serum transforming growth factor beta1 concentration and abdominal aortic aneurysm. *Br J Surg.* 2009 June;96(6):628-632.

103. Baas AF, Medic J, van 't SR et al. Association of the TGF-beta receptor genes with abdominal aortic aneurysm. *Eur J Hum Genet* 2009 August 12.
104. Schillinger M, Exner M, Mlekusch W, et al. Heme oxygenase-1 gene promoter polymorphism is associated with abdominal aortic aneurysm. *Thromb Res*. 2002 April 15;106(2):131-136.
105. Badger SA, Soong CV, O'Donnell ME, Mercer C, Young IS, Hughes AE. C-reactive protein (CRP) elevation in patients with abdominal aortic aneurysm is independent of the most important CRP genetic polymorphism. *J Vasc Surg*. 2009 January;49(1):178-184.
106. Badger SA, Soong C, Young I, McGinty A, Mercer C, Hughes A. The Influence of COX-2 Single Nucleotide Polymorphisms on Abdominal Aortic Aneurysm Development and the Associated Inflammation. *Angiology*. 2010 February;61(2):125-130.
107. Tromp G, Ogata T, Gregoire L, et al. HLA-DQA is associated with abdominal aortic aneurysms in the Belgian population. *Ann N Y Acad Sci*. 2006 November;1085:392-395.
108. Monux G, Serrano FJ, Vigil P, De la Concha EG. Role of HLA-DR in the pathogenesis of abdominal aortic aneurysm. *Eur J Vasc Endovasc Surg*. 2003 August;26(2):211-214.
109. Rasmussen TE, Hallett JW Jr, Metzger RL, et al. Genetic risk factors in inflammatory abdominal aortic aneurysms: polymorphic residue 70 in the HLA-DR B1 gene as a key genetic element. *J Vasc Surg*. 1997 February;25(2):356-364.
110. Schulz V, Hendig D, Schillinger M, et al. Analysis of sequence variations in the ABCC6 gene among patients with abdominal aortic aneurysm and pseudoxanthoma elasticum. *J Vasc Res*. 2005 September;42(5):424-432.
111. Atli FH, Manduz S, Katrancioglu N et al. eNOS G894T Polymorphism and Abdominal Aortic Aneurysms. *Angiology* 2009 July 27.
112. Fatini C, Sofi F, Sticchi E, et al. eNOS G894T polymorphism as a mild predisposing factor for abdominal aortic aneurysm. *J Vasc Surg*. 2005 September;42(3):415-419.
113. Kotani K, Shimomura T, Murakami F, et al. Allele frequency of human endothelial nitric oxide synthase gene polymorphism in abdominal aortic aneurysm. *Intern Med*. 2000 July;39(7):537-539.
114. Gotting C, Prante C, Schillinger M, et al. Xylosyltransferase I variants and their impact on abdominal aortic aneurysms. *Clin Chim Acta*. 2008 May;391(1–2):41-45.

Future of Cardiogenetics

28

Mohammad Hadi Zafarmand, K. David Becker, and Pieter A. Doevendans

I predict that comprehensive, genomics-based health care will become the norm, with individualized preventive medicine and early detection of illnesses.[1]

Elias A. Zerhouni, MD, NIH Director

During the past years, enormous efforts have been made in population genetics, testing common single nucleotide polymorphisms (SNPS) in large, collaborative, multicenter studies involving tens of thousands of subjects, which have revolutionized the production of genetic data. Genome-wide association studies (GWAS) as the current technology for analysis of complex disease genetics[2], have identified novel genes associated with disease and traits. In the field of cardiovascular diseases (CVDs) novel loci were recently found to be associated with coronary artery disease[3-5], early onset myocardial infarction[6], atrial fibrillation[7,8], left ventricular (LV) structure and function[9], stroke[10], blood pressure[11,12], and lipid levels.[13] Currently, there are a rapidly growing number of genetic associations that achieve very strong statistical support and are replicated in additional, independent studies. These discoveries increase expectations that genetic and genomic information will become an important component of personalized health care and disease prevention, and a great asset to translational research.[14-16]

Khoury and colleagues recently described a framework for the continuum of translational research that is needed to move genetic discoveries to clinical and public health applications.[17] This framework stipulates four stages of scientific evidence in multidisciplinary translational research and stresses the importance of developing evidence-based guidelines. The four phases of translational research include: (1) translation of basic genomics research into a possible health care application; (2) evaluation of the application for the development of evidence-based guidelines; (3) evaluation of the implementation and use of the application in health care practice; and (4) evaluation of impact of evidence-based recommendations and guidelines on real-world health outcomes.[17] Most research on genome-based applications for complex diseases including cardiovascular genetics is still in the first phase of the translational research framework. We briefly address these phases in cardiovascular genetics research.

28.1 Phase 1: Translation of Basic Cardiogenomics Research into a Candidate Health Care Application

Phase 1 of translational research, which starts after gene discovery, includes both observational studies and clinical trials, and aims to develop a potential application for clinical and public-health practice.[17] Such applications could be used in various aspects of health care, for instance, diagnostic testing, screening, predictive testing, or pharmacogenetics.

28.1.1 Diagnostic Testing

Diagnostic testing is performed on an individual with signs and/or symptoms of disease with the goal of

M.H. Zafarmand (✉)
Department of Cardiology and Julius Center
for Health Science and Primary Care, University Medical Center Utrecht, Utrecht, The Netherlands
e-mail: p.doevendans@umcutrecht.nl

establishing a precise diagnosis.[18] A genetic test can help establish a diagnosis using DNA of the patient, and can serve as the basis for treatment, anticipatory guidance, and genetic counseling. Genetic tests are widely used for the diagnosis and management of classical genetic disorders and characterized by a clear pattern of inheritance and high penetrance.[17] Despite the relatively simple mapping of the causal gene for a monogenic disease, the mapping procedure for the multiple susceptibility genes in common diseases and complex traits such as CAD is quite complex.[18] Diagnostic tests for inherited arrhythmias and cardiomyopathies such as the long QT syndromes, catecholaminergic polymorphic ventricular tachycardia, Brugada syndrome, hypertrophic cardiomyopathy, dilated cardiomyopathy, and arrhythmogenic right ventricular cardiomyopathy are available and advised, especially in families with affected members.[19] There is a spectrum of genetic tests sold to the public as being diagnostic, with some providing useful clinical information, while others based upon limited scientific information have not yet shown clinical utility. Tests with limited value can bring into question the potential clinical benefit of genetic tests that have substantial supporting evidence but are not yet accepted as standard of practice, and may raise the bar for acceptance to unreasonable levels. Genetic information provided directly to the public has the potential to lead to the misuse or physical or psychological harms to the public. The potential misuse of genetic information could occur in insurance enrollment, premium-setting, and employment decisions even though laws have been passed to provide protection against such misuse.[20] It has been suggested that increased risk predicted by genetic information from markers with small genetic effect may lead to increased psychological stress.[21] However, recent evidence suggests that knowledge of genetic predisposition to Alzheimer's disease based upon a marker that does not fully explain the genetic contribution to the disease may not increase anxiety levels.[22]

When genetic tests are performed, there have been calls for genetics professionals to help patients understand the health impact of their genetic information and there have also been requests for an adequate level of counseling.[23] There is a clear educational need for teaching different aspects of genetic counseling, genetic testing, and interpretation of molecular and clinical test results to health professionals and public health workers.[19] Along with efforts to guide healthcare professionals, it is also necessary to improve the education of patients and other consumers.[20]

28.1.2 Genetic Screening

Another application of the first phase of translational research is genetic screening of asymptomatic individuals, which can be done by using a combination of molecular and biochemical testing strategies.[24] Genetic screening programs, including newborn screening programs, carrier screening, and organized cascade screening of relatives of patients with genetic syndromes, are increasingly possible for a large number of disorders.[24] Currently, the most common settings for genetic screening are prenatal and newborn screening tests (such as tests for phenylketonuria, other inborn errors of metabolism, hemoglobinopathies, and congenital hypothyroidism), which began decades ago in North America.[25] Cascade screening, sometimes called presymptomatic genetic testing, of family members of patients with diagnosed genetic syndromes is a method for identifying people at risk of a genetic condition.[19] The test can be completed on request, or as part of an organized population cascade screening program. For instance, a population cascade screening program for familial hypercholesterolemia (FH) has been implemented in the Netherlands since 1994.[26] Cascade screening includes at least the first and second and, when possible, third-degree biological relatives.[24] In the first 5 years of the implementation of the cascade screening program for FH in the Netherlands, there was 90% compliance for screening.[27] Prior to screening, 39% of familial hypercholesterolemia patients were receiving lipid-lowering medication and 1 year after screening, 93% of these patients were taking such medication. An estimated 30% of patients with FH in the Netherlands have been detected in the first 5 years of cascade screening.[27,28] The implementation of genetic screening has several practical considerations such as the need for additional health professionals and laboratory staff, computer software for recording pedigrees, protocols for obtaining informed consent, frozen sample storage facilities, and an electronic database for storing results.[29]

28.1.3 Prediction of Risk

Identifying individuals at high risk for a disease is receiving a lot of attention, because it will help direct treatment to the subgroup that needs it most, and spending health budgets more efficiently. Rapid discoveries in genomics have fueled expectations about applications of predictive genetic tests in preventive and clinical health care.[30] Several researchers attempted to include genetic markers in algorithms predicting risk of CVD,[31–35] although the optimal set of risk increasing genotypes has yet to be identified.[36] To date, the genetic profiles that have been empirically studied included only a small number of susceptibility variants showing mostly small genetic effects (i.e., OR<1.5). In a simulation study, Pepe et al. estimated that either 150 genes each with odds ratio of 1.5 or 250 genes each with odds ratio of 1.25 will be required to identify substantial numbers of subjects at high risk.[37] Other simulation studies have shown that the predictive value of a larger number of independent genes can theoretically reach the same level as that of traditional risk factors predicting CVD, but it may not evidently become much better.[38,39] To date, whole genome scans have yielded many markers with small genetic effects that when combined do not account for a large portion of the heritability for complex conditions.[40] This has led to the suggestion that the known genetic variation is not useful in the calculation of risk.[38,39] However, in some cases, genetic markers may be added to traditional risk stratification information to improve assessment.[41] An amino acid change in the KIF6 gene increases the risk of heart attack and stroke by 55%.[42,43] Furthermore, this genetic marker can be used to guide therapy; carriers of the mutant allele have decreased risk of CVD-related events in response to statin treatment relative to controls.[44,45] Yet to be discovered genetic variants could be involved in unknown pathways or in pathways with unmeasurable intermediate factors, and these SNPS may also improve disease prediction beyond traditional risk factors. It is important to consider that genetic markers need not be required to be causative for use in risk prediction; adding to the clinical utility should be sufficient. However, it has been argued that gene discoveries may also identify novel etiological pathways and novel intermediate biomarkers, which consequently may be stronger predictors of disease than the genetic variant that led to its identification, the eventual predictive value of genetic markers beyond and above the nongenetic risk factors is still being debated.[38]

28.1.4 Pharmacogenetics

Pharmacogenetics is the study of genetic determinants of individual variation in response to drugs which affect drug metabolism, efficacy, and toxicity. For example, *CYP2D6* facilitates oxidative metabolism of CVD drugs, including flecainide, propafenone, and β-blockers.[46] Genomic approaches are currently being applied in academic and industry settings to address many of the translational research gaps through identification and validation of drug targets. A functional definition of a drug target is, as per Plump et al., "any molecular target which, when modified by a therapeutic agent, may result in a pharmacologic change associated with a clinical benefit."[47] Given the large number of CVD drugs available and the large number of patients eligible to receive these drugs, even small variations in drug efficacy and safety have important implications for clinical and public health, highlighting the importance and the potential value of pharmacogenomics research.[46,47] Pharmacogenomics has been viewed as one of the first applications of the Human Genome Project for medical science, by moving toward individualized pharmacotherapy or "finding the right medicine with right dosage for the right patient."[48–51] However, there is still a lack of research that provides the evidence to validate integration of pharmacogenetic testing into routine clinical practice. To overcome this gap, collaborative efforts of governments, academia, industries, patients, and payers will be required to provide advances in technology, information handling, and access to patients, data, and samples (including DNA).[47] Currently data and samples of hundreds of thousands patients have been collected in biobanks in Iceland, UK, Japan, and many other countries.[47,52–58]

28.2 Phase 2: Research Assesses the Value of a Genomic Application for Health Practice Leading to the Development of Evidence-Based Guidelines

The second phase of translational research initiates when there is convincing evidence after conducting a genetic test. In this phase, the clinical validity (the accuracy with which a test can predict the presence or absence of the phenotype or clinical disease) and clinical utility (the likelihood that the test will lead to an improved outcome) of genetic tests are measured in the population settings for which the tests are primarily intended.[17] Results from the second phase of translational research are systematic reviews and syntheses that will lead to evidence-based practice guidelines for clinical and public health.[17] When accumulating the evidence is difficult, especially for rare genetic diseases, making evidence-based guidelines can be time consuming.[17] Recently, guidelines for the diagnosis and management of different CVD disorders such of familial long QT syndrome, familial hypercholesterolemia, and cardiomyopathy have been provided.[59–61]

Assessment of social and behavioral issues linked to genetic tests is an important aspect of second phase of translation research.[62] Concerns about the implications of genetic information for privacy and autonomy have highlighted the importance of ethical, legal, and social issues (ELSI) in genetic screening.

28.3 Phase 3: Research Attempts to Move Evidence-Based Guidelines into Health Practice, Through Delivery, Dissemination, and Diffusion Research

The third phase of translational research addresses the spreading and incorporation of knowledge obtained through the second-phase research. The translation of evidence-based guidelines into clinical and public health practice is one of the most challenging problems in health care and prevention.[17] It needs participation of policymakers, researchers, and public and professional sectors for adoption of proven genomic applications.

28.4 Phase 4: Evaluation of Impact of Evidence-Based Recommendations and Guidelines on Real-World Health Outcomes

The fourth phase of translational research focuses on clinical and public health outcomes of adopted guidelines and includes surveillance of disease incidence, morbidity and mortality measures, quality-of-life indicators, clinical decision modeling, cost-effectiveness analysis, and monitoring quality of care studies.[17] Currently, there is limited data available on the implementation of the novel genomic applications into health practice. As most of translational research in cardiovascular genomics is still in the first phase, studies that assess the health impact are scarce.

It is estimated that up to 3% of published studies focused on phase 2 and beyond,[17] which means that massive research is still needed before their implementation can be considered. In a recent survey of 3,000 biomedical investigators, phase 2 researchers were more likely to have no funding compared with phase 1 or basic researchers.[63] However, the National Institute of Health (NIH) recently demonstrated its strong support for translational research through the funding of major US academic centers with five goals in mind: building a national clinical and translational research capacity, providing opportunities for training and career development, enhancing consortium-wide collaborations, improving community health through research, and advancing phase 1 researches to accelerate the translation of basic science discoveries to clinical testing.[64]

The etiology of complex diseases is essentially different from that of monogenic diseases, and hence translating the new emerging genomic knowledge into public health and medical care is one of the major challenges for the next decades.[17, 38, 65] Despite their success, a number of limitations of GWAS; (for instance, loci reported by GWAS explain only a small proportion of the observed phenotypic variation) underscore the potential strategies to overcome these. Alternative approaches that could account for the "missing heritability" are fine mapping and next-generation sequencing such as rapid sequencing of the human exome to further define functional variants within disease-associated loci and search for rarer variants.[2]

It has been stated that "no important health problem will be solved by clinical care or research alone, or by public health alone—but rather by all public and private sectors working together."[66] "This paradigm will certainly apply to almost all health problems in the genomics era," as Muin Khoury highlighted.[67]

References

1. Flock B. NIH seeks input on proposed repository for genetic information [news release]. August 30, 2006. Available at: http://www.nih.gov/news/pr/aug2006/od-30.htm. Accessed March 10, 2010.
2. Holmes MV, Shah SH, Angelakopoulou A, et al. A report on the Genetics of Complex Diseases meeting of the British Atherosclerosis Society, Cambridge, UK, 17–18 September 2009. *Atherosclerosis*. 2010 February;208(2):599-602.
3. McPherson R, Pertsemlidis A, Kavaslar N, et al. A common allele on chromosome 9 associated with coronary heart disease. *Science*. 2007 June 8;316(5830):1488-1491.
4. Samani NJ, Erdmann J, Hall AS, et al. Genomewide association analysis of coronary artery disease. *N Engl J Med*. 2007 August 2;357(5):443-453.
5. Schunkert H, Gotz A, Braund P, et al. Repeated replication and a prospective meta-analysis of the association between chromosome 9p21.3 and coronary artery disease. *Circulation*. 2008 April 1;117(13):1675-1684.
6. Kathiresan S, Voight BF, Purcell S, et al. Genome-wide association of early-onset myocardial infarction with single nucleotide polymorphisms and copy number variants. *Nat Genet*. 2009 March;41(3):334-341.
7. Pfeufer A, van Noord C, Marciante KD, et al. Genome-wide association study of PR interval. *Nat Genet*. 2010 42(2): 153-159.
8. Ellinor PT, Lunetta KL, Glazer NL, et al. Common variants in KCNN3 are associated with lone atrial fibrillation. *Nat Genet*. 2010 March;42(3):240-244.
9. Vasan RS, Glazer NL, Felix JF, et al. Genetic variants associated with cardiac structure and function: a meta-analysis and replication of genome-wide association data. *JAMA*. 2009 July 8;302(2):168-178.
10. Ikram MA, Seshadri S, Bis JC, et al. Genomewide association studies of stroke. *N Engl J Med*. 2009 April 23; 360(17):1718-1728.
11. Levy D, Ehret GB, Rice K et al. Genome-wide association study of blood pressure and hypertension. *Nat Genet*. 2009 May 10.
12. Newton-Cheh C, Johnson T, Gateva V et al. Genome-wide association study identifies eight loci associated with blood pressure. *Nat Genet*. 2009 41:666-676.
13. Kathiresan S, Willer CJ, Peloso GM, et al. Common variants at 30 loci contribute to polygenic dyslipidemia. *Nat Genet*. 2009 January;41(1):56-65.
14. Feero WG, Guttmacher AE, Collins FS. The genome gets personal–almost. *JAMA*. 2008 March 19;299(11):1351-1352.
15. Doevendans PA. Unravelling the pathophysiology of heart failure through human genomics. *Eur J Clin Invest*. 2001 May;31(5):378-379.
16. Doevendans PA. The human genome project is ready for us, but are we ready for the human genome project. *Cardiologie*. 1998;5:595-599.
17. Khoury MJ, Gwinn M, Yoon PW, Dowling N, Moore CA, Bradley L. The continuum of translation research in genomic medicine: how can we accelerate the appropriate integration of human genome discoveries into health care and disease prevention? *Genet Med*. 2007 October;9(10):665-674.
18. Robin NH, Tabereaux PB, Benza R, Korf BR. Genetic testing in cardiovascular disease. *J Am Coll Cardiol*. 2007 August 21;50(8):727-737.
19. Hofman N, van Langen, I, Wilde AA. Genetic testing in cardiovascular diseases. *Curr Opin Cardiol*. 2010 February 15.
20. US System of Oversight of Genetic Testing: A Response to the Charge of the Secretary of Health and Human Services, April 2008. Available at: http://oba.od.nih.gov/oba/SACGHS/reports/SACGHS_oversight_report.pdf. Accessed March 10, 2010.
21. Ransohoff DF, Khoury MJ. Personal genomics: information can be harmful. *Eur J Clin Invest*. 2010 January;40(1):64-68.
22. Green RC, Roberts JS, Cupples LA, et al. Disclosure of APOE genotype for risk of Alzheimer's disease. *N Engl J Med*. 2009 July 16;361(3):245-254.
23. Humphries SE, Ridker PM, Talmud PJ. Genetic testing for cardiovascular disease susceptibility: a useful clinical management tool or possible misinformation? *Arterioscler Thromb Vasc Biol*. 2004 April;24(4):628-636.
24. Grosse SD, Rogowski WH, Ross LF, Cornel MC, Dondorp WJ, Khoury MJ. Population screening for genetic disorders in the 21st century: evidence, economics, and ethics. *Public Health Genom*. 2010;13(2):106-115.
25. Therrell BL, Adams J. Newborn screening in North America. *J Inherit Metab Dis*. 2007 August;30(4):447-465.
26. Homsma SJ, Huijgen R, Middeldorp S, Sijbrands EJ, Kastelein JJ. Molecular screening for familial hypercholesterolaemia: consequences for life and disability insurance. *Eur J Hum Genet*. 2008 January;16(1):14-17.
27. Umans-Eckenhausen MA, Defesche JC, Sijbrands EJ, Scheerder RL, Kastelein JJ. Review of first 5 years of screening for familial hypercholesterolaemia in the Netherlands. *Lancet*. 2001 January 20;357(9251):165-168.
28. Leren TP. Cascade genetic screening for familial hypercholesterolemia. *Clin Genet*. 2004 December;66(6):483-487.
29. Herman K, van Heyningen C, Wile D. Cascade screening for familial hypercholesterolaemia and its effectiveness in the prevention of vascular disease. *Br J Diabet Vasc Dis*. 2009; 9:171-174.
30. Brand A, Brand H, Schulte in den BT. The impact of genetics and genomics on public health. *Eur J Hum Genet*. 2008 January;16(1):5-13.
31. Bare LA, Morrison AC, Rowland CM, et al. Five common gene variants identify elevated genetic risk for coronary heart disease. *Genet Med*. 2007 October;9(10):682-689.
32. Humphries SE, Cooper JA, Talmud PJ, Miller GJ. Candidate gene genotypes, along with conventional risk factor assessment, improve estimation of coronary heart disease risk in healthy UK men. *Clin Chem*. 2007 January;53(1):8-16.
33. Morrison AC, Bare LA, Chambless LE, et al. Prediction of coronary heart disease risk using a genetic risk score: the Atherosclerosis Risk in Communities Study. *Am J Epidemiol*. 2007 July 1;166(1):28-35.
34. Paynter NP, Chasman DI, Buring JE, Shiffman D, Cook NR, Ridker PM. Cardiovascular disease risk prediction with and without knowledge of genetic variation at chromosome 9p21.3. *Ann Intern Med*. 2009 January 20;150(2):65-72.
35. Talmud PJ, Cooper JA, Palmen J, et al. Chromosome 9p21.3 coronary heart disease locus genotype and prospective risk of CHD in healthy middle-aged men. *Clin Chem*. 2008 March;54(3):467-474.

36. Humphries SE, Yiannakouris N, Talmud PJ. Cardiovascular disease risk prediction using genetic information (gene scores): is it really informative? *Curr Opin Lipidol.* 2008 April;19(2):128-132.
37. Pepe MS, Gu JW, Morris DE. The potential of genes and other markers to inform about risk. *Cancer Epidemiol Biomarkers Prev.* 2010 March;19(3):655-665.
38. Janssens AC, van Duijn CM. Genome-based prediction of common diseases: advances and prospects. *Hum Mol Genet.* 2008 October 15;17(R2):R166-R173.
39. Janssens AC, Aulchenko YS, Elefante S, Borsboom GJ, Steyerberg EW, van Duijn CM. Predictive testing for complex diseases using multiple genes: fact or fiction? *Genet Med.* 2006 July;8(7):395-400.
40. Manolio TA, Collins FS, Cox NJ, et al. Finding the missing heritability of complex diseases. *Nature.* 2009 October 8;461(7265):747-753.
41. Gulcher J, Stefansson K. Genetic risk information for common diseases may indeed be already useful for prevention and early detection. *Eur J Clin Invest.* 2010 January;40(1):56-63.
42. Iakoubova O, Shepherd J, Sacks F. Association of the 719Arg variant of KIF6 with both increased risk of coronary events and with greater response to statin therapy. *J Am Coll Cardiol.* 2008 June 3;51(22):2195-2196.
43. Iakoubova OA, Sabatine MS, Rowland CM, et al. Polymorphism in KIF6 gene and benefit from statins after acute coronary syndromes: results from the PROVE IT-TIMI 22 study. *J Am Coll Cardiol.* 2008 January 29;51(4):449-455.
44. Iakoubova OA, Robertson M, Tong CH et al. KIF6 Trp719Arg polymorphism and the effect of statin therapy in elderly patients: results from the PROSPER study. *Eur J Cardiovasc Prev Rehabil.* 2010 March 2.
45. Shiffman D, Chasman DI, Zee RY, et al. A kinesin family member 6 variant is associated with coronary heart disease in the Women's Health Study. *J Am Coll Cardiol.* 2008 January 29;51(4):444-448.
46. Arnett DK, Baird AE, Barkley RA, et al. Relevance of genetics and genomics for prevention and treatment of cardiovascular disease: a scientific statement from the American Heart Association Council on Epidemiology and Prevention, the Stroke Council, and the Functional Genomics and Translational Biology Interdisciplinary Working Group. *Circulation.* 2007 June 5;115(22):2878-2901.
47. Plump AS, Lum PY. Genomics and cardiovascular drug development. *J Am Coll Cardiol.* 2009 March 31;53(13):1089-1100.
48. Guessous I, Gwinn M, Yu W, Yeh J, Clyne M, Khoury MJ. Trends in pharmacogenomic epidemiology: 2001–2007. *Public Health Genom.* 2009;12(3):142-148.
49. Evans WE, Relling MV. Moving towards individualized medicine with pharmacogenomics. *Nature.* 2004 May 27;429(6990):464-468.
50. Doevendans PA. Heart therapies may come from genetic studies. *Lancet.* 1998 January 23;353:300.
51. Doevendans PA, van Dantzig J, Meijer H, Schaap C. Molecular genetics of human cardiomyopathies. In: Peters R, Piek J, eds. *Molecular cardiology in clinical perspective.* 1st ed. Amsterdam: Knoll; 1997:33-53.
52. Fan CT, Lin JC, Lee CH. Taiwan Biobank: a project aiming to aid Taiwan's transition into a biomedical island. *Pharmacogenomics.* 2008 February;9(2):235-246.
53. Jaddoe VW, Bakker R, van Duijn CM, et al. The Generation R Study Biobank: a resource for epidemiological studies in children and their parents. *Eur J Epidemiol.* 2007;22(12):917-923.
54. Jayasinghe SR, Mishra A, Van DA, Kwan E. Genetics and cardiovascular disease: design and development of a DNA biobank. *Exp Clin Cardiol.* 2009;14(3):33-37.
55. Genetics Kaiser J. US hospital launches large biobank of children's DNA. *Science.* 2006 June 16;312(5780):1584-1585.
56. Ronningen KS, Paltiel L, Meltzer HM, et al. The biobank of the Norwegian Mother and Child Cohort Study: a resource for the next 100 years. *Eur J Epidemiol.* 2006;21(8):619-625.
57. Senior K. UK Biobank launched to mixed reception. *Lancet Neurol.* 2006 May;5(5):390.
58. Triendl R. Japan launches controversial Biobank project. *Nat Med.* 2003 August;9(8):982.
59. Hershberger RE, Lindenfeld J, Mestroni L, Seidman CE, Taylor MR, Towbin JA. Genetic evaluation of cardiomyopathy – a Heart Failure Society of America practice guideline. *J Card Fail.* 2009 March;15(2):83-97.
60. Sullivan D. Guidelines for the diagnosis and management of familial hypercholesterolaemia. *Heart Lung Circ.* 2007 February;16(1):25-27.
61. Skinner JR. Guidelines for the diagnosis and management of familial long QT syndrome. *Heart Lung Circ.* 2007 February;16(1):22-24.
62. Mihaescu R, Detmar SB, Cornel MC, van der Flier WM, Heutink P, Hol EM, et al. Translational research in genomics of Alzheimer's disease: a review of current practice and future perspectives. *J Alzheimers Dis* 2010; 20(4): 967-80
63. Zinner DE, Campbell EG. Life-science research within US academic medical centers. *JAMA.* 2009 September 2;302(9):969-976.
64. Lauer MS, Skarlatos S. Translational research for cardiovascular diseases at the national heart, lung, and blood institute: moving from bench to bedside and from bedside to community. *Circulation.* 2010 February 23;121(7):929-933.
65. Janssens AC, van Duijn CM. Genome-based prediction of common diseases: methodological considerations for future research. *Genome Med.* 2009 February 18;1(2):20.
66. Marks JS. Preventive care – the first step. *Manag Care.* 2005 September;14(9 Suppl):10-12.
67. Khoury MJ, Gwinn M, Burke W, Bowen S, Zimmern R. Will genomics widen or help heal the schism between medicine and public health? *Am J Prev Med.* 2007 October;33(4):310-317.

Appendix

HCM	MYH7		CSRP3
	MYBPC3		TCAP
	TNNT2		ABCC9
	TNNI3		PLN
	TPM1		ACTC1
	MYL2		TMPO
	MYL3		PSEN1
	ACTC1		PSEN2
	CSRP3		VCL
	TNNC1		FCMD
	MYH6		TPM1
	VCL		TNNC1
	TTN		ACTN2
	DES*		DSG2
	CRYAB*		NEXN
	GLA^		MYH6
	LAMP2^*		TTN
DCM	LMNA		CRYAB
	TNNT2		LAMP2^*
	MYH7	ARVD	PKP2
	TNNI3		DSG2
	MYBPC3		DSC2
	LDB3		RYR2?
	SCN5a		DSP
	DES		TMEM43
	EYA4		JUP @
	SGCD		TGF-β

(*continued*)

NCCM	LMNA	AF	KCNQ1
	LDB3		KCNE2
	TAZ		NPPA
	MYH7		KCNA5
	MYBPC3	IVF	SCN5A?
	TNNT2		DPP6 **
	TNNI3		KCNJ8**
	TPM1	ATS	KCJN2
	ACTC1	TS	CACNA1C
	CASQ2**	JLNS	KCNQ1 @
	PLN**		KCNE1 @
	LDB3	PFHB1a	SCN5A
LQTS	KCNQ1	PFH1b	TRPM4
	KCNH2	NSCCD	SCN1B
	SCN5A	CHD	GATA4
	ANK2		NKX2-5
	KCNE1		TNX20
	KCNE2		MYH6
	KCNJ2 &		TLL1
	CACNA1C +		ZIC3^^
	SCN4B		CFC1
	AKAP9		CRELD1
	SNTA1		GJA1
CPVT	RYR2		THRAP2
	CASQ2 @	AD CAD	LRP6
SQTS	KCNH2	FHC	APOB
	KCNQ1		ARH @
	KCNJ2		LDLR
BS	SCN5A		PCSK9
	GPD1L		
	CACNA1C		
	SCN1B		
	KCNE3		
	SCN3B		
	HCN4		

This table, which is not exhaustive, lists a number of genes that are associated with cardiac diseases in a more or less monogenic fashion. Some of these genes may be very infrequent causes of the disease <1%. The official gene symbols have been used (as in the OMIM – Online Mendelian Inheritance in Man database).

Appendix

Legenda

Disease Names

HCM hypertrophic cardiomyopathy, *DCM* dilated cardiomyopathy, *ARVD* arrhythmogenic right ventricular dysplasia, *NCCM* noncompaction cardiomyopathy, *LQTS* long QT syndromes, *CPVT* catecholaminergic polymorphic ventricular tachycardia, *SQTS* short QT syndrome, *BS* Brugada syndrome, *AF* atrial fibrillation, *IVF* idiopathic ventricular fibrillation, *ATS* Andersen-Tawil syndrome (cardiodysrythmic periodic paralysis/LQT7), *TS* Timothy syndrome (long QT syndrome with syndactyly an cognitive abnormalities/LQT8), *JLNS* Jervell-Lange-Nielsen syndrome (long QT syndrome with congenital deafness), *PFH1a* progressive familial heart block 1a, *PFH1b* progressive familial heart block 1b, *NSCCD* non-specific conduction defect, *CHD* congenital heart defect (since mutations in most of these genes can lead to various different HD, they are all grouped together): GATA4, NKX2-5 are mainly associated with ASD2, CRELD1 with AVSD, ZIC3 and CFC1 with laterality defects, THRAP2 with transposition of the great arteries, *AD CAD* autosomal dominant coronary artery disease, *FHC* familial hypercholesterolemia.

Symbols Used

* * = Skeletal muscle may also be affected
* @ = Recessive mutations in this gene cause the disease
* ? = Some discussion as to whether the disease caused by mutations in RyR2 cause real ARVD as defined by strict clinical criteria
* ** = Needs confirmation in separate studies
* & = LQT7 (some discussion as to whether Andersen–Tawil syndrome is a genuine long QT syndrome
* + = Causes LQT8 also known as Timothy syndrome or long QT syndactyly syndrome
* ^ = X-linked recessive inheritance

Index

A
ABCA1 deficiency
    clinical characteristics, 318–320
    diagnosis, 319
    genetics, 318
    management, 319
Abdominal aortic aneurysm (AAA)
    clinical diagnosis
        ruptured, 429
        unruptured, 428–429
    epidemiology, 423–424
    follow-up, 429–430
    molecular genetics
        association studies, 426–427
        segregation and linkage analysis, 425–426
    pathogenesis
        natural history, 424
        vascular pathology, 424–425
    therapy
        lifestyle, 430
        pharmacological treatment, 430
        surgical intervention, 430–431
Ablation therapy, 235
Acute coronary syndrome (ACS), 336, 338
Adenosine triphosphate (ATP) synthase, 123
AF. *See* Atrial fibrillation
American Heart Association (AHA), 97
Amyloidosis, 134–135
Anderson syndrome, 391
Angiotensin converting enzyme insertion/deletion, 360
ANP. *See* Atrial natriuretic peptide
Antiarrhythmic drug therapy, 235
Antisense oligonucleotides (ASOs), 313
Apolipoprotein AI deficiency
    clinical characteristics, 317
    diagnosis, 317
    genetics, 317
    management, 317–318
Apolipoprotein E, 359–360
Apoptosis, 377
Arrhythmias, 87, 147–148
Arrhythmogenic right ventricular dysplasia/cardiomyopathy (ARVD/C)
    clinical presentation, 84–85
    diagnosis, 85

    differential diagnosis, 90–91
    ECG criteria
        arrhythmias, 87
        depolarization abnormalities, 85–87
        diagnostics, 88–89
        endomyocardial biopsy, 88
        family history, 88
        global and regional dysfunction, 87–88
        modifications, 89–90
        nonclassical subtypes, 90
        repolarization abnormalities, 87
    epidemiology, 84
    molecular and genetic background
        autosomal dominant disease, 82–83
        autosomal recessive disease, 82
        desmosomal disease and dysfunction, 81–82
        desmosome function, 80–81
        nondesmosomal genes, 83–84
    molecular genetic analysis, 91
    prognosis and therapy, 91–93
Arterial thrombosis
    clinical relevance, 344–345
    coagulation
        factor V and prothrombin, 339
        factor VII, 337–338
        factor VIII, IX and XI, 338–339
        factor XII, 338
        tissue factor, 338
    factor XIII, 341
    fibrinogen, 340–341
    fibrinolysis, 341–342
    fibrinolytic inhibitors, 342
    hemostatic system, 332–335
    polymorphisms
        platelet receptors, 335–336
        Von Willebrand factor, 336–337
    protein C pathway, 339–340
Arterial tortuosity syndrome, 276–277
ARVD/C. *See* Arrhythmogenic right ventricular dysplasia/cardiomyopathy
Atherosclerotic plaque, 337, 338
Atrial fibrillation (AF)
    atrial natriuretic peptide, 218
    autonomic nervous system, 219–220
    clinical aspects, 220–222

connexins, 215–216
epidemiology, 207–208
gene associations
    ACE, 219
    GNB3, Enos, MMP-2, IL-10, 219
    genome wide association studies, 218–219
molecular background, 208–210
molecular genetics, 211
mutations, 215
pathophysiology, 210–211
potassium channel
    KCNA5, 214
    KCNE2 and KCNJ2, 212–213
    KCNQ1, 212
secondary hit hypothesis, 214–215
sodium channel
    gain-of-function mutations, 217–218
    loss-of-function mutations, 216–217
unknown loci, 218
Atrial natriuretic peptide (ANP), 218
Atrioventricular reentry tachycardia (AVRT)
clinical aspects
    symptoms, 246
    WPW syndrome, 246–247
electrocardiographic diagnosis, 247
epidemiology
    concealed bypasses, 244
    Mahaim tachycardia, 244
    Wolff–Parkinson–white syndrome, 243–244
genetics
    animal models, 249
    genes, 248–249
    Mahaim tachycardia, 249
pathogenesis
    accessory pathways (APs), 244–245
    arrhythmia mechanisms, 245–246
treatment, 247–248
Atrioventricular septal defect, 289
Autopsy, 230, 231, 402, 405–406
Autosomal recessive hypercholesterolemia (ARH), 313–314

B

Barth syndrome, 394
b-blocker therapy, 202
Becker muscular dystrophy, 387
Bentall procedure, 272, 273
Beta-adrenergic blockade, 158
Bicuspid aortic valves, 287–288
Bone morphogenetic proteins (BMPs), 249
Brugada syndrome (BS), 16, 404
cardiogenetics
    family screening, 182–183
    genetic diagnosis, 181–182
cellular and ionic mechanisms, 168–169
clinical manifestations, 169–173
diagnosis
    ECG findings, 173–174
    ECG modulators, 174–175
    prognosis and risk stratification, 177–179
    tools, 175–177
epidemiology, 165–166
genetics, 166–168
treatment
    ICD, 179–180
    pharmacological options, 180–181

C

CAD. *See* Coronary artery disease
Calcium-induced calcium release (CICR), 197, 198
Cardiac conduction defects (CCD)
Emery-Dreifus muscular dystrophy, 225
familial defects
    lamin A/C mutations, 257
    Lenègre disease, 255–256
    molecular screening, 258
    SCN5A overlap syndrome, 256
    secundum atrial septal defects, 257–258
    sodium channel beta1 subunit mutations, 256
    TRPM4 subfamily, 254–255
observations, 253–254
Cardiac conduction system (CCS), 244, 248, 249
Cardiac ion channels, 145
Cardiogenetics
ACE insertion/deletion, 360–362
adverse drug reactions, 360
apolipoprotein E, 359–360
atherosclerosis, 356
cholesterylesther transfer protein, 358–359
clinical and public health outcomes, 440
cytochrome P450, 356
delivery, dissemination, and diffusion research, 428
genomic application, 439–440
genotype–phenotype correlations, 114
health care
    diagnostic testing, 437–438
    genetic screening, 438
    pharmacogenetics, 439
    risk prediction, 439
HMG-CoA-reductase gene, 357–358
lipoprotein lipase, 356–357
molecular and cardiologic screening, 114
molecular strategies, 114–115
PCSK9, 358
pharmaceutics, 363
pharmacogenetics, 353–356, 364–365
polymorphisms, 363–364
thiopurine methyltransferase, 356
Cardiomyocytes, 80, 81, 88
Carnitine deficiency, 393
Carvajal syndrome, 90
Cascade screening, genetic test, 417
Catecholaminergic polymorphic ventricular tachycardia (CPVT), 404
clinical aspects
    diagnosis, 199
    epidemiology, 198–199
etiology
    molecular background, 197–198
    pathophysiology, 198

genetic diagnosis
    b-blocker therapy, 202
    cardiogenetics, 203–204
    differential diagnosis, 201
    follow-up, 203
    genotyping, 199–201
    lifestyle modifications, 202
    nonpharmacologic therapy, 203
    pharmacologic therapy, 202–203
    risk stratification/ indication, 203
Caveolinopathies, 388
CCD. *See* Cardiac conduction defects
CCS. *See* Cardiac conduction system
Cell cycle progression, 377
Central core disease, 392
Channelopathies, 166
Charcot-Marie-tooth disease, 394, 398
CHARGE syndrome, 295
Cholesteryl ester transfer protein (CETP), 358–359
    clinical characteristics, 320–321
    genetics, 320
Chromosomal abnormalities, 26–27
Clinical cardiogenetics
Clinical genetics
    De Novo mutations, 32–33
    founder mutations, 35
    genetic isolates
        consanguinity, 35–36
        counseling, 36–38
        testing, 36
    genotype–phenotype correlations, 34
    Hardy–Weinberg equilibrium, 34–35
    inherited disease, 23–24
    intake
        family history, 22
        inherited disease, 23–24
    Mendelian inheritance
        autosomal dominant, 28
        autosomal recessive, 28
        X-linked dominant, 29–30
        X-linked recessive, 28–29
    mitosis and meiosis
        chromosomal abnormalities, 26–27
        inheritance patterns, 27
        recombination, 26
    mosaicism, 33
    mutation–selection equilibrium, 35
    non-Mendelian inheritance
        maternal (mitochondrial), 31–32
        multifactorial inheritance, 30–31
    penetrance and variable expressivity, 33
    predictive testing
        adverse consequences, 39
        cardiogenetics, 43
        DNA, 38–39
        family members, 40–41
        minors, 39–40
        mutation testing, 42–43
        population genetic screening, 41–42
        prenatal diagnosis, 41

Congenital cardiac malformations, 106
Congenital heart defects, 105–107, 421
    atrial septal defect, 288–289
    atrioventricular septal defect, 289
    bicuspid aortic valves, 287–288
    genetic role, 285–287
    monogenic disease, 285
    research projects, 296–297
    syndromes
        CHARGE, 295
        Digeorge syndrome/velocardiofacial syndrome/22q11 deletion, 293–295
        down syndrome/trisomy 21, 292
        Holt-Oram, 295
        Noonan syndrome, 293
        Turner (Ullrich-Turner) syndrome, 292–293
    Williams syndrome, 295–296
Connective tissue and smooth muscle disorders
    arterial tortuosity syndrome, 276–277
    differential diagnosis, 270–272
    Ehlers–Danlos syndrome, 275–276
    genetic diagnosis, 269–270
    genotype–phenotype correlation, 270
    hereditary, 263
    Loeys–Dietz syndrome
        differential diagnosis, 274–275
        molecular background, 274
        treatment, 275
    Marfan syndrome
        cardiovascular system, 267–268
        clinical aspects, 265
        dural sac, 269
        molecular genetics, 264
        ocular system, 268
        pathophysiology, 264–265
        pulmonary system, 269
        skeletal system, 265–267
        skin and integuments, 269
    nonsyndromic aortic aneurysms and dissections, 277–278
    therapy and prognosis
        cardiovascular management, 272
        manifestations, 273
        pregnancy, 273
        surgical treatment, 272–274
Connexins, 215–216
Coronary artery disease (CAD)
    adhesion signaling, 377
    anatomy and pathology, 372
    apoptosis, 377
    cardiovascular system biology, 380
    cell cycle progression, 377
    chromosome 9p21.3
        pathophysiology, 374–375
        pleiotropic effects, 373–374
    chromosome 21q22, 378
    collagen processing, 376
    coronary calcification, 377
    EPC recruitment and inflammation, 376–377
    family history

familial forms, 371
  LRP6, 371–372
  MEF2A, 371
  predisposition, 369
  risk factors, 370
 genes affecting, 372–373
 genome-wide association study, 373
 genome-wide haplotype approach, 377–378
 heritability estimates, 372
 key players, 369
 LDL levels translates, 375–376
 LDL metabolism, 378
 risk prediction, 379–380
Coronary calcification, 377
CPVT. *See* Catecholaminergic polymorphic ventricular tachycardia
Cytoskeletal proteins
 desmin, 70–71
 dystrophin, 71
 sarcoglycans and glycoproteins, 71–72

D

Danon disease, 244, 249, 393
DCM. *See* Dilated cardiomyopathy
Death verification, 405
Denaturing high-performance liquid chromatography (DHPLC), 17
De Novo mutations, 32–33
Depolarization abnormalities, 85–87
Desmin, 70–71
Diabetic cardiomyopathy, 134
DiGeorge syndrome, 294
Dilated cardiomyopathy (DCM)
 epidemiology and prevalence
  diagnosis and clinical course, 64
  genes and mutations, 66–68
  genetic background, 64–66
 family screening, 74molecular pathophysiology
  actin, 68–69
  beta-myosin heavy chain, 69–70
  cardiac ATP-sensitive potassium channels, 72
  DCM and ventricular noncompaction cardiomyopathy, 72–73
  desmin, 70–71
  dystrophin, 71
  lamin A/C, 70
  metavincullin, 69
  phospholamban, 72
  sarcoglycans and glycoproteins, 71–72
  SCN5A, 72
  telethonin, 69
  thymopoietin, 70
  titin, 70
  tropomyosin and troponin C, 69
  troponin-I and troponin-T, 69
 multiple genetic defects, 74–75
 prognosis and risk stratification, 73–74
 therapy, 73
DNA storage, 406
Down syndrome/trisomy 21, 292
Doxycycline, 274

Duchenne muscular dystrophy, 386, 387
Dystrophin, 71
Dystrophin-associated glycoprotein complex cardiomyopathies, 387–388
Dystrophinopathic cardiomyopathy, 385–386

E

Ebstein's anomaly, 104, 106
Ehlers-Danlos syndrome, 275–276
Elastin (ELN), 296
Electrophysiological study (EPS), 177–179, 181, 184
Emery-Dreifuss muscular dystrophy (EDMD), 225, 389
Endocarditis prophylaxis, 273
Endomyocardial biopsy, 88
Endomyocardial fibrosis (EMB), 132
European Society of Cardiology (ESC), 97

F

Fabry disease, 136–137
Familial amyloid polyneuropathy, 394
Familial combined hyperlipidemia (FCH), 314
Familial defective apolipoprotein B (FDB), 315
Familial dilated cardiomyopathy (FDC), 257
Familial dysbetalipoproteinemia (FD), 323
Familial hypercholesterolemia (FH), 420, 438
Familial hypertriglyceridemia (FHTG)
 clinical characteristics, 323–324
 diagnosis, 324
 management, 324
Fibrinolysis, 341–342
Fibrinolytic inhibitors, 342
Fibrosis and myocyte disarray, 48
Filamin A, 241
FHTG. *See* Familial hypertriglyceridemia
Founder mutations, 35
Friedreich's ataxia, 124, 393–394
Fukuyama congenital muscular dystrophy (FCMD), 388–389

G

Gaucher disease (GD), 135
Genetic counseling
 cardiovascular diseases
  aortic dissections, 420–421
  congenital heart defects, 421
  familial hypercholesterolemia, 420
  multidisciplinary outpatient clinics, 418–420
 components, 414–415
 definition, 414
 evidence-based protocols and guidelines, 415
 in inherited cardiovascular disease, 417–418
 neurodegenerative disorders, 415–416
 presymptomatic/ predictive counseling, 415
 reproductive disorders, 415
 screening assessment criteria, 418
Genetic heterogeneity, 11–12
Genetic screening, 417, 438
Genetic testing
 cascade screening, 417
 diagnostic, 416
 genetic population screening, 417
 inherited cardiovascular disease, 417–418

Index 451

predictive, 416–417
SCD, 406–407
screening assessment criteria, 418
Genome-wide association studies (GWAs), 218–219, 297
Genotype-phenotype correlations, 34
Genotyping, 8–10
Ghent nosology, 265, 268–269
Glycogen storage disease type II, 392

H
Hardy-Weinberg equilibrium, 34–35
HCM. *See* Hypertrophic cardiomyopathy
Health practice, evidence-based guidelines
    delivery, dissemination, and diffusion research, 440
    genomic application, 439–440
    public health outcomes, 440
Hemostatic system
    phosphatidylserine, 332–333
    platelet activation and coagulation, 332
Hereditary neuromuscular diseases
    muscle disorders (*see* Muscle disorders)
    neuropathy (*see* Neuropathy)
HERG channel, 193
High molecular weight kininogen (HMWK), 334, 338
His-Purkinje system, 253
HMG-CoA-reductase gene, 357–358
Holter monitoring, 153
Holt-Oram syndrome, 295
Hydrops fetalis, 246
Hypereosinophylic syndrome, 137–138
Hypertrophic cardiomyopathy (HCM)
    disease penetrance, 51–52
    genetic counseling and testing, 56–57
    genetics, 54–56
    pathophysiology
        fibrosis and myocyte disarray, 48
        LVH, 50
        LVOTO, 51
        sarcomeric proteins, 49
    prevalence and diagnosis, 47
    sudden cardiac death and risk stratification, 53–54
    therapy, 52–53

I
Idiopathic dilated cardiomyopathy (IDC), 257, 258
Idiopathic ventricular fibrillation (IVF), 229, 232–236
    cardiac arrest (CA), 229–230
    diagnosis
        clinical, 230–231
        genetic, 232–233
        historical, 231–232
    therapy, follow-up and prognosis, 233–235
Implantable cardioverter defibrillator (ICD), 53, 54, 113–114

K
Kearns-Sayre syndrome, 126–127, 393

L
Lamin A/C mutations, 257
LDL-R adapting protein (LDLRAP), 313
Leber's hereditary optic neuropathy (LHON), 249
Left bundle branch block (LBBB), 255
Left cardiac sympathetic denervation (LCSD), 158, 159, 161, 203
Left dominant arrhythmogenic right ventricular dysplasia/cardiomyopathy (LDAC), 90
Left ventricular hypertrophy (LVH), 47, 49–50
Left ventricular outflow tract obstruction (LVOTO), 51, 53
Limb girdle muscular dystrophy (LGMD), 107
Lipoprotein lipase, 356–357
Lipoprotein metabolism
    ABCA1 deficiency
        clinical characteristics, 318–319
        diagnosis, 319
        genetics, 318
        management, 319
    apolipoprotein AI deficiency
        clinical characteristics, 317
        diagnosis, 317
        genetics, 317
        management, 317–318
    ARH, 313–314
    cholesteryl ester transfer protein (CETP)
        clinical characteristics, 320–321
        genetics, 320
    endogenous synthesis, 308
    exogenous and endogenous lipids, 306–308
    familial combined hyperlipidemia (FCH), 314
    familial defective apolipoprotein B (FDB), 313
    familial dysbetalipoproteinemia (FD)
        clinical characteristics, 323
        diagnosis, 323
        genetics, 323
        management, 323
    familial hypercholesterolemia (FH)
        clinical characteristics, 310
        diagnosis, 310, 312
        genetics, 310
        management, 311, 313
    familial hypertriglyceridemia (FHTG)
        clinical characteristics, 323–324
        diagnosis, 324
        management, 324
    familial LCAT deficiency and fish eye disease
        clinical characteristics, 319–320
        diagnosis, 320
        genetics, 319
        management, 320
    HDL metabolism, 308–309
    LPL-deficiency and Apo-CII deficiency, 322
    sitosterolemia, 315
    structure, 306
    Tangier disease
        clinical characteristics, 318–319
        diagnosis, 319
        genetics, 318
        management, 319
Loeys-Dietz syndrome
    differential diagnosis, 274–275
    molecular background, 274
    treatment, 275
Long QT syndrome (LQTS)

arrhythmia, 147–148
cardiac action potential, 143–145
cardiac ion channels, 145
clinical presentation, 148
diagnosis, 149–150
differential diagnosis, 154
epidemiology and prevalence, 146
epinephrine stress test, 153–154
exercise testing, 153
family screening, 160–161
genotype–phenotype correlations, 148–149
genotypes, 156
Holter monitoring, 153
molecular genetic diagnosis, 154–155
molecular genetics, 146–147
QT-dispersion, 153
QT-interval, 150–151
risk factors
    LQTS genotypes, 156
    mutation, 156
    postpartum period, 156
    symptoms, 157
risk stratification
    gender, 155
    QTc duration, 155
    time-dependent syncope, 155–156
sinus node dysfunction, 153
therapy and prognosis
    asymptomatic patients, 160
    beta-adrenergic blockade, 158
    general lifestyle measures, 157–158
    genotype specific measures, 159–160
    ICD, 158–159
    LCSD, 158
    pacemaker therapy, 158
T wave
    alternans, 153
    morphology, 151–152
Low-density-lipoprotein cholesterol (LDL-C) levels
clinical characteristics, 310
diagnosis, 310, 312
genetics, 310
management, 311, 313
LVH. *See* Left ventricular hypertrophy
LVOTO. *See* Left ventricular outflow tract obstruction
Lysosomal glycogenosis
danon disease, 393
Pompe's disease/glycogen storage disease type II, 392

M

Mahaim tachycardia, 244–245, 247, 249
Marfan syndrome, 239
cardiovascular system, 267–268
clinical aspects, 265
dural sac, 269
molecular genetics, 264
ocular system, 268
pathophysiology, 264–265
pulmonary system, 269
skeletal system, 265–267
skin and integuments, 269
MELAS syndrome, 125–126
Mendelian inheritance
autosomal dominant, 28
autosomal recessive, 28
X-linked dominant, 29–30
X-linked recessive, 28–29
Metallo-matrix proteinases (MMPs), 425
Metavincullin, 69
MFM. *See* Myofibrillar myopathies
Mitochondrial cardiomyopathy
Friedreich's ataxia, 124
heteroplasmy, 123
Kearns–Sayre syndrome, 126–127
MELAS syndrome, 125–126
mitochondrial diseases, 127
OXPHOS, 123
Mitochondrial disorders
Barth syndrome, 394
carnitine deficiency, 393
Friedreich's ataxia, 393–394
Kearns-Sayre syndrome, 393
Mitochondrial DNA mutations, 72
Mitral valve prolapse (MVP)
epidemiology, 239
genetic aspects, 241
pathophysiology and clinical aspects, 239–241
MLPA. *See* Multiplex ligation dependent probe amplification
MMPs. *See* Metallo-matrix proteinases
Molecular genetics
analysis, 91
DNA
    diagnostics, 12–15
    eukaryotic genes, 4
    genetic heterogeneity, 11
    genotyping, 8–10
    linkage and risk haplotype analysis, 15–16
    meiosis, 6
    mutations, 6–8
    Sanger sequencing, 16
    scanning methods, 16–17
    technologies, 17–19
genetic testing, 3
Mucopolysaccharidoses (MPS), 135–136
Multifactorial inheritance, 30–31
Multiple wavelet hypothesis, 210, 212, 214
Multiplex ligation dependent probe amplification (MLPA), 13, 15, 18
Muscular disorders
congenital myopathies
    CCD, 392
    myosin storage myopathy, 392
    nemaline rod myopathy, 392
enzymatic activity
    FCMD, 388–389
    LGMD2I/MDC1C, 388
inner nuclear membrane proteins
    autosomal dominant EDMD/LGMD1B, 390
    EDMD, 389
    X-linked emery dreifuss muscular dystrophy, 389–390

ion channel disorder
  Anderson syndrome, 391
metabolic disorders
  lysosomal glycogenosis, 392–393
  mitochondrial disorders, 393–394
myofibrillar myopathies (MFM), 391
nucleotide repeat disorders
  dystrophia myotonica, 390–391
  myotonic dystrophy type 2 (DM2) (*see* Proximal myotonic myopathy (PROMM))
sarcolemma-associated proteins
  Becker muscular, 387
  caveolinopathies, 388
  Duchenne muscular, 386, 387
  dystrophin-associated glycoprotein complex cardiomyopathies, 387–388
  dystrophinopathic cardiomyopathy, 385–386
  X-linked dilated cardiomyopathy, 387
Myofibrillar myopathies (MFM), 391
Myosin storage myopathy, 392
Myotonic dystrophy, 390–391

N

Naxos disease, 90
NCCM. *See* Noncompaction cardiomyopathy
NCLVM. *See* Noncompaction of the left ventricular myocardium
Nemaline rod myopathy, 392
Neuromuscular disease, 107–108
Neuropathy
  Charcot-Marie-tooth disease, 394, 398
  familial amyloid polyneuropathy, 394
  Refsum's disease, 398
Noncompaction cardiomyopathy (NCCM), 72–73
  asymptomatic disease, 115–116
  cardiogenetics
    genotype–phenotype correlations, 114
    molecular and cardiologic screening, 114
    molecular strategies, 114–115
  clinical aspects, 111–112
  coincidental etiologies, 111
  congenital heart disease, 105–107
  differential diagnosis, 112
  epidemiology, 102
  etiology and molecular genetics, 102–104
  implantable cardioverter defibrillator, 113–114
  isolation, 104–105
  mitochondrial disorders, 108–111
  neuromuscular disease, 107–108
  nonisolation, 105
  pathology
    macroscopy, 99–101
    microscopy, 101–102
  prognosis, 113
  syndromes, 108–111
  therapy and follow-up, 113
  trabeculations, 98–99
Noncompaction of the left ventricular myocardium (NCLVM), 97
Nondesmosomal genes, 83–84

Non-Mendelian inheritance
  maternal (mitochondrial), 31–32
  multifactorial inheritance, 30–31
Noonan syndrome, 293
Nuclear envelope proteins, 70

O

Ocular system, 268
Osteogenesis imperfecta, 279
Outpatient clinics
  genetic counseling (*see* Genetic counseling)
  genetic testing
    cascade screening, 417
    diagnostic, 416
    genetic population screening, 417
    predictive, 416–417
Oxidative phosphorylation (OXPHOS), 123, 125
OXPHOS. *See* Oxidative phosphorylation

P

Pacemaker therapy, 158
PCCD. *See* Progressive cardiac conduction defect
PCR. *See* Polymerase chain reaction
Permanent junctional reciprocating tachycardia (PJRT), 244, 245
Persistent ductus arteriosus (PDA), 285
Pharmacogenetics, 353–356, 364–365, 427
Phospholamban, 72
Poison polypeptide, 104
Polygenic threshold theory, 285
Polymerase chain reaction (PCR), 12–15
Pompe disease, 244, 249, 392
Potassium channel
  KCNA5, 214
  KCNE2 and KCNJ2, 212–213
  KCNQ1, 212
Preparticipation screening, 409
Progressive cardiac conduction defect (PCCD), 253, 256, 258
Protein C pathway, 339–340
Protein kinase A (PKA), 198
Proximal myotonic myopathy (PROMM), 391
Pseudoxanthoma elasticum (PXE), 134

Q

Quinidine, 169, 172, 182, 184

R

Radiofrequency catheter ablation (RFCA), 248, 250
Refsum's disease, 398
Repolarization abnormalities, 87
Restrictive cardiomyopathy (RCM), 97
  clinical aspects, 131
  diagnosis, 131–132
  differential diagnosis, 132
  endomyocardial causes
    endomyocardial fibrosis, 137
    hypereosinophylic syndrome, 137–138
  idiopathic and familial restrictive cardiomyopathy, 132–133
  infiltrative

cardiac amyloidosis, 134–135
    Gaucher disease, 135
    mucopolysaccharidoses, 135–136
  molecular background, 129–130
  non-infiltrative restrictive cardiomyopathy
    diabetic cardiomyopathy, 134
    pseudoxanthoma elasticum, 134
    scleroderma/systemic sclerosis, 133–134
  prognosis, 132
  storage diseases
    Fabry disease, 136–137
    glycogen storage disease, 137
    hemochromatosis, 136
  treatment, 132
RFCA. *See* Radiofrequency catheter ablation
Right bundle branch block (RBBB), 254, 255
RNA polymerase, 5

S

SAM. *See* Systolic anterior motion
Sanger sequencing, 16
Sarcoglycans and glycoproteins, 71–72
Sarcoidosis, 135
Sarcolemma-associated proteins
  Becker muscular, 387
  caveolinopathies, 388
  Duchenne muscular, 386, 387
  dystrophin-associated glycoprotein complex cardiomyopathies, 387–388
  dystrophinopathic cardiomyopathy, 385–386
  X-linked dilated cardiomyopathy, 387
Sarcomere proteins, 49
  actin, 68–69
  beta-myosin heavy chain, 69–70
  metavincullin, 69
  telethonin and troponin-T, 69
  titin, 70
  tropomyosin, 69
  troponin-I and troponin-C, 69
Sarcoplasmic reticulum calcium adenosine triphosphatase (SERCA), 198
SCD. *See* Sudden cardiac death
Scleroderma/Systemic sclerosis (SSc), 133–134
SCN5A overlap syndrome, 256
Secondary hit hypothesis, 214–215
Secundum atrial septal defects, 257–258
Short QT syndrome (SQTS)
  cardiogenetics, 195
  clinical presentation, 191
  diagnosis, 192–193
  differential diagnosis, 193
  epidemiology and prevalence, 190
  ICD therapy, 194
  molecular and genetic background, 189–190
  pathophysiology, 190–191
  pharmacologic therapy, 193–194
  risk stratification and indication, 194–195
Short QT syndrome pharmacologic therapy, 193–194
SIDS. *See* Sudden infant death syndrome
Single nucleotide polymorphisms (SNPs), 285, 287, 297, 426

Sinus node dysfunction, 153
Sitosterolemia, 315
Smooth muscle disorders. *See* Connective tissue and smooth muscle disorders
Sodium channel
  gain-of-function mutations, 217–218
  loss-of-function mutations, 216–217
Somatic mutation, 215–216
Southern blot analysis, 12, 15
SQTS. *See* Short QT syndrome
Stickler syndrome, 279
Sudden arrhythmic death syndrome, 402
Sudden cardiac death (SCD), 11
  arrhythmia, 54
  athletes screening, 408–409
  cardiogenetic clinic, 406
  causes, 404
  definitions, 401–403
  demographics, 404–405
  family history, 50
  first-degree relatives, 407–408
  genetic testing, 406–407
  incidence, 403
  postmortem diagnosis, 405–406
  preparticipation screening, 409
  subset, 53
Sudden infant death syndrome (SIDS), 401–403
Sudden unexplained death (SUD), 229–236, 402
Sudden unexplained death syndrome (SUDS), 166, 170
SUDS. *See* Sudden unexplained death syndrome
Supravalvular aortic stenosis (SVAS), 296
Supraventricular arrhythmias, 170, 174, 182
Supraventricular tachycardia (SVT), 243
Systolic anterior motion (SAM), 51

T

Tangier disease
  clinical characteristics, 318–319
  diagnosis, 319
  genetics, 318
  management, 319
Telethonin, 69
Tetralogy of Fallot (TOF), 294, 295
Thiopurine methyltransferase, 356
Thoracic aortic aneurysms and aortic dissections (TAAD)
  aortic dilatation, 277
  autosomal dominant manner, 277
  familial dilatation, 277
  genetic heterogeneity, 277
  Marfan-related disorders, 279
  mutations, 274
  phenotypic manifestations, 263
  thoracic aneurysms, 278
Thrombosis. *See* Arterial thrombosis
Thymopoietin, 70
Tissue factor, 338
Titin, 70
Transforming growth factor-b (TGF-b), 241
Tropomyosin and troponin C, 69

TRPM4 subfamily, 254–255
Turner (Ullrich-Turner) syndrome, 292–293

U
Unexplained cardiac arrest (UCA), 230, 232

V
Ventricular tachyarrhythmia, 195
Von Willebrand factor, 336–337

W
Williams syndrome, 278
Wolff–Parkinson–White syndrome, 243–244, 246–247

X
Xenopus laevis, 213, 217
X-linked dilated cardiomyopathy, 387
X-linked emery dreifuss muscular dystrophy, 389–390